The Hydraulics of Open Channel Flow

An introduction

Basic principles, sediment motion, hydraulic modelling, design of hydraulic structures

Hubert Chanson
University of Queensland, Australia

A member of the Hodder Headline Group
LONDON • SYDNEY • AUCKLAND
Copublished in North, Central and South America by
John Wiley & Sons Inc., New York • Toronto

First published in Great Britain in 1999 by
Arnold, a member of the Hodder Headline Group,
338 Euston Road, London NW1 3BH
http://www.arnoldpublishers.com

Copublished in North, Central and South America by
John Wiley & Sons Inc., 605 Third Avenue,
New York, NY 10158-0012

British Library Cataloguing in Publication Data
A catalogue record for this book is available from the British Library

Library of Congress Cataloging-in-Publication Data
A catalog record for this title is available from the Library of Congress

ISBN 0 340 74067 1
ISBN 0 470 36103 4 (Wiley)

1 2 3 4 5 6 7 8 9 10

Publisher: Eliane Wigzell
Production Editor: James Rabson
Production Controller: Iain McWilliams
Cover Design: Terry Griffiths

Typeset in 10/12 Times by Academic & Technical Typesetting, Bristol
Printed and bound in Great Britain by J W Arrowsmith Ltd, Bristol

What do you think about this book? Or any other Arnold title?
Please send your comments to feedback.arnold@hodder.co.uk

To Ya Hui

Pour Bernard et Nicky

Contents

Colour plates appear between pages xxxii and xxxiii

Preface		xi
Acknowledgements		xiii
About the author		xv
Glossary		xvi
List of symbols		xxxvi
Part 1	**Basic Principles of Open Channel Flows**	**1**
1.	Introduction	3
	1.1 Presentation	3
	1.2 Fluid properties	3
	1.3 Static fluids	4
	1.4 Open channel flow	6
	1.5 Exercises	8
2.	Fundamental equations	9
	2.1 Introduction	9
	2.2 The fundamental equations	9
	2.3 Exercises	20
3.	Applications of the Bernoulli equation to open channel flows	21
	3.1 Introduction	21
	3.2 Application of the Bernoulli equation – specific energy	21
	3.3 Froude number	39
	3.4 Properties of common open-channel shapes	46
	3.5 Exercises	48
4.	Applications of the momentum principle: hydraulic jump, surge and flow resistance in open channels	52
	4.1 Momentum principle and application	52
	4.2 Hydraulic jump	56
	4.3 Surges and bores	67
	4.4 Flow resistance in open channels	72

4.5	Flow resistance calculations in engineering practice	85
4.6	Exercises	91
5.	**Uniform flows and gradually varied flows**	98
5.1	Uniform flows	98
5.2	Non-uniform flows	105
5.3	Exercises	113
	Part 1 Revision exercises	116
	Revision exercise No. 1	116
	Revision exercise No. 2	117
	Revision exercise No. 3	118
	Revision exercise No. 4	119
	Revision exercise No. 5	120
	Revision exercise No. 6	121
	Revision exercise No. 7	122
	Appendices to Part 1	124
A1.1	Constants and fluid properties	124
A1.2	Unit conversions	128
A1.3	Mathematics	131
A1.4	Alternate depths in open channel flow	143
Part 2	**Introduction to Sediment Transport in Open Channels**	**147**
6.	**Introduction to sediment transport in open channels**	149
6.1	Introduction	149
6.2	Significance of sediment transport	149
6.3	Terminology	155
6.4	Structure of this section	156
6.5	Exercises	156
7.	**Sediment transport and sediment properties**	157
7.1	Basic concepts	157
7.2	Physical properties of sediments	161
7.3	Particle fall velocity	165
7.4	Angle of repose	171
7.5	Laboratory measurements	172
7.6	Exercises	173
8.	**Inception of sediment motion, occurrence of bed-load motion**	175
8.1	Introduction	175
8.2	Hydraulics of alluvial streams	175
8.3	Threshold of sediment bed motion	181
8.4	Exercises	188
9.	**Inception of suspended load motion**	190
9.1	Presentation	190
9.2	Initiation of suspension and critical bed shear stress	190

9.3   Onset of hyperconcentrated flow   192
9.4   Exercises   194

10. Sediment transport mechanisms 1. Bed-load transport   195
     10.1 Introduction   195
     10.2 Empirical correlations of bed-load transport rate   195
     10.3 Bed-load calculations   199
     10.4 Applications   202
     10.5 Exercises   209

11. Sediment transport mechanisms 2. Suspended load transport   210
     11.1 Introduction   210
     11.2 Advective diffusion of sediment suspension   212
     11.3 Suspended sediment transport rate   216
     11.4 Hyperconcentrated suspension flows   221
     11.5 Exercises   224

12. Sediment transport capacity and total sediment transport   225
     12.1 Introduction   225
     12.2 Total sediment transport rate (sediment transport capacity)   225
     12.3 Erosion, accretion and sediment bed motion   229
     12.4 Sediment transport in alluvial channels   233
     12.5 Applications   240
     12.6 Exercises   245

Part 2 Revision exercises   246
     Revision exercise No. 1   246
     Revision exercise No. 2   246
     Revision exercise No. 3   246
     Revision exercise No. 4   247

Appendix to Part 2   248
     A2.1 Some examples of reservoir sedimentation   248

Part 3   Hydraulic Modelling   **255**

13. Summary of basic hydraulic principles   257
     13.1 Introduction   257
     13.2 Basic principles   257
     13.3 Flow resistance   259

14. Physical modelling of hydraulics   261
     14.1 Introduction   261
     14.2 Basic principles   261
     14.3 Dimensional analysis   264
     14.4 Modelling fully-enclosed flows   270
     14.5 Modelling free-surface flows   272
     14.6 Design of physical models   277
     14.7 Summary   278
     14.8 Exercises   279

15. Numerical modelling: backwater computations   284
    15.1 Introduction   284
    15.2 Basic equations   284
    15.3 Backwater calculations   289
    15.4 Numerical integration   293
    15.5 Discussion   296
    15.6 Computer models   297
    15.7 Exercises   297

Part 3 Revision exercises   299
    Revision exercise No. 1   299
    Revision exercise No. 2   299

Appendices to Part 3   301
    A3.1 Physical modelling of movable boundary hydraulics   301
    A3.2 Extension of the backwater equation   305

Part 4 Design of Hydraulic Structures   **309**

16. Introduction to the design of hydraulic structures   311
    16.1 Introduction   311
    16.2 Structure of Part 4   311
    16.3 Professional design approach   311

17. Design of weirs and spillways   313
    17.1 Introduction   313
    17.2 Crest design   319
    17.3 Chute design   330
    17.4 Stilling basins and energy dissipators   333
    17.5 Design procedure   341
    17.6 Exercises   349

18. Design of drop structures and stepped cascades   355
    18.1 Introduction   355
    18.2 Drop structures   355
    18.3 Nappe flow on stepped cascades   362
    18.4 Exercises   363

19. Culvert design   365
    19.1 Introduction   365
    19.2 Basic features of a culvert   365
    19.3 Design of standard culverts   371
    19.4 Design of Minimum Energy Loss culverts   383
    19.5 Exercises   397

Part 4 Revision exercises   402
    Revision exercise No. 1   402
    Revision exercise No. 2 (hydraulic design of a new Gold Creek dam
    spillway)   402

Revision exercise No. 3 (hydraulic design of the Nudgee Road bridge
waterway)   406

Appendices to Part 4   410
  A4.1 Spillway chute flow calculations   410
  A4.2 Examples of Minimum Energy Loss weirs   417
  A4.3 Examples of Minimum Energy Loss culverts and waterways   421
  A4.4 Computer calculations of standard culvert hydraulics   430

Problems   **435**

P1. A study of the Marib dam and its sluice system (115 BC–575 AD)   437
  P1.1 Introduction   437
  P1.2 Hydraulics problem   441
  P1.3 Hydrological study: flood attenuation of the Marib reservoir   445

P2. A study of the Moeris reservoir, the Ha-Uar dam and the canal
     connecting the Nile river and Lake Moeris around 2900 BC to 230 BC   446
  P2.1 Introduction   446
  P2.2 Hydraulics problem   452
  P2.3 Hydrology of Egypt's Lake Moeris   455

P3. A study of the Moche River irrigation systems (Peru AD 200–AD 1532)   458
  P3.1 Introduction   458
  P3.2 Hydraulics problem   464
  P3.3 Hydrology of western Peru   467

References   470

Additional bibliography   481

Suggestion/correction form   483

Author index   485

Subject index   488

Preface

The book is an introduction to the hydraulics of open channel flows. The material is designed for undergraduate students in Civil, Environmental and Hydraulic Engineering. It will be assumed that the students have had an introductory course in fluid mechanics and are familiar with the basic principles of fluid mechanics: continuity, momentum, energy and Bernoulli principles.

The book will first develop the basic principles of fluid mechanics with applications to open channels. Open channel flow calculations are more complicated than pipe flow calculations because the location of the free-surface is often unknown *a priori* (i.e. beforehand). Later, the students are introduced to the basic concepts of sediment transport and hydraulic modelling (physical and numerical models). At the end of the course, the design of hydraulic structures is introduced. The book is designed to bring a basic understanding of the hydraulics of rivers, waterways and man-made canals (e.g. Plates 1 to 13) to the reader.

The lecture material is divided into four parts of increasing complexity.

Part 1. Introduction to the basic principles. Application of the fundamental fluid mechanics principles to open channels. The emphasis is on the application of the Bernoulli principle and momentum equation to open channel flows.

Part 2. Introduction to sediment transport in open channels. Basic definitions followed by simple applications. Occurrence of sediment motion in open channels. Calculations of sediment transport rate. Interactions between the sediment motion and the fluid motion.

Part 3. Modelling open channel flows. Physical modelling of open channel flows. Numerical modelling of open channel flows.

- Physical modelling: application of the basic principles of similitude and dimensional analysis to open channels.
- Numerical modelling: numerical integration of the energy equation; one-dimensional flow modelling.

Part 4. Introduction to the design of hydraulic structures for the storage and conveyance of water. Hydraulic design of dams, weirs and spillways. Design of drops and cascades. Hydraulic design of culverts: standard box culverts and minimum energy loss culvert.

Basic introduction to professional design of hydraulic structures. Application of the basic principles to real design situations. Analysis of complete systems.

Applications, tutorials and exercises are grouped into four categories: applications within the main text to illustrate the basic lecture material, exercises for each chapter, revision exercises using knowledge gained in several chapters within each Part, and major assignments (i.e. problems) involving expertise gained in several sections: e.g. typically Part I and one or two other sections. In the lecture material, complete and detailed solutions of the applications are given. Numerical solutions of some exercises, revision exercises and problems are available on the Internet (Publisher's site: http://www.arnoldpublishers.com/support/chanson).

A suggestion/correction form is placed at the end of the book. Comments, suggestions and criticisms are welcome and they will be helpful to improve the quality of future editions. Readers who find an error or mistake are welcome to record the error on the page and to send a copy to the author. 'Errare Humanum Est'[1].

---

[1] To err is human.

Acknowledgements

The author wants to thank especially Professor Colin J. Apelt, University of Queensland, for his help, support and assistance throughout the academic career of the writer.

He also expresses his gratitude to the following people who provided photographs and illustrations of interest.

The American Society of Mechanical Engineers
Mr and Mrs Michael and Linda Burridge, Brisbane QLD, Australia
Mr and Mrs Chanson, Paris, France
Ms Chou Ya-Hui, Brisbane QLD, Australia
Mr L. Stuart Davies, Welsh Water, United Kingdom
Dr Chris Fielding, Brisbane QLD, Australia
Mrs Jenny L.F. Hacker, Brisbane QLD, Australia
The Hydro-Electric Commission (HEC) of Tasmania, Australia
Mr D. Jeffery, Goulburn–Murray Water, Australia
Mr J. Mitchell, Brisbane QLD, Australia
Mr Pu Hua-Chih, Taipei, Taiwan, Republic of China
Dr R. Rankin, Rankin Publishers, Brisbane, Australia
Sinorama-Magazine, Taipei, Taiwan, Republic of China
US Geological Survey, USA

The author also thanks the following people who provided him with information and assistance.

Professor C.J. Apelt, Brisbane QLD, Australia
Mr Noel Bedford, Tamworth NSW, Australia
Dr D. Brady, Brisbane QLD, Australia
Mr and Mrs Chanson, Paris, France
Ms Chou Ya-Hui, Brisbane QLD, Australia
Concrete Pipe Association of Australasia, Sydney NSW, Australia
Mr Doug Davidson, Murwillumbah NSW, Australia
Mr L. Stuart Davies, Welsh Water, United Kingdom
Dr Michael R. Gourlay, Brisbane QLD, Australia
Dr D. Holloway, University of Tasmania, Australia

Mr Patrick James, Neutral Bay NSW, Australia
Mr D. Jeffery, Goulburn–Murray Water, Australia
Mr Steven Li, Hydro-Electric Commission of Tasmania, Australia
Ms Carolyn Litchfield, Brisbane QLD, Australia
Ms Joe McGregor, US Geological Survey (Photo Library), USA
Officine Maccaferri, Italy
Dr J. S. Montes, University of Tasmania, Australia
Professor N. Rajaratnam, University of Alberta, Canada
Mr Paul Royet, Cemagref, France
Mr Des Williamson, HydroTools, Canada.

And last but not the least, the author thanks all the people (including colleagues, former students, students, professionals) who gave him information, feedback and comments on his lecture material. In particular, he acknowledges the support of Professor C.J. Apelt in the preparation of the book. Some material on hydraulic structure design (more specifically spillway design and culvert design) is presented and derived from Professor Apelt's personal lecture notes.

About the Author

The author received a degree of 'Ingénieur Hydraulicien' from the Ecole Nationale Supérieure d'Hydraulique et de Mécanique de Grenoble (France) in 1983 and a degree of 'Ingénieur Génie Atomique' from the 'Institut National des Sciences et Techniques Nucléaires' in 1984. He worked in industry in France as a R&D engineer at the Atomic Energy Commission from 1984 to 1986, and as a computer professional in fluid mechanics for Thomson-CSF between 1989 and 1990. From 1986 to 1988, he studied at the University of Canterbury (New Zealand) as part of a PhD project.

The author has been a senior lecturer in fluid mechanics, hydraulics and environmental engineering at the University of Queensland since 1990. His research interests include design of hydraulic structures, experimental investigations of two-phase flows, water quality modelling in coastal and hydraulic structures, and environmental management and natural resources. He is the author of two books: '*Hydraulic Design of Stepped Cascades, Channels, Weirs and Spillways*' (Pergamon, 1995) and '*Air Bubble Entrainment in Free-Surface Turbulent Shear Flows*' (Academic Press, 1997). His publication record includes over 100 international refereed papers. The author has also been active as a consultant for both governmental agencies and private organizations.

He has been awarded four fellowships from the Australian Academy of Science. In 1995 he was a Visiting Associate Professor at National Cheng Kung University (Taiwan, Republic of China) and he was Visiting Research Fellow at Toyohashi University of Technology (Japan) in 1999.

Dr Chanson was the keynote lecturer at the 1998 ASME Fluids Engineering Symposium on Flow Aeration and at the Workshop on Flow Characteristics around Hydraulic Structures (Nihon University, Japan 1998). He has given several short courses in Australia and overseas.

Glossary

Abutment   part of the valley side against which the dam is constructed. Artificial abutments are sometimes constructed to take the thrust of an arch where there is no suitable natural abutment.

Académie des Sciences de Paris   The Académie des Sciences, Paris, is a scientific society, part of the Institut de France formed in 1795 during the French Revolution. The academy of sciences succeeded the Académie Royale des Sciences, founded in 1666 by Jean-Baptiste Colbert.

Accretion   increase of channel bed elevation resulting from the accumulation of sediment deposits.

Adiabatic   thermodynamic transformation occurring with neither loss nor gain of heat.

Advection   movement of a mass of fluid that causes change in temperature or in other physical or chemical properties of a fluid.

Aeration device (or *aerator*)   device used to artificially introduce air within a liquid. Spillway aeration devices are designed to introduce air into high-velocity flows. Such aerators basically include a deflector and air is supplied beneath the deflected waters. Downstream of the aerator, the entrained air can reduce or prevent cavitation erosion.

Afflux   rise of water level above normal level (i.e. natural flood level) on the upstream side of a culvert or an obstruction in a channel.

Aggradation   rise in channel bed elevation caused by deposition of sediment material. Another term is *accretion*.

Air   mixture of gases comprising the atmosphere of the Earth. The principal constituents are nitrogen (78.08%) and oxygen (20.95%). The remaining gases in the atmosphere include argon, carbon dioxide, water vapour, hydrogen, ozone, methane, carbon monoxide, helium, krypton and so on.

Air concentration   concentration of undissolved air defined as the volume of air per unit volume of air and water. It is also called the void fraction.

Alembert (d')   Jean le Rond d'Alembert (1717–1783) was a French mathematician and philosopher. He was a friend of Leonhard Euler and Daniel Bernoulli. In 1752 he published his famous d'Alembert's paradox for an ideal-fluid flow past a cylinder (Alembert 1752).

Alternate depth   in open channel flow, for a given flow rate and channel geometry, the relationship between the specific energy and flow depth indicates that, for a given specific energy, there is no real solution (i.e. no possible flow), one solution (i.e. critical flow) or two solutions for the flow depth. In the latter case, the two flow depths are called alternate depths. One corresponds to a subcritical flow, the second to a supercritical flow.

Analytical model   system of mathematical equations that are the algebraic solutions of the fundamental equations.

Apelt C.J. Apelt is an Emeritus Professor in civil engineering at the University of Queensland (Australia).

Apron the area at the downstream end of a weir to protect against erosion and scouring by water.

Aqueduct a conduit for conveying a large quantity of flowing waters. The conduit may include canals, siphons, pipelines.

Arch dam dam in plan dependent on arch action for its strength.

Arched dam gravity dam which is curved in plan. Alternatives include 'curved-gravity dam' and 'arch-gravity dam'.

Archimedes Greek mathematician and physicist. He lived between BC 290–280 and BC 212 (or 211). He spent most of his life in Syracuse (Sicily, Italy) where he played a major role in the defence of the city against the Romans. His treatise 'On floating bodies' is the first-known work on hydrostatics, in which he outlined the concept of buoyancy.

Aristotle Greek philosopher and scientist (384–322 BC), student of Plato. His work 'Meteorologica' is considered as the first comprehensive treatise on atmospheric and hydrological processes.

Armouring progressive coarsening of the bed material resulting from the erosion of fine particles. The remaining coarse material layer forms an armour, protecting further bed erosion.

Assyria land to the North of Babylon comprising, in its greatest extent, a territory between the Euphrates and the mountain slopes East of the Tigris. The Assyrian Kingdom lasted from about BC 2300 to BC 606.

Avogadro number number of elementary entities (i.e. molecules) in one mole of a substance: $6.022\,1367 \times 10^{23}$ mole^{-1}. Named after the Italian physicist Amedeo Avogadro.

Backwater in a tranquil flow motion (i.e. subcritical flow) the longitudinal flow profile is controlled by the downstream flow conditions: e.g. an obstacle, a structure, a change of cross-section. Any downstream control structure (e.g. bridge piers, weir) induces a backwater effect. More generally, the terms *backwater calculations* or *backwater profile* refer to the calculation of the longitudinal flow profile. The term is commonly used for both supercritical and subcritical flow motion.

Backwater calculation calculation of the free-surface profile in open channels. The first successful calculations were developed by the Frenchman J.B. Bélanger who used a finite difference step method for integrating the equations (Bélanger 1828).

Bagnold Ralph Alger Bagnold (1896–1990) was a British geologist and a leading expert on the physics of sediment transport by wind and water. During World War II, he founded the Long Range Desert Group and organized long-distance raids behind enemy lines across the Libyan desert.

Bakhmeteff Boris Alexandrovitch Bakhmeteff (1880–1951) was a Russian hydraulician. In 1912, he developed the concept of specific energy and an energy diagram for open channel flows.

Barrage French word for dam or weir, commonly used to describe a large dam structure in English.

Barré de Saint-Venant Adhémar Jean Claude Barré de Saint-Venant (1797–1886), French engineer of the 'Corps des Ponts-et-Chaussées', developed the equation of motion of a fluid particle in terms of the shear and normal forces exerted on it.

Barrel for a culvert, central section where the cross-section is minimum. Another term is the throat.

Bazin Henri Emile Bazin was a French hydraulician (1829–1917) and engineer, member of the French 'Corps des Ponts-et-Chaussées' and later of the Académie des Sciences de Paris. He worked as an assistant of Henri P.G. Darcy at the beginning of his career.

Bed form channel bed irregularity that is related to the flow conditions. Characteristic bed forms include ripples, dunes and antidunes.

Bed load sediment material transported by rolling, sliding and saltation motion along the bed.

Bélanger Jean-Baptiste Ch. Bélanger (1789–1874) was a French hydraulician and professor at the Ecole Nationale Supérieure des Ponts et Chaussées (Paris). He suggested first the application of the momentum principle to hydraulic jump flow (Bélanger 1828). In the same book, he presented the first 'backwater' calculation for open channel flow.

Bélanger equation momentum equation applied across a hydraulic jump in a horizontal channel (named after J.B.C Bélanger).

Belidor Bertrand Forêt de Belidor (1693–1761) was a teacher at the Ecole Nationale des Ponts et Chaussées. His treatise 'Architecture Hydraulique' (Belidor 1737–1753) was a well-known hydraulic textbook in Europe during the 18th and 19th centuries.

Bernoulli Daniel Bernoulli (1700–1782) was a Swiss mathematician, physicist and botanist who developed the Bernoulli equation in his 'Hydrodynamica, de viribus et motibus fluidorum' textbook (first draft in 1733, first publication in 1738, Strasbourg).

Bessel Friedrich Wilhelm Bessel (1784–1846) was a German astronomer and mathematician. In 1810 he computed the orbit of Halley's comet. As a mathematician he introduced the Bessel functions (or circular functions) which have found wide use in physics, engineering and mathematical astronomy.

Bidone Giorgio Bidone (1781–1839) was an Italian hydraulician. His experimental investigations on the hydraulic jump were published between 1820 and 1826.

Biesel Francis Biesel (1920–1993) was a French hydraulic engineer and a pioneer of computational hydraulics.

Blasius H. Blasius (1883–1970) was a German scientist, student and collaborator of L. Prandtl.

Boltzmann Ludwig Eduard Boltzmann (1844–1906) was an Austrian physicist.

Boltzmann constant ratio of the universal gas constant ($8.3143 \, \text{K J}^{-1}\text{mole}^{-1}$) to the Avogadro number ($6.022\,136\,7 \times 10^{23} \, \text{mole}^{-1}$). It equals: $1.380\,662 \times 10^{-23}$ J/K.

Borda Jean-Charles de Borda (1733–1799) was a French mathematician and military engineer. He achieved the rank of Capitaine de Vaisseau and participated in the US War of Independence with the French Navy. He investigated the flow through orifices and developed the Borda mouthpiece.

Borda mouthpiece a horizontal re-entrant tube in the side of a tank with a length such that the issuing jet is not affected by the presence of the walls.

Bore a surge of tidal origin is usually termed a bore (e.g. the Mascaret in the Seine river, France).

Bossut Abbé Charles Bossut (1730–1804) was a French ecclesiastic and experimental hydraulician, author of a hydrodynamic treatise (Bossut 1772).

Bottom outlet opening near the bottom of a dam for draining the reservoir and eventually flushing out reservoir sediments.

Boundary layer flow region next to a solid boundary where the flow field is affected by the presence of the boundary and where friction plays an essential part. A boundary layer flow is characterized by a range of velocities across the boundary layer region from zero at the boundary to the free-stream velocity at the outer edge of the boundary layer.

Boussinesq Joseph Valentin Boussinesq (1842–1929) was a French hydrodynamicist and Professor at the Sorbonne University (Paris). His treatise 'Essai sur la théorie des eaux courantes' (Boussinesq 1877) remains an outstanding contribution in hydraulics literature.

Boussinesq coefficient momentum correction coefficient named after J.V. Boussinesq who first proposed it (Boussinesq 1877).

Boussinesq–Favre wave an undular surge (see *Undular surge*).

Boys P.F.D. du Boys (1847–1924) was a French hydraulic engineer. He made a major contribution to the understanding of sediment transport and bed-load transport (Boys 1879).

Braccio ancient measure of length (from the Italian 'braccia'). One braccio equals 0.6096 m (or 2 ft).

Braided channel stream characterized by random interconnected channels separated by islands or bars. By comparison with islands, bars are often submerged at large flows.

Bresse Jacques Antoine Charles Bresse (1822–1883) was a French applied mathematician and hydraulician. He was Professor at the Ecole Nationale Supérieure des Ponts et Chaussées, Paris as successor of J.B.C Bélanger. His contribution to gradually-varied flows in open channel hydraulics is considerable (Bresse 1860).

Broad-crested weir a weir with a flat long crest is called a broad-crested weir when the crest length over the upstream head is greater than 1.5 to 3. If the crest is long enough, the pressure distribution along the crest is hydrostatic, the flow depth equals the critical flow depth $d_c = (q^2/g)^{1/3}$ and the weir can be used as a critical depth meter.

Buat Comte Pierre Louis George du Buat (1734–1809) was a French military engineer and hydraulician. He was a friend of Abbé C. Bossut. Du Buat is considered as the pioneer of experimental hydraulics. His textbook (Buat 1779) was a major contribution to flow resistance in pipes, open channel hydraulics and sediment transport.

Buoyancy tendency of a body to float, to rise or to drop when submerged in a fluid at rest. The physical law of buoyancy (or Archimedes' principle) was discovered by the Greek mathematician Archimedes. It states that any body submerged in a fluid at rest is subjected to a vertical (or buoyant) force. The magnitude of the buoyant force is equal to the weight of the fluid displaced by the body.

Buttress dam a special type of dam in which the water face consists of a series of slabs or arches supported on their air faces by a series of buttresses.

Byewash ancient name for a spillway: i.e. channel to carry waste waters.

Candela SI unit for luminous intensity, defined as the intensity in a given direction of a source emitting a monochromatic radiation of frequency 540×10^{12} Hz and which has a radiant intensity in that direction of 1/683 W per unit solid angle.

Carnot Lazare N.M. Carnot (1753–1823) was a French military engineer, mathematician, general and statesman who played a key role during the French Revolution.

Carnot Sadi Carnot (1796–1832), eldest son of Lazare Carnot, was a French scientist who worked on steam engines and described the Carnot cycle relating to the theory of heat engines.

Cartesian coordinate one of three coordinates that locate a point in space and measure its distance from one of three intersecting coordinate planes measured parallel to one of the three straight-line axes that are the intersections of the other two planes. It is named after the French mathematician René Descartes.

Cascade (1) a steep stream intermediate between rapids and a water fall. The slope is steep enough to allow a succession of small drops but not sufficient to cause the water to drop vertically (i.e. waterfall). (2) A man-made channel consisting of a series of steps: e.g. a stepped fountain, a staircase chute, a stepped sewer.

Cataract a series of rapids or waterfalls. It is usually coined for large rivers: e.g. the six cataracts of the Nile river between Karthum and Aswan.

Catena d'Acqua (Italian term for 'chain of water') variation of the cascade developed during the Italian Renaissance. Water is channelled down the centre of an architectural ramp contained on both sides by stone carved into a scroll pattern to give a chain-like appearance. Waters flow as a supercritical regime with regularly-spaced increases and decreases of channel width, giving a sense of continuous motion highlighted by shock wave patterns at the free-surface. One of the best examples is at Villa Lante, Italy. The stonework was carved into crayfish, the emblem of the owner, Cardinal Gambara.

Cauchy Augustin Louis de Cauchy (1789–1857) was a French engineer from the 'Corps des Ponts-et-Chaussées'. He devoted himself later to mathematics and he taught at Ecole Polytechnique, Paris, and at the Collège de France. He worked with Pierre-Simon Laplace and J. Louis Lagrange. In fluid mechanics, he contributed greatly to the analysis of wave motion.

Cavitation formation of vapour bubbles and vapour pockets within a homogeneous liquid caused by excessive stress (Franc *et al.* 1995). Cavitation may occur in low-pressure regions where the liquid has been accelerated (e.g. turbines, marine propellers, baffle blocks of dissipation basin). Cavitation modifies the hydraulic characteristics of a system, and it is characterized by damaging erosion, additional noise, vibrations and energy dissipation.

Celsius Anders Celsius (1701–1744) was a Swedish astronomer who invented the Celsius thermometer scale (or centigrade scale) in which the interval between the freezing and boiling points of water is divided into 100 degrees.

Celsius degree (or *degree centigrade*) temperature scale based on the freezing and boiling points of water: $0°$ and $100°$ Celsius respectively.

Chadar type of narrow sloping chute peculiar to Islamic gardens and perfected by the Mughal gardens in Northern India (e.g. at Nishat Bagh). These stone channels were used to carry water from one terrace garden down to another. A steep slope ($\theta \sim 20°$ to $35°$) enables sunlight to be reflected to the maximum degree. The chute bottom is very rough to enhance turbulence and free-surface aeration. The discharge per unit width is usually small, resulting in thin sheets of aerated waters.

Chézy Antoine Chézy (1717–1798) (or Antoine de Chézy) was a French engineer and member of the French 'Corps des Ponts-et-Chaussées'. He designed canals for the water supply of the city of Paris. In 1768 he proposed a resistance formula for open channel flows called the Chézy equation. In 1798, he became Director of the Ecole Nationale Supérieure des Ponts et Chaussées after teaching there for many years.

Chézy coefficient resistance coefficient for open channel flows first introduced by the Frenchman A. Chézy. Although it was thought to be a constant, the coefficient is a function of the relative roughness and Reynolds number.

Chimu Indian of a Yuncan tribe dwelling near Trujillo on the north-west coast of Peru. The Chimu empire lasted from AD 1250 to 1466. It was overrun by the Incas in 1466.

Choke in open channel flow, a channel contraction might obstruct the flow and induce the appearance of critical flow conditions (i.e. control section). Such a constriction is sometimes called a 'choke'.

Choking flow critical flow in a channel contraction. The term is used for both open channel flow and compressible flow.

Chord length (1) The chord or chord length of an airfoil is the straight line distance joining the leading and trailing edges of the foil. (2) The chord length of a bubble (or bubble chord length) is the length of the straight line connecting the two intersections of the air-bubble free-surface with the leading tip of the measurement probe (e.g. conductivity probe, conical hot-film probe) as the bubble is transfixed by the probe tip.

Clausius Rudolf Julius Emanuel Clausius (1822–1888) was a German physicist and thermodynamicist. In 1850 he formulated the second law of thermodynamics.

Clay earthy material that is plastic when moist and that becomes hard when baked or fired. It is composed mainly of fine particles of a group of hydrous alumino-silicate minerals (particle sizes usually less than 0.05 mm).

Clepsydra Greek name for water clock.

Cofferdam temporary structure enclosing all or part of the construction area so that construction can proceed in dry conditions. A diversion cofferdam diverts a stream into a pipe or channel.

Cohesive sediment sediment material of very small sizes (i.e. less than 50 μm) for which cohesive bonds between particles (e.g. intermolecular forces) are significant and affect the material properties.

Colbert Jean-Baptiste Colbert (1619–1683) was a French statesman. Under King Louis XIV, he was the Minister of Finances, the Minister of 'Bâtiments et Manufactures' (buildings and industries) and the Minister of the Marine.

Conjugate depth in open channel flow, another name for sequent depth.

Control considering an open channel, subcritical flows are controlled by the downstream conditions. This is called a 'downstream flow control'. Conversely, supercritical flows are controlled only by the upstream flow conditions (i.e. 'upstream flow control').

Control section in an open channel, the cross-section where critical flow conditions take place. The concept of 'control' and 'control section' are used with the same meaning.

Control surface is the boundary of a control volume.

Control volume refers to a region in space and is used in the analysis of situations where flow occurs into and out of the space.

Coriolis Gustave Gaspard Coriolis (1792–1843) was a French mathematician and engineer of the 'Corps des Ponts-et-Chaussées' who first described the Coriolis force (i.e. the effect of motion on a rotating body).

Coriolis coefficient kinetic energy correction coefficient named after G.G. Coriolis who first introduced the correction coefficient (Coriolis 1836).

Couette M. Couette was a French scientist who measured experimentally the viscosity of fluids with a rotating viscosimeter (Couette 1890).

Couette flow flow between parallel boundaries moving at different velocities, named after the Frenchman M. Couette. The most common Couette flows are the cylindrical Couette flow used to measure dynamic viscosity and the two-dimensional Couette flow between parallel plates.

Craya Antoine Craya was a French hydraulician and professor at the University of Grenoble.

Creager profile spillway shape developed from a mathematical extension of the original data of Bazin in 1886–1888 (Creager 1917).

Crest of spillway upper part of a spillway. The term 'crest of dam' refers to the upper part of an uncontrolled overflow.

Crib (1) framework of bars or spars for strengthening; (2) frame of logs or beams to be filled with stones, rubble or filling material and sunk as a foundation or retaining wall.

Crib dam gravity dam built up of boxes, cribs, crossed timbers or gabions, and filled with earth or rock.

Critical depth is the flow depth for which the mean specific energy is minimum.

Critical flow conditions in open channel flows, the flow conditions such that the specific energy (of the mean flow) is minimum are called the critical flow conditions. With commonly-used Froude number definitions, the critical flow conditions occur for $Fr = 1$. If the flow is critical, small changes in specific energy cause large changes in flow depth. In practice, critical flow over a long reach of a channel is unstable.

Culvert covered channel of relatively short length installed to drain water through an embankment (e.g. highway, railroad, dam).

Cyclopean dam gravity masonry dam made of very large stones embedded in concrete.

Danel Pierre Danel (1902–1966) was a French hydraulician and engineer. One of the pioneers of modern hydrodynamics, he worked from 1928 to his death for Neyrpic, known prior to 1948 as 'Ateliers Neyret-Beylier-Piccard et Pictet'.

Darcy Henri Philibert Gaspard Darcy (1805–1858) was a French civil engineer. He performed numerous experiments of flow resistance in pipes (Darcy 1858) and in open channels (Darcy and Bazin 1865), and of seepage flow in porous media (Darcy 1856). He gave his name to the Darcy–Weisbach friction factor and to the Darcy law in porous media.

Darcy law law of groundwater flow motion which states that the seepage flow rate is proportional to the ratio of the head loss over the length of the flow path. It was discovered by H.P.G. Darcy (1856) who showed that, for a flow of liquid through a porous medium, the flow rate is directly proportional to the pressure difference.

Darcy–Weisbach friction factor dimensionless parameter characterizing the friction loss in a flow. It is named after the Frenchman H.P.G. Darcy and the German J. Weisbach.

Debris debris comprises mainly large boulders, rock fragments, gravel-sized to clay-sized material, tree and wood material that accumulates in creeks.

Degradation lowering of channel bed elevation resulting from the erosion of sediments.

Descartes René Descartes (1596–1650) was a French mathematician, scientist and philosopher. He is recognized as the father of modern philosophy. He stated: 'cogito ergo sum' ('I think therefore I am').

Diffusion the process whereby particles of liquids, gases or solids intermingle as the result of their spontaneous movement caused by thermal agitation and, in dissolved substances, move from a region of higher concentration to one of lower concentration. The term turbulent diffusion is used to describe the spreading of particles caused by turbulent agitation.

Diffusion coefficient quantity of a substance that, in diffusing from one region to another, passes through each unit of cross-section per unit of time when the volume concentration is unity. The units of the diffusion coefficient are m^2/s.

Diffusivity another name for the diffusion coefficient.

Dimensional analysis organization technique used to reduce the complexity of a study, by expressing the relevant parameters in terms of numerical magnitude and associated units, and grouping them into dimensionless numbers. The use of dimensionless numbers increases the generality of the results.

Diversion channel waterway used to divert water from its natural course.

Diversion dam dam or weir built across a river to divert water into a canal. It raises the upstream water level of the river but does not provide any significant storage volume.

Drag reduction reduction of the skin friction resistance in fluids in motion.

Drainage layer layer of pervious material to relieve pore pressures and/or to facilitate drainage: e.g. drainage layer in an earthfill dam.

Drop (1) volume of liquid surrounded by gas in a free-fall motion (i.e. dropping); (2) by extension, small volume of liquid in motion in a gas; (3) a rapid change of bed elevation, also called a step.

Droplet small drop of liquid.

Drop structure single step structure characterized by a sudden decrease in bed elevation.

Du Boys (or *Duboys*) see *P.F.D. du Boys*.

Du Buat (or *Dubuat*) see *P.L.G. du Buat*.

Dupuit Arsène Jules Etienne Juvénal Dupuit (1804–1866) was a French engineer and economist. His expertise included road construction, economics, statics and hydraulics.

Earth dam massive earthen embankment with sloping faces and made watertight.

Ecole Nationale Supérieure des Ponts et Chaussées, Paris French civil engineering school founded in 1747. The direct translation is: 'National School of Bridge and Road Engineering'. Among the directors there were the famous hydraulicians A. Chézy and G. de Prony. Other famous professors included B.F. de Belidor, J.B.C Bélanger, J.A.C. Bresse, G.G. Coriolis and L.M.H. Navier.

Eddy viscosity another name for the momentum exchange coefficient. It is also called the 'eddy coefficient' by Schlichting (1979) (see *Momentum exchange coefficient*).

Embankment fill material (e.g. earth, rock) placed with sloping sides and with a length greater than its height.

Escalier d'Eau see *Water staircase*.

Euler Leonhard Euler (1707–1783) was a Swiss mathematician and physicist, and a close friend of Daniel Bernoulli.

Extrados upper side of a wing or exterior curve of a foil. The pressure distribution on the extrados must be smaller than that on the intrados to provide a positive lift force.

Face external surface which limits a structure: e.g. air face of a dam (i.e. downstream face), water face (i.e. upstream face) of a weir.

Favre H. Favre (1901–1966) was a Swiss professor at ETH-Zürich. He investigated, both experimentally and analytically, positive and negative surges. Undular surges are sometimes called Boussinesq–Favre waves. Later he worked on the theory of elasticity.

Fawer jump undular hydraulic jump.

Fick Adolf Eugen Fick was a 19th century German physiologist who developed the diffusion equation for neutral particle (Fick 1855).

Fixed-bed channel the bed and sidewalls are non-erodible. Neither erosion nor accretion occurs.

Flashboard a board or a series of boards placed on or at the side of a dam to increase the depth of water. Flashboards are usually lengths of timber, concrete or steel placed on the crest of a spillway to raise the upstream water level.

Flash flood flood of short duration with a relatively high peak flow rate.

Flashy term applied to rivers and streams whose discharge can rise and fall suddenly, and is often unpredictable.

Flettner Anton Flettner (1885–1961) was a German engineer and inventor. In 1924 he designed a rotor ship based on the Magnus effect. Large vertical circular cylinders were mounted on the ship. They were mechanically rotated to provide circulation and to propel the ship. More recently a similar system was developed for the ship 'Alcyone' of Jacques-Yves Cousteau.

Flip bucket a flip bucket, or ski-jump, is a concave curve at the downstream end of a spillway, to deflect the flow into an upward direction. Its purpose is to throw the water clear of the hydraulic structure and to induce the disintegration of the jet in air.

Fog small water droplets near ground level forming a cloud sufficiently dense to reduce visibility drastically. The term fog also refers to clouds of smoke particles or ice particles.

Forchheimer Philipp Forchheimer (1852–1933) was an Austrian hydraulician who contributed significantly to the study of groundwater hydrology.

Fortier André Fortier was a French scientist and engineer. He later became Professor at the Sorbonne, Paris.

Fourier Jean Baptiste Joseph Fourier (1768–1830) was a French mathematician and physicist known for his development of the Fourier series. In 1794 he was offered a professorship of mathematics at the Ecole Normale in Paris and was later appointed at the Ecole Polytechnique. In 1798 he joined the expedition to Egypt led by (then) General Napoléon Bonaparte. His research in mathematical physics culminated with the classical study 'Théorie Analytique de la Chaleur' (Fourier 1822) in which he enunciated his theory of heat conduction.

Free-surface interface between a liquid and a gas. More generally a free-surface is the interface between the fluid (at rest or in motion) and the atmosphere. In two-phase gas–liquid flow, the term 'free-surface' also includes the air–water interface of gas bubbles and liquid drops.

Free-surface aeration Natural aeration occurring at the free surface of high velocity flows is referred to as free-surface aeration or self-aeration.

French revolution (Révolution Française) revolutionary period that shook France between 1787 and 1799. It reached a turning point in 1789 and led to the destitution of the monarchy in 1791. The constitution of the First Republic was drafted in 1790 and adopted in 1791.

Frontinus Sextus Julius Frontinus (AD 35–103 or 104) was a Roman engineer and soldier. After AD 97, he was 'curator aquarum' in charge of the water supply system of Rome. He dealt with discharge measurements in pipes and canals. In his analysis he correctly related the proportionality between discharge and cross-section area. His book 'De Aquaeductu Urbis Romae' ('Concerning the Aqueducts of the City of Rome') described the operation and maintenance of Rome water supply system.

Froude William Froude (1810–1879) was an English naval architect and hydrodynamicist who invented the dynamometer and used it for the testing of model ships in towing tanks. He was assisted by his son Robert Edmund Froude who, after the death of his father, continued some of his work. In 1868, he used Reech's law of similarity to study the resistance of model ships.

Froude number   the Froude number is proportional to the square root of the ratio of the inertial forces over the weight of fluid. The Froude number is generally used for scaling free surface flows, open channels and hydraulic structures. Although the dimensionless number was named after William Froude, several French researchers used it before. Dupuit (1848) and Bresse (1860) highlighted the significance of the number to differentiate the open channel flow regimes. Bazin (1865a) confirmed the findings experimentally. Ferdinand Reech introduced the dimensionless number for testing ships and propellers in 1852. The number is called the Reech–Froude number in France.

Gabion   a gabion consists of rockfill material encased by a basket or a mesh. The word 'gabion' originates from the Italian 'gabbia' cage.

Gabion dam   crib dam built up of gabions.

Gas transfer   process by which gas is transferred into or out of solution: i.e. dissolution or desorption respectively.

Gate   valve or system for controlling the passage of a fluid. In open channels the two most common types of gates are the underflow gate and the overflow gate.

Gauckler   Philippe Gaspard Gauckler (1826–1905) was a French engineer and member of the French 'Corps des Ponts-et-Chaussées'. He re-analysed the experimental data of Darcy and Bazin (1865) and, in 1867, he presented a flow resistance formula for open channel flows (Gauckler–Manning formula) sometimes called improperly the Manning equation (Gauckler 1867). Later he became Directeur des Antiquités et des Beaux-Arts (Director of Anquities and Fine Arts) for the French Republic in Tunisia and he directed an extensive survey of Roman hydraulic works in Tunisia.

Gay-Lussac   Joseph-Louis Gay-Lussac (1778–1850) was a French chemist and physicist.

Ghaznavid (or *Ghaznevid*)   one of the Moslem dynasties (10th to 12th centuries) ruling south-western Asia. Its capital city was at Ghazni (Afghanistan).

G.K. formula   empirical resistance formula developed by the Swiss engineers E. Ganguillet and W.R. Kutter in 1869.

Gradually varied flow   is characterized by relatively small changes in velocity and pressure distributions over a short distance (e.g. long waterway).

Gravity dam   dam that relies on its weight for stability. Normally the term 'gravity dam' refers to a masonry or concrete dam.

Grille d'eau   (French for 'water screen') a series of water jets or fountains aligned to form a screen. An impressive example is 'les Grilles d'Eau' designed by A. Le Nôtre at Vaux-le Vicomte, France.

Hasmonean   the family or dynasty of the Maccabees, in Israel. The Hasmonean Kingdom was created following the uprising of the Jews in BC 166.

Helmholtz   Hermann Ludwig Ferdinand von Helmholtz (1821–1894) was a German scientist who made basic contributions to physiology, optics, electrodynamics and meteorology.

Hennin   Georg Wilhelm Hennin (1680–1750) was a young Dutchman hired by the tsar Peter the Great to design and build several dams in Russia (Danilveskii 1940, Schnitter 1994). He went to Russia in 1698 and stayed until his death in April 1750.

Hero of Alexandria   Greek mathematician (1st century AD) working in Alexandria, Egypt. He wrote at least 13 books on mathematics, mechanics and physics. He designed and experimented the first steam engine. His treatise 'Pneumatica' described Hero's fountain, siphons, steam-powered engines, a water organ, and hydraulic and mechanical water devices. It influenced directly the waterworks design during the Italian Renaissance. In his book 'Dioptra', Hero rightly stated the concept of continuity for incompressible flow: the discharge being equal to the area of the cross-section of the flow times the speed of the flow.

Himyarite   important Arab tribe of antiquity dwelling in southern Arabia (BC 700 to AD 550).

Hohokams Native Americans in south-west America (Arizona), they built several canal systems in the Salt River valley during the period BC 350 to AD 650. They migrated to northern Mexico around AD 900 where they built other irrigation systems.

Hokusai Katsushita Japanese painter and wood engraver (1760–1849). His 'Thirty-Six Views of Mount Fuji' (1826–1833) are known throughout the world..

Huang Chun-Pi one of the greatest masters of Chinese painting in modern China (1898–1991). Several of his paintings included mountain rivers and waterfalls: e.g. 'Red trees and waterfalls', 'The house by the water-falls', 'Listening to the sound of flowing waters', 'Water-falls'.

Hydraulic diameter is defined as the equivalent pipe diameter: i.e. four times the cross-sectional area divided by the wetted perimeter. The concept was first expressed by the Frenchman P.L.G. du Buat (Buat 1779).

Hydraulic fill dam embankment dam constructed of materials that are conveyed and placed by suspension in flowing water.

Hydraulic jump transition from a rapid (supercritical flow) to a slow flow motion (subcritical flow). Although the hydraulic jump was described by Leonardo da Vinci, the first experimental investigations were published by Giorgio Bidone in 1820. The present theory of the jump was developed by Bélanger (1828) and it has been verified experimentally by numerous researchers (e.g. Bakhmeteff and Matzke 1936).

Hyperconcentrated flow sediment-laden flow with large suspended sediment concentrations (i.e. typically more than 1% in volume). Spectacular hyperconcentrated flows are observed in the Yellow River basin (China) with volumetric concentrations larger than 8%.

Ideal fluid frictionless and incompressible fluid. An ideal fluid has zero viscosity: i.e. it cannot sustain shear stress at any point.

Idle discharge old expression for spill or waste water flow.

Inca South-American Indian of the Quechuan tribes of the highlands of Peru. The Inca civilization dominated Peru between AD 1200 and 1532. The domination of the Incas was terminated by the Spanish conquest.

Inflow (1) upstream flow; (2) incoming flow.

Inlet (1) upstream opening of a culvert, pipe or channel; (2) a tidal inlet is a narrow water passage between peninsulas or islands.

Intake any structure in a reservoir through which water can be drawn into a waterway or pipe. By extension, upstream end of a channel.

Interface surface forming a common boundary of two phases (e.g. gas–liquid interface) or two fluids.

International system of units see *Système international d'unités*.

Intrados lower side of a wing or interior curve of a foil.

Invert (1) lowest portion of the internal cross-section of a conduit; (2) channel bed of a spillway; (3) bottom of a culvert barrel.

Inviscid flow is a non-viscous flow.

Irrotational flow is defined as a zero vorticity flow. Fluid particles within a region have no rotation. If a frictionless fluid has no rotation at rest, any later motion of the fluid will be irrotational. In irrotational flow each element of the moving fluid undergoes no net rotation, with respect to chosen coordinate axes, from one instant to another.

Jet d'eau French expression for water jet. The term is commonly used in architecture and landscaping.

Jevons W.S. Jevons (1835–1882) was an English chemist and economist. His work on salt finger intrusions (Jevons 1858) was a significant contribution to the understanding of double-diffusive convection. He performed his experiments in Sydney, Australia, 23 years prior to Rayleigh's experiments (Rayleigh 1883).

JHRC Jump Height Rating Curve.

JHRL Jump Height Rating Level.

Karman Theodore von Karman (or von Kármán) (1881–1963) was a Hungarian fluid dynamicist and aerodynamicist who worked in Germany (1906 to 1929) and later in the USA. He was a student of Ludwig Prandtl in Germany. He gave his name to the vortex shedding behind a cylinder (Karman vortex street).

Karman constant (or von Karman constant) 'universal' constant of proportionality between the Prandtl mixing length and the distance from the boundary. Experimental results indicate that K = 0.40.

Kelvin (Lord) William Thomson (1824–1907), Baron Kelvin of Largs, was a British physicist. He contributed to the development of the second law of thermodynamics, the absolute temperature scale (measured in Kelvin), the dynamical theory of heat, fundamental work in hydrodynamics, etc.

Kelvin–Helmholtz instability instability at the interface of two ideal fluids in relative motion. The instability can be caused by a destabilizing pressure gradient of the fluid (e.g. clean-air turbulence) or free-surface shear (e.g. fluttering fountain). It is named after H.L.F. Helmholtz who solved first the problem (Helmholtz 1868) and Lord Kelvin (Kelvin 1871).

Keulegan Garbis Hovannes Keulegan (1890–1989) was an Armenian mathematician who worked as a hydraulician for the US Bureau of Standards since its creation in 1932.

Lagrange Joseph-Louis Lagrange (1736–1813) was a French mathematician and astronomer. During the 1789 Revolution, he worked on the committee to reform the metric system. He was Professor of mathematics at the École Polytechnique from the start in 1795.

Laminar flow is characterized by fluid particles moving along smooth paths in laminas or layers, with one layer gliding smoothly over an adjacent layer. Laminar flows are governed by Newton's law of viscosity which relates the shear stress to the rate of angular deformation: $\tau = \mu \times \partial V / \partial y$.

Laplace Pierre-Simon Laplace (1749–1827) was a French mathematician, astronomer and physicist. He is best known for his investigations into the stability of the solar system.

LDA velocimeter Laser Doppler Anemometer system.

Left abutment abutment on the left-hand side of an observer when looking downstream.

Left bank (left wall) looking downstream, the left bank or the left channel wall is on the left.

Leonardo da Vinci Italian artist (painter and sculptor) who extended his interest to medicine, science, engineering and architecture (AD 1452–1519).

Lining coating on a channel bed to provide water tightness, to prevent erosion or to reduce friction.

Lumber timber sawed or split into boards, planks or staves.

McKay Professor Gordon M. McKay (1913–1989) was Professor in Civil Engineering at the University of Queensland.

Mach Ernst Mach (1838–1916) was an Austrian physicist and philosopher. He established important principles of optics, mechanics and wave dynamics.

Mach number see *Sarrau–Mach number*.

Magnus H.G. Magnus (1802–1870) was a German physicist who investigated the so-called Magnus effect in 1852.

Magnus effect a rotating cylinder, placed in a flow, is subjected to a force acting in the direction normal to the flow direction: i.e. a lift force which is proportional to the flow velocity times the rotation speed of the cylinder. This effect, called the Magnus effect, has a wide range of applications (Swanson 1961).

Manning Robert Manning (1816–1897) was Chief Engineer of the Office of Public Works, Ireland. In 1889, he presented two formulae (Manning 1890). One was to become the so-called 'Gauckler–Manning formula', but Robert Manning did prefer to use the second formula that he gave in his paper. It must be noted that the Gauckler–Manning formula was proposed first by the Frenchman P.G. Gauckler (Gauckler 1867).

Mariotte Abbé Edme Mariotte (1620–1684) was a French physicist and plant physiologist. He was member of the Académie des Sciences de Paris and wrote a fluid mechanics treatise that was published after his death (Mariotte 1686).

Masonry dam dam constructed mainly of stone, brick or concrete blocks jointed with mortar.

Meandering channel alluvial stream characterized by a series of alternating bends (i.e. meanders) as a result of alluvial processes.

MEL culvert see *Minimum Energy Loss culvert*.

Metric system see Système métrique.

Minimum energy loss (MEL) culvert culvert designed with very smooth shapes to minimize energy losses. The design of a minimum energy loss culvert is associated with the concept of constant total head. The inlet and outlet must be streamlined in such a way that significant form losses are avoided (Apelt 1983).

Mixing length the mixing length theory is a turbulence theory developed by L. Prandtl, first formulated in 1925 (Prandtl 1925). Prandtl assumed that the mixing length is the characteristic distance travelled by a particle of fluid before its momentum is changed by the new environment.

Mochica (1) South American civilization (AD 200–1000) living in the Moche river valley, Peru along the Pacific coastline; (2) language of the Yuncas.

Mole mass numerically equal in grams to the relative mass of a substance (i.e. 12 g for Carbon 12). The number of molecules in one mole of gas is $6.022\,1367 \times 10^{23}$ (i.e. Avogadro number).

Momentum exchange coefficient in turbulent flows the apparent kinematic viscosity (or kinematic eddy viscosity) is analogous to the kinematic viscosity in laminar flows. It is called the momentum exchange coefficient, the eddy viscosity or the eddy coefficient. The momentum exchange coefficient is proportional to the shear stress divided by the strain rate. It was first introduced by the Frenchman J.V. Boussinesq (1877, 1896).

Moor (1) native of Mauritania, a region corresponding to parts of Morocco and Algeria; (2) Moslem of native North African races.

Morning-Glory spillway vertical discharge shaft, more particularly the circular hole form of a drop inlet spillway. The shape of the intake is similar to a Morning-Glory flower (American native plant, Ipomocea). It is sometimes called a Tulip intake.

Mud slimy and sticky mixture of solid material and water.

Mughal (or *Mughul* or *Mogul* or *Moghul*) name or adjective referring to the Mongol conquerors of India and to their descendants. The Mongols occupied India from 1526 up to the 18th century although the authority of the Mughal emperor became purely nominal after 1707. The fourth emperor, Jahangir (1569–1627), married a Persian princess Mehr-on Nesa who became known as Nur Jahan. His son Shah Jahan (1592–1666) built the famous Taj Mahal between 1631 and 1654 in memory of his favourite wife Arjumand Banu better known by her title: Mumtaz Mahal or Taj Mahal.

Nabataean habitant from an ancient kingdom to the east and south-east of Palestine that included the Neguev desert. The Nabataean kingdom lasted from around BC 312 to AD 106. The Nabataeans built a large number of soil-and-retention dams. Some are still in use today, as shown by Schnitter (1994).

Nappe flow flow regime on a stepped chute where the water bounces from one step to the next one as a succession of free-fall jets.

Navier Louis Marie Henri Navier (1785–1835) was a French engineer who primarily designed bridges but also extended Euler's equations of motion (Navier 1823).

Navier–Stokes equation momentum equation applied to a small control volume of incompressible fluid. It is usually written in vector notation. The equation was first derived by L. Navier in 1822 and S.D. Poisson in 1829 by a different method. It was derived later in a more modern manner by A.J.C. Barré de Saint-Venant in 1843 and G.G. Stokes in 1845.

Negative surge a negative surge results from a sudden change in flow that decreases the flow depth. It is a retreating wave front moving upstream or downstream.

Newton Sir Isaac Newton (1642–1727) was an English mathematician and physicist. His contributions in optics, mechanics and mathematics were fundamental.

Nikuradse J. Nikuradse was a German engineer who investigated experimentally the flow field in smooth and rough pipes (Nikuradse 1932,1933).

Non-uniform equilibrium flow the velocity vector varies from place to place at any instant: steady non-uniform flow (e.g. flow through an expanding tube at a constant rate) and unsteady non-uniform flow (e.g. flow through an expanding tube at an increasing flow rate).

Normal depth uniform equilibrium open channel flow depth.

Obvert roof of the barrel of a culvert. Another name is *soffit*.

One-dimensional flow neglects the variations and changes in velocity and pressure transverse to the main flow direction. An example of one-dimensional flow can be the flow through a pipe.

One-dimensional model model defined with one spatial coordinate, the variables being averaged in the other two directions.

Outflow downstream flow.

Outlet (1) downstream opening of a pipe, culvert or canal; (2) artificial or natural escape channel.

Pascal Blaise Pascal (1623–1662) was a French mathematician, physicist and philosopher. He developed the modern theory of probability. Between 1646 and 1648, he formulated the concept of pressure and showed that the pressure in a fluid is transmitted through the fluid in all directions. He also measured the air pressure both in Paris and on the top of a mountain overlooking Clermont-Ferrand (France).

Pascal unit of pressure named after the Frenchman, B. Pascal: one Pascal equals a Newton per square-metre.

Pelton turbine (or *wheel*) impulse turbine with one to six circular nozzles that deliver high-speed water jets into air, which then strike the rotor blades shaped like scoops and known as buckets. A simple bucket wheel was designed by Sturm in the 17th century. The American, Lester Allen Pelton, patented the actual double-scoop (or double-bucket) design in 1880.

Pervious zone part of the cross-section of an embankment comprising material of high permeability.

Pitot Henri Pitot (1695–1771) was a French mathematician, astronomer and hydraulician. He was a member of the French Académie des Sciences from 1724. He invented the Pitot tube to measure flow velocity in the Seine river (first presentation in 1732 at the Académie des Sciences de Paris).

Pitot tube device to measure flow velocity. The original Pitot tube consisted of two tubes, one with an opening facing the flow. L. Prandtl developed an improved design (e.g. Howe 1949) which provides the total head, piezometric head and velocity measurements. It is called a Prandtl–Pitot tube and more commonly a Pitot tube.

Pitting formation of small pits and holes on surfaces due to erosive or corrosive action (e.g. cavitation pitting).

Plato Greek philosopher (about BC 428–347) who greatly influenced Western philosophy.

Plunging jet liquid jet impacting (or impinging) into a receiving pool of liquid.

Poiseuille Jean-Louis Marie Poiseuille (1799–1869) was a French physician and physiologist who investigated the characteristics of blood flow. He carried out experiments and formulated first the expression of flow rates and friction losses in laminar fluid flow in circular pipes (Poiseuille 1839).

Poiseuille flow steady laminar flow in a circular tube of constant diameter.

Poisson Siméon Denis Poisson (1781–1840) was a French mathematician and scientist. He developed the theory of elasticity, a theory of electricity and a theory of magnetism.

Positive surge a positive surge results from a sudden change in flow that increases the depth. It is an abrupt wave front. The unsteady flow conditions may be solved as a quasi-steady flow situation.

Potential flow ideal-fluid flow with irrotational motion.

Prandtl Ludwig Prandtl (1875–1953) was a German physicist and aerodynamicist who introduced the concept of the boundary layer (Prandtl 1904) and developed the turbulent 'mixing length' theory. He was Professor at the University of Göttingen.

Prony Gaspard Clair François Marie Riche de Prony (1755–1839) was a French mathematician and engineer. He succeeded A. Chézy as director general of the Ecole Nationale Supérieure des Ponts et Chaussées, Paris during the French revolution.

Radial gate underflow gate for which the wetted surface has a cylindrical shape.

Rankine William J.M. Rankine (1820–1872) was a Scottish engineer and physicist. His contributions to thermodynamics and the steam-engine were important. In fluid mechanics, he developed the theory of sources and sinks, and used it to improve ship hull contours. One ideal-fluid flow pattern, the combination of uniform flow, source and sink, is named after him: i.e. flow past a Rankine body.

Rapidly varied flow is characterized by large changes over a short distance (e.g. sharp-crested weir, sluice gate, hydraulic jump).

Rayleigh John William Strutt, Baron Rayleigh (1842–1919) was an English scientist who made fundamental findings in acoustics and optics. His works are the basis of wave propagation theory in fluids. He received the Nobel Prize for Physics in 1904 for his work on the inert gas argon.

Reech Ferdinand Reech (1805–1880) was a French naval instructor who proposed first the Reech–Froude number in 1852 for the testing of model ships and propellers.

Rehbock Theodor Rehbock (1864–1950) was a German hydraulician and professor at the Technical University of Karlsruhe. His contribution to the design of hydraulic structures and physical modelling is important.

Renaissance period of great revival of art, literature and learning in Europe in the 14th, 15th and 16th centuries.

Reynolds Osborne Reynolds (1842–1912) was a British physicist and mathematician who expressed first the Reynolds number (Reynolds 1883) and later the Reynolds stress (i.e. turbulent shear stress).

Reynolds number dimensionless number proportional to the ratio of the inertial force over the viscous force.

Rheology science describing the deformation of fluid and matter.

Riblet series of longitudinal grooves. Riblets are used to reduce skin drag (e.g. on aircraft, ship hull). The presence of longitudinal grooves along a solid boundary modifies the bottom shear stress and the turbulent bursting process. Optimum groove width and depth are about 20 to 40 times the laminar sublayer thickness (i.e. about 10 to 20 μm in air, 1 to 2 mm in water).

Richelieu Armand Jean du Plessis (1585–1642), Duc de Richelieu and French Cardinal, was the Prime Minister of King Louis XIII of France from 1624 to his death.

Riemann Bernhard Georg Friedrich Riemann (1826–1866) was a German mathematician.

Right abutment abutment on the right-hand side of an observer when looking downstream.

Right bank (*right wall*) looking downstream, the right bank or the right channel wall is on the right.

Riquet Pierre Paul Riquet (1604–1680) was the designer and Chief Engineer of the Canal du Midi built between 1666 and 1681. The Canal provides an inland route between the Atlantic and the Mediterranean across southern France.

Rockfill material composed of large rocks or stones loosely placed.

Rockfill dam embankment dam in which more than 50% of the total volume comprises compacted or dumped pervious natural stones.

Roller in hydraulics, large-scale turbulent eddy: e.g. the roller of a hydraulic jump.

Roller Compacted Concrete (*RCC*) roller compacted concrete is defined as a no-slump consistency concrete that is placed in horizontal lifts and compacted by vibratory rollers. RCC has been commonly used as construction material of gravity dams since the 1970s.

Roll wave on steep slopes free-surface flows become unstable. The phenomenon is usually clearly visible at low flow rates. The waters flow down the chute in a series of wave fronts called roll waves.

Rouse Hunter Rouse (1906–1996) was an eminent hydraulician who was Professor and Director of the Iowa Institute of Hydraulic Research at the University of Iowa (USA).

Sabaen ancient name of the people of Yemen in Southern Arabia. Renowned for the visit of the Queen of Sabah (or Sheba) to the King of Israel around BC 950 and for the construction of the Marib dam (BC 115 to AD 575). The fame of the Marib dam was such that its final destruction in AD 575 was recorded in the Koran.

SAF St Anthony's Falls hydraulic laboratory at the University of Minnesota (USA).

Saltation (1) action of leaping or jumping; (2) in sediment transport, particle motion by jumping and bouncing along the bed.

Saint-Venant See Barré de Saint Venant.

Sarrau French Professor at Ecole Polytechnique, Paris, who first introduced the Sarrau–Mach number (Sarrau 1884).

Sarrau–Mach number dimensionless number proportional to the ratio of inertial forces over elastic forces. Although the number is commonly named after E. Mach who introduced it in 1887, it is often called the Sarrau number after Professor Sarrau who first highlighted the significance of the number (Sarrau 1884). The Sarrau–Mach number was once called the Cauchy number as a tribute to Cauchy's contribution to wave motion analysis.

Scalar a quantity that has a magnitude described by a real number and no direction. A scalar means a real number rather than a vector.

Scale effect discrepancy between model and prototype resulting when one or more dimensionless parameters have different values in the model and prototype.

Scour bed material removal caused by the eroding power of the flow.

Sediment any material carried in suspension by the flow or as bed load that would settle to the bottom in the absence of fluid motion.

Sediment load material transported by a fluid in motion.

Sediment transport transport of material by a fluid in motion.

Sediment transport capacity ability of a stream to carry a given volume of sediment material per unit time for given flow conditions. It is the sediment transport potential of the river.

Sediment yield total sediment outflow rate from a catchment, including bed-load and suspension.

Seepage interstitial movement of water that may take place through a dam, its foundation or abutments.

Sennacherib (or *Akkadian Sin-Akhkheeriba*) King of Assyria (BC 705–681), son of Sargon II (who ruled during BC 722–705). He built a huge water supply for his capital city Nineveh (near the actual Mossul, Iraq) in several stages. The latest stage comprised several dams and over 75 km of canals and paved channels.

Separation in a boundary layer, a deceleration of fluid particles leading to a reversed flow within the boundary layer is called a separation. The decelerated fluid particles are forced outwards and the boundary layer is separated from the wall. At the point of separation, the velocity gradient normal to the wall is zero:

$$\left(\frac{\partial V_x}{\partial y} \right)_{y=0} = 0$$

Separation point in a boundary layer, intersection of the solid boundary with the streamline dividing the separation zone and the deflected outer flow. The separation point is a stagnation point.

Sequent depth in open channel flow, the solution of the momentum equation at a transition between supercritical and subcritical flow gives two flow depths (upstream and downstream flow depths). They are called sequent depths.

Sewage refuse liquid or waste matter carried off by sewers. It may be a combination of water-carried wastes from residences and industries together with ground water, surface water and storm water.

Sewer an artificial subterranean conduit to carry off water and waste matter.

Shock waves with supercritical flows, a flow disturbance (e.g. change of direction, contraction) induces the development of shock waves propagating at the free-surface across the channel (e.g. Ippen and Harleman 1956, Hager 1992a). Shock waves are also called lateral shock waves, oblique hydraulic jumps, Mach waves, crosswaves, diagonal jumps.

Side-channel spillway a side-channel spillway consists of an open spillway (along the side of a channel) discharging into a channel running along the foot of the spillway and carrying the flow away in a direction parallel to the spillway crest (e.g. Arizona-side spillway of the Hoover Dam, USA).

Similitude correspondence between the behaviour of a model and that of its prototype, with or without geometric similarity. The correspondence is usually limited by scale effects.

Siphon pipe system discharging waters between two reservoirs or above a dam in which the water pressure becomes sub-atmospheric. The shape of a simple siphon is close to an omega (i.e. Ω-shape). Inverted-siphons carry waters between two reservoirs with a pressure larger than atmospheric. Their design follows approximately a U-shape. Inverted-siphons were commonly used by the Romans along their aqueducts to cross valleys.

Siphon-spillway device for discharging excess water in a pipe over the dam crest.

Skimming flow flow regime above a stepped chute for which the water flows as a coherent stream in a direction parallel to the pseudo-bottom formed by the edges of the steps. The same term is used to characterize the flow regime of large discharges above rockfill and closely-spaced large roughness elements.

Slope (1) side of a hill; (2) inclined face of a canal (e.g. trapezoidal channel); (3) inclination of the channel bottom from the horizontal.

Sluice gate underflow gate with a vertical sharp edge for stopping or regulating flow.

Soffit roof of the barrel of a culvert. Another name is *obvert*.

Specific energy quantity proportional to the energy per unit mass, measured with the channel bottom as the elevation datum, and expressed in metres of water. The concept of specific energy, first developed by B.A. Bakhmeteff in 1912, is commonly used in open channel flows.

Spillway opening built into a dam or the side of a reservoir to release (to spill) excess flood waters.

Splitter obstacle (e.g. concrete block, fin) installed on a chute to split the flow and to increase the energy dissipation.

Spray water droplets flying or falling through air: e.g. spray thrown up by a waterfall.

Stage-discharge curve relationship between discharge and free-surface elevation at a given location along a stream.

Stagnation point is defined as the point where the velocity is zero. When a streamline intersects itself, the intersection is a stagnation point. For irrotational flow a streamline intersects itself at right-angles at a stagnation point.

Staircase another adjective for 'stepped': e.g. a staircase cascade is a stepped cascade.

Stall aerodynamic phenomenon causing a disruption (i.e. separation) of the flow past a wing associated with a loss of lift.

Steady flow   occurs when conditions at any point of the fluid do not change with the time:

$$\frac{\partial V}{\partial t} = 0 \qquad \frac{\partial \rho}{\partial t} = 0 \qquad \frac{\partial P}{\partial t} = 0 \qquad \frac{\partial T}{\partial t} = 0$$

Stilling basin   structure for dissipating the energy of the flow downstream of a spillway, outlet work, chute or canal structure. In many cases, a hydraulic jump is used as the energy dissipator within the stilling basin.

Stokes   George Gabriel Stokes (1819–1903), British mathematician and physicist, known for his research in hydrodynamics and a study of elasticity.

Stop-logs   form of sluice gate comprising a series of wooden planks, one above the other, and held at each end.

Storm water   excess water running off the surface of a drainage area during and immediately following a period of rain. In urban areas, waters drained off a catchment area during or after a heavy rainfall are usually conveyed in man-made storm waterways.

Storm waterway   channel built for carrying storm waters.

Straub   L.G. Straub (1901–1963) was Professor and Director of the St Anthony Falls Hydraulics Laboratory at the University of Minnesota (USA).

Stream function   vector function of space and time which is related to the velocity field as: $V = -\text{curl}\,\varphi$. The stream function exists for steady and unsteady flow of incompressible fluid as it satisfies the continuity equation. The stream function was introduced by the French mathematician Lagrange.

Streamline   is the line drawn so that the velocity vector is always tangential to it (i.e. no flow across a streamline). When the streamlines converge the velocity increases. The concept of streamline was first introduced by the Frenchman J.C. de Borda.

Streamline maps   should be drawn so that the flow between any two adjacent streamlines is the same.

Stream tube   is a filament of fluid bounded by streamlines.

Subcritical flow   in open channel the flow is defined as subcritical if the flow depth is larger than the critical flow depth. In practice, subcritical flows are controlled by the downstream flow conditions.

Subsonic flow   compressible flow with a Sarrau–Mach number less than unity: i.e. the flow velocity is less than the sound celerity.

Supercritical flow   in an open channel, when the flow depth is less than the critical flow depth, the flow is supercritical and the Froude number is larger than one. Supercritical flows are controlled from upstream.

Supersonic flow   compressible flow with a Sarrau–Mach number larger than unity: i.e. the flow velocity is larger than the sound celerity.

Surface tension   property of a liquid surface displayed by its acting as if it were a stretched elastic membrane. Surface tension depends primarily upon the attraction forces between the particles within the given liquid and also upon the gas, solid or liquid that is in contact with it. The action of surface tension is to increase the pressure within a water droplet or within an air bubble. For a spherical bubble of diameter d_{ab}, the increase of internal pressure necessary to balance the tensile force caused by surface tension equals: $\Delta P = 4\sigma/d_{ab}$ where σ is the surface tension.

Surfactant (or *surface active agent*)   substance that, when added to a liquid, reduces its surface tension thereby increasing its wetting property (e.g. detergent).

Surge   a surge in an open channel is a sudden change of flow depth (i.e. abrupt increase or decrease in depth). An abrupt increase in flow depth is called a positive surge while a sudden decrease in depth is termed a negative surge. A positive surge is also called (improperly) a 'moving hydraulic jump' or a 'hydraulic bore'.

Surge wave   results from a sudden change in flow that increases (or decreases) the depth.

(a)

(b)

Plate 1 Mount Barney Creek in flood, February 1988, Queensland, Australia (courtesy of Dr R. Rankin). (a) Looking upstream; (b) view from the right bank, flow from the left to the right.

Plate 2 Meanders in Cuyama valley, California, February 1950 (by W.B. Hamilton, with permission of US Geological Survey, Ref. 316ct).

Plate 3 Lawn Hill Gorge in Queensland, Australia, on 4 June 1984 (courtesy of Mrs J. Hacker).

Plate 4 Flood plain of the South Alligator River in March 1998 (Kakadu National Park NT, Australia) (courtesy of Dr R. Rankin).

(a)

(b)

Plate 5 Man-made waterways: the Roman aqueducts. (a) Roman aqueduct at Tarragona, Spain, May 1997 (courtesy of Mr and Mrs Burridge). Aqueduct length 35 km, bridge length 249 m, bridge height 26 m. (b) Roman aqueduct at Pont-du-Gard, France, June 1998. Aqueduct length 49.5 km, bridge length 275 m, bridge height 48.8 m.

Plate 6 Remains of the Malpasset dam: a mistake not to repeat! Dam height 102.5 m, volume 50 Mm^3. Dam break on 2 December, 1959. Photograph taken in December 1981, view from upstream looking at the right abutment.

Plate 7 Dry reservoir of Koorawatha weir, railway dam completed in 1911. 9.1 m high thin arch wall, reservoir capacity 4×10^4 m^3. Photograph taken in December 1997, view from the right bank.

(a)

(b)

Plate 8 Antidune at Big Bend, Burdekin River, Australia, January 1996 (courtesy of Dr C. Fielding). (a) Bed forms left after a peak flow of 3200 m³/s, flow from left to right. Note the A5-size notepad on the left hand side, used for scaling. (b) Bed forms left after a peak flow of 3200 m³/s, looking downstream with tyre tracks in the foreground.

Plate 9 Dunes and runoff channels at Big Bend, Burdekin River, Australia, 20 January 1996 (courtesy of Dr C. Fielding). Bed forms left after a peak flow of 3200 m³/s.

Plate 10 Weir overflow at Moree, NSW, Australia, 16 December 1997, flow from left to right.

Plate 11 (a)

(b)

(c)

Plate 11 Spillway overflows. (a) (Opposite) Overflow on the Ternay dam spillway, June 1988. Dam height 41 m (completed in 1867), unlined rock stepped cascade. Refurbished in the 1980s. Small overflow $Q \sim 2\,m^3/s$. (b) Overflow on the Reece dam spillway (courtesy of the Hydro-Electric Commission, Tasmania). Design spillway capacity $4740\,m^3/s$, overflow event $365\,m^3/s$. (c) Dartmouth dam spillway (courtesy of Mr Jeffery, Goulburn-Murray Water). Dam height 180 m (completed in 1977), unlined rock stepped cascade. Design spillway capacity $2755\,m^3/s$. Small overflow in 1996 $Q \sim 225\,m^3/s$.

Plate 12 Dam hydraulics: an arch dam in a narrow valley, the Monteynard dam, June 1998. Dam height 155 m (completed in 1968), crest length 210 m, design spillway capacity 2500 m³/s.

Plate 13 When hydraulics, architecture and arts meet: les Jardins du Château de Versailles, France. Le Bosquet des Rocailles (ou Bosquet de la Salle de Bal), June 1988. Built between 1680 and 1683 by Jules Mansart, sculptures by Pierre Legros and Benoît Massou, modified in 1690. The fountain includes 20 water jets and 17 cascades.

Suspended load   transported sediment material maintained into suspension.

Système international d'unités   international system of units adopted in 1960 based on the metre-kilogram-second (MKS) system. It is commonly called the SI unit system. The basic seven units are: for length, the metre; for mass, the kilogram; for time, the second; for electric current, the ampere; for luminous intensity, the candela; for amount of substance, the mole; for thermodynamic temperature, the Kelvin. Conversion tables are given in Appendix B.

Système métrique   international decimal system of weights and measures which was adopted in 1795 during the French Révolution. Between 1791 and 1795, the Académie des Sciences de Paris prepared a logical system of units based on the metre for length and the kilogram for mass. The standard metre was defined as 1×10^{-7} times a meridional quadrant of the Earth. The gram was equal to the mass of $1\,cm^3$ of pure water at the temperature of its maximum density (i.e. 4°C) and 1 kilogram equalled 1000 grams. The litre was defined as the volume occupied by a cube of $1 \times 10^3\,cm^3$.

TWRC   Tail Water Rating Curve.

TWRL   Tail Water Rating Level.

Tainter gate   is a radial gate.

Tailwater depth   downstream flow depth.

Tailwater level   downstream free-surface elevation.

Total head   the total head is proportional to the total energy per unit mass and per gravity unit. It is expressed in metres of water.

Training wall   sidewall of chute spillway.

Trashrack   screen comprising metal or reinforced concrete bars located at the intake of a waterway to prevent the progress of floating or submerged debris.

Turbulence   flow motion characterized by its unpredictable behaviour, strong mixing properties and a broad spectrum of length scales (Lesieur 1994).

Turbulent flow   in turbulent flows the fluid particles move in very irregular paths, causing an exchange of momentum from one portion of the fluid to another. Turbulent flows have great mixing potential and involve a wide range of eddy length scales.

Turriano   Juanelo Turriano (1511–1585) was an Italian clockmaker, mathematician and engineer who worked for the Spanish Kings Charles V and later Philip II. It is reported that he checked the design of the Alicante dam for King Philip II.

Two-dimensional flow   all particles are assumed to flow in parallel planes along identical paths in each of these planes. There are no changes in flow normal to these planes. An example of two-dimensional flow can be an open channel flow in a wide rectangular channel.

Ukiyo-e (or *Ukiyoe*)   is a type of Japanese painting and colour woodblock prints during the period 1803–1867.

Undular hydraulic jump   hydraulic jump characterized by steady stationary free-surface undulations downstream of the jump and by the absence of a formed roller. The undulations can extend far downstream of the jump with decaying wavelengths, and the undular jump occupies a significant length of the channel. It is usually observed for $1 < Fr_1 < 1.5$ to 3 (Chanson 1995a). The first significant study of undular jump flow can be attributed to Fawer (1937) and undular jump flows should be called Fawer jumps in homage to Fawer's work.

Undular surge   positive surge characterized by a train of secondary waves (or undulations) following the surge front. Undular surges are sometimes called Boussinesq–Favre waves in homage to the contributions of J.B. Boussinesq and H. Favre.

Uniform equilibrium flow   occurs when the velocity is identically the same at every point, in magnitude and direction, for a given instant:

$$\frac{\partial V}{\partial s} = 0$$

in which time is held constant and ∂s is a displacement in any direction. That is, steady uniform flow (e.g. liquid flow through a long pipe at a constant rate) and unsteady uniform flow (e.g. liquid flow through a long pipe at a decreasing rate).

Universal gas constant (also called *molar gas constant* or *perfect gas constant*) fundamental constant equal to the pressure times the volume of gas divided by the absolute temperature for one mole of perfect gas. The value of the universal gas constant is $8.31441\,\mathrm{J\,K^{-1}mole^{-1}}$.

Unsteady flow the flow properties change with the time.

Uplift upward pressure in the pores of a material (interstitial pressure) or on the base of a structure. Uplift pressures led to the destruction of stilling basins and even to the failures of concrete dams (e.g. Malpasset dam break in 1959).

Upstream flow conditions flow conditions measured immediately upstream of the investigated control volume.

USACE United States Army Corps of Engineers.

USBR United States Bureau of Reclamation.

Validation comparison between model results and prototype data, to validate the model. The validation process must be conducted with prototype data that are different from those used to calibrate and to verify the model.

Vauban Sébastien Vauban (1633–1707) was Maréchal de France. He participated in the construction of several water supply systems in France, including the extension of the feeder system of the Canal du Midi between 1686 and 1687, and parts of the water supply system of the gardens of Versailles.

Velocity potential is defined as a scalar function of space and time such that its negative derivative with respect to any direction is the fluid velocity in that direction: $V = -\mathrm{grad}\,\Phi$. The existence of a velocity potential implies irrotational flow of an ideal-fluid. The velocity potential was introduced by the French mathematician J. Louis Lagrange (Lagrange 1781).

Vena contracta minimum cross-section area of the flow (e.g. jet or nappe) discharging through an orifice, sluice gate or weir.

Venturi meter in closed pipes, smooth constriction followed by a smooth expansion. The pressure difference between the upstream location and the throat is proportional to the velocity-square. It is named after the Italian physicist Giovanni Battista Venturi (1746–1822).

Villareal de Berriz Don Pedro Bernardo Villareal de Berriz (1670–1740) was a Basque nobleman. He designed several buttress dams, some of these being still in use (Smith 1971).

Viscosity fluid property which characterizes the fluid resistance to shear: i.e. resistance to a change in shape or movement of the surroundings.

Vitruvius Roman architect and engineer (BC 94–??). He built several aqueducts to supply the Roman capital with water. (Note: there is some uncertainty about his full name: 'Marcus Vitruvius Pollio' or 'Lucius Vitruvius Mamurra', Garbrecht 1987a.)

VNIIG Institute of Hydrotechnics Vedeneev in St Petersburg (Russia).

VOC Volatile Organic Compound.

Von Karman constant see Karman constant.

Wadi Arabic word for a valley that becomes a watercourse in rainy seasons.

Wake region the separation region downstream of the streamline that has separated from a boundary is called a wake or wake region.

Warrie Australian Aboriginal name for 'rushing water'.

Waste waterway old name for a spillway, particularly used in irrigation with reference to the waste of waters resulting from a spill.

Wasteweir a spillway. The name refers to the waste of hydroelectric power or irrigation water resulting from the spill. A 'staircase' wasteweir is a stepped spillway.

Water common name applied to the liquid state of the hydrogen–oxygen combination H_2O. Although the molecular structure of water is simple, the physical and chemical properties of H_2O are unusually complicated. Water is a colourless, tasteless, and odourless liquid at room temperature. One most important property of water is its ability to dissolve many other substances: H_2O is frequently called the universal solvent. Under standard atmospheric pressure, the freezing point of water is 0°C (273.16 K) and its boiling point is 100°C (373.16 K).

Water clock ancient device for measuring time by the gradual flow of water through a small orifice into a floating vessel. The Greek name is Clepsydra.

Waterfall abrupt drop of water over a precipice characterized by a free-falling nappe of water. The highest waterfalls are the Angel Falls (979 m) in Venezuela ('Churún Merú'), Tugel Falls (948 m) in South Africa, and Mtarazi (762 m) in Zimbabwe.

Water-mill mill (or wheel) powered by water.

Water staircase (or '*Escalier d'Eau*') is the common architectural name given to a stepped cascade with flat steps.

Weak jump a weak hydraulic jump is characterized by a marked roller, no free-surface undulation and low energy loss. It is usually observed after the disappearance of an undular hydraulic jump with increasing upstream Froude numbers.

Weber Moritz Weber (1871–1951) was a German Professor at the Polytechnic Institute of Berlin. The Weber number characterizing the ratio of inertial force over surface tension force was named after him.

Weber number Dimensionless number characterizing the ratio of inertial forces over surface tension forces. It is relevant in problems with gas–liquid or liquid–liquid interfaces.

Weir low river dam used to raise the upstream water level. Measuring weirs are built across a stream for the purpose of measuring the flow.

Weisbach Julius Weisbach (1806–1871) was a German applied mathematician and hydraulician.

Wen Cheng-Ming Chinese landscape painter (1470–1559). One of his famous works is the painting of 'Old trees by a cold waterfall'.

WES Waterways Experiment Station of the US Army Corps of Engineers.

WES standard spillway shape spillway shape developed by the US Army Corps of Engineers at the Waterways Experiment Station.

Wetted perimeter Considering a cross-section (usually selected normal to the flow direction), the wetted perimeter is the length of wetted contact between the flowing stream and the solid boundaries. For example, in a circular pipe flowing full, the wetted perimeter equals the circle perimeter.

Wetted surface in open channel, the term 'wetted surface' refers to the surface area in contact with the flowing liquid.

White waters non-technical term used to design free-surface aerated flows. The refraction of light by the entrained air bubbles gives the 'whitish' appearance to the free-surface of the flow.

White water sports include canoe, kayak and raft racing down swift-flowing turbulent waters.

Wing wall sidewall of an inlet or outlet.

Wood I.R. Wood is an Emeritus Professor in civil engineering at the University of Canterbury (New Zealand).

Yunca Indian of a group of South American tribes of which the Chimus and the Chinchas are the most important. The Yunca civilization developed a pre-Inca culture on the coast of Peru.

List of symbols

A	cross-section area (m^2)
A_s	particle cross-section area (m^2)
B	open channel free surface width (m)
B_{max}	inlet lip width (m) of MEL culvert (Chapter 19)
B_{min}	(1) minimum channel width (m) for onset of choking flow
	(2) barrel width (m) of a culvert (Chapter 19)
C	(1) celerity (m/s): e.g. celerity of sound in a medium, celerity of a small disturbance at a free-surface
	(2) dimensional discharge coefficient (Chapters 17 and 19)
c	discharge parameter of box culvert with submerged entrance
Ca	Cauchy number
$C_{Chézy}$	Chézy coefficient (m$^{1/2}$/s)
C_D	dimensionless discharge coefficient (SI units) (Chapter 17)
C_L	lift coefficient
C_d	(1) skin friction coefficient (also called drag coefficient)
	(2) drag coefficient
C_{des}	design discharge coefficient (SI units) (Chapter 17)
C_p	specific heat at constant pressure (J kg^{-1} K^{-1}): $C_p = \left(\dfrac{\partial h}{\partial T}\right)_P$
C_s	mean volumetric sediment concentration
$(C_s)_{mean}$	mean sediment suspension concentration
C_{sound}	sound celerity (m/s)
C_v	specific heat at constant volume (J kg^{-1} K^{-1})
C_s	sediment concentration
D	(1) circular pipe diameter (m)
	(2) culvert barrel height (m) (Chapter 19)
D_H	hydraulic diameter (m), or equivalent pipe diameter, defined as:

$$D_H = 4 \times \frac{\text{cross-sectional area}}{\text{wetted perimeter}} = \frac{4 \times A}{P_w}$$

D_s	sediment diffusivity (m^2/s)
D_1, D_2	characteristics of velocity distribution in turbulent boundary layer
d	flow depth (m) measured perpendicular to the channel bed

d_{ab}	air bubble diameter (m)
d_b	brink depth (m) (Chapter 18)
d_c	critical flow depth (m)
d_{charac}	characteristic geometric length (m)
d_{conj}	conjugate flow depth (m)
d_o	uniform equilibrium flow depth (m): i.e. normal depth
d_p	pool depth (m) (Chapter 18)
d_s	sediment size (m)
d_{tw}	tailwater flow depth (m)
d_{50}	median grain size (m) defined as the size for which 50% by weight of the material is finer
d_i	characteristic grain size (m), where $i = 10, 16, 50, 75, 84, 90$
d_*	dimensionless particle parameter: $d_* = d_s \sqrt[3]{(\rho_s/\rho - 1)g/\nu^2}$
e	internal energy per unit mass (J/kg)
E	mean specific energy (m) defined as: $E = H - z_o$
E	local specific energy (m) defined as: $E = \dfrac{P}{\rho g} + (z - z_o) + \dfrac{v^2}{2g}$
Eu	Euler number
E_b	bulk modulus of elasticity (Pa): $E_b = \rho \dfrac{\partial P}{\partial \rho}$
E_{co}	compressibility (1/Pa): $E_{co} = \dfrac{1}{\rho} \dfrac{\partial \rho}{\partial P}$
E_{min}	minimum specific energy (m)
E	total energy (J) of system
f	Darcy friction factor (also called head loss coefficient)
F	force (N)
$\mathbf{F}$	force vector
F_b	buoyant force (N)
F_d	drag force (N)
F_{fric}	friction force (N)
F_p	pressure force (N)
F_{visc}	viscous force (N/m^3)
F_{vol}	volume force per unit volume (N/m^3)
Fr	Froude number
F_r	ratio of prototype to model forces: $F_r = F_p/F_m$ (Chapter 14)
g	gravity constant (m/s^2) (see Appendix A); in Brisbane, Australia: $g = 9.80\,\text{m/s}^2$
$g_{centrif}$	centrifugal acceleration (m/s^2)
h	specific enthalpy (i.e. enthalpy per unit mass) (J/kg): $h = e + \dfrac{P}{\rho}$
h	(1) dune bed form height (m)
	(2) step height (m)
H	(1) mean total head (m): $H = d\cos\theta + z_o + \alpha \dfrac{V_{mean}^2}{2g}$ assuming a hydrostatic pressure distribution
	(2) depth-averaged total head (m) defined as: $H = \dfrac{1}{d} \displaystyle\int_0^d H\,\mathrm{d}y$
H_{des}	design upstream head (m) (Chapters 17 and 18)

H_{res}	residual head (m)
H_1	upstream total head (m)
H_2	downstream total head (m)
H	local total head (m) defined as: $H = \dfrac{P}{\rho g} + z + \dfrac{V^2}{2g}$
i	integer subscript
i	imaginary number: $i = \sqrt{-1}$
$JHRL$	jump height rating level (m R.L.)
k	permeability (m^2) of a soil
k_{Bazin}	Bazin resistance coefficient
$k_{Strickler}$	Strickler resistance coefficient (m$^{1/3}$/s)
k_s	equivalent sand roughness height (m)
K	hydraulic conductivity (m/s) of a soil
K	von Karman constant (K = 0.4)
K'	head loss coefficient: $K' = \Delta H/(0.5V^2/g)$
l	(1) dune bed form length (m)
	(2) step length (m)
L	length (m)
L_{crest}	crest length (m)
L_{culv}	culvert length (m) measured in the flow direction (Chapter 19)
L_d	drop length (m)
L_{inlet}	inlet length (m) measured in the flow direction (Chapter 19)
L_r	(1) length of roller of hydraulic jump (m) (Chapters 4 and 17)
	(2) ratio of prototype to model lengths: $L_r = L_p/L_m$ (Chapter 14)
M	momentum function (m^2)
Ma	Sarrau–Mach number
Mo	Morton number
M_r	ratio of prototype to model masses (Chapter 14)
M	total mass (kg) of system
$\dot{m}$	mass flow rate per unit width (kg s^{-1} m^{-1})
$\dot{m}_s$	sediment mass flow rate per unit width (kg s^{-1} m^{-1})
N	inverse of velocity distribution exponent
No	Avogadro constant: No = 6.0221367×10^{23} mole^{-1}
Nu	Nusselt number
N_{bl}	inverse of velocity distribution exponent in turbulent boundary layer (Appendix H)
$n_{Manning}$	Gauckler–Manning coefficient (s/m$^{1/3}$)
P	absolute pressure (Pa)
P_{atm}	atmospheric pressure (Pa)
$P_{centrif}$	centrifugal pressure (Pa)
P_r	ratio of prototype to model pressures (Chapter 14)
P_{std}	standard atmosphere (Pa) or normal pressure at sea level
P_v	vapour pressure (Pa)
P_w	wetted perimeter (m)
Po	porosity factor
q	discharge per meter width (m^2/s)
q_{des}	design discharge (m^2/s) per unit width (Chapters 17 to 19)

q_{max}	maximum flow rate per unit width (m²/s) in open channel for a constant specific energy
q_s	sediment flow rate per unit width (m²/s)
q_h	heat added to a system per unit mass (J/kg)
Q	total volume discharge (m³/s)
Q_{des}	design discharge (m³/s) (Chapters 17 to 19)
Q_{max}	maximum flow rate (m³/s) in open channel for a constant specific energy
Q_r	ratio of prototype to model discharges (Chapter 14)
Q_h	heat added to a system (J)
r	radius of curvature (m)
R	invert curvature radius (m)
R	fluid thermodynamic constant (J/kg/K) also called gas constant: $P = \rho R T$ perfect gas law (i.e. Mariotte law)
Re	Reynolds number
Re_*	shear Reynolds number
R_H	hydraulic radius (m) defined as: $R_H = \dfrac{\text{cross-sectional area}}{\text{wetted perimeter}} = \dfrac{A}{P_w}$
Ro	universal gas constant: $Ro = 8.3143\,\mathrm{J\,K^{-1}\,mole^{-1}}$
s	curvilinear coordinate (m) (i.e. distance measured along a streamline and positive in the flow direction)
$\mathbf{s}$	relative density of sediment: $\mathbf{s} = \rho_s/\rho$ (Chapters 7 to 12)
S	sorting coefficient of a sediment mixture: $S = \sqrt{d_{90}/d_{10}}$
S	specific entropy (i.e. entropy per unit mass) (J/K/kg): $dS = \left(\dfrac{dq_h}{T}\right)_{rev}$
S_c	critical slope
S_f	friction slope defined as: $S_f = -\dfrac{\partial H}{\partial s}$
S_o	bed slope defined as: $S_o = -\dfrac{\partial z_o}{\partial s} = \sin\theta$
S_t	transition slope for a multi-cell MEL culvert (Chapter 19)
t	time (s)
t_r	ratio of prototype to model times: $t_r = t_p/t_m$ (Chapter 14)
t_s	sedimentation time scale (s)
T	thermodynamic (or absolute) temperature (K)
T_o	reference temperature (K)
$TWRL$	tailwater rating level (m R.L.)
U	volume force potential (m²/s²)
V	(local) velocity (m/s)
v_s	particle volume (m³)
V	flow velocity (m/s)
V	velocity vector; in Cartesian coordinates the velocity vector equals: $V = (V_x, V_y, V_z)$
V_c	critical flow velocity (m/s)
V_{mean}	mean flow velocity (m/s): $V_{mean} = Q/A$
V_{max}	maximum velocity (m/s) in a cross-section; in a fully developed open channel flow, the velocity is maximum near the free-surface

V_r	ratio of prototype to model velocities: $V_r = V_p/V_m$ (Chapter 14)
V_s	average speed (m/s) of sediment motion
V_{srg}	surge velocity (m/s) as seen by an observer immobile on the channel bank
V_o	uniform equilibrium flow velocity (m/s)
V'	depth-averaged velocity (m/s): $V' = \dfrac{1}{d}\displaystyle\int_0^d V\,dy$
V_*	shear velocity (m/s) defined as: $V_* = \sqrt{\dfrac{\tau_o}{\rho}}$
w_o	(1) particle settling velocity (m/s) (2) fall velocity (m/s) of a single particle in a fluid at rest
w_s	settling velocity (m/s) of a suspension
w_s	work done by shear stress per unit mass (J/kg)
W	channel bottom width (m)
We	Weber number
W_p	work (J) done by the pressure force
W_s	work (J) done by the system by shear stress (i.e. torque exerted on a rotating shaft)
W_t	total work (J) done by the system
Wa	Coles wake function
X	horizontal coordinate (m) measured from spillway crest (Chapter 17)
X_r	ratio of prototype to model horizontal distances (Chapter 14)
x	Cartesian coordinate (m)
$\boldsymbol{x}$	Cartesian coordinate vector: $\boldsymbol{x} = (x, y, z)$
y	(1) distance (m) measured normal to the flow direction (2) distance (m) measured normal to the channel bottom (3) Cartesian coordinate (m)
$y_{channel}$	channel height (m)
y_s	characteristic distance (m) from channel bed
Y	vertical coordinate (m) measured from spillway crest (Chapter 17)
z	(1) altitude or elevation (m) measured positive upwards (2) Cartesian coordinate (m)
z_{apron}	apron invert elevation (m)
z_{crest}	spillway crest elevation (m)
z_o	(1) reference elevation (m) (2) bed elevation (m)
Z_r	ratio of prototype to model vertical distances (Chapter 14)

Greek symbols

α	Coriolis coefficient or kinetic energy correction coefficient
β	momentum correction coefficient (i.e. Boussinesq coefficient)
δ	(1) sidewall slope (2) boundary layer thickness (m)
δ_{ij}	identity matrix element
δ_s	bed-load layer thickness (m)
Δd	change in flow depth (m)

ΔE	change in specific energy (m)
ΔH	head loss (m): i.e. change in total head
ΔP	pressure difference (Pa)
Δq_s	change in sediment transport rate (m^2/s)
Δs	small distance (m) along the flow direction
ΔV	change in flow velocity (m/s)
Δz_o	change in bed elevation (m)
Δz_o	weir height (m) above natural bed level
ε_{ij}	velocity gradient element (m/s^2)
ϕ	sedimentological size parameter
ϕ	velocity potential (m^2/s)
ϕ_s	angle of repose
φ	stream function (m^2/s)
γ	specific heat ratio: $\gamma = C_p/C_v$
μ	dynamic viscosity (Pa s)
ν	kinematic viscosity (m^2/s): $\nu = \mu/\rho$
Π	wake parameter
π	$\pi = 3.141\,592\,653\,589\,793\,238\,462\,643$
θ	channel slope
ρ	density (kg/m^3)
ρ_r	density (kg/m^3)
ρ_s	ratio of prototype to model densities (Chapter 14)
ρ_{sed}	sediment mixture density (kg/m^3)
σ	surface tension (N/m)
σ_e	effective stress (Pa)
σ_{ij}	stress tensor element (Pa)
σ_g	geometric standard deviation of sediment size distribution: $\sigma_g = \sqrt{d_{84}/d_{16}}$
τ	shear stress (Pa)
τ_{ij}	shear stress component (Pa) of the i-momentum transport in the j-direction
τ_o	average boundary shear stress (Pa)
$(\tau_o)_c$	critical shear stress (Pa) for onset of sediment motion
τ_o'	skin friction shear stress (Pa)
τ_o''	bed form shear stress (Pa)
τ_1	yield stress (Pa)
τ_*	Shields parameter: $\tau_* = \dfrac{\tau_o}{\rho g(\rho_s/\rho - 1)\,d_s}$
$(\tau_*)_c$	critical Shields parameter for onset of sediment motion

Subscript

air	air
bl	bed-load
c	critical flow conditions
conj	conjugate flow property
des	design flow conditions
dry	dry conditions

exit	exit flow condition
i	characteristics of section $\{i\}$ (in the numerical integration process) (Chapter 15)
inlet	inlet flow condition
m	model
mean	mean flow property over the cross-section area
mixt	sediment–water mixture
model	model conditions
o	uniform equilibrium flow conditions
outlet	outlet flow condition
p	prototype
prototype	prototype conditions
r	ratio of prototype to model characteristics
s	(1) component in the s-direction
	(2) sediment motion
	(3) sediment particle property
sl	suspended load
tw	tailwater flow condition
w	water
wet	wet conditions
x	x-component
y	y-component
z	z-component
1	upstream flow conditions
2	downstream flow conditions

Abbreviations

CS	control surface
CV	control volume
D/S	downstream
GVF	gradually-varied flow
Hg	mercury
RVF	rapidly-varied flow
SI	Système international d'unités (International System of Units)
THL	total head line
U/S	upstream

Reminder

(1) At 20°C, the density and dynamic viscosity of water (at atmospheric pressure) are: $\rho_w = 998.2\,\text{kg/m}^3$ and $\mu_w = 1.005 \times 10^{-3}$ Pa s.

(2) Water at atmospheric pressure and 20.2°C has a kinematic viscosity of exactly $10^{-6}\,\text{m}^2/\text{s}$.

(3) Water in contact with air has a surface tension of about 0.0733 N/m at 20°C.

(3) At 20°C and atmospheric pressure, the density of air is about $1.2\,\text{kg/m}^3$.

Dimensionless numbers

Ca Cauchy number: $Ca = \dfrac{\rho V^2}{E_b}$

Note: the Sarrau–Mach number equals: $Ma \sim \sqrt{Ca}$

C_d (1) drag coefficient for bottom friction (i.e. friction drag):

$$C_d = \frac{\tau_o}{\frac{1}{2}\rho V^2} = \frac{\text{shear stress}}{\text{dynamic pressure}}$$

Note: another notation is C_f (e.g. Comolet 1976).

(2) drag coefficient for a structural shape (i.e. form drag):

$$C_d = \frac{F_d}{\frac{1}{2}\rho V^2 A} = \frac{\text{drag force per unit cross-section area}}{\text{dynamic pressure}}$$

where A is the projection of the structural shape (i.e. body) in the plane normal to the flow direction.

Eu Euler number is defined as: $Eu = \dfrac{V}{\sqrt{\Delta P/\rho}}$

Fr Froude number is defined as: $Fr = \dfrac{V}{\sqrt{gd_{\text{charac}}}} \propto \sqrt{\dfrac{\text{inertial force}}{\text{weight}}}$

Note: some authors use the notation:

$$Fr = \frac{V^2}{gd_{\text{charac}}} = \frac{\rho V^2 A}{\rho g A d_{\text{charac}}} \propto \frac{\text{inertial force}}{\text{weight}}$$

Ma Sarrau–Mach number: $Ma = \dfrac{V}{C}$

Mo Morton number is defined as: $Mo = \dfrac{g\mu_w^4}{\rho_w \sigma^3}$. The Morton number is a function only of fluid properties and gravity constant. If the same fluids (air and water) are used in both model and prototype, Mo may replace the Weber, Reynolds or Froude number as: $Mo = \dfrac{We^3}{Fr^2\,Re^4}$

Nu Nusselt number: $Nu = \dfrac{Hd_{\text{charac}}}{\lambda} \propto \dfrac{\text{heat transfer by convection}}{\text{heat transfer by conduction}}$

where H is the heat transfer coefficient ($\text{W m}^{-2}\,\text{K}^{-1}$) and λ is the thermal conductivity ($\text{W m}^{-1}\,\text{K}^{-1}$)

Re Reynolds number: $Re = \dfrac{Vd_{\text{charac}}}{\nu} \propto \dfrac{\text{inertial forces}}{\text{viscous forces}}$

Re_* shear Reynolds number: $Re_* = \dfrac{V_* k_s}{\nu}$

We Weber number: $We = \dfrac{V}{\sqrt{\dfrac{\sigma}{\rho d_{\text{charac}}}}} \propto \sqrt{\dfrac{\text{inertial force}}{\text{surface tension force}}}$

Note: some authors use the notation:

$$We = \frac{V^2}{\frac{\sigma}{\rho d_{charac}}} \propto \frac{\text{inertial force}}{\text{surface tension force}}$$

τ_* Shields parameter characterizing the onset of sediment motion:

$$\tau_* = \frac{\tau_o}{\rho g (\rho_s/\rho - 1) d_s} \propto \frac{\text{destabilizing force moment}}{\text{stabilizing moment of weight force}}$$

Notes

The variable d_{charac} is the characteristic geometric length of the flow field: e.g. pipe diameter, flow depth, sphere diameter. Some examples are listed below:

Flow	d_{charac}	Comments
Circular pipe flow	D	Pipe diameter
Flow in pipe of irregular cross-section	D_H	Hydraulic diameter
Flow resistance in open channel flow	D_H	Hydraulic diameter
Wave celerity in open channel flow	d	Flow depth
Flow past a cylinder	D	Cylinder diameter

Part 1 Basic Principles of Open Channel Flows

1

Introduction

Summary

This introductory chapter briefly reviews the basic fluid properties and some important results for fluids at rest. The concept of open channel flow is then defined and some applications are described.

1.1 Presentation

The term 'hydraulics' is related to the application of Fluid Mechanics principles to water engineering structures, and civil and environmental engineering facilities, especially hydraulic structures (e.g. canals, rivers, dams, reservoirs, water treatment plants).

In this book, we consider open channels in which liquid (i.e. water) flows with a free surface. Examples of open channels are natural streams and rivers. Man-made channels include irrigation and navigation canals, drainage ditches, sewer and culvert pipes running partially full, and spillways.

The primary factor in open channel flow analysis is the location of the free surface, which is unknown beforehand (i.e. '*a priori*'). The free surface rises and falls in response to perturbations to the flow (e.g. changes in channel slope or width). The main parameters of a hydraulic study are the geometry of the channel (e.g. width, slope, roughness), the properties of the flowing fluid (e.g. density, viscosity) and the flow parameters (e.g. velocity, flow depth).

1.2 Fluid properties

The *density* ρ of a fluid is defined as its mass per unit volume. All real fluids resist any force, tending to cause one layer to move over another, but this resistance is offered only while the movement is taking place. The resistance to the movement of one layer of fluid over an adjoining one is referred to as the *viscosity* of the fluid. Newton's law of viscosity postulates that, for the straight parallel motion of a given fluid, the tangential stress between two adjacent layers is proportional to the velocity gradient in a direction perpendicular to the layers:

$$\tau = \mu \frac{dV}{dy} \tag{1.1}$$

where τ is the shear stress between adjacent fluid layers, μ is the dynamic viscosity of the fluid, V is the velocity and y is the direction perpendicular to the fluid motion. Fluids that obey Newton's law of viscosity are called *Newtonian fluids*.

At the interface between a liquid and a gas, a liquid and a solid, or two immiscible liquids, a tensile force is exerted at the surface of the liquid and tends to reduce the area of this surface to the greatest possible extent. The *surface tension* is the stretching force required to form the film: i.e. the tensile force per unit length of the film in equilibrium.

The basic properties of air and water are detailed in Appendix A.

Notes
1. Isaac Newton (1642–1727) was an English mathematician.
2. The kinematic viscosity is the ratio of viscosity to mass density:

$$\nu = \frac{\mu}{\rho}$$

3. A Newtonian fluid is one in which the shear stress, in one-directional flow, is proportional to the rate of deformation as measured by the velocity gradient across the flow (i.e. equation (1.1)). The common fluids such as air, water and light petroleum oils, are Newtonian fluids. Non-Newtonian fluids will not be considered any further in Part I. They will be briefly mentioned in Part II (i.e. hyperconcentrated flows).

Application
At atmospheric pressure and 20°C the density and dynamic viscosity of water are:

$$\rho_w = 998.2 \, \text{kg/m}^3$$

$$\mu_w = 1.005 \times 10^{-3} \, \text{Pa s}$$

and the density of air is around:

$$\rho_{air} = 1.2 \, \text{kg/m}^3$$

Water in contact with air has a surface tension of about 0.0733 N/m at 20°C. Considering a spherical gas bubble (diameter d_{ab}) in a liquid, the increase of gas pressure required to balance the tensile force caused by surface tension equals: $\Delta P = 4\sigma/d_{ab}$.

1.3 Static fluids

Considering a fluid at rest (Fig. 1.1), the pressure at any point within the fluid follows Pascal's law. For any small control volume, there is no shear stress acting on the

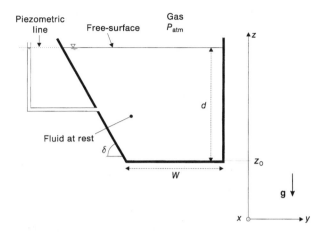

Fig. 1.1 Pressure variation in a static fluid.

control surface. The only forces acting on the control volume of the fluid are the gravity and the pressure[1] forces.

In a static fluid, the pressure at one point in the fluid has a unique value, independent of the direction. This is called Pascal's law. The pressure variation in a static fluid follows:

$$\frac{dP}{dz} = -\rho \times g \tag{1.2}$$

where P is the pressure, z is the vertical elevation positive upwards, ρ is the fluid density and g is the gravity constant (see Appendix A1.1).

For a body of fluid at rest with a free-surface (e.g. a lake) and with a constant density, the pressure variation equals:

$$P(x, y, z) = P_{atm} - \rho g(z - (z_o + d)) \tag{1.3}$$

where P_{atm} is the atmospheric pressure (i.e. air pressure above the free-surface), z_o is the elevation of the reservoir bottom and d is the reservoir depth (Fig. 1.1). $(d + z_o)$ is the free-surface elevation. Equation (1.3) implies that the pressure is independent of the horizontal coordinates (x, y). The term $\{-\rho g(z - (z_o + d))\}$ is positive within the liquid at rest. It is called the hydrostatic pressure.

The pressure force acting on a surface of finite area that is in contact with the fluid is distributed over the surface. The resultant force is obtained by integration:

$$F_p = \int P \, dA \tag{1.4}$$

where A is the surface area.

---

[1] By definition, the pressure always acts normal to a surface. That is, the pressure force has no component tangential to the surface.

Note

Blaise Pascal (1623–1662) was a French mathematician, physicist and philosopher. He developed the modern theory of probability. He also formulated the concept of pressure (between 1646 and 1648) and showed that the pressure in a fluid is transmitted through the fluid in all directions (i.e. Pascal's law).

Application

In Fig. 1.1, the pressure forces (per unit width) applied on the sides of the tank are:

$$F_p = \rho g d W \qquad \text{pressure force acting on the bottom per unit width}$$

$$F_p = \tfrac{1}{2}\rho g d^2 \qquad \text{pressure force acting on the right-wall per unit width}$$

For the left-wall, the pressure force acts in the direction normal to the wall. The integration of equation (1.4) yields:

$$F_p = \frac{1}{2}\rho g \frac{d^2}{\sin \delta} \qquad \text{pressure force acting on the left-wall per unit width}$$

1.4 Open channel flow

1.4.1 Definition

An open channel is a waterway, canal or conduit in which a liquid flows with a free-surface. Open channel flow describes the fluid motion in an open channel (Fig. 1.2). In most applications, the liquid is water and the air above the flow is usually at rest and at standard atmospheric pressure (see Appendix 1.1).

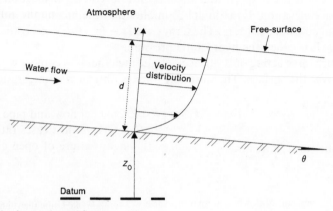

Fig. 1.2 Sketch of open channel flow.

Notes

1. In some practical cases (e.g. a closed conduit flowing partly full), the pressure of the air above the flow might become sub-atmospheric.
2. Next to the free-surface of an open channel flow, some air is entrained by friction at the free-surface. That is, the no-slip condition at the air–water interface induces the air motion. The term 'air boundary layer' is sometimes used to describe the atmospheric region where air is entrained through momentum transfer at the free-surface.
3. In a clear-water open channel flow, the free-surface is clearly defined: it is the interface between the water and the air. For an air–water mixture flow (called 'white waters'), the definition of the free-surface (i.e. the interface between the flowing mixture and the surrounding atmosphere) becomes somewhat complicated (e.g. Wood 1991, Chanson 1997).

1.4.2 Applications

Open channel flows are found in Nature as well as in man-made structures. In Nature, tranquil flows are observed in large rivers near their estuaries: e.g. the Nile river between Alexandria and Cairo, the Brisbane river in Brisbane. Rushing waters are encountered in mountain rivers, river rapids and torrents. Classical examples include the cataracts of the Nile river, the Zambesi rapids in Africa, and the Rhine waterfalls.

Man-made open channels can be water-supply channels for irrigation, power supply and drinking waters, conveyor channel in water treatment plants, storm waterways, some public fountains, culverts below roads and railways lines.

Open channel flows are observed in small-scale as well as large-scale situations. For example, the flow depth can be between a few centimetres in water treatment plants and over 10 m in large rivers. The mean flow velocity may range from less than 0.01 m/s in tranquil waters to above 50 m/s in high-head spillway. The range of total discharges[2] may extend from $Q \sim 0.001$ L/s in chemical plants to $Q > 10\,000\,\mathrm{m^3/s}$ in large rivers or spillways. In each flow situation, however, the location of the free-surface is unknown beforehand and it is determined by applying the continuity and momentum principles.

1.4.3 Discussion

There are characteristic differences between open channel flow and pipe flow (Table 1.1). In an open channel, the flow is driven by gravity in most cases rather than by pressure work, as with pipe flow. Another dominant feature of open channel flow is the presence of a free-surface:

---

[2] In the hydraulics of open channels, the water flow is assumed incompressible and the volume discharge is commonly used.

Table 1.1 Basic differences between pipe flow and open channel flow of an incompressible fluid

	Pipe flow	Open channel flow
Flow driven by	Pressure work	Gravity (i.e. potential energy)
Flow cross-section	Known (fixed by pipe geometry)	Unknown in advance because the flow depth is unknown beforehand
Characteristic flow parameters	Velocity deduced from continuity equation	Flow depth and velocity deduced by simultaneously solving the continuity and momentum equations
Specific boundary conditions		Atmospheric pressure at the flow free-surface

- the position of the free-surface is unknown '*a priori*',
- its location must be deduced by solving simultaneously the continuity and momentum equations, and
- the pressure at the free-surface is atmospheric.

1.5 Exercises

Give the values (*and* units) of the specified fluid and physical properties.

(a) Density of water at atmospheric pressure and 20°C.
(b) Density of air at atmospheric pressure and 20°C.
(c) Dynamic viscosity of water at atmospheric pressure and 20°C.
(d) Kinematic viscosity of water at atmospheric pressure and 20°C.
(e) Kinematic viscosity of air at atmospheric pressure and 20°C.
(f) Surface tension of air and water at atmospheric pressure and 20°C.
(g) Acceleration of gravity in Brisbane.

What is Newton's law of viscosity?

In a static fluid, express the pressure variation with depth.

2

Fundamental equations

Summary

In this chapter, the fundamental equations are developed and applied to open channel flows. It is shown that the momentum equation leads to the Bernoulli equation.

2.1 Introduction

In open channel flow the free surface is always at a constant absolute pressure (usually atmospheric) and the driving force of the fluid motion is gravity. In most practical situations, open channels contain waters. The general principles of open channel flow calculations developed in this chapter are, however, applicable to other liquids. Specific results (e.g. flow resistance) are based primarily upon experimental data obtained mostly with water.

2.2 The fundamental equations

2.2.1 Introduction

The law of conservation of mass states that the mass within a closed system remains constant with time (disregarding relativity effects):

$$\frac{\mathrm{D}M}{\mathrm{D}t} = 0 \qquad (2.1)$$

where M is the total mass and $\mathrm{D}/\mathrm{D}t$ is the absolute differential (see Appendix A1.3, section on Differential and differentiation). Equation (2.1) leads to the continuity equation.

The expression of Newton's second law of motion for a system is:

$$\sum F = \frac{\mathrm{D}}{\mathrm{D}t}(M \times V) \qquad (2.2)$$

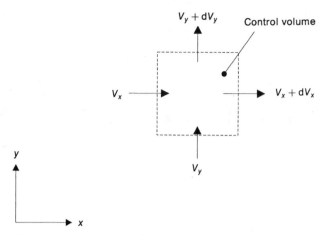

Fig. 2.1 Sketch of control volume for a two-dimensional flow.

where $\sum \boldsymbol{F}$ refers to the resultant of all external forces acting on the system including body forces such as gravity, and V is the velocity of the centre of mass of the system. The application of equation (2.2) is called the motion equation.

Equations (2.1) and (2.2) must be applied to a control volume. A *control volume* is a specific region of space selected for analysis. It refers to a region in space which fluid enters and leaves (e.g. Fig. 2.1). The boundary of a control volume is its *control surface*. The concept of a control volume used in conjunction with the differential form of the continuity, momentum and energy equations, is an open system.

All flow situations including open channel flows are subject to the following relationships (e.g. Streeter and Wylie 1981, pp. 88–89):

1. the first and second laws of thermodynamics,
2. the law of conservation of mass (i.e. the continuity relationship),
3. Newton's law of motion, and
4. the boundary conditions.

Other relations (such as a state equation, Newton's law of viscosity) may apply.

Note

A control volume may be either infinitesimally small or finite. It may move or remain fixed in space. It is an *imaginary* volume and *does not interfere* with the flow.

2.2.2 The continuity equation

The principle of conservation of mass states that the mass within a closed system remains constant with time:

$$\frac{\mathrm{D}M}{\mathrm{D}t} = \frac{\mathrm{D}}{\mathrm{D}t} \int_x \int_y \int_z \rho \, \mathrm{d}x \, \mathrm{d}y \, \mathrm{d}z = 0 \qquad (2.3)$$

where ρ is the fluid density and (x, y, z) are three components of a Cartesian system of coordinates. For an infinitesimal small control volume the continuity equation is:

$$\frac{\partial \rho}{\partial t} + \text{div}(\rho V) = 0 \tag{2.4a}$$

and in Cartesian coordinates:

$$\frac{\partial \rho}{\partial t} + \frac{\partial(\rho V_x)}{\partial x} + \frac{\partial(\rho V_y)}{\partial y} + \frac{\partial(\rho V_z)}{\partial z} = 0 \tag{2.4b}$$

where V_x, V_y and V_z are the velocity components in the x-, y- and z-directions respectively.

For an incompressible flow (i.e. $\rho = $ constant) the *continuity equation* becomes:

$$\text{div } V = 0 \tag{2.5a}$$

and in Cartesian coordinates:

$$\frac{\partial V_x}{\partial x} + \frac{\partial V_y}{\partial y} + \frac{\partial V_z}{\partial z} = 0 \tag{2.5b}$$

For an incompressible fluid (e.g. open channel flow), the inflow (i.e. amount of fluid entering into a control volume) equals the outflow.

Application
Considering an open channel flow with no flow across the side and bottom walls, equation (2.5) can be integrated between two cross-sections of areas A_1 and A_2. It yields:

$$Q = \int_{A_1} V \, dA = \int_{A_2} V \, dA$$

where Q is the total discharge (i.e. volume discharge). Integration of the equation leads to:

$$Q = V_1 A_1 = V_2 A_2$$

where V_1 and V_2 are the mean velocities across the cross-sections A_1 and A_2 respectively.

2.2.3 The momentum equation

The Navier–Stokes equation
Newton's law of motion is used as a basis for developing the momentum equation for a control volume:

$$\sum F = \frac{D}{Dt}(M \times V) = \frac{\partial}{\partial t}\left(\int_{CV} \rho V \, dVolume\right) + \int_{CS} \rho V V \, dArea \tag{2.6}$$

where CV and CS refer to the control volume and control surface respectively. Basically, it states that the change of momentum equals the sum of all forces applied to the control volume.

The forces acting on the control volume are: (1) the surface forces (i.e. pressure and shear forces) acting on the control surface and (2) the volume force (i.e. gravity) applied at the centre of mass of the control volume. For an infinitesimal small volume the momentum equation, applied to the i component of the vector equation is:

$$\frac{D(\rho V_i)}{Dt} = \left(\frac{\partial(\rho V_i)}{\partial t} + \sum_j V_j \frac{\partial(\rho V_i)}{\partial x_j}\right) = \rho F_{\text{vol}_i} + \sum_j \frac{\partial \sigma_{ij}}{\partial x_j} \tag{2.7}$$

where D/Dt is the absolute differential (or absolute derivative, see Appendix A1.3), V_i is the velocity component in the i direction, F_{vol} is the resultant of the volume forces (per unit volume) and σ_{ij} is the stress tensor (see notes below). The subscripts i and j refer to the Cartesian coordinate components (e.g. x, y). If the volume forces F_{vol} derives from a potential U, they can be rewritten as: $F_{\text{vol}} = -\text{grad } U$ (e.g. gravity force $F_{\text{vol}} = -\text{grad}(gz)$). Further, for a Newtonian fluid the stress forces are (1) the pressure forces and (2) the resultant of the viscous forces on the control volume. Hence, for a Newtonian fluid and for volume force deriving from a potential, the momentum equation becomes:

$$\frac{D(\rho \times V)}{Dt} = \rho F_{\text{vol}} - \text{grad } P + F_{\text{visc}} \tag{2.8}$$

where P is the pressure and F_{visc} is the resultant of the viscous forces (per unit volume) on the control volume.

In Cartesian coordinates (x, y, z):

$$\left(\frac{\partial(\rho V_x)}{\partial t} + \sum_j V_j \frac{\partial(\rho V_x)}{\partial x_j}\right) = \rho F_{\text{vol}_x} - \frac{\partial P}{\partial x} + F_{\text{visc}_x} \tag{2.9a}$$

$$\left(\frac{\partial(\rho V_y)}{\partial t} + \sum_j V_j \frac{\partial(\rho V_y)}{\partial x_j}\right) = \rho F_{\text{vol}_y} - \frac{\partial P}{\partial y} + F_{\text{visc}_y} \tag{2.9b}$$

$$\left(\frac{\partial(\rho V_z)}{\partial t} + \sum_j V_j \frac{\partial(\rho V_z)}{\partial x_j}\right) = \rho F_{\text{vol}_z} - \frac{\partial P}{\partial z} + F_{\text{visc}_z} \tag{2.9c}$$

where the subscript j refers to the Cartesian coordinate components (i.e. $j = x, y, z$). In equation (2.9), the term on the left-hand side is the sum of the momentum accumulation $\partial(\rho V)/\partial t$ plus the momentum flux $V\partial(\rho V)/\partial x$. The left-hand term is the sum of the forces acting on the control volume: the body force (or volume force) acting on the mass as a whole, and surface forces acting at the control surface.

For an incompressible flow (i.e. $\rho = \text{constant}$), for a Newtonian fluid and assuming that the viscosity is constant over the control volume, the motion equation becomes:

$$\rho\left(\frac{\partial V_x}{\partial t} + \sum_j V_j \frac{\partial V_x}{\partial x_j}\right) = \rho F_{\text{vol}_x} - \frac{\partial P}{\partial x} + F_{\text{visc}_x} \tag{2.10a}$$

$$\rho\left(\frac{\partial V_y}{\partial t} + \sum_j V_j \frac{\partial V_y}{\partial x_j}\right) = \rho F_{\text{vol}_y} - \frac{\partial P}{\partial y} + F_{\text{visc}_y} \tag{2.10b}$$

$$\rho\left(\frac{\partial V_z}{\partial t} + \sum_j V_j \frac{\partial V_z}{\partial x_j}\right) = \rho F_{\text{vol}_z} - \frac{\partial P}{\partial z} + F_{\text{visc}_z} \tag{2.10c}$$

where ρ, the fluid density, is assumed constant in time and space. Equations (2.10) are often called the *Navier–Stokes equation*.

Considering a *two-dimensional flow* in the (x, y) plane and for *gravity forces*, the Navier–Stokes equation becomes:

$$\rho\left(\frac{\partial V_x}{\partial t} + V_x \frac{\partial V_x}{\partial x} + V_y \frac{\partial V_x}{\partial y}\right) = -\rho g \frac{\partial z}{\partial x} - \frac{\partial P}{\partial x} + F_{\text{visc}_x} \tag{2.11a}$$

$$\rho\left(\frac{\partial V_y}{\partial t} + V_x \frac{\partial V_y}{\partial x} + V_y \frac{\partial V_y}{\partial y}\right) = -\rho g \frac{\partial z}{\partial y} - \frac{\partial P}{\partial y} + F_{\text{visc}_y} \tag{2.11b}$$

where z is aligned along the vertical direction and is positive upward. Note that the x- and y-directions are perpendicular to each other and are independent of (and not necessarily orthogonal to) the vertical direction.

Notes

1. For gravity force the volume force potential U is:

$$U = \mathbf{g} \times \mathbf{x}$$

where $\mathbf{g}$ is the gravity acceleration vector and $\mathbf{x} = (x, y, z)$, z being the vertical direction positive upward. It yields that the gravity force vector equals:

$$F_{\text{vol}} = -\text{grad}(gz)$$

2. A Newtonian fluid is characterized by a linear relation between the magnitude of shear stress τ and the rate of deformation $\partial V/\partial y$ (equation (1.1)), and the stress tensor is:

$$\sigma_{ij} = -P\delta_{ij} + \tau_{ij}$$

$$\tau_{ij} = -\frac{2\mu}{3}\varepsilon\delta_{ij} + 2\mu\varepsilon_{ij}$$

where P is the static pressure, τ_{ij} is the shear stress component of the i momentum transported in the j direction, δ_{ij} is the identity matrix element: $\delta_{ii} = 1$ and $\delta_{ij} = 0$ (for i different from j),

$$\varepsilon_{ij} = \frac{1}{2}\left(\frac{\partial V_i}{\partial x_j} + \frac{\partial V_j}{\partial x_i}\right) \qquad \text{and} \qquad \varepsilon = \text{div } V = \sum_i \frac{\partial V_i}{\partial x_i}$$

3. The vector of the viscous forces is:

$$F_{\text{visc}_i} = \text{div } \tau_i = \sum_j \frac{\partial \tau_{ij}}{\partial x_j}$$

For an incompressible flow the continuity equation gives: $\varepsilon = \text{div } V = 0$. And the viscous force per unit volume becomes:

$$F_{\text{visc}_i} = \sum_j \mu \frac{\partial^2 V_i}{\partial x_j \, \partial x_j}$$

where μ is the dynamic viscosity of the fluid. Substituting this into equations (2.11), yields:

$$\rho \left(\frac{\partial V_x}{\partial t} + V_x \frac{\partial V_x}{\partial x} + V_y \frac{\partial V_x}{\partial y} \right) = -\rho g \frac{\partial z}{\partial x} - \frac{\partial P}{\partial x} + \sum_j \mu \frac{\partial^2 V_x}{\partial x_j \, \partial x_j} \qquad (2.12a)$$

$$\rho \left(\frac{\partial V_y}{\partial t} + V_x \frac{\partial V_y}{\partial x} + V_y \frac{\partial V_y}{\partial y} \right) = -\rho g \frac{\partial z}{\partial y} - \frac{\partial P}{\partial y} + \sum_j \mu \frac{\partial^2 V_y}{\partial x_j \, \partial x_j} \qquad (2.12b)$$

Equation (2.12) is the *original* Navier–Stokes equation.
4. Equation (2.12) was derived by Navier in 1822 and Poisson in 1829 by entirely different methods. Equations (2.10) to (2.12) were derived later in a manner similar to the above by Barré de Saint-Venant in 1843 and Stokes in 1845.
5. Louis Navier (1785–1835) was a French engineer who primarily designed bridges but also extended Euler's equations of motion. Siméon Denis Poisson (1781–1840) was a French mathematician and scientist. He developed the theory of elasticity, a theory of electricity and a theory of magnetism. Adhémar Jean Claude Barré de Saint-Venant (1797–1886), a French engineer, developed the equations of motion of a fluid particle in terms of the shear and normal forces exerted on it. George Gabriel Stokes (1819–1903), a British mathematician and physicist, is known for his research in hydrodynamics and a study of elasticity.

Application

Considering on open channel flow in a rectangular channel (Fig. 2.2), we assume a one-dimensional flow, with uniform velocity distribution, a constant channel slope θ and a constant channel width B. The Navier–Stokes equation in the s-direction is:

$$\rho \left(\frac{\partial v}{\partial t} + v \frac{\partial v}{\partial s} \right) = -\rho g \frac{\partial z}{\partial s} - \frac{\partial P}{\partial s} + F_{\text{visc}}$$

where v is the velocity along a streamline. Integrating the Navier–Stokes equation over the control volume, the forces acting on the control volume shown in Fig. 2.2 in the s-direction are:

$$\int_{\text{CV}} -\rho g \frac{dz}{ds} = +\rho g A \Delta s \sin \theta \qquad \text{Volume force (i.e. weight)}$$

$$\int_{\text{CV}} -\frac{dP}{ds} = -\rho g d \Delta d B \cos \theta$$

$$\text{Pressure force (assuming hydrostatic pressure distribution)}$$

$$\int_{\text{CV}} F_{\text{vis}} = -\tau_0 P_w \Delta s \qquad \text{Friction force (i.e. boundary shear)}$$

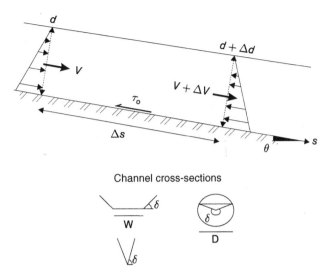

Fig. 2.2 Control volume for an open channel flow.

where A is the cross-section area (i.e. $A = Bd$ for a rectangular channel), d is the flow depth, Δs is the length of the control volume, τ_o is the average bottom shear stress and P_w is the wetted perimeter.

Assuming a *steady* flow, the change in momentum equals:

$$\int_{CV} \rho v \frac{dv}{ds} = \rho A \Delta s V \frac{\Delta V}{\Delta s}$$

where V is the mean flow velocity. The term (ρV) is the momentum per unit volume.

The integration of the Navier–Stokes equation for a one-dimensional steady open channel flow yields:

$$\rho A V \Delta V = +\rho g A \Delta s \sin\theta - \rho g d \Delta dB \cos\theta - \tau_o P_w \Delta s$$

The term on the left is the gradient of momentum flux: $\Delta(\frac{1}{2}(\rho V)(V A))$ (i.e. the rate of change of momentum). On the right-hand side of the equation, $(+\rho g A \sin\theta \Delta s)$ is the gravity force (potential energy), $(-\rho g d \Delta dB \cos\theta)$ is the pressure term (work of the flow) and $(-\tau_o P_w \Delta s)$ is the friction force (losses).

Application of the momentum equation

Considering an arbitrary control volume, it is advantageous to select a volume with a control surface perpendicular to the flow direction denoted s. For a steady and incompressible flow the forces acting on the control volume in the s-direction are equal to the rate of change in the flow momentum (i.e. no momentum accumulation). The momentum equation gives:

$$\sum F_s = \rho_2 A_2 V_2 V_{s2} - \rho_1 A_1 V_1 V_{s1} \tag{2.13a}$$

where $(\sum F_s)$ is the resultant of all the forces in the s-direction, the subscripts 1 and 2 refer to the upstream and downstream cross-sections respectively and $(V_{si})_{i=1,2}$ is the velocity component in the s-direction. Combining with the continuity equation for a steady and incompressible flow, it yields:

$$\sum F_s = \rho Q (V_{s2} - V_{s1}) \tag{2.13b}$$

In simple terms: the momentum equation states that the change in momentum flux is equal to the sum of all forces (volume and surface forces) acting on the control volume.

Note

For a steady incompressible flow, the momentum flux equals $(\rho Q V)$.

Application: hydraulic jump

In open channels, the transition from a rapid flow to a slow flow is called a hydraulic jump (Fig. 2.3). The transition occurs suddenly and is characterized by a sudden rise of the liquid surface. The forces acting on the control volume are the hydrostatic pressure forces at each end of the control volume, the gravity force (i.e. the weight of water), the invert reaction force and the bottom friction. Considering a horizontal rectangular open channel of constant channel width B, and neglecting the shear stress at the channel bottom, the resultant of the forces acting in the s-direction are the result of hydrostatic pressure at the ends of the control volume. The continuity equation and momentum equations are respectively:

$$V_1 d_1 B = V_2 d_2 B \tag{2.14}$$

$$\rho Q (V_2 - V_1) = (\tfrac{1}{2}\rho g d_1^2 - \tfrac{1}{2}\rho g d_2^2) B \tag{2.15}$$

where B is the channel width (assumed constant) and Q is the total discharge (i.e. $Q = V d B$).

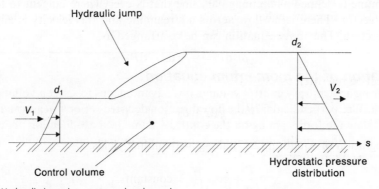

Fig. 2.3 Hydraulic jump in a rectangular channel.

Notes

If the velocity distribution is not uniform over the cross-section, the average velocity can be used by introducing the momentum correction factor β defined as:

$$\beta = \frac{\int_A \rho v^2 \, dA}{\rho V^2 A}$$

where A is the cross-sectional area (normal to the flow direction) and V is the mean flow velocity ($V = Q/A$).

Let us consider a practical case. For a steady flow in a horizontal open channel of rectangular cross-section, and assuming a hydrostatic pressure distribution, the momentum equation (2.15) becomes:

$$\rho Q(\beta_1 V_2 - \beta_2 V_1) = (\tfrac{1}{2}\rho g d_1^2 B_1) - (\tfrac{1}{2}\rho g d_2^2 B_2)$$

where Q is the discharge (m^3/s), d is the flow depth, B is the channel width and β the momentum correction coefficient. If the shape of the velocity distribution does not change substantially between cross-sections 1 and 2, the variations of momentum correction coefficient are negligible. The momentum equation becomes:

$$\rho \beta Q(V_2 - V_1) = (\tfrac{1}{2}\rho g d_1^2 B_1) - (\tfrac{1}{2}\rho g d_2^2 B_2)$$

The Bernoulli equation

The 'local form' of the Bernoulli equation can be deduced from the Navier–Stokes equation.

Considering {H1} the flow is along a *streamline*, assuming that {H2} the fluid is *frictionless* (i.e. $F_{\text{visc}} = 0$), {H3} the volume force *potential* (i.e. *gravity*) is *independent of time* (i.e. $\partial U/\partial t = 0$), for {H4} a *steady* flow (i.e. $\partial V/\partial t = 0$) and {H5} an *incompressible* flow (i.e. $\rho = $ constant), the Navier–Stokes equation (2.10) along the streamline becomes:

$$\rho v \frac{dv}{ds} = -\rho g \frac{dz}{ds} - \frac{dP}{ds} \tag{2.16}$$

where v is the velocity along the streamline, s is the direction along the streamline. A streamline is defined as an imaginary line that is everywhere tangent to the fluid velocity vector. There is no flow across a streamline and the velocity is aligned in the s-direction. The above equation can be re-arranged as:

$$\rho v \, dv = -\rho g \, dz - dP$$

and it can be rewritten as:

$$\frac{dP}{\rho} + g \, dz + d\left(\frac{v^2}{2}\right) = 0 \tag{2.17}$$

The integration of equation (2.17) along a streamline yields:

$$\frac{P}{\rho} + gz + \frac{v^2}{2} = \text{constant} \tag{2.18}$$

Equation (2.18) is the *Bernoulli equation*. Equation (2.17) is called the differential form of the Bernoulli equation. Each term of the Bernoulli equation may be interpreted by analogy as a form of energy:

1. P/ρ is analogous to the flow work per unit of mass of flowing fluid (net work done by the fluid element on its surroundings while it is flowing),
2. $U = gz$ is similar to the potential energy per unit mass, and
3. $v^2/2$ is related to the kinetic energy per unit mass.

If there are no friction losses, the sum of the fluid's potential energy, kinetic energy and pressure work is a constant. Along a streamline the flow 'energy' may be rearranged between kinetic energy (i.e. velocity), potential energy (i.e. altitude) or pressure work (i.e. flow depth) but the sum of all the terms must remain constant.

Notes

1. The Bernoulli equation is named after the Swiss mathematician Daniel Bernoulli (1700–1782) who developed the equation in his 'Hydrodynamica, de viribus et motibus fluidorum' textbook (first draft in 1733, first publication in 1738, Strasbourg) (Carvill 1981, Garbrecht 1987a, pp. 245–258).
2. Under particular conditions each of the assumptions underlying the Bernoulli equation may be abandoned.
   (a) For a gas flow such that the change of pressure is only a small fraction of the absolute pressure (i.e. less than 5%), the gas may be considered incompressible. Equation (2.18) may be applied with an average density ρ.
   (b) For unsteady flow with gradually changing conditions (i.e. slow emptying of a reservoir) the Bernoulli equation may be applied without noticeable error.
   (c) For a real fluid, the Bernoulli equation is used by first neglecting viscous shear to obtain theoretical results. The resulting equation may then be affected by an experimental coefficient to correct the analytical solution so that it conforms to the actual physical case.
3. Liggett (1993) developed a superb discussion of the complete integration process of the Navier–Stokes equation leading to the Bernoulli equation.

2.2.4 The energy equation

The first law of thermodynamics for a system states that the net energy (e.g. heat, potential energy) supplied to the system equals the increase in energy of the system plus the energy that leaves the system as work is done:

$$\frac{DE}{Dt} = \frac{\Delta Q_h}{\Delta t} - \frac{\Delta W_t}{\Delta t} \qquad (2.19)$$

where E is the total energy of the system, ΔQ_h the heat transferred *to* the system and ΔW_t the work done *by* the system. The energy of the system is the sum of: (1) the potential energy term (gz); (2) the kinetic energy term $(V^2/2)$; and (3) the internal energy e.

The work done by the system on its surroundings includes the work done by the pressure forces:

$$\Delta W_p = \Delta t \int_{CS} PV \, dA$$

and the work done by shear forces (i.e. on a rotating shaft) ΔW_s.

For a steady and one-dimensional flow through a control volume the first law of thermodynamics becomes:

$$\frac{\Delta Q_h}{\Delta t} + \left(\frac{P_1}{\rho_1} + gz_1 + \frac{V_1^2}{2} \right) \rho_1 V_1 A_1 = \frac{\Delta W_s}{\Delta t} + \left(\frac{P_2}{\rho_2} + gz_2 + \frac{V_2^2}{2} \right) \rho_2 V_2 A_2 \qquad (2.20)$$

where the subscripts 1 and 2 refer to the upstream and downstream flow conditions respectively.

Since the flow is steady the conservation of mass implies:

$$\rho_1 V_1 A_1 = \rho_2 V_2 A_2 \qquad (2.21)$$

and dividing the first law of thermodynamics by $(\rho V A)$ the energy equation in differential form becomes:

$$dq_h - dw_s = d\left(\frac{P}{\rho} \right) + (g \, dz + V \, dV + de) \qquad (2.22)$$

where e is the internal energy per unit mass, q_h is the heat added to the system per unit mass and w_s is the work done (by shear forces) by the system per unit mass. For a frictionless fluid (reversible transformation) the first law of thermodynamics may be written in terms of the entropy S as:

$$de = T \, dS - P d\left(\frac{1}{\rho} \right) \qquad (2.23)$$

The Clausius inequality states that:

$$dS > \frac{dq_h}{T}$$

Replacing the internal energy by the above equation and calling 'Losses' the term $(T \, dS - dq_h)$, the energy equation becomes:

$$\left(\frac{dP}{\rho} + g \, dz + V \, dV \right) + dw_s + d(\text{Losses}) = 0 \qquad (2.24)$$

In the absence of work of shear forces (i.e. $w_s = 0$) this equation differs from the differentiation of the Bernoulli equation (2.17) by the loss term only.

Notes

1. For non-steady flows the energy equation (for two particular cases) becomes:
   (a) for perfect and frictionless gas without heat added to the system, the energy equation is:

   $$\rho C_p \frac{dT}{dt} = \frac{dP}{dt}$$

(b) for an incompressible and undilatable fluid the energy equation is:

$$\rho C_{\mathrm{p}} \frac{\mathrm{d}T}{\mathrm{d}t} = \kappa \Delta T + \Psi + \Phi$$

where C_{p} is the specific heat at constant pressure, κ is the thermal diffusivity, Ψ is the volume density of heat added to the system and Φ is the dissipation rate. The knowledge of the density ρ (from the continuity equation) and the pressure distribution (from the Navier–Stokes equation) enables the calculation of the temperature distribution.

2. If work is done on the fluid in the control volume (i.e. pump) the work done by shear forces w_{s} is negative.

2.3 Exercises

Considering a circular pipe (diameter 2.2 m), the total flow rate is 1600 kg/s. The fluid is slurry (density 1360 kg/m³). What is the mean flow velocity?

For a two-dimensional steady incompressible flow, write the Navier–Stokes equation in Cartesian coordinates.

Considering a sluice gate in a horizontal smooth rectangular channel, write the momentum and Bernoulli equations as functions of the flow rate, channel width, upstream and downstream depths and the force of the gate onto the fluid only.

3

Applications of the Bernoulli equation to open channel flows

Summary

Applications of the Bernoulli equation to open channel flow situations (neglecting the flow resistance) are developed. In this chapter it is assumed that the fluid is frictionless. The concept of specific energy is introduced for free surface flows and critical flow conditions are defined. The Froude number is introduced.

3.1 Introduction

The Bernoulli equation is derived from the Navier–Stokes equation, considering the flow along a streamline, assuming that the volume force potential is independent of the time, for a frictionless and incompressible fluid, and for a steady flow. For gravity forces the differential form of the Bernoulli equation is:

$$\left(\frac{\mathrm{d}P}{\rho} + g\,\mathrm{d}z + V\,\mathrm{d}V \right) = 0 \tag{2.17}$$

where V is the velocity, g is the gravity constant, z is the altitude, P is the pressure and ρ is the fluid density.

This chapter describes hydraulic applications in which the Bernoulli equation is valid: e.g. smooth channel, short transitions, sluice gate.

3.2 Application of the Bernoulli equation – specific energy

3.2.1 Bernoulli equation

Summary
The Bernoulli equation (2.18) was obtained for a frictionless steady flow of incompressible fluid. It states that, along a streamline,

$$\frac{V^2}{2} + gz + \frac{P}{\rho} = \text{constant} \tag{2.18}$$

Dividing by the gravity constant, the Bernoulli equation becomes:

$$\frac{v^2}{2g} + z + \frac{P}{\rho g} = \text{constant} \tag{3.1}$$

where the constant is expressed in metres.

Along any streamline, the *total head* is defined as:

$$H = \frac{v^2}{2g} + z + \frac{P}{\rho g} \qquad \text{total head along a streamline} \tag{3.2a}$$

where v, z, P are the local fluid velocity, altitude and pressure. The term $(P/(\rho g) + z)$ is often called the *piezometric head*.

Note

If the flow is accelerated the following equation must be used:

$$\frac{D}{Dt}\left(\frac{v^2}{2}\right) + \frac{DU}{Dt} + \frac{1}{\rho}\frac{DP}{Dt} = \frac{1}{\rho}\frac{\partial P}{\partial t}$$

where U is the volume force potential (i.e. gravity force: $U = gz$).

If the velocity is not uniform, the total head H is a function of the distance y from the channel:

$$H(y) = \frac{V(y)^2}{2g} + z(y) + \frac{P(y)}{\rho g} \tag{3.2b}$$

where y is measured perpendicular to the channel bottom. Assuming a hydrostatic pressure gradient (see next section), the local elevation $z(y)$ and pressure $P(y)$ can be transformed:

$$z(y) = z_o + y\cos\theta$$
$$P(y) = P_{\text{atm}} + \rho g(d - y)\cos\theta$$

where z_o is the bottom elevation, θ is the channel slope, d is the flow depth (Fig. 3.1). With uniform or non-uniform velocity distributions, the piezometric head is a constant *assuming a hydrostatic pressure distribution*:

$$\left(\frac{P}{\rho g} + z\right) = \frac{P_{\text{atm}}}{\rho g} + z_o + d\cos\theta \qquad \text{piezometric head} \tag{3.3}$$

In summary: at a given cross-section, although the total head varies with respect to the distance y from the channel bottom, the piezometric head in an open channel is constant if the pressure is hydrostatic.

Application of the Bernoulli equation

Hydrostatic pressure distribution in open channel flow

If the flow is not accelerated (i.e. $dv = 0$) the differential form of the Bernoulli equation (2.17) gives the pressure distribution:

$$\frac{dP}{\rho} + g\,dz = 0$$

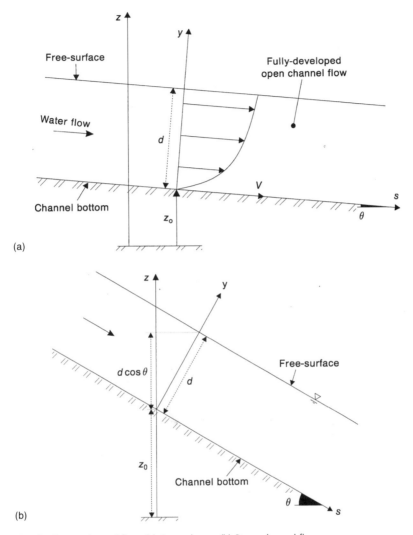

Fig. 3.1 Sketch of open channel flow. (a) General case; (b) Steep channel flow.

For a horizontal channel it leads to:

$$P = P_{\text{atm}} - \rho g(z - (z_o + d)) \tag{3.4}$$

where z is along the vertical direction and positive upward, z_o is the bottom elevation and d is the flow depth. Equation (3.4) is simply a rewriting of the hydrostatic pressure distribution in static fluid (i.e. (1.3)).

Pressure distribution in open channel flow
In open channel flow, the assumption of hydrostatic pressure distribution is valid in gradually varied flow along a flat slope. Equation (3.4) is not valid if the flow acceleration is important, if there is a marked acceleration perpendicular to the flow, if the slope is steep or if the streamline curvature is pronounced. For the flow

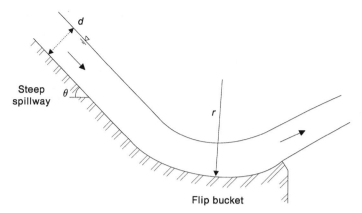

Fig. 3.2 Open channel flow in a curved channel.

along a curved streamline, the centrifugal acceleration equals:

$$g_{centrif} = \pm \frac{V^2}{r} \tag{3.5}$$

where $g_{centrif}$ is the centrifugal acceleration due to the streamline curvature, r is the radius of curvature of the streamline, $(+)$ is used for concave streamline curvature, $(-)$ is used for convex curvature.

Considering a curved channel (Fig. 3.2), the centrifugal pressure equals approximately:

$$P_{centrif} = \pm \frac{\rho V^2 d}{r} \tag{3.6}$$

where $P_{centrif}$ is the pressure rise due to the channel bottom curvature (e.g. flip bucket at spillway toe, Henderson 1966, p. 189), V is the mean flow velocity, r is the radius of curvature of the invert, $(+)$ is used for concave boundary curvature, $(-)$ is used for convex curvature.

Considering now an open channel flow down a steep slope θ (e.g. Fig. 3.1(b)), the pressure at the channel bottom is (in the absence of channel curvature):

$$P = P_{atm} + \rho g d \cos \theta \tag{3.7}$$

where d is the flow depth measured perpendicular to the channel bottom and θ the channel slope.

In the general case, the pressure at the channel bottom equals:

$$P = P_{atm} + \rho d \left(g \cos \theta \pm \frac{V^2}{r} \right) \tag{3.8}$$

where θ is the bed slope and r is the radius of curvature of the invert.

Mean total head

In an open channel flow, it is often convenient to consider the depth-averaged total head H defined as:

$$H = \frac{1}{d} \int_0^d H(y) \, dy = \frac{1}{d} \int_0^d \left(\frac{V(y)^2}{2g} + z(y) + \frac{P(y)}{\rho g} \right) dy$$

Replacing the pressure P and elevation z by their respective expressions, the mean total head becomes:

$$H = \frac{V^2}{2g} + z_0 + \frac{d}{g}\left(g\cos\theta \pm \frac{V^2}{r}\right) + P_{\text{atm}} \qquad (3.9a)$$

where z_0 is the bottom elevation, and a uniform velocity distribution, (i.e. $v(y) = V$) is assumed.

In open channel hydraulics the local pressure is often taken as the pressure relative to the atmospheric pressure P_{atm} and the mean total head is rewritten as:

$$H = \frac{V^2}{2g} + z_0 + \frac{d}{g}\left(g\cos\theta \pm \frac{V^2}{r}\right) \qquad (3.9b)$$

where V is the mean flow velocity, z_0 is the bed elevation, d is the flow depth, θ is the bed slope.

Notes
1. An important and practical result to remember is that: *the pressure gradient is not hydrostatic if the curvature of the streamlines is important.* In the absence of additional information, the free-surface curvature and/or the bottom curvature can be used to give some indication of the streamline curvature. The free-surface and the channel bottom are both streamlines (i.e. no flow across the free-surface nor the bottom). Any rapid change of flow stage (i.e. a short transition) is always accompanied with a change of streamline curvature, under which circumstances it is seldom possible to assume hydrostatic pressure distribution. Rouse (1938, pp. 319–321) discussed in detail one particular example: the broad-crested weir.
2. A flip bucket (e.g. Fig. 3.2) or ski-jump is a deflector located at the end of a spillway and designed to deflect the flow away from the spillway foundation. The waters take off at the end of the deflector as a free-fall flow and impact further downstream in a dissipation pool (e.g. Fig. 17.3(b)).

Pitot tube
A direct application of the Bernoulli equation is a device for measuring fluid velocity called the Pitot tube. Aligned along a streamline, the Pitot tube enables one to measure the total head H and the pressure head (i.e. $z + P/(\rho g)$) denoted H_1 and H_2 respectively in Fig. 3.3. The fluid velocity is then deduced from the total head difference.

Notes
1. The Pitot tube is named after the Frenchman H. Pitot. The first presentation of the concept of the Pitot tube was made in 1732 at the French Academy of Sciences by Henri Pitot.
2. The original Pitot tube included basically a total head reading. L. Prandtl improved the device by introducing a pressure reading. The modified Pitot tube (illustrated in Fig. 3.3) is sometimes called a Pitot–Prandtl tube.

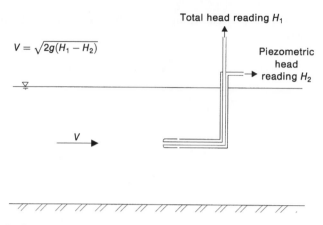

$$V = \sqrt{2g(H_1 - H_2)}$$

Total head reading H_1

Piezometric head reading H_2

V

Fig. 3.3 Sketch of a Pitot tube.

3. The Pitot tube is also called the Pitot static tube.
4. For many years, aeroplanes used Pitot tubes to estimate their relative velocity.

Application: smooth transition

Considering a smooth flow transition (e.g. a step, a weir), for a steady, incompressible, frictionless flow, between an upstream section (1) and a downstream section (2), the continuity and Bernoulli equations applied to the open channel flow become:

$$V_1 A_1 = V_2 A_2 = Q \tag{3.10a}$$

$$z_{o1} + d_1 \cos \theta_1 + \frac{V_1^2}{2g} = z_{o2} + d_2 \cos \theta_2 + \frac{V_2^2}{2g} \tag{3.11}$$

assuming a hydrostatic pressure distribution and where z_{o_1} and z_{o_2} are the bottom elevations, d_1 and d_2 are the flow depths measured normal to the channel bottom (Fig. 3.4). For a rectangular channel, the continuity

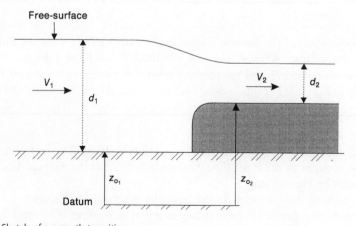

Free-surface

V_1

d_1

V_2 d_2

z_{o_1} z_{o_2}

Datum

Fig. 3.4 Sketch of a smooth transition.

equations becomes:

$$V_1 d_1 B_1 = V_2 d_2 B_2 = Q \qquad (3.10b)$$

If the discharge Q, the bed elevations (z_{o_1}, z_{o_2}), channel widths (B_1, B_2) and channel slopes (θ_1, θ_2) are known, the downstream flow conditions (d_2, V_2) can be deduced from the upstream flow conditions (d_1, V_1) using equations (3.10) and (3.11).

3.2.2 Influence of the velocity distribution

Introduction

The above developments were obtained assuming (or implying) uniform velocity distributions: i.e. the velocity was assumed constant over the entire cross-section. In practice, the velocity distribution is not uniform because of the bottom and sidewall friction.

It will be shown that most results are still valid using the mean flow velocity (i.e. $V = Q/A$) instead of the uniform velocity by introducing velocity correction coefficients: the Boussinesq or momentum correction coefficient, and the Coriolis or kinetic energy correction coefficient.

Velocity distribution

The velocity distribution in fully-developed turbulent open channel flows is given approximately by Prandtl's power law:

$$\frac{V}{V_{\text{max}}} = \left(\frac{y}{d}\right)^{1/N} \qquad (3.12)$$

where the exponent $1/N$ varies from $1/4$ down to $1/12$ depending upon the boundary friction and cross-section shape. The most commonly-used power law formulae are the one-sixth power $(1/6)$ and the one-seventh power $(1/7)$ formula.

Notes

1. Chen (1990) developed a complete analysis of the velocity distribution in an open channel and pipe flow with reference to flow resistance. Moreover, he showed that the Gauckler–Manning formula (see Chapter 4) implies a $1/6$ power formula while the $1/7$ power formula derives from the Blasius resistance formula (i.e. for smooth turbulent flow). In uniform equilibrium flows, the velocity distribution exponent is related to the flow resistance:

$$N = K\sqrt{\frac{8}{f}}$$

where f is the Darcy friction factor and K is the von Karman constant (K = 0.4).

2. In practical engineering applications, N can range from 4 (for shallow waters in wide rough channels) up to 12 (smooth narrow channel). A value $N = 6$ is reasonably representative of open channel flows in smooth-concrete channels. However, it must be remembered that N is a function of the flow resistance.

3. For a wide rectangular channel, the relationship between the mean flow velocity V and the free-surface velocity V_{max} derives from the continuity equation:

$$q = Vd = \int_0^d v \, dy = \frac{N}{N+1} V_{max} d$$

Velocity coefficients

Momentum correction coefficient

If the velocity distribution is not uniform over the cross-section, a correction coefficient must be introduced in the momentum equation if the average velocity is used. The momentum correction factor β is defined as (e.g. Streeter and Wylie 1981, p. 116):

$$\beta = \frac{\int_A \rho v^2 \, dA}{\rho V^2 A} \tag{3.13}$$

where V is the mean velocity over a cross-section (i.e. $V = Q/A$). The momentum correction coefficient is sometimes called the Boussinesq coefficient.

Considering a steady flow in a horizontal channel and assuming a hydrostatic pressure distribution (e.g. Fig. 2.3), the momentum equation is often rewritten as:

$$\rho Q(\beta_2 V_2 - \beta_1 V_1) = (\tfrac{1}{2}\rho g d_1^2 B_1) - (\tfrac{1}{2}\rho g d_2^2 B_2)$$

where Q is the discharge (m³/s), d is the flow depth, B is the channel width and β the momentum correction coefficient.

Notes

1. The Boussinesq coefficient is named after J. Boussinesq, French mathematician (1842–1929) who proposed it first (Boussinesq 1877).
2. The momentum correction coefficient is always larger than unity. ($\beta = 1$) implies a uniform velocity distribution.
3. For a $1/N$ power law velocity distribution (equation (3.12)), the momentum correction coefficient equals:

$$\beta = \frac{(N+1)^2}{N(N+2)}$$

Kinetic energy correction coefficient

If the velocity varies across the section, the mean of the velocity head $(v^2/(2g))_{mean}$ is not equal to $V^2/(2g)$, where the subscript mean refers to the mean value over the cross-section. The ratio of these quantities is called the *Coriolis coefficient*, denoted

α and defined as:

$$\alpha = \frac{\int_A \rho v^3 \, \mathrm{d}A}{\rho V^3 A} \tag{3.14}$$

If the energy equation is rewritten in terms of the *mean total head*, the later must be transformed as:

$$H = \alpha \frac{V^2}{2g} + z_\mathrm{o} + d \cos \theta$$

assuming a hydrostatic pressure distribution.

Notes

1. The Coriolis coefficient is named after G.G. Coriolis, a French engineer, who first introduced the correction coefficient (Coriolis 1836).
2. The Coriolis coefficient is also called the kinetic energy correction coefficient.
3. α is equal to or larger than 1 but rarely exceeds 1.15 (see the discussion in Li and Hager 1991). For a uniform velocity distribution: $\alpha = 1$.
4. For a $1/N$ power law velocity distribution (equation (3.12)), the kinetic energy coefficient equals:

$$\alpha = \frac{(N+1)^3}{N^2(N+3)}$$

Correction coefficient in the Bernoulli equation

As shown in Chapter 2, the Bernoulli equation derives from the Navier–Stokes equation: i.e. from the conservation of momentum. Application of the Bernoulli equation to open channel flow (within the frame of relevant assumptions) yields:

$$\beta \frac{V'^2}{2g} + z_\mathrm{o} + d = \text{constant} \tag{3.15}$$

where V' is the depth-averaged velocity:

$$V' = \frac{1}{d} \int_0^d v \, \mathrm{d}y$$

Note that V' usually differs from V.

Equation (3.15) is the depth-averaged Bernoulli equation in which the correction factor is the Boussinesq coefficient.

Notes

1. Some textbooks introduce (incorrectly) the kinetic energy correction coefficient in the Bernoulli equation. This is *incorrect*.
2. The reader is referred to the excellent discussion by Liggett (1993) for further details.

3.2.3 Specific energy

Definition

The *specific energy* E is defined as:

$$E(y) = \frac{P(y)}{\rho g} + \frac{V(y)^2}{2g} + (z(y) - z_o) \tag{3.16}$$

where P is the pressure, V is the velocity, z is the elevation, z_o is the bed elevation. The specific energy is similar to the energy per unit mass, measured with the channel bottom as the datum. The specific energy changes along a channel because of changes of the bottom elevation and energy losses (e.g. friction loss).

The mean specific energy is defined as:

$$E = H - z_o \tag{3.17a}$$

where H is the mean total head. For a *flat channel*, assuming a *hydrostatic pressure distribution* (i.e. $P = \rho g d$) and a uniform velocity distribution, it yields:

$$E = d + \frac{V^2}{2g} \qquad \text{(flat channel and hydrostatic pressure distribution)} \tag{3.17b}$$

where d is the flow depth and V is the mean flow velocity.

Notes

It must be emphasized that, even if the total head H is constant, the specific energy varies with the bed elevation z_o (or bed altitude) as:

$$H = E + z_o$$

For a non-uniform velocity distribution, assuming a hydrostatic pressure distribution, the *mean specific energy* as used in the energy equation becomes:

$$E = d \cos \theta + \alpha \frac{V^2}{2g}$$

where θ is the channel slope, α is the Coriolis coefficient and V is the mean flow velocity: $V = Q/A$.

For a horizontal channel, the Bernoulli equation implies that the specific energy is constant. This statement is true only within the assumptions of the Bernoulli equation: i.e. for an incompressible, frictionless and steady flow along a streamline.

For a rectangular channel it is convenient to combine the continuity equation and the specific energy definition. Using the total discharge Q, the expression for the specific energy becomes (for a flat channel):

$$E = d + \frac{Q^2}{2gd^2B^2} \tag{3.17c}$$

where B is the free surface width.

Analysis of the specific energy

Flow depth versus specific energy

The specific energy is usually studied as a function of the flow depth d. It is convenient to plot the relationship: $d = f(E)$ as shown in Fig. 3.5. In rectangular channels there is only one specific energy–flow depth curve for a given discharge per unit width Q/B.

For a tranquil and slow flow, the velocity is small and the flow depth is large. The kinetic energy term $V^2/(2g)$ is very small and the specific energy tends to the flow depth d (i.e. asymptote $E = d$).

For a rapid flow (e.g. a torrent), the velocity is large and, by continuity, the flow depth is small. The pressure term $P/(\rho g)$ (i.e. flow depth) is small compared with the kinetic energy term. The specific energy term tends to an infinite value when d tends to zero (i.e. asymptote $d = 0$).

At any cross-section, the specific energy has a unique value. For a given value of specific energy and a given flow rate, there may be zero, one or two possible flow depths (Fig. 3.5, Appendix A1.4).

Critical flow conditions

For a constant discharge Q and a given cross-section, the relationship $E = f(d)$ indicates the existence of a minimum specific energy (Fig. 3.5). The flow conditions,

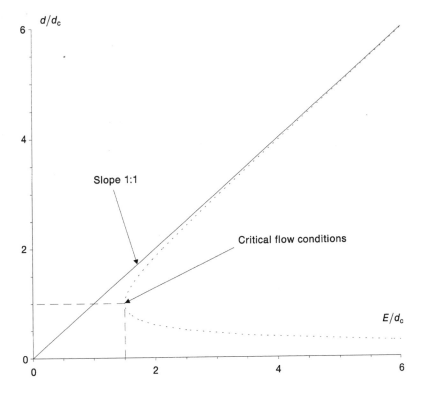

Fig. 3.5 Dimensionless specific energy curve for a flat rectangular channel (equation (3.20)).

(d_c, V_c), such that the mean specific energy is minimum, are called the *critical flow conditions*. They take place for:

$$\left(\frac{\partial E}{\partial d}\right)_{(Q\text{ constant})} = 0$$

The critical flow conditions may be expressed in terms of the discharge Q and geometry of the channel (cross-section A, free surface width B) after transformation of equation (3.16). For a rectangular flat channel of constant width, the minimum specific energy E_{min} and the critical flow depth are, respectively:

$$E_{min} = \tfrac{3}{2}d_c \tag{3.18}$$

$$d_c = \sqrt[3]{\frac{Q^2}{gB^2}} \tag{3.19}$$

The specific energy can be rewritten in dimensionless terms (for a flat channel) as:

$$\frac{E}{d_c} = \frac{d}{d_c} + \frac{1}{2}\left(\frac{d_c}{d}\right)^2 \tag{3.20}$$

Equation (3.20) is plotted in Fig. 3.5. Note that equation (3.20) is a unique curve: it is valid for any discharge.

The relationship specific energy versus flow depth indicates two trends (Fig. 3.5). When the flow depth is greater than the critical depth (i.e. $d/d_c > 1$), an increase in specific energy causes an increase in depth. For $d > d_c$, the flow is termed *subcritical*. When the flow depth is less than the critical depth, an increase in specific energy causes a decrease in depth. The flow is *supercritical*.

If the flow is critical (i.e. the specific energy is minimum), small changes in specific energy cause large changes in depth. In practice, critical flow over a long reach of channel is unstable. The definition of specific energy and critical flow conditions are summarized in the following table for flat channels of irregular cross-section (area A) and channels of rectangular cross-section (width B):

Variable	Channel of irregular cross-section	Rectangular channel
Specific energy E	$d + \dfrac{Q^2}{2gA^2}$	$d + \dfrac{q^2}{2gd^2}$
$\dfrac{\partial E}{\partial d}$	$1 - \dfrac{Q^2 B}{gA^3}$	$1 - \dfrac{q^2}{gd^3}$
$\dfrac{\partial^2 E}{\partial d^2}$	$\dfrac{Q^2}{gA^3}\left(\dfrac{3B^2}{A} - \dfrac{\partial B}{\partial d}\right)$	$\dfrac{3q^2}{gd^4}$
$\lim E_{d\to 0^+}$ $\lim E_{d\to +\infty}$	infinite d	infinite d
Critical depth d_c		$\sqrt[3]{\dfrac{q^2}{g}}$
Critical velocity V_c	$\sqrt{g\dfrac{A_c}{B_c}}$	$\sqrt{gd_c}$

Variable	Channel of irregular cross-section	Rectangular channel
Minimum specific energy E_{min}	$d_c + \dfrac{1}{2}\dfrac{A_c}{B_c}$	$\dfrac{3}{2}d_c$
Froude number Fr	$\dfrac{V}{\sqrt{g\dfrac{A}{B}}}$	$\dfrac{V}{\sqrt{gd}}$
$\dfrac{\partial E}{\partial d}$	$1 - Fr^2$	$1 - Fr^2$

Notes: A = cross-section area; B = free-surface width; Q = total discharge; q = discharge per unit width.

Notes
1. In open channel flow, the Froude number is defined such that it equals 1 for critical flow conditions. The corollary is that critical flow conditions are reached if $Fr = 1$.
2. A general dimensionless expression of the specific energy is:

$$\frac{E}{d_c} = \frac{d}{d_c}\left(\cos\theta + \tfrac{1}{2}Fr^2\right)$$

For a flat channel it yields:

$$\frac{E}{d_c} = \frac{d}{d_c}\left(1 + \tfrac{1}{2}Fr^2\right)$$

3. The ratio A/B of the cross-sectional area over the free-surface width is sometimes called the mean depth.

Application of the specific energy

For a frictionless flow in a horizontal channel, the specific energy is constant along the channel. The specific energy concept can be applied to predict the flow under a sluice gate as a function of the gate operation (Fig. 3.6). For a given gate opening, a specific discharge $(q = Q/B)$ takes place and there is only one specific energy/flow depth curve.

The upstream and downstream values of the specific energy are equal (by application of the Bernoulli equation). For this value of specific energy, there are two possible values of the flow depth: one subcritical depth (i.e. upstream depth) and one supercritical depth (i.e. downstream depth). If the value of the specific energy is known, the upstream and downstream flow depths are deduced from the specific energy curve (Fig. 3.6(a)). If the specific energy is not known *a priori*, the knowledge of the upstream flow depth and of the flow rate fixes the specific energy, and the downstream flow depth is deduced by a graphical solution. If the discharge remains constant but for a larger gate opening (Fig. 3.6(b)), the upstream depth is smaller and the downstream flow depth is larger.

In Fig. 3.6(c), the specific energy is the same as in Fig. 3.6(a) but the flow rate is larger. The upstream and downstream flow depths are located on a curve corresponding to the larger flow rate. The graphical solution of the specific energy/flow depth

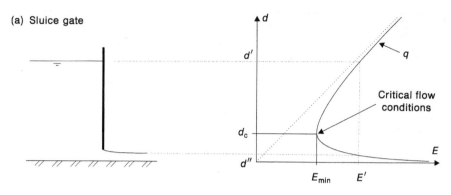

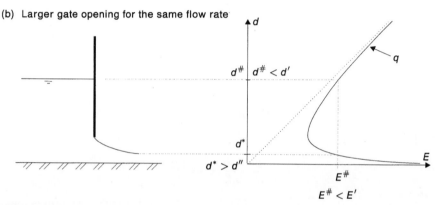

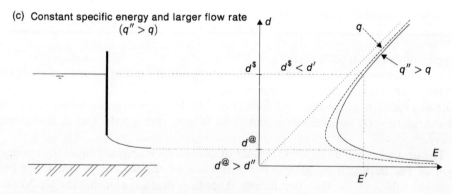

Fig. 3.6 Specific energy diagrams for flow under a sluice gate.

relationship indicates that the upstream flow depth is smaller and the downstream depth is larger than for the case in Fig. 3.6(a). This implies that the gate opening in Fig. 3.6(c) must be larger than in Fig. 3.6(a) to compensate for the larger flow rate.

Note

A sluice gate is a device used for regulating flow in open channels.

Discussion

Change in specific energy associated with a fixed discharge

Considering a rectangular channel, Figs 3.5 and 3.6 show the relationship flow depth versus specific energy. For a specific energy such as $E > E_{min}$, the two possible depths are called the *alternate depths* (Appendix A1.4).

Let us consider now a change in bed elevation. For a constant total head H, the specific energy decreases when the channel bottom rises. For a tranquil flow (i.e. $Fr < 1$ or $d > d_c$), a decrease in specific energy implies a decrease in flow depth as:

$$\frac{\partial E}{\partial d} = 1 - Fr^2 > 0 \tag{3.21}$$

For rapid flow (i.e. $Fr > 1$ or $d < d_c$), the relation is inverted: a decrease of specific energy implies an increase of flow depth. The results are summarized in the following table.

Flow depth	Fr	Type of flow	Bed raised	Bed lowered
$d > d_c$	<1	slow, tranquil, fluvial *Subcritical*	d decreases	d increases
$d < d_c$	>1	fast, shooting, torrential *Supercritical*	d increases	d decreases

Figure 3.7 shows practical applications of the specific energy concept to flow transitions in an open channel. For a change of bed elevation, the downstream flow conditions (E_2, d_2) can be deduced directly from the specific energy/flow depth curve. They are functions of the upstream flow conditions (E_1, d_1) and the change of bed elevation Δz_o.

Notes

1. Examples of applications of the relationship E versus d are shown in Figs 3.6 and 3.7: sluice gate, transition with bed elevation.
2. Rouse (1946, p. 139) illustrated with nice photographs the effects of a change in bed elevation for subcritical $(d > d_c)$ and supercritical $(d < d_c)$ flows.

Case of fixed specific energy

For a fixed specific energy E, the relationship flow rate versus flow depth in a rectangular flat channel is:

$$Q = Bd\sqrt{2g(E - d)} \tag{3.22a}$$

where d is the flow depth and B is the channel width. In dimensionless terms, it becomes:

$$\frac{Q}{B\sqrt{gE^3}} = \frac{d}{E}\sqrt{2\left(1 - \frac{d}{E}\right)} \tag{3.22b}$$

Equation (3.22b) is plotted in Fig. 3.8.

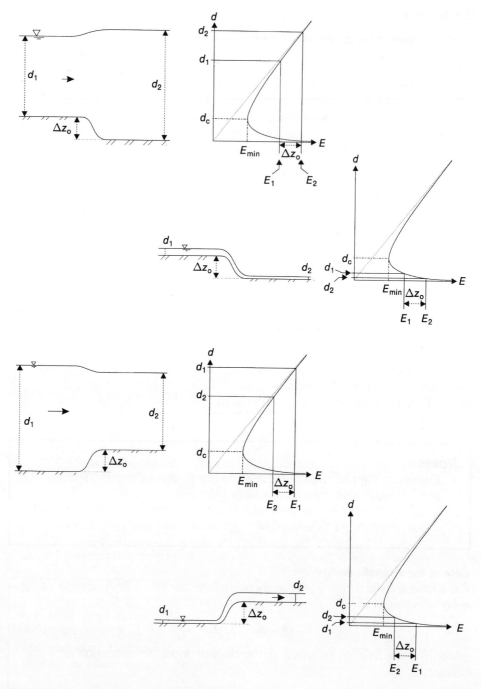

Fig. 3.7 Specific energy and transition problem.

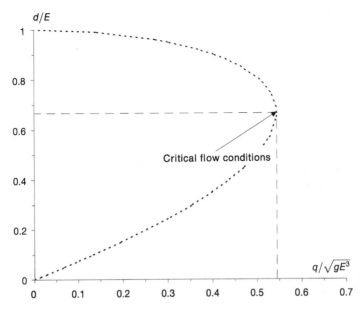

Fig. 3.8 Dimensionless discharge-depth curve for given specific energy in a horizontal rectangular channel (equation (3.22b)).

Figure 3.8 indicates that there is a maximum discharge Q_{max} such that the problem has a real solution. It can be shown that, for a fixed specific energy E, the maximum discharge is obtained for critical flow conditions (i.e. $d = d_c$) (see notes below). The results are detailed in the following table.

E constant	General section	Rectangular channel
Critical depth d_c	$d_c = E - \dfrac{A_c}{2B_c}$	$d_c = \frac{2}{3} E$
Maximum discharge Q_{max}	$Q_{max} = \sqrt{g \dfrac{A_c^3}{B_c}}$	$Q_{max} = B\sqrt{\left(\frac{2}{3}\right)^3 g E^3}$

Notes

1. The expression of the maximum discharge for a given total head was first formulated by the Frenchman J.B. Bélanger for the flow over a broad-crested weir (Bélanger 1849). He showed in particular that maximum flow rate is achieved for $\partial Q / \partial d = 0$ (if the streamlines are parallel to the weir crest).

2. Assuming a rectangular channel of constant width, equation (3.17) can be rewritten as:

$$Q = Bd\sqrt{2g(E - d)} \qquad (3.22a)$$

For a constant specific energy, the maximum discharge is obtained for $\partial Q / \partial d = 0$, i.e.

$$\frac{2gE - 3gd}{\sqrt{2g(E - d)}} = 0$$

or

$$E = \tfrac{3}{2}d$$

Such a relationship between specific energy and flow depth is obtained only at critical flow conditions (i.e. equation (3.18), Fig. 3.5). The result implies that $E = E_{\min}$ and $d = d_c$ (critical flow depth) for $Q = Q_{\max}$.

3.2.4 Limitations of the Bernoulli equation

The application of the Bernoulli equation is valid only within the range of assumptions (i.e. steady frictionless flow of incompressible fluid).

For *short and smooth transitions* the energy losses are negligible and the Bernoulli equation may be applied quite successfully. Short transitions may be gates (e.g. sluice or radial gates), weirs and steps (e.g. Figs 3.6, 3.7 and 3.9).

If energy losses (e.g. friction loss) occur, they must be taken in account in the energy equation. The Bernoulli equation (i.e. equation (3.1)) is no longer valid. For example, the Bernoulli equation is not valid at a hydraulic jump, where turbulent energy losses are significant.

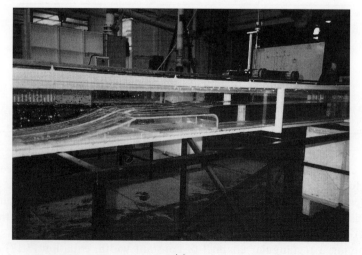

(a)

Fig. 3.9 Overflow above a broad-crested weir (weir height: 0.067 m, channel width: 0.25 m, crest length: 0.42 m). (a) $q = 0.0112\,\mathrm{m^2/s}$, upstream depth: 0.097 m, flow from the right to the left.

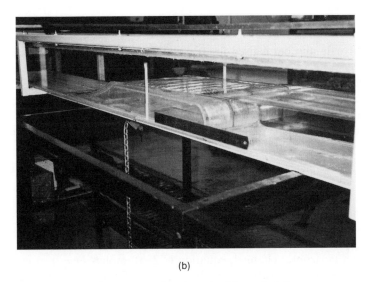

(b)

Fig. 3.9 (b) Undular flow above the weir crest, flow from the right to the left.

There is, however, another limitation of the use of the Bernoulli equation. At short transitions the velocity and flow depth may *not* be *uniform* across a section (e.g. flow through bridge piers) and the complete energy equation must be used (Henderson 1966, pp. 49–50).

Note

At short transitions (e.g. gate, weir intake), there is a rapid change in velocity and pressure distributions. It is often observed that the assumptions of uniform velocity distributions and hydrostatic pressure distributions are not valid.

3.3 Froude number

3.3.1 Definition

The *Froude number* is a dimensionless number proportional to the square root of the ratio of the inertial forces over the weight of fluid:

$$Fr = \frac{V}{\sqrt{gd_{\text{charac}}}} = \sqrt{\frac{\rho V^2 A}{\rho g A L}} \propto \sqrt{\frac{\text{inertial force}}{\text{weight}}}$$

where d_{charac} is a characteristic dimension (see notes below). For open channel flows, the relevant characteristic length is the flow depth d. For a rectangular channel, the Froude number is defined as:

$$Fr = \frac{V}{\sqrt{gd}} \qquad \text{rectangular channel} \qquad (3.23a)$$

> **Notes**
> 1. The Froude number is used for scaling open channel flows and free-surface flows.
> 2. The length d_{charac} is the characteristic geometric dimension: internal diameter of the pipe for pipe flows, flow depth for open channel flow in a rectangular channel, ship draught, hydrofoil thickness, etc.
> 3. For a channel of irregular cross-section shape, the Froude number is usually defined as:
>
> $$Fr = \frac{V}{\sqrt{g\dfrac{A}{B}}} \tag{3.23b}$$
>
> where V is the mean flow velocity, A is the cross-section area and B is the free-surface width.
>
> For a horizontal channel, equation (3.23b) implies $Fr = 1$ at critical flow conditions (see the section on the analysis of the specific energy, in this chapter).
> 4. Some authors use the notation:
>
> $$Fr = \frac{V^2}{gd_{charac}} = \frac{\rho V^2 A}{\rho g A d_{charac}} \propto \frac{\text{inertial force}}{\text{weight}}$$

3.3.2 Similarity and Froude number

In problems of fluid flow there are always at least four parameters involved: pressure difference ΔP, velocity V, length d and fluid density ρ. In open channel hydraulics another important parameter is the gravity acceleration g.

Considering a closed pipe circuit, and provided that the liquid remains liquid (i.e. no cavitation), the piezometric pressure $(P + \rho g z)$ is merely affected by the gravity effect as the pressure P is free to take any value at all. In an open channel, through the existence of a free surface where the pressure must take a prescribed value P_{atm}, the gravity can influence the flow pattern. For P_{atm} fixed, the gravity determines the piezometric pressure and hence the flow properties. In any flow situations where the gravity effects are significant, the appropriate dimensionless number (i.e. the Froude number) must be considered. In practice, model studies of open channel flows and hydraulic structures are performed using a Froude similitude. That is, the Froude number is the same for the model and the prototype (see Part III).

> **Notes**
> 1. For open channel flow (one-phase and one-component flow) the Froude number plays a very important part in the appropriate flow equations. This has been already noted at critical flow conditions defined as $d = d_c$ for which $Fr = 1$.

2. For two-phase (e.g. ice–water flows) or two-component (e.g. air–water flows) open channel flows, the Froude number might not be a significant number.
3. If the viscosity μ plays an effective part, the Reynolds number is another relevant parameter in dimensional analysis (see Part III).

Application

A prototype channel is 1000 m long and 12 m wide. The channel is rectangular and the flow conditions are: $d = 3$ m and $Q = 15 \, \text{m}^3/\text{s}$. What would be the size and discharge of a 1/25 scale model using a Froude similitude?

Solution

On the 1/25 scale model, all the geometric dimensions (including the flow depth) are scaled by a factor of 1/25:

Length	$L_{model} = L_{prototype}/25$
Width	$B_{model} = B_{prototype}/25$
Flow depth	$d_{model} = d_{prototype}/25$

The flow velocity on the model is deduced by a Froude similitude. The Froude similitude implies that the Froude number is the same on both model and prototype:

Froude similitude $Fr_{model} = Fr_{prototype}$

It yields:

Velocity $V_{model} = V_{prototype}/\sqrt{25}$

For the discharge, the continuity equation $Q = VdB$ provides the scaling ratio:

Discharge $Q_{model} = Q_{prototype}/(25)^{5/2}$

As a result the model characteristics are:

Length	$L_{model} = 40 \, \text{m}$
Width	$B_{model} = 0.25 \, \text{m}$
Flow depth	$d_{model} = 0.12 \, \text{m}$
Velocity	$V_{model} = 0.083 \, \text{m/s}$
Discharge	$Q_{model} = 0.0048 \, \text{m}^3/\text{s}$
Froude similitude	$Fr_{model} = Fr_{prototype} = 0.077$

3.3.3 Critical conditions and wave celerity

Considering an oscillatory wave[1], the velocity C or celerity[2] of waves of length l is given by:

$$C^2 = \frac{gl}{2\pi} \tanh\left(\frac{2\pi d}{l}\right) \tag{3.24}$$

---

[1] The term progressive wave can also be used (e.g. Liggett 1994).
[2] The speed of a disturbance is called celerity: e.g. sound celerity, wave celerity.

where d is the flow depth and tanh is the hyperbolic tangent function (Henderson 1966, p. 37; Liggett 1994, p. 394). For a long wave of small amplitude (i.e. $d/l \ll 1$) this expression becomes:

$$C^2 = gd$$

A similar result can be deduced from the continuity equation applied to a weak free-surface gravity wave of height $\Delta d \ll d$ in an open channel (Streeter and Wylie 1981, p. 514). The wave celerity equals:

$$C = \sqrt{gd} \qquad (3.25)$$

where C is the celerity of a small free-surface disturbance in open channel flow.

The Froude number may be written in a form analogous to the Sarrau–Mach number (see Section 3.3.4) in gas flow, as the ratio of the flow velocity divided by the celerity of small disturbances:

$$Fr = \frac{V}{C} \qquad (3.26)$$

where $C = \sqrt{gd}$ is the celerity of small waves (or small disturbance) at the free-surface in open channels.

For Froude or Sarrau–Mach numbers less than unity, disturbances at a point are propagated to all parts of the flow; however, for Froude and Sarrau–Mach numbers of greater than unity, disturbances propagate downstream[3] only. In other words, in a supercritical flow, small disturbances propagate only in the downstream flow direction, while in subcritical flows, small waves can propagate in both the upstream and downstream flow directions.

Note

Oscillatory waves (i.e. progressive waves) are characterized by no net mass transfer.

3.3.4 Analogy with compressible flow

In compressible flows, the pressure and the fluid density depend on the velocity magnitude relative to the celerity of sound in the fluid C_{sound}. The compressibility effects are often expressed in terms of the Sarrau–Mach number $Ma = V/C_{sound}$. Both the Sarrau–Mach number and the Froude number are expressed as the ratio of the fluid velocity over the celerity of a disturbance (celerity of sound and celerity of small wave respectively). Dimensional analysis shows that dynamic similarity in compressible flows is achieved with equality of both the Sarrau–Mach and Reynolds numbers, and equal value of the specific heat ratio.

The propagation of pressure waves (i.e. sound waves) in a compressible fluid is comparable with the movement of small amplitude waves on the surface of an

---

[3] Downstream means in the flow direction.

open channel flow. It was shown (e.g. Thompson 1972, Liggett 1994) that the combination of the motion equation for two-dimensional compressible flow with the state equation produces the same basic equation as for open channel flow (of an incompressible fluid), in which the gas density is identified with the flow depth (i.e. free-surface position). Such a result is obtained however assuming: an inviscid flow, a hydrostatic pressure gradient (and zero channel slope), and the ratio of specific heat γ must equal 2.

The formal analogy and correspondence of flow parameters are summarized in the following table.

	Open channel flow	Compressible flow
Basic parameters	flow depth d	gas density ρ
	velocity V	velocity V
	d^γ	absolute pressure P
	$d^{(\gamma-1)}$	absolute temperature T
Other parameters	$\sqrt{gd}$	sound celerity C
	Froude number	Sarrau–Mach number
	gravity acceleration g	$\dfrac{1}{2}\dfrac{P}{\rho^\gamma}$
	channel width B	flow area A
Flow analogies	hydraulic jump	normal shock wave
	oblique shock wave[4]	oblique shock wave
Basic assumptions	inviscid flow	$\gamma = 2$
	hydrostatic pressure gradient	

Application

The study of two-dimensional supercritical flow in an open channel is very similar to the study of supersonic gas flow. Liggett (1994) developed the complete set of flow equations. The analogy was applied with some success during the early laboratory studies of supersonic flows.

Notes

1. At the beginning of high-speed aerodynamics (i.e. the first half of the 20th century), compressible flows were investigated experimentally in open channels using water. For example, the propagation of oblique shock waves in supersonic (compressible) flows was deduced from the propagation of oblique shock waves at the free-surface of supercritical open channel flows. Interestingly, the celerity C in open channel flow is slow (compared with the sound celerity) and it can be easily observed.

   With the development of high-speed wind tunnels in the 1940s and 1950s, some compressible flow experimental results were later applied to open channel flow situations. Nowadays the analogy is seldom applied because of limitations.

2. The main limitations of the compressible flow/open channel flow analogy are:
   (a) The ratio of specific heat must equal 2. For real gases the maximum possible value for γ is 5/3 (see Appendix A1.1). For air, $\gamma = 1.4$. The

---

[4] Also called oblique jump or diagonal jump.

difference in specific heat ratio (between the analogy and real gases) implies that the analogy can only be approximate.

(b) The accuracy of free-surface measurements is disturbed by surface tension effects and the presence of capillary waves at the free-surface.

(c) Other limitations of the analogy include the hydraulic jump case. The hydraulic jump is an analogue to a normal shock wave. Both processes characterize a flow discontinuity with energy dissipation (i.e. irreversible energy loss). But in a hydraulic jump, the ratio of the sequent depths (i.e. upstream and downstream depth) is not identical to the density ratio across a normal shock wave (except for $Fr = 1$).

3.3.5 Critical flows and controls

Occurrence of critical flow – control section

For an open channel flow, the basic equations are the continuity equation (i.e. conservation of mass), and the motion equation or the Bernoulli equation which derives from the Navier–Stokes equation. The occurrence of critical flow conditions provides one additional equation:

$$V = \sqrt{gd} \qquad \text{(flat rectangular channel)} \qquad (3.27)$$

in addition to the continuity equation and the Bernoulli equation. These three conditions fix all the flow properties at the location where critical flow occurs, called the *control section.*

For a given discharge, the flow depth and velocity are fixed at a control section, independently of the upstream and downstream flow conditions. For a rectangular channel, it yields:

$$d = d_c = \sqrt[3]{\frac{Q^2}{gB^2}} \qquad (3.28)$$

$$V = V_c = \sqrt[3]{g\frac{Q}{B}} \qquad (3.29)$$

where B is the channel width.

Corollary: at a control section the discharge can be calculated once the depth is known: e.g. the critical depth meter.

Hydraulic structures that cause critical flow (e.g. gates, weirs) are used as control sections for open channel flows. Examples of control sections include: spillway crest, weir, gate, overfall, etc. (e.g. Henderson 1966, Chapter 6). Some hydraulic structures are built specifically to create critical flow conditions (i.e. the critical depth meter). Such structures provide a means to record the flow rates simply by measuring the critical flow depth: e.g. gauging stations, broad-crested weir, sharp-crested weirs (e.g. Rouse 1938, pp. 319–326; Henderson 1966, pp. 210–214; Bos 1976; Bos *et al.* 1991). A control section 'controls' the upstream flow if it is subcritical and also controls the downstream flow if it is supercritical. A classical example is the sluice gate (Fig. 3.6).

Upstream and downstream controls

In subcritical flow a disturbance travelling at a celerity C can move upstream and downstream because the wave celerity C is larger than the flow velocity V (i.e. $V/C < 1$). A control mechanism (e.g. sluice gate) can make its influence on the flow upstream of the control. Any small change in the downstream flow conditions affects tranquil (subcritical) flows. Therefore, subcritical flows are controlled by downstream conditions. This is called a *downstream flow control*. Conversely, a disturbance cannot travel upstream in a supercritical flow because the celerity is less than the flow velocity (i.e. $V/C > 1$). Hence, supercritical flows can only be controlled from upstream (i.e. *upstream flow control*). All rapid flows are controlled by the upstream flow conditions.

Discussion

All supercritical flow computations (i.e. 'backwater' calculations) must be started at the upstream end of the channel (e.g. gate, critical flow). Tranquil flow computations are started at the downstream end of the channel and are carried upstream.

Considering a channel with a steep slope upstream followed by a mild slope downstream (see definitions in Chapter 5), critical flow conditions occur at the change of slope. Computations must proceed at the same time from the upstream end (supercritical flow) and from the downstream end (subcritical flow). Near the break in grade, there is a transition from a supercritical flow to a subcritical flow. This transition is called a hydraulic jump. It is a flow discontinuity that is solved by applying the momentum equation across the jump (see Chapter 4).

Application: influence of the channel width

For an incompressible open channel flow, the differential form of the continuity equation ($Q = VA = $ constant) along a streamline in the s-direction gives:

$$V\frac{\partial A}{\partial s} + A\frac{\partial V}{\partial s} = 0 \tag{3.30a}$$

where A is the cross-section area. Equation (3.30a) is valid for any shape of cross-section. For a channel of rectangular cross-section, it becomes:

$$q\frac{\partial B}{\partial s} + B\frac{\partial q}{\partial s} = 0 \tag{3.30b}$$

where B is the channel surface width and q is the discharge per unit width ($q = Q/B$).

For a rectangular and horizontal channel (z_0 constant, H fixed) the differentiation of the Bernoulli equation is:

$$\frac{\partial d}{\partial s} - \frac{q^2}{gd^3}\frac{\partial d}{\partial s} + \frac{q}{gd^2}\frac{\partial q}{\partial s} = 0 \tag{3.31}$$

Introducing the Froude number and using the continuity equation (3.30) it yields:

$$(1 - Fr^2)\frac{\partial d}{\partial s} = Fr^2\frac{d}{B}\frac{\partial B}{\partial s} \tag{3.32}$$

In a horizontal channel, equation (3.32) provides a means to predict the flow depth variation associated with an increase or a decrease in channel width.

Considering an upstream subcritical flow (i.e. $Fr < 1$) the flow depth decreases if the channel width B decreases: i.e. $\partial d/\partial s < 0$ for $\partial B/\partial s < 0$. As a result the Froude number increases and when $Fr = 1$ critical flow conditions occur for a channel width B_{min}.

Considering an upstream supercritical flow (i.e. $Fr > 1$), a channel contraction (i.e. $\partial B/\partial s < 0$) induces an increase of flow depth. As a result, the Froude number decreases and when $Fr = 1$ critical flow conditions occur for a particular downstream channel width B_{min}.

With both types of upstream flow (i.e. sub- and supercritical flow), a constriction that is severe enough may induce critical flow conditions at the throat. The characteristic channel width B_{min} for which critical flow conditions occur is deduced from the Bernoulli equation:

$$B_{min} = \frac{Q}{\sqrt{\frac{8}{27}gE_1^3}} \tag{3.33}$$

where E_1 is the upstream specific energy. Note that equation (3.33) is valid for a horizontal channel of rectangular cross-section. B_{min} is the minimum channel width of the contracted section for the appearance of critical flow conditions. For $B > B_{min}$, the channel contraction does not induce critical flow and it is not a control section. With $B < B_{min}$ critical flow takes place, and the flow conditions at the control section may affect (i.e. modify) the upstream flow conditions.

Notes

1. At the location of critical flow conditions, the flow is sometimes referred to as 'choking'.
2. Considering a channel contraction such that the critical flow conditions are reached (i.e. $B = B_{min}$) the flow downstream of that contraction will tend to be subcritical if there is a downstream control, or supercritical in the absence of downstream control (Henderson 1966, pp. 47–49).
3. The symbol used for the channel width may be:
   B channel free surface width (m) (e.g. Henderson 1966),
   W channel bottom width (m), also commonly used for rectangular channel width.

3.4 Properties of common open-channel shapes

3.4.1 Properties

In practice, natural and man-made channels do not often have a rectangular cross-section. The most common shapes of artificial channels are circular (i.e. made of pipes) or trapezoidal.

The geometrical characteristics of the most common open channel shapes are summarized in the following table.

Shape	Flow depth d	Free-surface width B	Cross-section area A	Wetted perimeter P_w	Hydraulic diameter D_H
Trapezoidal		$W + 2d\cot\delta$	$d(W + d\cot\delta)$	$W + \dfrac{2d}{\sin\delta}$	$4\dfrac{A}{P_w}$
Triangular		$2d\cot\delta$	$d^2\cot\delta$	$\dfrac{2d}{\cot\delta}$	$2d\cos\delta$
Circular	$\dfrac{D}{2}\left(1 - \cos\dfrac{\delta}{2}\right)$	$D\sin\dfrac{\delta}{2}$	$\dfrac{D^2}{8}(\delta - \sin\delta)$	$\delta\dfrac{D}{2}$	$D\left(1 - \dfrac{\sin\delta}{\delta}\right)$
Rectangular		W	Wd	$W + 2d$	$4\dfrac{Bd}{B + 2d}$

Notes: A = cross-section area; B = free surface width; D = pipe diameter (circular pipe); D_H = hydraulic diameter; d = flow depth on the channel centre-line; P_w = wetted perimeter.

Note
The hydraulic diameter (also called equivalent pipe diameter) D_H is defined as:

$$D_H = 4\frac{A}{P_w}$$

where A is the cross-section area and P_w is the wetted perimeter.

3.4.2 Critical flow conditions

Critical flow conditions are defined as the flow conditions for which the mean specific energy is minimum. In a horizontal channel and assuming hydrostatic pressure distribution, critical flow conditions imply (Section 3.3.3):

$$\frac{A_c^3}{B_c} = \frac{Q^2}{g} \tag{3.34a}$$

where A_c and B_c are, respectively, the cross-sectional area and free surface width at critical flow conditions. Typical results are summarized in the following table.

Channel cross-section shape	Critical flow depth d_c	Minimum specific energy E_{min}
General shape	$\dfrac{A_c^3}{B_c} = \dfrac{Q^2}{g}$	$d_c + \dfrac{1}{2}\dfrac{A_c}{B_c}$
Trapezoidal	$\dfrac{(d_c(W + d_c\cot\delta))^3}{W + 2d_c\cot\delta} = \dfrac{Q^2}{g}$	$d_c + \dfrac{1}{2}\dfrac{d_c(W + d_c\cot\delta)}{W + 2d_c\cot\delta}$
Triangular	$d_c^5 = 2\dfrac{Q^2}{g(\cot\delta)^2}$	$\frac{5}{4}d_c$
Rectangular	$d_c^3 = \dfrac{Q^2}{gB^2}$	$\frac{3}{2}d_c$

For a rectangular channel the critical flow depth is:

$$d_c = \sqrt[3]{\frac{Q^2}{gB^2}} \qquad \text{rectangular channel} \qquad (3.34b)$$

3.5 Exercises

In a 3.5 m wide rectangular channel, the flow rate is $14\,\text{m}^3/\text{s}$. Compute the flow properties in the three following cases:

	Case 1	Case 2	Case 3	Units
Flow depth	0.8	1.15	3.9	m
Cross-section area				
Wetted perimeter				
Mean flow velocity				
Froude number				
Specific energy				

Considering a two-dimensional flow in a rectangular channel, the velocity distribution follows a power law:

$$\frac{V}{V_{max}} = \left(\frac{y}{d}\right)^{1/N}$$

where d is the flow depth and V_{max} is a characteristic velocity.

(a) Develop the relationship between the maximum velocity V_{max} and the mean flow velocity $Q/(dB)$.
(b) Develop the expression of the Coriolis coefficient and the momentum correction coefficient as functions of the exponent N only.

Explain in words the meaning and significance of the momentum correction coefficient. Develop a simple flow situation for which you write the fundamental equation(s) using the momentum correction coefficient. Using this example, discuss in words the correction resulting from taking into account the momentum correction coefficient.

Considering water discharging in a wide inclined channel of rectangular cross-section, the bed slope is $30°$ and the flow is a fully-developed turbulent shear flow. The velocity distribution follows a 1/6th power law:

$$\text{i.e. } V(y) \text{ is proportional to } \left(\frac{y}{d}\right)^{1/6}$$

where y is the distance from the channel bed measured perpendicular to the bottom, $V(y)$ is the velocity at a distance y normal from the channel bottom and d is the flow depth. The discharge per unit width in the channel is $2\,\text{m}^2/\text{s}$. At a gauging station (in the gradually varied flow region), the observed flow depth equals $0.9\,\text{m}$. At that location, compute and give the values (and units) of the specified quantities in the following list: (a) velocity at $y = 0.1\,\text{m}$; (b) velocity at the free-surface; (c) momentum correction coefficient; (d) Coriolis coefficient; (e) bottom pressure; (f) specific energy.

Considering a smooth transition in a rectangular open channel, compute the downstream flow properties as functions of the upstream conditions in the following cases:

	Upstream conditions	D/S conditions	Units
$d =$	5		m
$V =$	0.5		m/s
$B =$	20	18	m
$Q =$			
$z_o =$	6	6	m
$E =$			

	Upstream conditions	D/S conditions	Units
$d =$	0.1		m
$V =$			
$B =$	300	300	m
$Q =$	12	12	m^3/s
$z_o =$	0	-1	m
$E =$			

	Upstream conditions	D/S conditions	Units
$d =$	1		m
$V =$	2		m/s
$B =$	0.55	0.65	m
$Q =$			m^3/s
$z_o =$	3	2.2	m
$E =$			

Considering a horizontal channel of rectangular cross-section, develop the relationship between E, Q and d. Demonstrate that, for a fixed specific energy E, the maximum discharge (in a horizontal channel) is obtained if and only if the flow conditions are critical.

Considering an un-gated spillway crest, the reservoir free-surface elevation above the crest is 0.17 m. The spillway crest is rectangular and 5 m long. Assuming a smooth spillway crest, calculate the water discharge into the spillway.

Considering a broad-crested weir, draw a sketch of the weir. What is the main purpose of a broad-crest weir? A broad-crested weir is installed in a horizontal and smooth channel of rectangular cross-section. The channel width is 10 m. The bottom of the weir is 1.5 m above the channel bed. The water discharge is 11 m^3/s. Compute the depth of flow upstream of the weir, above the sill of the weir and downstream of the weir (in the absence of downstream control), assuming that critical flow conditions take place at the weir crest.

A prototype channel is 1000 m long and 12 m wide. The channel is rectangular and the flow conditions are: $d = 3$ m and $Q = 15$ m^3/s. Calculate the size (length, width, flow depth) and discharge of a 1/25 scale model using a Froude similitude.

Considering a rectangular and horizontal channel (z_o constant, H fixed), investigate the effects of change in channel width. Using continuity and Bernoulli equations, deduce the relationship between the longitudinal variation of flow depth ($\Delta d/\Delta s$), the variation in channel width ($\Delta B/\Delta s$) and the Froude number.
  Application: The upstream flow conditions being: $d_1 = 0.05$ m, $B_1 = 1$ m, $Q = 10$ L/s and the downstream channel width being $B_2 = 0.8$ m, compute the downstream flow depth and downstream Froude number. Is the downstream flow sub-, super- or critical?

A rectangular channel is 23 m wide and the water flows at 1.2 m/s. The channel contracts smoothly to 17.5 m width without energy loss.

- If the flow rate is 41 m^3/s, what is the depth and flow velocity in the channel contraction?
- If the flow rate is 0.16 m^3/s, what is the depth and flow velocity in the channel contraction?

A measuring flume is a short channel contraction designed to induce critical flow conditions and to measure (at the throat) the discharge. Considering the measuring flume sketched in Fig. E.3.1, the inlet, throat and outlet sections are rectangular. The channel bed is smooth and it is horizontal at Sections 1, 2 and 3. The channel width equals $B_1 = B_3 = 15$ m, $B_2 = 8.5$ m. The weir height is: $\Delta z_o = 1.1$ m. The total flow rate is 88 m^3/s. We shall assume that critical flow conditions (and hydrostatic pressure distribution) occur at the weir crest (i.e. Section 2). We shall investigate the channel flow conditions (assuming a supercritical downstream flow).

(a) At Section 2, compute: flow depth, specific energy.
(b) In Fig. E.3.1, five water surface profiles are labelled a, b, c, d and e. Which is the correct one?

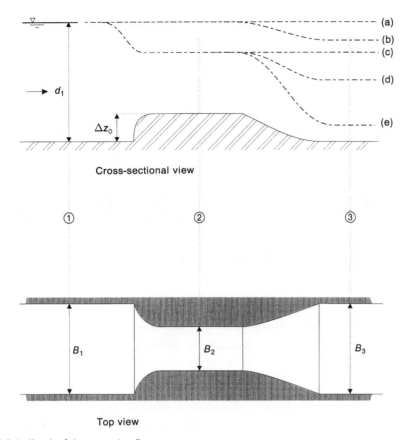

Cross-sectional view

Fig. E.3.1 Sketch of the measuring flume.

(c) At Section 1, compute the following properties: specific energy, flow depth, flow velocity.

(d) At Section 3, compute the specific energy, flow depth, flow velocity.

4

Applications of the momentum principle: hydraulic jump, surge and flow resistance in open channels

Summary

In this chapter the momentum equation is applied to open channel flows. The momentum principle is always used for hydrodynamic force calculations: e.g. force acting on a gate, flow resistance in uniform equilibrium flow. Other applications include the hydraulic jump, surge and bore. Hydraulic jump and positive surge calculations are developed for frictionless flow. Later, the momentum equation is combined with flow resistance formulations. Several calculations of friction loss in open channel are discussed with analogy to pipe flow. Head losses calculations are developed with the Darcy, Chézy, Manning and Strickler coefficients.

4.1 Momentum principle and application

4.1.1 Introduction

In the Chapter 3, subcritical and supercritical flows, controls and control sections were introduced. The basic results may be summarized as:

(a) critical flow occurs for the minimum specific energy; at critical flow conditions the Froude number is unity;
(b) at a control section critical flow conditions take place and this fixes a unique relationship between depth and discharge in the vicinity (e.g. sluice gate, weir);
(c) subcritical flows are controlled from downstream (e.g. reservoir) while supercritical flows have upstream control (e.g. spillway, weir);
(d) a control influences both the flows upstream and downstream of the control section: i.e. downstream flow control and upstream flow control respectively.

There are, however, flow situations for which the use of Bernoulli equation is not valid (e.g. significant fiction loss and energy loss) or cannot predict the required parameters (e.g. force exerted by the fluid onto a structure). In such cases the momentum principle may be used in conjunction with the continuity equation.

4.1.2 Momentum principle

The momentum principle states that, for a selected control volume, the rate of change in momentum flux equals the sum of the forces acting on the control volume. For a horizontal rectangular open channel, the momentum equation applied in the flow direction leads to:

$$\rho Q V_2 - \rho Q V_1 = \tfrac{1}{2}\rho g d_1^2 B - \tfrac{1}{2}\rho g d_2^2 B - F_{\text{fric}} \qquad (4.1)$$

where V and d are, respectively, the mean velocity and flow depth, the subscripts 1 and 2 refer to the upstream and downstream cross-sections respectively (e.g. Fig. 4.1), ρ is the fluid density, B is the channel width, Q is the total flow rate and F_{fric} is the boundary friction force.

In addition, the continuity equation gives:

$$Q = V_1 d_1 B = V_2 d_2 B \qquad (4.2)$$

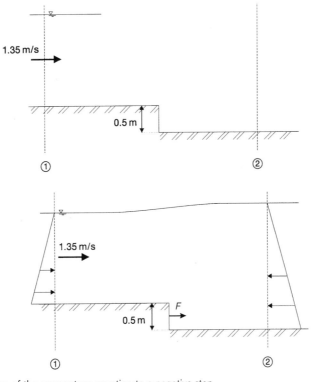

Fig. 4.1 Application of the momentum equation to a negative step.

Discussion

The total momentum flux across a section equals: $\rho VVA = \rho QV$ (see Section 2.2.3). The rate of change in momentum flux is then: $(\rho QV_2 - \rho QV_1)$ in a horizontal channel. The forces acting on a control volume include the pressure forces, friction force and gravity force. For a horizontal rectangular channel with hydrostatic distribution, the pressure force at a section is: $\rho g d^2 B / 2$.

Further discussions on the momentum principle have been developed in Section 2.2.3. If the velocity distribution is not uniform, the momentum correction coefficient or Boussinesq coefficient must be taken into account (see Section 3.2.2). The momentum flux is then expressed as: $\beta \rho VVA = \beta \rho QV$ where β is the Boussinesq coefficient.

Application

The backward-facing step, sketched in Fig. 4.1, is in a 5 m wide channel of rectangular cross-section. The total discharge is $55 \, \text{m}^3/\text{s}$. The fluid is water at 20°C. The bed of the channel, upstream and downstream of the step, is horizontal and smooth. Compute the pressure force acting on the vertical face of the step and indicate the direction of the force (i.e. upstream or downstream).

Solution

First we sketch a free-surface profile between sections 1 and 2 (we might have to modify it after further calculations).

Secondly we must compute the upstream and downstream flow properties. Applying the continuity equation, the upstream flow depth equals: $d_1 = 8.15 \, \text{m}$. Assuming a hydrostatic pressure distribution, the upstream specific energy equals: $E_1 = 8.24 \, \text{m}$. Note that the upstream flow is subcritical ($Fr_1 = 0.15$).

The downstream specific energy is deduced from the Bernoulli principle assuming a smooth transition:

$$E_1 + \Delta z_o = E_2 \qquad \text{Bernoulli equation}$$

where Δz_o is the drop height (0.5 m). It gives $E_2 = 8.74 \, \text{m}$. The downstream flow depth is deduced from the definition of the specific energy (in a rectangular channel assuming a hydrostatic pressure distribution):

$$E_2 = d_2 + \frac{Q^2}{2gB^2 d_2^2}$$

where B is the channel width. The solution of the equation is: $d_2 = 8.66 \, \text{m}$. Note that the drop in bed elevation yields an increase of flow depth. Indeed the upstream flow being subcritical ($d_1 = 8.15 \, \text{m} > d_c = 2.31 \, \text{m}$), the relationship specific energy versus flow depth (Fig. 3.5) predicts such an increase in flow depth for an increasing specific energy.

Now let us consider the forces acting on the control volume contained between sections 1 and 2: the pressure forces at sections 1 and 2, the weight of the control volume, the reaction force of the bed (equal, in magnitude, to the weight in the absence of vertical fluid motion) and the pressure force F exerted by the vertical face of the step on the fluid. The friction force is zero as the channel is assumed smooth (i.e. frictionless).

The momentum equation as applied to the control volume between sections 1 and 2 is:

$$\rho Q V_2 - \rho Q V_1 = \tfrac{1}{2}\rho g d_1^2 B - \tfrac{1}{2}\rho g d_2^2 B + F \qquad \text{Momentum principle}$$

in the horizontal direction.

The solution of the momentum equation is: $F = +205\,\text{kN}$ (remember: F is the force exerted by the step onto the fluid). In other words, the pressure force exerted by the fluid onto the vertical face of the step acts in the upstream direction and equals 205 kN.

4.1.3 Momentum function

The momentum function in a rectangular channel is defined as (e.g. Henderson 1966, pp. 67–70):

$$M = \frac{\rho Q V + F_p}{\rho g B} \qquad (4.3a)$$

where F_p is the pressure force acting in the flow direction on the channel cross-section area (i.e. $F_p = 1/2\rho g d^2 B$, in a horizontal channel). For a flat rectangular channel and assuming a hydrostatic pressure distribution, the momentum function becomes:

$$M = \frac{d^2}{2} + \frac{q^2}{gd} \qquad (4.3b)$$

where q is the discharge per unit width. In dimensionless terms, it yields:

$$\frac{M}{d_c^2} = \frac{1}{2}\left(\frac{d}{d_c}\right)^2 + \frac{d_c}{d} \qquad (4.3c)$$

Application

The momentum equation (4.1) can be rewritten in terms of the momentum function:

$$M_2 - M_1 = -\frac{F_{\text{fric}}}{\rho g}$$

Note

The momentum function is minimum at critical flow conditions. For a momentum function such as $M > M_{\text{min}}$ there are two possible depths called the *sequent* or *conjugate depths*. For a non-rectangular channel the momentum function M becomes (Henderson 1966, pp. 72–73):

$$M = \bar{d}A + \frac{Q^2}{gA}$$

where $\bar{d}$ is the depth from the surface to the centroid of the section.

4.2 Hydraulic jump

4.2.1 Presentation

An open channel flow can change from subcritical to supercritical in a relatively 'low-loss' manner at gates or weirs. In these cases the flow regime evolves from subcritical to supercritical with the occurrence of critical flow conditions associated with relatively small energy loss (e.g. broad-crested weir). The transition from supercritical to subcritical flow is, on the other hand, characterized by a strong dissipative mechanism. It is called a hydraulic jump (Figs 2.3, 4.2 and 4.3).

Note
An example of subcritical-to-supercritical transition is the flow at the crest of a steep spillway. Upstream of the channel intake, the flow motion in the reservoir is tranquil. Along the steep chute, the flow is supercritical. Critical flow conditions are observed near the crest of the spillway. In practice, it is reasonable to assume critical flow at the spillway crest although the pressure distribution is not hydrostatic at the crest depending upon the crest shape (e.g. see discussions in Rouse 1938, p. 325; Creager *et al.* 1945, vol. II, p. 371).

Definition
Considering a channel with both an upstream control (e.g. sluice gate) leading to a supercritical flow (downstream of the control section) and a downstream control (e.g. reservoir) imposing a subcritical flow at the downstream end – the channel conveys an upstream supercritical flow, a downstream subcritical flow and a transition flow.

Fig. 4.2 Hydraulic jump in a natural waterway: Bald Rock Creek at the Junction QLD, Australia (9 November 1997). Flow from the left to the right.

The transition from a supercritical flow to a subcritical flow is called a *hydraulic jump*. A hydraulic jump is extremely turbulent. It is characterized by the development of large-scale turbulence, surface waves and spray, energy dissipation and air entrainment (Fig. 4.3). The large-scale turbulence region is usually called the 'roller'. A hydraulic jump is a region of rapidly-varied flow.

The flow within a hydraulic jump is extremely complicated (Hager 1992b) and it is not required usually to consider its details. To evaluate the basic flow properties and energy losses in such a region, the momentum principle is used.

4.2.2 Basic equations

For a *steady flow* in a *horizontal rectangular* channel of *constant channel width*, the three fundamental equations become:

(A) *Continuity equation*

$$Q = V_1 d_1 B = V_2 d_2 B \qquad (4.2)$$

where V_1 and d_1 are respectively the velocity and flow depth at the upstream cross-section (Fig. 4.4), V_2 and d_2 are defined at the downstream cross-section, B is the channel width and Q is the total flow rate.

(B) *Momentum equation (Bélanger equation)*[1]
The momentum equation states that the sum of all the forces acting on the control volume equals the change in momentum flux across the control volume. For a hydraulic jump, it yields

$$\left(\tfrac{1}{2}\rho g d_1^2 - \tfrac{1}{2}\rho g d_2^2\right) B - F_{\text{fric}} = \rho Q (V_2 - V_1) \qquad (4.3)$$

where F_{fric} is a drag force exerted by the channel roughness on the flow (Fig. 4.4).

(C) *Energy equation*
The energy equation (2.24) can be transformed as:

$$H_1 = H_2 + \Delta H \qquad (4.4a)$$

where ΔH is the energy loss (or head loss) at the jump, and H_1 and H_2 are upstream and downstream total heads respectively. Assuming a hydrostatic pressure distribution and taking the channel bed as the datum, equation (4.4a) becomes:

$$\frac{V_1^2}{2g} + d_1 = \frac{V_2^2}{2g} + d_2 + \Delta H \qquad (4.4b)$$

Note that equations (4.1) to (4.4) were developed assuming hydrostatic pressure distributions at both the upstream and downstream ends of the control volume (Fig. 4.4). Furthermore, the upstream and downstream velocity distributions were assumed uniform for simplicity.

---

[1] Jean-Baptiste Bélanger (1789–1874) was the first to suggest the application of the momentum principle to the hydraulic jump flow (Bélanger 1828). The momentum equation applied across a hydraulic jump is often called the Bélanger equation.

(a)

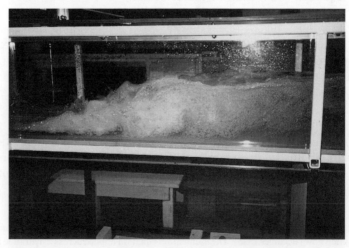

(b)

(c)

Fig. 4.3 Photographs of hydraulic jump in a rectangular channel. (a) Undular hydraulic jump: $Fr_1 = 1.95$, $d_1 = 0.014$ m, $B = 0.5$ m, flow from left to right (side and top views). (b) Steady/strong jump: $Fr_1 = 9.4$, $d_1 = 0.013$ m, $B = 0.5$ m, flow from left to right. (c) Strong jump: $Fr_1 = 13.2$, $d_1 = 0.013$ m, $B = 0.5$ m, flow from left to right.

Notes

1. In the momentum equation, (ρV) is the momentum per unit volume.
2. In the simple case of uniform velocity distribution, the term (ρVV) is the momentum flux per unit area across the control surface. (ρVVA) is the total momentum flux across the control surface.

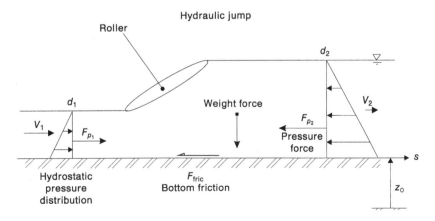

Fig. 4.4 Application of the momentum equation to a hydraulic jump.

Neglecting the drag force on the fluid, the continuity and momentum equations provide a relationship between the upstream and downstream flow depths as:

$$d_2 = \sqrt{\left(\frac{d_1}{2}\right)^2 + \frac{2Q^2}{gd_1 B^2}} - \frac{d_1}{2} \tag{4.5a}$$

or in dimensionless terms:

$$\frac{d_2}{d_1} = \frac{1}{2}\left(\sqrt{1 + 8Fr_1^2} - 1\right) \tag{4.5b}$$

where Fr_1 is the upstream Froude number: $Fr_1 = V_1/\sqrt{gd_1}$. It must be noted that $Fr_1 > 1$. The depths d_1 and d_2 are referred to as *conjugate depths* (or sequent depths). Using equation (4.5) the momentum equation yields:

$$Fr_2 = \frac{2^{3/2}Fr_1}{\left(\sqrt{1 + 8Fr_1^2} - 1\right)^{3/2}} \tag{4.6}$$

where Fr_2 is the downstream Froude number.

The energy equation gives the head loss:

$$\Delta H = \frac{(d_2 - d_1)^3}{4d_1 d_2} \tag{4.7a}$$

and in dimensionless terms:

$$\frac{\Delta H}{d_1} = \frac{\left(\sqrt{1 + 8Fr_1^2} - 3\right)^3}{16\left(\sqrt{1 + 8Fr_1^2} - 1\right)} \tag{4.7b}$$

Equations (4.5) to (4.7) are summarized on Fig. 4.5. Figure 4.5 provides means to estimate rapidly the jump properties as functions of the upstream Froude number. For example, for $Fr_1 = 5$, we can deduce: $Fr_2 \sim 0.3$, $d_2/d_1 \sim 6.5$, $\Delta H/d_1 \sim 7$.

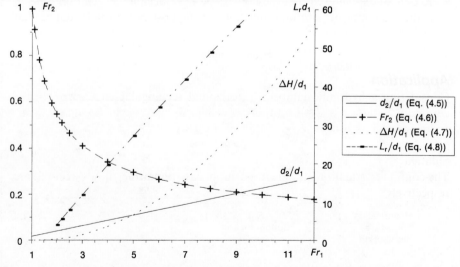

Fig. 4.5 Flow properties downstream of a hydraulic jump in a rectangular horizontal channel.

Notes

1. In a hydraulic jump, the downstream flow depth d_2 is always larger than the upstream flow depth d_1.
2. The main parameter of a hydraulic jump is its upstream Froude number Fr_1.
3. If only the downstream flow conditions are known, the solution of the continuity and momentum equations is given by:

$$\frac{d_1}{d_2} = \tfrac{1}{2}\left(\sqrt{1 + 8Fr_2^2} - 1\right)$$

4. A hydraulic jump is a very effective way of dissipating energy. It is also an effective mixing device. Hydraulic jumps are commonly used at the end of a spillway or in a dissipation basin to destroy much of the kinetic energy of the flow. The hydraulic power dissipated in a jump equals: $\rho g Q \Delta H$, where ΔH is computed using equation (4.7).
5. Hydraulic jumps are characterized by air entrainment. Air is trapped at the impingement points of the supercritical inflow with the roller. The rate of air entrainment may be approximated as:

$$\frac{Q_{air}}{Q} \approx 0.018(Fr_1 - 1)^{1.245} \qquad \text{(Rajaratnam 1967)}$$

$$\frac{Q_{air}}{Q} \approx 0.014(Fr_1 - 1)^{1.4} \qquad \text{(Wisner 1965)}$$

Wood (1991) and Chanson (1997) presented complete reviews of these formulae.

If the jump is in a closed duct then, as the air is released, it could accumulate on the roof of the duct. This phenomena is called 'blowback' and has caused failures in some cases (Falvey 1980).

6. Recent studies showed that the flow properties (including air entrainment) of hydraulic jumps are not only functions of the upstream Froude number but also of the upstream flow conditions: i.e. partially-developed or fully-developed inflow. The topic is currently under investigation.

Application

Considering a hydraulic jump in a horizontal rectangular channel, write the continuity equation and momentum principle. Neglecting the boundary shear, deduce the relationships: $d_2/d_1 = f(Fr_1)$ and $Fr_2 = f(Fr_1)$.

Solution

The continuity equation and the momentum equation (in the flow direction) are respectively:

$$q = V_1 d_1 = V_2 d_2 \qquad \qquad \text{[C]}$$

$$\tfrac{1}{2}\rho g d_1^2 - \tfrac{1}{2}\rho g d_2^2 = \rho q(V_2 - V_1) \qquad \qquad \text{[M]}$$

where q is the discharge per unit width.

[C] implies: $V_2 = V_1 d_1/d_2$. Substituting [C] into [M], it yields:

$$\tfrac{1}{2}\rho g d_1^2 - \tfrac{1}{2}\rho g d_2^2 = \rho V_1^2 d_1 \left(\frac{d_1}{d_2}\right) - \rho V_1^2 d_1$$

Dividing by $(\rho g d_1^2)$ it becomes:

$$\tfrac{1}{2} - \tfrac{1}{2}\left(\frac{d_2}{d_1}\right)^2 = Fr_1^2 \left(\frac{d_1}{d_2}\right) - Fr_1^2$$

After transformation we obtain a polynomial equation of degree three in terms of d_2/d_1:

$$\tfrac{1}{2}\left(\frac{d_2}{d_1}\right)^3 - \left(\tfrac{1}{2} + Fr_1^2\right)\left(\frac{d_2}{d_1}\right) + Fr_1^2 = 0$$

or

$$\tfrac{1}{2}\left(\frac{d_2}{d_1} - 1\right)\left(\left(\frac{d_2}{d_1}\right)^2 + \left(\frac{d_2}{d_1}\right) - 2Fr_1^2\right) = 0$$

The solutions of the momentum equation are the obvious solution $d_2 = d_1$ and the solutions of the polynomial equation of degree two:

$$\left(\frac{d_2}{d_1}\right)^2 + \left(\frac{d_2}{d_1}\right) - 2Fr_1^2 = 0$$

The (only) meaningful solution is:

$$\frac{d_2}{d_1} = \tfrac{1}{2}\left(\sqrt{1 + 8Fr_1^2} - 1\right)$$

Using the continuity equation in the form: $V_2 = V_1 d_1/d_2$, and dividing by $\sqrt{g d_2}$, yields:

$$Fr_2 = \frac{V_2}{\sqrt{g d_2}} = \frac{V_1}{\sqrt{g d_1}}\left(\frac{d_1}{d_2}\right)^{3/2} = Fr_1 \frac{2^{3/2}}{\left(\sqrt{1 + 8Fr_1^2} - 1\right)^{3/2}}$$

Application

A hydraulic jump takes place in a 0.4 m wide laboratory channel. The upstream flow depth is 20 mm and the total flow rate is 31 L/s. The channel is horizontal, rectangular and smooth. Calculate the downstream flow properties and the energy dissipated in the jump. If the dissipated power could be transformed into electricity, how many 100 W bulbs could be lighted with the jump?

Solution

The upstream flow velocity is deduced from the continuity equation

$$V_1 = \frac{Q}{B d_1} = \frac{(31 \times 10^{-3})}{0.4 \times (20 \times 10^{-3})} = 3.875 \, \text{m/s}$$

The upstream Froude number equals: $Fr_1 = 8.75$ (i.e. supercritical upstream flow). The downstream flow properties are deduced from the above equation:

$$\frac{d_2}{d_1} = \frac{1}{2}\left(\sqrt{1 + 8Fr_1^2} - 1\right) = 11.9$$

$$Fr_2 = Fr_1 \frac{2^{3/2}}{\left(\sqrt{1 + 8Fr_1^2} - 1\right)^{3/2}} = 0.213$$

Hence: $d_2 = 0.238\,\text{m}$ and $V_2 = 0.33\,\text{m/s}$.

The head loss in the hydraulic jump equals:

$$\Delta H = \frac{(d_2 - d_1)^3}{4d_1 d_2} = 0.544\,\text{m}$$

The hydraulic power dissipated in the jump equals:

$$\rho g Q \Delta H = 998.2 \times 9.8 \times 31 \times 10^{-3} \times 0.544 = 165\,\text{W}$$

where ρ is the fluid density (kg/m^3), Q is in m^3/s and ΔH is in m. In the laboratory flume, the dissipation power equals 165 W: one 100 W bulb and one 60 W bulb could be lighted with the jump power.

4.2.3 Discussion

Types of hydraulic jump

Experimental observations highlighted different types of hydraulic jumps, depending upon the upstream Froude number Fr_1 (e.g. Chow 1973, p. 395). For hydraulic jumps in rectangular horizontal channels, Chow's classification is commonly used and it is summarized in the following table.

Fr_1	Definition (Chow 1973)	Remarks
1	Critical flow	No hydraulic jump.
1 to 1.7	Undular jump	Free-surface undulations developing downstream of jump over considerable distances. Negligible energy losses. Also called Fawer jump in homage to the work of Fawer (1937).
1.7 to 2.5	Weak jump	Low energy loss.
2.5 to 4.5	Oscillating jump	Wavy free surface. Unstable oscillating jump. Production of large waves of irregular period. Each irregular oscillation produces a large wave which may travel far downstream, damaging and eroding the banks. *To be avoided* if possible.
4.5 to 9	Steady jump	45 to 70% of energy dissipation. Steady jump. Insensitive to downstream conditions (i.e. tailwater depth). *Best economical design.*
>9	Strong jump	Rough jump. Up to 85% of energy dissipation. *Risk of channel bed erosion.* To be avoided.

> **Notes**
> 1. The above classification must be considered as *rough* guidelines.
>   For example, experiments performed at the University of Queensland (Australia) showed the existence of undular hydraulic jumps for an upstream Froude number between 1 to 3 (Chanson 1995a). Further, the inflow conditions (uniform velocity distribution, partially developed or fully-developed) modify substantially the flow properties and affect the classification of jumps. Also, the shape of the channel cross-section affects the hydraulic jump characteristics (Hager 1992b). Note that the above table is given for a rectangular cross-section only.
> 2. A hydraulic jump is a very unsteady flow. Experimental measurements of bottom pressure fluctuations indicated that the mean pressure is quasi-hydrostatic below the jump *but large pressure fluctuations* are observed (see reviews in Hager 1992b and Chanson 1995b). The re-analysis of bottom pressure fluctuation records below hydraulic jumps over long periods of time indicates that the extreme minimum pressures might become negative (i.e. below atmospheric pressure) and could lead to uplift pressures on the channel bottom. The resulting uplift loads on the channel bed might lead to substantial damage, erosion and destruction of the channel.

Length of the roller

Chow (1973) proposed some guidelines to estimate the length of the roller of the hydraulic jump as a function of the upstream flow conditions. Hager *et al.* (1990) reviewed a broader range of data and correlations. For a wide channel (i.e. $d_1/B < 0.10$), they proposed the following correlation:

$$\frac{L_r}{d_1} = 160 \tanh\left(\frac{Fr_1}{20}\right) - 12 \qquad 2 < Fr_1 < 16 \tag{4.8}$$

where L_r is the length of the roller. Equation (4.8) is valid for rectangular horizontal channels with $2 < Fr_1 < 16$. Such a correlation can be used when designing energy dissipation basins (see Fig. 4.5).

> **Notes**
> 1. The hyperbolic tangent tanh is defined as:
>
> $$\tanh(x) = \frac{e^x - e^{-x}}{e^x + e^{-x}}$$
>
> 2. Equation (4.8) is an empirical correlation based upon model and prototype data. It fits reasonably well experimental data for hydraulic jumps with partially-developed inflow conditions (in rectangular channels).

Application: energy dissipation basin

Hydraulic jumps are known for their energy dissipation performances. Prior to the late 19th century, designers tried to avoid hydraulic jumps whenever possible to minimize the risks of channel destruction. Since the beginning of the 20th century and with the introduction of high-resistance materials (e.g. reinforced concrete), hydraulic jumps are used to dissipate flow energy downstream of supercritical flow structures (e.g. spillways, bottom outlets) (e.g. Henderson 1966, pp. 221–225). In practice, energy dissipation structures[2] are designed to induce a steady jump or a strong jump. The lowest design (inflow) Froude number must be above 4.5. The selection of a strong jump requires a careful analysis of the risks of bed erosion.

Application

Considering a dissipation basin at the downstream end of a spillway, the total discharge is $Q = 2000 \, \text{m}^3/\text{s}$. The energy dissipation structure is located in a horizontal rectangular channel (25 m wide). The flow depth at the downstream end of the spillway is 2.3 m. Compute the energy dissipation in the basin

Solution

The upstream flow conditions of the jump are:

$$d_1 = 2.3 \, \text{m}$$

$$V_1 = 34.8 \, \text{m/s}$$

$$Fr_1 = 7.3 \qquad \text{(i.e. steady jump)}$$

The downstream flow conditions are:

$$d_2 = 22.707 \, \text{m}$$

$$V_2 = 3.52 \, \text{m/s}$$

$$Fr_2 = 0.236$$

The head loss across the jump equals:

$$\Delta H = 40.7 \, \text{m}$$

The power dissipated in the jump is:

$$\rho g Q \Delta H = 796 \times 10^6 \, \text{W} \qquad \text{(i.e. nearly 800 MW!)}$$

In a dissipation basin with a flat horizontal bottom, the location of the jump may change as a function of the upstream and downstream flow conditions. That is, with a change of (upstream or downstream) flow conditions, the location of the jump changes in order to satisfy the Bélanger equation (i.e. equation (4.1)). The new location might not be suitable and could require a very long and uneconomical structure.

In practice, design engineers select desirable features to make the jump stable and as short as possible. Abrupt rise and abrupt drop, channel expansion and channel contraction, and chute blocks introduce additional flow resistance and tend to promote the jump formation. Hager (1992b) described a wide range of designs.

---

[2] Energy dissipation structures are also called stilling basins, transition structures or energy dissipators.

Application

Considering a hydraulic jump in a horizontal rectangular channel located immediately upstream of an abrupt rise (Fig. 4.6), estimate the downstream flow depth d_3 (see Fig. 4.6) for the design flow conditions: $d_1 = 0.45\,\text{m}$, $V_1 = 10.1\,\text{m/s}$. The step height equals: $\Delta z_o = 0.5\,\text{m}$.

Solution

First, we will assume that the complete jump is located upstream of the bottom rise as sketched in Figure 4.6.

The continuity equation between sections 1 and 3 is:

$$q = V_1 d_1 = V_3 d_1 \qquad \text{continuity equation}$$

The momentum equation between section 1 (upstream flow) and section 2 (i.e. immediately *upstream* of the abrupt rise) is:

$$\left(\tfrac{1}{2}\rho g d_1^2 - \tfrac{1}{2}\rho g d_2^2\right) = \rho q(V_2 - V_1) \qquad \text{momentum equation}$$

where q is the discharge per unit width. The momentum equation implies hydrostatic pressure distribution at section 2.

The momentum equation between section 2 (taken immediately *downstream* of the bottom rise) and section 3 is:

$$\left(\tfrac{1}{2}\rho g (d_2 - \Delta z_o)^2 - \tfrac{1}{2}\rho g d_3^2\right) = \rho q(V_3 - V_2) \qquad \text{momentum equation}$$

The solution of the non-linear system of equations is:

$$q = 4.54\,\text{m}^2/\text{s}$$
$$Fr_1 = 4.8$$
$$d_2/d_1 = 6.32 \qquad \text{(equation (4.5))}$$
$$d_2 = 2.84\,\text{m}$$
$$V_2 = 1.6\,\text{m/s} \qquad \text{(continuity equation)}$$
$$d_3 = 2.26\,\text{m}$$
$$V_3 = 2.0\,\text{m/s}$$
$$Fr_3 = 0.43$$

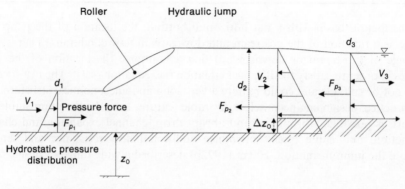

Fig. 4.6 Sketch of hydraulic jump at an abrupt bottom rise.

Comments

1. $(d_2 - \Delta z_o)$ is not equal to d_3. At section 2, the velocity V_2 is indeed slower than that at section 3.
2. The hydraulic jump remains confined upstream of the abrupt rise as long as $d_3 > d_c$ where d_c is the critical flow depth (in this case $d_c = 1.28\,\text{m}$).
3. Experimental data showed that the pressure distribution at section 2 is not hydrostatic. In practice, the above analysis is somehow oversimplified although it provides a good order of magnitude.

4.3 Surges and bores

4.3.1 Introduction

A *surge* wave results from a sudden change in flow (e.g. a partial or complete closure of a gate) that increases the depth: such a flow situation is called a positive surge (see section also 4.3.4). In such a case the application of the momentum principle to unsteady flow is simple: the unsteady flow conditions are solved as a quasi-steady flow situation using the momentum equation developed for the hydraulic jump.

Notes

1. A positive surge is an abrupt wave front (see section 4.3.4).
2. When the surge is of tidal origin it is usually termed a *bore*. The difference of name does not mean a difference in principle. Hydraulic bores results from the propagation of tides into tidal estuaries and rivers. Tricker (1965) presented numerous photographs of interest. Classical examples are described in Table 4.1.

Table 4.1 Examples of positive surges (bores) in estuaries

River	Country	Name of the bore	Reference – comments
Amazon river	Brazil	'Pororoca'	Bazin (1865b, p. 624). $\Delta d = 3.7$ to $5\,\text{m}$.
Chien Tang river	China	Hang-chow bore	Chow (1973, p. 558), Tricker (1965). $V_{srg} = 4.1$ to 5.7 m/s, $\Delta d = 1$ to $3.7\,\text{m}$.
Severn river	England (near Gloucester)		Tricker (1965). $V_{srg} = 3.1$ to 6 m/s, $\Delta d = 1$ to $1.8\,\text{m}$.
Seine river	France	'Mascaret'	Bazin (1865b, p. 623), Tricker (1965). $V_{srg} = 2$ to 10 m/s, $\Delta d = 2$ to $7.3\,\text{m}$.
Ganges river	India	Hoogly bore	Bazin (1865b, p. 623). $V_{srg} = 10\,\text{m/s}$, $\Delta d = 3.7$ to $4.5\,\text{m}$.
Petitcodiac river	USA (New Brunswick)		Tricker (1965). $\Delta d = 1$ to $1.5\,\text{m}$.

Notes: V_{srg} = velocity of the surge; Δd = height of the wave front.

4.3.2 Equations

Considering a positive surge in a rectangular channel (Fig. 4.7), the surge is an unsteady flow situation for an observer standing on the bank (Fig. 4.7 left). But the surge is seen by an observer travelling at the surge speed V_{srg} as a steady-flow case called a *quasi-steady hydraulic jump*.

For a rectangular horizontal channel and considering a control volume across the front of the surge travelling at a velocity V_{srg} (Fig. 4.7a), the continuity equation is:

$$(V_1 + V_{srg})d_1 = (V_2 + V_{srg})d_2 \qquad (4.9a)$$

where V_{srg} is the surge velocity, as seen by an observer immobile (standing) on the channel bank, the subscript 1 refers to the initial flow conditions and the subscript 2 refers to the new flow conditions.

By analogy with the hydraulic jump (Fig. 4.4), the momentum equation for the control volume, *neglecting friction loss*, yields (see section 4.2.2):

$$Fr_2 = \frac{2^{3/2} Fr_1}{\left(\sqrt{1 + 8Fr_1^2} - 1\right)^{3/2}} \qquad (4.6)$$

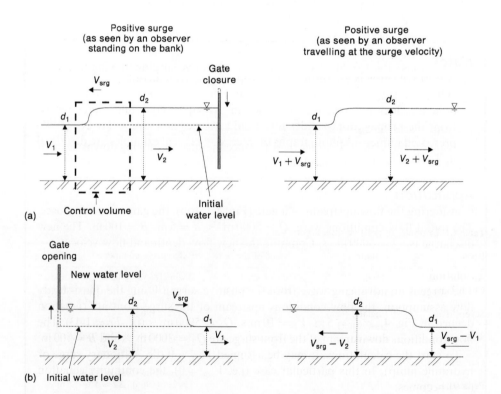

Fig. 4.7 Positive surges and wave front propagation. (a) Advancing front moving upstream. (b) Advancing front moving downstream.

where the Froude numbers Fr_1 and Fr_2 are defined as (Fig. 4.7a):

$$Fr_1 = \frac{V_1 + V_{srg}}{\sqrt{gd_1}}$$

$$Fr_2 = \frac{V_2 + V_{srg}}{\sqrt{gd_2}}$$

This is a system of two equations (4.9a) and (4.6) with five variables (i.e. d_1, d_2, V_1, V_2, V_{srg}). Usually the upstream conditions V_1, d_1 are known and the new flow rate Q_2/Q_1 is determined by the rate of closure of the gate (e.g. complete closure: $Q_2 = 0$).

Note that the continuity equation provides an estimate of the velocity of the surge:

$$V_{srg} = \frac{Q_1 - Q_2}{(d_2 - d_1)B} \tag{4.10}$$

Equations (4.9a) and (4.6) can be solved graphically or numerically to provide the new flow depth d_2 and velocity V_2, and the surge velocity V_{srg} as functions of the initial flow conditions (i.e. d_1, V_1) and the new flow rate Q_2 (e.g. Henderson 1966, pp. 75–77).

Notes
1. A stationary surge (i.e. $V_{srg} = 0$) is a hydraulic jump.
2. A surge can be classified as for a hydraulic jump as a function of its 'upstream' Froude number Fr_1. As an example, a surge with $Fr_1 = (V_1 + V_{srg})/\sqrt{gd_1} = 1.4$ is called an undular surge.
3. Equations (4.9a) and (4.10) are valid for a positive surging moving upstream (Fig. 4.7a). For a positive surge moving downstream (Fig. 4.7b), equation (4.9a) becomes:

$$(V_1 - V_{srg})d_1 = (V_2 - V_{srg})d_2 \tag{4.9b}$$

Note that V_{srg} must be larger than V_1 since the surge is moving downstream in the direction of the initial flow.

Application
Considering the flow upstream of a gate (Fig. 4.7, top), the gate suddenly closes. The initial flow conditions were: $Q = 5000\,\text{m}^3/\text{s}$, $d = 5\,\text{m}$, $B = 100\,\text{m}$. The new discharge is: $Q = 3000\,\text{m}^3/\text{s}$. Compute the new flow depth and flow velocity.

Solution
The surge is an advancing wave front (i.e positive surge). Using the quasi-steady flow assumption, the flow conditions upstream of the surge front are (notation defined in Fig. 4.7): $d_1 = 5\,\text{m}$, $V_1 = 10\,\text{m/s}$, $Q_1 = 5000\,\text{m}^3/\text{s}$ (i.e. $Fr = 1.43$). The flow conditions downstream of the front surge are $Q_2 = 3000\,\text{m}^3/\text{s}$ and $B = 100\,\text{m}$.

To start the calculations, it may be assumed $V_{srg} = 0$ (i.e. stationary surge or hydraulic jump). In this particular case (i.e. $V_{srg} = 0$), the continuity equation (4.9) becomes:

$$V_1 d_1 = V_2 d_2$$

Using the definition of the Froude number, it yields:

$$Fr_1\sqrt{gd_1^3} = Fr_2\sqrt{gd_2^3}$$

For an initialization step where $V_{srg} = 0$, the above equation is more practical than equation (4.10).

Notation Equation	V_{srg}	Fr_1	d_2 Eq. (4.10)	Fr_2 Eq. (4.6)	V_2 Definition of Fr_2	$V_2 d_2 B$ Continuity equation
1st iteration	0.0	1.43	23.4[a]	0.72	10.9	2540
2nd iteration	2	1.71	15	0.62	5.5	8229
3rd iteration	3	1.86	11.7	0.58	3.2	3730
Solution	3.26	1.89	11.1	0.57	2.70	3000

Note: [a] initialization step: $d_2 = d_c$.

Comments

It must be noted that the initial flow conditions are supercritical (i.e. $V_1/\sqrt{gd_1} = 1.43$). The surge is a large disturbance travelling upstream against a supercritical flow. After the passage of the surge, the flow becomes subcritical (i.e. $V_2/\sqrt{gd_2} = 0.3$). The surge can be characterized as a weak surge ($Fr_1 = 1.89$). For a surge flow, engineers should not be confused between the surge Froude numbers $(V_1 + V_{srg})/\sqrt{gd_1}$ and $(V_2 + V_{srg})/\sqrt{gd_2}$, and the initial and new channel Froude numbers ($V_1/\sqrt{gd_1}$ and $V_2/\sqrt{gd_2}$ respectively). Positive surge calculations are performed with the surge Froude numbers.

4.3.3 Discussion

Considering the simple case of a positive surge travelling upstream of a sluice gate (after the gate closure), the flow sketch is sketched in Fig. 4.7 top. Several important results derive from the basic equations and they are summarized below.

(a) For an observer travelling with the flow upstream of the surge front (i.e. at a velocity V_1), the celerity of the surge (relative to the upstream flow) is:

$$V_1 + V_{srg} = \sqrt{gd_1}\sqrt{\frac{1}{2}\frac{d_2}{d_1}\left(1+\frac{d_2}{d_1}\right)} \qquad (4.11)$$

Note that if $d_2 > d_1$ then $(V_1 + V_{srg}) > \sqrt{gd_1}$. For a small wave (i.e. $d_2 = d_1 + \Delta d$), the term $V_1 + V_{srg}$ tends to the celerity of a small disturbance $\sqrt{gd_1}$ (see section 3.3.3, critical conditions and wave celerity).

(b) As $(V_1 + V_{srg}) > \sqrt{gd_1}$ a surge can move upstream even if the upstream (initial) flow is supercritical. Earlier it was stated that a *small* disturbance celerity C cannot move upstream in a supercritical flow. However, a large disturbance can make its way against supercritical flow provided it is *large enough* and in so doing the flow becomes subcritical (see the example in the previous section). A positive surge is a large disturbance.

(c) Considering the flow upstream of the surge front: for an observer moving at the same speed as the upstream flow, the celerity of a small disturbance travelling upstream is $\sqrt{gd_1}$. For the same observer, the celerity of the surge (i.e. a large disturbance) is $(V_1 + V_{srg}) > \sqrt{gd_1}$. As a result, the surge travels faster than a small disturbance. The surge overtakes and absorbs any small disturbances that may exist at the free-surface of the upstream water (i.e. in front of the surge).

Relative to the downstream water the surge travels more slowly than small disturbances as $(V_2 + V_{srg}) < \sqrt{gd_2}$. Any small disturbance, downstream of the surge front and moving upstream toward the wave front, overtakes the surge and is absorbed into it. Consequently, the wave absorbs random disturbances on both sides of the surge and this makes the positive surge *stable* and *self-perpetuating*.

(d) A lower limit of the surge velocity V_{srg} is set by the fact that $(V_1 + V_{srg}) > \sqrt{gd_1}$. It may be used as the initialization step of the iterative process for solving the equations. An upper limit of the surge celerity exists but it is a function of d_2 and V_2, the unknown variables.

Note

A positive surge can travel over very long distance without losing much energy because it is self-perpetuating. In natural and artificial channels, observations have shown that the wave front may travel over dozens of kilometres.

In water supply channels, brusque operation of controls (e.g. gates) may induce large surge which might overtop the channel banks, damaging and eroding the channel. In practice rapid operation of gates and controls must be avoided.

4.3.4 Positive and negative surges

Definitions

Positive surges

A positive surge is an *advancing wave front* resulting from an increase of flow depth moving upstream (i.e. closure of a downstream gate) or downstream (i.e. opening of an upstream gate, dam break) (Fig. 4.7).

Negative surges

A negative surge is a retreating wave front resulting from a decrease in flow depth moving upstream (e.g. opening of a downstream gate) or downstream (e.g. closure of an upstream gate) (Fig. 4.8) (e.g. Henderson 1966, p. 294; Streeter and Wylie 1981, p. 516).

Discussion

Positive surge flows are solved using the quasi-steady flow analogy (Section 4.3.2). For negative surges, the flow is unsteady and no quasi-steady flow analogy exists.

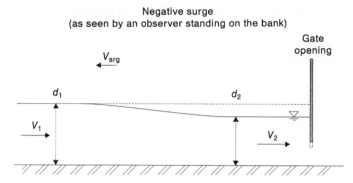

Fig. 4.8 Negative surge.

The complete unsteady analysis is necessary (e.g. Henderson 1966, pp. 294–304; Streeter and Wylie 1981, pp. 516–519; Liggett 1994, pp. 282–289).

For practising engineers, it is important to recognize between positive and negative surge cases. The table below summarizes the four possible cases:

Surge wave:	Positive surge		Negative surge	
Moving	downstream	upstream	downstream	upstream
Front of	deep water	deep water	shallow water	shallow water
Energy balance	loss of energy	loss of energy	gain of energy	gain of energy
Wave front stability	Stable	Stable	Unstable	Unstable
Analysis	quasi-steady	quasi-steady	unsteady	unsteady

> **Notes**
> 1. A positive surge is characterized by a steep advancing front. It is easy to recognize. Negative surges are more difficult to notice as the free-surface curvature is very shallow.
> 2. *For a negative surge, the quasi-steady analysis is not valid.*

4.4 Flow resistance in open channels

4.4.1 Presentation and definitions

Introduction

In a real fluid flow situation, energy is continuously dissipated. In open channel flows, flow resistance can be neglected over a short transition[3] as a first approximation, and the continuity and Bernoulli equations can be applied to estimate the downstream flow properties as functions of the upstream flow conditions and boundary

---

[3] In open channel hydraulics, the flow below a sluice gate, or a hydraulic jump; or the flow above a weir or at an abrupt drop or rise, may be considered as a short transition.

conditions. However, the approximation of frictionless flow is no longer valid for long channels. Considering a water supply canal extending over several kilometres, the bottom and sidewall friction retards the fluid, and, at equilibrium, the friction force counterbalances exactly the weight force component in the flow direction.

The laws of flow resistance in open channels are essentially the same as those in closed pipes (Henderson 1966). In an open channel, the calculations of the boundary shear stress are complicated by the existence of the free surface and the wide variety of possible cross-sectional shapes. Another difference is the propulsive force acting in the direction of the flow. In closed pipes, the flow is driven by a pressure gradient along the pipe whereas, in open channel flows, the fluid is propelled by the weight of the flowing water resolved down a slope.

Head loss

For open channel flow as for pipe flow, the head loss ΔH over a distance Δs (along the flow direction) is given by the *Darcy equation*:

$$\Delta H = f \frac{\Delta s}{D_H} \frac{V^2}{2g} \tag{4.12}$$

where f is the Darcy coefficient,[4] V is the mean flow velocity and D_H is the hydraulic diameter or equivalent pipe diameter. In open channels and assuming hydrostatic pressure distribution, the energy equation can be conveniently rewritten as:

$$d_1 \cos \theta_1 + z_{o_1} + \alpha_1 \frac{V_1^2}{2g} = d_2 \cos \theta_2 + z_{o_2} + \alpha_2 \frac{V_2^2}{2g} + \Delta H \tag{4.13}$$

where the subscripts 1 and 2 refer to the upstream and downstream cross-section of the control volume, d is the flow depth measured normal to the channel bottom, θ is the channel slope, z_o is the bed elevation, V is the mean flow velocity and α is the kinetic energy correction coefficient (i.e. Coriolis coefficient).

Notes
1. Henri P.G. Darcy (1805–1858) was a French civil engineer. He gave his name to the Darcy–Weisbach friction factor.
2. The hydraulic diameter and hydraulic radius are defined respectively as:

$$D_H = 4 \frac{\text{cross-sectional area}}{\text{wetted perimeter}} = \frac{4A}{P_w}$$

$$R_H = \frac{\text{cross-sectional area}}{\text{wetted perimeter}} = \frac{A}{P_w} \quad \text{(e.g. Henderson 1966, p. 91)}$$

where the subscript H refers to the hydraulic diameter or radius, A is the cross-section area and P_w is the wetted perimeter.

The hydraulic diameter is also called the equivalent pipe diameter. Indeed it is noticed that:

$$D_H = \text{pipe diameter } D \text{ for a circular pipe}$$

---

[4] Also called the Darcy–Weisbach friction factor or head loss coefficient.

The hydraulic radius was used in the early days of hydraulics as a mean flow depth. We note that:

R_H = flow depth d for an open channel flow in a wide rectangular channel

The author of the present textbook, believes that it is preferable to use the hydraulic diameter rather than the hydraulic radius, as the friction factor calculations are done with the hydraulic diameter (and not the hydraulic radius).

Bottom shear stress and shear velocity

The average shear stress on the wetted surface or *boundary shear stress* equals:

$$\tau_o = C_d \frac{1}{2} \rho V^2 \tag{4.14a}$$

where C_d is the skin friction coefficient[5] and V is the mean flow velocity. In open channel flow, it is common practice to use the Darcy friction factor f, which is related to the skin friction coefficient by:

$$f = 4C_d$$

It yields:

$$\tau_o = \frac{f}{8} \rho V^2 \tag{4.14b}$$

The *shear velocity* V_* is defined as (e.g. Henderson 1966, p. 95):

$$V_* = \sqrt{\frac{\tau_o}{\rho}} \tag{4.15}$$

where τ_o is the boundary shear stress and ρ is the density of the flowing fluid. The shear velocity is a measure of shear stress and velocity gradient near the boundary.

As for pipe flows, the flow regime in open channels can be either laminar or turbulent. In industrial applications, it is commonly accepted that the flow becomes turbulent for Reynolds numbers larger than 2000 to 3000, the *Reynolds number* being defined for pipe and open channel flows as:

$$Re = \rho \frac{V D_H}{\mu} \tag{4.16}$$

where μ is the dynamic viscosity of the fluid, D_H is the hydraulic diameter and V is the mean flow velocity.

Most open channel flows are turbulent. There are three types of turbulent flows: smooth, transition and fully rough. Each type of turbulent flow can be distinguished as a function of the *shear Reynolds number* defined as:

$$Re_* = \frac{V_* k_s}{\nu} \tag{4.17}$$

---

[5] Also called drag coefficient or Fanning friction factor (e.g. Liggett 1994).

where k_s is the average surface roughness (e.g. Henderson 1966, p. 95–96). For turbulent flows, the transition between smooth turbulence and fully-rough turbulence is approximately defined as follows.

Flow situation (Reference)	Open channel flow (Henderson 1966)	Pipe flow (Schlichting 1979)
Smooth turbulent	$Re_* < 4$	$Re_* < 5$
Transition	$4 < Re_* < 100$	$5 < Re_* < 75$
Fully rough turbulent	$100 < Re_*$	$75 < Re_*$

Notes
1. The shear velocity being a measure of shear stress and velocity gradient near the boundary, a large shear velocity V_* implies large shear stress and large velocity gradient. The shear velocity is commonly used in sediment-laden flows to calculate the sediment transport rate.
2. The shear velocity may be rewritten as:

$$\frac{V_*}{V} = \sqrt{\frac{f}{8}}$$

where V is the mean flow velocity.

Friction factor calculation
For open channel flow, the effect of turbulence becomes sensible for $Re > 2000$ to 3000 (e.g. Comolet 1976). In most practical cases, open channel flows are turbulent and the friction factor (i.e. Darcy coefficient) may be estimated from the *Colebrook–White formula* (Colebrook 1939):

$$\frac{1}{\sqrt{f}} = -2.0 \log_{10}\left(\frac{k_s}{3.71 D_H} + \frac{2.51}{Re\sqrt{f}}\right) \tag{4.18}$$

where k_s is the equivalent sand roughness height, D_H is the hydraulic diameter and Re is the Reynolds number defined as:

$$Re = \rho \frac{V D_H}{\mu} \tag{4.16}$$

Equation (4.18) is a non-linear equation in which the friction factor f is present on both the left- and right-sides. A graphical solution of the Colebrook–White formula is the Moody diagram (Moody 1944) given in Fig. 4.9.

Notes
1. The Colebrook–White formula is valid *only* for turbulent flows. It can be applied to any type of turbulent flows (i.e. smooth, transition, fully-rough).
2. More generally, the Darcy friction factor of open channel flows can be estimated as for pipe flows:

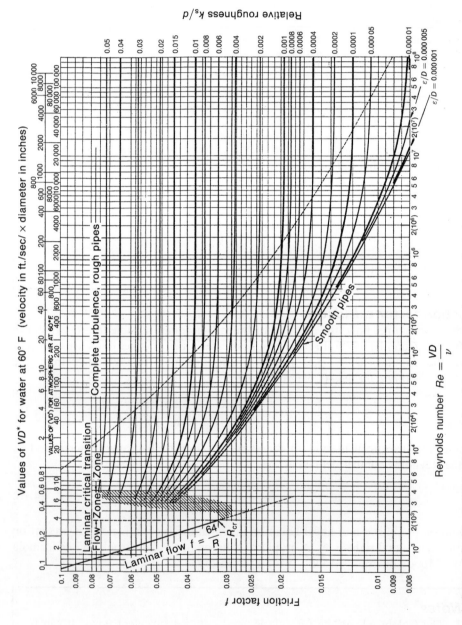

Fig. 4.9 Moody diagram (after Moody 1944, with permission of the American Society of Mechanical Engineers).

Laminar flow

$$f = \frac{64}{Re} \qquad Re < 2000 \text{ (e.g. Streeter and Wylie 1981, p. 238)}$$

Smooth turbulent flow

$$f = \frac{0.3164}{Re^{1/4}} \qquad Re < 1 \times 10^5 \text{ (Blasius' formula)}$$

(e.g. Schlichting 1979, Streeter and Wylie 1981, p. 236)

$$\frac{1}{\sqrt{f}} = 2.0 \log_{10}(Re\sqrt{f}) - 0.8 \qquad Re > 1 \times 10^5 \text{ (Karman–Nikuradse's formula)}$$

(e.g. Henderson 1966)

Turbulent flow

$$\frac{1}{\sqrt{f}} = -2.0 \log_{10}\left(\frac{k_s}{3.71 D_H} + \frac{2.51}{Re\sqrt{f}}\right)$$

Colebrook–White's formula (Colebrook 1939)

A less-accurate formula which can be used to initialize the calculation with the Colebrook–White formula is the Altsul's formula:

$$f = 0.1\left(1.46\frac{k_s}{D_H} + \frac{100}{Re}\right)^{1/4} \qquad \text{Altsul's formula (Idelchik 1969, 1986)}$$

Fully rough turbulent flow

$$\frac{1}{\sqrt{f}} = 2.0 \log_{10}\left(\frac{D_H}{k_s}\right) + 1.14 \qquad \text{deduced from the Colebrook's formula}$$

3. Roughness height
   Typical roughness heights are:

k_s	Material
0.01 to 0.02 mm	PVC (plastic)
0.02 mm	painted pipe
1 to 10 mm	riveted steel
0.25 mm	cast iron (new)
1 to 1.5 mm	cast iron (rusted)
0.3 to 3 mm	concrete
3 to 10 mm	untreated shot-concrete
0.6 to 2 mm	planed wood
5 to 10 mm	rubble masonry
3 mm	straight uniform earth channel

4. The calculation of the friction factor in turbulent flow is an iterative process. In practice the Moody diagram (Fig. 4.9) is often used to estimate the friction factor as:

$$f = f\left(\rho\frac{V D_H}{\mu}; \frac{k_s}{D_H}\right)$$

5. For calculations of head losses, Darcy coefficients and roughness heights, a recommended reference textbook for professional engineers is Idelchik (1969, 1986).

Application

Considering an open channel flow in a rectangular channel, the total discharge is $2.4 \, m^3/s$ and the flow depth is 3.1 m. The channel width is 5 m and the bottom and sidewalls are made of smooth concrete. Estimate the Darcy friction factor and the average boundary shear stress. The fluid is water at 20°C.

Solution

The flow is subcritical ($Fr = 0.03$). For smooth concrete, it is reasonable to assume $k_s = 1 \, mm$. The flow velocity, hydraulic diameter, relative roughness and Reynolds number are respectively:

$$V = 0.155 \, m/s$$

$$D_H = 5.54 \, m$$

$$k_s/D_H = 1.81 \times 10^{-4} \quad \text{relative roughness}$$

$$Re = 8.51 \times 10^5 \quad \text{Reynolds number (equation (4.16))}$$

The flow is turbulent ($Re > 2000$ to 3000).

Using the Colebrook–White's formula, the friction factor and bottom shear stress are:

$$f = 0.015 \quad \text{(equation (4.18))}$$

$$\tau_o = 0.044 Pa \quad \text{(equation (4.14))}$$

Comments

The shear velocity and shear Reynolds number are:

$$V_* = 0.0066 \, m/s \quad \text{(equation (4.15))}$$

$$Re_* = 6.6 \quad \text{(equation (4.17))}$$

The flow is turbulent in the transition region between smooth-turbulent and rough-turbulent.

Historically, flow resistance in open channels was investigated earlier than in pipe flows. The first successful empirical resistance formulae were obtained in the 18th century (Chézy formula) and 19th century (Gauckler–Manning formula). Such empirical formulae are inaccurate!

In the following section the momentum equation for uniform equilibrium open channel flows is developed. The flow resistance is analysed as for pipe flows. Well-known empirical resistance formulae are then presented. The correspondence between the empirical coefficients and the Darcy friction factor is also detailed.

4.4.2 Flow resistance of open channel flows

Momentum equation in steady uniform equilibrium open channel flow

The fundamental problem of steady uniform equilibrium flow is determining the relation between the flow velocity, the uniform flow depth, the channel slope and the channel geometry.

For a *steady* and *uniform equilibrium* flow the flow properties are independent of time and of position along the flow direction. That is:

$$\frac{\partial V}{\partial t} = 0$$

and

$$\frac{\partial V}{\partial s} = 0$$

where t is the time and s is the coordinate in the flow direction.

The momentum equation along a streamline states the exact balance between the shear forces and the gravity component. Considering a control volume as shown in Fig. 4.10, the momentum equation yields:

$$\tau_o P_w \Delta s = \rho g A \Delta s \sin \theta \qquad (4.19a)$$

where τ_o is the bottom shear stress, P_w is the wetted perimeter, Δs is the length of the control volume, A is the cross-section area and θ the channel slope. Replacing the bottom shear stress by its expression (equation (4.14)), the momentum equation for uniform equilibrium flows becomes:

$$V_o = \sqrt{\frac{8g}{f}} \sqrt{\frac{(D_H)_o}{4} \sin \theta} \qquad (4.19b)$$

where V_o is the uniform (equilibrium) flow velocity and $(D_H)_o$ is the hydraulic diameter of uniform equilibrium flows.

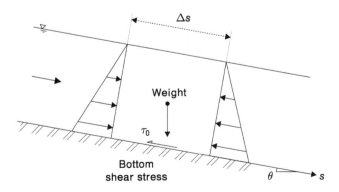

Fig. 4.10 Application of the momentum equation to uniform equilibrium open channel flow.

The momentum equation for steady uniform open channel flow (4.19a) is rewritten usually as:

$$S_f = S_o \qquad (4.19c)$$

where S_f is called the friction slope and S_o is the channel slope defined as:

$$S_f = -\frac{\partial H}{\partial s} = \frac{4\tau_o}{\rho g D_H} \qquad (4.20)$$

$$S_o = -\frac{\partial z_o}{\partial s} = \sin\theta \qquad (4.21)$$

where H is the mean total head and z_o is the bed elevation. Note that the definitions of the friction and bottom slope are general and apply to both uniform equilibrium and gradually-varied flows.

Notes

1. In sediment transport calculations, the momentum equation is usually expressed as equation (4.19a). Equation (4.19b) is more often used in clear-water flows.
2. For a flat channel, the channel slope S_o (or bed slope) is almost equal to the slope tangent:

$$S_o = -\frac{\partial z_o}{\partial s} = \sin\theta \sim \tan\theta$$

   S_o might be denoted 'i' (Comolet 1976).
3. For uniform equilibrium and gradually-varied flows the shear stress τ_o can be expressed as:

$$\tau_o = \rho g \frac{D_H}{4} S_f$$

4. Equations (4.19) to (4.21) are valid for any shape of channel cross-section.

Combining the definitions of bottom shear stress (equation (4.14)) and of friction slope, the momentum equation can be rewritten as:

$$V = \sqrt{\frac{8g}{f}} \sqrt{\frac{D_H}{4} S_f} \qquad (4.22)$$

This relationship (i.e. (4.22)) is valid for both uniform equilibrium and gradually varied flows.

Chézy coefficient

The dependence of the flow velocity on the channel slope and hydraulic radius can be deduced from equations (4.14) and (4.19). Replacing τ_o by its expression (equation (4.14)), the momentum equation (4.19b) can be transformed to give the Chézy equation:

$$V = C_{\text{Chézy}} \sqrt{\frac{D_H}{4} \sin\theta} \qquad (4.23)$$

where $C_{\text{Chézy}}$ is the Chézy coefficient (units: $m^{1/2}/s$), D_H is the hydraulic diameter and θ is the channel slope (e.g. Henderson 1966, pp. 91–96; Streeter and Wylie 1981, p. 229).

The Chézy equation (4.23) was first introduced in 1768 as an empirical correlation. Equation (4.23) is defined for uniform equilibrium and non-uniform gradually varied flows. The Chézy coefficient ranges typically from $30\,m^{1/2}/s$ (small rough channel) up to $90\,m^{1/2}/s$ (large smooth channel). Equations (4.19b) and (4.23) look similar. But it must be emphasized that equation (4.19b) was deduced from the momentum equation for uniform equilibrium flows. Equation (4.19b) is not valid for non-uniform equilibrium flows for which equation (4.22) should be used.

Notes

1. The Chézy equation was introduced in 1768 by the French engineer A. Chézy when designing canals for the water supply of the city of Paris.
2. The Chézy equation applies for *turbulent* flows. Although Chézy gave several values for $C_{\text{Chézy}}$, several researchers assumed later that $C_{\text{Chézy}}$ was independent of the flow conditions. Further studies clearly showed its dependence upon the Reynolds number and channel roughness.
3. The Chézy equation is valid for uniform equilibrium and non-uniform (gradually varied) turbulent flows.
4. For *uniform equilibrium flows* the boundary shear stress τ_o can be rewritten in terms of the Chézy coefficient as:

$$\tau_o = \rho g \frac{V^2}{(C_{\text{Chézy}})^2}$$

The Chézy equation in uniform equilibrium flows can be rewritten in terms of the shear velocity as:

$$\frac{V}{V_*} = \frac{C_{\text{Chézy}}}{\sqrt{g}}$$

and the Chézy coefficient and the Darcy friction factor are related by

$$C_{\text{Chézy}} = \sqrt{\frac{8g}{f}}$$

5. At uniform equilibrium and in fully-rough turbulent flows (e.g. natural streams), the Chézy coefficient becomes:

$$C_{\text{Chézy}} = 17.7 \log_{10}\left(\frac{D_H}{k_s}\right) + 10.1$$

(Uniform equilibrium fully-rough turbulent flow)

This expression derives from the Colebrook–White formula and is very close to the Keulegan formula (see below).

6. In *non-uniform gradually-varied flows* the combination of (4.22) and (4.23) indicates that the Chézy coefficient and the Darcy friction factor are related by:

$$C_{\text{Chézy}} = \sqrt{\frac{8g}{f}}\sqrt{\frac{S_f}{\sin\theta}}$$

7. Empirical estimates of the Chézy coefficient include the Bazin and Keulegan formulae:

$$C_{\text{Chézy}} = \frac{87}{1 + \dfrac{k_{\text{Bazin}}}{\sqrt{\dfrac{D_H}{4}}}} \qquad \text{Bazin formula}$$

where $k_{\text{Bazin}} = 0.06$ (very smooth boundary: cement, smooth wood), 0.16 (smooth boundary: wood, brick, freestone), 0.46 (stonework), 0.85 (fine earth), 1.30 (raw earth), 1.75 (grassy boundary, pebble bottom, grassy channel) (e.g. Comolet 1976); and

$$C_{\text{Chézy}} = 18.0 \log_{10}\left(\frac{D_H}{k_s}\right) + 8.73 \qquad \text{Keulegan formula}$$

which is valid for fully-rough turbulent flows and where k_s (in mm) = 0.14 (cement), 0.5 (planed wood), 1.2 (brick), 10 to 30 (gravel) (Keulegan 1938).

Interestingly Keulegan formula was validated with Bazin's data (Bazin 1865a).

The Gauckler–Manning coefficient

Natural channels have irregular channel bottoms, and information on the channel roughness is not easy to obtain. An empirical formulation, called the Gauckler–Manning formula, was developed for turbulent flows in rough channels.

The Gauckler–Manning formula is deduced from the Chézy equation by setting:

$$C_{\text{Chézy}} = \frac{1}{n_{\text{Manning}}} \left(\frac{D_H}{4}\right)^{1/6} \tag{4.24}$$

and equation (4.23) becomes:

$$V = \frac{1}{n_{\text{Manning}}} \left(\frac{D_H}{4}\right)^{2/3} \sqrt{\sin\theta} \tag{4.25}$$

where n_{Manning} is the *Gauckler–Manning coefficient* (units: $s/m^{1/3}$), D_H is the hydraulic diameter and θ is the channel slope.

The Gauckler–Manning coefficient is an *empirical* coefficient, found to be a characteristic of the surface roughness alone. Such an approximation might be reasonable as long as the water is not too shallow nor the channel too narrow.

Notes

1. Equation (4.25) is improperly called the 'Manning formula'. In fact it was first proposed by Gauckler (1867) based upon the re-analysis of experimental data obtained by Darcy and Bazin (1865).
2. Philippe Gaspard Gauckler (1826–1905) was a French engineer and member of the French 'Corps des Ponts-et-Chaussées'.

3. Robert Manning (1816–1897) was chief-engineer at the Office of Public Works, Ireland. He presented two formula in 1890 in his paper 'On the flow of water in open channels and pipes' (Manning 1890). One of the formulae was the 'Gauckler–Manning' formula (equation (4.25)) but Robert Manning preferred to use the second formula that he gave in the paper. Further information on the 'history' of the Gauckler–Manning formula was reported by Dooge (1991).
4. The Gauckler–Manning equation is valid for uniform equilibrium and non-uniform (gradually varied) flows.
5. Equation (4.25) is written in SI units. The units of the Gauckler–Manning coefficient n_{Manning} are $\text{s/m}^{1/3}$. A main criticism of the (first) Manning formula (equation (4.25)) is its dimensional aspect: i.e. n_{Manning} is not dimensionless (Dooge 1991).
6. The Gauckler–Manning equation applies for *fully rough turbulent* flows and *water* flows. It is an *empirical* relationship but has been found reasonably reliable.
7. Typical values of n_{Manning} (in SI Units) are:

$$
n_{\text{Manning}} =
\begin{array}{ll}
0.010 & \text{glass, plastic} \\
0.012 & \text{planed wood} \\
0.013 & \text{unplaned wood} \\
0.012 & \text{finished concrete} \\
0.014 & \text{unfinished concrete} \\
0.025 & \text{earth} \\
0.029 & \text{gravel} \\
0.05 & \text{flood plain (light brush)} \\
0.15 & \text{flood plain (trees)}
\end{array}
$$

Yen (1991b) proposed an extensive list of values for a wide range of open channels.

The Strickler coefficient

In Europe, the Strickler equation is used by defining:

$$
C_{\text{Chézy}} = k_{\text{Strickler}} \left(\frac{D_{\text{H}}}{4} \right)^{1/6}
\tag{4.26}
$$

and equation (4.23) becomes Strickler's equation:

$$
V = k_{\text{Strickler}} \left(\frac{D_{\text{H}}}{4} \right)^{2/3} \sqrt{\sin \theta}
\tag{4.27}
$$

where $k_{\text{Strickler}}$ is only a function of the surfaces.

Notes

1. The Gauckler–Manning and Strickler coefficients are related as:

$$
k_{\text{Strickler}} = 1/n_{\text{Manning}}
$$

2. The Strickler equation is used for pipes, gallery and channel carrying water. This equation is preferred to the Gauckler–Manning equation in Europe.
3. The coefficient $k_{Strickler}$ varies from 20 (rough stone, rough surface) to $80\,\text{m}^{1/3}/\text{s}$ (smooth concrete, cast iron).
4. The Strickler equation is valid for uniform and non-uniform (gradually varied) flows.

Particular flow resistance approximations

In Nature, rivers and streams do not exhibit a regular uniform bottom roughness. The channel bed consists often of unsorted sand, gravels and rocks. Numerous researchers have attempted to relate the equivalent roughness height k_s to a characteristic grain size (e.g. median grain size d_{50}). The analysis of numerous experimental data suggested that:

$$k_s \propto d_{50} \qquad (4.28)$$

where the constant of proportionality k_s/d_{50} ranges from 1 to well over 6 (see Table 12.2)! Obviously it is extremely difficult to relate grains size distributions and bed forms to a single parameter (i.e. k_s). For gravel-bed streams, Henderson (1966) produced a relationship between the Gauckler–Manning coefficient and the gravel size:

$$n_{Manning} = 0.038 d_{75}^{1/6} \qquad (4.29)$$

where the characteristic grain size d_{75} is in metres and $n_{Manning}$ is in $\text{s/m}^{1/3}$. Equation (4.29) was developed for $k_s/D_H < 0.05$.

For flood plains the vegetation may be regarded as a kind of roughness. Chow (1973) presented several empirical formulations for grassed channels as well as numerous photographs to assist in the choice of a Gauckler–Manning coefficient.

Notes
1. Strickler (1923) proposed the following empirical correlation for the Gauckler–Manning coefficient of rivers:

$$n_{Manning} = 0.041 d_{50}^{1/6}$$

where d_{50} is the median grain size (in m).
2. In torrents and mountain streams the channel bed might consist of gravel, stones and boulders with a size of the same order of magnitude as the flow depth. In such cases, the overall flow resistance results from a combination of skin friction drag, form drag and energy dissipation in hydraulic jumps behind large boulders. Neither the Darcy friction factor nor the Gauckler–Manning coefficient should be used to estimate the friction losses. Experimental investigations should be performed to estimate an overall Chézy coefficient (for each discharge).

4.5 Flow resistance calculations in engineering practice

4.5.1 Introduction

The transport of real fluids is associated with energy losses and friction losses. With pipe systems, pumps or high-head intakes are needed to provide the required energy for the fluid transport (kinetic energy, potential energy) and the associated energy loss (flow resistance). Two types of flow regime are encountered: laminar flow at low Reynolds numbers, and turbulent flows. In turbulent flows, the head loss can be estimated from the Darcy equation (4.12) in which the friction factor f is a function of the Reynolds number Re and relative roughness k_s/D_H. An extremely large number of experiments were performed in pipe flow systems to correlate f with Re and k_s/D_H. Usually the friction factor in turbulent flows is calculated with the Colebrook–White formula (eq. (4.18)) or from the Moody diagram (Fig. 4.9).

Flow resistance calculations in open channels
In open channels, the Darcy equation is valid using the hydraulic diameter as an equivalent pipe diameter. It is the only sound method to estimate the energy loss.

For various (mainly historical) reasons, empirical resistance coefficients (e.g. the Chézy coefficient) were and are still used. The Chézy coefficient was introduced in 1768 while the Gauckler–Manning coefficient was first presented in 1865: i.e. well before the classical pipe flow resistance experiments in the 1930s. Historically both the Chézy and the Gauckler–Manning coefficients were expected to be constant and functions of the roughness only. But it is now well recognized that these coefficients are only constant for a range of flow rates (e.g. Chow 1973, Chen 1990, Yen 1991a). Most friction coefficients (except perhaps the Darcy friction factor) are estimated 100%-empirically and they apply only to fully-rough turbulent water flows.

In practice, the Chézy equation is often used to compute open channel flow properties. The Chézy equation can be related to the Darcy equation using equations (4.22) and (4.23). As a lot of experimental data are available to estimate the friction factor f (or Darcy coefficient), the accurate estimate of the Darcy friction factor and Chézy coefficient is possible for standard geometry and material. But the data do not apply to natural rivers with vegetation, trees, large stones, boulders and complex roughness patterns (e.g. Fig. 4.11) and with movable boundaries (see also Chapter 12).

4.5.2 Selection of a flow resistance formula

Flow resistance calculations in open channels must be performed in terms of the *Darcy friction factor*. First the type of flow regime (laminar or turbulent) must be determined. Then the friction factor is estimated using the classical results (Section 4.3.1).

In turbulent flows, the choice of the boundary equivalent roughness height is important. Hydraulic handbooks (e.g. Idelchik 1969,1986) provide a selection of appropriate roughness heights for standard materials.

(a)

(b)

Fig. 4.11 Examples of natural rivers and flood plains. (a) Lance Creek and its flood plain, looking upstream (by H.E. Malde, with permission of US Geological Survey, Ref. 3994ct). Note the driftwood on the flood terrace that is flooded about every 10 years. (b) Burdekin River QLD, Australia (10 July 1983) (courtesy of Mrs J. Hacker). From Herveys Development Road Bridge, looking South.

The main limitations of the Darcy equation for turbulent flows are:

- the friction factor can be estimated for relative roughness k_s/D_H less than 0.05;
- classical correlations for f were validated for uniform-size roughness and regular roughness patterns.

In simple words, the Darcy equation cannot be applied to complex roughness patterns: e.g. vegetation and trees (in flood plains), shallow waters over rough channels.

(c)

Fig. 4.11 (c) Flood plain with trees of the South Alligator River, Kakadu National Park, Northern Territory, Australia in March 1998 (courtesy of Dr R. Rankin).

For *complex channel bed roughness*, practising engineers might estimate the flow resistance by combining the Chézy equation (4.23) with an 'appropriate' Gauckler–Manning or Chézy coefficient. Such an approximation is valid only for fully-rough turbulent flows.

Discussion

Great care must be taken when using the Gauckler–Manning equation (or Strickler equation) as the values of the coefficient are empirical. It is well known to river engineers that the estimate of the Gauckler–Manning (or Strickler) coefficient is a most difficult choice. Furthermore, it must be emphasized that the Gauckler–Manning formula is valid *only* for fully-rough turbulent flows of water. It should not be applied to smooth (or transition) turbulent flows. It is not valid for fluids other than water. The Gauckler–Manning equation was developed and 'validated' for clear-water flows only.

In practice, it is recommended to calculate the flow resistance using the Darcy friction factor. Empirical correlations such as the Bazin, Gauckler–Manning and/or Strickler formula could be used to check the result. If there is substantial discrepancy between the sets of results, experimental investigations must be considered.

Note

Several computer models of river flows using the unsteady flow equations are based on the Gauckler–Manning equation. Professionals (engineers, designers, managers) must not put too much confidence in the results of these models as long as the resistance coefficients have not been checked and verified with experimental measurements.

Applications

1. In a rectangular open channel (boundary roughness: PVC), the uniform equilibrium flow depth equals 0.5 m. The channel width is 10 m and the channel slope is 0.002°. Compute the discharge. The fluid is water at 20°C.

Solution

The problem must be solved by iterations. First we will assume that the flow is turbulent (we will need to check this assumption later) and we assume $k_s = 0.01$ mm (PVC).

An initial velocity (e.g. 0.1 m/s) is assumed to estimate the Reynolds number and hence the Darcy friction factor at the first iteration. The mean flow velocity is calculated using the momentum equation (or the Chézy equation). The full set of calculations are summarized in the following table. The total flow rate equals 1.54 m³/s. The results indicate that the flow is subcritical ($Fr = 0.139$) and turbulent ($Re = 5.6 \times 10^6$). The shear Reynolds number Re_* equals 0.12: i.e. the flow is smooth-turbulent.

Note: as the flow is not fully-rough turbulent, the Gauckler–Manning equation must not be used.

Iteration	V (initialization) (m/s)	f Eq. (4.18)	$C_{\text{Chézy}}$ (m$^{1/2}$/s) Eq. (4.23)	V (m/s) Eq. (4.23)
1	0.1	0.015	69.8	0.28
2	0.28	0.012	76.6	0.305
3	0.305	0.012	77.2	0.31
4	0.31	0.012	77.3	0.31

2. Considering a uniform equilibrium flow in a rectangular concrete channel ($B = 2$ m), the total discharge is 10 m³/s. The channel slope is 0.02°. Compute the flow depth. The fluid is water at 20°C.

Solution

The problem must be solved again by iterations. We will assume a turbulent flow and $k_s = 1$ mm (concrete).

At the first iteration, we need to assume a flow depth d (e.g. $d = 0.5$ m) to estimate the relative roughness k_s/D_H, the mean velocity (by continuity) and the Reynolds number. We deduce then the friction factor, the Chézy coefficient and the mean flow velocity (Chézy equation). The new flow depth is deduced from the continuity equation.

The iterative process is repeated until convergence.

Iteration	d (initialization) (m)	f Eq. (4.18)	$C_{\text{Chézy}}$ (m$^{1/2}$/s) Eq. (4.23)	V (m/s) Eq. (4.23)	d (m)
1	0.5	0.018	65.4	0.705	7.09
2	7.1	0.0151	72.1	1.26	4.0
3	4.0	0.0153	71.7	1.20	4.18
4	4.2	0.0152	71.7	1.20	4.15
5	4.15	0.0152	71.7	1.20	4.16
6	4.16	0.0152	71.7	1.20	4.16

The calculations indicate that the uniform equilibrium flow depth equals 4.16 m. The flow is subcritical ($Fr = 0.19$) and turbulent ($Re = 3.8 \times 10^6$). The shear Reynolds number equals 52.1: i.e. the flow is at transition between smooth-turbulent and fully-rough-turbulent.

Note: as the flow is not fully-rough turbulent, the Gauckler–Manning equation must not be used.

3. Considering a rectangular concrete channel ($B = 12$ m), the flow rate is 23 m³/s. The channel slope is 1°. Estimate the uniform equilibrium flow depth using both the Darcy friction factor and the Gauckler–Manning coefficient (if the flow is fully-rough turbulent). Compare the results. Investigate the sensitivity of the results upon the choice of the roughness height and Gauckler–Manning coefficient.

Solution
The calculations are performed in a similar manner to the previous example. We will detail the effects of roughness height and Gauckler–Manning coefficient upon the flow depth calculation.

For concrete, the equivalent roughness height varies from 0.3 to 3 mm for finished concrete and from 3 to 10 mm for rough concrete. For damaged concrete the equivalent roughness height might be greater than 10 mm. The Gauckler–Manning coefficient for concrete can be between 0.012 (finished concrete) and 0.014 s/m$^{1/3}$ (unfinished concrete). The results of the calculation are summarized in the following table.

Calculation	Surface	k_s (mm)	f	$n_{Manning}$ (s/m$^{1/3}$)	$C_{Chézy}$ (m$^{1/2}$/s)	d (m)
Darcy friction factor	Finished	0.3	0.0143		74.1	0.343
		1	0.0182		65.7	0.373
	Unfinished	3	0.0223		58.0	0.406
		10	0.032		49.5	0.452
Gauckler–Manning formula	Finished			0.012	69.6	0.359
	Unfinished			0.014	74.8	0.341

First the Reynolds number is typically within the range 7×10^6 to 7.2×10^6 (i.e. turbulent flow). The shear Reynolds number is between 70 and 2700. That is, the flow is fully-rough turbulent and within the validity range of the Gauckler–Manning formula. Secondly the flow depth increases with increasing roughness height. The increase in flow depth results from a decrease of flow velocity with increasing flow resistance. Thirdly, let us observe the discrepancy of results for unfinished concrete between the Darcy friction factor calculations and the Gauckler–Manning formula. Further, note that the calculations (using the Darcy friction factor) are sensitive to the choice of roughness height.

4.5.3 Flow resistance in a flood plain

A practical problem is the flow resistance calculations of a river channel and the adjacent flood plain (Figs 4.11 and 4.12). Usually the flood plain is much rougher than the river channel and the flow depth is much smaller. The energy slope of both portions must be the same. For uniform equilibrium flow, this implies that the bed slope S_o is the same in the channel and plain.

Two practical applications of such calculations are:

1. assuming that the total discharge is known, estimate the flow depths in the river d_1 and in the flood plain d_2, or
2. assuming known flow depths (d_1 and d_2) find the total discharge in the flood plain and channel.

In practice the complete hydraulic calculations are an iterative process. The main equations are summarized in the table below:

	River channel	Flood plain
Flow depth	d_1	d_2
Flow velocity	V_1	V_2
Width	B_1	B_2
Bed altitude	z_{o_1}	$z_{o_2} = z_{o_1} + y_{\text{channel}}$
Wetted perimeter	$P_{w_1} = B_1 + d_1 + y_{\text{channel}}$	$P_{w_2} = B_2 + d_2$
Cross-section area	$A_1 = d_1 B_1$	$A_2 = d_2 B_2$
Hydraulic diameter	$D_{H_1} = 4\dfrac{A_1}{P_{w_1}}$	$D_{H_2} = 4\dfrac{A_2}{P_{w_2}}$
Equivalent roughness height	k_{s_1}	k_{s_2}
Darcy friction factor	$f_1 = f\left(\dfrac{k_{s_1}}{D_{H_1}}; \dfrac{V_1 D_{H_1}}{\nu}\right)$	$f_2 = f\left(\dfrac{k_{s_2}}{D_{H_2}}; \dfrac{V_2 D_{H_2}}{\nu}\right)$
Momentum equation	$V_1 = \sqrt{\dfrac{8g}{f_1}}\sqrt{\dfrac{D_{H_1}}{4}}\sqrt{\sin\theta}$	$V_2 = \sqrt{\dfrac{8g}{f_2}}\sqrt{\dfrac{D_{H_2}}{4}}\sqrt{\sin\theta}$
Continuity equation	$Q_1 = V_1 A_1$	$Q_2 = V_2 A_2$

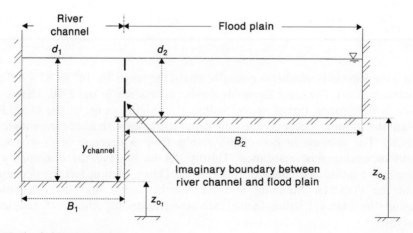

Fig. 4.12 Flood plain cross-section.

For a given flow depth the total discharge equals:

$$Q = Q_1 + Q_2 \qquad\qquad [C]$$

The flow depths are deduced by iterative calculations.

Notes
1. The longitudinal bed slope ($\sin\theta$) is the same for each portion (i.e. the main channel and flood plain).
2. The flow depths d_1 and d_2 are related as: $d_1 = d_2 + y_{channel}$.
3. Note that friction (and energy loss) is assumed zero at the interface between the river channel flow and the flood plain flow. In practice, turbulent energy losses and secondary currents are observed at the transition between the channel flow and the much slower flood plain flow.

4.6 Exercises

Numerical solutions to some of these exercises are available from the Web at www.arnoldpublishers.com/support/chanson

Momentum equation

The backward-facing step, sketched on Fig. E.4.1, is in a 5 m wide channel of rectangular cross-section. The total discharge is 55 m³/s. The fluid is water at 20°C. The bed of the channel, upstream and downstream of the step, is horizontal and smooth. Compute the pressure force acting on the vertical face of the step and indicate the direction of the force (i.e. upstream or downstream).

Solution: Applying the continuity equation, the upstream flow depth equals: $d_1 = 8.15$ m. Assuming a hydrostatic pressure distribution, the upstream specific energy equals: $E_1 = 8.24$ m. Note that the upstream flow is subcritical ($Fr_1 = 0.15$).

The downstream specific energy is deduced from the Bernoulli principle assuming a smooth transition:

$$E_1 + \Delta z_o = E_2 \qquad \text{(Bernoulli equation)}$$

where Δz_o is the drop height (0.1 m). It gives $E_2 = 8.34$ m. The downstream flow depth is deduced from the definition of the specific energy:

$$E_2 = d_2 + \frac{Q^2}{2gB^2 d_2^2}$$

where B is the channel width. The solution is: $d_2 = 8.25$ m.

Considering the forces acting on the control volume contained between Sections 1 and 2, the momentum equation as applied to the control volume between Sections 1

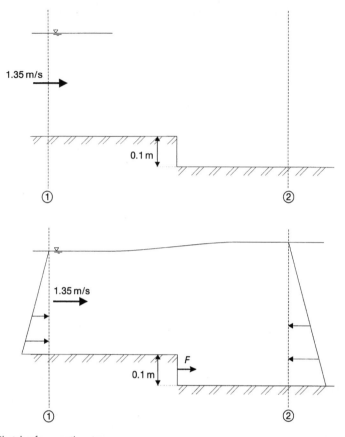

Fig. E.4.1 Sketch of a negative step.

and 2 is:

$$\rho Q V_2 - \rho Q V_1 = \tfrac{1}{2}\rho g d_1^2 B - \tfrac{1}{2}\rho g d_2^2 B + F \qquad \text{(Momentum principle)}$$

in the horizontal direction. Note that the friction is zero as the bed is assumed smooth.

The solution of the momentum equation is: $F = +40.3\,\text{kN}$ (remember: F is the force exerted by the step onto the fluid). In other words, the pressure force exerted by the fluid onto the vertical face of the step acts in the upstream direction and equals 40.3 kN.

Considering a broad-crested weir laboratory model (0.25 m wide rectangular channel), the upstream flow conditions (measured upstream of the weir) are: $d_1 = 0.18\,\text{m}$, $V_1 = 0.36\,\text{m/s}$. The bed of the channel is horizontal and smooth (both upstream and downstream of the weir).

Assuming critical flow conditions at the crest and supercritical flow downstream of the weir, compute the following quantities: (a) flow depth downstream of the weir and (b) the horizontal force acting on the weir (i.e. sliding force).

Considering a sluice gate in a rectangular channel ($B = 0.80$ m), the observed upstream and downstream flow depths are respectively: $d_1 = 0.450$ m, $d_2 = 0.118$ m. (a) Derive the expression of the flow rate in terms of the upstream and downstream depths, and of the channel width. (b) Derive the expression of the force on the gate in terms of the upstream and downstream depths, the channel width and the flow rate. (c) Compute the flow rate. (d) Compute the force on the gate.

Assume a smooth horizontal channel.

Hydraulic jump

Considering a hydraulic jump, are the following statements true or not?

Statement	Yes	No	Correct answer
The upstream Froude number is less than 1			
The downstream velocity is less than the upstream velocity			
The energy loss term is always positive			
The upstream depth is larger than the downstream flow depth			
Hydraulic jump can occur in closed-conduit flow			
Air entrainment occurs in a hydraulic jump			
Hydraulic jump is dangerous			
If the upstream flow is such as $Fr_1 = 1$ can hydraulic jump occur?			
What then is the downstream Froude number?	–	–	

What is the definition of: alternate depths, conjugate depths and sequent depths? In each case, explain your answer in words and use appropriate example(s) if necessary.

Considering a hydraulic jump in a rectangular horizontal channel, sketch the flow and write the following basic equations across the jump: (a) continuity equation, (b) momentum equation and (c) energy equation. Deduce the expression of the downstream flow depth as a function of the upstream flow depth and upstream Froude number only.

Application: The upstream flow depth is 1.95 m. The channel width is 2 m. The flow rate is 70 m³/s. Compute the downstream flow depth and the head loss across the jump.

Considering a hydraulic jump in a horizontal channel of trapezoidal cross-section. The channel cross-section is symmetrical around the channel centreline, the angle between the sidewall and the horizontal being δ. Write the continuity and momentum equations for the hydraulic jump. Use the subscript 1 to refer to the upstream flow conditions and the subscript 2 for the downstream flow conditions. Neglect the bottom friction.

A hydraulic jump flow takes place in a horizontal rectangular channel. The upstream flow conditions are: $d = 1$ m, $q = 11.2$ m²/s. Calculate the downstream flow depth, downstream Froude number, head loss in the jump and roller length.

Considering the upstream flow: $Q = 160\,\text{m}^3/\text{s}$, $B = 40\,\text{m}$, design a hydraulic jump dissipation structure. Select an appropriate upstream flow depth for an optimum energy dissipation. What would be the upstream and downstream Froude numbers?

Considering a hydraulic jump in a horizontal rectangular channel of constant width and neglecting the friction force, apply the continuity, momentum and energy equations to the following case: $Q = 1500\,\text{m}^3/\text{s}$, $B = 50\,\text{m}$. The energy dissipation in the hydraulic jump is $\Delta H = 5\,\text{m}$. Calculate the upstream and downstream flow conditions (d_1, d_2, Fr_1, Fr_2). Discuss the type of hydraulic jump.

Considering a rectangular horizontal channel downstream of a radial gate, a stilling basin (i.e. energy dissipation basin) is to be designed immediately downstream of the gate. Assume no flow contraction at the gate. The design conditions are: $Q = 65\,\text{m}^3/\text{s}$, $B = 5\,\text{m}$. Select an appropriate gate opening for an optimum design of a dissipation basin. Explain in words what gate opening you would choose while designing this hydraulic structure to dissipate the energy in a hydraulic jump. Calculate the downstream Froude number. What is the minimum length of the dissipation basin? What hydraulic power would be dissipated in the jump at design flow conditions? How many 100 W bulbs could be powered (ideally) with the hydraulic power dissipated in the jump?

Surges and bores

What are the differences between a surge, a bore and a wave? Give examples of a bore.

Considering a flow downstream of a gate in a horizontal and rectangular channel, the initial flow conditions are: $Q = 500\,\text{m}^3/\text{s}$, $B = 15\,\text{m}$ and $d = 7.5\,\text{m}$. The gate suddenly opens and provides the new flow conditions: $Q = 700\,\text{m}^3/\text{s}$. (a) Can the flow situation be described using the quasi-steady flow analogy? (Sketch the flow and justify your answer in words.) (b) Assuming a frictionless flow, compute the surge velocity and the new flow depth. (c) Are the new flow conditions supercritical? (d) Is the wave stable? (If a small disturbance starts from the gate (i.e. gate vibrations) after the surge wave what will happen? Would the wave become unstable?)

Considering a flow upstream of a gate in a horizontal and rectangular channel, the initial flow conditions are: $Q = 100\,\text{m}^3/\text{s}$, $B = 9\,\text{m}$ and $d = 7.5\,\text{m}$. The gate suddenly opens and provides the new flow conditions: $Q = 270\,\text{m}^3/\text{s}$. Characterize the surge: i.e. positive or negative. Is the wave stable?

Considering the flow upstream of a gate, the gate suddenly closes. The initial flow conditions were: $Q = 5000\,\text{m}^3/\text{s}$, $d = 5\,\text{m}$, $B = 100\,\text{m}$. The new discharge is: $Q = 3000\,\text{m}^3/\text{s}$. Compute the new flow depth and flow velocity.

Solution: The surge is an advancing wave front (i.e positive surge). Using the quasi-steady flow assumption, the initial flow conditions are the flow conditions upstream of the surge front: $d_1 = 5\,\text{m}$, $V_1 = 10\,\text{m/s}$, $Fr = 1.43$. To start the calculations, $V_s = 0$ may be assumed (i.e. stationary surge also called hydraulic jump). In this particular case (i.e. $V_{srg} = 0$), the continuity equation becomes: $V_1 d_1 = V_2 d_2$.

Notation Equation	V_{srg}	Fr_1	d_2 [C]	Fr_2 [M]	V_2 Def. Fr_2	$V_2 d_2 B$ Q_2
1st iteration	0.0	1.43	23.4	0.72	10.9	2540
2nd iteration	2	1.71	15	0.62	5.5	8229
Solution	3.26	1.89	11.1	0.57	2.70	3000

Notes
1. For an initialization step (where $V_{srg} = 0$), the continuity equation can be rewritten as: $Fr_1 \sqrt{gd_1^3} = Fr_2 \sqrt{gd_2^3}$ using the definition of the Froude number. The above equation is more practical than the general continuity equation for surge.
2. It must be noted that the initial flow conditions are supercritical. The surge is a large disturbance travelling upstream against a supercritical flow.
3. The surge can be classified as a hydraulic jump using the (surge) upstream Froude number Fr_1. For this example, the surge is a weak surge.

Flow resistance

Considering a uniform equilibrium flow down a rectangular channel, develop the momentum equation. For a wide rectangular channel, deduce the expression of the normal depth as a function of the Darcy friction factor, discharge per unit width and bed slope.

For a uniform equilibrium flow down an open channel: (a) Write the Chézy equation. Define clearly all your symbols. (b) What are the SI units of the Chézy coefficient? (c) Give the expression of the Chézy coefficient as a function of the Darcy–Weisbach friction factor. (d) Write the Gauckler–Manning equation. Define clearly your symbols. (e) What are the SI units of the Gauckler–Manning coefficient?

Considering a gradually-varied flow in a rectangular open channel, the total discharge is 2.4 m³/s and the flow depth is 3.1 m. The channel width is 5 m, and the bottom and sidewalls are made of smooth concrete. Estimate: (a) the Darcy friction factor and (b) the average boundary shear stress. The fluid is water at 20°C.

In a rectangular open channel (boundary roughness: PVC), the uniform equilibrium flow depth equals 0.9 m. The channel is 10 m wide and the bed slope is 0.0015°. The fluid is water at 20°C. Calculate: (a) the flow rate, (b) the Froude number, (c) the Reynolds number, (d) the relative roughness, (e) the Darcy friction factor and (f) the mean boundary shear stress.

Considering a uniform equilibrium flow in a trapezoidal grass waterway (bottom width: 15 m, sidewall slope: 1 V:5 H), the flow depth is 5 m and the longitudinal bed slope is 3 m/km. Assume a Gauckler–Manning coefficient of 0.05 s/m$^{1/3}$ (flood

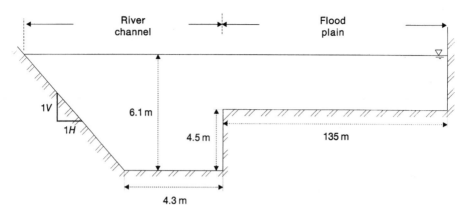

Fig. E.4.2 Sketch of a flood plain.

plain, light brush). The fluid is water at 20°C. Calculate: (a) the discharge, (b) the critical depth, (c) the Froude number, (d) the Reynolds number and (e) the Chézy coefficient.

Considering a uniform equilibrium flow in a trapezoidal concrete channel (bottom width: 2 m, sidewall slope: 30°), the total discharge is $10 \, \text{m}^3/\text{s}$. The channel slope is 0.02°. The fluid is water at 20°C. Estimate: (a) the critical depth, (b) the normal depth, (c) the Froude number, (d) the Reynolds number and (e) the Darcy friction factor.

Considering the flood plain, sketched on Fig. E.4.2, the mean channel slope is 0.05°. The river channel is lined with concrete and the flood plain is riprap material (equivalent roughness height: 8 cm). The fluid is water with a heavy load of suspended sediment (fluid density: $1080 \, \text{kg/m}^3$). The flow is assumed to be in uniform equilibrium. Compute and give the values (and units) of the following quantities: (a) volume discharge in the river channel; (b) volume discharge in the flood plain; (c)

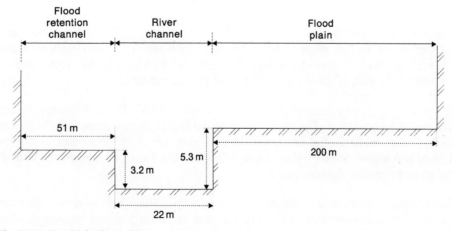

Fig. E.4.3 Sketch of a flood plain.

total volume discharge (river channel + flood plain); (d) total mass flow rate (river channel + flood plain); (e) is the flow subcritical or supercritical? Justify your answer clearly. (Assume no friction (and energy loss) at the interface between the river channel flow and the flood plain flow.)

Considering a river channel with a flood plain in each side (Fig. E.4.3), the river channel is lined with finished concrete. The lowest flood plain is liable to flooding (it is land used as a flood water retention system) and its bed consists of gravel ($k_s = 20\,\text{mm}$). The right bank plain is a grassed area (centipede grass, $n_{\text{Manning}} \sim 0.06\,\text{SI units}$). The longitudinal bed slope of the river is 2.5 m/km. For the 1-in-50-years flood ($Q = 500\,\text{m}^3/\text{s}$), compute: (a) the flow depth in the main channel, (b) the flow depth in the right flood plain, (c) the flow rate in the main channel, (d) the flow rate in the flood water retention system and (e) the flow rate in the right flood plain. (Assume no friction (and energy loss) at the interface between the river channel flow and the flood plain flows.)

5

Uniform flows and gradually varied flows

Summary

First, the properties of uniform equilibrium open channel flows are described (for steady flows). Then the gradually varied flow assumptions are detailed. Later, the energy equation is applied to gradually varied flows, and backwater calculations are introduced.

5.1 Uniform flows

5.1.1 Presentation

Definition

Uniform equilibrium open channel flows are characterized by a constant depth and constant mean flow velocity:

$$\frac{\partial d}{\partial s} = 0$$

and

$$\frac{\partial V}{\partial s} = 0$$

where s is the coordinate in the flow direction. Uniform equilibrium open channel flows are commonly called 'uniform flows' or 'normal flows'. The expression should not be confused with 'uniform velocity distribution flows'. In open channels, the uniform equilibrium flow regime is 'equivalent' to a fully developed pipe flow regime.

Uniform equilibrium flow can occur only in a straight channel with a constant channel slope and cross-sectional shape, and a constant discharge. The depth corresponding to uniform flow in a particular channel is called the *normal depth* or uniform flow depth.

Basic equations

For a uniform open channel flow, the shear forces (i.e. flow resistance) exactly balance the gravity force component. Considering a control volume as shown in Fig. 4.9, the momentum equation becomes (see Section 4.4):

$$\tau_o P_w \Delta s = \rho g A \Delta s \sin \theta \tag{4.19a}$$

where τ_o is the average shear stress on the wetted surface, P_w is the wetted perimeter, θ is the channel slope, A is the cross-section area and Δs is the length of the control volume.

For steady uniform open channel flow, this equation is equivalent to the energy equation:

$$\frac{\partial H}{\partial s} = \frac{\partial z_o}{\partial s} \tag{5.1}$$

where H is the mean total head and z_o is the bed elevation. This is usually rewritten as:

$$S_f = S_o$$

where S_o is the *bed slope* and S_f the *friction slope*.

Notes
1. For a flat channel, the bed slope is:

$$S_o = \sin \theta = -\frac{\partial z_o}{\partial s} \sim \tan \theta$$

2. For uniform or non-uniform flows the *friction slope* S_f is defined as:

$$S_f = -\frac{\partial H}{\partial s} = \frac{4 \tau_o}{\rho g D_H}$$

3. At any cross-section the relationship between the Darcy coefficient and the friction slope leads to:

$$S_f = f \frac{1}{D_H} \frac{V^2}{2g} = \frac{Q^2 P_w f}{8g A^3}$$

4. The uniform flow depth or *normal depth* is denoted d_o.

5.1.2 Discussion

Mild and steep slopes
A channel slope is usually 'classified' by comparing the uniform flow depth d_o with the critical flow depth d_c. When the uniform flow depth is larger than the critical flow depth, the uniform equilibrium flow is tranquil and subcritical. The slope is called a *mild slope*. For $d_o < d_c$, the uniform flow is supercritical and the slope is *steep*.

$d_o > d_c$	mild slope	Uniform flow: $Fr_o < 1$ (subcritical flow)
$d_o = d_c$	critical slope	Uniform flow: $Fr_o = 1$ (critical flow)
$d_o < d_c$	steep slope	Uniform flow: $Fr_o > 1$ (supercritical flow)

> ### Note
> For a wide rectangular channel, the ratio d_o/d_c can be rewritten as:
>
> $$\frac{d_o}{d_c} = \sqrt[3]{\frac{f}{8\sin\theta}}$$
>
> where f is the friction factor and θ is the channel slope. The above result shows that the notion of steep and mild slopes is not only a function of the bed slope but also a function of the flow resistance: i.e. of the flow rate and roughness height.

Critical slope

A particular case is the situation where $d_o = d_c$: i.e. the uniform equilibrium flow is critical. The channel slope, for which the uniform flow is critical, is called the *critical slope* and is denoted S_c.

Critical slopes are seldom found in nature because critical flows and near-critical flows[1] are unstable. They are characterized by flow instabilities (e.g. free-surface undulations and waves) and the flow becomes rapidly unsteady and non-uniform.

> ### Note
> In the general case, the critical slope satisfies:
>
> $$S_c = \sin\theta_c = S_f = f\frac{V_o^2}{2g(D_H)_o}$$
>
> where V_o and $(D_H)_o$ are the uniform flow velocity and hydraulic diameter respectively which must also satisfy: $V_o = V_c$ and $(D_H)_o = (D_H)_c$.
> For a wide rectangular channel, the critical slope satisfies:
>
> $$S_c = \sin\theta_c = \frac{f}{8}$$

Application: most efficient cross-section shape

Uniform flows seldom occur in Nature. Even in artificial channels of uniform section (e.g. Figs 5.1 and 5.2), the occurrence of uniform flows is not frequent because of the existence of controls (e.g. weirs, sluice gates) which rule the relationship between depth and discharge. However, uniform equilibrium flow is of importance as a reference. Most channels are analysed and designed for uniform equilibrium flow conditions.

During the design stages of an open channel, the channel cross-section, roughness and bottom slope are given. The objective is to determine the flow velocity, depth and flow rate, given any one of them. The design of channels involves selecting the channel shape and bed slope to convey a given flow rate with a given flow depth. For a given

---

[1] Near-critical flows are characterized by a specific energy only slightly greater than the minimum specific energy and by a Froude number close to unity (i.e. $0.7 < Fr < 1.5$ typically). Such flows are unstable, as any small change in specific energy (e.g. bed elevation, roughness) induces a large variation of flow depth.

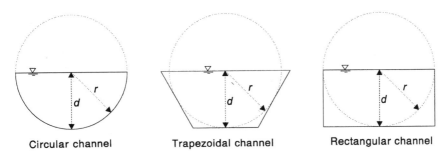

| Circular channel | Trapezoidal channel | Rectangular channel |

Fig. 5.1 Cross-sections of maximum flow rate: i.e. 'optimum design'.

discharge, slope and roughness, the designer aims to minimize the cross-sectional area A in order to reduce construction costs (e.g. Henderson 1966, p. 101).

The most 'efficient' cross-sectional shape is determined for uniform flow conditions. Considering a given discharge Q, the velocity V is maximum for the minimum cross-section A. According to the Chézy equation (4.23) the hydraulic diameter is then maximum. It can be shown that:

1. the wetted perimeter is also minimum (e.g. Streeter and Wylie 1981, pp. 450–452); and
2. the semi-circular section (the semi-circle having its centre in the surface) is the best hydraulic section (e.g. Henderson 1966, pp. 101–102).

For several types of channel cross-sections the 'best' design is shown in Fig. 5.1 and summarized in the following table.

Cross-section	Optimum width B	Optimum cross-section area A	Optimum wetted perimeter P_{w}	Optimum hydraulic diameter D_{H}
Rectangular	$2d$	$2d^2$	$4d$	$2d$
Trapezoidal	$\dfrac{2}{\sqrt{3}}d$	$\sqrt{3}d^2$	$2\sqrt{3}d$	$2d$
Semi-circle	$2d$	$\dfrac{\pi}{2}d^2$	πd	$2d$

5.1.3 Uniform flow depth in non-rectangular channels

For any channel shape, the uniform flow conditions are deduced from the momentum equation (4.19a). The uniform flow depth must satisfy:

$$\frac{A_{\mathrm{o}}^3}{(P_{\mathrm{w}})_{\mathrm{o}}} = \frac{fQ^2}{8g\sin\theta} \tag{5.2a}$$

where the cross-section area A_{o} and wetted perimeter $(P_{\mathrm{w}})_{\mathrm{o}}$ for uniform flow conditions are functions of the normal flow depth d_{o}. Typical examples of relationships between A, P_{w} and d have been detailed in Section 3.4.

(a)

(b)

Fig. 5.2 Examples of waterways. (a) Storm waterway in Canberra ACT, Australia (December 1997). Trapezoidal concrete-lined channel and artifical flood plains. (b) The McPartlan canal connecting Lake Pedler to Lake Gordon, Tasmania (20 December 1992). Trapezoidal concrete-lined channel. Flow from the left to the top background

For a wide rectangular channel equation (5.2a) yields:

$$d_o = \sqrt[3]{\frac{q^2 f}{8g \sin \theta}} \qquad \text{wide rectangular channel} \qquad (5.2b)$$

where q is the discharge per unit width. Equation (5.2b) applies only to rectangular channels such as $B > 10d_o$.

Notes
1. The calculation of the normal depth (equation (5.2)) is an iterative process (see the next application). In practice, the iterative method converges rapidly. The Chézy equation (i.e. (4.23)) may also be used (see Sections 4.4.2 and 4.5.2).
2. Calculations of the normal flow depth always converge to a unique solution for either laminar or turbulent flows. But near the transition between laminar and turbulent flow regimes, the calculations might not converge. Indeed flow conditions near the laminar–turbulent flow transition are naturally unstable.

Applications
1. Considering a rectangular channel, the channel slope is 0.05°. The bed is made of bricks ($k_s = 10$ mm) and the width of the channel is 3 m. The channel carries water (at 20°C). For a 2 m³/s flow rate at uniform equilibrium flow conditions, is the flow super- or sub-critical?

Solution
We must compute both the normal and critical depths to answer the question. For a rectangular channel, the critical depth equals (Chapter 3):

$$d_c = \sqrt[3]{\frac{Q^2}{gB^2}}$$

It yields: $d_c = 0.357$ m.

The uniform equilibrium flow depth (i.e. normal depth) is calculated by an iterative method. At the first iteration, we assume a flow depth. For that flow depth, we deduce the velocity (by continuity), the Reynolds number, relative roughness and Darcy friction factor. The flow velocity is computed from the momentum equation:

$$V_o = \sqrt{\frac{8g}{f}} \sqrt{\frac{(D_H)_o}{4} \sin \theta} \qquad (4.19b)$$

Note that the Chézy equation (4.23) may also be used. The flow depth d_o is deduced by continuity:

$$A_o = \frac{Q}{V_o}$$

where A_o is the cross-sectional area for normal flow conditions (i.e. $A_o = Bd_o$ for a rectangular cross-sectional channel).

The process is repeated until convergence:

Iteration	d initialization (m)	Re	k_s/D_H	f	V_o (m/s)	d_o (m)
1	0.357	2.14×10^6	0.0087	0.0336	0.738	0.903
2	0.903	1.65×10^6	0.0044	0.0284	1.146	0.581
Solution		1.84×10^6	0.0054	0.0299	1.005	0.663

The normal flow is a fully-rough turbulent flow ($Re = 1.8 \times 10^6$, $Re_* = 622$). This observation justifies the calculation of the Darcy friction factor using the Colebrook–White correlation. The normal depth is 0.663 m.

As the normal depth is greater than the critical depth, the uniform equilibrium flow is subcritical.

2. For the same channel and the same flow rate as above, compute the critical slope S_c.

Solution
The question can be re-worded as: for $Q = 2\,\mathrm{m}^3/\mathrm{s}$ in a rectangular ($B = 3\,\mathrm{m}$) channel made of bricks ($k_s = 10\,\mathrm{mm}$), what is the channel slope θ for which the uniform equilibrium flow is critical?

The critical depth equals 0.357 m (see previous example). For that particular depth, we want to determine the channel slope such that $d_o = d_c = 0.357\,\mathrm{m}$. Equation (4.19) can be transformed as:

$$\sin \theta_c = \frac{V_o^2}{\dfrac{8g}{f} \dfrac{(D_H)_o}{4}}$$

in which $d_o = d_c = 0.357\,\mathrm{m}$ and $V_o = V_c = 1.87\,\mathrm{m/s}$.
It yields: $S_c = \sin \theta_c = 0.00559$. The critical slope is 0.32°.

Comments
For wide rectangular channels, the critical slope can be approximated as: $S_c = f/8$. In our application, calculations indicate that, for the critical slope, $f \sim 0.036$ leading to: $S_c \sim 0.0045$ and $\theta_c \sim 0.259°$. That is, the approximation of a wide channel induces an error of about 25% on the critical slope S_c.

3. For a triangular channel (26.6° wall slope, i.e. 1 V:2 H) made of concrete, compute the (centreline) flow depth for uniform flow conditions. The flow rate is $2\,\mathrm{m}^3/\mathrm{s}$ and the channel slope is 0.05°. Is the channel slope mild or steep?

Solution
The uniform flow depth is calculated by successive iterations. We assume $k_s \sim 1\,\mathrm{mm}$ in the absence of further information on the quality of the concrete:

Iteration	d initialization (m)	Re	k_s/D_H	f	V_o (m/s)	d_o (m)
1	0.7	2.5×10^6	0.0008	0.0187	1.07	0.97
2	0.97	9.2×10^5	0.0006	0.0177	1.29	0.88
Solution		9.9×10^5	0.00062	0.0180	1.62	0.899

The normal (centreline) depth equals: $d_o = 0.899$ m. The uniform flow is turbulent ($Re = 9.9 \times 10^5$, $Re = 58.2$). To characterize the channel slope as mild or steep, we must first calculate the critical depth. For a non-rectangular channel, the critical depth must satisfy a minimum specific energy. It yields (Chapter 3):

$$\frac{A^3}{B} = \frac{Q^2}{g} \qquad (3.34a)$$

where A is the cross-section area and B is the free-surface width. For a triangular channel, $A = 0.5Bd$ and $A = 2d \cot \delta$ where δ is the wall slope ($\delta = 26.6°$). Calculations indicate that the critical depth equals: $d_c = 0.728$ m. As the normal depth is larger than the critical depth, the channel slope is *mild*.

5.2 Non-uniform flows

5.2.1 Introduction

In most practical cases, the cross-section, depth and velocity in a channel vary along the channel and the uniform flow conditions are not often reached. In this section, we analyse steady gradually-varied flows in open channels. Flow resistance, and changes in bottom slope, channel shape, discharge and surface conditions are taken into account. Examples are shown on Fig. 5.3.

Figure 5.3 illustrates typical longitudinal free-surface profiles. Upstream and downstream control can induce various flow patterns. In some cases, a hydraulic jump might take place. A jump is a rapidly-varied flow and calculations were developed in Chapter 4 (Section 4.4.2). However, it is also a control section and it affects the free-surface profile upstream and downstream.

It is usual to denote d the actual depth (i.e. the non-uniform flow depth), d_o the normal depth (i.e. uniform flow depth) and d_c the critical depth.

Note

A non-uniform flow is such that along a stream line in the s-direction:

$$\frac{\partial V}{\partial s} \neq 0 \qquad \text{or} \qquad \frac{\partial d}{\partial s} \neq 0$$

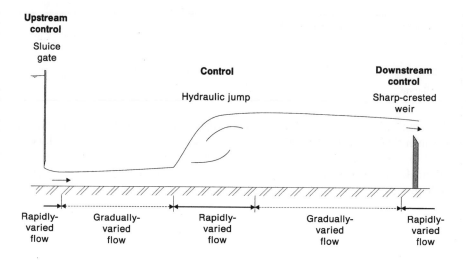

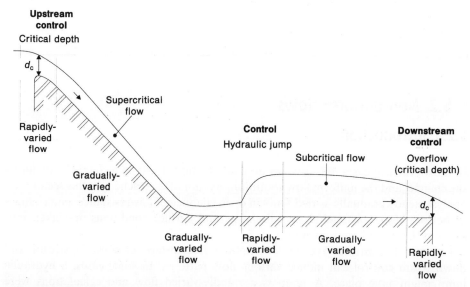

Fig. 5.3 Examples of non-uniform flows.

5.2.2 Equations for gradually-varied flow: backwater calculation

For subcritical flows the flow situation is governed (controlled) by the downstream flow conditions. A downstream hydraulic structure (e.g. bridge piers, gate) will increase the upstream depth and create a 'backwater' effect. The term 'backwater calculations' refers more generally to the calculations of the longitudinal free-surface profile for both sub- and super-critical flows.

The backwater calculations are developed assuming:

[H1] a *non-uniform* flow,
[H2] a *steady* flow,
[H3] that the flow is *gradually varied*, and
[H4] that, at a given section, the *flow resistance is the same as for a uniform flow* for the same depth and discharge, regardless of trends of the depth.

The gradually-varied flow (GVF) calculations do not apply to uniform equilibrium flows, nor to unsteady flows, nor to rapidly-varied flows (RVF). The last assumption [H4] implies that the Darcy, Chézy or Gauckler–Manning equation may be used to estimate the flow resistance, although these equations were originally developed for uniform equilibrium flows only.

Along a streamline in the *s*-direction, the energy equation, written in terms of mean total head becomes:

$$\frac{\partial H}{\partial s} = -f \frac{1}{D_H} \frac{V^2}{2g} \tag{5.3a}$$

where V is the fluid velocity, f is the Darcy friction factor, D_H is the hydraulic diameter. Using the definition of the friction slope, it yields:

$$\frac{\partial H}{\partial s} = -S_f \tag{5.3b}$$

where S_f is the friction slope defined as:

$$S_f = f \frac{1}{D_H} \frac{V^2}{2g}$$

Assuming a hydrostatic pressure gradient, equation (5.3a) can be transformed as:

$$\frac{\partial H}{\partial s} = \frac{\partial}{\partial s} \left(z_o + d \cos\theta + \frac{V^2}{2g} \right) = -S_f \tag{5.3c}$$

where z_o is the bed elevation, d is the flow depth measured normal to the channel bottom and θ is the channel slope.

Introducing the mean specific energy (i.e. $H = E + z_o$), equation (5.3) can be rewritten:

$$\frac{\partial E}{\partial s} = S_o - S_f \tag{5.4}$$

where S_o is the bed slope ($S_o = \sin\theta$). For any cross-sectional shape it was shown that the differentiation of the mean specific energy E with respect to d equals (see Section 3.2.3):

$$\frac{\partial E}{\partial d} = 1 - Fr^2 \tag{5.5}$$

where Fr is the Froude number (defined such that $Fr = 1$ for critical flow conditions). Hence the energy equation leads to:

$$\frac{\partial d}{\partial s} (1 - Fr^2) = S_o - S_f \tag{5.6}$$

Notes

1. The first backwater calculations were developed by Bélanger (1828) who used a finite difference step method of integration.
2. The assumption: '*the flow resistance is the same as for a uniform flow for the same depth and discharge*' implies that the Darcy equation, the Chézy equation or the Gauckler–Manning equation may be used to estimate the flow resistance, although these equations were initially developed for normal flows only. For rapidly varied flows, these equations are no longer valid.

   In practice, the friction factor and friction slope are calculated as in uniform equilibrium flow but using the non-uniform flow depth d.

   It is well recognized that accelerating flows are characterized by lesser flow resistance than normal flows for the same flow properties (V, d). This phenomenon is sometimes called drag reduction in accelerating flow. For decelerating flows, the flow resistance is larger than that for uniform equilibrium flow with identical flow properties.
3. For non-uniform flows we have:

$$d > d_o \quad \Rightarrow \quad S_f < S_o$$

$$d = d_o \quad \Rightarrow \quad S_f = S_o$$

$$d < d_o \quad \Rightarrow \quad S_f > S_o$$

   and at the same time:

   $d > d_c$ subcritical flow $Fr < 1$

   $d = d_c$ critical flow $Fr = 1$

   $d < d_c$ supercritical flow $Fr > 1$

4. For uniform or non-uniform flows the bed slope S_o and the friction slope S_f are defined respectively as:

$$S_f = f \frac{1}{D_H} \frac{V^2}{2g} \tag{4.20}$$

$$S_o = \sin \theta = -\frac{\partial z_o}{\partial s} \tag{4.21}$$

5.2.3 Discussion

Singularity of the energy equation

The energy equation (5.3), (5.4) or (5.6) can be applied to non-uniform (gradually-varied) flow situations as long as the friction slope differs from the bed slope. Indeed equation (5.6) may be transformed as:

$$\frac{\partial d}{\partial s} = \frac{S_o - S_f}{1 - Fr^2} \tag{5.7}$$

where S_f and Fr are functions of the flow depth d for a given value of Q. Equation (5.7) emphasizes three possible singularities of the energy equation. Their physical meaning is summarized below:

Case	Singularity condition	Physical meaning
[1]	$\dfrac{\partial d}{\partial s} = 0$	Uniform equilibrium flow: i.e. $S_f = S_o$.
[2]	$Fr = 1$	Critical flow conditions. $S_f = S_c$.
[3]	$\dfrac{\partial d}{\partial s} = 0$ and $Fr = 1$	Uniform equilibrium flow conditions and critical slope: i.e. $S_f = S_o = S_c$. This case is seldom encountered because critical (and near-critical) flows are unstable.

Free-surface profiles

With the aid of equations (5.6) it is possible to establish the behaviour of $(\partial d/\partial s)$ as a function of the magnitudes of the flow depth d, the critical depth d_c and the normal depth d_o. The sign of $(\partial d/\partial s)$ and the direction of depth change can be determined by the actual depth relative to both the normal and critical flow depths. Altogether, there are 12 different types of free-surface surface profiles (of non-uniform gradually varied flows), excluding uniform flow. They are listed in the table below.

S_o	$\dfrac{d}{d_c}$	$\dfrac{d_o}{d_c}$	$\dfrac{d}{d_o}$	Fr	$\dfrac{S_f}{S_o}$	d, d_o, d_c	Name[a]
(1)	(2)	(3)	(4)	(5)	(6)	(7)	(8)
>0	<1	<1	<1	>1	>1	$d < d_o < d_c$	S3
			>1	>1	<1	$d_o < d < d_c$	S2
		>1	<1	>1	>1	$d < d_c < d_o$	M3
			>1	>1	<1	Not possible	
	>1	<1	<1	<1	>1	Not possible	
			>1	<1	<1	$d_o < d_c < d$	S1
		>1	<1	<1	>1	$d_c < d < d_o$	M2
			>1	<1	<1	$d_c < d_o < d$	M1
= 0	<1	N/A	N/A	>1	N/A	$d < d_c$	H3
	>1	N/A	N/A	<1	N/A	$d > d_c$	H2
<0	<1	N/A	N/A	>1	N/A	$d < d_c$	A3
	>1	N/A	N/A	<1	N/A	$d > d_c$	A2

Remarks: [a] Name of the free-surface profile where the letter is descriptive for the slope: H for horizontal, M for mild, S for steep and A for adverse (negative slope) (e.g. Chow 1973).

For a given free-surface profile, the relationships between the flow depth d, the normal depth d_o and the critical depth d_c give the shape of the longitudinal profile. The relationship between d and d_c enables prediction of the Froude number, and the relationship between d and d_o gives the sign of $(S_o - S_f)$. Combining with the differential form of the mean specific energy (i.e. (5.6)), we can determine how the behaviour of $(\Delta d/\Delta s)$ (i.e. the longitudinal variation of flow depth) is affected by the relative magnitude of d, d_o and d_c (see table below):

Name[a]	Case	$S_o - S_f$	Fr	$\dfrac{\Delta d}{\Delta s}$	Remarks
(1)	(2)	(3)	(4)	(5)	(6)
M1	$d > d_o > d_c$	$S_o > S_f$	$Fr < 1$	Positive	The water surface is asymptotic to a horizontal line (backwater curve behind a dam).
M2	$d_o > d > d_c$	$S_o < S_f$	$Fr < 1$	Negative	S_f tends to S_o and we reach a transition from subcritical to critical flow (i.e. overfall).
M3	$d_o > d_c > d$	$S_o < S_f$	$Fr > 1$	Positive	Supercritical flow that tends to a hydraulic jump.
S1	$d > d_c > d_o$	$S_o > S_f$	$Fr < 1$	Positive	
S2	$d_c > d > d_o$	$S_o > S_f$	$Fr > 1$	Negative	
S3	$d_c > d_o > d$	$S_o < S_f$	$Fr > 1$	Positive	
H2	$d > d_c$	$S_o = 0 < S_f$	$Fr < 1$	Negative	
H3	$d_c > d$	$S_o = 0 < S_f$	$Fr > 1$	Positive	
A2	$d > d_c$	$S_o < 0 < S_f$	$Fr < 1$	Negative	
A3	$d_c > d$	$S_o < 0 < S_f$	$Fr > 1$	Positive	

Remarks:
[a] Name of the free-surface profile where the letter is descriptive for the slope: H for horizontal, M for mild, S for steep and A for adverse.
$\Delta d/\Delta s$ is negative for gradually-accelerating flow. For $\Delta d/\Delta s > 0$ the flow is decelerating by application of the continuity principle.

Notes

1. The classification of backwater curves was developed first by the French professor J.A.C. Bresse (Bresse 1860). Bresse originally considered only wide rectangular channels.
2. The friction slope may be re-formulated as:

	Expression	Comments
Friction slope	$S_f = \dfrac{Q^2 P_w f}{8g A^3}$	f is the Darcy friction factor.
	$S_f = \dfrac{Q^2 P_w}{(C_{\text{Chézy}})^2 A^3}$	$C_{\text{Chézy}}$ is the Chézy coefficient.

3. The Froude number can be re-formulated as:

	Expression	Comments
Froude number	$Fr = \dfrac{Q}{\sqrt{g\dfrac{A^3}{B}}}$	General case (i.e. any cross-sectional shape).
	$Fr = \dfrac{Q}{\sqrt{gd^3 B^2}}$	Rectangular channel (channel width B).

Application

Considering a mild slope channel, discuss all possible cases of free-surface profiles (non-uniform flows). Give some practical examples.

Solution

Considering a gradually-varied flow down a mild slope channel, the normal depth must be larger than the critical depth (i.e. the definition of a mild slope). There are three cases of non-uniform flows:

Case	d_o/d_c	d/d_c	Fr	S_f	$\Delta d/\Delta s$	Remarks
$d > d_o > d_c$	>1	>1	<1	$< S_o$	>0	The water surface is asymptotic to a horizontal line (e.g. backwater curve behind a dam).
$d_o > d > d_c$	>1	>1	<1	$> S_o$	<0	S_f tends to S_o and we reach a transition from subcritical to critical flow (e.g. overfall).
$d_o > d_c > d$	>1	<1	>1	$> S_o$	>0	Supercritical flow that tends to a hydraulic jump (e.g. flow downstream of sluice gate with a downstream control).

Discussion

The *M1-profile* is a subcritical flow. It occurs when the downstream of a long mild channel is submerged in a reservoir of greater depth than the normal depth d_o of the flow in the channel.

The *M2-profile* is also a subcritical flow ($d > d_c$). It can occur when the bottom of the channel at the downstream end is submerged in a reservoir to a depth less than the normal depth d_o.

The *M3-profile* is a supercritical flow. This profile occurs usually when a supercritical flow enters a mild channel (e.g. downstream of a sluice gate, downstream of a steep channel).

The *S1-profile* is a subcritical flow in a steep channel.

The *S2-profile* is a supercritical flow. Examples are the profiles formed on a channel as the slope changes from steep to steeper and downstream of an enlargement of a step channel.

The *S3-profile* is a supercritical flow. An example is the flow downstream of a sluice gate in a steep slope when the gate opening is less than the normal depth.

5.2.4 Backwater computations

The backwater profiles are obtained by combining the differential equation (5.6), the flow resistance calculations (i.e. friction slope) and the boundary conditions. The solution is then obtained by numerical computations or graphical analysis.

Discussion

To solve equation (5.6) it is *essential* to determine correctly the *boundary conditions*. By boundary conditions we mean the flow conditions imposed upstream, downstream and along the channel. These boundary conditions are, in practice:

(a) *control devices*: sluice gate, weir or reservoir, that impose flow conditions for the depth, the discharge, a relationship between the discharge and the depth, etc., and

(b) *geometric characteristics*: bed altitude, channel width, local friction factor, etc. (e.g. Fig. 5.3).

The backwater profile is calculated using equation (5.6):

$$\frac{\partial d}{\partial s}(1 - Fr^2) = S_o - S_f \qquad (5.6)$$

It must be emphasized that the results depend critically on the *assumed friction factor f*. As discussed earlier, a great uncertainty applies to the assumed friction factor.

Notes

1. Some textbooks suggest using equation (5.7) to compute the backwater profile. But equation (5.7) presents a numerical singularity for critical flow conditions (i.e. $Fr = 1$). In practice, equation (5.7) *should not* be used, unless it is known '*a priori*' (i.e. beforehand) that the critical flow condition does not take place anywhere along the channel.
   It is suggested *very strongly* to compute the backwater curves using equation (5.6).

2. Note that equation (5.6) has a solution for: $S_f = S_o$ (i.e. uniform flow).

Several integration methods exist: analytical solutions are rare but numerous numerical integration methods exist. For simplicity, we propose to develop the standard step method (distance computed from depth). This method is simple, extremely reliable and very stable. It is *strongly* recommended to practising engineers who are not necessarily hydraulic experts.

Standard step method (distance calculated from depth)

Equation (5.4) or (5.6) can be rewritten in a finite difference form as:

$$\Delta s = \frac{\Delta E}{(S_o - S_f)_{\text{mean}}} \qquad (5.8)$$

where the subscript 'mean' indicates the *mean value over an interval* and ΔE is the change in mean specific energy along the distance Δs for a given change of flow depth Δd.

In practice, calculations are started at a location of known flow depth d (e.g. a control section). The computations can be performed either in the upstream or the downstream flow direction. A new flow depth $d + \Delta d$ is selected. The flow properties for the new flow depth are computed. We can deduce the change in specific energy ΔE and the average difference $(S_o - S_f)$ over the control volume (i.e. interval). The longitudinal position of the new flow depth is deduced from equation (5.8).

A practical computation example is detailed below to explain the step method. The flow conditions are: flow discharge $Q = 100 \, \text{m}^3/\text{s}$, channel slope $\theta = 5°$ (i.e. $S_o = 0.0875$), roughness $k_s = 5 \, \text{mm}$, rectangular channel, and known flow depth: $d = 4 \, \text{m}$ at $s = 0$.

d (m)	B (m)	A (m²)	P_w (m)	D_H (m)	f	V (m/s)	Fr	S_f	S_o	$S_o - S_f$	ΔE (m)	Δs (m)	s (m)
4.0	10.0	40.0	18.0	8.89	0.015	2.50	0.399	5.38×10^{-4}	8.75×10^{-2}				0.0
										0.08689	0.0829	0.9544	
4.1	9.0	36.9	17.2	8.58	0.015	2.71	0.427	6.54×10^{-4}	8.75×10^{-2}				0.954
										0.08685	0.0824	0.9485	
4.2	9.0	37.8	17.4	8.69	0.015	2.65	0.412	6.16×10^{-4}	8.75×10^{-2}				1.903

where:
A is the channel cross-section area ($A = dB$ for a rectangular channel),
B is the channel free surface width,
d is the flow depth,
D_H is the hydraulic diameter ($D_H = 4A/P_w$),
Fr is the Froude number defined in terms of the flow depth d ($Fr = V/\sqrt{gd}$ for a rectangular channel),
f is the Darcy coefficient, computed as a function of the Reynolds number Re and the relative roughness k_s/D_H,
P_w is the wetted perimeter ($P_w = B + 2d$ for a rectangular channel),
Re is the Reynolds number defined in terms of the average flow velocity and hydraulic diameter D_H ($Re = VD_H/\nu$),
V is the mean flow velocity ($V = Q/A$),
V_* is the shear velocity ($V_* = \sqrt{fV^2/8}$),
Re_* is the shear Reynolds number ($Re_* = V_* k_s/\nu$),
S_f is the friction slope ($S_f = (Q^2 P_w f)/(8gA^3)$),
S_o is the bed slope ($S_o = \sin\theta$),
$\Delta E = \Delta d(1 - Fr^2)$ (mean value over the interval), and s is the curvilinear coordinate.

Warning

This step method applies only to the energy equation and does not necessarily converge (e.g. uniform flow conditions). Do not use this numerical method for any other equations.

Notes

1. The above method is called the 'step method/distance calculated from depth' (e.g. Henderson 1966, pp. 126–130). The method, based on equation (5.8), is numerically valid for subcritical, critical and supercritical flows. It is not valid for rapidly varied flows (e.g. hydraulic jumps).
2. Another method is based on equation (5.7) as:

$$\Delta d = \frac{S_o - S_f}{1 - Fr^2} \Delta s$$

It is called the 'step method/depth calculated from distance' (e.g. Henderson 1966, pp. 136–140). This method has the great disadvantage to be singular (i.e. unsolvable) at critical flow conditions.

5.3 Exercises

What are the basic assumptions for gradually-varied flow calculations?

What is the definition of the normal depth?

What is the definition of a steep slope? Does the notion of steep slope depend only on the bed slope? Discuss your answer. A discharge of $100 \, \text{m}^3/\text{s}$ flows in a 12 m wide concrete channel. The channel slope is $5\,\text{m/km}$. Calculate: (a) normal depth and (b) critical depth; (c) indicate the slope type: steep or mild.

A smooth-concrete channel carries a water flow. The channel cross-section is rectangular with a $2\,\text{m}$ bottom width. The longitudinal bed slope is $1.3\,\text{m/km}$ (i.e. $S_o = 0.0013$). The uniform equilibrium flow depth equals $3.1\,\text{m}$.
(a) Compute the following flow properties: flow velocity, flow rate, average bed shear stress and critical flow depth.
(b) For the uniform equilibrium flow, where would be the optimum location of a control?
Assume $k_s = 1\,\text{mm}$ for the finished concrete lining.

The Cotter dam (Canberra ACT) is a $27\,\text{m}$ high un-gated weir. The spillway is $60\,\text{m}$ wide with a $50°$ slope. At the toe of the spillway, the slope changes from $50°$ to $0.05°$. The weir is followed by a $20\,\text{km}$ long channel with a $60\,\text{m}$ width. The channel slope is $0.05°$. Assuming that the spillway and the channel are made of smooth concrete, compute the free-surface profiles above the spillway and in the channel for a discharge of $500\,\text{m}^3/\text{s}$.

Solution: In the reservoir upstream of the weir, the flow motion is tranquil. The flow above the spillway is expected to be rapid (to be checked). Hence a transition from slow to rapid flow motion occurs at the spillway crest: i.e. critical flow conditions. The channel downstream of the weir has a long section of constant slope, constant roughness, constant cross-section shape and constant shape. Uniform flow conditions will be obtained at the end of the channel. Note: if the uniform flow conditions are subcritical, a transition from supercritical to subcritical flow is expected between the end of the spillway and the end of the channel.

Critical and uniform flow calculations
There is only one discharge and channel shape, and it yields only one set of critical flow conditions: $d_c = 1.92\,\text{m}$, $V_c = 4.34\,\text{m/s}$.

Assuming a roughness height $k_s = 1\,\text{mm}$, the uniform flow conditions for the spillway and the downstream channel are:

	d_o (m)	V_o (m/s)	Fr_o
Spillway	0.28	29.7	17.9
Channel	2.27	3.67	0.78

Hydraulic controls
The preliminary calculations indicate that: (1) the spillway slope is *steep*, and (2) the channel slope is *mild*. Hence, critical flow conditions occur at two locations: (A) at the spillway crest, transition from subcritical to supercritical flow, and (B) in the channel, transition from supercritical to subcritical flow (i.e. hydraulic jump).

Backwater calculations
The easiest location to start the backwater calculations is the spillway crest. The flow is critical and hence we know the flow depth and flow velocity. By applying the energy equation, we can deduce the free-surface location at any position along the spillway.

In the channel downstream of the spillway, the first calculations predict the occurrence of a hydraulic jump. In such a case, we can assume that the flow depth downstream of the hydraulic jump equals the uniform flow depth. We can deduce then the flow depth immediately upstream of the jump ($d = 1.61$ m, $Fr = 1.30$) by applying the momentum and continuity equations across the hydraulic jump.

The location of the hydraulic jump is deduced from the backwater calculations. Starting at the bottom of the spillway, the flow is decelerated along the channel. When the flow depth equals the sequent depth (i.e. 1.61 m), the hydraulic jump takes place.

Considering a 20 m high un-gated weir with a stepped chute spillway, the spillway is 20 m wide and the slope is 45°. At the toe of the spillway, the slope changes from 45° to 0.1°. The spillway is followed by a dissipation basin (10 m long, 20 m wide) consisting of concrete baffle blocks and then by an 800 m long concrete channel with a 25 m width. The channel slope is 0.1°.

Assume that the friction factor of the stepped spillway is 1.0, and the friction factor of the baffle block lining is $f = 2.5$. Compute the free-surface profiles above the spillway and in the channel for a discharge of 160 m^3/s. Is the dissipation basin operating properly?

Notes

1. A dissipation basin or stilling basin is a short length of paved channel placed at the foot of a spillway or any other source of supercritical flow. The aim of the designer is to make a hydraulic jump within the basin so that the flow is converted to subcritical before it reaches the exposed riverbed or channel downstream (Henderson 1966).
2. Baffle blocks or chute blocks consist of reinforced concrete blocks installed to increase the resistance to the flow and to dissipate part of the kinetic energy of the flow.

Part 1 Revision exercises

Revision exercise No. 1

The undershot sluice gate sketched in Fig. R.1 is in a channel 5 m wide and the discharge, Q, is $4\,\text{m}^3/\text{s}$. The upstream flow depth is 1.2 m. The bed of the channel is horizontal and smooth.

- Sketch on Fig. R.1 the variation of the pressure with depth at sections 1 and 2, and on the upstream face of the sluice gate. Sections 1 and 2 are located far enough from the sluice gate for the velocity to be essentially horizontal and uniform.

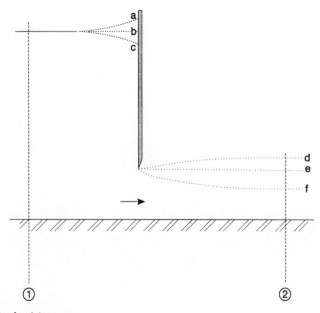

Fig. R.1 Sketch of a sluice gate.

- Show on Fig. R.1 the forces acting on the control volume contained between sections 1 and 2. Show also your choice for the positive direction of distance and of force.
- Write the momentum equation as applied to the control volume between sections 1 and 2, using the sign convention you have chosen. Show on Fig. R.1 the forces and velocities used in the momentum equation.
- Three water surface profiles on the upstream side of the sluice gate are labelled a, b, c. Which is the correct one? Three water surface profiles downstream of the sluice gate are labelled d, e, f. Which is the correct one?
- Compute and give the values (and units) of the specified quantities in the following list. (a) Velocity of flow at section 1. (b) Specific energy at section 1. (c) Specific energy at section 2. (d) Assumption used in answer (c). (e) Depth of flow at section 2. (f) Velocity of flow at section 2. (g) The force acting on the sluice gate. (h) The direction of the force in (g): i.e. upstream or downstream. (i) The critical depth for the flow in Fig. R.1. (j) What is the maximum possible discharge per unit width for a flow with specific energy entered at (b). (This question is a general question, not related to the sluice gate.)

Revision exercise No. 2

A channel step, sketched in Fig. R.2, is in a 12 m wide channel and the discharge, Q, is $46\,m^3/s$. The flow depth at section 1 is 1.6 m and the step height is 0.4 m. The bed of the channel, upstream and downstream of the step, is horizontal and smooth.

- Sketch on Fig. R.2 the free-surface profile between sections 1 and 2. (You might have to modify it after your calculations in the following questions.)
- Sketch on Fig. R.2 the variation of the pressure with depth at sections 1 and 2, and on the vertical face of the step. Sections 1 and 2 are located far enough from the step for the velocity to be essentially horizontal and uniform.
- Show on Fig. R.2 the forces acting on the control volume contained between sections 1 and 2. Show also your choice for the positive direction of distance and of force.
- Write the momentum equation as applied to the control volume between sections 1 and 2, using the sign convention you have chosen. Show on Fig. R.2 the forces and velocities used in the momentum equation.

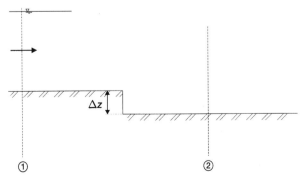

Fig. R.2 Sketch of a stepped transition.

- Compute and give the values (and units) of the specified quantities in the following list. (a) Velocity of flow at section 1. (b) Specific energy at section 1. (c) Specific energy at section 2. (d) Assumption used in answer (c). (e) Depth of flow at section 2. (f) Plot the correct free-surface profile on Fig. R.2 using (e). (g) Velocity of flow at section 2. (h) The force acting on the step. (i) The direction of the force in (h): i.e. upstream or downstream. (j) The critical depth for the flow in Fig. R.2. (k) What is maximum possible discharge per unit width for a flow with specific energy entered at (b). (This question is a general question, not related to the stepped channel.)

Revision exercise No. 3

A channel step, sketched in Fig. R.3, is in a 5 m wide channel of rectangular cross-section. The total discharge is 64 000 kg/s. The fluid is slurry (density: 1200 kg/m³). The downstream flow depth (i.e. at section 2) is 1.5 m and the step height is 0.8 m. The bed of the channel, upstream and downstream of the step, is horizontal and smooth. The flow direction is from section 1 to section 2.

- Compute and give the values (and units) of the following quantities. (a) Velocity of flow at section 2. (b) Froude number at section 2. (c) Specific energy at section 2.
- For the flow at section 2, where would you look for the hydraulic control?
- Sketch on Fig. R.3 the free-surface profile between sections 1 and 2. You might have to modify it after your calculations in the following questions.
- Show on Fig. R.3 *all* the forces acting on the control volume contained between sections 1 and 2. Show also your choice for the positive direction of distance and of force.
- Compute and give the values (and units) of the specified quantities in the following list. (a) Specific energy at section 1. (b) Depth of flow at section 1. (c) Velocity of flow at section 1. (d) Froude number at section 1. (e) Plot the correct free-surface profile on Fig. R.3 using (b). (f) The force acting on the step. (g) The

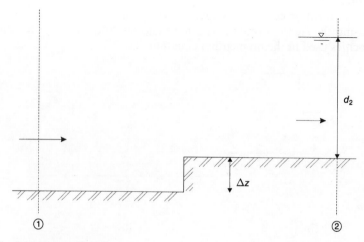

Fig. R.3 Sketch of an upward-step.

direction of the force in (f): i.e. upstream or downstream. (h) The critical depth for the flow in Fig. R.3.

Revision exercise No. 4

The horizontal rectangular channel sketched in Fig. R.4 is used as a stilling basin in which energy dissipation takes place in a hydraulic jump. The channel is equipped with four baffle blocks. The width of the channel is 20 m. The bed of the channel is horizontal and smooth. The inflow conditions are: $d_1 = 4.1$ m, $V_1 = 22$ m/s, and the *observed* downstream flow depth is $d_2 = 19.7$ m.

- Sketch on Fig. R.4 an appropriate control volume between the upstream and downstream flow locations (on the cross-sectional view).
- Sketch on Fig. R.4 (cross-sectional view) the variation of the pressure with depth at sections 1 and 2. Sections 1 and 2 are located far enough from the hydraulic jump for the velocity to be essentially horizontal and uniform.

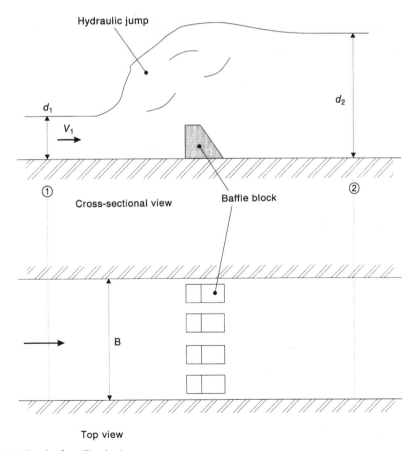

Fig. R.4 Sketch of a stilling basin.

- Show on Fig. R.4 *all* the forces acting on the control volume contained between sections 1 and 2. Show also your choice for the positive direction of distance and of force.
- Write the momentum equation as applied to the control volume between sections 1 and 2, using the sign convention that you have chosen.
- Compute and give the values (and units) of the specified quantities in the following list. (a) Total flow rate at section 1. (b) Specific energy at section 1. (c) Froude number at section 1. (d) Total force acting on the baffle blocks. (e) Velocity of flow at section 2. (f) Specific energy at section 2. (g) Energy loss between section 1 and 2. (h) Force acting on a single block. (i) The direction of the force in (h) and (g): i.e. upstream or downstream.

Revision exercise No. 5

The 'Venturi' flume sketched on Fig. R.5 is in a rectangular channel and the discharge is $2 \, \text{m}^3/\text{s}$. The channel bed upstream and downstream of the throat is horizontal and smooth. The channel width is $B_1 = B_3 = 5 \, \text{m}$. The throat characteristics are: $\Delta z_o = 0.5 \, \text{m}$, $B_2 = 2.5 \, \text{m}$. The upstream flow depth is $d_1 = 1.4 \, \text{m}$.

- Sketch on Fig. R.5 a free-surface profile in the Venturi flume between sections 1 and 3. You might have to modify it after your calculations in the following questions.

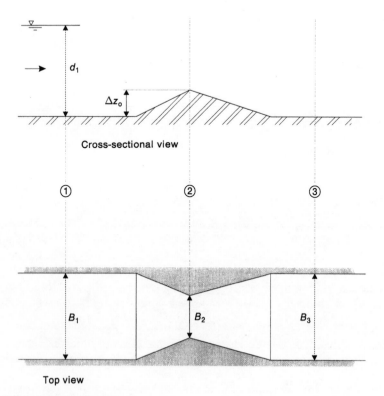

Fig. R.5 Sketch of a Venturi flume.

- Sketch on Fig. R.5 the variation of the pressure with depth at sections 1 and 3. Sections 1 and 2 are located far enough from the throat for the velocity to be essentially horizontal and uniform.
- Compute and give the values (and units) of the specified quantities in the following list. (a) Velocity of flow at section 1. (b) Specific energy at section 1. (c) Specific energy at section 2 (i.e. at the throat). (d) Assumption used in answer (c). (e) Depth of flow at section 2 (i.e. at the throat). (f) Velocity of flow at section 2 (i.e. at the throat). (g) Depth of flow at section 3. (h) Velocity of flow at section 3. (i) Specific energy at section 3. (j) The critical depth for the flow at section 2 (i.e. at the throat). (k) What is maximum possible discharge per unit width for a flow with specific energy entered at (c) (i.e. at the throat)? (This question is a general question, not related to the sluice gate.)

Revision exercise No. 6

An (undershot) *sluice gate* is in a rectangular channel 0.8 m wide. The flow depth upstream of the gate is 0.45 m and, downstream of the sluice gate, the free-surface elevation is 0.118 m above the channel bed. The bed of the channel is horizontal and smooth.

- Sketch the sluice gate and the free-surface profiles upstream and downstream of the sluice gate.
- Sketch on your figure the variation of the pressure with depth at a section 1 (upstream of the gate) and a section 2 (downstream of the gate), and on the upstream face of the sluice gate. Sections 1 and 2 are located far enough from the sluice gate for the velocity to be essentially horizontal and uniform.
- Compute and give the values (and units) of the following quantities. (a) Flow rate. (b) Assumption used in answer (a). (c) Velocity of flow at section 1. (d) Specific energy at section 1. (e) Specific energy at section 2. (f) Velocity of flow at section 2.
- Write the momentum equation as applied to the control volume between sections 1 and 2, using a consistent sign convention. Show on your figure the forces and velocities used in the momentum equation.
- Compute and give the values (and units) of the specified quantities in the following list. (a) The force acting on the sluice gate. (b) The direction of the force in (a): i.e. upstream or downstream. (c) The critical depth for the flow.

A *broad-crested weir* is to be constructed downstream of the sluice gate. The purpose of the gate is to force the dissipation of the kinetic energy of the flow, downstream of the sluice gate, between the gate and the weir: i.e. to induce a (fully-developed) hydraulic jump between the gate and the weir. The channel between the gate and the weir is horizontal and smooth. The weir crest will be set at a vertical elevation Δz above the channel bed, to be determined for the flow conditions in the above questions.

- Sketch the channel, the sluice gate, the hydraulic jump and the broad-crested weir.
- Write the momentum equation between section 3 (located immediately upstream of the jump and downstream of the gate) and section 4 (located immediately downstream of the jump and upstream of the weir).
- For the above flow conditions, compute and give the values (and units) of the specified quantities in the following list. (a) Flow depth upstream of the hydraulic

jump (section 3). (b) Specific energy upstream of the jump (section 3). (c) Flow depth downstream of the jump (section 4). (d) Specific energy downstream of the jump (section 4). (e) Flow depth above the crest. (f) Specific energy of the flow at the weir crest. (g) Weir crest elevation Δz. (h) Assumptions used in answer (g).

Revision exercise No. 7

A settling basin is a deep and wide channel in which heavy sediment particles may settle, to prevent siltation of the downstream canal. Settling basins are commonly located at the upstream end of an irrigation canal. In this section we shall investigate the flow characteristics of the outlet of the settling basin (Fig. R.6).

The outlet section is a rectangular channel sketched on Fig. R.6. The channel bed is smooth and, at sections 1, 2 and 3, it is horizontal. The channel width equals $B_1 = 12 \, \text{m}$, $B_2 = B_3 = 4.2 \, \text{m}$. The change in bed elevation is: $\Delta z_0 = 0.4 \, \text{m}$. The

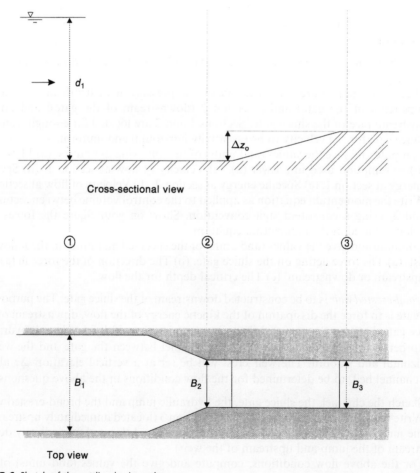

Fig. R.6 Sketch of the settling basin outlet.

total flow rate is $48\,200\,\mathrm{kg/s}$. The fluid is slurry (density $1405\,\mathrm{kg/m^3}$).The inflow conditions are: $d_1 = 4\,\mathrm{m}$.

(a) Compute and give the value of velocity of flow at section 1. (b) Compute the specific energy at section 1. (c) calculate the specific energy at section 2. (d) In answer (c), what basic principle did you *not* use? (e) Compute the depth of flow at section 2. (f) Calculate the velocity of flow at section 2. (g) Compute the depth of flow at section 3. (h) Compute the velocity of flow at section 3. (i) Calculate the specific energy at section 3. (j) Compute the critical depth for the flow at section 2. (k) What is the maximum possible discharge per unit width for a flow with specific energy entered at question (i)? (This question is a general question, not related to the settling basin.)

Appendices to Part 1

A1.1 Constants and fluid properties

Acceleration of gravity

Standard acceleration of gravity

The standard acceleration of gravity equals:

$$g = 9.80665 \, \text{m/s}^2 \qquad (A1.1)$$

This value is roughly equal to that at sea level and at 45° latitude. The gravitational acceleration varies with latitude and elevation owing to the form and rotation of the earth and may be estimated as:

$$g = 9.806056 - 0.025027 \cos(2\text{Latitude}) - 3 \times 10^{-6} z \qquad (A1.2)$$

where z is the altitude with the sea level as the origin, and the latitude is in degrees (Rouse 1938).

Altitude z (m) (1)	g (m/s^2) (2)
−1000	9.810
0 (Sea level)	9.807
1000	9.804
2000	9.801
3000	9.797
4000	9.794
5000	9.791
6000	9.788
7000	9.785
10 000	9.776

Absolute gravity values

Gravity also varies with the local geology and topography. Measured values of g are reported in the following table.

Location (1)	g (m/s^2) (2)	Location (1)	g (m/s^2) (2)	Location (1)	g (m/s^2) (2)
Addis Ababa, Ethiopia	9.7743	Helsinki, Finland	9.81090	Québec, Canada	9.80726
Algiers, Algeria	9.79896	Kuala Lumpur, Malaysia	9.78034	Quito, Ecuador	9.7726
Anchorage, USA	9.81925	La Paz, Bolivia	9.7745	Sapporo, Japan	9.80476
Ankara, Turkey	9.79925	Lisbon, Portugal	9.8007	Reykjavik, Iceland	9.82265
Aswan, Egypt	9.78854	Manila, Philippines	9.78382	Taipei, Taiwan	9.7895
Bangkok, Thailand	9.7830	Mexico City, Mexico	9.77927	Teheran, Iran	9.7939
Bogota, Colombia	9.7739	Nairobi, Kenya	9.77526	Thule, Greenland	9.82914
Brisbane, Australia	9.794	New Delhi, India	9.79122	Tokyo, Japan	9.79787
Buenos Aires, Argentina	9.7949	Paris, France	9.80926	Vancouver, Canada	9.80921
Christchurch, NZ	9.8050	Perth, Australia	9.794	Ushuaia, Argentina	9.81465
Denver, USA	9.79598	Port-Moresby, PNG	9.782		
Guatemala, Guatemala	9.77967	Pretoria, South Africa	9.78615		

Reference: Morelli (1971).

Properties of water

Temperature (Celsius) (1)	Density ρ_w (kg/m^3) (2)	Dynamic viscosity μ_w (Pa s) (3)	Surface tension σ (N/m) (4)	Vapour pressure P_v (Pa) (5)	Bulk modulus of elasticity E_b (Pa) (6)
0	999.9	1.792×10^{-3}	0.0762	0.6×10^3	2.04×10^9
5	1000.0	1.519×10^{-3}	0.0754	0.9×10^3	2.06×10^9
10	999.7	1.308×10^{-3}	0.0748	1.2×10^3	2.11×10^9
15	999.1	1.140×10^{-3}	0.0741	1.7×10^3	2.14×10^9
20	998.2	1.005×10^{-3}	0.0736	2.5×10^3	2.20×10^9
25	997.1	0.894×10^{-3}	0.0726	3.2×10^3	2.22×10^9
30	995.7	0.801×10^{-3}	0.0718	4.3×10^3	2.23×10^9
35	994.1	0.723×10^{-3}	0.0710	5.7×10^3	2.24×10^9
40	992.2	0.656×10^{-3}	0.0701	7.5×10^3	2.27×10^9

Reference: Streeter and Wylie (1981).

Gas properties

Basic equations

The *state equation* of perfect gas is:

$$P = \rho R T \tag{A1.3}$$

where P is the absolute pressure (in Pascal), ρ is the gas density (in kg/m^3), T is the absolute temperature (in Kelvin) and R is the gas constant (in J/kg K) (see table below). For a perfect gas, the *specific heat* at constant pressure C_p and the specific heat at constant volume C_v are related to the gas constant as:

$$C_p = \frac{\gamma}{\gamma - 1} R \tag{A1.4a}$$

$$C_p = C_v R \tag{A1.4b}$$

where γ is the specific heat ratio (i.e. $\gamma = C_p/C_v$).

During an *isentropic transformation* of perfect gas, the following relationships hold:

$$\frac{P}{\rho^\gamma} = \text{constant} \tag{A1.5a}$$

$$TP^{(1-\gamma)/\gamma} = \text{constant} \tag{A1.5b}$$

Physical properties

Gas	Formula	Gas constant R (J/kg K)	Specific heat		Specific heat ratio γ
			C_p (J/kg K)	C_v (J/kg K)	
(1)	(2)	(3)	(4)	(5)	(6)
Perfect gas					
Mono-atomic gas	(e.g. He)		$\frac{5}{2}R$	$\frac{3}{2}R$	$\frac{5}{3}$
Di-atomic gas	(e.g. O_2)		$\frac{7}{2}R$	$\frac{5}{2}R$	$\frac{7}{5}$
Poly-atomic gas	(e.g. CH_4)		$4R$	$3R$	$\frac{4}{3}$
Real gas[a]					
Air		287	1.004	0.716	1.40
Helium	He	2077.4	5.233	3.153	1.67
Nitrogen	N_2	297	1.038	0.741	1.40
Oxygen	O_2	260	0.917	0.657	1.40
Water vapour	H_2O	462	1.863	1.403	1.33

Note: [a] At low pressures and at 299.83 K (Streeter and Wylie 1981).

Compressibility and bulk modulus of elasticity

The compressibility of a fluid is a measure of the change in volume and density when the fluid is subjected to a change of pressure. It is defined as:

$$E_{co} = \frac{1}{\rho}\frac{\partial \rho}{\partial P} \tag{A1.6}$$

The reciprocal function of the compressibility is called the bulk modulus of elasticity:

$$E_b = \rho\frac{\partial P}{\partial \rho} \tag{A1.7}$$

For a perfect gas, the bulk modulus of elasticity equals:

$$E_b = \gamma P \quad \text{adiabatic transformation for a perfect gas} \tag{A1.7b}$$

$$E_b = P \quad \text{isothermal transformation for a perfect gas} \tag{A1.7c}$$

Celerity of sound

Introduction

The celerity of sound in a medium is:

$$C_{sound} = \sqrt{\frac{\partial P}{\partial \rho}} \tag{A1.8}$$

where P is the pressure and ρ is the density. It may be rewritten in terms of the bulk modulus of elasticity E_b:

$$C_{sound} = \sqrt{\frac{E_b}{\rho}} \tag{A1.9}$$

Equation (A1.7) applies to both liquids and gases.

Sound celerity in gas

For an isentropic process and a perfect gas, equation (A1.9) yields:

$$C_{sound} = \sqrt{\gamma R T} \tag{A1.10}$$

where γ and R are the specific heat ratio and gas constant respectively (see above).

The dimensionless velocity of a compressible fluid is called the Sarrau–Mach number:

$$Ma = \frac{V}{C_{sound}} \tag{A1.11}$$

Classical values

Celerity of sound in water at 20°C: 1485 m/s
Celerity of sound in dry air at sea level and 20°C: 343 m/s

Atmospheric parameters

Air pressure

The standard atmosphere or normal pressure at sea level equals:

$$P_{std} = 1\,atm = 360\,mm\ of\ Hg = 101\,325\,Pa \tag{A1.12}$$

where Hg is the chemical symbol of mercury. Unit conversion tables are provided in Appendix A1.2. The atmospheric pressure varies with the elevation above sea level (i.e. altitude). For dry air, the atmospheric pressure at altitude z equals

$$P_{atm} = P_{std} \times \exp\left(-\int_0^z \frac{0.0034841g}{T}\,dz\right) \tag{A1.13}$$

where T is the absolute temperature in Kelvin and equation (A1.13) is expressed in SI units.

Reference: Miller (1971).

Air temperature

In the troposphere (i.e. $z < 10\,000$ m), the air temperature decreases with altitude, on average, at a rate of 6.5×10^{-3} K/m (i.e. 6.5 Kelvin per km).

Table A1.1 presents the distributions of average air temperatures (Miller 1971) and corresponding atmospheric pressures with altitude (equation (A1.13)).

Table A1.1 Distributions of air temperature and air pressure as functions of the altitude (for dry air and standard acceleration of gravity)

Altitude z (m) (1)	Mean air temperature (K) (2)	Atmospheric pressure (Eq. (A1.13)) (Pa) (3)	Atmospheric pressure (Eq. (A1.13)) (atm) (4)
0	288.2	1.013×10^5	1.000
500	285.0	9.546×10^4	0.942
1000	281.7	8.987×10^4	0.887
1500	278.4	8.456×10^4	0.834
2000	275.2	7.949×10^4	0.785
2500	272.0	7.468×10^4	0.737
3000	268.7	7.011×10^4	0.692
3500	265.5	6.576×10^4	0.649
4000	262.2	6.164×10^4	0.608
4500	259.0	5.773×10^4	0.570
5000	255.7	5.402×10^4	0.533
5500	252.5	5.051×10^4	0.498
6000	249.2	4.718×10^4	0.466
6500	246.0	4.404×10^4	0.435
7000	242.8	4.106×10^4	0.405
7500	239.5	3.825×10^4	0.378
8000	236.3	3.560×10^4	0.351
8500	233.0	3.310×10^4	0.327
9000	229.8	3.075×10^4	0.303
9500	226.5	2.853×10^4	0.282
10 000	223.3	2.644×10^4	0.261

Viscosity of air

Viscosity and density of air at 1.0 atm:

Temperature (K) (1)	μ_{air} (Pa s) (2)	ρ_{air} (kg/m^3) (3)
300	18.4×10^{-6}	1.177
400	22.7×10^{-6}	0.883
500	26.7×10^{-6}	0.705
600	29.9×10^{-6}	0.588

The viscosity of air at standard atmosphere is commonly fitted by the Sutherland formula (Sutherland 1893):

$$\mu_{air} = 17.16 \times 10^{-6} \left(\frac{T}{273.1}\right)^{3/2} \frac{383.7}{T + 110.6} \tag{A1.14}$$

A simpler correlation is:

$$\frac{\mu_{air}(T)}{\mu_{air}(T_o)} = \left(\frac{T}{T_o}\right)^{0.76} \tag{A1.15}$$

where μ_{air} is in Pa s, and the temperature T and reference temperature T_o are expressed in Kelvin.

A1.2 Unit conversions

Introduction

The systems of units derived from the metric system have gradually been replaced by a single system, called the Système International d'Unités (SI unit system, or International System of Units). The present book is presented in SI Units.

Some countries continue to use British and American units, and conversion tables are provided in this appendix. Basic references in unit conversions include Degremont (1979) and ISO (1979).

Principles and rules

Unit symbols are written in small letters (i.e. m for metre, kg for kilogramme) but a capital is used for the first letter when the name of the unit derives from a surname (e.g. Pa after Blaise Pascal, N after Isaac Newton).

Multiples and submultiples of SI units are formed by adding one prefix to the name of the unit: e.g. km for kilometre, cm for centimetre, dam for decametre, μm for micrometre (or micron).

Multiple/submultiple factor	Prefix	Symbol
1×10^9	giga	G
1×10^6	mega	M
1×10^3	kilo	k
1×10^2	hecto	d
1×10^1	deca	da
1×10^{-1}	deci	d
1×10^{-2}	centi	c
1×10^{-3}	milli	m
1×10^{-6}	micro	μ
1×10^{-9}	nano	n

The basic SI units are the metre, kilogramme, second, Ampere, Kelvin, mole and candela. Supplementary units are the radian and the steradian. All other SI units derive from these basic units.

Units and conversion factors

Quantity (1)	Unit (symbol) (2)	Conversion (3)	Comments (4)
Length	1 inch (in)	$= 25.4 \times 10^{-3}$ m	Exactly
	1 foot (ft)	$= 0.3048$ m	Exactly
	1 yard (yd)	$= 0.9144$ m	Exactly
	1 mil	$= 25.4 \times 10^{-6}$ m	1/1000 inch
	1 mile	$= 1.609.344$ m	Exactly
Area	1 square inch (in^2)	$= 6.4516 \times 10^{-4}$ m^2	Exactly
	1 square foot (ft^2)	$= 0.092\,903\,06$ m^2	Exactly
Volume	1 Litre (L)	$= 1.0 \times 10^{-3}$ m^3	Exactly. Previous symbol: l.
	1 cubic inch (in^3)	$= 16.387\,064 \times 10^{-6}$ m^3	Exactly
	1 cubic foot (ft^3)	$= 28.3168 \times 10^{-3}$ m^3	Exactly
	1 gallon UK (gal UK)	$= 4.546\,09 \times 10^{-3}$ m^3	
	1 gallon US (gal US)	$= 3.785\,41 \times 10^{-3}$ m^3	
	1 barrel US	$= 158.987 \times 10^{-3}$ m^3	For petroleum, etc
Velocity	1 foot per second (ft/s)	$= 0.3048$ m/s	Exactly

Quantity (1)	Unit (symbol) (2)	Conversion (3)	Comments (4)
Velocity	1 mile per hour (mph)	$= 0.447\,04\,\text{m/s}$	Exactly
Acceleration	1 foot per second squared (ft/s^2)	$= 0.3048\,\text{m/s}^2$	Exactly
Mass	1 pound (lb or lbm)	$= 0.453\,592\,37\,\text{kg}$	Exactly
	1 ton UK	$= 1016.05\,\text{kg}$	
	1 ton US	$= 907.185\,\text{kg}$	
Density	1 pound per cubic foot (lb/ft^3)	$= 16.0185\,\text{kg/m}^3$	
Force	1 kilogram-force (kgf)	$= 9.806\,65\,\text{N (exactly)}$	Exactly
	1 pound force (lbf)	$= 4.448\,221\,615\,260\,5\,\text{N}$	
Moment of force	1 foot pound force (ft lbf)	$= 1.355\,82\,\text{N m}$	
Pressure	1 Pascal (Pa)	$= 1\,\text{N/m}^2$	
	1 standard atmosphere (atm)	$= 101\,325\,\text{Pa}$ $= 760\,\text{mm of Mercury at}$ normal pressure (i.e. mm of Hg)	Exactly
	1 bar	$= 10^5\,\text{Pa}$	Exactly
	1 torr	$= 133.322\,\text{Pa}$	
	1 conventional metre of water (m of H_2O)	$= 9.806\,65 \times 10^3\,\text{Pa}$	Exactly
	1 conventional metre of Mercury (m of Hg)	$= 1.333\,224 \times 10^5\,\text{Pa}$	
	1 Pound per Square Inch (PSI)	$= 6.894\,757\,2 \times 10^3\,\text{Pa}$	
Temperature	T (Celsius)	$= T\ \text{(Kelvin)} - 273.16$	0 Celsius is 0.01 K below the temperature of the triple point of water
	T (Fahrenheit)	$= T\ \text{(Celsius)}\frac{9}{5} + 32$	
	T (Rankine)	$= \frac{9}{5}T\ \text{(Kelvin)}$	
Dynamic viscosity	1 Pa s	$= 0.006\,720\,\text{lbm/ft/s}$	
	1 Pa s	$= 10\,\text{Poises}$	Exactly
	1 N s/m^2	$= 1\,\text{Pa s}$	Exactly
	1 Poise (P)	$= 0.1\,\text{Pa s}$	Exactly
	1 milliPoise (mP)	$= 1.0 \times 10^{-4}\,\text{Pa s}$	Exactly
Kinematic viscosity	1 square foot per second (ft^2/s)	$= 0.092\,903\,0\,\text{m}^2/\text{s}$	
	1 m^2/s	$= 10.7639\,\text{ft}^2/\text{s}$	
	1 m^2/s	$= 10^4\,\text{Stokes}$	
Surface tension	1 dyne/cm	$= 0.99987 \times 10^{-3}\,\text{N/m}$	
	1 dyne/cm	$= 5.709 \times 10^{-6}\,\text{pound/inch}$	
Work energy	1 Joule (J)	$= 1\,\text{N m}$	
	1 Joule (J)	$= 1\,\text{W s}$	
	1 Watt hour (W h)	$= 3.600 \times 10^3\,\text{J}$	Exactly
	1 electronvolt (eV)	$= 1.60219 \times 10^{-19}\,\text{J}$	
	1 Erg	$= 10^{-7}\,\text{J}$	Exactly
	1 foot pound force (ft lbf)	$= 1.355\,82\,\text{J}$	
Power	1 Watt (W)	$= 1\,\text{J/s}$	
	1 foot pound force per second (ft lbf/s)	$= 1.355\,82\,\text{W}$	
	1 horsepower (hp)	$= 745.700\,\text{W}$	

A1.3 Mathematics

Introduction

References

Beyer, W. H. (1982) *CRC Standard Mathematical Tables* (CRC Press: Boca Raton, Florida, USA).

Korn, G. A., and Korn, T. M. (1961) *Mathematical Handbook for Scientist and Engineers* (McGraw-Hill: New York, USA).

Spiegel, M. R. (1968) *Mathematical Handbook of Formulas and Tables* (McGraw-Hill: New York, USA).

Notation

x, y, z	Cartesian coordinates
r, θ, z	polar coordinates
$\dfrac{\partial}{\partial x}$	partial differentiation with respect to the x-coordinate
$\dfrac{\partial}{\partial y}, \dfrac{\partial}{\partial z}$	partial differential (Cartesian coordinate)
$\dfrac{\partial}{\partial r}, \dfrac{\partial}{\partial \theta}$	partial differential (polar coordinate)
$\dfrac{\partial}{\partial t}$	partial differential with respect to time t
$\dfrac{D}{Dt}$	absolute derivative
δ_{ij}	identity matrix element: $\delta_{ii} = 1$ and $\delta_{ij} = 0$ (for i different from j)
$N!$	N-factorial: $N! = 1 \times 2 \times 3 \times 4 \times \cdots \times (N-1) \times N$

Constants

e	constant such as $\ln(e) = 1$: $e = 2.718\,281\,828\,459\,045\,235\,360\,287$
π	$\pi = 3.141\,592\,653\,589\,793\,238\,462\,643$
$\sqrt{2}$	$\sqrt{2} = 1.414\,213\,562\,373\,095\,0488$
$\sqrt{3}$	$\sqrt{3} = 1.732\,050\,807\,568\,877\,2935$

Vector operations

Definitions

Considering a three-dimensional space, the coordinates of a point M or of a vector A can be expressed in a Cartesian system of coordinates or a cylindrical (or polar) system of coordinates as:

	Cartesian system of coordinates	Cylindrical system of coordinates
Point M	(x, y, z)	(r, θ, z)
Vector A	(A_x, A_y, A_z)	(A_r, A_θ, A_z)

The relationship between the cartesian coordinates and the polar coordinates of any point M are:

$$r^2 = x^2 + y^2$$

$$\tan\theta = \frac{y}{x}$$

Vector operations
Scalar product of two vectors

$$A \times B = |A| \times |B| \times \cos(A, B)$$

where $|A| = \sqrt{A_x^2 + A_y^2 + A_z^2}$. Two non-zero vectors are perpendicular to each other if and only if their scalar product is null.

Vector product

$$A \wedge B = i(A_y \times B_z - A_z \times B_y) + j(A_z \times B_x - A_x \times B_z) + k(A_x \times B_y - A_y \times B_x)$$

where i, j and k are the unity vectors in the x-, y- and z-directions respectively.

Differentials and differentiation

Absolute differential

The absolute differential D/Dt of a scalar $\Phi(r)$ along the curve: $r = r(t)$ is, at each point (r) of the curve, the rate of change of $\Phi(r)$ with respect to the parameter t as r varies as a function of t:

$$\frac{D\Phi}{Dt} = \left(\frac{Dr}{Dt} \times \nabla\right)\Phi = \frac{\partial x}{\partial t} \times \frac{\partial \Phi}{\partial x} + \frac{\partial y}{\partial t} \times \frac{\partial \Phi}{\partial y} + \frac{\partial z}{\partial t} \times \frac{\partial \Phi}{\partial z}$$

If Φ depends explicitly on t $[\Phi = \Phi(r, t)]$ then:

$$\frac{D\Phi}{Dt} = \frac{\partial \Phi}{\partial t} + \left(\frac{Dr}{Dt} \times \nabla\right)\Phi = \frac{\partial \Phi}{\partial t} + \frac{\partial x}{\partial t} \times \frac{\partial \Phi}{\partial x} + \frac{\partial y}{\partial t} \times \frac{\partial \Phi}{\partial y} + \frac{\partial z}{\partial t} \times \frac{\partial \Phi}{\partial z}$$

and this may be rewritten as:

$$\frac{D\Phi}{Dt} = \frac{\partial \Phi}{\partial t} + V \times \nabla\Phi = \frac{\partial \Phi}{\partial t} + V_x \times \frac{\partial \Phi}{\partial x} + V_y \times \frac{\Phi}{\partial y} + V_z \times \frac{\partial \Phi}{\partial z}$$

In the above equations ∇ is the nabla differential operator. It is considered as a vector:

$$\nabla = i\left(\frac{\partial}{\partial x}\right) + j\left(\frac{\partial}{\partial y}\right) + k\left(\frac{\partial}{\partial z}\right)$$

Differential operators
Gradient

$$\text{grad }\Phi(x, y, z) = \nabla\Phi(x, y, z) = i\frac{\partial \Phi}{\partial x} + j\frac{\partial \Phi}{\partial y} + k\frac{\partial \Phi}{\partial z} \qquad \text{Cartesian coordinate}$$

Divergence

$$\operatorname{div} F(x,y,z) = \nabla F(x,y,z) = \frac{\partial F_x}{\partial x} + \frac{\partial F_y}{\partial y} + \frac{\partial F_z}{\partial z}$$

Curl

$$\operatorname{curl} F(x,y,z) = \nabla \wedge F(x,y,z) = \mathbf{i}\left(\frac{\partial F_z}{\partial y} - \frac{\partial F_y}{\partial z}\right) + \mathbf{j}\left(\frac{\partial F_x}{\partial z} - \frac{\partial F_z}{\partial x}\right) + \mathbf{k}\left(\frac{\partial F_y}{\partial x} - \frac{\partial F_x}{\partial y}\right)$$

Laplacian operator

$$\nabla\Phi(x,y,z) = \nabla \times \nabla\Phi(x,y,z)$$

$$= \operatorname{div}\operatorname{grad}\Phi(x,y,z) = \frac{\partial^2\Phi}{\partial x^2} + \frac{\partial^2\Phi}{\partial y^2} + \frac{\partial^2\Phi}{\partial z^2} \qquad \text{Laplacian of scalar}$$

$$\nabla F(x,y,z) = \nabla \times \nabla F(x,y,z) = \mathbf{i}\Delta F_x + \mathbf{j}\Delta F_y + \mathbf{k}\Delta F_z \qquad \text{Laplacian of vector}$$

Polar coordinates

$$\operatorname{grad}\Phi(r,\theta,z) = \left(\frac{\partial\Phi}{\partial r} ; \frac{1}{r} \times \frac{\partial\Phi}{\partial\theta} ; \frac{\partial\Phi}{\partial z}\right)$$

$$\operatorname{div} F(r,\theta,z) = \frac{1}{r}\left(\frac{\partial(r \times F_r)}{\partial r} + \frac{\partial F_\theta}{\partial\theta} + r \times \frac{\partial F_z}{\partial z}\right)$$

$$\operatorname{curl} F(r,\theta,z) = \left(\frac{1}{r}\frac{\partial F_z}{\partial\theta} - \frac{\partial F_\theta}{\partial z} ; \frac{\partial F_r}{\partial z} - \frac{\partial F_z}{\partial r} ; \frac{1}{r}\frac{\partial(r \times F_\theta)}{\partial r} - \frac{1}{r} \times \frac{\partial F_r}{\partial\theta}\right)$$

$$\Delta\Phi(r,\theta,z) = \frac{1}{r}\frac{\partial}{\partial r}\left(r\frac{\partial\Phi}{\partial r}\right) + \frac{1}{r^2}\frac{\partial^2\Phi}{\partial\theta^2} + \frac{\partial^2\Phi}{\partial z^2}$$

Operator relationship

Gradient

$$\operatorname{grad}(f + g) = \operatorname{grad} f + \operatorname{grad} g$$

$$\operatorname{grad}(f \times g) = g \times \operatorname{grad} f + f \times \operatorname{grad} g$$

where f and g are scalars.

Divergence

$$\operatorname{div}(F + G) = \operatorname{div} F + \operatorname{div} G$$

$$\operatorname{div}(f \times F) = f \times \operatorname{div} F + F \times \operatorname{grad} f$$

where f is a scalar.

$$\operatorname{div}(F \wedge G) = G \times \operatorname{curl} F - F \times \operatorname{curl} G$$

Curl

$$\operatorname{curl}(F + G) = \operatorname{curl} F + \operatorname{curl} G$$

$$\operatorname{curl}(f \times F) = f \times \operatorname{curl} F - G \wedge \operatorname{grad} f$$

where f is a scalar.

$$\text{curl}(\text{grad}\, f) = 0$$

$$\text{div}(\text{curl}\, F) = 0$$

Laplacian

$$\Delta f = \text{div}\, \text{grad}\, f$$

$$\Delta F = \text{grad}\, \text{div}\, F - \text{curl}(\text{curl}\, F)$$

where f is a scalar.

$$\Delta(f + g) = \Delta f + \Delta g$$

$$\Delta(F + G) = \Delta F + \Delta G$$

$$\Delta(f \times g) = g \times \Delta f + f \times \Delta g + 2 \times \text{grad}\, f \times \text{grad}\, g$$

where f and g are scalars.

Trigonometric functions

Definitions

The basic definitions may be stated in terms of right-angled triangle geometry:

$$\sin(x) = \frac{\text{opposite}}{\text{hypotenuse}}$$

$$\cos(x) = \frac{\text{adjacent}}{\text{hypotenuse}}$$

$$\tan(x) = \frac{\sin(x)}{\cos(x)} = \frac{\text{opposite}}{\text{adjacent}}$$

$$\cot(x) = \frac{\cos(x)}{\sin(x)} = \frac{\text{adjacent}}{\text{opposite}}$$

The power-series expansions of these functions are:

$$\sin(x) = x - \frac{x^3}{3!} + \frac{x^5}{5!} - \frac{x^7}{7!} + \cdots \qquad \text{for any value of } x$$

$$\cos(x) = 1 - \frac{x^2}{2!} + \frac{x^4}{4!} - \frac{x^6}{6!} + \cdots \qquad \text{for any value of } x$$

$$\tan(x) = x + \frac{x^3}{3} + \frac{2}{15}x^5 + \frac{17}{315}x^7 + \cdots \qquad \text{for } -\pi/2 < x < \pi/2$$

$$\cot(x) = \frac{1}{x} - \frac{x}{3} - \frac{1}{45}x^3 - \frac{2}{945}x^5 + \cdots \qquad \text{for } 0 < \text{Abs}(x) < \pi$$

where $n! = n \times (n - 1) \times (n - 2) \times \cdots \times 1$.

Relationships

$$\tan(x) = \frac{\sin(x)}{\cos(x)}$$

$$\cot(x) = \frac{1}{\tan(x)}$$

$$\sin^2(x) + \cos^2(x) = 1$$

$$\frac{1}{\sin^2(x)} - \cot^2(x) = 1$$

$$\frac{1}{\cos^2(x)} - \tan^2(x) = 1$$

$$\sin(-x) = -\sin(x)$$

$$\cos(-x) = \cos(x)$$

$$\tan(-x) = -\tan(x)$$

$$\cot(-x) = -\cot(x)$$

$$\sin(x) = \cos\left(\frac{\pi}{2} - x\right) \qquad \text{for } 0 < x < \pi/2$$

$$\cos(x) = \sin\left(\frac{\pi}{2} - x\right) \qquad \text{for } 0 < x < \pi/2$$

$$\tan(x) = \cot\left(\frac{\pi}{2} - x\right) \qquad \text{for } 0 < x < \pi/2$$

$$\cot(x) = \tan\left(\frac{\pi}{2} - x\right) \qquad \text{for } 0 < x < \pi/2$$

$$\sin(x + y) = \sin(x)\cos(y) + \cos(x)\sin(y)$$

$$\sin(x - y) = \sin(x)\cos(y) - \cos(x)\sin(y)$$

$$\cos(x + y) = \cos(x)\cos(y) - \sin(x)\sin(y)$$

$$\cos(x - y) = \cos(x)\cos(y) + \sin(x)\sin(y)$$

$$\tan(x + y) = \frac{\tan(x) + \tan(y)}{1 - \tan(x)\tan(y)}$$

$$\tan(x - y) = \frac{\tan(x) - \tan(y)}{1 + \tan(x)\tan(y)}$$

$$\cot(x + y) = \frac{\cot(x)\cot(y) - 1}{\cot(x) + \cot(y)}$$

$$\cot(x - y) = \frac{\cot(x)\cot(y) + 1}{\cot(x) - \cot(y)}$$

$$\sin(x) + \sin(y) = 2\sin\left(\frac{x+y}{2}\right)\cos\left(\frac{x-y}{2}\right)$$

$$\sin(x) - \sin(y) = 2\cos\left(\frac{x+y}{2}\right)\sin\left(\frac{x-y}{2}\right)$$

$$\cos(x) + \cos(y) = 2\cos\left(\frac{x+y}{2}\right)\cos\left(\frac{x-y}{2}\right)$$

$$\cos(x) - \cos(y) = -2\sin\left(\frac{x+y}{2}\right)\sin\left(\frac{x-y}{2}\right)$$

$$\sin(x)\sin(y) = \tfrac{1}{2}(\cos(x-y) - \cos(x+y))$$

$$\cos(x)\cos(y) = \tfrac{1}{2}(\cos(x-y) + \cos(x+y))$$

$$\sin(x)\cos(y) = \tfrac{1}{2}(\sin(x-y) + \sin(x+y))$$

Derivatives

$$d(\sin(x)) = \cos(x)\,dx$$

$$d(\cos(x)) = -\sin(x)\,dx$$

$$d(\tan(x)) = \frac{1}{\cos^2(x)}dx$$

$$d(\cot(x)) = \frac{-1}{\sin^2(x)}dx$$

$$d\left(\frac{1}{\cos(x)}\right) = \frac{\tan(x)}{\cos(x)}dx$$

$$d\left(\frac{1}{\sin(x)}\right) = \frac{-\cot(x)}{\sin(x)}dx$$

Inverse trigonometric functions

The inverse trigonometric functions are expressed as $\sin^{-1}$, $\cos^{-1}$, $\tan^{-1}$ and $\cot^{-1}$.
The power-series expansions of these functions are:

$$\sin^{-1}(x) = x + \frac{1}{2}\times\frac{x^3}{3} + \frac{1\times 3}{2\times 4}\times\frac{x^5}{5} + \frac{1\times 3\times 5}{2\times 4\times 6}\times\frac{x^7}{7} + \cdots \quad \text{for } -1 < x < 1$$

$$\cos^{-1}(x) = \frac{\pi}{2} - \left(x + \frac{1}{2}\times\frac{x^3}{3} + \frac{1\times 3}{2\times 4}\times\frac{x^5}{5} + \frac{1\times 3\times 5}{2\times 4\times 6}\times\frac{x^7}{7} + \cdots\right) \quad \text{for } -1 < x < 1$$

$$\tan^{-1}(x) = x - \frac{x^3}{3} + \frac{x^5}{5} - \frac{x^7}{7} + \cdots \quad \text{for } -1 < x < 1$$

$$\cot^{-1}(x) = \frac{\pi}{2} - \left(x - \frac{x^3}{3} + \frac{x^5}{5} - \frac{x^7}{7} + \cdots\right) \quad \text{for } -1 < x < 1$$

The following relationships can be established:

$$\sin^{-1}(-x) = -\sin^{-1}(x)$$

$$\cos^{-1}(-x) = \pi - \cos^{-1}(x)$$

$$\tan^{-1}(-x) = -\tan^{-1}(x)$$

$$\cot^{-1}(-x) = \pi - \cot^{-1}(x)$$

$$\sin^{-1}(x) + \cos^{-1}(x) = \frac{\pi}{2}$$

$$\tan^{-1}(x) + \cot^{-1}(x) = \frac{\pi}{2}$$

$$\sin^{-1}\left(\frac{1}{x}\right) + \cos^{-1}\left(\frac{1}{x}\right) = \frac{\pi}{2}$$

$$\sin^{-1}(x) = \tan^{-1}\left(\frac{1}{\sqrt{1-x^2}}\right)$$

$$\cot^{-1}(x) = \frac{\pi}{2} - \tan^{-1}(x^2)$$

$$\sin^{-1}(x) + \sin^{-1}(y) = \sin^{-1}(x\sqrt{1-y^2} + y\sqrt{1-x^2})$$

$$\sin^{-1}(x) - \sin^{-1}(y) = \sin^{-1}(x\sqrt{1-y^2} - y\sqrt{1-x^2})$$

$$\cos^{-1}(x) + \cos^{-1}(y) = \cos^{-1}(xy - \sqrt{1-x^2}\sqrt{1-y^2})$$

$$\cos^{-1}(x) - \cos^{-1}(y) = \cos^{-1}(xy + \sqrt{1-x^2}\sqrt{1-y^2})$$

$$\tan^{-1}(x) + \tan^{-1}(y) = \tan^{-1}\left(\frac{x+y}{1-xy}\right)$$

$$\tan^{-1}(x) - \tan^{-1}(y) = \tan^{-1}\left(\frac{x-y}{1+xy}\right)$$

Derivatives

$$d(\sin^{-1}(x)) = \frac{1}{\sqrt{1-x^2}}dx \qquad \text{for } -\pi/2 < \sin^{-1}(x) < \pi/2$$

$$d(\cos^{-1}(x)) = \frac{-1}{\sqrt{1-x^2}}dx \qquad \text{for } 0 < \cos^{-1}(x) < \pi$$

$$d(\tan^{-1}(x)) = \frac{1}{1+x^2}dx \qquad \text{for } -\pi/2 < \tan^{-1}(x) < \pi/2$$

$$d(\cot^{-1}(x)) = \frac{-1}{1+x^2}dx \qquad \text{for } 0 < \sin^{-1}(x) < \pi$$

Hyperbolic functions

Definitions

There are six hyperbolic functions that are comparable to the trigonometric functions. They are designated by adding the letter h to the trigonometric abbreviations: sinh, cosh, tanh, coth, sech, csch.

The basic definitions may be stated in terms of exponentials:

$$\sinh(x) = \frac{1}{\operatorname{csch}(x)} = \frac{e^x - e^{-x}}{2}$$

$$\cosh(x) = \frac{1}{\operatorname{sech}(x)} = \frac{e^x + e^{-x}}{2}$$

$$\tanh(x) = \frac{1}{\coth(x)} = \frac{e^x - e^{-x}}{e^x + e^{-x}}$$

The power-series expansion of these functions are:

$$\sinh(x) = x + \frac{x^3}{3!} + \frac{x^5}{5!} + \frac{x^7}{7!} + \cdots \qquad \text{for any value of } x$$

$$\cosh(x) = 1 + \frac{x^2}{2!} + \frac{x^4}{4!} + \frac{x^6}{6!} + \cdots \qquad \text{for any value of } x$$

$$\tanh(x) = x - \frac{x^3}{3} + \frac{2}{15}x^5 - \frac{17}{315}x^7 + \cdots \qquad \text{for } -\pi/2 < x < \pi/2$$

$$\coth(x) = \frac{1}{x} + \frac{x}{3} - \frac{1}{45}x^3 + \frac{2}{945}x^5 + \cdots \qquad \text{for } 0 < \text{Abs}(x) < \pi$$

where $n! = n \times (n-1) \times (n-2) \times \cdots \times 1$.

Relationships

$$\sinh(-x) = -\sinh(x)$$

$$\cosh(-x) = \cosh(x)$$

$$\cosh^2(x) - \sinh^2(x) = 1$$

$$1 - \tanh^2(x) = \operatorname{sech}^2(x)$$

$$1 - \coth^2(x) = -\operatorname{csch}^2(x)$$

$$\sinh(x + y) = \sinh(x)\cosh(y) + \cosh(x)\sinh(y)$$

$$\sinh(x - y) = \sinh(x)\cosh(y) - \cosh(x)\sinh(y)$$

$$\cosh(x + y) = \cosh(x)\cosh(y) + \sinh(x)\sinh(y)$$

$$\cosh(x - y) = \cosh(x)\cosh(y) - \sinh(x)\sinh(y)$$

$$\tanh(x+y) = \frac{\tanh(x)+\tanh(y)}{1+\tanh(x)\tanh(y)}$$

$$\tanh(x-y) = \frac{\tanh(x)-\tanh(y)}{1-\tanh(x)\tanh(y)}$$

$$\sinh(2x) = 2\sinh(x)\cosh(x)$$

$$\cosh(2x) = \cosh^2(x)+\sinh^2(x) = 2\cosh^2(x)-1 = 1+2\sinh 2(x)$$

$$\tanh(2x) = \frac{2\tanh(x)}{1+\tanh^2(x)}$$

$$\sinh\left(\frac{x}{2}\right) = \sqrt{\tfrac{1}{2}(\cosh(x)-1)}$$

$$\cosh\left(\frac{x}{2}\right) = \sqrt{\tfrac{1}{2}(\cosh(x)+1)}$$

$$\tanh\left(\frac{x}{2}\right) = \frac{\cosh(x)-1}{\sinh(x)} = \frac{\sinh(x)}{\cosh(x)+1}$$

Derivatives

$$d(\sinh(x)) = \cosh(x)\,dx$$

$$d(\cosh(x)) = \sinh(x)\,dx$$

$$d(\tanh(x)) = \operatorname{sech}^2(x)\,dx$$

$$d(\coth(x)) = -\operatorname{csch}^2(x)\,dx$$

$$d(\operatorname{sech}(x)) = -\operatorname{sech}(x)\tanh(x)\,dx$$

$$d(\operatorname{csch}(x)) = -\operatorname{csch}(x)\coth(x)\,dx$$

Inverse hyperbolic functions

The inverse hyperbolic functions are expressible in terms of logarithms.

$$\sinh^{-1}(x) = \ln(x+\sqrt{x^2+1}) \qquad \text{for any value of } x$$

$$\cosh^{-1}(x) = \ln(x+\sqrt{x^2-1}) \qquad \text{for } x > 1$$

$$\tanh^{-1}(x) = \frac{1}{2}\ln\left(\frac{1+x}{1-x}\right) \qquad \text{for } x^2 < 1$$

$$\coth^{-1}(x) = \frac{1}{2}\ln\left(\frac{x+1}{x-1}\right) \qquad \text{for } x^2 > 1$$

$$\operatorname{sech}^{-1}(x) = \ln\left(\frac{1}{x}+\sqrt{\frac{1}{x^2}-1}\right) \qquad \text{for } x < 1$$

$$\operatorname{csch}^{-1}(x) = \ln\left(\frac{1}{x}+\sqrt{\frac{1}{x^2}+1}\right) \qquad \text{for any value of } x$$

The following relationships can be established:

$$\sinh^{-1}(-x) = \sinh^{-1}(x)$$

$$\tanh^{-1}(-x) = -\tanh^{-1}(x)$$

$$\coth^{-1}(-x) = -\coth^{-1}(x)$$

$$\coth^{-1}(x) = \tanh^{-1}\left(\frac{1}{x}\right)$$

Derivatives

$$d(\sinh^{-1}(x)) = \frac{dx}{\sqrt{x^2 + 1}}$$

$$d(\cosh^{-1}(x)) = \frac{dx}{\sqrt{x^2 - 1}}$$

$$d(\tanh^{-1}(x)) = \frac{dx}{1 - x^2} = d(\coth^{-1}(x))$$

$$d(\operatorname{sech}^{-1}(x)) = \frac{-dx}{x\sqrt{1 - x^2}}$$

$$d(\operatorname{csch}^{-1}(x)) = \frac{-dx}{x\sqrt{1 + x^2}}$$

Complex numbers

Definition

A complex number z consists of two distinct scalar (i.e. real) parts a and b, and is written in the form:

$$z = a + ib$$

where $i = \sqrt{-1}$. The first part a is called the real part and the second part b is called the imaginary part of the complex number.

The modulus (i.e. absolute value) of a complex number is designated r and defined as:

$$r = \sqrt{x^2 + y^2}$$

The argument θ of the complex number is the position vector measured from the positive x-axis in an anti-clockwise direction:

$$\theta = \tan^{-1}\left(\frac{y}{x}\right)$$

An alternative mode of expressing a complex number is:

$$z = x + iy = r(\cos\theta + i\sin\theta) = r\,e^{i\theta}$$

Properties

Various operations involving complex numbers are

addition

$$z_1 + z_2 = (x_1 + x_2) + i(y_1 + y_2)$$

multiplication

$$z_1 z_2 = r_1 r_2 \, e^{i(\theta_1 + \theta_2)}$$

power

$$z^n = r^n \, e^{in\theta}$$

division

$$\frac{z_1}{z_2} = \frac{r_1}{r_2} e^{i(\theta_1 + \theta_2)}$$

multiplication by i

$$iz = -y + ix = r \, e^{i(\theta + \pi/2)}$$

logarithm

$$\ln(z) = \ln(r) + i\theta$$

Conjugate number

If a complex number is: $z = x + iy$, the conjugate number is defined as:

$$\bar{z} = x - iy = r \, e^{i(\theta - \pi/2)}$$

The main properties of a conjugate are:

$$z\bar{z} = r^2 \qquad \text{(real)}$$

$$z + \bar{z} = 2x \qquad \text{(real)}$$

$$z - \bar{z} = 2iy \qquad \text{(imaginary)}$$

Polynomial equations

Presentation

A polynomial equation of degree n is:

$$a_0 + a_1 x + a_2 x^2 + \cdots + a_n x^n = 0$$

This polynomial equation (with real coefficients) has at least $(n - 2)$ real solutions.

Polynomial equation of degree two

A polynomial equation of degree two has the form:

$$a_0 + a_1 x + a_2 x^2 = 0$$

And it has two, one or zero real solutions depending upon the sign of the discriminant:

$$\Delta = a_1^2 - 4a_2 a_0$$

There is no real solution for $\Delta < 0$, one solution for $\Delta = 0$ and two real solutions for $\Delta > 0$:

$$x_1 = \frac{-a_1 + \sqrt{\Delta}}{2a_2}$$

$$x_2 = \frac{-a_1 - \sqrt{\Delta}}{2a_2}$$

Polynomial equation of degree three

A polynomial equation of degree three has at least one real solution.

$$a_0 + a_1 x + a_2 x^2 + a_3 x^3 = 0$$

Denoting A_1 and A_2 the following variables:

$$A_1 = \frac{1}{3}\frac{a_1}{a_3} - \frac{1}{9}\left(\frac{a_2}{a_3}\right)^2$$

$$A_2 = \frac{1}{6}\frac{a_2 a_1}{a_3^2} - \frac{1}{2}\frac{a_0}{a_3} - \frac{1}{27}\left(\frac{a_2}{a_3}\right)^3$$

the discriminant of the polynomial equation of degree three is:

$$\Delta = R_1^3 + A_2^2$$

If $\Delta > 0$ there is one real solution. If $\Delta = 0$ there are two real solutions and there are three real solutions when $\Delta < 0$.

Note

In practice, if an obvious solution is known (i.e. $x = -c_0$), the polynomial equation may be rewritten as:

$$a_0 + a_1 x + a_2 x^2 + a_3 x^3 = (x + c_0)(b_2 x^2 + b_1 x + b_0) = 0$$

The following relationships hold:

$$b_2 = a_3$$

$$c_0 b_2 + b_1 = a_2$$

$$c_0 b_1 + b_0 = a_1$$

$$c_0 b_0 = a_0$$

The solutions of the polynomial equation of degree three are $\{x = -c_0\}$ and the solutions of the polynomial equation of degree two:

$$b_2 x^2 + b_1 x + b_0 = 0$$

A1.4 Alternate depths in open channel flow

Presentation

Considering an open channel flow in a rectangular channel, the continuity and Bernoulli equations state:

$$q = Vd \tag{A1.16}$$

$$H - z_o = d + \frac{V^2}{2g} \tag{A1.17}$$

where q is the discharge per unit width, d is the flow depth, V is the flow velocity, H is the mean total head, z_o is the bed elevation and g is the gravity constant. The Bernoulli equation (A1.17) is developed assuming a flat horizontal channel, hydrostatic pressure distribution and uniform velocity distribution.

For a given specific energy $(E = H - z_o)$ and flow rate q_w, the system of two equations (A1.16) and (A1.17) has:

- no solution (i.e. no flow) for $H - z_o < 3/2\sqrt[3]{q^2/g}$,
- one solution (i.e. critical flow conditions) for $H - z_o = 3/2\sqrt[3]{q^2/g}$, or
- two (meaningful) solutions.

In the latter case, the two possible flow depths d' and d'' are called alternate depths (Fig. A1.1). One corresponds to a subcritical flow (i.e. $Fr = q/\sqrt{gd^3} < 1$) and the other to a supercritical flow.

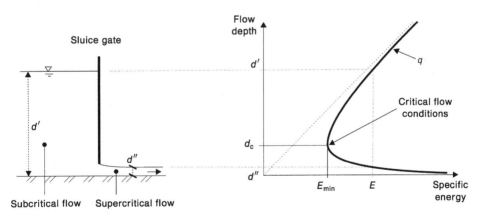

Fig. A1.1 Definition sketch: alternate depths across a sluice gate.

Discussion

For a given specific energy and known discharge, the Bernoulli equation yields:

$$\left(\frac{d}{d_c}\right)^3 - \frac{H - z_o}{d_c}\left(\frac{d}{d_c}\right)^2 + \frac{1}{2} = 0 \tag{A1.18}$$

where $d_c = \sqrt[3]{q^2/g}$ is the critical flow depth. Equation (A1.18) is a polynomial equation of degree 3 in terms of the dimensionless flow depth d/d_c and it has three real solutions for $(H - z_0)/d_c \geq 3/2$.

The solutions of equation (A1.18) (for $(H - z_0)/d_c \geq 3/2$) are:

$$\frac{d^{(1)}}{d_c} = \frac{H - z_0}{d_c}\left(\frac{1}{3} + \frac{2}{3}\cos\left(\frac{\Gamma}{3}\right)\right) \qquad \text{Subcritical flow} \qquad (A1.19a)$$

$$\frac{d^{(2)}}{d_c} = \frac{H - z_0}{d_c}\left(\frac{1}{3} + \frac{2}{3}\cos\left(\frac{\Gamma}{3} + \frac{2\pi}{3}\right)\right) \qquad \text{Negative solution} \qquad (A1.19b)$$

$$\frac{d^{(3)}}{d_c} = \frac{H - z_0}{d_c}\left(\frac{1}{3} + \frac{2}{3}\cos\left(\frac{\Gamma}{3} + \frac{4\pi}{3}\right)\right) \qquad \text{Supercritical flow} \qquad (A1.19c)$$

where

$$\cos\Gamma = 1 - \frac{27}{4}\left(\frac{d_c}{H - z_0}\right)^3$$

Equations (A1.19a) and (A1.19c) are plotted as functions of $(H - z_0)/d_c$ in Fig. A1.2.

Note that equation (A1.19a) may be rewritten in terms of the Froude number as:

$$Fr = \left(\frac{d}{d_c}\right)^{-3/2} \qquad (A1.20)$$

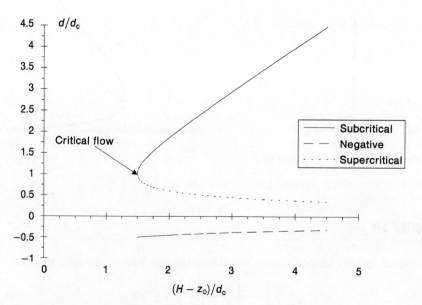

Fig. A1.2 Flow depth solutions as functions of $(H - z_0)/d_c$.

It yields:

$$Fr^{(1)} = \left(\frac{H - z_o}{d_c} \left(\frac{1}{3} + \frac{2}{3}\cos\left(\frac{\Gamma}{3}\right) \right) \right)^{-3/2} \qquad \text{Subcritical flow} \qquad \text{(A1.21a)}$$

$$Fr^{(2)} = \left(\frac{H - z_o}{d_c} \left(\frac{1}{3} + \frac{2}{3}\cos\left(\frac{\Gamma}{3} + \frac{2\pi}{3}\right) \right) \right)^{-3/2} \qquad \text{Complex number} \qquad \text{(A1.21b)}$$

$$Fr^{(3)} = \left(\frac{H - z_o}{d_c} \left(\frac{1}{3} + \frac{2}{3}\cos\left(\frac{\Gamma}{3} + \frac{4\pi}{3}\right) \right) \right)^{-3/2} \qquad \text{Supercritical flow} \qquad \text{(A1.21c)}$$

Part 2 Introduction to Sediment Transport in Open Channels

6

Introduction to sediment transport in open channels

6.1 Introduction

Waters flowing in natural streams and rivers have the ability to scour channel beds, to carry particles (heavier than water) and to deposit materials, hence changing the bed topography. This phenomenon (i.e. *sediment transport*) is of great economic importance: e.g. to predict the risks of scouring of bridges, weirs, channel banks; to estimate the siltation of a reservoir upstream of a dam wall; to predict the possible bed form changes of rivers and estuaries.

Numerous failures have resulted from the inability of engineers to predict sediment motion: e.g. bridge collapse (pier foundation erosion), formation of sand bars in estuaries and navigable rivers, destruction of banks and levees.

In this book, we shall introduce the basic concepts of sediment transport in open channels.

6.2 Significance of sediment transport

6.2.1 Sediment transport in large alluvial streams

The transport of sediment is often more visible in mountain streams, torrents and creeks (e.g. Fig. 6.1). However larger rivers are also famous for their capacity to carry sediment load: e.g. the Nile river, the Mississippi river and the Yellow River.

The Nile river is about 6700 km long and the mean annual flow is about $97.8 \times 10^9 \, m^3$ (i.e. $3100 \, m^3/s$). Large floods take place each year from July to October with maximum flow rates over $8100 \, m^3/s$ during which the river carries a large sediment load. In Ancient Egypt, peasants and farmers expected the Nile floods to deposit fertile sediments in the 20 km wide flood plains surrounding the main channel.

The Mississippi river is about 3800 km long. Its mean annual discharge is about $536 \times 10^9 \, m^3$ (or $17\,000 \, m^3/s$). On average, the river transports about $7000 \, kg/s$ of sediment to the Gulf of Mexico. During flood periods much larger sediment transport rates are observed. McMath (1883) reported maximum daily sediment transport rates of $4.69 \times 10^9 \, kg$ (i.e. $54\,300 \, kg/s$) in the Mississippi river at Fulton (on 10 July 1880)

Fig. 6.1 Aerial view of Camas Creek flowing into the North Fork of the Flathead River, Montana, USA (12 July 1981) (by P. Carrara, with permission of US Geological Survey, Ref. 356ct).

and over 5.97×10^9 kg (or 69 110 kg/s) in the Missouri river[1] at St Charles (on 3 July 1879). During the 1993 flood, the river at the Nebraska Station carried 11.9×10^9 kg (4580 kg/s) of sediment load during July. Between 12 and 28 July 1993, the Missouri river at Kansas City scoured the bed by 4.5 m and the sediment load reached 8700 kg/s between 26 June and 14 September 1993 at Hermann gauging site (Bhowmik 1996).

The Yellow River (or Huang Ho river) flows 5460 km across China and its catchment area is 745 000 km^2. The annual mean (water) flow is 48.3×10^9 m^3 (i.e. 1530 m^3/s) and the average annual sediment load is 1.6 billion tons which comes primarily from the Middle reach regions. In the lower reach the river bed was subjected to numerous changes, and the location of the river estuary varied by as much as 600 km for the past 3000 years.

6.2.2 Failures caused by sediment transport processes

Moore Creek dam, Tamworth, Australia

The Moore Creek dam was completed in 1898 to supply water to the town of Tamworth, NSW (325 km North of Sydney) (Fig. 6.2). At the time of construction,

---

[1] The Missouri river is the longest tributary (3726 km long) of the Mississippi River and it is sometimes nicknamed 'Big Muddy' because of the amount of sediment load.

(a)

(b)

Fig. 6.2 Photographs of the Moore Creek dam. (a) Old photograph (shortly after completion). (b) Recent photograph (14 June 1997).

the 18.6 m high dam was designed with advanced structural features: i.e. thin single-arch wall (7.7 m thick at base, 0.9 m thick at crest), vertical downstream wall and battered upstream face, made of Portland cement concrete. The volume of the reservoir was 220 000 m^3 and the catchment area is 51 km^2.

Between 1898 and 1911, the reservoir was filled with 85 000 m^3 of sediment.[2] In 1924 the dam was out of service because the reservoir was fully silted (25 years after its construction). The dam is still standing today. Moore Creek dam must be considered as an engineering failure. It failed because the designers understood neither the basic concepts of sediment transport nor the catchment erosion mechanisms.

Old Quipolly dam, Werris Creek, Australia

Completed in 1932, the Old Quipolly dam[3] is a concrete single-arch dam: it is 19 m high with a 184 m long crest (1 m thickness at crest) (Fig. 6.3). The reservoir was built to supply water to the town of Werris Creek, NSW. The catchment area is 70 km^2 and the original storage capacity was 860 000 m^3.

In 1941, 130 000 m^3 of sediment had accumulated. In 1943 the siltation volume amounted to 290 000 m^3. By 1952, more than half of the initial storage had disappeared. The reservoir was disused in 1955. Nowadays the reservoir is fully silted. The dam is still standing and the wall is in excellent conditions (author's inspection in 1997). The reservoir acts presently as a sediment trap for the new Quipolly dam built in 1955 and located 3 km downstream.

The Old Quipolly dam is nevertheless another engineering failure caused by reservoir sedimentation. It is probably the most extreme siltation record recorded in Australia, although several dams also suffered siltation in south-east Australia (Table 6.1).

Mount Isa railway bridges, Queensland, Australia

In the early 1960s, the reconstruction of the railway line to Mount Isa (mining town, in the Australian desert) required the construction of several bridges across creeks. Several bridges failed because of erosion of the pier foundations: e.g. bridge 235 across Julia Creek, bridge across Eastern creek (Fig. 6.4), bridge across Corella Creek (Nonda).

The bridges collapsed because of the inability of (overseas) engineers to understand the local hydrology and associated sediment motion.

Shihmen dam, Taiwan

The Shihmen dam (Taoyuan County, Taiwan, Republic of China) is a 133 m high dam built between 1958 and 1964. The maximum reservoir capacity was more than 60 000 000 m^3 and the catchment area is 763 km^2. Although the dam was inaugurated in 1964, the reservoir began filling in May 1963. In September 1963, 20 000 000 m^3 of

---

[2] For the period 1898–1911 observations suggested that most siltation took place during the floods of February 1908 and January 1910.

[3] Also known as Coeypolly Creek dam No. 1 (International Commission on Large Dams 1984). A second dam was built in 1955, 3 km downstream of the old dam.

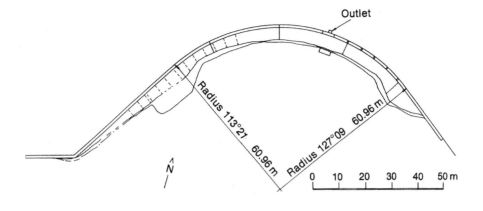

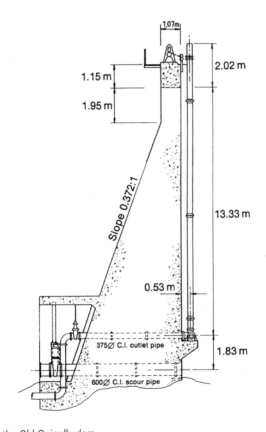

Fig. 6.3 Sketch of the Old Quipolly dam.

silt accumulated during cyclonic flood (typhoon Gloria). Between 1985 and 1995, the reservoir was dredged and over 10 000 000 m³ of sediment were removed. However, during that period, the volume of sediment flowing into the reservoir amounted to about 15 000 000 m³ (over 10 years). More than 100 sediment-trap dams were built

Table 6.1 Examples of reservoir siltation in south-east Australia

Dam (1)	Year of completion (2)	Catchment area (km²) (3)	Original reservoir capacity (m³) (4)	Cumulative siltation volume (m³) (5)	Date of record (6)
Corona dam, Broken Hill NSW, Australia	1890?	15	120 000⁺	120 000 fully-silted	1910?
Stephens Creek dam, Broken Hill NSW, Australia	1892ᶜ	510	24 325 000	1 820 000 2 070 000 4 500 000 de-silting*	1907 1912 1958 1971
Junction Reefs dam, Lyndhurst NSW, Australia	1896	–	200 000⁺	fully-silted	1997
Moore Creek dam, Tamworth NSW, Australia	1898	51	220 000⁺	85 000 ~200 000 fully-silted	1911 1924 1985
Koorawatha dam, Koorawatha NSW, Australia	1901	–	40 500	fully-silted	1997
Gap weir, Werris Creek NSW, Australia	1902	160	⁺	fully-silted	1924
Cunningham Creek dam, Harden NSW, Australia	1912	820	–⁺	216 000 258 600 379 000 522 600 758 000	1916 1917 1920 1922 1929
Illalong Creek dam, Binalong NSW, Australia	1914	130	260 000⁺	75% siltation	1997
Umberumberka dam, Broken Hill NSW, Australia	1915	420	13 197 000	3 600 000 4 500 000 5 013 000 5 700 000	1941 1961 1964 1982
Korrumbyn Creek dam, Murwillumbah NSW, Australia	1919	3	27 300⁺	fully-silted	1985
Borenore Creek dam, Orange NSW, Australia	1928ᵃ	22	230 000	150 000	1981
Tenterfield Creek dam, Tenterfield NSW, Australia	1930ᵇ	38	1 500 000	110 000	1951
Old Quipolly dam, Werris Creek NSW, Australia	1932	70	860 000⁺	130 000 290 000 ~430 000 fully-silted	1941 1943 1952 1985
Inverell dam, Inverell NSW, Australia	1939	600	153 000 000⁺	fully-silted	1982
Illawambra dam, Bega Valley NSW, Australia	1959	19	4500⁺	3000 fully-silted	1976 1985

Notes: ᵃ dam raised in 1943; ᵇ dam raised in 1974; ᶜ spillway crest raised in 1909; ⁺ fully-silted nowadays; * extensive desilting in 1971.

upstream of Shihmen reservoir to trap incoming sediments. In 1996 only one sediment-trap dam was still functioning, all the others being filled up. It is believed that the maximum depth of water in the Shihmen reservoir was less than 40 m in 1997.

The Shihmen reservoir (Fig. 6.5) was designed to operate for at least 70 years Thirty years after completion, it has become, in fact, a vast sediment trap with an

Fig. 6.4 Sketch of a collapsed railway bridge across Eastern Creek (after an original photograph). Collapse caused by pier scour.

inappropriate storage capacity to act as a flood control or water supply reservoir (Chang 1996).

6.3 Terminology

Classical (clear-water) hydraulics is sometimes referred to as *'fixed boundary hydraulics'*. It can be applied to most man-made channels (e.g. concrete-lined channels, rock tunnels, rockfill-protected channels) and to some extent to grassed waterways. However most natural streams are characterized by some sediment

Fig. 6.5 The Shihmen reservoir during a dry season in summer 1993 (courtesy of H.C. Pu and Sinorama).

transport: i.e. the channel boundaries are movable. '*Movable boundary hydraulics*' applies to streams with gravel or sand beds, estuaries (i.e. silt or sand beds), sandy coastlines and man-made canals in earth, sand or gravel. Movable boundary hydraulics is characterized by variable boundary roughness and variable channel dimensions. Strong interactive processes take place between the water flow and the bed form changes.

6.4 Structure of this section

In Part II, the lecture material is regrouped as a series of definitions (Chapter 7), the basic concepts of sediment motion (Chapters 8 and 9), the calculations of the sediment transport capacity (Chapters 10 and 11) and the applications to natural alluvial streams, including the concepts of erosion, accretion and bed form motion (Chapter 12). Further examples of reservoir sedimentation are discussed in Appendix A2.1.

6.5 Exercises

Name two rivers that are world-famous for their sediment transport capacity.

Does sediment transport occur in fixed-boundary hydraulics?

Are the boundary roughness and channel dimensions fixed in movable boundary hydraulics?

7

Sediment transport and sediment properties

7.1 Basic concepts

7.1.1 Definitions

'*Sediment transport*' is the general term used for the transport of material (e.g. silt, sand, gravel, boulders) in rivers and streams. The transported material is called the '*sediment load*'. Distinction is made between the '*bed load*' and the '*suspended load*'. The bed load characterizes grains rolling along the bed while suspended load refers to grains maintained in suspension by turbulence. The distinction is, however, sometimes arbitrary when both loads are of the same material.

Note
The word 'sediment' commonly refers to fine materials that settle to the bottom. Technically, however, the term sediment transport includes the transport of both fine and large materials (e.g. clay, silt, gravel, boulders).

7.1.2 Bed formation

In most practical situations, the sediments behave as a non-cohesive material (e.g. sand, gravel) and the fluid flow can distort the bed into various shapes. The bed form results from the drag force exerted by the bed on the fluid flow as well as the sediment motion induced by the flow onto the sediment grains. This interactive process is quite complex.

In a simple approach, the predominant parameters that affect the bed form are the bed slope, the flow depth and velocity, the sediment size and particle fall velocity. At low velocities, the bed does not move. With increasing flow velocities, the inception of bed movement is reached and the sediment bed begins to move. The basic bed forms which may be encountered are the *ripples* (usually of heights less than 0.1 m), *dunes*, *flat bed*, *standing waves*, and *antidunes*. At high flow velocities (e.g. mountain streams, torrents), chutes and step-pools may form. They consist of a succession of chutes and

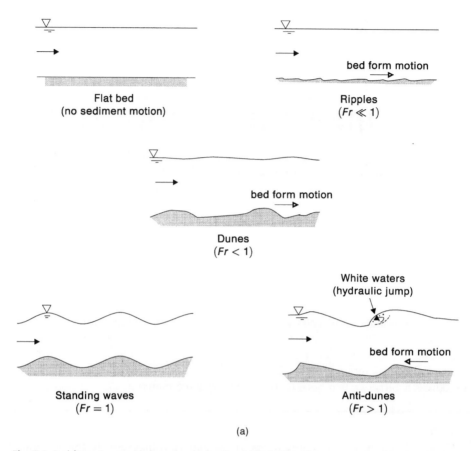

(a)

Fig. 7.1 Bed forms in movable boundary hydraulics. (a) Typical bed forms.

free-falling nappes (i.e. supercritical flow) connected by pools where the flow is usually subcritical.

The typical bed forms are summarized in Fig. 7.1 and Table 7.1. In Table 7.1, column 3 indicates the migration direction of the bed forms. Ripples and dunes move in the downstream direction. Antidunes and step-pools are observed with supercritical flows and they migrate in the upstream flow direction. Field observations are illustrated in Fig. 7.2.

Notes
1. Both ripples and dunes are observed with wind-blown sand and in open channel flow.
2. Note that, in alluvial rivers, dunes form with subcritical flow conditions only. Antidunes are associated with supercritical flow while standing waves are characteristics of near-critical flow conditions.
3. The transition between dune and standing-wave bed forms occurs with a flat bed. The flat-bed is an unstable bed pattern, often observed at Froude

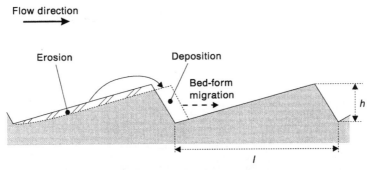

Dune migration (in the downstream direction)

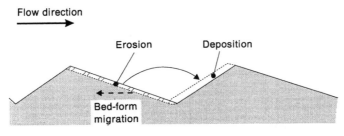

Anti-dune migration (in the upstream direction)

(b)

Fig. 7.1 (b) Bed form motion.

Table 7.1 Basic bed forms in alluvial channels (classification by increasing flow velocities)

Bed form	Flow	Bed form motion	Comments
(1)	(2)	(3)	(4)
Flat bed	No Flow (or $Fr \ll 1$)	NO	No sediment motion.
Ripples	$Fr \ll 1$	D/S	Three-dimensional forms. Observed also with air flows (e.g. sand ripples in a beach caused by wind).
Dunes	$Fr < 1$	D/S	Three-dimensional forms. Sand dunes can also be caused by wind.
Flat bed	$Fr \leq 1$	NO	Observed also with wind flow.
Standing waves	$Fr = 1$	NO	Critical flow conditions. Bed standing waves in phase with free-surface standing waves.
Antidunes	$Fr > 1$	U/S	Supercritical flow with tumbling flow and hydraulic jump upstream of antidune crests.
Chute-pools	$Fr > 1$	U/S	Very active antidunes.
Step-pools	$Fr > 1$	–	Cascade of steps and pools. Steps are often caused by rock bed.

References: Henderson (1966), Graf (1971).
Notes: D/S = in downstream flow direction; Fr = Froude number; U/S = in upstream flow direction.

(a)

(b)

(c)

numbers slightly below unity: e.g. $Fr = 0.77$ (Kennedy 1963), 0.83 (Laursen 1958) and 0.57 to 1.05 (Julien and Raslan 1998).

Discussion

Ripples are associated with the presence of a laminar boundary layer. Their size is independent of the flow depth d. Usually their characteristic dimensions (length l, height h) satisfy: $l \ll 1000d$ and $h < 100d$.

Dunes are associated with a turbulent boundary layer. In rivers, their size is about proportional to the flow depth (see also Table 12.3). In open channels, dunes take place in subcritical flow.

With standing wave and antidune bed forms, the free-surface profile is in phase with the bed form profile. In natural streams, antidunes and standing waves are seldom seen because the bed forms are often not preserved during the receding stages of a flood. Kennedy (1963) investigated standing wave bed-forms in the laboratory while Alexander and Fielding (1997) presented superb photographs of gravel antidunes.

7.2 Physical properties of sediments

7.2.1 Introduction

Distinction is made between two categories of sediment: cohesive material (e.g. clay, silt) and non-cohesive material (e.g. sand, gravel). In this section, the basic properties of both types are developed. In practice, however, we will primarily consider non-cohesive materials in this introductory course.

7.2.2 Property of single particles

The *density* of quartz and clay minerals is typically: $\rho_s = 2650 \, \text{kg/m}^3$. Most natural sediments have densities similar to that of quartz.

The *relative density* of a sediment particle equals:

$$\mathbf{s} = \frac{\rho_s}{\rho} \tag{7.1}$$

where ρ is the fluid density. For a quartz particle in water, $\mathbf{s} = 2.65$ but a quartz particle in air has a relative density: $\mathbf{s} = 2200$.

Fig. 7.2 (opposite) Field observations of bed forms. (a) Sand dunes in Cape River after the flood, 270 km north of Clermont QLD, Australia (8 July 1983) (courtesy of Mrs J. Hacker). (b) Gravel antidune at Brigalow Bend, Burdekin River, Australia in August 1995 (courtesy of Dr C. Fielding). Bed forms left after a peak flow of 8000 m³/s. Flow from the left to the right. (c) Gravel antidune at Brigalow Bend, Burdekin River, Australia in August 1995 (courtesy of Dr C. Fielding). Bed forms left after a peak flow of 8000 m³/s. Looking downstream, with an observation trench in the foreground.

Notes
1. In practice, a dense particle is harder to move than a light one.
2. Let us keep in mind that heavy minerals (e.g. iron, copper) have larger values of density than quartz.
3. The relative density is dimensionless. It is also called the specific gravity. It is sometimes denoted γ_s.

The most important property of a sediment particle is its characteristic size. It is termed the *diameter* or *sediment size*, and denoted d_s. In practice, natural sediment particles are not spherical but exhibit irregular shapes. Several definitions of sediment size are available:

- the sieve diameter,
- the sedimentation diameter, and
- the nominal diameter.

The *sieve diameter* is the size of particle that passes through a square mesh sieve of given size but not through the next smallest size sieve: e.g. $1\,\text{mm} < d_s < 2\,\text{mm}$.

The *sedimentation diameter* is the size of a quartz sphere that settles down (in the same fluid) with the same settling velocity as the real sediment particle. The *nominal diameter* is the size of the sphere of same density and same mass as the actual particle.

The sediment size may also be expressed as a function of the sedimentological size parameter ϕ (or Phi-scale) defined as:

$$d_s = 2^{-\phi} \tag{7.2a}$$

or

$$\phi = -\frac{\ln(d_s)}{\ln(2)} \tag{7.2b}$$

where d_s is in mm.

A typical sediment size classification is shown in Table 7.2.

Notes
1. Large particles are harder to move than small ones.
2. Air can move sand (e.g. sand dunes formed by wind action). Water can move sand, gravel, boulders or breakwater armour blocks (weighing several tonnes).
3. The sedimentation diameter is also called the standard fall diameter.

Table 7.2 Sediment size classification

Class name (1)	Size range (mm) (2)	Phi-scale ϕ (3)	Remarks (4)
Clay	$d_s < 0.002$ to $0.004\,\text{mm}$	$+8$ to $+9 < \phi$	
Silt	0.002 to $0.004 < d_s < 0.06\,\text{mm}$	$+4 < \phi < +8$ to $+9$	
Sand	$0.06 < d_s < 2.0\,\text{mm}$	$-1 < \phi < +4$	Silica
Gravel	$2.0 < d_s < 64\,\text{mm}$	$-6 < \phi < -1$	Rock fragments
Cobble	$64 < d_s < 256\,\text{mm}$	$-8 < \phi < -6$	Original rocks
Boulder	$256 < d_s$	$\phi < -8$	Original rocks

7.2.3 Properties of sediment mixture

The density of a dry sediment mixture equals:

$$(\rho_{sed})_{dry} = (1 - Po)\rho_s \tag{7.3}$$

where Po is the porosity factor.

The density of wet sediment is:

$$(\rho_{sed})_{wet} = Po\rho + (1 - Po)\rho_s \tag{7.4}$$

The porosity factor ranges basically from 0.26 to 0.48. In practice, Po is typically about 0.36 to 0.40.

Notes

1. The density of sediments may also be expressed as a function of the void ratio: i.e. the ratio of volume of voids (or pores) to volume of solids. The void ratio is related to the porosity as:

$$\text{Void ratio} = Po/(1 - Po)$$

2. Another characteristic of a porous medium is the permeability. For a one-dimensional flow through the pores of the sediment bed, the velocity of seepage is given by the Darcy law:

$$V = -K\frac{dH}{dx}$$

where K is the hydraulic conductivity (or coefficient of permeability, in m/s) and H is the piezometric head (in m). The hydraulic conductivity not only depends on the permeability of the soil by also on the properties of the fluid and dimensional analysis yields (Raudkivi and Callander 1976, p. 15):

$$K = k\frac{\rho g}{\mu}$$

where k is the permeability (in m^2), ρ is the fluid density, g is the gravity constant and μ is the fluid dynamic viscosity. Typical values of the hydraulic conductivity are (Raudkivi and Callander 1976, p. 19):

Soil type	Fine sand	Silty sand	Silt
K (m/s)	5×10^{-4} to 1×10^{-5}	2×10^{-5} to 1×10^{-6}	5×10^{-6} to 1×10^{-7}

3. Discussion: conversion between parts per million (ppm) and kilograms per cubic metre (kg/m^3).

For suspended sediment, the sediment concentration may be expressed in kg/m^3 and it is calculated as the ratio of dry sediment mass to the volume of a water–sediment mixture. It can also be expressed as a volume concentration (dimensionless). Another unit, parts per million (ppm), is sometimes used. It is defined as the ratio of the weight of sediment to the weight of the water–sediment mixture times one million.

The conversion relationships are:

$$\text{mass concentration} = \rho_s C_s$$

$$\text{concentration in parts per million by weight} = \frac{1 \times 10^6 s C_s}{1 + (s-1)C_s}$$

where C_s is the volumetric sediment concentration and $s = \rho_s/\rho$.

Application
Calculate the dry and wet densities of a sand mixture with a 38% porosity.

Solution
Assuming a quartz sand ($\rho_s = 2650 \, \text{m}^3/\text{s}$), the dry and wet densities of the sediment mixture are estimated using equations (7.3) and (7.4):

$$(\rho_{\text{sed}})_{\text{dry}} = (1 - Po)\rho_s = 1643 \, \text{kg/m}^3$$

$$(\rho_{\text{sed}})_{\text{wet}} = Po\rho + (1 - Po)\rho_s = 2022 \, \text{kg/m}^3$$

7.2.4 Particle size distribution

Natural sediments are mixtures of many different particle sizes and shapes. The *particle size distribution* is usually represented by a plot of the weight percentage of the total sample, which is smaller than a given size plotted as a function of the particle size. A cumulative curve fitted to data points is shown in Fig. 7.3.

The characteristic *sediment size* d_{50} is defined as the size for which 50% by weight of the material is finer. Similarly the characteristic sizes d_{10}, d_{75} and d_{90} are values of grain sizes for which 10%, 75% and 90% of the material weight is finer, respectively.

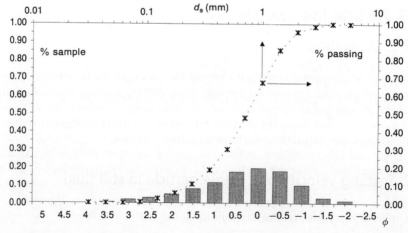

Fig. 7.3 Typical particle size distribution curve. (a) Percentage sampling as a function of sedimentological size parameter ϕ (linear scale). (b) Cumulative percentage passing as a function of the particle size d_s in mm (semi-logarithmic scale).

d_{50} is commonly used as the characteristic grain size and the range of particle size is often expressed in terms of a sorting coefficient S:

$$S = \sqrt{\frac{d_{90}}{d_{10}}} \qquad (7.5)$$

Another descriptor is the geometric standard deviation based upon a log-normal distribution of grain sizes σ_g:

$$\sigma_g = \sqrt{\frac{d_{84}}{d_{16}}} \qquad (7.6)$$

Small values of S and σ_g imply a nearly uniform sediment size distribution. A large value of S means a broad sediment size distribution.

Notes
1. The sediment size d_{50} is called the *median grain size*.
2. Other definitions may be used to characterize the range of particles sizes. For example:

$$\text{Gradation coefficient} = \frac{1}{2}\left(\frac{d_{84}}{d_{50}} + \frac{d_{50}}{d_{16}}\right) \qquad \text{(Julien 1995)}$$

3. The size distribution may be recorded using a settling tube (see next sections).

Comments
The size distribution of cohesive sediments (e.g. clay, silt) may vary with the environmental conditions to which the sediments have been subjected and also the procedures that are used to determine their size distribution. In the following, we shall consider primarily non-cohesive sediments.

7.3 Particle fall velocity

7.3.1 Presentation

In a fluid at rest, a suspended particle (heavier than water) falls: i.e. it has a downward (vertical) motion. The *terminal fall velocity* is the particle velocity at equilibrium, the sum of the gravity force, buoyancy force and fluid drag force being equal to zero.

In an open channel flow, the particle fall velocity is further affected by the flow turbulence and the interactions with surrounding particles.

7.3.2 Settling velocity of a single particle in still fluid

For a spherical particle settling in a still fluid, the terminal fall velocity w_o equals:

$$w_o = -\sqrt{\frac{4gd_s}{3C_d}(s-1)} \qquad \text{Spherical particle} \qquad (7.7)$$

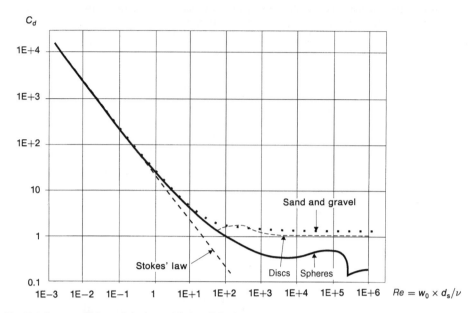

Fig. 7.4 Drag coefficient of single particle in still fluid.

where d_s is the particle diameter, C_d is the drag coefficient and $s = \rho_s/\rho$. The negative sign indicates a downward motion (for $s > 1$).

Dimensional analysis implies that the drag coefficient is a function of Reynolds number and particle shape:

$$C_d = f\left(\rho\frac{w_o d_s}{\mu}; \text{ particle shape}\right) \qquad (7.8)$$

where ρ and μ are the fluid density and dynamic viscosity respectively.

At low particle Reynolds numbers ($w_o d_s/\nu < 1$), the flow around the particle is laminar. At large Reynolds numbers ($w_o d_s/\nu > 1000$), the flow around the spherical particle is turbulent and the drag coefficient is nearly constant. Typical drag coefficient values for spherical particles are presented in Fig. 7.4.

Sediment particles have irregular shapes and the drag coefficient differs from that for spherical particles. Their shape is often angular, sometimes disc-shaped, and the drag coefficient can be expected to be larger than that of spheres. For sands and gravels, a simple approximation of the drag coefficient is:

$$C_d = \frac{24\mu}{\rho|w_o|d_s} + 1.5 \qquad Re < 1 \times 10^4 \qquad (7.9)$$

where $|w_o|$ is the absolute value of the particle fall velocity (Fig. 7.4).

Combining equations (7.7) and (7.9), an estimate of the terminal fall velocity of a sediment particle is:

$$w_o = -\sqrt{\frac{4gd_s}{3\left(\dfrac{24\mu}{\rho|w_o|d_s} + 1.5\right)}(s-1)} \qquad (7.10)$$

Table 7.3 Computed settling velocity of sediment particles in still water (equation (7.10))

d_s (m):	5×10^{-5}	0.0001	0.0002	0.0005	0.001	0.002	0.005	0.01	0.02	0.05	0.1	0.2	0.5
w_o (m/s):	2.2×10^{-3}	8.5×10^{-3}	0.027	0.071	0.11	0.17	0.27	0.39	0.54	0.85	1.2	1.7	2.7

Notes: $s = 2.69$; w_o = terminal fall velocity of single particle in water at 20°C.

Computed values of settling velocity of sediment particles are reported in Table 7.3. Experimental observations are presented in Table 7.4.

Table 7.4 Terminal settling velocity of sediment particle in still water (observations)

d_s [a] (mm) (1)	w_o [a] (m/s) (2)	$w_o d_s / \nu$ (3)	C_d [a] (4)	Comment (5)
0.089	0.005	0.44	55	Sand grains
0.147	0.013	1.9	15	
0.25	0.028	7.0	6	
0.42	0.050	21	3	
0.76	0.10	75	1.8	
1.8	0.17	304	1.5	

Notes: w_o = terminal fall velocity of single particle in water at 20°C; [a] data from Engelund and Hansen (1972).

Application
For a spherical particle, derive the basic relationship between the settling velocity, drag coefficient and particle characteristics.

Solution
Considering a spherical particle in still water, the forces acting on the settling particle are:

the drag force $0.5 C_d \rho w_o^2 A_s$,

the weight force $\rho_s g v_s$ and

the buoyant force F_b

where A_s is the area of the particle in the z-direction, z being the vertical axis positive upwards, C_d is the drag coefficient, g is the gravity constant, ρ is the water density, ρ_s is the particle density, w_o is the terminal fall velocity in still water and v_s is the volume of the particle.

The buoyant force on a submerged body is the difference between the vertical component of pressure force on its underside and the vertical pressure component on its upper side. To illustrate the concept of buoyancy, let us consider a diver in a swimming pool. The pressure force exerted on the diver equals the weight of water above him/her. As the pressure below him/her is larger than that immediately above, a reaction force (i.e. the buoyant force) is applied to the diver in the vertical direction. The buoyant force counteracts the pressure force and equals the weight of displaced liquid. For a sediment particle subjected to a hydrostatic pressure gradient, the buoyant force equals:

$$F_b = \rho g v_s$$

assuming a constant pressure gradient over the particle height d_s.

In the force balance, the drag force is opposed to the particle motion direction and the buoyant force is positive (upwards). At equilibrium the balance of the forces yields:

$$+\frac{1}{2}C_d\rho w_o^2 A_s - \rho_s g v_s + \rho g v_s = 0$$

The settling velocity equals:

$$w_o^2 = \frac{2g}{C_d A_s}\left(\frac{\rho_s}{\rho} - 1\right)v_s$$

Note that the settling velocity is negative (i.e. downwards).

Spherical particle
For a spherical particle, the cross-sectional area and volume of the particle are respectively:

$$A_s = \frac{\pi}{4}d_s^2$$

$$v_s = \frac{\pi}{6}d_s^3$$

And the settling velocity becomes

$$w_o = -\sqrt{\frac{4gd_s}{3C_d}\left(\frac{\rho_s}{\rho} - 1\right)} \qquad \text{Spherical particle}$$

Remarks
Note that, on Earth, the buoyant force is proportional to the liquid density ρ and to the gravity acceleration g. The buoyancy is larger in denser liquids: e.g. a swimmer floats better in the water of the Dead Sea than in fresh water. In gravitationless water (e.g. a waterfall) the buoyant force is zero.

Comments
1. *Drag coefficient of spherical particles*
   Considering the flow around a sphere, the flow is everywhere laminar at very small Reynolds numbers ($Re < 1$). The flow of a viscous incompressible fluid around a sphere was solved by Stokes (1845, 1851). His results imply that the drag coefficient equals:

$$C_d = \frac{24}{Re} \qquad \text{Laminar flow around a spherical particle (Stokes' law)}$$

   which is defined in terms of the particle size and settling velocity: $Re = \rho w_o d_s/\mu$.
   In practice, Stokes' law is valid for $Re < 0.1$. A better estimate of the drag coefficient, which includes the inertial terms of the vorticity transport is

(Julien 1995):

$$C_d = \frac{24}{Re}\left(1 + \frac{3}{16}Re - \frac{19}{1280}Re^2 + \frac{71}{20480}Re^3 + \dots\right)$$

Laminar flow around a spherical particle ($Re < 0.1$)

At large Reynolds numbers ($Re > 1000$), the flow around the spherical particle is turbulent and the drag coefficient is nearly constant:

$$C_d \approx 0.5$$

Turbulent flow around a spherical particle ($1 \times 10^3 < Re < 1 \times 10^5$)

2. *Drag coefficient of natural sediment particles*
   For natural sand and gravel particles, experimental values of drag coefficient were measured by Engelund and Hansen (1967) (Table 7.4). Their data are best fitted by:

$$C_d = \frac{24}{Re} + 1.5$$

   based on experimental values obtained for sand and gravel ($Re < 1 \times 10^4$)

   Another empirical correlation was recently proposed (Cheng 1997):

$$C_d = \left(\left(\frac{24}{Re}\right)^{2/3} + 1\right)^{3/2}$$

   based on experimental values for natural sediment particles ($Re < 1 \times 10^4$)

3. George Gabriel Stokes (1819–1903) was a British physicist and mathematician, known for his studies of the behaviour of viscous fluids (Stokes 1845, 1851).
4. *Terminal settling velocity*
   Gibbs *et al.* (1971) provided an empirical formula for the settling velocity of a spherical particle ($50\,\mu m < d_s < 5\,mm$). Their results can be expressed in SI units as:

$$w_o = 10\frac{-30\nu + \sqrt{900\nu^2 + gd_s^2(s-1)(0.003869 + 2.480d_s)}}{0.011607 + 7.4405d_s}$$

Spherical particles

   where ν is the kinematic viscosity of the fluid (i.e. $\nu = \mu/\rho$).
5. Note that the temperature affects the fall velocity as the fluid viscosity changes with temperature.

Discussion: settling velocity of sediment particles

Observed values of terminal settling velocity are reported in Table 7.4. First, note that large-size particles fall faster than small particles. For example, the terminal fall velocity of a 0.01 mm particle (i.e. silt) is about 0.004 cm/s while a 10 mm gravel settles at about 34 cm/s. Practically, fine particles (e.g. clay, silt) settle in a laminar flow motion ($w_o d_s/\nu < 1$) while large particles (e.g. gravel, boulders) fall in a turbulent flow motion ($w_o d_s/\nu > 1000$).

Further, at the limits, the fall velocity and the particle size satisfy:

$$w_o \propto d_s^2 \qquad \text{Laminar flow motion } (w_o d_s/\nu < 1) \qquad (7.11a)$$

$$w_o \propto \sqrt{d_s} \qquad \text{Turbulent flow motion } (w_o d_s/\nu > 1000) \qquad (7.11b)$$

Application

The size distribution of a sandy mixture is recorded using a settling tube in which the settling time in water over a known settling distance is measured. The results are:

Settling rate (cm/s)	32	16	10.7	8	5.33	4	3.2	2.7
Mass (g)	0.4	1.20	4.31	7.00	15.98	31.60	38.12	42.59

Settling rate (cm/s)	2.29	2	1.78	1.6	1.45	1.3	1.23	1.14
Mass (g)	44.80	48.04	49.07	49.60	49.81	49.87	49.95	50.00

Deduce the median grain size and the sorting coefficient $S = \sqrt{d_{90}/d_{10}}$.

Solution

First, we deduce the median settling velocity $(w_o)_{50}$ and the settling velocities $(w_o)_{10}$ and $(w_o)_{90}$ for which 10% and 90% by weight of the material settle faster respectively. The equivalent sedimentation diameters can then be determined.

For the measured data, it yields: $(w_o)_{50} = 4.56$ cm/s, $(w_o)_{10} = 10.0$ cm/s and $(w_o)_{90} = 2.27$ cm/s.

The equivalent sedimentation diameter may be deduced from equation (7.7):

$$d_s = \frac{3C_d}{4g(s-1)} w_o^2$$

in which the drag coefficient may be estimated using equation (7.9):

$$C_d = \frac{24\mu}{\rho|w_o|d_s} + 1.5 \qquad \text{for sand and gravels}$$

For a known settling velocity, the equivalent sedimentation diameter may be derived from the above equations by iterative calculations or using Table 7.3.

The equivalent sedimentation diameters for the test are: $d_{50} = 0.31$ mm, $d_{10} = 0.18$ mm, $d_{90} = 0.83$ mm and the sorting coefficient is: $S = 2.15$.

Remarks

Fine particles settle more slowly than heavy particles. As a result, the grain sizes d_{10} and d_{90} are deduced from the settling velocities $(w_o)_{90}$ and $(w_o)_{10}$ for which 90% and 10% by weight of the material settle faster respectively.

7.3.3 Effect of sediment concentration

The settling velocity of a single particle is modified by the presence of surrounding particles. Experiments have shown that thick homogeneous suspensions have a slower fall velocity than that of a single particle. Furthermore, the fall velocity of the suspension decreases with increasing volumetric sediment concentration. This effect, called *hindered settling*, results from the interaction between the downward fluid motion induced by each particle on the surrounding fluid and the return flow (i.e. upward fluid motion) following the passage of a particle. As a particle settles down, a volume of fluid equal to the particle volume is displaced upwards. In thick sediment suspension, the drag on each particle tends to oppose the upward fluid displacement.

Notes
1. The fall velocity of a suspension w_s may be estimated as:

$$w_s = (1 - 2.15C_s)(1 - 0.75C_s^{0.33})w_o$$

Fall velocity of a fluid–sediment suspension

where w_o is the terminal velocity of a single particle and C_s is the volumetric sediment concentration. Van Rijn (1993) recommended this formula, derived from experimental work, for sediment concentrations up to 35%.
2. It must be noted that a very dense cloud of particles settling in clear water may fall faster than an individual particle. A very dense cloud tends to behave as a large particle rather than as a suspension.

7.3.4 Effect of turbulence on the settling velocity

In turbulent flows, several researchers discussed the effects of turbulence upon the sediment settling velocity. Graf (1971) presented a comprehensive review of the effects of turbulence on suspended solid particles. Recently, Nielsen (1993) suggested that the fall velocity of sediment particles increases or decreases depending upon turbulence intensity, the particle density, and the characteristic length scale and time scale of the turbulence. Although the subject is not yet fully understood, it is agreed that turbulence may affect drastically the particle settling motion.

7.4 Angle of repose

Considering the stability of a single particle in a horizontal plane, the threshold condition (for motion) is achieved when the centre of gravity of the particle is vertically above the point of contact. The critical angle at which motion occurs is called the *angle of repose* ϕ_s.

Angle of repose

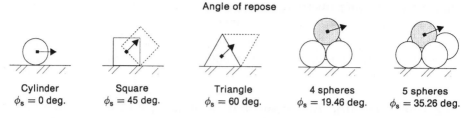

| Cylinder | Square | Triangle | 4 spheres | 5 spheres |
| $\phi_s = 0$ deg. | $\phi_s = 45$ deg. | $\phi_s = 60$ deg. | $\phi_s = 19.46$ deg. | $\phi_s = 35.26$ deg. |

Fig. 7.5 Examples of angle of repose.

The angle of repose is a function of the particle shape and, on a flat surface, it increases with angularity. Typical examples are shown in Fig. 7.5. For sediment particles, the angle of repose usually ranges from 26° to 42°. For sands, ϕ_s is typically between 26° and 34°. Van Rijn (1993) recommended using more conservative values (i.e. larger values) for the design of stable channels (Table 7.5).

Table 7.5 Angle of repose for stable channel design

| d_s | ϕ_s Rounded particles | ϕ_s Angular particles | Comments |
| (mm) | (degrees) | (degrees) | |
(1)	(2)	(3)	(4)
< 1	30	35	
5	32	37	Gravel
10	35	40	Gravel
50	37	42	Gravel
> 100	40	45	e.g. cobble, boulders

Note: Silicate material, see van Rijn (1993).

Note

For a two-dimensional polygon, the angle of repose equals 180° divided by the number of sides of the polygon. For example: $\phi_s = 60°$ for an equilateral triangle, $\phi_s = 0$ for a circle (Fig. 7.5).

7.5 Laboratory measurements

7.5.1 Particle size distribution

In the laboratory, particle size distributions may be determined by direct measurements or indirect methods.

Direct measurements include immersion and displacement volume measurements, direct measurements of particle diameter, and semi-direct measurements of particle sizes using sieves.

Indirect methods of particle size measurements relate fall velocity measurements to particle size. The most common methods are the visual accumulation tube (VAT), the bottom withdrawal tube (BWT) and the pipette. The former (VAT) is used only for sands. The last two methods are used only for silts and clays.

7.5.2 Concentration of suspended sediments

Suspended sediment concentrations may be measured from 'representative' samples of the sediment-laden flow. The sampling techniques may be: instantaneous sampling, point sampling or depth-integrated sampling. Graf (1971) and Julien (1995) reviewed various techniques.

7.6 Exercises

Numerical solutions to some of these exercises are available on the Web at www.arnoldpublishers.com/support/chanson

Bed forms

What is the difference between ripples and dunes? Can dune formation occur with wind-blown sands?

In what direction do antidunes migrate? Can antidunes be observed with wind-blown sands?

In a natural stream, can dunes form in supercritical flows?

Considering a natural stream, the water discharge is 6.4 m²/s and the observed flow depth is 4.2 m. What is the most likely type of bed form with a movable bed?

In a natural stream, the flow velocity is 4.1 m/s and the observed flow depth is 0.8 m. What is the most likely type of bed form with a movable bed? In what direction will the bed forms migrate (i.e. upstream or downstream)?

Considering a 2.3 m wide creek, the water discharge is 1.5 m³/s and the observed flow depth is 0.35 m. What is the most likely type of bed form with a movable bed? In what direction will the bed forms migrate (i.e. upstream or downstream)?

Sediment properties

The dry density of a sand mixture is 1655 kg/m³. Calculate (and give units): (a) the sand mixture porosity and (b) the wet density of the mixture.

The characteristic grain sizes of a sediment mixture are: $d_{10} = 0.1$ mm, $d_{50} = 5.55$ mm, $d_{90} = 1.1$ mm. Indicate the type of sediment material. Calculate: (a) the sedimentological size parameters ϕ corresponding to d_{10}, d_{50} and d_{90}, (b) the sorting coefficient and (c) the dry sediment mixture density.

Settling velocity

Considering a spherical particle (density ρ_s) settling in still water, list all the forces acting on the particle at equilibrium. Write the motion equation for the settling particle at equilibrium. Deduce the analytical expression of the settling velocity.

Considering a 1.1 mm sediment particle settling in water at 20°C, calculate the settling velocity using (a) the formula of Gibbs *et al.* (1971) and (b) the semi-analytical expression derived in the present book:

$$w_0 = 10 \frac{-30\nu + \sqrt{900\nu^2 + gd_s^2(s-1)(0.003869 + 2.480d_s)}}{0.011607 + 7.4405d_s} \qquad \text{Gibbs } et\ al.\ (1971)$$

$$w_0 = -\sqrt{\frac{4gd_s}{3\left(\dfrac{24\mu}{\rho|w_0|d_s} + 1.5\right)}(s-1)} \qquad \text{This book}$$

The size distribution of a sandy mixture is recorded using a settling tube in which the settling time in water over a known settling distance is measured. The results are:

Settling rate (cm/s)	35	12.0	10.0	8.0	5.0	4.0	3.0	2.8
Mass (g)	1.22	2.50	9.20	15.10	25.98	55.60	72.81	81.09

Settling rate (cm/s)	2.39	2.1	1.92	1.8	1.65	1.5	1.43	1.34
Mass (g)	89.80	93.01	96.03	96.90	98.01	99.28	99.90	100.00

Deduce the median grain size and the sorting coefficient $S = \sqrt{d_{90}/d_{10}}$.

Considering a light cloud of sediment particles settling in still water, the particle size is 0.45 mm and the volumetric sediment concentration is 5%. Calculate: (a) the terminal velocity of a single particle and (b) the fall velocity of the suspension.

Inception of sediment motion, occurrence of bed-load motion

8.1 Introduction

In this chapter, the conditions for the onset of sediment transport are reviewed. First, the threshold of bed load motion is described: under what conditions do sediment particles start to roll along the bed? In the next chapter the inception of sediment suspension will be presented.

Accurate estimate of the onset of sediment motion is required: (1) to prevent erosion of channel bed; (2) to predict the risks of scouring below foundations (e.g. bridge pier foundations); and (3) to select rock armour material.

> **Note**
> In practice, bed-load transport starts occurring at lower velocities than sediment suspension, for a given bed geometry and particle size distribution.

8.2 Hydraulics of alluvial streams

8.2.1 Introduction

In most cases, river and stream flows behave as steady flows and uniform equilibrium flow conditions are reached. Application of the momentum equation provides an expression of the mean flow velocity V (at equilibrium):

$$V = \sqrt{\frac{8g}{f}} \sqrt{\frac{D_H}{4} \sin \theta} \qquad \text{Uniform equilibrium flow} \qquad (8.1)$$

where f is the Darcy friction factor, θ is the bed slope and D_H is the hydraulic diameter.

For alluvial streams, the knowledge of the mean flow velocity is insufficient to predict accurately the occurrence of sediment motion and associated risks of scouring.

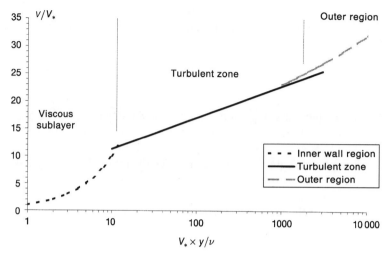

Fig. 8.1 Velocity distribution in turbulent flows.

The knowledge of the velocity profile and more specifically of the velocity next to the channel bed is required.

8.2.2 Velocity distributions in turbulent flows

Considering a turbulent flow along a 'smooth' boundary, the flow field can be divided into three regions (Fig. 8.1): the *inner wall region* (i.e. 'viscous sublayer') next to the wall where the turbulent stress is negligible and the viscous stress is large, the '*outer region*' where the turbulent stress is large and the viscous stress is small, and an overlap region, sometimes called the '*turbulent zone*'.

The thickness of the inner wall region is about $(10\nu/V_*)$, where ν is the fluid kinematic viscosity (i.e. $\nu = \mu/\rho$) and V_* is the shear velocity.

Notes

1. The shear velocity is defined as: $V_* = \sqrt{\tau_o/\rho}$ where τ_o is the mean bed shear stress and ρ is the fluid density.

2. In uniform equilibrium flow down an open channel, the average shear velocity equals:

$$V_* = \sqrt{g\frac{D_H}{4}\sin\theta} \qquad \text{Uniform equilibrium flow}$$

where D_H is the hydraulic diameter and θ is the bed slope.

Application: velocity profile in a turbulent boundary layer

For a turbulent boundary layer flow along a smooth boundary with zero pressure gradient, the velocity distribution follows (e.g. Schlichting 1979):

$$\frac{V}{V_*} = \frac{V_* y}{\nu} \qquad \text{Viscous sublayer:} \quad \frac{V_* y}{\nu} < 5$$

$$\frac{V}{V_*} = \frac{1}{K}\ln\left(\frac{V_* y}{\nu}\right) + D_1 \qquad \text{Turbulent zone: } 30 \text{ to } 70 < \frac{V_* y}{\nu} \text{ and } \frac{y}{\delta} < 0.1 \text{ to } 0.15$$

$$\frac{V_{max} - V}{V_*} = -\frac{1}{K}\ln\left(\frac{y}{\delta}\right) \qquad \text{Outer region: } \frac{y}{\delta} > 0.1 \text{ to } 0.15$$

where V is the velocity at a distance y measured normal to the boundary, δ is the boundary layer thickness, V_{max} is the maximum velocity at the outer edge of the boundary layer (i.e. free-stream velocity), K is the von Karman constant (K = 0.40) and D_1 is a constant ($D_1 = 5.5$, Schlichting 1979).

The turbulent zone equation is called the logarithmic profile or the 'law of the wall'. The outer region equation is called the 'velocity defect law' or 'outer law'. Coles (1956) extended the logarithmic profile to the outer region by adding a 'wake law' term to the right-hand side term:

$$\frac{V}{V_*} = \frac{1}{K}\ln\left(\frac{V_* y}{\nu}\right) + D_1 + \frac{\Pi}{K}Wa\left(\frac{y}{\delta}\right)$$

$$\text{Turbulent zone and outer region: } 30 \text{ to } 70 < \frac{V_* y}{\nu}$$

where Π is the wake parameter, and Wa is Coles' wake function, originally estimated as (Coles 1956):

$$Wa\left(\frac{y}{\delta}\right) = 2\sin^2\left(\frac{\pi}{2}\frac{y}{\delta}\right)$$

Roughness effects
Surface roughness has an important effect on the flow in the wall-dominated region (i.e. inner wall region and turbulent zone). Numerous experiments showed that, for a turbulent boundary layer along a rough plate, the 'law of the wall' follows:

$$\frac{V}{V_*} = \frac{1}{K}\ln\left(\frac{V_* y}{\nu}\right) + D_1 + D_2 \qquad \text{Turbulent zone: } \frac{y}{\delta} < 0.1 \text{ to } 0.15$$

where D_2 is a function of the roughness height, of roughness shape and spacing (e.g. Schlichting 1979, Schetz 1993). For smooth turbulent flows, D_2 equals zero.

In the turbulent zone, the roughness effect (i.e. $D_2 < 0$) implies a 'downward shift' of the velocity distribution (i.e. law of the wall). For large roughness, the laminar sublayer (i.e. inner region) disappears. The flow is said to be 'fully-rough' and D_2 may be estimated as (Schlichting 1979, p. 620):

$$D_2 = 3 - \frac{1}{K}\ln\left(\frac{k_s V_*}{\nu}\right) \qquad \text{Fully-rough turbulent flow}$$

where k_s is the equivalent roughness height. After transformation, the velocity distribution in the turbulent zone for fully-rough turbulent flow becomes:

$$\frac{V}{V_*} = \frac{1}{K}\ln\left(\frac{y}{k_s}\right) + 8.5$$

$$\text{turbulent zone: } \frac{y}{\delta} < 0.1 \text{ to } 0.15 \text{ for fully-rough turbulent flow}$$

Discussion

In a turbulent boundary layer, the velocity distribution may be approximated by a power law function:

$$\frac{V}{V_{\text{max}}} = \left(\frac{y}{\delta}\right)^{1/N}$$

where N is a function of the boundary roughness.

For smooth turbulent flows, $N = 7$ (e.g. Schlichting 1979). For uniform equilibrium flow in open channels, $N = K\sqrt{8/f}$, where f is the Darcy friction factor (Chen 1990).

8.2.3 Velocity profiles in alluvial streams

Most river flows are turbulent and the velocity profile is fully-developed in rivers and streams: i.e. the boundary layer thickness equals the flow depth.

For an alluvial stream, the bed roughness effect might be substantial (e.g. in a gravel-bed stream) and, hence, the complete velocity profile (and the bottom shear stress) is affected by the ratio of the sediment size to the inner wall region thickness: i.e. $d_s/(10\nu/V_*)$.

If the sediment size is small compared with the sub-layer thickness (i.e. $V_* d_s/\nu < 4$ to 5), the flow is *smooth turbulent*. If the sediment size is much larger than the sub-layer thickness (i.e. $V_* d_s/\nu > 75$ to 100), the flow is called *fully-rough turbulent*. For 4 to 5 $< V_* d_s/\nu <$ 75 to 100, the turbulent flow regime is a *transition* regime.

Note

The law of flow resistance is related to the type of turbulent flow regime. For smooth turbulent flows, the friction factor is independent of the boundary roughness and it is a function of the mean flow Reynolds number only.

For fully-rough turbulent flows, the flow resistance is independent of the Reynolds number and the friction factor is a function only of the relative roughness height.

Applications

1. For a wide river, calculate the bed shear stress and shear velocity for a steady uniform flow with a 1.5 m depth and a bed slope of 0.0003.

Solution

In uniform equilibrium flows, the shear velocity equals:

$$V_* = \sqrt{g\frac{D_{\text{H}}}{4}\sin\theta} \qquad \text{Uniform equilibrium flow}$$

where D_{H} is the hydraulic diameter. For a wide channel the hydraulic diameter is about four times the flow depth d and the shear velocity becomes:

$$V_* \approx \sqrt{gd\sin\theta} \qquad \text{Uniform equilibrium flow}$$

For the given flow conditions, $V_* = 0.066$ m/s and the mean bed shear stress equals: $\tau_o = 4.40$ Pa.

2. Considering an alluvial stream, the longitudinal bed slope is 0.001 and the river width is about 65 m. The flow in the river is fully-rough turbulent. Velocity measurements were performed at various distances from the bed and showed that the velocity profile is approximately logarithmic. The data are:

Elevation y (m)   0.02 0.03 0.04 0.05 0.07 0.10 0.20 0.30 0.40 0.50 0.60
Velocity V (m/s) 1.32 1.47 1.54 1.63 1.75 1.85 2.10 2.23 2.33 2.41 2.47

Plot the velocity profile as $V = f(\ln(y))$. Deduce the shear velocity, the characteristic grain size and the flow rate.

In a first approximation, the characteristic grain size may be assumed to be the equivalent roughness height.

Solution
First, plot the velocity distribution.

For a fully-rough turbulent flow, the logarithmic velocity profile may be estimated as (Schlichting 1979, p. 619):

$$\frac{V}{V_*} = \frac{1}{K} \ln\left(\frac{y}{k_s}\right) + 8.5 \qquad \text{turbulent zone and fully-rough turbulent flow}$$

where V is the local velocity, k_s is the equivalent roughness height and K is the von Karman constant (K $= 0.4$).

The relationship $V = f(\ln(y))$ is a straight line, and the slope and constant of the linear relationship are respectively:

$$\text{Slope: } \frac{V_*}{K}$$

$$\text{Constant: } V_*\left(8.5 - \frac{\ln(k_s)}{K}\right)$$

For the experiment data, the slope and constant of the relationship $V = f(\ln(y))$ give the shear velocity $V_* = 0.135$ m/s and the equivalent roughness height $k_s = 0.012$ m.

The estimate of the flow rate requires knowledge of the mean flow velocity and flow depth. Both characteristics may be deduced from the shear velocity and the equivalent roughness height.

In uniform equilibrium flow and for a wide river, the shear velocity and flow depth are related by:

$$V_* \approx \sqrt{gd \sin\theta} \qquad \text{Uniform equilibrium flow}$$

Hence, the flow depth equals: $d = 1.86$ m. Note that the wide channel assumption is justified as $d \ll 65$ m (channel width). Furthermore, the mean flow velocity V and the shear velocity V_* are related by:

$$\frac{V}{V_*} = \sqrt{\frac{8}{f}}$$

where f is the Darcy friction factor, which is a function of the relative roughness height: $k_s/D_H \approx k_s/(4d)$ (for a wide channel). Using the Moody diagram or the Colebrook–White formula, $f = 0.022$ for $k_s/D_H = 0.0016$ and the mean velocity equals: $V = 2.6\,\text{m/s}$. The total flow rate is: $Q = 310\,\text{m}^3/\text{s}$.

8.2.4 Forces acting on a sediment particle

For an open channel flow with a movable bed, the forces acting on each sediment particle are (Fig. 8.2):

- the gravity force $\rho_s g v_s$,
- the buoyancy force $F_b = \rho g v_s$,
- the drag force $C_d \rho A_s V^2/2$,
- the lift force $C_L \rho A_s V^2/2$, and
- the reaction forces of the surrounding grains,

where v_s is the volume of the particle, A_s is a characteristic particle cross-section area, C_d and C_L are the drag and lift coefficients respectively, and V is a characteristic velocity next to the channel bed.

The gravity force and the buoyancy force both act in the vertical direction while the drag force acts in the flow direction and the lift force in the direction perpendicular to the flow direction (Fig. 8.2). The inter-granular forces are related to the grain disposition and packing.

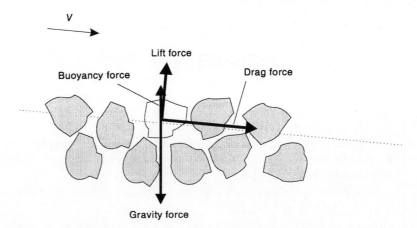

Fig. 8.2 Forces acting on a sediment particle (inter-granular forces not shown for the sake of clarity).

Notes
1. The drag force and the lift force on an object are respectively the integration of the longitudinal component and normal component of the pressure distribution acting on the body. These expressions are often rewritten in terms of

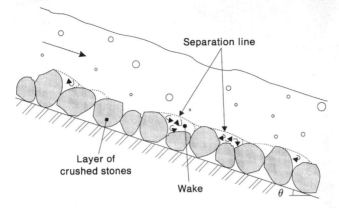

Fig. 8.3 Sketch of turbulent flow patterns around alluvial bed elements.

the mean flow velocity, fluid density and dimensionless coefficients: i.e. the drag and lift coefficients.

2. For an ideal-fluid flow, the drag and lift forces acting on a sphere in a uniform flow are zero (i.e. d'Alembert's paradox).

3. The pressure distribution on a grain is related to the flow pattern around the particle. In turbulent flows, several processes may take place at the particle wall: e.g. wake, flow separation (Fig. 8.3). Hence the pressure distribution around the particle differs from the hydrostatic pressure distribution.

4. For alluvial streams, the characteristic velocity near the sediment bed is the shear velocity V_*.

5. Warning: in many alluvial channels, the shear stress (hence V_*) is not uniform around the wetted perimeter and inertia force takes place wherever the fluid is accelerated.

8.3 Threshold of sediment bed motion

8.3.1 Introduction

The term *threshold of sediment motion* describes the flow conditions and boundary conditions for which the transport of sediment starts to occur. The threshold of sediment motion cannot be defined with an exact (absolute) precision but most experimental observations have provided reasonably accurate and consistent results.

8.3.2 Simple dimensional analysis

The relevant parameters for the analysis of sediment transport threshold are: the bed shear stress τ_o, the sediment density ρ_s, the fluid density ρ, the grain diameter d_s, the gravity acceleration g and the fluid viscosity μ:

$$f_1(\tau_o, \rho, \rho_s, \mu, g, d_s) = 0 \qquad (8.2a)$$

In dimensionless terms, it yields:

$$f_2\left(\frac{\tau_o}{\rho g d_s}; \frac{\rho_s}{\rho}; \frac{d_s\sqrt{\rho\tau_o}}{\mu}\right) = 0 \qquad (8.2b)$$

Discussion

The ratio of the bed shear stress to fluid density is homogeneous (in units) with a velocity squared. Introducing the *shear velocity* V_* defined as:

$$V_* = \sqrt{\frac{\tau_o}{\rho}} \qquad (8.3)$$

equation (8.2a) can be transformed as:

$$f_3\left(\frac{V_*}{\sqrt{gd_s}}; \frac{\rho_s}{\rho}; \rho\frac{d_s V_*}{\mu}\right) = 0 \qquad (8.2c)$$

The first term is a form of Froude number. The second is the relative density (also called specific gravity). The last term is a Reynolds number defined in terms of the grain size and shear velocity. It is often denoted Re_* and called the *shear Reynolds number* or *particle Reynolds number*.

Note that the shear Reynolds number was previously introduced in relation to the velocity distribution in turbulent flows.

Notes

1. The boundary shear stress equals:

$$\tau_o = C_d \tfrac{1}{2}\rho V^2$$

where C_d is the skin friction coefficient and V is the mean flow velocity. In open channel flow, it is common practice to use the Darcy friction factor f, which is related to the skin friction coefficient by: $f = 4C_d$. It yields:

$$\tau_o = \frac{f}{8}\rho V^2$$

2. The shear velocity is a measure of shear stress and velocity gradient near the boundary. It may be rewritten as:

$$\frac{V_*}{V} = \sqrt{\frac{f}{8}}$$

where V is the mean flow velocity and f is the Darcy friction factor.

3. In uniform equilibrium flow down an open channel, the shear velocity equals:

$$V_* = \sqrt{g\frac{D_H}{4}\sin\theta} \qquad \text{Uniform equilibrium flow}$$

where D_H is the hydraulic diameter and θ is the bed slope. And the mean boundary shear stress equals:

$$\tau_o = \rho g \frac{D_H}{4} \sin\theta \qquad \text{Uniform equilibrium flow}$$

For a wide rectangular channel, it yields:

$$V_* \approx \sqrt{gd \sin\theta} \qquad \text{Uniform equilibrium flow in wide rectangular channel}$$

$$\tau_o \approx \rho g d \sin\theta \qquad \text{Uniform equilibrium flow in wide rectangular channel}$$

8.3.3 Experimental observations

Particle movement occurs when the moments of the destabilizing forces (i.e. drag, lift, buoyancy), with respect to the point of contact, become larger than the stabilizing moment of the weight force. The resulting condition is a function of the angle of repose (e.g. van Rijn 1993, p. 4.1).

Experimental observations highlighted the importance of the stability parameter τ_* (which may be derived from dimensional analysis) defined as:

$$\tau_* = \frac{\tau_o}{\rho(s-1)gd_s} \tag{8.4}$$

A critical value of the stability parameter may be defined at the inception of bed motion: i.e. $\tau_* = (\tau_*)_c$. Shields (1936) showed that $(\tau_*)_c$ is primarily a function of the shear Reynolds number $V_* d_s/\nu$ (Fig. 8.4).

Bed load motion occurs for:

$$\tau_* > (\tau_*)_c \qquad \text{Bed load motion} \tag{8.5}$$

In summary: the initiation of bed load transport occurs when the bed shear stress τ_o is larger than a critical value:

$$(\tau_o)_c = \rho(s-1)gd_s(\tau_*)_c$$

Experimental observations showed that $(\tau_*)_c$ is primarily a function of $(d_s V_*/\nu)$ (Fig. 8.4).

Notes
1. The stability parameter τ_* is called commonly the *Shields parameter* after A. Shields who introduced it first (Shields 1936). It is a dimensionless parameter. Note that the Shields parameter is sometimes denoted θ.
2. The stability parameter may be rewritten as:

$$\tau_* = \frac{V_*^2}{(s-1)gd_s} = \frac{\tau_o}{\rho(s-1)gd_s}$$

3. $(\tau_*)_c$ is commonly called the critical Shields parameter.

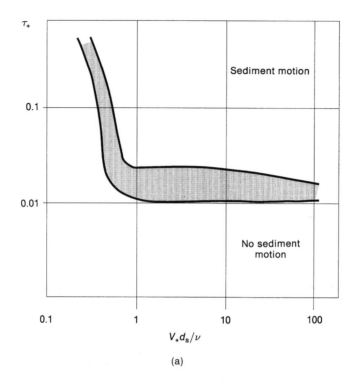

(a)

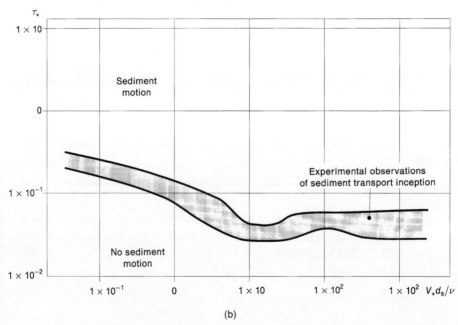

(b)

Fig. 8.4 Threshold of bed load motion (Shields diagram). (a) Shields parameter as a function of the particle Reynolds number for sand in air. (b) Shields parameter as a function of the particle Reynolds number for sediment in water (experimental data reviewed by Yalin and Karahan 1979).

Comments

First let us note that, for given fluid and sediment properties (i.e. ν, ρ, s) and given bed shear stress τ_0, the Shields parameter τ_* decreases with increasing sediment size: i.e. $\tau_* \propto 1/d_s$. Hence, for given flow conditions, sediment motion may occur for small particle sizes while no particle motion occurs for large grain sizes.

It is worth noting also that, for sediment particles in water (Fig. 8.4b), the Shields diagram exhibits different trends corresponding to different turbulent flow regimes:

- the smooth turbulent flow $(Re_* < 4 \text{ to } 5)$ $0.035 < (\tau_*)_c$,
- the transition regime $(4 \text{ to } 5 < Re_* < 75 \text{ to } 100)$ $0.03 < (\tau_*)_c < 0.04$,
- the fully-rough turbulent flow $(75 \text{ to } 100 < Re_*)$ $0.03 < (\tau_*)_c < 0.06$.

For fully-rough turbulent flows, the critical Shields parameter $(\tau_*)_c$ is nearly constant, and the critical bed shear stress for bed load motion becomes linearly proportional to the sediment size:

$$(\tau_0)_c \propto d_s \qquad \text{fully-rough turbulent flow} \qquad (8.6)$$

Notes

1. On the Shields diagram (Fig. 8.4), the Shields parameter and the particle Reynolds number are both related to the shear velocity and the particle size. Some researchers proposed a modified diagram: τ_* as a function of a particle parameter $d_* = (Re_*^2/\tau_*)^{1/3}$.

2. The critical Shields parameter may be estimated as (Julien 1995):

$$(\tau_*)_c = 0.5 \tan \phi_s \qquad \text{for } d_* < 0.3$$

$$(\tau_*)_c = 0.25 d_*^{-0.6} \tan \phi_s \qquad \text{for } 0.3 < d_* < 19$$

$$(\tau_*)_c = 0.013 d_*^{0.4} \tan \phi_s \qquad \text{for } 19 < d_* < 50$$

$$(\tau_*)_c = 0.06 \tan \phi_s \qquad \text{for } 50 < d_*$$

where ϕ_s is the angle of repose and d_* is the dimensionless particle parameter defined as:

$$d_* = d_s \sqrt[3]{\frac{(s-1)g}{\nu^2}}$$

and ν is the fluid kinematic viscosity.

Applications

1. Considering a stream with a flow depth of 1.7 m and a bed slope $\sin \theta = 0.002$, indicate whether a 5 mm gravel bed will be subjected to bed load motion. Find out what is the critical particle size for bed load motion in the stream.

Solution

Assuming a wide rectangular channel and assuming that the stream flow is nearly uniform equilibrium, the shear velocity equals:

$$V_* = \sqrt{g \frac{D_H}{4} \sin \theta} \approx \sqrt{gd \sin \theta} = 0.18 m/s$$

where d is the flow depth. The mean bed shear stress equals: $\tau_0 = 33.3 \, \text{Pa}$.

For a 5 mm gravel bed, the Shields parameter is:

$$\tau_* = \frac{V_*^2}{(s-1)gd_s} = 0.41$$

assuming $s = 2.65$. And the particle Reynolds number equals 910. For these values, Fig. 8.4(b) indicates bed load motion.

To estimate the critical sediment size for bed load motion in the stream, let us assume a gravel bed (i.e. fully-rough turbulent flow). For a rough turbulent flow, the critical sediment size for initiation of bed motion satisfies:

$$(\tau_0)_c = (\tau_*)_c(\rho_s - \rho)gd_s = \tau_0 = 33.3\,\text{Pa}$$

Assuming $(\tau_*)_c = 0.06$, it yields $d_s = 34\,\text{mm}$ and $Re_* = 6200$. Note that the value of Re_* corresponds to a fully-rough turbulent flow and hence equation (8.6) may be used to estimate the critical particle size.

Conclusion: the largest particle size subjected to bed load motion in the stream is 34 mm. For $d_s > 34\,\text{mm}$ the particles will not move.

2. During a storm, the wind blows over a sandy beach (0.5 mm sand particles). The wind boundary layer is about 100 m high at the beach and the free-stream velocity at the outer edge of the wind boundary layer is 35 m/s. Estimate the risk (or not) of beach sand erosion.

Solution
First we must compute the bed shear stress or the shear velocity.

The Reynolds number $V\delta/\nu$ of the boundary layer is 2.3×10^8. The flow is turbulent. As the sand diameter is very small compared with the boundary layer thickness, let us assume that the boundary layer flow is smooth turbulent. In a turbulent boundary layer along a smooth boundary, the mean bed shear stress equals:

$$\tau_0 = 0.0225\rho V^2 \left(\frac{\nu}{V\delta}\right)^{1/4}$$

where ρ is the fluid density, ν is the kinematic viscosity, δ is the boundary layer thickness and V is the free-stream velocity (at the outer edge of the boundary layer) (e.g. Schlichting 1979, p. 637, Streeter and Wylie 1981, p. 215).

For the beach, it yields: $\tau_0 = 0.27\,\text{Pa}$. And the Shields parameter equals: $\tau_* = 0.021$. The particle Reynolds number equals: $Re_* = 16$.

For these values, Fig. 8.4(a) indicates that the flow conditions are at the onset of sediment motion. Sediment transport on the beach will be small.

8.3.4 Discussion

Several parameters may affect the inception of bed load motion: particle size distribution, bed slope, bed forms, material cohesion. Van Rijn (1993) presented a comprehensive review.

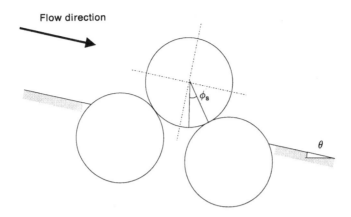

Fig. 8.5 Angle of repose for granular material on steep slope.

(a) The particle size distribution has an effect when the size range is wide because the fine particles will be shielded by the larger particles. After an initial erosion of the fine particles, the coarser particles will form an armour layer preventing further erosion. This process is called *bed armouring*.

(b) On steep channels (Fig. 8.5), the bed slope assists in destabilizing the particles and bed motion occurs at lower bed shear stresses than in flat channels. At the limit, when the bed slope becomes larger than the repose angle (i.e. $\theta > \phi_s$), the grains roll even in the absence of flow: i.e. the bed slope is unstable.

(c) When bed forms (e.g. ripples, dunes) develop, the critical bed shear stress for initiation of bed motion becomes different from that for a flat bed. Indeed, the bed shear stress above a bed form includes a skin friction component plus a form drag component which is related to the non-uniform pressure distribution in the flow surrounding the bed form (see Chapter 12).

(d) For clay and silty sediment beds, the cohesive forces between sediment particles may become important. This causes a substantial increase of the bed resistance to scouring.

A related effect is the existence of seepage through the sediment bed. When the seepage pressure forces exerted on the sediment particles become larger than the submerged weight, the grains will be subjected to some motion. This process is called *bed fluidization*.

Notes

1. For steep channels, van Rijn (1993) suggested defining the Shields parameter as: $[(\sin(\phi_s - \theta)/\sin\phi_s)\tau_*]$, where ϕ_s is the angle of repose, θ is the bed slope and τ_* is the Shields parameter for flat channels.

2. Recently Chiew and Parker (1994) demonstrated that the critical shear velocity for bed motion on steep slopes must be estimated as:

$$\sqrt{\cos\phi_s\left(1 - \frac{\tan\phi_s}{\tan\theta}\right)}\,V_*$$

where V_* is the critical shear velocity for a flat bed. Note that the above formula is also valid for adverse slopes $(\theta < 0)$.

Application: bed fluidization

Considering a seepage flow through a particle bed, the forces exerted on each particle (at equilibrium) are the reaction forces of the surrounding grains, the submerged weight (i.e. weight minus buoyancy) and the pressure force induced by the seepage flow.

At the initiation of fluidization, the submerged weight equals the pressure force component in the vertical direction. For a spherical particle, it yields:

$$-(\rho_s - \rho)gv_s - \frac{dP}{dz}v_s = 0 \qquad \text{Bed fluidization condition}$$

where z is the vertical axis positive upwards, g is the gravity constant, ρ is the water density, ρ_s is the particle density, and v_s is the volume of the particle. The vertical pressure gradient can be derived from Darcy's law:

$$\frac{dP}{dz} = -\rho g \left(1 + \frac{V_z}{K}\right)$$

where K is the hydraulic conductivity (in m/s) and V_z is the vertical component of the seepage velocity.

Bed fluidization occurs for:

$$V_z \geq K(\mathbf{s} - 1) \qquad \text{Bed fluidization}$$

8.4 Exercises

Numerical solutions for some of these exercises are available from the Web at www.arnoldpublishers.com/support/chanson

Considering a smooth turbulent boundary layer, experimental observations suggest that the shear velocity is about 0.21 m/s and the boundary layer thickness equals 0.18 m. Calculate: (a) the viscous sublayer thickness and (b) the range of validity of the outer region velocity profile (i.e. 'velocity defect law'). (The fluid is air at 20°C.)

Considering a stream with a flow depth of 2 m and a bed slope $\sin \theta = 0.0003$, the median grain size of the channel bed is 2.5 mm. Calculate: (a) mean flow velocity, (b) shear velocity, (c) bed shear stress, (d) Reynolds number, (e) shear Reynolds number and (f) Shields parameter. (g) Will sediment motion take place? (Assume that the equivalent roughness height of the channel bed equals the median grain size.)
Solution: Yes (bed-load motion and saltation).

The wind blows over a sandy beach. The wind boundary layer is 200 m high and the free-stream velocity (i.e. at outer edge of boundary layer) is 42 m/s. Deduce the critical particle size for beach sand erosion.

Considering a stream with a flow depth of 0.95 m and a bed slope $\sin \theta = 0.0019$: (a) indicate whether a 15 mm gravel bed will be subjected to bed load motion; (b) find out the critical particle size for bed load motion in the stream.

Solution: particles with $d_{50} = 15$ mm will move: bed-load motion and saltation. Bed load motion occurs for $d_s < 25$–30 mm.

A stream carries a discharge of 81 m³/s. The channel is 42 m wide and the longitudinal bed slope is 15 m per km. The bed consists of a mixture of fine sands ($d_{50} = 0.9$ mm). Assume that uniform equilibrium flow conditions are achieved. (a) Compute the flow depth. (b) Compute the flow velocity. (c) For such a flow, where would you locate a control: i.e. upstream or downstream? (Justify your answer.) (d) Predict the occurrence of bed motion. (e) If bed motion occurs, what is the type of bed form?

Considering the natural stream ($\sin \theta = 0.0009$, $d_{50} = 3$ mm) the flow rate per unit width is 12 m²/s. Will the channel bed sediments be subjected to sediment motion. In the affirmative, indicate whether the sediment motion is bed-load, saltation or suspension. (Assume that the equivalent roughness height of the channel bed equals the median grain size.)

Solution: Yes (bed-load motion)

9

Inception of suspended load motion

9.1 Presentation

In this chapter, the inception of sediment suspension is presented. As discussed in the previous chapter, the inception of sediment motion is related to the shear velocity (or to the bed shear stress). Considering a given channel and bed material, no sediment motion is observed at very low bed shear stress until τ_o exceeds a critical value. For τ_o larger than the critical value, bed load motion takes place. The grain motion along the bed is not smooth, and some particles bounce and jump over the others. With increasing shear velocities, the number of particles bouncing and rebounding increases until the cloud of particles becomes a suspension.

The onset of sediment suspension is not a clearly defined condition.

Note

When particles are moving along the bed with jumps and bounces, the mode of sediment transport is sometimes called '*saltation*' (Fig. 10.1 later).

9.2 Initiation of suspension and critical bed shear stress

Considering a particle in suspension, the particle motion in the direction normal to the bed is related to the balance between the particle fall velocity component ($w_o \cos \theta$) and the turbulent velocity fluctuation in the direction normal to the bed. Turbulence studies (e.g. Hinze 1975, Schlichting 1979) suggested that the turbulent velocity fluctuation is of the same order of magnitude as the shear velocity. With this reasoning, a simple criterion for the initiation of suspension (which does not take into account the effect of bed slope) is:

$$\frac{V_*}{w_o} > \text{critical value} \tag{9.1a}$$

Table 9.1 Criteria for suspended load motion

Reference (1)	Criterion for suspension (2)	Remarks (2)
Bagnold (1966)	$\dfrac{V_*}{w_o} > 1$	as given by van Rijn (1993).
van Rijn (1984b)		Deduced from experimental investigations.
	$\dfrac{V_*}{w_o} > \dfrac{4}{\sqrt[3]{\dfrac{(s-1)g}{\nu^2}} d_s}$	for $1 < \sqrt[3]{\dfrac{(s-1)g}{\nu^2}} d_s \leq 10$ where $d_s = d_{50}$
	$\dfrac{V_*}{w_o} > 0.4$	for $\sqrt[3]{\dfrac{(s-1)g}{\nu^2}} d_s > 10$ where $d_s = d_{50}$
Raudkivi (1990)	$\dfrac{V_*}{w_o} > 0.5$	Note: 'rule of thumb' (Raudkivi 1990, p. 142)! Inception of suspension (i.e. saltation).
	$\dfrac{V_*}{w_o} > 1.2$	Dominant suspended load (i.e. suspension).
Julien (1995)	$\dfrac{V_*}{w_o} > 0.2$	Turbulent water flow over rough boundaries. Inception of suspension.
	$\dfrac{V_*}{w_o} > 2.5$	Dominant suspended load.
Sumer *et al.* (1996)	$\dfrac{V_*^2}{(s-1)gd_s} > 2$	Experimental observations in sheet-flow. Sediment size: $0.13 < d_s < 3\,\text{mm}$.

Notes: $s = \rho_s/\rho$; V_* = shear velocity; w_o = terminal settling velocity.

Several researchers proposed criteria for the onset of suspension (Table 9.1). In a first approximation, suspended sediment load occurs for:

$$\frac{V_*}{w_o} > 0.2 \text{ to } 2 \tag{9.1b}$$

Application

Considering a stream with a flow depth of 3.2 m and a bed slope $\sin\theta = 0.001$, indicate whether a 3 mm gravel bed will be subjected to sediment motion. In the affirmative, indicate whether the sediment motion is bed-load, saltation or suspension.

Solution

Assuming a wide rectangular channel and assuming that the stream flow is nearly uniform equilibrium, the shear velocity equals:

$$V_* = \sqrt{g\,\frac{D_H}{4}\sin\theta} \approx \sqrt{gd\sin\theta} = 0.18\,\text{m/s}$$

where d is the flow depth.

For a 3 mm gravel bed, the Shields parameter is:

$$\tau_* = \frac{V_*^2}{(s-1)gd_s} = 0.67$$

assuming $s = 2.65$. The particle Reynolds number Re_* equals 540. For these values, the Shields diagram predicts sediment motion (see Chapter 8).

Let us estimate the settling velocity. For a single particle in still fluid, the fall velocity equals:

$$w_o = - \sqrt{\frac{4gd_s}{3\left(\dfrac{24\mu}{\rho|w_o|d_s} + 1.5\right)}(s-1)}$$

It yields: $w_o = 0.20\,\text{m/s}$. Hence the ratio V_*/w_o is about unity.

Sediment transport takes place with bed load and saltation. The flow conditions are near the initiation of sediment suspension (equation (9.1b)).

9.3 Onset of hyperconcentrated flow

9.3.1 Definition

Hyperconcentrated flows are sediment-laden flows with large suspended sediment concentrations (i.e. typically more than 1% in volume). Spectacular hyperconcentrated flows are observed in the Yellow River basin (China) with volumetric concentrations larger than 8%: i.e. the sediment mass flow rate being more than 25% of the water mass flow rate.

Hyperconcentrated flows exhibit different flow properties from clear-water flows. For example, the fluid viscosity can no longer be assumed to be that of water (see Chapter 11). In practice, numerous researchers observed that the properties of hyperconcentrated flows differ from those of 'classical' sediment-laden flows when the volumetric sediment concentration exceeds 1 to 3%.

Notes

1. A 'hyperconcentrated flow' is a suspension flow with a large volume sediment concentration (i.e. over 1 to 3%). Sometimes the term might also include 'mud flow' and 'debris flow', but the rheology of these flows differs significantly from that of hyperconcentrated flows. In mud and debris flow, the volumetric sediment concentration might be larger than 50%!
2. Mud is a sticky mixture of fine solid particles and water (i.e. soft wet earth).
3. Debris comprise mainly large boulders, rock fragments, gravel-sized to clay-sized material, tree and wood material that accumulate in creeks. *Debris flow* characterizes *sediment laden flow* in which roughly more than half of the solid volume is coarser than sand. Both flows are often encountered in steep mountainous areas: e.g. Western parts of Canada, east coast of Taiwan, Austria.
4. Hyperconcentrated flows behave as non-Newtonian fluids. The relationship between the fluid shear stress and the velocity gradient is non-linear. The rheology of hyperconcentrations is currently the topic of active research (see also Chapter 11).

(a)

(b)

Fig. 9.1 Examples of extreme land erosion associated with hyperconcentrated suspension flows. (a) 'Moon walk' in Kaohsiung County, Taiwan, Republic of China (September 1995). (b) Draix catchment, France (June 1998). Black marl basin in the Durance catchment.

9.3.2 Discussion

Hyperconcentrated flows are often associated with severe land degradation in the catchment. Two dramatic examples are illustrated in Fig. 9.1. In each catchment, hyperconcentrated flows occur during floods.

Notes

1. In the Yellow River basin (China), the hyperconcentrated flows are characterized by up to 10% of the sediment material being clay and the remaining
   materials consisting of silt and fine sand.
2. In the Durance basin (France), hyperconcentrated sediment concentrations
   may exceed 15% (in volume), particularly in black marl[1] catchments. Meunier
   (1995) observed sediment generation by miniature debris flows in small size
   parts of the slope.

9.4 Exercises

Numerical solutions to these exercises are available from the Web at
www.arnoldpublishers.com/support/chanson

During a storm, the wind blows over a beach (0.2 mm sand particles). The velocity at
the outer edge of the boundary layer is 110 m/s and the boundary layer is 45 m thick.
Calculate: (a) shear velocity, (b) settling velocity of 0.2 mm sand particles in air and (c)
occurrence of suspended-load motion. (Assume that the equivalent roughness height
of the channel bed equals the median grain size.)

Considering a stream with a flow depth of 2.3 m and a bed slope $\sin \theta = 0.002$, the
median grain size of the channel bed is 1.1 mm. Calculate: (a) mean flow velocity,
(b) shear velocity, (c) settling velocity, (d) occurrence of bed-load motion and (e)
occurrence of suspension. (Assume that the equivalent roughness height of the channel bed equals the median grain size.)

---

[1] Also called badlands (in English), 'Marnes Noires' or 'Terres Noires' (in French).

10

Sediment transport mechanisms 1. Bed-load transport

10.1 Introduction

When the bed shear stress exceeds a critical value, sediments are transported in the form of bed-load and suspended load. For bed-load transport, the basic modes of particle motion are *rolling* motion, *sliding* motion and *saltation* motion (Fig. 10.1).

In this chapter, formulations to predict the bed load transport rate are presented. Figure 10.2 shows a natural stream subjected to significant bed-load transport.

Notes
1. Saltation refers to the transport of sediment particles in a series of irregular jumps and bounces along the bed (Fig. 10.1(a)).
2. In this section, predictions of bed-load transport are developed for plane bed. Bed form motion and bed form effects on bed-load transport are not considered in this section (see Chapter 12).

Definitions

The sediment transport rate may be measured by weight (units: N/s), by mass (units: kg/s) or by volume (units: m^3/s). In practice the sediment transport rate is often expressed per unit width and is measured either by mass or by volume. These are related by:

$$\dot{m}_s = \rho_s q_s \qquad (10.1)$$

where $\dot{m}_s$ is the mass sediment flow rate per unit width, q_s is the volumetric sediment discharge per unit width and ρ_s is the specific mass of sediment.

10.2 Empirical correlations of bed-load transport rate

10.2.1 Introduction

Bed load transport occurs when the bed shear stress τ_o exceeds a critical value $(\tau_o)_c$. In

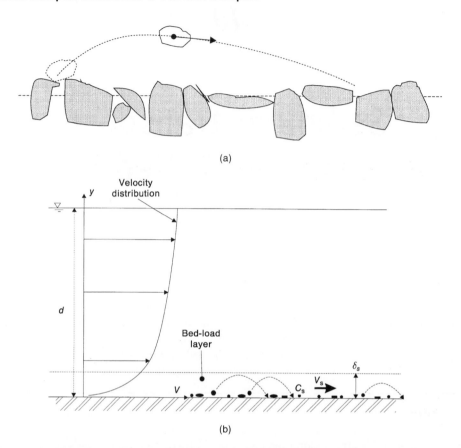

(a)

(b)

Fig. 10.1 Bed-load motion. (a) Sketch of saltation motion. (b) Definition sketch of the bed-load layer.

dimensionless terms, the condition for bed-load motion is:

$$\tau_* > (\tau_*)_c \qquad \text{Bed-load transport} \qquad (10.2)$$

where τ_* is the Shields parameter (i.e. $\tau_* = \tau_0/(\rho(s-1)gd_s)$) and $(\tau_*)_c$ is the critical Shields parameter for initiation of bed load transport (Fig. 8.4).

10.2.2 Empirical bed load transport predictions

Many researchers attempted to predict the rate of bed load transport. The first successful development was proposed by P.F.D. du Boys in 1879. Although his model of sediment transport was incomplete, the proposed relationship for bed-load transport rate (Table 10.1) proved to be in good agreement with a large amount of experimental measurements.

Subsequently, numerous researchers proposed empirical and semi-empirical correlations. Some are listed in Table 10.1. Graf (1971) and van Rijn (1993) discussed their applicability. The most notorious correlations are the Meyer-Peter and Einstein formulae.

(a)

(b)

Fig. 10.2 Bed-load transport in natural streams. (a) South Korrumbyn Creek NSW, Australia looking upstream (25 April 1997). Stream bed 2 km upstream of the Korrumbyn Creek weir, fully-silted today (Appendix A2.1). (b) Cedar Creek QLD, Australia looking downstream (9 December 1992) (courtesy of Mrs J. Hacker). Note the coarse material left after the flood.

Table 10.1 Empirical and semi-empirical correlations of bed load transport

Reference (1)	Formulation (2)	Range (3)	Remarks (4)
Boys (1879)	$q_s = \lambda \tau_o (\tau_o - (\tau_o)_c)$		λ was called the characteristic sediment coefficient.
	$\lambda = \dfrac{0.54}{(\rho_s - \rho)g}$ Schoklitsch (1914)		Laboratory experiments with uniform grains of various kinds of sand and porcelain.
	$\lambda \propto d_s^{-3/4}$ Straub (1935)	$0.125 < d_s < 4\,\text{mm}$	Based upon laboratory data.
Schoklitsch (1930)	$q_s = \lambda'(\sin\theta)^k (q - q_c)$ $q_c = 1.944 \times 10^{-2} d_s (\sin\theta)^{-4/3}$	$0.305 < d_s < 7.02\,\text{mm}$	Based upon laboratory experiments.
Shields (1936)	$\dfrac{q_s}{q} = 10 \dfrac{\sin\theta}{s} \dfrac{\tau_o - (\tau_o)_c}{\rho g(s-1)d_s}$	$1.06 < s < 4.25$ $1.56 < d_s < 2.47\,\text{mm}$	
Einstein (1942)	$\dfrac{q_s}{\sqrt{(s-1)gd_s^3}} =$ $2.15 \exp\left(-0.391\dfrac{\rho(s-1)gd_s}{\tau_o}\right)$	$\dfrac{q_s}{\sqrt{(s-1)gd_s^3}} < 0.4$ $1.25 < s < 4.25$ $0.315 < d_s < 28.6\,\text{mm}$	Laboratory experiments. Weak sediment transport formula for sand mixtures. Note: $d_s \approx d_{35}$ to d_{45}.
Meyer-Peter (1949, 1951)	$\dfrac{\dot{m}^{2/3}\sin\theta}{d_s} - 9.57(\rho g(s-1))^{10/9} =$ $0.462(s-1)\dfrac{\left(\rho g(\dot{m}_s)^2\right)^{2/3}}{d_s}$	$1.25 < s < 4.2$	Laboratory experiments. Uniform grain size distribution.
	$\dfrac{q_s}{\sqrt{(s-1)gd_s^3}} =$ $\left(\dfrac{4\tau_o}{\rho(s-1)gd_s} - 0.188\right)^{3/2}$		Laboratory experiments. Particle mixtures. Note: $d_s \approx d_{50}$.
Einstein (1950)	Design chart $\dfrac{q_s}{\sqrt{(s-1)gd_s^3}} = f\left(\dfrac{\rho(s-1)gd_s}{\tau_o}\right)$	$\dfrac{q_s}{\sqrt{(s-1)gd_s^3}} < 10$ $1.25 < s < 4.25$ $0.315 < d_s < 28.6\,\text{mm}$	Laboratory experiments. For sand mixtures. Note: $d_s \approx d_{35}$ to d_{45}.
Schoklitsch (1950)	$\dot{m}_s = 2500(\sin\theta)^{3/2}(q - q_c)$ $q_c = 0.26(s-1)^{5/3}d_{40}^{3/2}(\sin\theta)^{-7/6}$		Based upon laboratory experiments and field measurements (Danube and Aare rivers).
Nielsen (1992)	$\dfrac{q_s}{\sqrt{(s-1)gd_s^3}} =$ $\left(\dfrac{12\tau_o}{\rho(s-1)gd_s} - 0.05\right)\sqrt{\dfrac{\tau_o}{\rho(s-1)gd_s}}$	$1.25 < s < 4.22$ $0.69 < d_s < 28.7\,\text{mm}$	Re-analysis of laboratory data.

Notes: $\dot{m}$ = mass water flow rate per unit width; $\dot{m}_s$ = mass sediment flow rate per unit width; q = volumetric water discharge; q_s = volumetric sediment discharge per unit width; $(\tau_o)_c$ = critical bed shear stress for initiation of bed load.

Note

P.F.D. du Boys (1847-1924) was a French hydraulic engineer. In 1879, he proposed a bed load transport model, assuming that sediment particles move in sliding layers (Boys 1879).

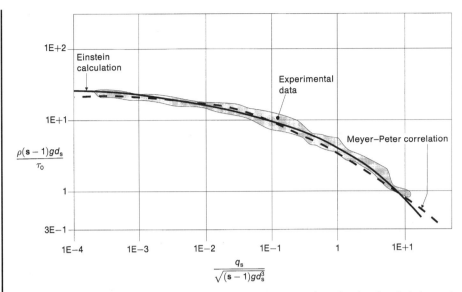

Fig. 10.3 Bed-load transport rate: comparison between Meyer-Peter formula, Einstein calculation and laboratory data (Meyer-Peter *et al.* 1934, Gilbert 1914, Chien 1954).

Discussion

The correlation of Meyer-Peter (1949, 1951) has been popular in Europe. It is considered most appropriate for wide channels (i.e. large width to depth ratios) and coarse material. Einstein's (1942, 1950) formulations derived from physical models of grain saltation, and they have been widely used in America. Both the Meyer-Peter and Einstein correlations give close results (e.g. Graf 1971, p. 150), usually within the accuracy of the data (Fig. 10.3).

It must be noted that empirical correlations should not be used outside of their domain of validity. For example, Engelund and Hansen (1972) indicated explicitly that Einstein's (1950) bed-load transport formula differs significantly from experimental data for large amounts of bed load (i.e. $q_s / \sqrt{(s-1)gd_s^3} > 10$).

10.3 Bed-load calculations

10.3.1 Presentation

Bed-load transport is closely associated with inter-granular forces. It takes place in a thin region of fluid close to the bed (sometimes called the *bed-load layer* or saltation layer) (Fig. 10.1, 10.4). Visual observations suggest that the bed-load particles move within a region of less than 10 to 20 particle-diameter heights.

During the bed-load motion, the moving grains are subjected to hydrodynamic forces, gravity force and inter-granular forces. Conversely the (submerged) weight of the bed load is transferred as a normal stress to the (immobile) bed grains. The

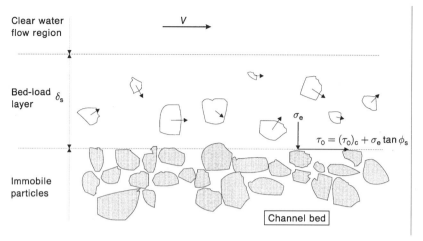

Fig. 10.4 Sketch of bed-load motion at equilibrium.

normal stress σ_e exerted by the bed load on the immobile bed particles is called the *effective stress* and it is proportional to:

$$\sigma_e \propto \rho(s-1)g\cos\theta C_s\delta_s \tag{10.3}$$

where δ_s is the bed-load layer thickness, C_s is the volumetric concentration of sediment in the bed-load layer and θ is the longitudinal bed slope.

The normal stress increases the frictional strength of the sediment bed and the boundary shear stress applied to the top layer of the immobile grains becomes:

$$\tau_0 = (\tau_0)_c + \sigma_e \tan\phi_s \tag{10.4}$$

where $(\tau_0)_c$ is the critical bed shear stress for initiation of bed load and ϕ_s is the angle of repose.

Notes
1. The concept of effective stress and associated bed shear stress (as presented above) derives from the work of Bagnold (1956, 1966).
2. For sediment particles, the angle of repose ranges usually from 26° to 42° and hence $0.5 < \tan\phi_s < 0.9$. For sands, it is common to choose: $\tan\phi_s \approx 0.6$.

10.3.2 Bed load transport rate

The bed load transport rate per unit width may be defined as:

$$q_s = C_s\delta_s V_s \tag{10.5}$$

where V_s is the average sediment velocity in the bed-load layer (Fig. 10.1(b)).

Physically the transport rate is related to the characteristics of the bed-load layer: its mean sediment concentration C_s, its thickness δ_s which is equivalent to the average saltation height measured normal to the bed (Fig. 10.1 and 10.4) and the average speed V_s of sediment moving along the plane bed.

Notes
1. A steady sediment transport in the bed-load layer is sometimes called a (no-suspension) *sheet-flow*.
2. Note that the volumetric sediment concentration has a maximum value. For rounded grains, the maximum sediment concentration is 0.65.

10.3.3 Discussion

Several researchers have proposed formulae to estimate the characteristics of the bed-load layer (Table 10.2). Figure 10.5 presents a comparison between two formulae. Overall the results are not very consistent. In practice there is still great uncertainty on the prediction of bed load transport.

Note that the correlations of van Rijn (1984a) are probably more accurate to estimate the saltation properties (i.e. C_s, δ_s/d_s and V_s/V_*) (within their range of validity).

Table 10.2 Bed-load transport rate calculations

Reference (1)	Bed-load layer characteristics (2)	Remarks (3)
Fernandez-Luque and van Beek (1976)	$\dfrac{V_s}{V_*} = 9.2\left(1 - 0.7\sqrt{\dfrac{(\tau_*)_c}{\tau_*}}\right)$	Laboratory data $1.34 \le s \le 4.58$ $0.9 \le d_s \le 3.3\,\mathrm{mm}$ $0.08 \le d \le 0.12\,\mathrm{m}$
Nielsen (1992)	$C_s = 0.65$ $\dfrac{\delta_s}{d_s} = 2.5(\tau_* - (\tau_*)_c)$ $\dfrac{V_s}{V_*} = 4.8$	Simplified model.
Van Rijn (1984a, 1993)	$C_s = \dfrac{0.117}{d_s}\left(\dfrac{\nu^2}{(s-1)g}\right)^{1/3}\left(\dfrac{\tau_*}{(\tau_*)_c} - 1\right)$ $\dfrac{\delta_s}{d_s} = 0.3\left(d_s\left(\dfrac{(s-1)g}{\nu^2}\right)^{1/3}\right)^{0.7}\sqrt{\dfrac{\tau_*}{(\tau_*)_c} - 1}$ $\dfrac{V_s}{V_*} = 9 + 2.6\log_{10}\left(d_s\left(\dfrac{(s-1)g}{\nu^2}\right)^{1/3}\right) - 8\sqrt{\dfrac{(\tau_*)_c}{\tau_*}}$	For $\dfrac{\tau_*}{(\tau_*)_c} < 2$ and $d_s = d_{50}$. Based on laboratory data $0.2 \le d_s \le 2\,\mathrm{mm}$ $d > 0.1\,\mathrm{m}$ $Fr < 0.9$
	$C_s = \dfrac{0.117}{d_s}\left(\dfrac{\nu^2}{(s-1)g}\right)^{1/3}\left(\dfrac{\tau_*}{(\tau_*)_c} - 1\right)$ $\dfrac{\delta_s}{d_s} = 0.3\left(d_s\left(\dfrac{(s-1)g}{\nu^2}\right)^{1/3}\right)^{0.7}\sqrt{\dfrac{\tau_*}{(\tau_*)_c} - 1}$ $\dfrac{V_s}{V_*} = 7$	$d_s = d_{50}$. Based on laboratory data $0.2 \le d_s \le 2\,\mathrm{mm}$ $d > 0.1\,\mathrm{m}$ $Fr < 0.9$

Notes: V_* = shear velocity; $(\tau_*)_c$ = critical Shields parameter for initiation of bed load.

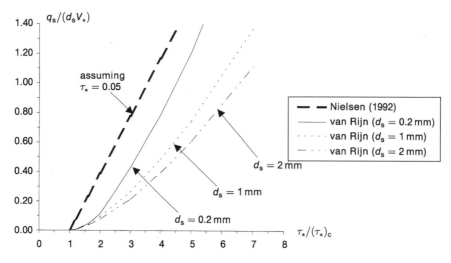

Fig. 10.5 Dimensionless bed-load transport rate $q_s/(d_s V_*)$ as a function of the dimensionless Shields parameter $\tau_*/(\tau_*)_c$ (Table 10.2).

Notes
1. The calculations detailed in Table 10.2 apply to flat channels (i.e. $\sin \theta < 0.001$ to 0.01) and in absence of bed forms (i.e. plane bed only).
2. For steep channels several authors showed a strong increase of bed-load transport rate. It is believed that the longitudinal bed slope affects the transport rate because the threshold conditions (i.e. initiation of bed load) are affected by the bed slope, the sediment motion is changed with steep bed slope and the velocity distribution near the bed is modified.

Discussion
The prediction of bed-load transport rate is *not* an accurate prediction. One researcher (van Rijn 1984a) stated explicitly that:

> the overall inaccuracy [...] may not be less than a factor 2

10.4 Applications

10.4.1 Application No. 1

The bed-load transport rate must be estimated for the Danube river (Central Europe) at a particular cross-section. The known hydraulic data are: flow rate of about 530 m³/s, flow depth of 4.27 m, bed slope being about 0.0011. The channel bed is a sediment mixture with a median grain size of 0.012 m and the channel width is about 34 m.

Predict the sediment-load rate using the Meyer-Peter correlation, the Einstein formula, and equation (10.5) using both Nielsen and van Rijn coefficients.

First calculations

Assuming a wide channel, the mean shear stress and shear velocity equals respectively:

$$T_o = \rho g d \sin \theta = 998.2 \times 9.81 \times 4.27 \times 0.0011 = 46.0 \, \text{Pa}$$

$$V_* = \sqrt{gd \sin \theta} = 0.215 \, \text{m/s}$$

The Shields parameter equals:

$$T_* = \frac{T_o}{\rho(s-1)gd_s} = \frac{46.0}{998.2 \times 1.65 \times 9.81 \times 0.012} = 0.237$$

assuming $s = 2.65$ (quartz particles). And the particle Reynolds number is:

$$Re_* = \frac{V_* d_s}{\nu} = \frac{0.215 \times 0.012}{1.007 \times 10^{-6}} = 2558$$

For these flow conditions (T_*, Re_*), the Shields diagram predicts sediment motion:

$$T_* = 0.237 > (T_*)_c \approx 0.05$$

Note that V_*/w_o is small and the flow conditions are near the initiation of suspension. In the first approximation, the suspended sediment transport will be neglected.

Approach No. 1: Meyer-Peter correlation

For the hydraulic flow conditions, the dimensionless parameter used for the Meyer-Peter formula is:

$$\frac{\rho(s-1)gd_s}{T_o} = \frac{998.2 \times 1.65 \times 9.81 \times 0.012}{46} = 4.215$$

Application of the Meyer-Peter formula (for a sediment mixture) leads to:

$$\frac{q_s}{\sqrt{(s-1)gd_s^3}} = 0.66$$

Hence: $q_s = 0.0035 \, \text{m}^2/\text{s}$.

Approach No. 2: Einstein function

For the hydraulic flow conditions, the dimensionless parameter used for the Einstein formula is:

$$\frac{\rho(s-1)gd_{35}}{T_o}$$

In the absence of information on the grain size distribution, it will be assumed: $d_{35} \approx d_{50}$. It yields:

$$\frac{\rho(s-1)gd_{35}}{T_o} \approx 4.215$$

For a sediment mixture, the Einstein formula gives:

$$\frac{q_s}{\sqrt{(s-1)gd_s^3}} = 0.85$$

Hence: $q_s = 0.0045 \, \text{m}^2/\text{s}$.

Approach No. 3: bed-load calculation (equation (10.5))

The bed load transport rate per unit width equals:

$$q_s = C_s \delta_s V_s \tag{10.5}$$

Using Nielsen's (1992) simplified model, it yields:

$$q_s = 0.65 \times 2.5(\tau_* - (\tau_*)_c)d_s 4.8 V_*$$

$$= 0.65 \times 2.5 \times (0.237 - 0.05) \times 0.012 \times 4.8 \times 0.215 = 0.0038 \, \text{m}^2/\text{s}$$

With the correlation of van Rijn (1984a), the saltation properties are:

$$C_s = 0.00145$$

$$\frac{\delta_s}{d_s} = 31.6$$

$$\frac{V_s}{V_*} = 7$$

and the sediment transport rate is: $q_s = 0.00083 \, \text{m}^2/\text{s}$.

Summary

	Meyer-Peter	Einstein	Eq. (10.5) (Nielsen)	Eq. (10.5) (van Rijn)
Q_s (m^3/s)	0.119	0.153	0.128	0.0281 (?)
q_s (m^2/s)	0.0035	0.0045	0.0038	0.00083 (?)
Mass sediment rate (kg/s)	314	405	339	74.6 (?)

Four formulae were applied to predict the sediment transport rate by bed-load. Three formulae give reasonably close results. Let us review the various formulae.

Graf (1971) commented that the Meyer-Peter formula 'should be used carefully at [...] high mass flow rate', emphasizing that most experiments with large flow rates used by Meyer-Peter *et al.* (1934) were performed with light sediment particles (i.e. lignite breeze, s = 1.25). Graf stated that one advantage of the Meyer-Peter formula is that 'it has been tested with large grains'.

The Einstein formula has been established with more varied experimental data than the Meyer-Peter formula. The present application is within the range of validity of the data (i.e. $q_s/\sqrt{(s-1)gd_s^3} = 0.85 \ll 10$).

Equation (10.5) gives reasonably good overall results using Nielsen's (1992) simplified model. In the present application, the grain size (0.012 m) is large compared with the largest grain size used by van Rijn (1984a) to validate his formulae (i.e. $d_s \leq 0.002$ m). Hence it is understandable that equation (10.5) with van Rijn's formulae is inaccurate (and not applicable).

For the present application, it might be recommended to consider the Meyer-Peter formula, which was developed and tested in Europe.

Note

All bed-load formulae would predict the maximum bed-load transport rate for a stream in equilibrium. This transport capacity may or may not be equal to the actual bed-load if the channel is subjected to degradation or aggradation (see Chapter 12).

10.4.2 Application No. 2

A wide stream has a depth of 0.6 m and the bed slope is 0.0008. The bed consists of a mixture of heavy particles ($\rho_s = 2980 \, \text{kg/m}^3$) with a median particle size $d_{50} = 950 \, \mu\text{m}$.

Compute the bed-load transport rate using the formulae of Meyer-Peter and Einstein, and equation (10.5) for uniform equilibrium flow conditions.

First calculations

Assuming a wide channel, the mean shear stress and shear velocity equals respectively:

$$\tau_o = \rho g d \sin \theta = 998.2 \times 9.80 \times 0.6 \times 0.0008 = 4.70 \, \text{Pa}$$

$$V_* = \sqrt{gd \sin \theta} = 0.069 \, \text{m/s}$$

The Shields parameter equals:

$$\tau_* = \frac{\tau_o}{\rho(s-1)gd_s} = \frac{4.70}{998.2 \times 1.98 \times 9.80 \times 0.00095} = 0.255$$

The particle Reynolds number is:

$$Re_* = \frac{V_* d_s}{\nu} = \frac{0.069 \times 0.00095}{1.007 \times 10^{-6}} = 65.1$$

For these flow conditions (τ_*, Re_*), the Shields diagram predicts sediment motion:

$$\tau_* = 0.255 > (\tau_*)_c \approx 0.05$$

Note that V_*/w_o is less than 0.7. In the first approximation, the suspended sediment transport is negligible.

Approach No. 1: Meyer-Peter correlation

For the hydraulic flow conditions, the dimensionless parameter used for the Meyer-Peter formula is:

$$\frac{\rho(s-1)gd_s}{\tau_o} = \frac{998.2 \times 1.98 \times 9.80 \times 0.00095}{4.7} = 3.91$$

Application of the Meyer-Peter formula (for a sediment mixture) leads to:

$$\frac{q_s}{\sqrt{(s-1)gd_s^3}} = 0.76$$

Hence: $q_s = 9.82 \times 10^{-5} \, \text{m}^2/\text{s}$.

Approach No. 2: Einstein function

For the hydraulic flow conditions, the Einstein formula is based on the d_{35} grain size. In the absence of information on the grain size distribution, it will be assumed: $d_{35} \approx d_{50}$.

For a sediment mixture, the Einstein formula gives:

$$\frac{q_s}{\sqrt{(s-1)gd_s^3}} \approx 1$$

Hence: $q_s = 1.29 \times 10^{-4}\,\mathrm{m^2/s}$.

Approach No. 3: bed-load calculation (equation (10.5))

The bed-load transport rate per unit width equals:

$$q_s = C_s \delta_s V_s \tag{10.5}$$

Using Nielsen's (1992) simplified model, it yields:

$$q_s = 0.65 \times 2.5 \times (\tau_* - (\tau_*)_c)d_s 4.8 V_*$$

$$= 0.65 \times 2.5 \times (0.255 - 0.05) \times 0.00095 \times 4.8 \times 0.069 = 1.05 \times 10^{-4}\,\mathrm{m^2/s}$$

With the correlation of van Rijn (1984a), the saltation properties are:

$$C_s = 0.019$$

$$\frac{\delta_s}{d_s} = 5.848$$

$$\frac{V_s}{V_*} = 7$$

and the sediment transport rate is: $q_s = 0.5 \times 10^{-4}\,\mathrm{m^2/s}$.

Summary

	Meyer-Peter	Einstein	Eq. (10.5) (Nielsen)	Eq. (10.5) (van Rijn)
q_s (m²/s)	0.98×10^{-5}	1.2×10^{-4}	1.05×10^{-4}	0.5×10^{-4}
Mass sediment rate (kg/s/m)	0.29	0.38	0.31	0.15

All the formulae give consistent results (within the accuracy of the calculations!).

For small-size particles, (i.e. $d_{50} < 2\,\mathrm{mm}$), the formulae of van Rijn are recommended because they were validated with over 500 laboratory and field data.

Note, however, that 'the overall inaccuracy of the predicted [bed-load] transport rates may not be less than a factor 2' (van Rijn 1984a, p. 1453).

10.4.3 Application No. 3

The North Fork Toutle river flows on the north-west slopes of Mount St. Helens (USA), which was devastated in May 1980 by a volcanic eruption. Since 1980 the river has carried a large volume of sediment.

Measurements were performed on 28 March 1989 at the Hoffstadt Creek bridge. At that location the river is 18 m wide. Hydraulic measurements indicated that the flow depth was 0.83 m, the depth-averaged velocity was 3.06 m/s and the bed slope was $\sin\theta = 0.0077$. The channel bed is a sediment mixture with a median grain size of 15 mm and $d_{84} = 55\,\mathrm{mm}$.

Predict the sediment-load rate using the Meyer-Peter correlation, the Einstein formula, and equation (10.5) using both Nielsen and van Rijn coefficients.

First calculations

Assuming a wide channel ($d = 0.83\,\text{m} \ll 18\,\text{m}$), the mean shear stress and shear velocity equals respectively:

$$\tau_o = \rho g d \sin\theta = 998.2 \times 9.81 \times 0.83 \times 0.0077 = 62.6\,\text{Pa}$$

$$V_* = \sqrt{g d \sin\theta} = 0.25\,\text{m/s}$$

The Shields parameter equals:

$$\tau_* = \frac{\tau_o}{\rho(s-1)g d_s} = 0.258$$

assuming $s = 2.65$ (quartz particles) and using $d_s = d_{50}$. And the particle Reynolds number is:

$$Re_* = \frac{V_* d_s}{\nu} = 3725$$

For these flow conditions (τ_*, Re_*), the Shields diagram predicts sediment motion:

$$\tau_* = 0.258 > (\tau_*)_c \approx 0.05$$

Note that V_*/w_o is small ($V_*/w_o \approx 0.5$) and the flow conditions are near the initiation of suspension. In the first approximation, the suspended sediment load will be neglected.

Approach No. 1: Meyer-Peter correlation

For the hydraulic flow conditions, the dimensionless parameter used for the Meyer-Peter formula is:

$$\frac{\rho(s-1)g d_s}{\tau_o} = 3.87$$

using $d_s = d_{50}$. Application of the Meyer-Peter formula (for a sediment mixture) would lead to:

$$\frac{q_s}{\sqrt{(s-1)g d_s^3}} = 0.78$$

Hence: $q_s = 0.0057\,\text{m}^2/\text{s}$.

Approach No. 2: Einstein function

For the hydraulic flow conditions, the dimensionless parameter used for the Einstein formula is:

$$\frac{\rho(s-1)g d_{35}}{\tau_o}$$

In the absence of information on the grain size distribution, we assume: $d_{35} \approx d_{50}$.
For a sediment mixture, the Einstein formula gives:

$$\frac{q_s}{\sqrt{(s-1)g d_s^3}} \sim 40$$

But note that the flow conditions are outside of the range of validity of the formula. That is, the Einstein formula should not be used.

Approach No. 3: bed-load calculation (equation (10.5))

The bed-load transport rate per unit width equals:

$$q_s = C_s \delta_s V_s \qquad (10.5)$$

Using Nielsen's (1992) simplified model, it yields:

$$q_s = 0.65 \times 2.5 \times (\tau_* - (\tau_*)_c) d_s 4.8 V_* = 0.0061 \, \text{m}^2/\text{s}$$

With the correlation of van Rijn (1984a), the saltation properties are:

$$C_s = 0.00129$$

$$\frac{\delta_s}{d_s} = 38.97$$

$$\frac{V_s}{V_*} = 7$$

And the sediment transport rate is: $q_s = 0.0013 \, \text{m}^2/\text{s}$.

Summary

	Meyer-Peter	Einstein	Eq. (10.5) (Nielsen)	Eq. (10.5) (van Rijn)	Data
Q_s (m^3/s)	0.10	N/A	0.11	0.024	
q_s (m^2/s)	0.0057	N/A	0.0061	0.0023	
Mass sediment rate (kg/s)	274	N/A	290	63	205.2

Pitlick (1992) described in-depth the field study performed at the Hoffstadt Creek bridge on the North Fork Toutle river. On 28 March 1989, the main observations were:

$$d = 0.83 \, \text{m}, \quad V = 3.06 \, \text{m/s}, \quad \sin\theta = 0.0077, \quad f = 0.054, \quad \tau_o = 63 \, \text{N/m}^2,$$

$$C_s = 0.031, \quad \dot{m}_s = 11.4 \, \text{kg/m/s}$$

The channel bed was formed in dunes (up to 0.16 m high).

Discussion

First let us note that two methods of calculations are incorrect: the Einstein formula and equation (10.5) using van Rijn's correlation. The flow conditions and sediment characteristics are outside of the range of applicability of Einstein's formula as $q_s/\sqrt{(s-1)gd_s^3} > 10$. In addition the grain size (0.015 m) is larger than the largest grain size used by van Rijn (1984a) to validate his formulae (i.e. $d_s \le 0.002$ m). Secondly it is worth noting that the Meyer-Peter formula and equation (10.5) using Nielsen's correlations give reasonable predictions.

This last application is an interesting case: it is well documented. The river flow is characterized by heavy bed-load transport and the bed-forms are a significant feature of the channel bed.

10.5 Exercises

Numerical solutions to some of these exercises are available from the Web at
www.arnoldpublishers.com/support/chanson

Considering a 20 m wide channel, the bed slope is 0.00075 and the observed flow
depth is 3.2 m. The channel bed is sandy ($d_s = 0.008$ m). Calculate: (a) mean velocity,
(b) mean boundary shear stress, (c) shear velocity, (d) Shields parameter and (e)
occurrence of bed-load motion. If bed-load motion occurs, calculate: (f) bed-load
layer sediment concentration, (g) bed-load layer thickness, (h) average sediment
velocity in bed-load layer and (i) bed-load transport rate. (Assume that the equivalent
roughness height of the channel bed equals the median grain size. Use the Nielsen
simplified model.)

Considering a wide channel, the discharge is 20 m²/s, the observed flow depth is
4.47 m and the bed slope is 0.001. The channel bed consists of a sand mixture
($d_{50} = 0.011$ m). Calculate the bed-load transport rate using: (a) Meyer-Peter correlation, (b) Einstein function, (c) Nielsen simplified model and (d) van Rijn correlations.
(Assume that the equivalent roughness height of the channel bed equals the median
grain size.)

A 25 m wide channel has a bed slope of 0.0009. The bed consists of a mixture of light
particles ($\rho_s = 2350$ kg/m³) with a median particle size $d_{50} = 1.15$ mm. The flow rate
is 7.9 m³/s. Calculate the bed-load transport rate at uniform equilibrium flow conditions using the formulae of Meyer-Peter, Einstein, Nielsen and van Rijn.
(Assume that the equivalent roughness height of the channel bed equals the median
grain size.)

11

Sediment transport mechanisms 2. Suspended load transport

11.1 Introduction

Sediment suspension can be described as the motion of sediment particles during which the particles are surrounded by fluid. The grains are maintained within the mass of fluid by turbulent agitation without (frequent) bed contact. Sediment suspension takes place when the flow turbulence is strong enough to balance the particle weight (i.e. $V_*/w_o > 0.2$ to 2, Chapter 9).

The amount of particles transported by suspension is called the *suspended load*. In this section, the mechanisms of suspended load transport are described. Then methods to predict the suspended load transport rate are also presented.

Notes
1. In natural streams suspension is more likely to occur with fine particles (e.g. silt, fine sands). Figure 11.1 illustrates an example of suspension material left after a flood.
2. The term *wash load* describes an inflow of fine-particles in suspension which do not interact with the bed material (bed-load and suspension) and remains in suspension (Fig. 12.1 later).

Mechanisms of suspended load transport

The transport of suspended matter occurs by a combination of advective turbulent diffusion and convection. *Advective diffusion* characterizes the random motion and mixing of particles through the water depth superimposed to the longitudinal flow motion. In a stream with particles heavier than water, the sediment concentration is larger next to the bottom and turbulent diffusion induces an upward migration of the grains to regions of lower concentrations. A time-averaged balance between settling and diffusive flux derives from the continuity equation for sediment matter:

$$D_s \frac{dc_s}{dy} = -w_o c_s \tag{11.1}$$

Fig. 11.1 Suspension material left after the flood in 1963 (by E.D. McKee, with permission of US Geological Survey, Ref. 79ct). Silt deposit in the flood plain of the Colorado River, Grand Canyon National Park, USA – note the shrinkage cracks.

where c_s is the local sediment concentration at a distance y measured normal to the channel bed, D_s is the sediment diffusivity and w_o is the particle settling velocity.

Sediment motion by *convection* occurs when the turbulent mixing length is large compared with the sediment distribution length scale. Convective transport may be described as the entrainment of sediments by very-large-scale vortices: e.g. at bed drops, in stilling basins and hydraulic jumps (Fig. 11.2).

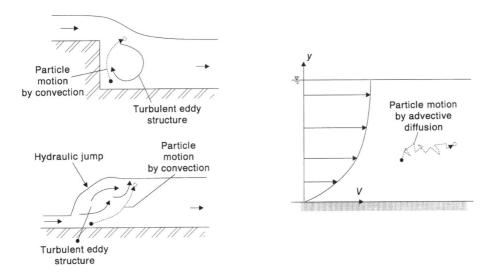

Fig. 11.2 Suspended sediment motion by convection and diffusion processes.

Notes
1. The advective diffusion process is sometimes called gradient diffusion.
2. Equation (11.1) derives from Fick's law. It is valid at equilibrium: i.e. when the turbulent diffusion balances exactly the particle settling motion.
3. The sediment diffusivity D_s is also called the sediment mixing coefficient or sediment diffusion coefficient.

11.2 Advective diffusion of sediment suspension

11.2.1 Introduction

For a constant sediment diffusion coefficient, the solution of the continuity equation for sediment (equation (11.1)) is:

$$C_s = (C_s)_{y=y_s} \exp\left(-\frac{w_o}{D_s}(y - y_s)\right) \qquad \text{Constant diffusivity coefficient} \qquad (11.2)$$

where $(C_s)_{y=y_s}$ is the sediment concentration at a reference location $(y = y_s)$. Equation (11.2) is valid for uniform turbulence distribution at steady-state conditions: e.g. it can be applied to suspension in tanks with oscillating and rotating stirring devices.

In natural (flowing) streams, the turbulence is generated by boundary friction: it is stronger close to the channel bed than near the free-surface. Hence, the assumption of constant sediment diffusivity (i.e. D_s = constant) is not realistic and equation (11.2) should not be applied to natural streams.

11.2.2 Sediment concentration in streams

In flowing streams the sediment diffusivity may be assumed to be nearly equal to the turbulent diffusion coefficient (i.e. the 'eddy viscosity'). The eddy viscosity is a coefficient of momentum transfer and it expresses the transfer of momentum from points where the momentum per unit volume (ρv) is high to points where it is lower. It is also called the turbulent mixing coefficient.

In open channel flows, the eddy viscosity and hence the sediment diffusivity D_s may be estimated as:

$$D_s \approx KV_*(d - y)\frac{y}{d} \qquad (11.3)$$

where d is the flow depth, V_* is the shear velocity and K is the von Karman constant (K = 0.4).

For a parabolic diffusivity law (equation (11.3)), the integration of the continuity equation for sediment (equation (11.1)) gives the distribution of sediment

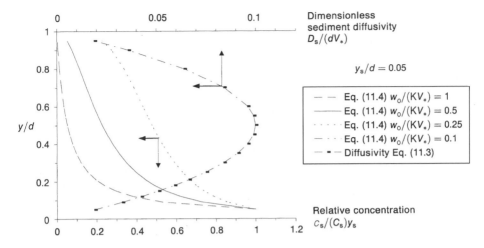

Fig. 11.3 Sediment concentration distribution (equation (11.4)) and sediment diffusivity profile (equation (11.3)).

concentration across the flow depth:

$$C_s = (C_s)_{y=y_s} \left(\frac{\dfrac{d}{y} - 1}{\dfrac{d}{y_s} - 1} \right)^{w_o/(KV_*)} \tag{11.4}$$

where $(C_s)_{y=y_s}$ is the reference sediment concentration at the reference location $(y = y_s)$.

Equation (11.4) was first developed by Rouse (1937) and it was successfully verified with laboratory and field data (e.g. Vanoni 1946). Typical plots are presented in Fig. 11.3.

Notes
1. The dimensionless number $(w_o/(KV_*))$ is sometimes called the *Rouse number* (e.g. Julien 1995) after H. Rouse.
2. Equation (11.4) is not valid very close to the wall. It predicts an infinite bed concentration (i.e. at $y = 0$). Typically it is used for $y > y_s$.

11.2.3 Discussion

Sediment motion and flow regions in sediment-laden flows
In a sediment-laden flow, the basic flow regions consist of:

- the bed-load layer next to the channel bottom (if $\tau_* > (\tau_*)_c$), and
- a suspension region above (if sediment suspension occurs) (Fig. 11.4).

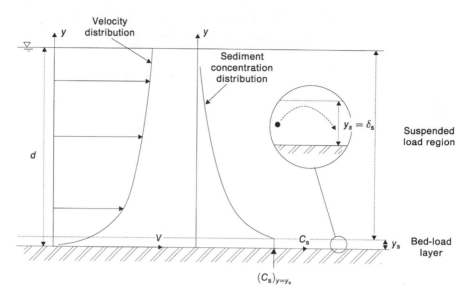

Fig. 11.4 Sketch of sediment-laden flows.

The type of sediment motion is a function of the sediment properties, bed slope and flow conditions. The results may be summarized in a modified Shields diagram shown in Fig. 11.5. Figure 11.5 presents the Shields parameter τ_* as a function of a dimensionless particle parameter $d_* = d_s^3 \sqrt{(s-1)g/\nu^2}$. The critical Shields parameter for initiation of sediment motion is plotted as a solid line, and a criterion for initiation of suspension (i.e. $V_*/w_0 = 1$) is also shown.

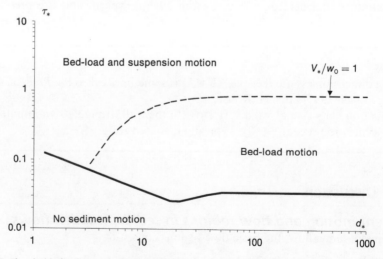

Fig. 11.5 Threshold of sediment motion (bed-load and suspension): Shields parameter $\tau_* = \tau_0/(\rho g(s-1)d_s)$ as a function of the dimensionless particle parameter $d_* = d_s \sqrt[3]{(s-1)g/\nu^2}$ (for water at 20°C and $s = 2.69$).

Reference concentration and elevation

Several researchers showed that the accurate prediction of sediment concentration distributions (i.e. equation (11.3)) depends critically upon the selection of the reference concentration $(C_s)_{y=y_s}$ and elevation y_s.

Very close to the bed (i.e. in the bed-load layer), sediment motion occurs by bed-load transport (rolling, sliding and saltation). Suspension takes place above the bed-load layer (Fig. 11.4). Hence a logical choice for y_s is the outer edge of the bed-load layer:

$$y_s = (\delta_s)_{bl} \qquad \text{bed-load layer thickness} \qquad (11.5)$$

and the reference concentration must be taken as the bed-load layer sediment concentration:

$$(C_s)_{y=y_s} = (C_s)_{bl} \qquad \text{average bed-load layer sediment concentration} \qquad (11.6)$$

where the subscript bl refers to the bed-load layer properties (Table 11.1) (van Rijn 1984b).

Sediment diffusivity

Although equation (11.4) was successfully compared with numerous data, its derivation implies a parabolic distribution of the sediment diffusivity (i.e. equation (11.3)). The re-analysis of model and field data (Anderson 1942, Coleman 1970) shows that the sediment diffusivity is not zero next the free-surface (as predicted by equation (11.3)). The distribution of a sediment diffusion coefficient is better estimated by a semi-parabolic profile:

$$D_s \approx K V_*(d-y)\frac{y}{d} \qquad y/d < 0.5 \qquad (11.7a)$$

$$D_s \approx 0.1 \qquad \text{Outer region } (y/d > 0.5) \qquad (11.7b)$$

Table 11.1 Dimensionless sediment diffusivities $D_s/(dV_*)$ in suspended-sediment flows and for salt diffusion in open channel flows

Reference (1)	$D_s/(dV_*)$ (2)	Comments (3)
Sediment-laden flows		
Lane and Kalinske (1941)[a]	0.067[b]	
Anderson (1942)[c]	0.08 to 0.3[d]	Enoree river (USA). $0.9 < d < 1.52$ m
Coleman (1970)	0.05 to 0.4[d]	Model data $q_w \leq 0.2\,\text{m}^2/\text{s}$
Matter diffusion in open channel		
Elder (1959)	0.228	Model data. Lateral (transverse) diffusion of permanganate solution. $d = 0.01$ to 0.15 m
Fischer *et al.* (1979)	0.067	Vertical mixing
	0.15	Transverse mixing in straight channels. Average of experimental measurements $(0.09 \leq D_s/(V_*d) \leq 0.26)$
	0.4 to 0.8	Transverse mixing in curved channels and irregular sides. Experimental data $(0.36 \leq D_s/(V_*d) \leq 3.4)$

Notes: [a] as reported in Graf (1971); [b] depth-averaged value; [c] see also Coleman (1970); [d] sediment diffusion coefficient in the outer flow region $(y/d > 0.5)$.

Comments

1. The above developments (equations (11.3) and (11.7)) imply that the diffusion of sediment particles has the same properties as the diffusion of small coherent fluid structures (i.e. $D_s = \nu_T$).

   In practice, the sediment diffusivity D_s does not equal exactly the turbulent mixing coefficient (i.e. eddy viscosity). First, the diffusion of solid particles differs from the diffusion of fluid particles. Secondly, the solid sediment particles may interfere with the flow turbulence and, hence, the presence of solid particles might affect the turbulent mixing coefficient. Finally, the assumption of $D_s = \nu_T$ does not hold well for all grain sizes.
2. Table 11.1 summarizes a number of experimental data from which values of dimensionless diffusivity $D_s/(dV_*)$ were estimated. The results are compared with values of dimensionless diffusivity for salt diffusion in streams.
3. A recent study (Chanson 1997) developed an analogy between the diffusion of suspended particles and that of entrained air bubbles in open channel supercritical flows. Interestingly, the analysis showed the same order of magnitude for the mean (dimensionless) diffusion coefficient in both types of flows (air–water and sediment–water flows).

11.3 Suspended sediment transport rate

11.3.1 Presentation

Considering sediment motion in an open channel (Fig. 11.4), the suspended load transport rate equals:

$$q_s = \int_{\delta_s}^{d} C_s V \, dy \tag{11.8}$$

where q_s is the volumetric suspended-load transport rate per unit width, C_s is the sediment concentration (equation (11.4)), V is the local velocity at a distance y measured normal to the channel bed, d is the flow depth and δ_s is the bed-load layer thickness.

Notes

1. Equation (11.8) implies that the longitudinal velocity component of the suspended load equals that of the water flow in which the particles are convected.
2. In practice, field measurements are often presented as:

$$q_s = \int_{\delta_s}^{d} C_s V \, dy = (C_s)_{mean} q = (C_s)_{mean} V d$$

where q is the water discharge per unit width, V is the mean flow velocity, d is the flow depth and $(C_s)_{mean}$ is the depth-averaged sediment concentration (or mean suspended-sediment concentration).

11.3.2 Calculations

The suspended-load transport rate may be computed using equation (11.8) in which C_s is computed using equation (11.4), and δ_s and $(C_s)_{bl}$ are deduced from the bed-load layer characteristics (Chapter 10):

$$C_s = (C_s)_{bl} \left(\frac{\dfrac{d}{y} - 1}{\dfrac{d}{(\delta_s)_{bl}} - 1} \right)^{w_o/(KV_*)} \tag{11.4}$$

$$(C_s)_{bl} = \frac{0.117}{d_s} \left(\frac{\nu^2}{(s-1)g} \right)^{1/3} \left(\frac{\tau_*}{(\tau_*)_c} - 1 \right) \qquad \text{(van Rijn 1984a)} \tag{11.6}$$

$$(\delta_s)_{bl} = 0.3 d_s \left(d_s \left(\frac{(s-1)g}{\nu^2} \right)^{1/3} \right)^{0.7} \sqrt{\frac{\tau_*}{(\tau_*)_c} - 1} \qquad \text{(van Rijn 1984a)} \tag{11.7}$$

where $\tau_* = \tau_o/(\rho(s-1)gd_s)$, $(\tau_*)_c$ is the critical Shields parameter for bed-load initiation, d_s is the sediment size, τ_o is the bed shear stress, $s = \rho_s/\rho$ and ν is the kinematic viscosity of the fluid.

For a rough-turbulent flow (e.g. gravel-bed streams), the velocity distribution may be estimated as:

$$\frac{V}{V_*} = \frac{1}{K} \ln \left(\frac{y}{k_s} \right) + 8.5 \tag{11.9}$$

where V_* is the shear velocity, K is the von Karman constant ($K = 0.4$) and k_s is the equivalent roughness height.

Notes
1. The sediment concentration in the bed-load layer has an upper limit of 0.65 (for rounded grains). Equation (11.6) should be used in the form:

$$(C_s)_{bl} = \text{Min} \left(\frac{0.117}{d_s} \left(\frac{\nu^2}{(s-1)g} \right)^{1/3} \left(\frac{\tau_*}{(\tau_*)_c} - 1 \right); \ 0.65 \right) \tag{11.6b}$$

2. Equation (11.9) was obtained for the turbulent zone (i.e. $y/\delta < 0.1$ to 0.15 of a fully-rough turbulent boundary layer. In fully-developed open channel flow, it is valid for $y/d < 0.1$ to 0.15.
3. Some researchers (e.g. Graf 1971, Julien 1995) proposed extending equation (11.9) (in first approximation) to the entire flow field. Such an approximation is reasonable with heavy particles as most of the sediment flux (i.e. $C_s V$) occurs in the lowest flow region (i.e. in or close to the turbulent zone).
4. For (fully-developed) turbulent open channel flows, the velocity distribution may also be expressed as:

$$\frac{V}{V_{\text{max}}} = \left(\frac{y}{d} \right)^{1/N}$$

where V_{max} is the free-surface velocity. By continuity, the free-surface velocity equals:

$$V_{max} = \frac{N+1}{N}\frac{q}{d} = \frac{N+1}{N}V_*\sqrt{\frac{8}{f}} \qquad \text{for a wide rectangular channel}$$

and q is the discharge per unit width and f is the Darcy friction factor.

For uniform equilibrium flow, Chen (1990) showed that the exponent N is related to the Darcy friction factor:

$$N = K\sqrt{\frac{8}{f}}$$

11.3.4 Application

A mixture of fine sands ($d_{50} = 0.1\,\text{mm}$) flows on a steep slope ($\sin\theta = 0.03$). The water discharge is $q = 1.5\,\text{m}^2/\text{s}$.

(a) Predict the occurrence of bed-load motion and/or suspension.
(b) If bed-load motion occurs, estimate the characteristics of the bed-load layer using the formulae of van Rijn (1984a) and calculate the bed-load transport rate.
(c) If suspension occurs, plot the sediment concentration distribution and velocity distribution on the same graph (plot v/V_* and C_s as functions of y/d).
(d) Calculate the suspended-load transport rate.
(e) Calculate the total sediment transport rate.

Note: perform the calculations for a flat bed and assuming uniform equilibrium flow. The bed roughness height k_s will be assumed to be the median grain size.

Solution
First, the basic flow properties (mean velocity, flow depth, Darcy friction factor) must be computed. Finally the shear velocity and mean bottom shear stress are deduced. Finally, the fall velocity of the sediment particle is calculated.

The mean flow velocity, flow depth and Darcy friction factor are the solutions of a system of three equations (with three unknowns): the continuity equation (for water), the momentum equation for the uniform equilibrium flow and the friction factor formula. These are:

$$Q = VA \qquad\qquad\qquad [\text{C}]$$

$$V = \sqrt{\frac{8g}{f}}\sqrt{\frac{D_H}{4}\sin\theta} \qquad\qquad [\text{M}]$$

$$f = \mathscr{F}\left(\frac{VD_H}{\nu}; \frac{k_s}{D_H}\right) \qquad \text{Darcy friction factor calculation}$$

where V is the mean flow velocity, D_H is the hydraulic diameter, Q is the water discharge and A is the cross-sectional area of the flow (in a plane normal to the flow direction).

For a wide rectangular channel and assuming turbulent flow, the above equations become:

$$q \approx Vd \qquad \text{wide rectangular channel}$$

$$V \approx \sqrt{\frac{8g}{f}} \sqrt{d \sin\theta} \qquad \text{uniform equilibrium flow in wide rectangular channel}$$

$$\frac{1}{\sqrt{f}} = -2.0 \log_{10}\left(\frac{k_s}{3.71 D_H} + \frac{2.51\nu}{V D_H \sqrt{f}}\right)$$

$$\text{turbulent flow (Colebrook–White formula)}$$

Iterative calculations yield:

$$V = 6.62 \,\text{m/s}$$

$$f = 0.012$$

$$d = 0.227 \,\text{m}$$

The shear velocity and bed shear stress are:

$$V_* = V\sqrt{\frac{f}{8}} = 0.258 \,\text{m/s}$$

$$\tau_0 = \rho V_*^2 = 66.6 \,\text{Pa}$$

For a 0.1 mm sediment particle, the settling velocity is:

$$w_o = 0.0081 \,\text{m/s}$$

(a) Let us predict the occurrence of bed-load motion using the Shields diagram. The Shields parameter and particle Reynolds number are:

$$\tau_* = \frac{\tau_0}{\rho(s-1)g d_s} = 41.15$$

$$Re_* = \frac{V_* d_s}{\nu} = 25.66$$

The Shields diagram indicates: $(\tau_*)_c \approx 0.08$. Hence bed-load motion occurs (because $\tau_* > (\tau_*)_c$).

Now let us consider the occurrence of suspension. The ratio V_*/w_o equals 32. Hence suspension takes place.

(b) The characteristics of the bed-load layer are computed using the formulae of van Rijn (1984a):

$$(C_s)_{bl} = \text{Min}\left(\frac{0.117}{d_s}\left(\frac{\nu^2}{(s-1)g}\right)^{1/3}\left(\frac{\tau_*}{(\tau_*)_c} - 1\right); \, 0.65\right) \qquad (11.6b)$$

$$(\delta_s)_{bl} = 0.3 d_s\left(d_s\left(\frac{(s-1)g}{\nu^2}\right)^{1/3}\right)^{0.7}\sqrt{\frac{\tau_*}{(\tau_*)_c} - 1} \qquad (11.7)$$

Calculations yield:

$$(C_s)_{bl} = 0.65 \qquad (\delta_s)_{bl} = 0.0013 \, \text{m}$$

The bed-load transport rate equals:

$$(q_s)_{bl} = C_s \delta_s V_s \tag{11.5}$$

Using the formulae of Nielsen (1992), the transport rate per unit width equals:

$$(q_s)_{bl} = 0.0083 \, \text{m}^2/\text{s}$$

(c) The flow Reynolds number VD_H/ν equals 1.5×10^3 and the shear Reynolds number of the flow $k_s V_*/\nu$ equals 25.7. The flow is turbulent, in the transition (between smooth and fully-rough). Hence, the above calculation of the Darcy friction factor is correct. However, the entire velocity distribution cannot be estimated by equation (11.9) (valid for fully-rough turbulent flow only). Instead, the velocity distribution may be described by:

$$\frac{V}{V_{\text{max}}} = \left(\frac{y}{d}\right)^{1/N}$$

where V_{max} is the free-surface velocity:

$$V_{\text{max}} = \frac{N+1}{N} V_* \sqrt{\frac{8}{f}} = 7.3 \, \text{m/s} \qquad \text{for a wide rectangular channel}$$

For uniform equilibrium flow, Chen (1990) showed that the exponent N is related to the Darcy friction factor:

$$N = K\sqrt{\frac{8}{f}} = 10.2$$

The volumetric sediment concentration distribution can be estimated using the Rouse profile:

$$C_s = (C_s)_{bl} \left(\frac{\dfrac{d}{y} - 1}{\dfrac{d}{(\delta_s)_{bl}} - 1} \right)^{w_o/(KV_*)} \tag{11.4}$$

where K is the von Karman constant (K = 0.4).

Figure 11.6 presents the distributions of sediment concentration C_s, of dimensionless velocity V/V_* and of the dimensionless suspended sediment flux $C_s V/V_*$.

(d) The suspended-load transport rate is calculated as

$$q_s = \int_{y_s}^{d} C_s V \, dy \tag{11.8}$$

Numerical integration gives:

$$q_s = 0.62 \, \text{m}^2/\text{s} \qquad \text{Suspension transport rate}$$

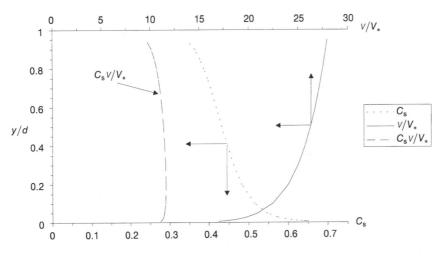

Fig. 11.6 Sediment concentration distribution, dimensionless velocity profile v/V_* and dimensionless suspended sediment flux $c_s v/V_*$.

(e) The total sediment transport rate equals:

$$q_s = 0.625\,\mathrm{m}^2/\mathrm{s}$$

Note that most sediment transport occur by suspension. The contribution of bed-load transport is negligible.

11.4 Hyperconcentrated suspension flows

11.4.1 Presentation

The last example showed a sediment-laden flow with a large amount of suspension. The depth-averaged sediment concentration was about 30%. With such an amount of sediment:

- Are the (sediment–water) fluid properties the same as those of clear-water?
- What are the interactions between suspended sediment and turbulence?
- Can the flow properties be computed as for those of clear-water flows (e.g. Chézy equation, Colebrook–White formula)?

Applications
Wan and Wang (1994) described numerous examples of hyperconcentrated flows in the Yellow River basin (China).

During one hyperconcentrated flood event, the Yellow River bed was eroded by 8.8 m in 72 hours (Longmen, Middle reach of Yellow River, August 1970, $Q = 13\,800\,\mathrm{m}^3/\mathrm{s}$). During another flood event, a wash load (sediment concentration: $964\,\text{kg/m}^3$) was transported over $50\,\text{km}$ in narrow and deep canals.

'Clogging' of open channels might also result during the recession of a hyper-concentrated flood. A whole river might stop flowing sometimes (e.g. Heihe river, July 1967).

11.4.2 Fluid properties

In a homogeneous suspension, the basic fluid properties (e.g. density and viscosity) differ from those of clear-water. The density of a particle–water mixture equals:

$$\rho_{\text{mixt}} = \rho(1 + C_{\text{s}}(\text{s} - 1)) \tag{11.10}$$

where ρ is the water density, C_{s} is the volumetric sediment concentration and $\text{s} = \rho_{\text{s}}/\rho$. The dynamic viscosity of the mixture equals:

$$\mu_{\text{mixt}} = \mu(1 + 2.5C_{\text{s}}) \tag{11.11}$$

where μ is the water dynamic viscosity (Einstein 1906, 1911).

Note
More sophisticated formulations of the mixture viscosity have been proposed. For example:

$$\mu_{\text{mixt}} = \mu(1 + 2.5C_{\text{s}} + 6.25C_{\text{s}}^2) \quad \text{(Graf 1971)}$$

$$\mu_{\text{mixt}} = \mu(1 - 1.4C_{\text{s}})^{-2.5} \quad \text{(Wan and Wang 1994)}$$

Rheology
For a Newtonian fluid (e.g. clear-water), the shear stress acting in a direction (e.g. the x-direction) is proportional to the velocity gradient in the normal direction (e.g. the y-direction):

$$\tau = \mu \frac{dv}{dy} \tag{11.12}$$

where μ is the dynamic viscosity of the fluid (see Chapter 1).

Hyperconcentrated suspensions do not usually satisfy equation (11.12). They are called non-Newtonian fluids. For homogenous suspension of fine particles, the constitutive law of the fluid is:

$$\tau = \tau_1 + \mu_{\text{mixt}} \frac{dv}{dy} \tag{11.13}$$

where μ_{mixt} is the dynamic viscosity of the sediment–water mixture and τ_1 is a constant, called the *yield stress*.

Notes
1. Equation (11.12) is also called Newton's law of viscosity.
2. Fluids satisfying equation (11.13) are called *Bingham plastic fluids*.
3. The Bingham plastic model is well suited to sediment-laden flows with large concentrations of fine particles (Wan and Wang 1994, Julien 1995).
4. The yield stress of a hyperconcentrated mixture is generally formulated as a function of sediment concentration. Julien (1995) proposed:

$$\tau_1 \approx 0.1 \exp(3(C_s - 0.05)) \qquad \text{for sands}$$

$$\tau_1 \approx 0.1 \exp(13(C_s - 0.05)) \qquad \text{for silts (95\%) and clays (5\%)}$$

$$\tau_1 \approx 0.1 \exp(23(C_s - 0.05)) \qquad \text{for silts (70\%) and clays (30\%)}$$

11.4.3 Discussion

The characteristics of hyperconcentrated flows are not yet fully understood. It is clear that the fluid and flow properties of hyperconcentration cannot be predicted, as can those of clear-water flows. Furthermore, the interactions between bed-load and suspended-load motions are not known, and there are substantial differences between fine-particle hyperconcentrated flows and large-particle debris flows.

Fig. 11.7 Extreme reservoir siltation caused by hyperconcentrated flows: Le Saignon dam (France), June 1998. Dam height: 14.5 m (completed in 1961), reservoir capacity: 1.4×10^5 m^3, catchment area: 3.5 km^2. Fully-silted reservoir after two years of operation. View from the right dam abutment: the dam wall being on the right edge of the photograph, and the silted intake tower in the centre.

Currently, the topic is under active research investigations. Wan and Wang(1994) presented a solid review of the Chinese expertise, in particular with fine-particle hyperconcentrations. Takahashi (1991) reviewed the Japanese knowledge in large-particle flows and mountain debris flows. Meunier (1995) described a series of field investigations in France, including studies of the catchment hydrology, soil erosion, hydraulics and sediment transport. Julien (1995) presented a brief summary of the American experience.

Practically, hyperconcentrated flows must be closely monitored. They could lead to severe soil erosion, a disastrous reservoir siltation and channel clogging with associated upstream flooding. For example, the reservoir shown in Fig. 11.7 became fully-silted in less than two years, despite the presence of upstream debris dams in the catchment.

11.5 Exercises

Numerical solutions to some of these exercises are available from the Web at www.arnoldpublishers.com/support/chanson

Sediment suspension is observed in a hydraulic jump stilling basin. Does sediment motion occur by advective diffusion or convection?
   Solution: convection.

A natural stream has a bed slope of 0.008 and the observed flow depth is 0.87 m. The bed consists of very fine sands ($d_{50} = 0.18$ mm). Experimental observations indicate that the bed material is entrained in suspension and that the sediment concentration at a distance: $y_s = 0.1$ m from the bed is 15%. (a) Calculate the Rouse number $w_o/(KV_*)$. (b) Plot the sediment concentration distribution between $y_s = 0.1$ mm and the free-surface. (Assume that the sediment concentration distribution follows the Rouse profile.)

Fine sands ($d_{50} = 0.2$ mm) in suspension flow down a 3 m wide steep mountain river ($\sin \theta = 0.032$). The normal depth equals 0.23 m. Calculate: (a) water discharge, (b) mean boundary shear stress, (c) shear velocity, (d) settling velocity and (e) Shields parameter. (f) Predict the occurrence of bed-load motion and/or suspension. If sediment motion occurs, calculate: (g) bed-load layer sediment concentration, (h) bed-load layer thickness, (i) average sediment velocity (in bed-load layer), (j) bed-load transport rate, (k) suspension transport rate and (l) total sediment transport rate. (Assume that the sediment concentration distribution follows the Rouse profile.)

12

Sediment transport capacity and total sediment transport

12.1 Introduction

In this chapter we review the calculations of the total sediment transport rate. They are based upon previous results (Chapters 10 and 11) obtained for a flat-bed, and they can be used to predict the maximum sediment transport capacity. The effects of sediment erosion, accretion and bed-form motion are then discussed. Later the effects of bed-forms on flow resistance are presented.

12.2 Total sediment transport rate (sediment transport capacity)

12.2.1 Presentation

The total sediment discharge is the total volume of sediment particles in motion per unit time. It includes the sediment transport by bed load motion and by suspension as well as the wash load (Fig. 12.1).

Numerous formulae were proposed to predict the bed-load transport (Chapter 10), the suspension discharge (Chapter 11) and the total transport rate (e.g. Table 12.1). Despite significant progress during the past few decades, none of these can accurately predict the 'real' sediment motion in natural streams.

Most importantly, let us remember that existing formulae (e.g. Chapters 10 and 11, Table 12.1) predict only the *sediment transport capacity* of a known bed sediment mixture: i.e. the maximum sediment transport rate. They do not take into account the sediment inflow nor erosion and accretion.

12.2.2 Calculation of the sediment transport capacity

The sediment transport capacity (i.e. maximum total sediment transport rate) equals the sum of the bed load transport rate and suspension transport rate:

$$q_s = (q_s)_{bl} + (q_s)_s \tag{12.1}$$

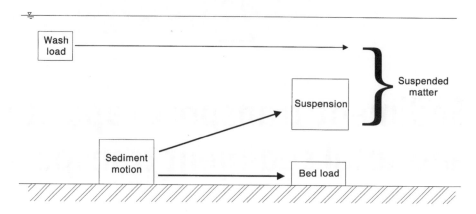

Fig. 12.1 Sediment transport classification.

Table 12.1 Empirical and semi-empirical correlations of total sediment transport

Reference (1)	Formulation (2)	Range (3)	Remarks (4)
Einstein (1950)	$q_s = (q_s)_{bl}\left(1 + I_1 \ln\left(\dfrac{30d}{d_s}\right) + I_2\right)$ where I_1 and I_2 are two integrals deduced from design charts	Small suspension transport rate	
Engelund and Hansen (1967)	$q_s = 0.4 f \dfrac{\tau_o}{\rho}\sqrt{\dfrac{d_s}{(s-1)g}}$	$0.19 < d_s < 0.93$ mm	Based on laboratory data by Guy *et al.* (1966) ($B = 2.44$ m). Validated for $d_{50} > 0.15$ mm and $\sqrt{d_{90}/d_{10}} < 1.6$.
Graf (1971)	$\dfrac{q_s}{q} = 10.39\dfrac{VD_H/4}{\sqrt{(s-1)gd_{50}^3}}$ $\times\left(\dfrac{(s-1)d_{50}}{\sin\theta D_H/4}\right)^{-2.52}$	$0.1 < \dfrac{(s-1)d_{50}}{\sin\theta\dfrac{D_H}{4}} < 15$	Based on experimental data in open channel and pipe flows.
Van Rijn (1984c)	$\dfrac{q_s}{q} = 0.005\left(\dfrac{V-(V)_c}{\sqrt{(s-1)gd_{50}}}\right)^{2.4}\left(\dfrac{d_{50}}{d}\right)^{1.2}$ $+ 0.012\left(\dfrac{V-(V)_c}{\sqrt{(s-1)gd_{50}}}\right)^{2.4}\left(\dfrac{d_{50}}{d}\right)$ $\times\left(d_{50}\sqrt[3]{\dfrac{(s-1)g}{\nu^2}}\right)^{-0.6}$ $(V)_c = 0.19 d_{50}^{0.1}\log_{10}\left(\dfrac{D_H}{d_{90}}\right)$ for $0.1 < d_{50} < 0.5$ mm $(V)_c = 8.5 d_{50}^{0.6}\log_{10}\left(\dfrac{D_H}{d_{90}}\right)$ for $0.5 < d_{50} < 2$ mm	$0.1 < d_s < 2$ mm $d_{84}/d_{16} = 4$ $1 < d < 20$ m $0.5 < V < 2.5$ m/s	Regression analysis of calculations (at 15°C).

Notes: B = channel width; D_H = hydraulic diameter; q_s = total volumetric sediment discharge per unit width.

(a)

(b)

Fig. 12.2 Sediment deposits after floods. (a) Cattle Creek at Gargett bridge in August 1977. Large bank of gravel left after the flood (courtesy of Mrs J. Hacker). Major tributary of Pioneer River, Inland of Mackay QLD, Australia. (b) Flinders River on 11 June 1984. Sediment deposit on bank after the flood (courtesy of Mrs J. Hacker). At Molesworth QLD, Australia. River subject to very large floods during tropical cyclones.

where:

$$(q_s)_{bl} = (C_s)_{bl}(\delta_s)_{bl}(V_s)_{bl} \qquad \text{bed-load transport rate} \qquad (12.2)$$

$$(q_s)_s = \int_{\delta_s}^{d} c_s v \, dy = (C_s)_{mean} q \qquad \text{suspended-sediment transport rate} \qquad (12.3)$$

and δ_s is the bed-load layer thickness, d is the flow depth, q is the water discharge per unit width and the subscript bl refers to the bed load layer. Equations (12.2) and (12.3) have been detailed in Chapters 10 and 11 respectively (for flat-bed channels).

(c)

(d)

Fig. 12.2 (Continued) (c) Bridge across the Flinders River at Molesworth QLD, Australia on 11 June 1984 (courtesy of Mrs J. Hacker). Note the silt deposit on the bridge. The bridge road had to be re-opened (i.e. cleared) by bulldozer. (d) Mirani Road Bridge looking upstream in August 1977 (courtesy of Mrs J. Hacker). Debris of old bridge piers in the foreground.

12.2.3 Discussion

Equations (12.1) to (12.3) give an estimate of the (maximum) sediment transport capacity of the flow for a given channel configuration and for a flat movable bed. 'Reasonable' predictions might be obtained for straight prismatic channels with a relatively wide cross-section formed with uniform bed material, but not in natural streams.

Limitations of the calculations include:

- non-uniformity of the flow, presence of secondary currents,
- bends, channel irregularities, bank shape, formation of bars, presence of bed forms (e.g. Figs 7.2 and 12.2),
- change of flow regime (e.g. change in bed slope),
- transition regime (e.g. between dunes and flat bed, Fig. 12.4 later).

12.3 Erosion, accretion and sediment bed motion

12.3.1 Presentation

Erosion or *accretion* of the channel bed is not only a function of the sediment transport capacity (equation (12.1)) but also of the inflow conditions (i.e. discharge, sediment inflow). Let us consider some examples.

(1) *Change of bed slope*
When the bed slope decreases, the mean velocity in the downstream reach is lower and hence the lower reach has a smaller sediment transport capacity than the upstream reach for a given discharge and channel cross-sectional shape. As a result, the sediment inflow is larger than the sediment transport capacity and accretion usually takes place in the downstream reach.

(2) *Increase of channel width*
When the channel width increases, lower flow velocities take place in the downstream section: the sediment transport capacity is smaller in the downstream reach and accretion may take place. This geometry is well suited to the design of settling basins: e.g. at the intake of an irrigation channel, a settling basin will trap sediment materials to prevent siltation of the irrigation system.

(3) *Change in flood flow*
A change in river discharge (e.g. receding flood) affects the flow properties: e.g. flow depth, mean velocity, boundary shear stress. Often erosion is observed during the rising stage of a flood and accretion occurs during the falling stage (e.g. Fig. 12.2). As a result, massive sediment movement may be observed: e.g. formation of sand bars in rivers, foundation scouring (houses, bridge piers), or shift of the main river course.

Erosion and accretion must be predicted using the continuity equation for the bed sediment material.

Note
Accretion is the increase of channel bed elevation resulting from the accumulation of sediment deposits. It is called also deposition and aggradation.

12.3.2 Continuity principle for bed material

By continuity the longitudinal change in sediment transport rate $\partial q_s / \partial s$ must be equal to the change in bed elevation, taking into account the porosity of the bed material. Considering a small control volume (Fig. 12.3(a)), the continuity equation for sediment states that the change in sediment transport rate Δq_s during a time Δt equals:

$$\Delta q_s \Delta t = -(1 - Po)\Delta z_o \cos \theta \Delta s \qquad (12.4a)$$

where Po is the bed porosity, Δz_o is the change in bed elevation during the time Δt, Δs is the length of the control volume in the flow direction and θ is the longitudinal bed slope. Equation (12.4a) may be re-arranged:

$$\frac{\Delta q_s}{\Delta s} = -(1 - Po)\frac{\Delta z_o}{\Delta t}\cos \theta \qquad (12.4b)$$

An increase in sediment transport rate (i.e. $\Delta q_s / \Delta s > 0$) implies a drop in bed elevation and hence erosion. Conversely a decrease in sediment transport rate (i.e. $\Delta q_s / \Delta s < 0$) implies accretion.

Notes
1. The continuity equation for bed material may be rewritten in a differential form:

$$\frac{\partial q_s}{\partial s} = -(1 - Po)\cos \theta \frac{\partial z_o}{\partial t} \qquad (12.4c)$$

assuming no lateral sediment inflow and where s is the longitudinal coordinate in the flow direction.
2. The term $(1 - Po)$ is sometimes called the *packing factor*.
3. For the calculation of erosion and accretion, the volumes of material deduced from sediment transport rates must be computed taking into account the porosity (i.e. the voids) of the sediment bed (i.e. equation (12.4)).

12.3.3 Applications

Practical applications of the continuity equation for movable bed materials are presented below: a movable bed downstream of a fixed-bed and the migration of dunes above a fixed bed (Fig. 12.3), the erosion rate of a stream reach.

(A) Considering a movable bed downstream of a fixed bed, the channel is horizontal and the bed is initially flat (i.e. $t = 0$). Let us compute the downstream sediment discharge q_s assuming a movable bed erosion such that the bed elevation $z_o(s, t)$ varies linearly with time and that $q_s(s = 0, t) = 0$ (i.e. no sediment inflow).

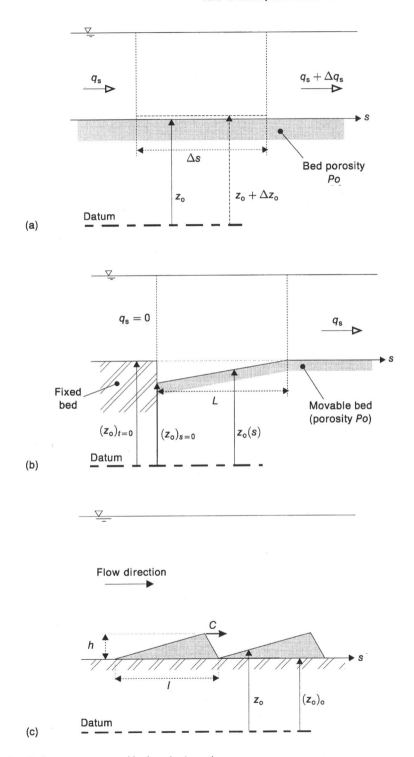

Fig. 12.3 Sediment transport and bed erosion/accretion.

Solution

The elevation of the movable bed ($s > 0$) equals:

$$z_o(s, t) = (z_o)_{t=0} - \Delta(t)\left(1 - \frac{s}{L}\right)t$$

where t is the time, s is the longitudinal coordinate, $\Delta(t) = (z_o)_{t=0} - (z_o)_{s=0}$, $(z_o)_{t=0}$ is the initial bed elevation and $(z_o)_{s=0}$ is the movable bed elevation at $s = 0$ (Fig. 12.3(b)).

The integration of the continuity equation for movable bed material (equation (12.4)) yields:

$$q_s(s, t) = \int_0^s \frac{\partial q_s}{\partial s}\, ds = \int_0^s (1 - Po)\frac{\Delta(t)}{t}\left(1 - \frac{s}{L}\right) ds = \frac{(1 - Po)\Delta(t)}{t}s\left(1 - \frac{s}{2L}\right)$$

At $s = L$, the sediment transport rate equals:

$$q_s(L, t) = \frac{(1 - Po)\Delta(t)L}{t}$$

Note that, in this particular case, the calculations are independent of the flow conditions because the variations of the bed elevation $z_o(s, t)$ are defined.

(B) Considering the migration of two-dimensional bed forms with constant shape (Fig. 12.3(c)), let us compute the local sediment transport rate and the mean sediment transport rate assuming no sediment suspension. We shall assume that the bed forms migrate over a horizontal fixed bed.

Solution

If the bed forms move at a constant velocity C without any change of shape, the channel bed elevation may be described as:

$$z_o = (z_o)_o + F(s - Ct)$$

where s is the longitudinal coordinate, t is the time, $(z_o)_o$ is the fixed bed elevation and F is a function describing the bed form shape.

The integration of the continuity equation for sediment material (equation (12.4)) yields:

$$q_s(s, t) = (1 - Po)CF(s - Ct)$$

assuming no sediment suspension.

Discussion

For dune bed forms (Fig. 12.3(c), $C > 0$), the above result implies that $q_s(s, t)$ is zero at the troughs and it is maximum at the dune crest. This indicates that the bed shear stress at the surface of the dunes (i.e. effective shear stress) varies from a negligible value at the trough (i.e. less then $(\tau_o)_c$) to a maximum value at the crest.

For anti-dunes, the same equation may apply but the bed form velocity C is negative. Hence $q_s(s, t)$ is maximum at the wave trough and minimum at the wave crest.

Overall, the mean sediment transport rate (i.e. bed-load transport) of two-dimensional triangular dune bed forms (Fig. 12.3(c)) equals:

$$q_s = \frac{1}{2}(1 - Po)Ch$$

Note that this result has a similar form to equation (12.2). In alluvial streams, however, the dune pattern is often very irregular, three-dimensional and continuously changing.

(C) A 1 km long reach of a river has a longitudinal slope $\sin\theta = 0.0014$. The water discharge is $8.5\,\text{m}^2/\text{s}$ and the mean sediment concentration of the inflow is 0.25%. The sediment transport capacity of the reach is $0.13\,\text{m}^2/\text{s}$. Deduce the rate of erosion (or accretion) of the channel bed for the 1 km long reach.

Solution
Let us consider the 1 km long section of a river with a bed slope $\theta = 0.08°$. The sediment inflow is:

$$(q_s)_o = [(C_s)_{mean}]_o q = 0.0025 \times 8.5 = 0.021\,\text{m}^2/\text{s}$$

The maximum sediment transport rate of the reach (i.e. the sediment transport capacity) is larger than the sediment inflow:

$$q_s = 0.13\,\text{m}^2/\text{s} > (q_s)_o = 0.021\,\text{m}^2/\text{s}$$

Hence, bed erosion will take place along the reach. The continuity equation for the reach is:

$$\frac{\Delta q_s}{\Delta s} = \frac{q_s - (q_s)_o}{L} = -(1 - Po)\frac{\Delta z_o}{\Delta t}\cos\theta = -(1 - Po)\frac{\partial z_o}{\partial t}\cos\theta$$

assuming a uniform erosion rate along the river reach and no lateral sediment inflow, and where L is the reach length ($L = 1000\,\text{m}$). It yields:

$$\frac{\partial z_o}{\partial t} = -\frac{1}{(1 - 0.4) \times 0.9999}\frac{0.13 - 0.021}{1000} = -0.00018\,\text{m/s}$$

assuming a 40% porosity of the movable bed.

In summary: the channel bed is eroded and the bed elevation drops at a rate of 0.65 m per hour (15.7 m per day!). The erosion rate will remain constant until the modifications of the channel bed profile change the sediment transport capacity of the reach.

12.4 Sediment transport in alluvial channels

12.4.1 Introduction

The above developments are based upon results (Chapters 10 and 11) obtained with a flat bed. In practice, bed forms are observed (Chapter 7, Figs 7.1 and 7.2).

Movable hydraulics is characterized by a dual interaction between bed forms and flow resistance. The type of bed form and the sediment transport rate are both related to the discharge, which is, in turn, a function of the bed form resistance.

12.4.2 Influence of bed forms on flow resistance

In an alluvial channel, the boundary friction is related to the skin friction (or grain-related friction) and to the form losses caused by the bed forms. Figure 12.4 presents a typical relationship between the mean boundary shear stress and the flow velocity. It indicates that the effect of bed forms is particularly substantial with ripples and dunes.

In alluvial streams the mean bed shear stress τ_o may be divided into:

$$\tau_o = \tau'_o + \tau''_o \tag{12.5}$$

where τ'_o is the skin friction shear stress and τ''_o is the form-related shear stress.

Skin friction shear stress
The *skin friction shear stress* equals:

$$\tau'_o = \frac{f}{8}\rho V^2 \tag{12.6}$$

where ρ is the fluid density, V is the mean flow velocity and f is the Darcy–Weisbach friction factor. For turbulent flows, the friction factor may be calculated using the Colebrook–White formula:

$$\frac{1}{\sqrt{f}} = -2.0\log_{10}\left(\frac{k_s}{3.71D_H} + \frac{2.51}{Re\sqrt{f}}\right) \qquad \text{Turbulent flow} \tag{12.7}$$

where D_H is the hydraulic diameter, $Re = VD_H/\nu$, ν is the fluid kinematic viscosity and k_s is the equivalent roughness height, which is related to the sediment size (Table 12.2).

The skin friction shear stress τ'_o is hence a function of the grain size d_s, fluid properties (ρ, ν) and flow characteristics (V, D_H).

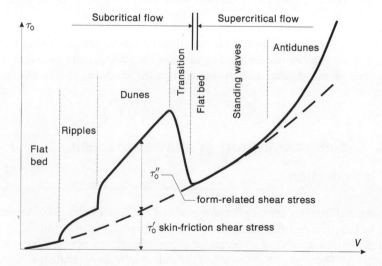

Fig. 12.4 Bed shear stress as a function of the mean flow velocity.

Table 12.2 Equivalent roughness height k_s of sediment materials (flat bed channel)

Reference (1)	k_s (2)	Range (3)	Remarks (4)
Engelund (1966)	$2.5d_{65}$		Experimental data.
Kamphuis (1974)	1.5 to $2.5d_{90}$	$d/d_{90} > 10$, $B/d > 3$	Laboratory data ($B = 0.38$ m). $d_{84}/d_{16} \sim 1.73$.
	$\approx 2d_{90}$		Large values of d/d_{90}.
Maynord (1991)	$2d_{90}$	$d/d_{90} > 2$, $B/d \geq 5$, $Re_* \geq 100$	Re-analysis of laboratory data: $0.0061 < d_{50} < 0.076$ m $0.007 < d_{90} < 0.071$ m
Van Rijn (1984c)	$3d_{90}$		Re-analysis of flume and field data.
Sumer *et al.* (1996)	$1.5d_{50}$ to $21.1d_{50}$		Experimental data in sheet-flow ($B = 0.3$ m). $0.00013 < d_{50} < 0.003$ m $1 \leq d_{84}/d_{16} \leq 2.3$

Notes: B = channel width; k_s = equivalent roughness height (grain friction related).

Notes
1. τ_o' is also called the *effective bed shear stress*.
2. It must be emphasized that there is no definite correlation between the grain size and the equivalent roughness height (e.g. Table 12.2). Usually k_s is larger than the median grain size and a first approximation might be: $k_s = 2d_{90}$.
3. In practice, the shear velocity (and hence the friction factor) might be deduced from velocity distribution measurements (on a flat bed). Another method consists of deducing the friction factor from the energy slope (i.e. from the slope of the total head line).

Bed-form shear stress

The *bed-form shear stress* τ_o'' is related to the fluid pressure distribution on the bed form and to the form loss. Fundamental experiments in irregular open channels (Kazemipour and Apelt 1983, Fig. 12.5) showed that the form losses could account for up to 92% of the total loss.

Next to the surface of a bed form, the fluid is accelerating immediately upstream of the bed form crest and decelerating downstream of the crest, associated sometimes with separation and recirculation (Fig. 12.6). The form loss may be crudely analysed as a sudden expansion downstream of the bed form crest. For a two-dimensional dune element, it yields:

$$\tau_o'' = \frac{1}{8}\rho V^2 \frac{h^2}{ld} \tag{12.8}$$

where h and l are respectively the height and length of the dune (Fig. 12.6). Experimental observations of dune dimensions are summarized in Table 12.3.

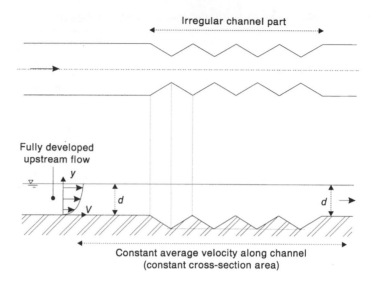

Fig. 12.5 Sketch of the experimental apparatus of Kazemipour and Apelt (1983).

Notes

1. At a sudden channel expansion, the head loss equals:

$$\Delta H = k \frac{V_1^2}{2g} \left(1 - \frac{A_1}{A_2} \right)^2 \qquad \text{Borda–Carnot formula}$$

where A_1 and A_2 are, respectively, the upstream and downstream cross-sectional areas of the flow ($A_1 < A_2$), V_1 is the upstream flow velocity and k is a head loss coefficient, function of the ratio A_1/A_2. For A_1/A_2 close to unity, $k = 1$. This formula is called the Borda–Carnot formula after the French scientists J.C. de Borda (1733–1799) and S. Carnot (1796–1832).

2. The Borda–Carnot formula may be applied to a wide rectangular channel with two-dimensional dunes (Fig. 12.6):

$$\Delta H = k \frac{V^2}{2g} \left(1 - \frac{1}{1 + h/d} \right)^2$$

where V is the mean flow velocity at the bed form crest and d is the flow depth measured from the bed form crest (Fig. 12.6). For $h/d \ll 1$, it becomes:

$$\Delta H \approx \frac{V^2}{2g} \left(\frac{h}{d} \right)^2 \qquad h \ll d$$

That is, the head loss along a dune element (length l) equals:

$$\Delta H \approx \left(\frac{h^2 D_H}{d^2 l} \right) \frac{l}{D_H} \frac{V^2}{2g} \qquad h \ll d$$

For a wide channel, it yields equation (12.8) (see Chapter 4).

3. Equation (12.8) was first applied to alluvial streams by Engelund (1966).
4. Several researchers investigated the characteristic sizes of bed forms. Van Rijn (1993) presented a comprehensive review. Typical dimensions of dune bed form are presented in Table 12.3.

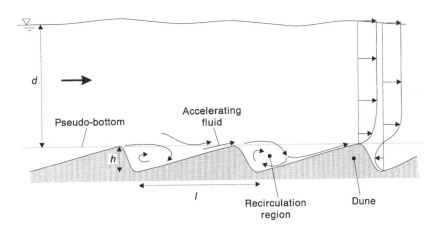

Fig. 12.6 Flow pattern over a dune.

Important

The bed-load transport must be related to the effective shear stress (skin friction shear stress) only and not to the form roughness. In natural rivers, the Shields parameter and bed-load layer characteristics must be calculated using the skin friction bed shear stress.

In other words, the onset of sediment motion must be predicted using the Shields parameter defined as:

$$\tau_* = \frac{\tau'_o}{\rho g (s - 1) d_s}$$

Comments

All characteristic parameters of bed-load transport and bed-load layer must be computed in terms of the *effective bed shear stress* τ'_o . This includes the bed-load layer thickness $(\delta_s)_{bl}$, sediment concentration in bed-load layer $(C_s)_{bl}$, bed-load velocity $(V_s)_{bl}$, bed-load transport rate $(q_s)_{bl}$.

For the suspended material, the main parameters of the sediment concentration distribution $c_s(y)$ and velocity distribution $v(y)$ are related to the *total bed shear stress*. That is, the Rouse number $(w_o/(KV_*))$, the shear velocity V_* and the mean flow velocity V are calculated in terms of the mean boundary shear stress $\tau_o (\tau_o = \tau'_o + \tau''_o)$.

Table 12.3 Characteristic dimensions of bed forms

Reference (1)	Dune dimensions (2)	Comments (3)
Dunes Laursen (1958)	$\dfrac{l}{d} = 4$ to 7.1	Model data: $d_{50} = 0.1\,\text{mm}$, $0.076 \le d \le 0.305\,\text{m}$, $0.28 \le Fr \le 0.47$, $4 \times 10^{-4} \le \sin\theta \le 0.0018$.
Yalin (1964)	$\dfrac{h}{d} = \dfrac{1}{6}\left(1 - \dfrac{(\tau_o)_c}{\tau_o}\right)$ $\dfrac{l}{d} \approx 5$	Based on model and field data: $0.137 \le d_{50} \le 2.45\,\text{mm}$ $0.013 \le d \le 28.2\,\text{m}$ $1 \times 10^{-5} \le \sin\theta \le 0.014$
Engelund and Hansen (1967)	$\dfrac{l}{h} = \dfrac{1.88}{f}$	Based on laboratory data by Guy *et al.* (1966).
Van Rijn (1984c)	$\dfrac{h}{d} = 0.11\left(\dfrac{d_{50}}{d}\right)^{0.3}$ $\times\left(1 - \exp\left(-0.5\left(\dfrac{\tau_o}{(\tau_o)_c} - 1\right)\right)\right)$ $\times\left(25 - \left(\dfrac{\tau_o}{(\tau_o)_c} - 1\right)\right)$ $\dfrac{l}{d} = 7.33$	Based on model and field data: $0.19 \le d_{50} \le 3.6\,\text{mm}$ $0.11 \le d \le 16\,\text{m}$ $0.34 \le V \le 1.55\,\text{m/s}$ Note: τ_o and $(\tau_o)_c$ are effective bed shear stresses.
Standing Waves Tison (1949)	$\dfrac{h}{d} = 0.2$ $\dfrac{h}{l} \approx 0.09$ to 0.14	Laboratory data: $d_{50} = 0.25\,\text{mm}$ $0.035 \le d \le 0.05\,\text{m}$ $0.42 \le Fr \le 0.68$
Kennedy (1963)	$\dfrac{l}{d} = 2\pi Fr^2$ $\dfrac{h}{l} \approx 0.014$	Ideal fluid flow calculations validated with laboratory data.
Present study	$\dfrac{l}{d} \approx 11.1 Fr^{3.5}\ \left(\text{i.e. } 3.4 < \dfrac{l}{d} < 8.3\right)$	Based on undular jump free-surface data (Chanson 1995a): $0.7 \le Fr < 0.9$
Antidunes Alexander and Fielding (1997)	$l = 8$ to $19\,\text{m}$ $h = 0.25$ to $1\,\text{m}$ $h/l \le 0.13$ typically (observed maximum h/l: 0.28)	Gravel antidunes in Burdekin river (Australia) for $Q = 7700\,\text{m}^3/\text{s}$ and $3200\,\text{m}^3/\text{s}$: $0.15 \le d_{50} \le 2.8\,\text{mm}$ $2 \le d \le 4\,\text{m}$

Note: f = Darcy friction factor; h = bed form height; l = bed form length (Fig. 12.6).

12.4.3 Design chart

The above considerations have highlighted the complexity of the interactions between sediment transport, bed form and flow properties.

For pre-design calculations, Engelund and Hansen (1967) developed a design chart that regroups all the relevant parameters (Fig. 12.7):

- a Froude number $(Vd)/\sqrt{g(\mathbf{s} - 1)d_{\text{s}}^3}$,
- a dimensionless sediment transport rate $q_{\text{s}}/\sqrt{g(\mathbf{s} - 1)d_{\text{s}}^3}$,

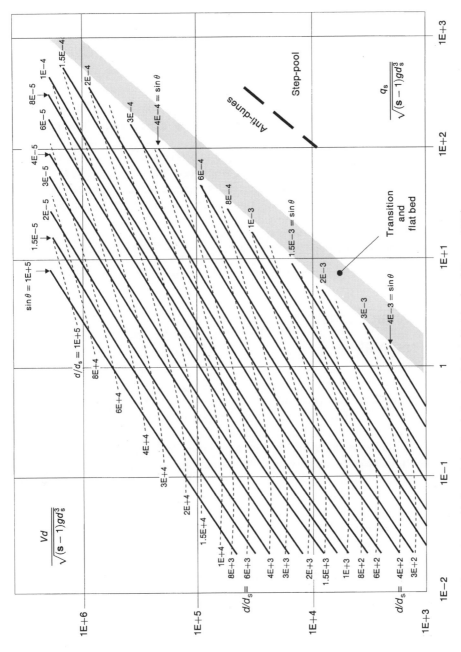

Fig. 12.7 Open channel flow with movable boundaries. Pre-design calculations (Engelund and Hansen 1967).

- the bed slope $\sin \theta$,
- the dimensionless flow depth d/d_s, and
- the bed form.

Figure 12.7 takes into account the effect of bed forms and it may also be used to predict the type of bed form. In the region for which dunes occur, only two of the first four relevant parameters are required to deduce the entire flow properties.

Notes

1. The design chart of Engelund and Hansen (1967) was developed for fully-rough turbulent flows. It was validated with the experimental data of Guy *et al.* (1966).
2. The calculations are not valid for ripple bed forms with $d_s V_*/\nu < 12$. The chart does not predict the presence of ripple bed forms.
3. For the calculation of erosion and accretion, the sediment volumes, deduced from the sediment transport rates, must be computed, taking into account the porosity (i.e. the voids) of the sediment bed.
4. Engelund and Hansen (1967) used the median grain size $d_s = d_{50}$ for their design chart (Fig. 12.7).

12.5 Applications

12.5.1 Presentation

Practically, the most important variables in designing alluvial channels are the water discharge Q, the sediment transport rate Q_s and the sediment size d_s.

The complete calculations of the flow properties and sediment transport are deduced by an iterative process. In the first stage, simplified design charts (e.g. Engelund and Hansen 1967, Fig. 12.7) may be used to have a 'feel' of the type of bed forms. In the second stage, complete calculations must be developed to predict the hydraulic flow conditions (V, d), the type of bed forms and the sediment transport capacity. The continuity equation for sediment material may then be used to assess the rate of erosion (or accretion).

Complete calculations (for a prismatic section of alluvial channel) must proceed in several successive steps.

(A) Determination of the channel characteristics

These include the bed slope $\sin \theta$, the shape of the cross-section and channel width, the movable bed properties (ρ_s, P_o, particle size distribution), and eventually the presence of bed forms (e.g. resulting from past flood events).

(B) Selection of the inflow conditions

The inflow conditions are the total water discharge Q and the sediment inflow (i.e. $(C_s)_{mean}$).

(C) Calculations of sediment-laden flow properties

[1] Preliminary calculations

Assuming a flat bed, uniform equilibrium flow and $k_s = d_s$, deduce a first estimate of the flow properties (V and d) and the bed shear stress τ_o:

$$V = \sqrt{\frac{8g}{f}}\sqrt{\frac{D_H}{4}\sin\theta} \qquad \text{Uniform equilibrium flow}$$

$$\tau_o = \rho g \frac{D_H}{4}\sin\theta \qquad \text{Uniform equilibrium flow}$$

Then assess the occurrence of sediment motion (i.e. Shields diagram).

[2] Pre-design calculations to assess the type of bed form

Using the preliminary calculations, apply the results to the design chart of Engelund and Hansen (1967) (Fig. 12.7).

The known parameters are usually:

the Froude number $(Vd)/\sqrt{g(s-1)d_{50}^3}$
the bed slope $\sin\theta$
the dimensionless ratio d/d_{50}

[3] Complete flow calculations

The calculations are iterative until the type of bed form, the bed resistance and the flow conditions satisfy the continuity and momentum equations:

$$Q = VA \qquad\qquad \text{[C]}$$

$$\tau_o = \rho g \frac{D_H}{4}\sin\theta \qquad \text{Uniform equilibrium flow} \qquad \text{[M]}$$

where V is the mean flow velocity, A is the flow cross-section area and D_H is the hydraulic diameter.

(a) Select a mean flow velocity V.
(b) Compute the skin friction shear stress τ_o':

$$\tau_o' = \frac{f}{8}\rho V^2 \qquad \text{skin friction shear stress}$$

where the roughness height might be assumed $k_s = 2d_{90}$.
(c) In the presence of bed forms, calculate the bed form shear stress:

$$\tau_o'' = \frac{1}{8}\rho V^2 \frac{h^2}{ld} \qquad \text{bed form shear stress}$$

where h and l are bed form height and length (Table 12.3).
(d) Compare $\tau_o' + \tau_o''$ and τ_o

If $(\tau_o' + \tau_o'') > \tau_o$, select a lower flow velocity to satisfy the momentum equation. Iterative calculations proceed until $[(\tau_o' + \tau_o'') = \tau_o]$ within a reasonable accuracy.

[4] Sediment transport calculations

First the sediment transport capacity is calculated:

$$Q_s = (Q_s)_{bl} + (Q_s)_s$$

then the continuity equation for the sediment material is applied to the channel reach to predict erosion or accretion.

Comments
A series of calculations is valid for a prismatic section of the channel. Each time that one channel characteristic changes (e.g. slope, sediment size, cross-section shape), the calculations must be performed again.

12.5.2 Application No. 1

An alluvial stream in which the flow depth is 1.8 m has a longitudinal bed slope of $\sin \theta = 0.001$. The characteristics of the bed material are: $d_{50} = 0.8$ mm, $d_{90} = 2.1$ mm. Predict the occurrence of sediment motion, the type of bed form and the mean flow velocity.

Solution
The calculations will proceed in three steps.
[1] Preliminary calculations (assuming flat bed and $k_s = d_{50}$).
[2] Determination of type of bed form (using Fig. 12.7).
[3] Complete calculations (taking into account the grain roughness and the type of bed form).

Step [1]: Assuming a wide channel and uniform equilibrium flow, the bed shear stress equals:

$$\tau_o = \rho g \frac{D_H}{4} \sin \theta \approx \rho g d \sin \theta = 17.6 \, \text{Pa} \qquad \text{Uniform equilibrium flow} \qquad [\text{M}]$$

Assuming a flat bed and $k_s = d_{50}$, the mean velocity is:

$$V = \sqrt{\frac{8g}{f}} \sqrt{\frac{D_H}{4} \sin \theta} \qquad \text{Uniform equilibrium flow and flat bed}$$

The (iterative) calculations give: $V = 3.36$ m/s and $f = 0.0125$.
    The Shield parameter equals:

$$\tau_* = \frac{\tau_o}{\rho(s-1)gd_{50}} = 1.36$$

and the particle Reynolds number $(d_{50} V_*/\nu)$ equals 106. For these values, the Shields diagram (Fig. 8.4) predicts *sediment motion*.

Step [2]: Using the preliminary results, let us use the design chart of Engelund and Hansen (1967) (Fig. 12.7). The known parameters are:

$(Vd)/\sqrt{g(s-1)d_{50}^3} = 6.7 \times 10^4$
the bed slope: $\sin \theta = 0.001$
the dimensionless flow depth: $d/d_{50} = 2.25 \times 10^3$

Figure 12.7 indicates that the bed form is either transition (from dune to flat-bed) or flat bed. Note that the preliminary calculations (step [1]) predicted a flow Froude number: $Fr = 0.8$. According to Table 7.1 and Fig. 7.1, the type of bed form is the *flat bed*.

Step [3]: The complete flow calculations are developed. The total bed shear stress must satisfy the momentum equation:

$$\tau_o = \tau_o' + \tau_o'' \approx \rho g d \sin \theta = 17.6 \, \text{Pa}$$

where τ_o' is the skin friction shear stress and τ_o'' is the form-related shear stress. The skin friction shear stress equals:

$$\tau_o' = \frac{f}{8}\rho V^2$$

assuming $k_s = 2 d_{90}$ and the bed form shear stress is calculated as:

$$\tau_o'' = \frac{1}{8}\rho V^2 \frac{h^2}{ld}$$

For a flat bed, $h = 0$ and $\tau_o'' = 0$. Hence it yields:

$$\tau_o = \tau_o' + 0 = 17.6 \, \text{Pa}$$

Application of the continuity and momentum equations gives the mean flow velocity, friction factor and Reynolds number: $V = 2.84 \, \text{m/s}$ and $Re = 5 \times 10^6$ (fully-rough turbulent flow).

Note: the mean velocity (computed using τ_o) differs from the value deduced from the preliminary calculations (step [1]).

12.5.3 Application No. 2

An alluvial stream in which the flow depth is 1.35 m has a longitudinal bed slope of $\sin \theta = 0.0002$. Characteristics of the sediment mixture are: $d_{50} = 1.5 \, \text{mm}$, $d_{90} = 2.5 \, \text{mm}$. Predict the sediment transport capacity (taking into account the bed form).

Solution
The calculations proceed in four steps.

Step [1]
Assuming a wide channel and uniform equilibrium flow, the bed shear stress equals:

$$\tau_o = \rho g \frac{D_H}{4} \sin \theta \approx \rho g d \sin \theta = 2.64 \, \text{Pa} \qquad \text{Uniform equilibrium flow} \qquad \text{[M]}$$

Assuming a flat bed and $k_s = d_{50}$ (i.e. $k_s/D_H \approx 2.8 \times 10^{-4}$), the mean velocity is:

$$V = \sqrt{\frac{8g}{f}}\sqrt{\frac{D_H}{4}\sin \theta} \qquad \text{Uniform equilibrium flow and flat bed}$$

The (iterative) calculations give: $V = 1.19 \, \text{m/s}$ and $f = 0.015$.

The Shield parameter equals:

$$\tau_* = \frac{\tau_o}{\rho(s-1)gd_{50}} = 0.109$$

and the particle Reynolds number $(d_{50}V_*/\nu)$ equals 77. For these values, the Shields diagram (Fig. 8.4) predicts *sediment motion*.

Step [2]
Using the preliminary results, let us use the design chart of Engelund and Hansen (1967) (Fig. 12.7). The known parameters are:

$(Vd)/\sqrt{g(s-1)d_{50}^3} = 6.89 \times 10^3$
the bed slope: $\sin\theta = 2 \times 10^{-4}$
the dimensionless flow depth: $d/d_{50} = 6.75 \times 10^3$

Figure 12.7 indicates that the bed forms are *dunes*. Note that Fig. 12.7 suggests: $q_s \sim 0.05\sqrt{(s-1)gd_s^3}$.

Step [3]
The complete flow calculations are developed. The total bed shear stress must satisfy the momentum equation:

$$\tau_o = \tau_o' + \tau_o'' \approx \rho gd \sin\theta = 2.64\,\text{Pa} \qquad [\text{M}]$$

where τ_o' is the skin friction shear stress and τ_o'' is the form-related shear stress. The skin friction shear stress equals:

$$\tau_o' = \frac{f}{8}\rho V^2$$

assuming $k_s = 2d_{90}$ and the bed form shear stress is calculated as:

$$\tau_o'' = \frac{1}{8}\rho V^2 \frac{h^2}{ld}$$

where the dune dimensions may be calculated using the correlations of van Rijn (1984c):

$$\frac{h}{d} = 0.11\left(\frac{d_{50}}{d}\right)^{0.3}\left(1 - \exp\left(-0.5\left(\frac{\tau_o}{(\tau_o)_c} - 1\right)\right)\right)\left(25 - \left(\frac{\tau_o}{(\tau_o)_c} - 1\right)\right)$$

$$\frac{l}{d} = 7.33$$

The calculations are iterative:

1. assume V,
2. compute h, l, τ_o' and τ_o'',
3. check the momentum equation by comparing $(\tau_o' + \tau_o'')$ and τ_o (= 2.64 Pa),
4. if $(\tau_o' + \tau_o'') > \tau_o$, select a smaller flow velocity V; if $(\tau_o' + \tau_o'') < \tau_o$, select a larger flow velocity,
5. repeat the calculations until $(\tau_o' + \tau_o'') = \tau_o$ with an acceptable accuracy.

Complete calculations give: $V = 0.9\,\text{m/s}$, $\tau_o' = 1.95\,\text{Pa}$, $\tau_o'' = 0.70\,\text{Pa}$, $h = 0.15\,\text{m}$, $l = 9.9\,\text{m}$.

Step [4]

The sediment transport capacity is the sum of the bed-load transport rate and suspension transport rate.

The bed-load transport rate (see Chapter 10) is calculated using flow properties based upon the effective shear stress τ_o': i.e. V_*', τ_*', $(\tau_*)_c'$, $(C_s)_{bl}$, $(d_s)_{bl}$, $(V_s)_{bl}$.

Complete calculations confirm bed-load motion, and the bed-load transport rate equals $2.48 \times 10^{-5}\,\mathrm{m^2/s}$.

The suspension transport rate (see Chapter 11) must be computed using the overall bed shear stress τ_o. For the present application, the ratio V_*/w_o equals 0.36, where $V_* = \sqrt{\tau_o/\rho} = 0.051\,\mathrm{m/s}$.

The flow conditions are near the onset of suspension and the suspension transport rate may be assumed negligible. Overall the sediment transport capacity of the alluvial stream with dune bed forms is: $q_s = 2.48 \times 10^{-5}\,\mathrm{m^2/s}$ (i.e. $0.066\,\mathrm{kg/s/m}$).

12.6 Exercises

Numerical solutions to some of these exercises are available from the Web at www.arnoldpublishers.com/support/chanson.

Continuity equation for sediment material

A 5 km long reach of a river has a longitudinal slope 1.4 m/km and the channel bed consists of fine sands. The river is 110 m wide and the flow rate is a water discharge of $920\,\mathrm{m^3/s}$. The mean sediment concentration of the inflow is 0.22%. Preliminary calculations indicate that the sediment transport capacity of the reach is $0.0192\,\mathrm{m^2/s}$. (a) Calculate the rate of erosion (or accretion) of the channel bed for the 5 km long reach. (b) Compute the change in bed elevation (positive upwards) during one day.

Considering a 300 m long reach of a stream (bed slope 0.0004), the inflow sediment transport rate is $0.00034\,\mathrm{m^3/s}$. During a 2 hour flood event, the river bed was scoured by 0.5 m, on average, over the reach. Compute the sediment transport capacity of the reach during the 2 hour event. Assume a sand and gravel river bed.

Total sediment transport capacity

An alluvial stream in which the flow depth is 2.2 m has a bed slope of 0.0015. The bed material characteristics are: $d_{50} = 0.95\,\mathrm{mm}$, $d_{90} = 3.2\,\mathrm{mm}$, $\rho_s = 2480\,\mathrm{kg/m^3}$. Predict: (a) occurrence of sediment motion, (b) type of bed-form, (c) mean flow velocity and (d) total sediment transport capacity. (Take into account the bed form. Assume $k_s = 2d_{90}$.)

Considering a 1500 m reach of an alluvial channel (85 m wide), the median grain size of the movable bed is 0.5 mm, $d_{90} = 3.15\,\mathrm{mm}$ and the longitudinal bed slope is $\sin\theta = 0.00027$. The observed flow depth is 1.24 m and the mean sediment concentration of the inflow is 2.1%. Calculate: (a) discharge, (b) effective boundary shear stress, (c) type of bed form, (d) bed form shear stress, (e) total shear stress, (f) total sediment transport capacity and (g) rate of erosion (or accretion). (Assume uniform equilibrium flow conditions. Take into account the bed form and use the design chart of Engelund and Hansen (1967) to predict the type of bed form. Assume $k_s = 2d_{90}$.)

Part 2 Revision exercises

Revision exercise No. 1

Considering a 52 m wide stream with a flow depth of 0.85 m and a bed slope $\sin\theta = 0.012$, the median grain size of the channel bed is 0.35 mm. (a) Compute the water discharge. (b) Predict the occurrence of bed-load motion and/or suspension. (c) If bed load motion occurs, calculate the bed-load transport rate. (d) If suspension occurs, plot the sediment concentration distribution (Rouse concentration distribution) and velocity distribution on the same graph (plot v/V_*) and C_s as functions of y/d. (e) Calculate the suspended-load transport rate. (f) Calculate the total sediment transport rate.

Note: Perform the calculations for a flat bed and assuming a uniform equilibrium flow. The bed roughness height k_s will be assumed to be the median grain size.

Revision exercise No. 2

A channel of trapezoidal cross-section (bottom width 2 m, sidewall slope 1 V : 3 H) has a longitudinal slope of 0.012. The channel bed and sloping sidewall consist of a mixture of fine sands ($d_s = 0.2$ mm). The flow rate is $Q = 4\,\text{m}^3/\text{s}$ and normal flow conditions are achieved. (a) Compute the flow depth. (b) Will bed-load motion take place? (c) Will sediment suspension occur? (d) Calculate the total boundary shear stress. (e) What is the type of bed forms? (Assume uniform equilibrium flow conditions. Take into account the bed form and use the design chart of Engelund and Hansen (1967) to predict the type of bed form. Assume $k_s = d_s$. The fluid is water at 20°C and the sediment grains are quartz particles.)

Revision exercise No. 3

Considering a 50 m reach of an alluvial channel (25 m wide), the bed material characteristics are: $d_{50} = 0.5$ mm, $d_{90} = 1.95$ mm and $\sin\theta = 0.00025$. The water discharge is 16.5 m^3/s and the mean sediment concentration of the inflow is 0.8%. Predict the rate of erosion (or accretion) that will take place. (Assume uniform equilibrium flow conditions. Take into account the bed form and use the design chart of Engelund and Hansen (1967) to predict the type of bed form. Assume $k_s = 2d_{90}$.)

Revision exercise No. 4

An artificial canal carries a discharge of $25\,\text{m}^3/\text{s}$. The channel cross-section is trapezoidal and symmetrical, with a 0.5 m bottom width and 1 V:2.5 H sidewall slopes. The longitudinal bed slope is 8.5 per km. The channel bottom and sidewall consist of a mixture of fine sands ($d_{50} = 0.3\,\text{mm}$).

A gauging station is set at a bridge crossing the waterway. The observed flow depth, at the gauging station, is 2.2 m. (a) Compute the flow velocity, Darcy friction factor, boundary shear stress, shear velocity. (b) At the gauging station, is the flow laminar, smooth turbulent, transition-turbulent or fully-rough turbulent? (c) Compute the critical flow depth. (d) From where is the flow controlled? (e) Predict the occurrence of bed motion and the type of bed-load motion. (f) If bed motion occurs, what is the likely type of bed form? (g) If bed-load motion occurs, compute the bed-load transport rate (using Nielsen's formula).

Use the Moody and Shields diagrams. Assume that the equivalent roughness height equals the median sediment size.

Appendix to Part 2

A2.1 Some examples of reservoir sedimentation

Introduction

Natural streams and rivers have the ability to scour channel beds and to carry a large amount of solid particles. When a dam is built across a river, it often acts as a sediment trap. After several years of use, a reservoir might become full of sediments and cease to provide water storage.

In this appendix, some examples of reservoir siltation are described.

Reservoir siltation in Australia

The Umberumberka dam, Broken Hill, NSW

Built between 1913 and 1915, the Umberumberka dam was completed in 1915 to supply water to the town of Broken Hill, NSW (900 km west-north-west of Sydney). The 41 m high dam was designed as a concrete arched dam. The original volume of the reservoir was $13\,197\,000\,m^3$ and the catchment area is $420\,km^2$ (Wasson and Galloway 1986).

The dam is located on the Umberumberka Creek, at a scarp immediately downstream of an alluvial fan. The average rainfall over the catchment is about 220 mm (for the period 1939–1972). The dam traps almost all the incoming sediment material. Estimates of the volume of sediment trapped in the reservoir have been made regularly since the dam construction.

The volume of trapped sediment is summarized in Table A2.1 and compared with other dams in the same region (i.e. Corona dam and Stephens Creek dam). Note that most siltation occurred between 1915 and 1941 and the rate of sedimentation has since decreased drastically (Wasson and Galloway 1986). The Umberumberka reservoir is still in use today.

The Korrumbyn Creek dam, Murwillumbah, NSW

The Korrumbyn Creek dam was built between 1917 and 1918, and completed in late 1918 to supply water to the town of Murwillumbah, NSW (150 km south of

Table A2.1 Reservoir siltation in the Broken Hill region (north-east area of New South Wales, Australia)

Dam	Year of completion	Catchment area (km²)	Original reservoir capacity (m³)	Cumulative siltation volume (m³)	Date of record
(1)	(2)	(3)	(4)	(5)	(6)
Corona dam, Broken Hill, NSW, Australia	1890?	15	120 000[a]	120 000 fully-silted	1910?
Stephens Creek dam, Broken Hill, NSW, Australia	1892	510	24 325 000	4 500 000 de-silting[b]	1958 1971
Umberumberka dam, Broken Hill, NSW, Australia	1915	420	13 197 000	3 600 000 4 500 000 5 013 000 5 700 000	1941 1961 1964 1982

Notes: [a] fully-silted nowadays; [b] extensive desilting in 1971.

Brisbane). The 14.1 m high dam was designed with advanced structural features: i.e. single arch wall (1.1 m thick at crest, 5.2 m thick at base, 61 m radius in plan) with the left-abutment tangent of gravity cross-section, made of Portland cement concrete (Fig. A2.1). The cross-section of the arch wall has a vertical upstream wall and a battered downstream face. The dam was equipped with two bottom outlets (one pipe outlet valve and one scour valve) and with an overfall spillway (at elevation 217.9 m R.L.). The volume of the reservoir was 27 300 m³ and the catchment area is only 3 km² (Chanson and James 1998).

The reservoir was rapidly abandoned because a log jammed the scour pipe entrance during a flood and could not be removed: i.e. the dam could no longer be scoured to remove the sediments. Prior to the incident, the water level used to drop during dry periods, and the water would turn green and foul as it warmed up, making it unfit for use. Further the catchment is very steep (bed slope > 7.6°) and the channel bed contains a wide range of sediment materials. As a result the reservoir silted up very rapidly by bed-load. For these reasons the life of the reservoir was very short (less than 20 years).

The dam still stands today, the reservoir being occupied by an overgrown tropical rainforest (Fig. A2.2).

The Cunningham Creek dam, Harden, NSW

The Cunningham Creek dam, located near Harden, NSW, was completed in 1912 to supply water to the railway. The 13 m high dam is a curved arched dam and the catchment area is 850 km². The reservoir was over 1.3 km long and about 110 m wide, and the reservoir area at dam crest level is 150 000 m².

The dam was equipped with a bottom outlet (pipe outlet), and a crane was installed at the dam crest to facilitate the outlet operation and to assist in clearing the outlet (Hellström 1941).

The reservoir silted up very rapidly (Table A2.2) and this was well documented (Hellström 1941). Attempts to flush sediments were not successful. During the first ten years of operation, the silt was deposited at a nearly constant rate of 49 000 m³ per annum (60 m³/km²/year). The dam was nearly full (~ 90%) by 1929. The reservoir was filled primarily by suspended-load. Another railway dam (Illalong

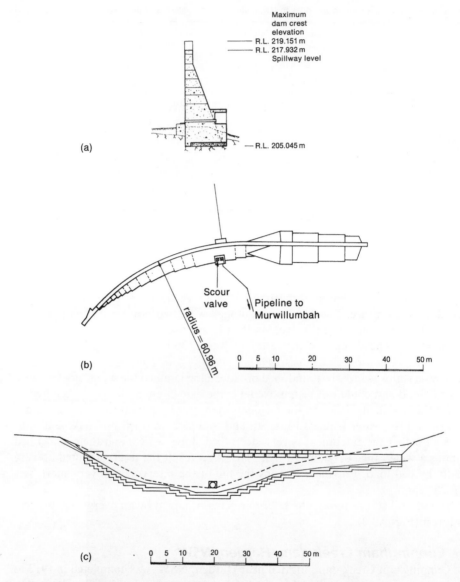

Fig. A2.1 Sketch of the Korrumbyn Creek dam after original drawings. (a) Cross-section of the wall at the bottom outlets. (b) Top view. (c) View from downstream.

Creek dam), located less than 30 km from Cunningham Creek dam, suffered a similar sedimentation problem (Table A2.2).

Extreme reservoir siltation

Reservoir sedimentation results from soil erosion in the catchment (wind, rainfall and man-made erosion), surface run-off, sediment transport in the creeks, streams and

(a)

(b)

Fig. A2.2 Recent photographs of the Korrumbyn Creek dam taken on 25 April 1997. (a) View of the sub-tropical forest occupying the reservoir. Note the dam wall in the bottom right corner. (b) Dam wall and outlet system.

Table A2.2 Siltation of Cunningham Creek and Illalong Creek reservoirs

Dam (1)	Year of completion (2)	Catchment area (km^2) (3)	Original reservoir capacity (m^3) (4)	Cumulative siltation volume (m^3) (5)	Date of record (6)
Cunningham Creek dam, Harden, NSW, Australia	1912	820	–	216 000 258 600 379 000 522 600 758 000 Fully-silted	1916 1917 1920 1922 1929 1998
Illalong Creek dam, Binalong, NSW, Australia	1914	130	2.6×10^5	Fully-silted	1998

Table A2.3 Examples of extreme reservoir sedimentation rates

Reservoir (1)	Sedimentation rate $(m^3/km^2/year)$ (2)	Study period (3)	Catchment area (km^2) (4)	Annual rainfall (mm) (5)
Asia				
Wu-Sheh (Taiwan) (S)	10 838	1957–1958	205	–
	9959	1959–1961	205	–
	7274	1966–1969	205	–
Shihmen (Taiwan) (S)	4366	1958–1964	763	> 2000
Tsengwen (Taiwan) (S)	6300	1973–1983	460	3000
Muchkundi (India)	1165	1920–1930?	67	–
North Africa				
El Ouldja (Algeria) (W)	7960 (F)	1948–1949	1.1	1500
El Fodda (Algeria) (W)	5625 (F)	1950–1952	800	555
	3060 (F)	1932–1948	800	555
Hamiz (Algeria) (W)	1300	1879–1951	139	–
El Gherza (Algeria)	615	1951–1967	1300	35
	577	1986–1992	1300	35
North America				
Sweetwater (USA)	10 599	1894–1895	482	240
White Rock (USA)	570	1923–1928	295	870
Zuni (USA) (*)	546	1906–1927	1290	250 to 400
Roosevelt (USA)	438	1906–1925	14 900	–
Europe				
Saignon (France) (*)	25 714	1961–1963	3.5	–
Saifnitz (Austria) (*)	6820	1876	4	–
Monte Reale (Italy) (*)	1927	1904–1905	436	–
Wetzmann (Aust.) (*)	1852	1883–1884	324	–
Pont-du-Loup (France) (*)	1818	1927–1928	750	–
Pontebba (Austria)	1556	1862–1880	10	–
Lavagnina (Italy)	784	1884–1904	26	1800
Roznov (Poland) (S)	398	1958–1961	4885	600
Cismon (Italy)	353	1909–1919	496	1500
Abbeystead (UK) (*)	308	1930–1948	49	1300 to 1800
Porabka (Poland) (S)	288	1958–1960	1082	600
Australia				
Quipolly (*)	1143	1941–1943	70	686
Pykes Creek	465	1911–1945	125	–
Umberumberka	407	1961–1964	420	220
Corona (*)	400	1890–1910	15	–
Eildon	381	1939–1940	3885	–
Moore Creek (*)	174	1911–1924	51	674
Pekina Creek (*)	174	1911–1944	136	340 to 450
Korrumbyn Creek (*)	*1400 (?)*	1918–1924?	3	1699

Notes: (S) Summer rainfall climate; (W) Winter rainfall climate; (F) Important flushing; (*) fully-silted today; (–) data not available.

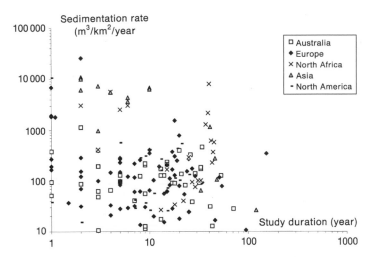

Fig. A2.3 Extreme reservoir sedimentation rates.

rivers, and sediment trapping in the reservoir. Altogether reservoir sedimentation is a very complex process. Practically, the primary consequence is the reduction of the reservoir capacity as well as its economic and strategic impact.

Extreme reservoir sedimentation has been observed all around the world (Table A2.3). Usually the sedimentation rate is defined as: (sediment inflow − sediment flushing-dredging)/year/km^2 of catchment. Prototype observations are shown in Fig. A2.3 where all the reported data must be considered as extreme events. In some cases, extreme siltation rates might lead to the reservoir siltation and its disuse (Table A2.3).

Part 3 Hydraulic Modelling

13

Summary of basic hydraulic principles

13.1 Introduction

For design engineers, it is essential to predict accurately the behaviour of hydraulic structures under the design conditions (e.g. maximum discharge capacity), the operation conditions (e.g. gate operation) and the emergency situations (e.g. Probable Maximum Flood for a spillway). During design stages, the engineers need a reliable prediction 'tool' to compare the performances of various design options.

In fluid mechanics and hydraulics, many problems are insoluble by theory or by reference to empirical data. In practice, a proper design of hydraulic structures derives from a model: a model being defined as a system that operates similarly to other systems and which provides accurate prediction of other systems' characteristics.

In hydraulics, two types of models are commonly used: computational models and physical models. Numerical models are computer software that solves the basic fluid mechanics equations: e.g. numerical integration of the backwater equation to predict the longitudinal free-surface profile of a gradually-varied open channel flow. Their application is restricted, however, to simple flow situations and boundary conditions for which the basic equations can be numerically integrated and are meaningful. The calibration and validation of computational models is extremely difficult, and most computer models are applicable only in a very specific range of flow conditions and boundary conditions. Most often, physical models must be used, including for the validation of computational models.

In this section, we consider open channels in which water flows with a free surface. First, the basic principles are summarized. Then physical and numerical modelling of hydraulic flows are discussed.

13.2 Basic principles

13.2.1 Introduction

The basic fluid mechanics principles are the continuity equation (i.e. conservation of mass), the momentum principle (or conservation of momentum) and the energy equation. A related principle is the Bernoulli equation which derives from the motion equation (e.g. Chapter 2 and Liggett 1993).

13.2.2 Basic equations in open channel flows

In an open channel flow the free surface is always a streamline and it is all the time at a constant absolute pressure (usually atmospheric). The driving force of the fluid motion is gravity and the fluid is water in most practical situations.

For an incompressible fluid (e.g. water) and for zero lateral inflow (i.e. Q constant), the *continuity equation* yields:

$$Q = V_1 A_1 = V_2 A_2 \tag{13.1}$$

where Q is the total discharge (i.e. volume discharge), V is the mean flow velocity across the cross-section A, and the subscripts 1 and 2 refer to the upstream and downstream locations respectively.

The *momentum equation* states that the change of momentum flux equals the sum of all the forces applied to the fluid. Considering an arbitrary control volume, it is advantageous to select a volume with a control surface perpendicular to the flow direction, denoted s. For a steady and incompressible flow, the momentum equation gives:

$$F_s = \rho Q (V_{s2} - V_{s1}) \tag{13.2}$$

where F_s is the resultant of the forces in the s-direction, ρ is the fluid density and $(V_{si})_{i=1,2}$ is the velocity component in the s-direction. In hydraulics, the *energy equation* is often written as:

$$H_1 = H_2 + \Delta H \tag{13.3}$$

where H is the mean total head and ΔH is the head loss.

Notes
Along any streamline, the total head is defined as:

$$H = \frac{V^2}{2g} + z + \frac{P}{\rho g} \qquad \text{total head along a streamline}$$

where V, z, P are the local fluid velocity, altitude and pressure.

Assuming a hydrostatic pressure distribution, the mean total head (in a cross-section normal to the flow direction) equals:

$$H = \alpha \frac{V^2}{2g} + z_o + d \cos \theta \qquad \text{mean total head}$$

where V is the mean flow velocity (i.e. $V = Q/A$), z_o is the bed elevation (positive upwards), θ is the bed slope and α is the kinetic energy correction coefficient (i.e. Coriolis coefficient).

13.2.3 The Bernoulli equation

The Bernoulli equation derives from the Navier–Stokes equation. The derivation is developed along a *streamline*, for a *frictionless* fluid (i.e. no viscous effect), for the volume force potential (i.e. gravity) being independent of the time, and for a *steady* and *incompressible* flow.

Along a streamline, the Bernoulli equation implies:

$$\frac{P}{\rho} + gz + \frac{V^2}{2} = \text{constant} \tag{13.4}$$

Application: smooth transition

The Bernoulli equation may be applied to smooth transitions (e.g. a step, a weir) for a steady, incompressible, frictionless flow. Assuming a hydrostatic pressure distribution and a uniform velocity distribution, the continuity equation and Bernoulli equation applied to a horizontal channel give:

$$Q = VA = \text{constant}$$

$$H = z_o + d + \frac{V^2}{2g} = \text{constant}$$

13.3 Flow resistance

In open channel flows, bottom and sidewall friction resistance can be neglected over a short transition[1] as a first approximation. However, the approximation of frictionless flow is no longer valid for long channels. Considering a water supply canal extending over several kilometres, the bottom and sidewall friction retards the fluid, and at equilibrium the friction force counterbalances exactly the weight force component in the flow direction.

The laws of flow resistance in open channels are essentially the same as those in closed pipes (Henderson 1966). The head loss ΔH over a distance Δs (along the flow direction) is given by the Darcy equation:

$$\Delta H = f \frac{\Delta s}{D_H} \frac{V^2}{2g} \tag{13.5}$$

where f is the Darcy coefficient,[2] V is the mean flow velocity and D_H is the hydraulic diameter or equivalent pipe diameter.

In open channels and assuming hydrostatic pressure distribution, the energy equation may be conveniently rewritten as:

$$d_1 \cos\theta_1 + z_{o1} + \alpha_1 \frac{V_1^2}{2g} = d_2 \cos\theta_2 + z_o2 + \alpha_2 \frac{V_2^2}{2g} + \Delta H \tag{13.6}$$

---

[1] In open channel hydraulics, the flow below a sluice gate, a hydraulic jump, the flow above a weir, or at an abrupt drop or rise may be considered as short transitions.
[2] Also called the Darcy–Weisbach friction factor or head loss coefficient.

where the subscripts 1 and 2 refer to the upstream and downstream cross-section of the control volume, d is the flow depth measured normal to the channel bottom and V is the mean flow velocity.

Notes

1. The hydraulic diameter is defined as:

$$D_H = 4 \frac{\text{cross-sectional area}}{\text{wetted perimeter}} = \frac{4A}{P_w}$$

The hydraulic diameter is also called the equivalent pipe diameter.

2. The Darcy friction factor of open channel flows can be estimated as for pipe flows:

$$f = \frac{64}{Re} \qquad \text{Laminar flow } (Re < 2000)$$

$$\frac{1}{\sqrt{f}} = -2.0 \log_{10} \left(\frac{k_s}{3.71 D_H} + \frac{2.51}{Re\sqrt{f}} \right)$$

Turbulent flow (Colebrook–White's formula, Colebrook 1939)

where k_s is the equivalent roughness height and Re is the Reynolds number: $Re = \rho V D_H / \mu$. A less-accurate formula may be used to initialize the calculation with the Colebrook–White formula:

$$f = 0.1 \left(1.46 \frac{k_s}{D_H} + \frac{100}{Re} \right)^{1/4}$$

Turbulent flow (Altsul's formula, Idelchik 1969, 1986)

14

Physical modelling of hydraulics

14.1 Introduction

Definition: the physical hydraulic model

A physical model is a scaled representation of a hydraulic flow situation. Both the boundary conditions (e.g. channel bed, sidewalls), the upstream flow conditions and the flow field must be scaled in an appropriate manner (Fig. 14.1).

Physical hydraulic models are commonly used during design stages to *optimize* a structure and to ensure a safe operation of the structure. They have an important further role to assist non-engineering people during the 'decision-making' process. A hydraulic model may help the decision-makers to visualize and to picture the flow field, before selecting a 'suitable' design.

In civil engineering applications, a physical hydraulic model is usually a smaller-size representation of the prototype (i.e. the *full-scale* structure) (e.g. Fig. 14.2). Other applications of model studies (e.g. water treatment plant, flotation column) may require the use of models larger than the prototype. In any case the model is investigated in a laboratory under controlled conditions.

Discussion

Hydraulic modelling cannot be disassociated from the basic theory of fluid mechanics. To be efficient and useful, experimental investigations require theoretical guidance which derives primarily from the basic principles (see Chapter 13) and the theory of similarity (see the next subsection).

In the present section, we will consider the physical modelling of hydraulic flows: i.e. the use of laboratory models (with controlled flow conditions) to predict the behaviour of prototype flow situations.

14.2 Basic principles

In a physical model, the flow conditions are said to be similar to those in the prototype if the model displays similarity of form (*geometric similarity*), similarity of motion (*kinematic similarity*) and similarity of forces (*dynamic similarity*).

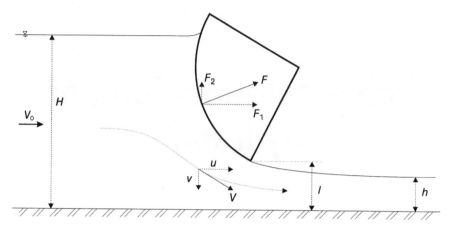

Fig. 14.1 Basic flow parameters.

Basic scale ratios

Geometric similarity implies that the ratios of prototype characteristic lengths to model lengths are equal:

$$L_r = \frac{l_p}{l_m} = \frac{d_p}{d_m} = \frac{H_p}{H_m} \qquad \text{Length} \tag{14.1}$$

where the subscripts p and m refer to prototype (full-scale) and model parameters respectively, and the subscript r indicates the ratio of prototype-to-model quantity. Length, area and volume are the parameters involved in geometric similitude.

Kinematic similarity implies that the ratios of prototype characteristic velocities to model velocities are the same:

$$V_r = \frac{V_p}{V_m} = \frac{(V_1)_p}{(V_1)_m} = \frac{(V_2)_p}{(V_2)_m} \qquad \text{Velocity} \tag{14.2}$$

Dynamic similarity implies that the ratios of prototype forces to model forces are equal:

$$F_r = \frac{F_{1_p}}{F_{1_m}} = \frac{F_{2_p}}{F_{2_m}} \qquad \text{Force} \tag{14.3}$$

Work and power are other parameters involved in dynamic similitude.

Notes
1. Geometric similarity is not enough to ensure that the flow patterns are similar in both model and prototype (i.e. kinematic similarity).
2. The combined geometric and kinematic similarities give the prototype-to-model ratios of time, acceleration, discharge, angular velocity.

(a)

(b)

Fig. 14.2 Example of physical model: breakwater jetty for the Changhua Reclamation area, along the north-west coastline of Taiwan, Republic of China (January 1994). (a) Prototype breakwater jetty. (b) Model breakwater jetty in a wave flume (Tainan Hydraulic Laboratory).

Subsequent scale ratios

The quantities L_r, V_r and F_r (defined in equations (14.1) to (14.3)) are the basic scale ratios. Several scale ratios can be deduced from equations (14.1) to (14.3):

$$M_r = \rho_r L_r^3 \qquad \text{Mass} \tag{14.4}$$

$$t_r = \frac{L_r}{V_r} \qquad \text{Time} \tag{14.5}$$

$$Q_r = V_r L_r^2 \qquad \text{Discharge} \tag{14.6}$$

$$P_r = \frac{F_r}{L_r^2} \qquad \text{Pressure} \tag{14.7}$$

where ρ is the fluid density. Further scale ratios may be deduced in particular flow situations.

Application

In open channel flows, the presence of the free-surface means that gravity effects are important. The Froude number ($Fr = V/\sqrt{gL}$) is always significant. Secondary scale ratios can be derived from the constancy of the Froude number[1] which implies:

$$V_r = \sqrt{L_r} \qquad \text{Velocity}$$

Other scale ratios are derived from the Froude similarity (e.g. Henderson 1966):

$$Q_r = V_r L_r^2 = L_r^{5/2} \qquad \text{Discharge}$$

$$F_r = \frac{M_r L_r}{T_r^2} = \rho_r L_r^3 \qquad \text{Force}$$

$$P_r = \frac{F_r}{L_r^2} = \rho_r L_r \qquad \text{Pressure}$$

14.3 Dimensional analysis

14.3.1 Basic parameters

The basic relevant parameters needed for any dimensional analysis (Fig. 14.1) may be grouped into the following groups.

(a) Fluid properties and physical constants (see Appendix A1.1). These consist of the density of water ρ (kg/m^3), the dynamic viscosity of water μ (N s/m^2), the surface tension of air and water σ (N/m), the bulk modulus of elasticity of water E_b (Pa), and the acceleration of gravity g (m/s^2).

---

[1] It is assumed that the gravity acceleration is identical in both the model and the prototype.

(b) Channel (or flow) geometry. These may consist of the characteristic length(s) L (m).
(c) Flow properties. These consist of the velocity(ies) V (m/s) and the pressure difference(s) ΔP (Pa).

14.3.2 Dimensional analysis

Taking into account all basic parameters, dimensional analysis yields:

$$\mathscr{F}_1(\rho, \mu, \sigma, E_b, g, L, V, \Delta P) = 0 \qquad (14.8)$$

There are eight basic parameters and the dimensions of these can be grouped into three categories: mass (M), length (L) and time (T). The Buckingham Π-theorem (Buckingham 1915) implies that the quantities can be grouped into five ($5 = 8 - 3$) independent dimensionless parameters:

$$\mathscr{F}_2 \left(\frac{V}{\sqrt{gL}}; \frac{\rho V^2}{\Delta P}; \frac{\rho VL}{\mu}; \frac{V}{\sqrt{\dfrac{\sigma}{\rho L}}}; \frac{V}{\sqrt{\dfrac{E_b}{\rho}}} \right) \qquad (14.9a)$$

$$\mathscr{F}_2(Fr; Eu; Re; We; Ma) \qquad (14.9b)$$

The first ratio is the Froude number Fr, characterizing the ratio of the inertial force to gravity force. Eu is the Euler number, proportional to the ratio of inertial force to pressure force. The third dimensionless parameter is the Reynolds number Re which characterizes the ratio of inertial force to viscous force. The Weber number We is proportional to the ratio of inertial force to capillary force (i.e. surface tension). The last parameter is the Sarrau–Mach number, characterizing the ratio of inertial force to elasticity force.

Notes
1. The Froude number is used generally for scaling free surface flows, open channels and hydraulic structures. Although the dimensionless number was named after William Froude (1810–1879), several French researchers used it before: e.g. Bélanger (1828), Dupuit (1848), Bresse (1860), Bazin (1865a). Ferdinand Reech (1805–1880) introduced the dimensionless number for testing ships and propellers in 1852, and the number should really be called the Reech–Froude number.
2. Leonhard Euler (1707–1783) was a Swiss mathematician and physicist, and a close friend of Daniel Bernoulli.
3. Osborne Reynolds (1842–1912) was a British physicist and mathematician who expressed first the 'Reynolds number' (Reynolds 1883).
4. The Weber number characterizing the ratio of inertial force over surface tension force was named after Moritz Weber (1871–1951), German Professor at the Polytechnic Institute of Berlin.
5. The Sarrau–Mach number is named after Professor Sarrau who first highlighted the significance of the number (Sarrau 1884) and E. Mach who

introduced it in 1887. The Sarrau–Mach number was once called the Cauchy number as a tribute to Cauchy's contribution to wave motion analysis.

Discussion

Any combination of the dimensionless numbers involved in equation (14.9) is also dimensionless and may be used to replace one of the combinations. It can be shown that one parameter can be replaced by the Morton number $Mo = (g\mu^4)/(\rho\sigma^3)$, also called the liquid parameter, since:

$$Mo = \frac{We^3}{Fr^2 Re^4} \tag{14.10}$$

The Morton number is a function only of fluid properties and the gravity constant. For the same fluids (air and water) in both model and prototype, Mo is a constant (i.e. $Mo_p = Mo_m$).

14.3.3 Dynamic similarity

Traditionally model studies are performed using geometrically similar models. In a geometrically similar model, true dynamic similarity is achieved if and only if each dimensionless parameter (or Π-terms) has the same value in both model and prototype:

$$Fr_p = Fr_m; \quad Eu_p = Eu_m; \quad Re_p = Re_m; \quad We_p = We_m; \quad Ma_p = Ma_m \tag{14.11}$$

Scale effects will exist when one or more Π-terms have different values in the model and prototype.

Practical considerations

In practice, hydraulic model tests are performed under controlled flow conditions. The pressure difference ΔP may usually be controlled. This enables ΔP to be treated as a dependent parameter. Further compressibility effects are small in clear-water flows[2] and the Sarrau–Mach number is usually very small in both model and prototype. Hence, dynamic similarity in most hydraulic models is governed by:

$$\frac{\Delta P}{\rho V^2} = \mathscr{F}_3 \left(\frac{V}{\sqrt{gL}}; \frac{\rho V L}{\mu}; \frac{V}{\sqrt{\dfrac{\sigma}{\rho L}}} \right) \tag{14.12a}$$

$$Eu = \mathscr{F}_3(Fr; Re; We) \qquad \text{Hydraulic model tests} \tag{14.12b}$$

There are a multitude of phenomena that might be important in hydraulic flow situations: e.g. viscous effects, surface tension, gravity effect. The use of the same fluid on both prototype and model prohibits simultaneously satisfying the Froude, Reynolds and Weber number scaling criteria (equation (14.12)) because the Froude

---

[2] This statement is not true in air–water flows (e.g. free-surface aerated flows) as the sound celerity may decrease to about 20 m/s for 50% volume air content (e.g. Cain 1978, Chanson 1997).

number similarity requires $V_r = \sqrt{L_r}$, the Reynolds number scaling implies that $V_r = 1/L_r$ and the Weber number similarity requires: $V_r = 1/\sqrt{L_r}$.

In most cases, only the most dominant mechanism is modelled. Hydraulic models commonly use water and/or air as flowing fluid(s). In *fully-enclosed flows* (e.g. pipe flows), the pressure losses are basically related to the Reynolds number *Re*. Hence, a Reynolds number scaling is used: i.e. the Reynolds number is the same in both model and prototype. In *free-surface flows* (i.e. flows with a free surface), gravity effects are always important and a Froude number modelling is used (i.e. $Fr_m = Fr_p$) (e.g. Fig. 14.2).

Discussion

When inertial and surface tension forces are dominant, a Weber number similarity must be selected. Studies involving air entrainment in flowing waters (i.e. white waters), de-aeration in shaft or bubble plumes are often based upon a Weber number scaling.

The Euler number is used in practice for the scaling of models using air rather than water: e.g. hydraulic models in wind tunnels, or a manifold system with water flow which is scaled at a smaller size with an air flow system.

14.3.4 Scale effects

Scale effects may be defined as the distortions introduced by effects (e.g. viscosity, surface tension) other than the dominant one (e.g. gravity in free-surface flows). They take place when one or more dimensionless parameters (see Section 14.3.3) differ between model and prototype.

Scale effects are often small but they are not always negligible altogether. Considering an overflow above a weir, the fluid is subjected to some viscous resistance along the invert. However the flow above the crest is not significantly affected by resistance, the viscous effects are small and the discharge–head relationship can be deduced as for ideal-fluid flow.

In free-surface flows, the gravity effect is dominant. If the same fluid (i.e. water) is used in both the model and the prototype, it is impossible to keep both the Froude and Reynolds numbers in the model and full-scale. Indeed it is elementary to show that a Froude similitude implies $(Re)_r = L_r^{3/2}$, and the Reynolds number becomes much smaller in the model than in the prototype (if $L_r < 1$).

Note that different fluids may be used to have the same Reynolds and Froude numbers in both the model and prototype, but this expedient is often not practical nor economical.

Some examples of scale effects

Example No. 1
Considering the drag exerted on two-dimensional bodies, Fig. 14.3 shows the effects of the Reynolds number on the drag coefficient. Dynamic similarity

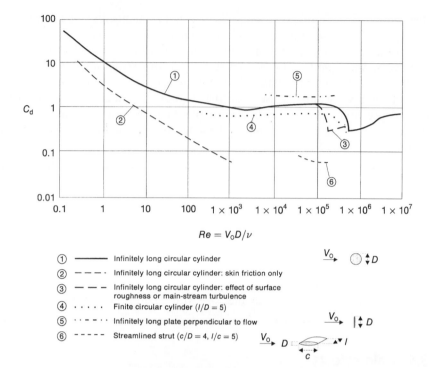

Fig. 14.3 Drag coefficient on two-dimensional bodies.

(equation (14.12)) requires the drag coefficient to be the same in the model and prototype. If the Reynolds number is smaller in the model than at full-scale (in most practical cases), Fig. 14.3 suggests that the model drag coefficient would be larger than that of the prototype and dynamic similarity could not be achieved. Moreover, the drag force comprises the form drag and the surface drag (i.e. skin friction). In small-size models, the surface drag might become predominant, particularly if the model flow is not fully-rough turbulent or the geometrical scaling of roughness height is not achievable.

In practice, an important rule, in model studies is that the model Reynolds number Re_m should be kept as large as possible, so that the model flow is fully-rough turbulent (if prototype flow conditions are fully-rough turbulent).

Example No. 2
Another example is the effect of the corner radius on the drag force on two-dimensional bodies (Fig. 14.4). Figure 14.4 shows significant differences in the Reynolds number–drag coefficient relationships depending upon the relative radius r/D. When the corner radius on the prototype is small and L_r is large, it is impossible to have the same ratio of corner radius to body size in the model and prototype because the model cannot be manufactured with the required accuracy. In such cases, the drag force is not scaled adequately.

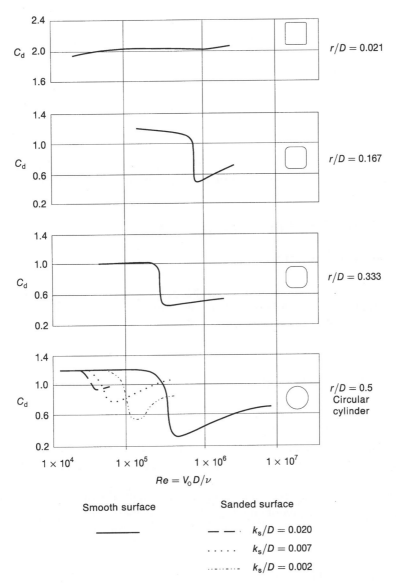

Fig. 14.4 Effect of corner radius and surface roughness on the drag coefficient of two-dimensional bodies.

Example No. 3
A different example is the flow resistance of bridge piers. Henderson (1966) showed that the resistance to flow of normal bridge pier shapes is such that the drag coefficient is about or over unity, implying that the form drag is a significant component of the total drag. In such a case, the viscous effects are relatively small, and dynamic similarity is achievable, provided that model viscous effects remain negligible.

Discussion

If scale effects would become significant in a model, a smaller prototype-to-model scale ratio L_r should be considered to minimize the scale effects. For example, in a 100:1 scale model of an open channel, the gravity effect is predominant but viscous effects might be significant. A geometric scale ratio of 50:1 or 25:1 may be considered to reduce or eliminate viscous scale effects.

Another example is the entrainment of air bubbles in free-surface flows. Gravity effects are predominant but it is recognized that surface tension scale effects can take place for $L_r > 10$ to 20 (or $L_r < 0.05$ to 0.1) (e.g. Wood 1991, Chanson 1997).

At the limit, no scale effect is observed at full-scale (i.e. $L_r = 1$) as all the Π-terms (equation (14.11)) have the same values in the prototype and model when $L_r = 1$.

14.4 Modelling fully-enclosed flows

14.4.1 Reynolds models

Fully-enclosed flow situations include pipe flows, turbomachines and valves. For such flow situations, viscosity effects on the solid boundaries are important. Physical modelling is usually performed with a Reynolds similitude: i.e. the Reynolds number is kept identical in both the model and prototype:

$$Re_p = Re_m \qquad (14.13)$$

If the same fluid is used in both the model and prototype, equation (14.13) implies:

$$V_r = 1/L_r \qquad \text{(Reynolds similitude)}$$

For $L_r > 1$, the model velocity must be larger than that in the prototype.

Discussion

For example, if the model scale is 10:1 (i.e. $L_r = 10$), the velocity in the model must be ten times that in the prototype. By using a different fluid in the model, the ratio (μ_r/ρ_r) becomes different from unity and V_m can be reduced.

14.4.2 Discussion

Flow resistance in pipe flows

For pipe flows, the Darcy equation relates the pressure losses to the pipe geometry (diameter D, length L) and to the flow velocity V:

$$\Delta P = f \frac{L}{D} \frac{\rho V^2}{2} \qquad (14.14)$$

where f is the Darcy–Weisbach friction factor. After transformation and combining with equation (14.10), it leads:

$$\frac{fL}{2D} = Eu = \mathscr{F}_4(Fr; Re; We; Ma; \ldots) \qquad (14.15)$$

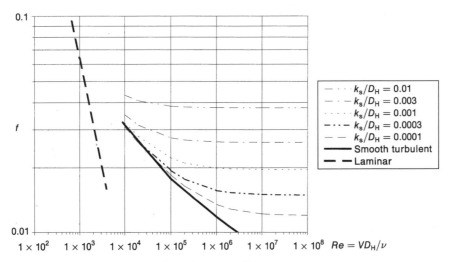

Fig. 14.5 Friction factor versus Reynolds number in pipe flows.

In pipe flows, gravity and surface tension have no effect on the pressure losses. For steady liquid flows, the compressibility effects are negligible. The roughness height k_s is, however, an additional characteristic length. For a uniformly distributed-roughness, equation (14.15) becomes:

$$\frac{fL}{2D} = \mathscr{F}_4\left(Re;\ \frac{k_s}{D}\right) \tag{14.16}$$

Equation (14.16) expresses the dimensionless relationship between friction losses in pipes, the Reynolds number and relative roughness. An illustration is the Moody diagram (Fig. 14.5).

Skin friction and form drag

Considering the drag on a body (e.g. Figs 14.3 and 14.4), the pressure losses associated with the modification of the flow field caused by the presence of the body are usually expressed in terms of the drag force F_d on the body. The Euler number is rewritten as: $Eu = F_d/(\rho V^2 A)$, where A is the projection of the body in the plane normal to the flow direction. F_d/A is equivalent to the pressure loss ΔP.

Equations (14.10) and (14.15) may be combined to relate the drag coefficient C_d to the Π-terms:

$$C_d = \frac{Eu}{2} = \frac{F_d}{\dfrac{\rho}{2} V^2 A} = \mathscr{F}_5(Fr;\ Re;\ We;\ Ma;\ldots) \tag{14.17}$$

In equation (14.17), the Reynolds number Re is related to the *skin friction drag* due to viscous shear as well as to *form drag* resulting from the separation of the flow streamlines from the body.

14.4.3 Practical considerations in modelling fully-enclosed flows

The flow regime in pipes is either laminar or turbulent. In industrial applications, it is commonly accepted that the flow becomes turbulent for Reynolds numbers larger than 2000 to 3000, the Reynolds number being defined in terms of the equivalent pipe diameter D and of the mean flow velocity V.

For turbulent flows, the flow regime can be sub-divided into three categories: smooth, transition and fully rough. Each category can be distinguished as a function of a shear Reynolds number defined as:

$$Re_* = \rho \frac{V_* k_s}{\mu} \tag{14.18}$$

where V_* is the shear velocity. The transition between smooth turbulence and fully-rough turbulence is approximately defined as:

Flow situation Ref.	Open channel flow (Henderson 1966)	Pipe flow (Schlichting 1979)
Smooth turbulent	$Re_* < 4$	$Re_* < 5$
Transition	$4 < Re_* < 100$	$5 < Re_* < 75$
Fully rough turbulent	$100 < Re_*$	$75 < Re_*$

Dynamic similarity of fully-enclosed flows implies the same resistance coefficient in both the model and the prototype. This can be achieved with the Reynolds number being the same in the model and prototype, or with both flows in the model and prototype being fully-rough turbulent (Fig. 14.5).

If the full-scale flow is turbulent, it is extremely important to ensure that the model flow is also turbulent. Laminar and turbulent flows exhibit very important basic differences. In most cases, turbulent flows should not be scaled with laminar flow models.

Note

The Reynolds number can be kept constant by changing the flowing fluid. For example the atmospheric wind flow past a tall building could be modelled in a small-size water tunnel at the same Reynolds number.

14.5 Modelling free-surface flows

14.5.1 Presentation

In free-surface flows (e.g. rivers, wave motion), gravity effects are predominant. Model-prototype similarity is performed usually with a Froude similitude:

$$Fr_p = Fr_m \tag{14.19}$$

If the gravity acceleration is the same in both the model and prototype, a Froude number modelling implies:

$$V_r = \sqrt{L_r} \qquad \text{(Froude similitude)}$$

Note that the model velocity is less than that in the prototype for $L_r > 1$ and the time scale equals $t_r = \sqrt{L_r}$.

Remarks

Froude number modelling is typically used when friction losses are small and the flow is highly turbulent: e.g. spillways, overflow weirs, flow past bridge piers. It is also used in studies involving large waves: e.g. breakwater or ship models.[3]

A main concern is the potential for scale effects induced by viscous forces. Scale effects caused by surface tension effects are another concern, in particular when free-surface aeration (i.e. air entrainment) takes place.

14.5.2 Modelling hydraulic structures and wave motion

In hydraulic structures and for wave motion studies (Fig. 14.2), the gravity effect is usually predominant in the prototype. The flow is turbulent, and hence viscous and surface tension effects are negligible in the prototype if the flow velocity is reasonably small. In such cases a Froude similitude must be selected.

The most economical strategy is:

1. to choose a geometric scale ratio L_r such as to keep the model dimensions small, and
2. to ensure that the model Reynolds number Re_m is large enough to make the flow turbulent at the smallest test flows.

14.5.3 Modelling rivers and flood plains

In river modelling, gravity effects and viscous effects are basically of the same order of magnitude. For example, in uniform equilibrium flows (i.e. normal flows), the gravity force component counterbalances exactly the flow resistance and the flow conditions are deduced from the continuity and momentum equations.

In practice, river models are scaled with a Froude similitude (equation (14.19)) and viscous scale effects must be minimized. The model flow must be turbulent, and possibly fully-rough turbulent with the same relative roughness as for the prototype:

$$Re_m > 5000 \qquad (14.20)$$

$$(k_s)_r = L_r \qquad (14.21)$$

---

[3] The testing of ship models is very specialized. Interestingly, F. Reech and W. Froude were among the first to use the Froude similitude for ship modelling.

where the Reynolds number is defined in terms of the hydraulic diameter (i.e. $Re = \rho V D_{H}/\mu$).

Distorted models

A distorted model is a physical model in which the geometric scale is different between each main direction. For example, river models are usually designed with a larger scaling ratio in the horizontal directions than in the vertical direction: $X_r > Z_r$. The scale distortion does not distort seriously the flow pattern and it usually gives good results.

A classical example of a distorted model is that of the Mississippi river, built by the US Army Corps of Engineers. The Mississippi basin is about $3\,100\,000\,km^2$ and the river is nearly 3800 km long. An outdoor model was built with a scale of 2000:1. If the same scaling ratio was applied to both the vertical and horizontal dimensions, prototype depths of about 6 m would imply model depths of about 3 mm. With such small flow depths, surface tension and viscous effects would be significant. The Mississippi model was built, in fact, with a distorted scale: $Z_r = 100$ and $X_r = 2000$. Altogether the model size is about 1.5 km per 2 km!

A distorted model of rivers is designed with a Froude similitude:

$$Fr_p = Fr_m \tag{14.19}$$

where the Froude number scaling ratio is related to the vertical scale ratio:

$$Fr_r = \frac{V_r}{\sqrt{Z_r}} \tag{14.22}$$

As for an undistorted model, the distorted model flow must be turbulent (equation (14.20)), and preferably fully-rough turbulent with the same relative roughness as for the prototype:

$$(k_s)_r = Z_r \tag{14.23}$$

The Froude similitude (equation (14.22)) implies:

$$V_r = \sqrt{Z_r} \qquad \text{Velocity} \tag{14.24}$$

$$Q_r = V_r X_r Z_r = Z_r^{3/2} X_r \qquad \text{Discharge} \tag{14.25}$$

$$t_r = \frac{X_r}{V_r} = \frac{X_r}{\sqrt{Z_r}} \qquad \text{Time} \tag{14.26}$$

$$(\tan\theta)_r = \frac{Z_r}{X_r} \qquad \text{Longitudinal bed slope} \tag{14.27}$$

where θ is the angle between the channel bed and the horizontal.

With a distorted scale model, it is possible to select small physical models (i.e. X_r large). In addition to the economical and practical benefits, distorted models also have the following advantages compared with non-distorted models:

- the flow velocities and turbulence in the model are larger (equation (14.24)),
- the time scale is reduced (equation (14.26)),
- the model Reynolds number is larger, improving the prototype-to-model dynamic similarity, and

• the larger vertical scale (i.e. $Z_r < X_r$) allows a greater accuracy on the flow depth measurements.

Discussion

Practically it is recommended that the model distortion (i.e. the ratio X_r/Z_r) should be less than 5 to 10. Some disadvantages of distorted models may be mentioned for completeness: the velocity directions are not always reproduced correctly, and some observers might be distracted unfavourably by the model distortion leading to inaccurate or incorrect judgements.

Movable-bed models

Movable-bed hydraulic models are some of the most difficult types of models and they often give unsatisfactory results.

The primary difficulty is to scale both the sediment movement and the fluid motion. Furthermore, the bed roughness becomes a function of the bed geometry and of the sediment transport. Early movable bed model studies on the River Mersey (England) and Seine River (France) in the 1880s showed that the time scale governing the fluid flow differs from the time scale governing sediment motion (see Appendix A3.1).

A detailed analysis of sediment transport modelling is developed in Appendix A3.1. Several authors (e.g. Henderson 1996, pp. 497–508, Graf 1971, pp. 392–398) also discussed various methods for 'designing' a movable-bed model.

The most important point is the need to verify and to calibrate a movable-bed model before using it as a prediction tool.

14.5.4 Resistance scaling

The modelling of flow resistance is not a simple matter. Often the geometric similarity of roughness height and spacing is not enough. For example, it is observed sometimes that the model does not reproduce the flow patterns in the prototype because the model is too 'smooth' or too 'rough'. In some cases (particularly with a large scale ratio L_r), the model flow is not as turbulent as the prototype flow. A solution is to use roughness elements (e.g. mesh, wire, vertical rods) to enhance the model flow turbulence, hence to simulate more satisfactorily the prototype flow pattern.

Another aspect is the scaling of the resistance coefficient. The flow resistance can be described in terms of the Darcy friction factor or an empirical resistance coefficient (e.g. Chézy or Gauckler–Manning coefficients).

In uniform equilibrium flows, the momentum equation implies:

$$V_r = \sqrt{L_r} = \sqrt{\frac{(D_H)_r(\sin\theta)_r}{f_r}} \tag{14.28}$$

For an undistorted model, a Froude similitude (equation (14.19) and (14.28)) implies that the model flow resistance will be similar to that in the prototype:

$$f_r = 1 \tag{14.29}$$

Most prototype flows are fully-rough turbulent and the Darcy friction factor is primarily a function of the relative roughness.

Another approach is based upon the Gauckler–Manning coefficient. The Chézy equation implies that, in gradually-varied and uniform equilibrium flows, the following scaling relationship holds:

$$V_r = \sqrt{L_r} = \frac{1}{(n_{\text{Manning}})_r}((D_H)_r)^{2/3}\sqrt{(\sin\theta)_r} \tag{14.30}$$

For an undistorted scale model, equation (14.30) becomes:

$$(n_{\text{Manning}})_r = L_r^{1/6} \tag{14.31}$$

Equation (14.31) indicates that the notion of complete similarity is applied both to the texture of the surface and to the shape of its general outline (Henderson 1966). In practice, the lowest achievable value of n_{Manning} is about 0.009 to $0.010\,\text{s/m}^{1/3}$ (i.e. for glass). With such a value, the prototype resistance coefficient $(n_{\text{Manning}})_p$ and the Gauckler–Manning coefficient similarity $(n_{\text{Manning}})_r$ could limit the maximum geometrical similarity ratio L_r. If L_r is too small (typically less than 40), the physical model might not be economical nor convenient.

In summary, a physical model (based upon a Froude similitude) has proportionally more resistance than the prototype. If the resistance losses are small (e.g. at a weir crest), the resistance scale effects are not considered. In the cases of river and harbour modelling, resistance is significant. The matter may be solved using distorted models.

Distorted models

With a distorted scale model, equations (14.28) and (14.30) become respectively:

$$V_r = \sqrt{Z_r} = \sqrt{\frac{(D_H)_r(\sin\theta)_r}{f_r}} \tag{14.32}$$

$$V_r = \sqrt{Z_r} = \frac{1}{(n_{\text{Manning}})_r}((D_H)_r)^{2/3}\sqrt{(\sin\theta)_r} \tag{14.33}$$

For a wide channel (i.e. $(D_H)_r = Z_r$) and a flat slope (i.e. $(\sin\theta)_r = (\tan\theta)_r$), the scaling of flow resistance in distorted models implies:

$$f_r = \frac{Z_r}{X_r} \qquad \text{wide channel and flat slope} \tag{14.34}$$

$$(n_{\text{Manning}})_r = \frac{Z_r^{2/3}}{\sqrt{X_r}} \qquad \text{wide channel and flat slope} \tag{14.35}$$

Discussion

In practice $Z_r/X_r < 1$ and equation (14.34) would predict a model friction factor lower than that in the prototype. But equation (14.35) could imply a model resistance coefficient larger or smaller than that in the prototype depending upon the ratio $Z_r^{2/3}/\sqrt{X_r}$!

14.6 Design of physical models

14.6.1 Introduction

Before building a physical model, engineers must have the appropriate topographic and hydrological field information. The type of model must then be selected, and a question arises:

> Which is the dominant effect: e.g. viscosity, gravity or surface tension?

14.6.2 General case

In the general case, the engineer must choose a proper geometric scale. The selection procedure is an iterative process.

Step 1. Select the smallest geometric scale ratio L_r to fit within the constraints of the laboratory.

Step 2. For L_r, and for the similitude criterion (e.g. Froude or Reynolds), check if the discharge can be scaled properly in the model, based upon the maximum model discharge $(Q_m)_{max}$.

> For L_r and the similitude criterion, is the maximum model discharge large enough to model the prototype flow conditions?

Step 3. Check if the flow resistance scaling is achievable in the model.

> Is it possible to achieve the required f_m or $(n_{Manning})_m$ in the model?

Step 4. Check the model Reynolds number Re_m for the smallest test flow rate.

> For $(Q_p)_{min}$, what are the flow conditions in the model: e.g. laminar or turbulent, smooth-turbulent or fully-rough-turbulent? If the prototype flow is turbulent, model flow conditions must be turbulent (i.e. typically $Re_m > 5000$).

Step 5. Choose the convenient scale.

When a simple physical model is not feasible, more advanced modelling techniques can be used: e.g. a two-dimensional model (e.g. spillway flow), a distorted scale model (e.g. river flow).

14.6.3 Distorted scale models

For a distorted scale model, the engineer must select two (or three) geometric scales. The model design procedure is again an iterative process:

Step 1. Select the smallest horizontal scale ratio X_r to fit within the constraints of the laboratory.

Step 2. Determine the possible range of vertical scale Z_r such as:
+ the smallest scale $(Z_r)_1$ is that which gives the limit of the discharge scaling ratio, based upon the maximum model discharge $(Q_m)_{max}$,
+ the largest scale $(Z_r)_2$ is that which gives the feasible flow resistance coefficient (i.e. feasible f_m or $(n_{Manning})_m$), and
+ check the distortion ratio X_r/Z_r.
$(X_r/Z_r$ should be less than 5 to 10.)

Step 3. Check the model Reynolds number Re_m for the smallest test flow rate. This might provide a new largest vertical scale ratio $(Z_r)_3$.
+ check the distortion ratio X_r/Z_r.

Step 4. Select a vertical scale ratio which satisfies: $(Z_r)_1 < Z_r < \text{Min}[(Z_r)_2, (Z_r)_3]$. If this condition cannot be satisfied, a smaller horizontal scale ratio must be chosen.
+ check the distortion ratio X_r/Z_r.
In practice it is recommended that X_r/Z_r should be less than 5 to 10.

Step 5. Choose the convenient scales (X_r, Z_r).

14.7 Summary

Physical hydraulic modelling is a design technique used by engineers to optimize the structure design, to ensure the safe operation of the structure and/or to facilitate the decision-making process.

In practice, most hydraulic models are scaled with either a Froude or a Reynolds similitude: i.e. the selected dimensionless number is the same in the model and in the prototype (i.e. full-scale).

The most common fluids are air and water. Free-surface flow modelling is most often performed with the same fluid (e.g. water) in full-scale and the model. Fully-enclosed flow modelling might be performed with water in the prototype and air in the model. The selection of fluid in the model and the prototype fixes the density scale ratio ρ_r.

Table 14.1 summarizes the scaling ratios for the Froude and Reynolds similitudes.

Table 14.1 Scaling ratios for Froude and Reynolds similitudes (undistorted model)

Parameter	Unit	Scale ratio with		
		Froude law[a]	Froude law[a] (distorted model)	Reynolds law
(1)	(2)	(3)	(4)	(5)
Geometric properties				
Length	m	L_r	X_r, Z_r	L_r
Area	m^2	L_r^2	–	L_r^2
Kinematic properties				
Velocity	m/s	$\sqrt{L_r}$	$\sqrt{Z_r}$	$1/L_r \times \mu_r/\rho_r$
Discharge per unit width	m^2/s	$L_r^{3/2}$	$\sqrt{Z_r}X_r$	μ_r/ρ_r
Discharge	m^3/s	$L_r^{5/2}$	$Z_r^{3/2}X_r$	$L_r\mu_r/\rho_r$
Time	s	$\sqrt{L_r}$	$X_r/\sqrt{Z_r}$	$L_r^2\rho_r/\mu_r$
Dynamic properties				
Force	N	$\rho_r L_r^3$	–	μ_r^2/ρ_r
Pressure	Pa	$\rho_r L_r$	$\rho_r Z_r$	$\mu_r^2/\rho_r \times 1/L_r^2$
Density	kg/m^3	ρ_r	ρ_r	ρ_r
Dynamic viscosity	Pa s	$L_r^{3/2}\sqrt{\rho_r}$	–	μ_r
Surface tension	N/m	L_r^2	–	$\mu_r^2/\rho_r \times 1/L_r$

Note: [a] assuming identical gravity acceleration in model and prototype.

14.8 Exercises

Numerical solutions to some of these exercises are available from the Web at www.arnoldpublishers.com/support/chanson

A butterfly valve is to be tested in a laboratory to determine the discharge coefficient for various openings of the disc. The prototype size will be 2.2 m in diameter and it will be manufactured from cast steel with machined inside surfaces (roughness height estimated to be about 0.5 mm). The maximum discharge to be controlled by the valve is 15 m^3/s. The laboratory model is a 5:1 scale model.

(a) What surface condition is required in the model? What model discharge is required to achieve complete similarity with the prototype, if water is used in both? (b) Can these conditions be achieved? (c) If the maximum flow available for model tests is 200 L/s, could you accurately predict prototype discharge coefficients from the results of the model tests?

Summary sheet:

(a) $(k_s)_m =$	$Q_m =$
(b) Yes/No	Reasons:
(c) $Re_p =$	Re_m
Discussion:	

The inlet of a Francis turbine is to be tested in a laboratory to determine the performances for various discharges. The prototype size of the radial flow rotor will be:

inlet diameter $= 0.6$ m, width $= 0.08$ m, inlet crossflow area $= \pi \times$ diameter $\times$ width. It will be manufactured from cast steel with machined inside surfaces (roughness height estimated to be about 0.3 mm). The maximum discharge to be turbined (by the Francis wheel) is 1.4 m³/s. The laboratory model is a 5:1 scale model.

(a) What surface condition is required in the model? What model discharge is required to achieve complete similarity with the prototype, if water is used in both?

(b) Can these conditions be achieved? (Compute the minimum required model total head and flow rate. Compare these with the pump performances of a typical hydraulic laboratory: $H \sim 10$ m, $Q \sim 100$ L/s.)

(c) If the maximum flow available for model tests is 150 L/s, would you be able to predict accurately prototype performances from the results of the model tests? (Justify your answer.)

Summary sheet:

(a) $(k_s)_m =$	$Q_m =$
(b) Yes/No	Reasons:
(c) $Re_p =$	Re_m
Discussion:	

An overflow spillway is to be designed with an uncontrolled crest followed by a stepped chute and a hydraulic jump dissipator. The maximum spillway capacity will be 4300 m³/s. The width of the crest, chute and dissipation basin will be 55 m. A 50:1 scale model of the spillway is to be built. Discharges ranging between the maximum flow rate and 10% of the maximum flow rate are to be reproduced in the model.

(a) Determine the maximum model discharge required. (b) Determine the minimum prototype discharge for which negligible scale effects occur in the model. (Comment on your result.) (c) What will be the scale for the force ratio?

Laboratory tests indicate that operation of the basin may result in unsteady wave propagation downstream of the stilling basin with a model wave amplitude of about 0.05 m and model wave period of 47 seconds. Calculate: (d) the prototype wave amplitude and (e) the prototype wave period.

Summary sheet:

(a) Maximum $Q_m =$	
(b) Minimum $Q_p =$	Why?
(c) $Force_r =$	
(d) $A_p =$	
(e) $t_p =$	

A 35.5:1 scale model of a concrete overfall spillway and stilling basin is to be built. The prototype discharge will be 200 m³/s and the spillway crest length is 62 m. (a) Determine the maximum model discharge required and the minimum prototype discharge for which negligible scale effects occur in the model. (b) In tests involving baffle blocks for stabilizing the hydraulic jump in the stilling basin, the force measured on each block was 9.3 N. What is the corresponding prototype force? (c) The channel downstream of the stilling basin is to be lined with rip-rap (angular blocks of rock) approximately 650 mm in size. The velocity measured near the rip-rap is as low as 0.2 m/s. Check whether the model Reynolds number is large enough for the drag coefficient of the model rocks to be the same as in the prototype. What will be the scale for the force ratio?

Summary sheet:

(a) Maximum $Q_m =$	
Minimum $Q_p =$	Why?
(b) Force$_p =$	
(c) $Re_m =$	Comment:
Force ratio $=$	

A sluice gate will be built across a 25 m wide rectangular channel. The maximum prototype discharge will be 275 m³/s and the channel bed will be horizontal and con-crete-lined (i.e. smooth). A 35:1 scale model of the gate is to be built for laboratory tests.

(a) What similitude should be used? Calculate: (b) model width and (c) maximum model flow rate.

For one particular gate opening and flow rate, the *laboratory* flow conditions are: upstream flow depth of 0.2856 m, downstream flow depth of 0.0233 m. (d) Compute the model discharge. State the basic principle(s) involved. (e) Compute the model force acting on the sluice gate. State the basic principle(s) involved. (f) What will be the corresponding prototype discharge and force on the gate? (g) What will be the scale for the force ratio?

Gate operation may result in unsteady flow situations. If a prototype gate operation has the following characteristics: gate opening duration = 15 minutes, initial discharge = 180 m³/s, new discharge = 275 m³/s, calculate: (h) gate opening duration and (i) discharges to be used in the model tests.

Summary sheet:

(a) Similitude:	Why?
(b) $B_m =$	
(c) $(Q_m)_{max} =$	
(d) $Q_m =$	Principle(s):
(e) $F_m =$	Principle(s):
(f) $Q_p =$	
Force$_p =$	
(g) Force$_r =$	
(h) $t_m =$	
(i) $Q_m =$ (before)	$Q_m =$ (after)

A hydraulic jump stilling basin, equipped with baffle blocks, is to be tested in a labora-tory to determine the dissipation characteristics for various flow rates. The maximum prototype discharge will be $220\,\mathrm{m^3/s}$ and the rectangular channel will be 10 m wide. (Assume the channel bed to be horizontal and concrete-lined, i.e. smooth.) A 40:1 scale model of the stilling basin is to be built. Discharges ranging between the maximum flow rate and 10% of the maximum flow rate are to be reproduced in the model.

(a) What similitude should be used? (Justify your selection.) (b) Determine the maximum model discharge required. (c) Determine the minimum prototype discharge for which negligible scale effects occur in the model. (Comment on your result.)

For one particular inflow condition, the *laboratory* flow conditions are: upstream flow depth of 0.019 m, upstream flow velocity of 2.38 m/s, downstream flow depth of 0.122 m. (d) Compute the model force exerted on the baffle blocks. (State the basic principle(s) involved.) (e) What is the direction of force in (d): i.e. upstream or downstream? (f) What will be the corresponding prototype force acting on the blocks? (g) Compute the prototype head loss.

Operation of the basin may result in unsteady wave propagation downstream of the stilling basin. (h) What will be the scale for the time ratio?

Tests will be made on a model sea wall of 1/18 prototype size. (a) If the prototype wave climate is: wave period = 12 seconds, wave length = 20 m, wave amplitude = 2.1 m, what wave period, wave length and wave amplitude should be used in the model tests? (b) If the maximum force exerted by a wave on the model sea wall is 95 N, what corresponding force will be exerted on the prototype?

Summary sheet:

(a) $t_{\mathrm{m}} =$ $L_m =$	
$A_{\mathrm{m}} =$	
(b) Force$_{\mathrm{p}} =$	

A fixed bed model is to be built of a certain section of a river. The maximum full-scale discharge is $2750\,\mathrm{m^3/s}$, the average width of the river is 220 m and the bed slope is 0.16 m per kilometre. The Gauckler–Manning coefficient for the prototype is estimated at $0.035\,\mathrm{s/m^{1/3}}$. Laboratory facilities limit the scale ratio to 200:1 and maximum model discharge is 45 L/s. Note that the smoothest model surface feasible has a Gauckler–Manning coefficient of about $0.014\,\mathrm{s/m^{1/3}}$. Discharges ranging between the maximum flow rate and 15% of the maximum flow rate are to be reproduced in the model.

Determine the acceptable maximum and minimum values of the vertical scale ratio Z_r. Select a suitable scale for practical use, and calculate the corresponding model values of the Gauckler–Manning coefficient, maximum discharge and normal depth (at maximum discharge). (It can be assumed that the river channel is wide (i.e. $D_H \sim 4d$) in both the model and prototype for all flows. Assume that uniform equilibrium flow is achieved in the model and prototype.)

Summary sheet:

Minimum Z_r =	Why?
Maximum Z_r =	Why?
Alternative Max. Z_r =	Why?
Allowable range for Z_r =	
Your choice for Z_r =	Why?
Corresponding values of Q_r =	$(n_{Manning})_r$ =
$(n_{Manning})_r$ =	Maximum Q_m =
d_m =	at maximum flow rate

An artificial concrete channel model is to be built. Laboratory facilities limit the scale ratio to 50:1 and the maximum model discharge is 50 L/s. The maximum full-scale discharge is 150 m³/s, the cross-section of the channel is approximately rectangular (50 m bottom width) and the bed slope is 0.14 m per kilometre. (Note: The roughness height of the prototype is estimated as 3 mm while the smoothest model surface feasible has a Darcy friction factor of about $f = 0.03$.) Discharges ranging between the maximum flow rate and 10% of the maximum flow rate are to be reproduced in the model.

For an undistorted model: (a) what would be the model discharge at maximum full-scale discharge? (b) what would be the Darcy coefficient of the model flow? (c) what would be the Darcy coefficient of the prototype channel? (d) comment and discuss your findings. (Assume normal flow conditions.)

A distorted model is to be built. (e) Determine the acceptable maximum and minimum values of the vertical scale ratio Z_r. (f) Select a suitable scale for practical use. Calculate the corresponding model values of: (g) Darcy coefficient, (h) maximum discharge and (i) normal depth (at maximum discharge).

A fixed bed model is to be made of a river with a surface width of 80 m. The Gauckler–Manning coefficient for the river is estimated at 0.026 s/m$^{1/3}$. Scale ratios of $X_r = 150$ and $Z_r = 25$ have been selected. (a) Find the required model values of the Gauckler–Manning coefficient corresponding to prototype depths of water of 2.0 and 5.0 m, if the cross-sectional shape is assumed to be rectangular. (b) What material would you recommend to use in the laboratory model for a prototype depth of 2.0 m?

15

Numerical modelling: backwater computations

15.1 Introduction

Numerical models of hydraulic flows are computer programs designed to solve the basic fluid mechanics equations. They are applied typically to simple flows with uncomplicated boundary conditions for which the basic equations can be numerically integrated and are meaningful. The calibration and validation of computational models is extremely difficult, and most computer models are applicable only to a very specific range of flow conditions and boundary conditions.

In this chapter, we present an example of hydraulic modelling: the numerical integration of the energy equation for gradually-varied flows in an open channel. It is one of the simplest forms of a numerical model: i.e. the numerical integration of a steady one-dimensional flow.

15.2 Basic equations

15.2.1 Presentation

Considering a steady open channel flow and along a streamline in the s-direction, the energy equation written in terms of mean total head is:

$$\frac{\partial H}{\partial s} = \frac{\partial}{\partial s}\left(z + \frac{P}{\rho g} + \frac{V^2}{2g}\right)_{\text{mean}} = -f\frac{1}{D_\text{H}}\frac{V^2}{2g} \qquad (15.1a)$$

where s is the distance along the channel bed (Fig. 15.1), z is the elevation (positive upwards), P is the pressure, V is the fluid velocity along a streamline, f is the Darcy friction factor, D_H is the hydraulic diameter, V is the mean flow velocity (i.e. $V = Q/A$) and the subscript 'mean' refers to the mean total head over the cross-sectional area. In the following development, we shall consider the mean flow properties only.

The energy equation (i.e. equation (15.1)) is also called the *backwater equation*. It is usually rewritten as:

$$\frac{\partial H}{\partial s} = -S_\text{f} \qquad (15.1b)$$

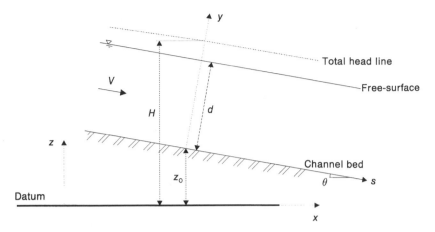

Fig. 15.1 Sketch of open channel flow.

where S_f is the friction slope defined as:

$$S_f = \frac{\tau_o}{\rho g D_H} = f \frac{1}{D_H} \frac{V^2}{2g} \qquad (15.2)$$

where τ_o is the boundary shear stress and D_H is the hydraulic diameter.

Notes
1. S_f is called the friction slope or energy slope. That is, it is the slope of the total energy line (or total head line – THL) (Fig. 15.1).
2. Assuming a hydrostatic pressure gradient, equation (15.1a) can be transformed as:

$$\frac{\partial H}{\partial s} = \frac{\partial}{\partial s}\left(z_o + d\cos\theta + \alpha\frac{V^2}{2g}\right) = -S_f$$

   where z_o is the bed elevation, d is the flow depth measured normal to the channel bottom, θ is the longitudinal bed slope and α is the kinetic energy correction coefficient (i.e. Coriolis coefficient).
3. Equation (15.1a) is the basic backwater equation, first developed by J.B. Bélanger (Bélanger 1828). It is a one-dimensional model of gradually-varied steady flows in an open channel.
4. Further developments of the backwater equation are presented in Appendix A3.2.

15.2.2 Basic assumptions

Equation (15.1) may be applied to natural and artificial channels. It is important to remember the basic assumptions (Chapter 5).

The backwater calculations are developed assuming [H1] a non-uniform flow, [H2] a steady flow, [H3] that the flow is gradually varied, and [H4] that, at a given section, the flow resistance is the same as for a uniform flow for the same depth and discharge, regardless of trends of the depth.

The gradually-varied flow (GVF) calculations (i.e. backwater calculations) do not apply to uniform equilibrium flows, nor to unsteady flows nor to rapidly-varied flows (RVF). Furthermore, the last assumption [H4] implies that the Darcy, Chézy or Gauckler–Manning equations may be used to estimate the flow resistance, although these equations were originally developed for uniform flows only (see Section 15.2.4).

Note

In gradually varied flows, the friction slope and Darcy coefficient are calculated as in uniform equilibrium flow (see Chapter 4) but using the local non-uniform equilibrium flow depth.

15.2.3 Applications

In natural streams, the cross-sections are irregular and variable, and the location of the free-surface elevation in gradually-varied flows is predicted using equation (15.1). Furthermore, the bed slope is usually small in natural systems (i.e. typically $\tan \theta < 0.01$). For such small slopes, the following approximation holds: $\tan \theta \approx \sin \theta \approx \theta$, where θ is in radians. The flow depth (measured normal to the channel bed) approximately equals the free-surface elevation above the channel bed within 0.005%. These variables are used interchangeably.

At hydraulic structures and spillways, the flow is often supercritical. The channel slope may range from zero to steep (over 75°). Negative slopes might also take place (e.g. flip bucket). With such structures, it is extremely important to assess first whether gradually varied flow calculations (i.e. backwater calculations) can be applied or not (see Section 15.2.2). In the affirmative, it is frequent that the velocity distribution is not uniform and the pressure distribution might not be hydrostatic.

In man-made channels, simpler forms of the backwater equation are available. Assuming a hydrostatic pressure distribution, the backwater equation may be expressed in term of the flow depth:

$$\frac{\partial d}{\partial s}(\cos \theta - \alpha Fr^2) = S_o - S_f \tag{15.3}$$

where $S_o = \sin \theta$, α is the Coriolis coefficient (or kinetic energy correction coefficient) and Fr is the Froude number defined as:

$$Fr = \frac{Q}{\sqrt{g \dfrac{A^3}{B}}} \tag{15.4}$$

A is the flow cross-section area and B is the open channel free surface width.

Notes

1. If the velocity varies across the section, the mean of the velocity head $(V^2/(2g))_{mean}$ is not equal to $V^2/(2g)$, where the subscript 'mean' refers to the mean value over a cross-section. The ratio of these quantities is called the Coriolis coefficient, denoted α and defined as:

$$\alpha = \frac{\int_A \rho v^3 \, dA}{\rho V^3 A}$$

2. In artificial channels of simple shape, equation (15.3) can be integrated analytically (e.g. Chow 1973, pp. 237–242) or semi-analytically (e.g. Henderson 1966, pp. 130–136, Chow 1973, pp. 252–262).
3. Equation (15.4) is a common definition of the Froude number for a channel of irregular cross-sectional shape.
4. For a rectangular channel, the Froude number equals: $Fr = V/\sqrt{gd}$.

15.2.4 Discussion: flow resistance calculations

For channels extending over several kilometres, the boundary shear opposes the fluid motion and retards the flow. The flow resistance and gravity effects are of the same order of magnitude and it is essential to predict accurately the flow resistance. In fact, the friction force exactly equals and opposes the weight force component in the flow direction at equilibrium (i.e. normal flow conditions).

The flow resistance caused by friction is given by the Darcy equation:

$$\Delta H = f \frac{\Delta s}{D_H} \frac{V^2}{2g} \tag{15.5a}$$

where ΔH is the total head loss over a distance Δs (in the flow direction), f is the Darcy coefficient, V is the mean flow velocity and D_H is the hydraulic diameter or equivalent pipe diameter. Using the definition of the friction slope (equation (15.2)), equation (15.5a) can be rewritten:

$$V = \sqrt{\frac{8g}{f}} \sqrt{\frac{D_H}{4} S_f} \tag{15.5b}$$

Equation (15.5b) is sometimes called the Chézy equation. It is valid for gradually varied flows and for uniform equilibrium flows.

Discussion

Note the similarity between equations (15.1a) and (15.5a). Equation (15.5a) may be rewritten as:

$$\frac{\Delta H}{\Delta s} = f \frac{1}{D_H} \frac{V^2}{2g} \qquad \text{Head loss gradient}$$

where ΔH is the head loss (always positive).

Empirical resistance coefficients

The Chézy equation was originally introduced (in 1768) as an empirical correlation by the French engineer A. Chézy:

$$V = C_{\text{Chézy}}\sqrt{\frac{D_{\text{H}}}{4}\ S_{\text{f}}} \tag{15.6}$$

where $C_{\text{Chézy}}$ is the Chézy coefficient (units: $\text{m}^{1/2}/\text{s}$), ranging typically from $30\,\text{m}^{1/2}/\text{s}$ (small rough channel) up to $90\,\text{m}^{1/2}/\text{s}$ (large smooth channel).

An empirical formulation (called the Gauckler–Manning formula) was deduced from the Chézy equation:

$$V = \frac{1}{n_{\text{Manning}}}\left(\frac{D_{\text{H}}}{4}\right)^{2/3}\sqrt{\sin\theta} \tag{15.7}$$

where n_{Manning} is the Gauckler–Manning coefficient (units: $\text{s/m}^{1/3}$). The Gauckler–Manning coefficient was assumed to be a characteristic of the surface roughness alone. Such an approximation might be reasonable as long as the water is not too shallow nor the channel too narrow. However, the assumption is not strictly correct. It can be shown that, for a given channel, the Gauckler–Manning coefficient depends upon the discharge, flow depth and relative roughness.

Notes
1. The Chézy equation was introduced in 1768 by A. Chézy when designing canals for the water supply of the city of Paris. Some researchers assumed wrongly that $C_{\text{Chézy}}$ would be a constant, but the original study never implied such an assumption.
2. The Gauckler–Manning formula was proposed first in 1867 (Gauckler 1867). In 1889, R. Manning proposed two formulae: one became known as the 'Gauckler–Manning formula' but Manning preferred to use the second formula that he presented in his paper (Manning 1890)!

Discussion

The friction slope may be expressed in terms of the Darcy, Chézy or Gauckler–Manning coefficients:

$$S_{\text{f}} = f\frac{1}{D_{\text{H}}}\frac{V^2}{2g} \tag{15.2}$$

$$S_{\text{f}} = \frac{1}{C_{\text{Chézy}}^2}\frac{V^2}{\frac{D_{\text{H}}}{4}} \tag{15.8}$$

$$S_{\text{f}} = n_{\text{Manning}}^2\frac{V^2}{\left(\frac{D_{\text{H}}}{4}\right)^{4/3}} \tag{15.9}$$

where V is the mean flow velocity.

In practice, equation (15.2) must be the preferred friction slope definition. In the case of complex bed forms, equations (15.8) or (15.9) might be used if some information on $C_{\text{Chézy}}$ or n_{Manning} is available.

15.3 Backwater calculations

15.3.1 Presentation

In the next sections, we will develop a basic case of backwater calculations. First a general method is described. Then calculations are developed and commented upon. At each step, a practical case is shown. The same test case is considered: i.e. a reservoir discharging into a channel, followed by an overfall (Fig. 15.2).

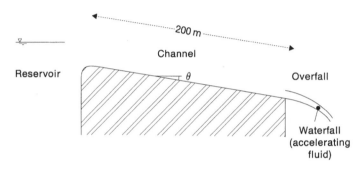

Fig. 15.2 Sketch of a long channel carrying water from a reservoir to an overfall.

15.3.2 Method

First: a sketch of the longitudinal profile of the channel must be drawn. Control structures (e.g. gate, weir) must be included.

Secondly: the location of 'obvious' control sections[1] must be highlighted as at these locations the flow depth is known. Classical control sections include: sluice gate, overflow gate, broad-crested weir, hydraulic jump, overfall, bottom slope change from very-flat to very-steep (or from very-steep to very-flat).

Thirdly: the uniform flow properties must be computed at each position along the channel. In practice, uniform flow conditions must be re-computed after *each* change of bottom slope, total discharge, cross-section shape (e.g. from rectangular to trapezoidal), cross-section characteristics (e.g. bottom width), boundary roughness (i.e. bottom and sidewalls).

Then: plot the uniform flow properties on the longitudinal sketch of the channel. Highlight each change of flow regime (i.e. sub- to super-critical flow, super- to sub-critical flow): these are control sections.

Lastly: sketch the composite free-surface profiles.

[1] A control section is defined as the location where critical flow occurs (see Chapter 3).

This first analysis is *uppermost important*. An incorrect estimate of the composite pro-file may lead to wrong or false results that are not necessarily detected by computer programs nor 'computer experts'. It is the responsibility of the hydraulic engineer to state clearly and accurately the correct boundary conditions.

Backwater calculations start from positions of known flow depth. Calculations may be performed in the upstream and downstream flow directions up to the next control sections. At a control section, the type of control affects the flow computation through the control section. That is, the Bélanger equation (i.e. momentum equation) must be applied for a hydraulic jump; the Bernoulli equation (and specific energy con-cept) is applied for a gate (or a weir); for a smooth change from sub- to supercritical flow (e.g. spillway crest), calculations continue downstream of the critical flow depth section.

Application

Let us consider the outflow from a reservoir of constant free-surface elevation into a long channel (200 m length). The channel ends by a free overfall (at the downstream end) (Fig. 15.2). The channel has a rectangular cross-section (width: 3.5 m). The channel bed is made of smooth concrete and the equivalent roughness height of bed and sidewalls is $k_s = 2$ mm. Assuming that the flow rate in the channel is 5 m^3/s, sketch the composite profile for each of the following channel slopes.

(A) Case 1: $\theta = 0.02°$.

(B) Case 2: $\theta = 1.4°$.

Solution

From the first sketch (Fig. 15.2), we note two regions where the flow character-istics are obvious.

1. In the reservoir, the fluid is quiescent (still). The flow must be subcritical.
2. Downstream of the overfall crest, the free-falling flow is accelerating. The flow must be supercritical.

First result: there must be (at least) one control section between the reservoir and the waterfall as the flow regime must change from subcritical in the reservoir to supercritical at the overfall.

For a constant-slope (and constant-width) channel, one set of uniform flow conditions must be computed. The normal flow computations give:

(A) Case 1: $\theta = 0.02°$, $d_o = 1.33$ m, $V_o = 1.07$ m/s, $Fr_o = 0.30$.

(B) Case 2: $\theta = 1.4°$, $d_o = 0.307$ m, $V_o = 4.65$ m/s, $Fr_o = 2.68$.

In Case 1, the channel flow will tend towards subcritical normal flow conditions (i.e. $Fr_o = 0.3$) while in Case 2 the waters will accelerate towards supercritical equilibrium flow conditions ($Fr_o = 2.7$).

Conclusions

In Case 1, the flow is subcritical all along the channel. At the downstream end of the channel, the free-falling water is accelerated and becomes supercritical. Critical flow conditions are observed only near the end of the channel (see the discussion in Henderson 1966, pp. 28–29 and 191–202). The complete composite profile is plotted in Fig. 15.3(a). Note that the flow in the reservoir is also sub-critical: i.e. there is no control section at the upstream end of the channel.

In Case 2, the channel flow is supercritical. At the channel intake, there is a change of flow regime from subcritical (in the reservoir) to supercritical. This is a control section (Fig. 15.3(b)). At the downstream of the channel, the flow is accelerated and remain supercritical: i.e. critical flow conditions are not observed at the overfall.

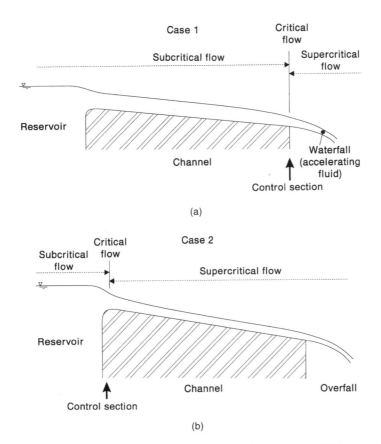

Fig. 15.3 Sketch of the composite profiles. (a) Case 1: channel slope $\theta = 0.02°$. (b) Case 2: channel slope $\theta = 1.4°$.

15.3.3 Calculations

First, the calculations must start from positions of known flow depth. Secondly, the reader must remember that sub-critical flows are controlled from downstream while super-critical flows are controlled from upstream. In practice, it is strongly advisable to start backwater calculations of subcritical flow from the downstream control and to proceed in the upstream direction. Similarly, it is recommended to start backwater computations of supercritical flow from the upstream control and to proceed in the flow direction.

The calculations must be conducted as detailed in Chapter 5. An example is detailed below.

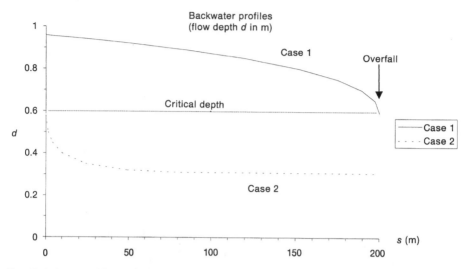

Fig. 15.4 Computed free-surface profiles.

Application

Let us consider the same example: the outflow from a reservoir of constant free-surface elevation into a long channel (200 m length). The channel ends by a free overfall at the downstream end (Fig. 15.2). The channel has a rectangular cross-section (width: 3.5 m). The equivalent roughness height of the channel bed is $k_s = 2\,\text{mm}$.

Assuming that the flow rate in the channel is $5\,\text{m}^3/\text{s}$, compute the backwater profile for the following two different channel slopes:
(A) Case 1: $\theta = 0.02°$.
(B) Case 2: $\theta = 1.4°$.
Subsidiary question: in each case, what is the reservoir free-surface elevation above the channel intake (crest)?

Solution

Case 1. In Case 1, a control section is located at the downstream end of the channel (Fig. 15.3(a)). At that location (i.e. $s = 200\,\text{m}$) the flow depth equals the critical flow depth $d_c = 0.593\,\text{m}$ (to a first approximation). Backwater calculations are performed from the control section in the upstream flow direction up to the channel end. Results are summarized in Fig. 15.4. Note that, near the upstream end of the channel, the flow depth equals 0.96 m: i.e. normal flow conditions do not take place because the channel is too short.

The flow depth in the reservoir is deduced by assuming a smooth transition (i.e. frictionless) between the reservoir and the channel. The application of the

Bernoulli equation indicates that the total head in the reservoir is the same as at the channel intake (where $d = 0.96$ m, $V = 1.49$ m/s). As the velocity in the reservoir is negligible, the reservoir free-surface elevation must be 1.07 m above the channel bed (at intake).

Case 2. In Case 2, the control section is located at the upstream end of the channel (Fig. 15.3(b)). At that location (i.e. $s = 0$ m) the flow depth equals the critical flow depth $d_c = 0.593$ m. Backwater calculations are performed from the control section into the downstream flow direction up to the downstream channel end. Results are summarized in Fig. 15.4. Near the downstream end, the flow depth equals 0.31 m: i.e. the flow is almost uniform equilibrium.

  The flow depth in the reservoir is deduced by assuming a smooth flow transition (frictionless) between the reservoir and the channel. The application of the Bernoulli equation indicates that the total head in the reservoir is the same as at the channel intake (where $d = d_c = 0.593$ m, $V = V_c = 2.4$ m/s). As the velocity in the reservoir is negligible, the free-surface elevation above the channel bed (at intake) equals: 0.889 m.

15.3.4 Comments

The above application is a simple case with only one control section and two distinctive boundary conditions (upstream reservoir and downstream overfall) (Fig. 15.3). Usually, backwater profiles include more control sections. Particular controls might be required to apply the Bernoulli equation (e.g. gate) or the momentum equation (e.g. hydraulic jump) to continue the calculations. If the longitudinal channel profile includes a relatively long prismatic channel, uniform equilibrium flow conditions may also be attained.

  Remember that uniform equilibrium flows, hydraulic jumps and control sections are some form of singularity. Altogether, backwater calculations are a difficult exercise, and they require practice and experience. Note that, in the previous application (Figs 15.2 to 15.4), two regions of rapidly varied flow exist, although the channel flow is gradually varied: at the channel intake and at the overfall. In the region of rapidly varied flow, the backwater calculations cannot predict the flow properties accurately.

15.4 Numerical integration

15.4.1 Introduction

The backwater equation (equation (15.1a)) is a non-linear equation that cannot be solved analytically, but it can be integrated numerically. One of the most common integration methods is the standard step method.

  The method developed here (i.e. step method/depth calculated from distance) is used in several numerical models (e.g. HEC, Table 15.1).

Table 15.1 Examples of computational models for open channel flow calculations

Model (1)	Description (2)	Remarks (3)
HEC-1 and HEC-2	One-dimensional model for steady subcritical flow down flat slope (assuming hydrostatic pressure distribution). Flow resistance computed using the Gauckler–Manning formula.	Developed by USACE. Standard step method/depth calculated from distance.
WES	One-dimensional model for supercritical flow in curved channels. Flow resistance calculated using the Colebrook–White formula.	Developed by USACE-WES.
MIKE-11	One-dimensional model.	Developed by DHI.
GENFLO2D	Two-dimensional model (analytical solution of Navier–Stokes equation). Steady quasi-uniform flows in complex compound channels.	Developed by Water Solutions. Mixing length turbulence model.
TELEMAC	Two-dimensional model (finite element method): Saint-Venant equations. Flow resistance computed using the Strickler formula.	Developed by SOGREAH.
TABS-2/RAM-2	Two-dimensional model (finite element method). RAM-2 is the steady flow hydraulic model.	Developed by USACE-WES.

Comments

There are two integration step methods: the 'step method/distance calculated from depth' and the 'step method/depth calculated from distance'. Henderson (1966) discussed each one in detail.

Note that the 'step method/depth calculated from distance' (developed below and used in several software) assumes a mild slope with subcritical flow or a steep slope with supercritical flow. It is indeterminate for critical flow conditions!

15.4.2 Standard step method (depth calculated from distance)

The basic equation is:

$$\frac{\partial H}{\partial s} = -S_f \tag{15.1b}$$

where H is the mean total head over the cross-section and S_f is the friction slope. Assuming a hydrostatic pressure gradient, the mean total head equals:

$$H = z_o + d \cos \theta + \alpha \frac{V^2}{2g} \tag{15.10}$$

where z_o is the bed elevation, d is the flow depth measured normal to the channel bottom, θ is the channel slope, α is the Coriolis coefficient, V is the mean velocity and g is the gravity constant (Fig. 15.1).

Considering the gradually varied flow sketched in Fig. 15.5, the backwater equation can be integrated between two stations denoted $\{i\}$ and $\{i+1\}$:

$$\frac{H_{i+1} - H_i}{s_{i+1} - s_i} \approx -\tfrac{1}{2}(S_{f_{i+1}} + S_{f_i}) \tag{15.11a}$$

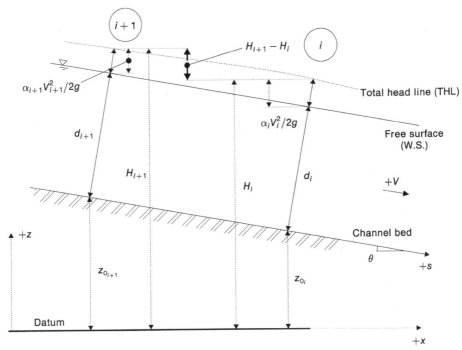

Fig. 15.5 Backwater calculations: definition sketch.

where s is the streamwise coordinate, and the subscripts i and $i+1$ refer to stations $\{i\}$ and $\{i+1\}$ respectively. It yields:

$$H_{i+1} - H_i \approx -\tfrac{1}{2}(S_{f_{i+1}} + S_{f_i})(s_{i+1} - s_i) \qquad (15.11b)$$

If the flow conditions at station $\{i\}$ are known, the total head at station $\{i+1\}$ equals:

$$H_{i+1} \approx H_i - \tfrac{1}{2}(S_{f_{i+1}} + S_{f_i})(s_{i+1} - s_i) \qquad (15.11c)$$

Notes
1. The elevation (i.e. altitude) is positive upwards.
2. By convention, the downstream direction is the positive direction.
3. The calculations are solved iteratively because the friction slope at station $\{i+1\}$ is unknown beforehand.

15.4.3 Computational algorithm

The calculations can be performed only if the total flow rate Q is known.
    The calculation steps are:

Step 1. At station $\{i\}$ (i.e. $s = s_i$), the water level and hence the flow depth are known. The following variables can be deduced:

Cross-section area $\quad A_i$

Wetted perimeter $\quad P_{w_i}$

Hydraulic diameter $\quad D_{H_i} = 4A_i/P_{w_i}$

Flow velocity $\quad V_i = Q/A_i$ $\quad\quad$ (Continuity equation)

Friction slope $\quad S_{f_i}$ $\quad\quad\quad\quad$ (equations (15.2), (15.8) or (15.9))

Total head $\quad H_i = z_{o_i} + d_i \cos\theta_i + \alpha_i V_i^2/(2g)$

Step 2. Estimate (i.e. guess) the water level (or flow depth $d_{i+1}^{(1)}$) at station $\{i+1\}$ (i.e. $s = s_{i+1}$).

Step 3. Calculate then:

Cross-section area $\quad A_{i+1}^{(1)}$

Wetted perimeter $\quad P_{w_{i+1}}^{(1)}$

Hydraulic diameter $\quad D_{H_{i+1}}^{(1)}$

Flow velocity $\quad V_{i+1}^{(1)}$ $\quad\quad$ (Continuity equation)

Friction slope $\quad S_{f_{i+1}}^{(1)}$ $\quad\quad\quad$ (equations (15.2), (15.8) or (15.9))

Total head $\quad H_{i+1}^{(1)} = z_{o_{i+1}} + d_{i+1}^{(1)} \cos\theta_{i+1} + \alpha_{i+1}^{(1)}(V_{i+1}^{(1)})^2/(2g)$

$$\text{equation (15.10)}$$

Step 4. Check the calculations by calculating $H_{i+1}^{(*)}$ using equation (15.11c):

$$H_{i+1}^{(*)} \approx H_i - \tfrac{1}{2}(S_{f_{i+1}}^{(1)} + S_{f_i})(s_{i+1} - s_i) \qquad\qquad \text{equation (15.11c)}$$

Step 5. If $H_{i+1}^{(*)} = H_{i+1}^{(1)}$ with an acceptable accuracy, the estimate of $d_{i+1}^{(1)}$ is acceptable, and $d_{i+1} = d_{i+1}^{(1)}$. And we can then proceed to the next interval. In the negative, we must repeat the Steps 2, 3, 4 and 5 until an acceptable accuracy is obtained.

Note

For a full-scale channel, an acceptable accuracy might be: $H_{i+1}^{(*)} - H_{i+1}^{(1)} < 1\,\text{mm}$.

15.5 Discussion

To solve the backwater equation (15.1), it is essential to determine correctly the boundary conditions: i.e. the flow conditions imposed upstream, downstream and along the channel. These boundary conditions are, in practice: (a) control devices (e.g. sluice gate, weir, reservoir) that impose flow conditions for the depth, the discharge or a relationship between the discharge and the depth, and (b) geometric characteristics (e.g. bed altitude, channel width, local roughness).

The computations proceed from a hydraulic control or from a position of known flow characteristics. Usually the integration proceeds in the upstream direction if the flow is subcritical, and in the downstream flow direction for supercritical flows.

It must be emphasized that the calculations depend critically on the assumed flow resistance coefficient. In practice, a great uncertainty applies to the assumed friction factor. Note also that the above calculations are applicable to both subcritical and supercritical flows within the GVF calculation assumptions (Section 15.2.2). The only other assumption is that of hydrostatic pressure distribution.

In practice, design engineers must remember that backwater calculations are not valid for rapidly-varied flow situations (e.g. gate, weir crest) including across a hydraulic jump. A hydraulic jump is a rapidly-varied flow situation. The relationship between the upstream and downstream flow conditions derives from the momentum equation (i.e. the Bélanger equation). Furthermore, the backwater calculations should not be applied to unsteady flow situations.

15.6 Computer models

Several computer models have been developed to compute backwater profiles (e.g. Table 15.1). The simplest are one-dimensional models based on the backwater equation (i.e. equations (15.1a or 15.1b)). Most of these use a step method/depth calculated from distance (see Section 15.4.2): e.g. HEC-1 and HEC-2. More complicated hydraulic models are based on the Saint-Venant equations and they are two-dimensional or three-dimensional.

The use of computer models requires a sound understanding of the basic equations and a good knowledge of the limitations of the models. It is of utmost importance that computer program users have a through understanding of the hydraulic mechanisms and of the key physical processes that take place in the system (i.e. river and catchment). The writer knows too many 'computer users' who have little idea of the model limitations.

Practically, hydraulic engineers must consider both physical and numerical modelling. Physical modelling assists in understanding the basic flow mechanisms as well as visualizing the flow patterns. In addition, physical model data may be used to calibrate and to verify a numerical model. Once validated, the computational model is a valuable tool to predict a wide range of flow conditions.

15.7 Exercises

Numerical solutions to these exercises are available from the Web at www.arnoldpublishers.com/support/chanson

A waterway has a trapezoidal cross-section with a 7 m bottom width, a vertical sidewall and a battered (1 V : 5 H) sidewall. The longitudinal bed slope is 6 m per km. The channel bottom and sidewalls are concrete lined. The waterway carries a discharge of $52 \, \text{m}^3/\text{s}$. (a) Compute the critical flow depth. (b) Compute the normal flow depth.

A gauging station is set at a bridge crossing the waterway. The observed flow depth at the gauging station is 2.2 m. (c) From where is the flow controlled: i.e. upstream or downstream? (Justify your answer.) (d) Compute the average bed shear stress at the gauging station. (e) Upstream of the gauging station, will the flow depth be larger than, equal to or less than that at the gauging station? (Justify your answer.)

Write a calculation sheet to integrate the backwater equation in the form:

$$\frac{\partial H}{\partial s} = -S_f$$

using the standard step method ('distance calculated from depth'). In order to limit the task, use the following specifications: [1] the cross section shape is limited to a symmetrical trapezoidal and specified bottom width, angle of side slope, elevation of bed level above datum and distance of section from a reference location (positive in the downstream direction); [2] in the iterative procedure required to calculate the correct total head and distance which constitute the result of the integration, you may use any depth increment provided that the changes in friction slope remain small. (For engineering purposes, agreement would be satisfactory if $\Delta S_f/(S_o - S_f) < 5\%$ and $\Delta S_f/S_f < 5\%$.)

In order to validate your calculation sheet, use it to compute the following backwater profile: trapezoidal channel: 6 m bottom width, 1 V:1.5 H side slope, bed slope $S_o = 0.001$, $k_s = 25$ mm, $Q = 3.84$ m^3/s.

Calculate the free-surface elevation upstream of a gauging station, where the known depth is 1.05 m and the bed elevation is R.L. 779.8 m, at the following locations: $s = 0$ m, $s = -20$ m, $s = -89$ m, $s = -142$ m.

Prepare (1) a brief description of the calculations sheet, (2) a printout of the complete calculation sheet, (3) the results of the validation test, and (4) the attached summary sheet.

Summary sheet:

$Q =$

s (m)	Free-surface elevation (m R.L.)	V (m/s)
0		
−20 m		
−89 m		
−142 m		

Uniform equilibrium flow conditions in the channel:

$$d_o =$$

$$Fr_o =$$

Type of free-surface profile observed:

Part 3 Revision exercises

Revision exercise No. 1

An undershot sluice gate is to be tested in a laboratory to determine the discharge characteristics for various opening heights. The maximum prototype discharge will be $220\,\mathrm{m^3/s}$ and the gate is installed in a rectangular channel 20 m wide. Note that the channel bed will be horizontal and concrete-lined (i.e. smooth). A 50:1 scale model of the gate is to be built. What similitude should be used? (Justify in words your selection.)

Determine: (a) the maximum model discharge required and (b) the minimum prototype discharge for which negligible scale effects occur in the model.

For one particular gate opening, the *laboratory* flow conditions are: upstream flow depth of 0.21 m, downstream flow depth of 0.01 m. (c) Compute the model discharge. State the basic principle(s) involved. (d) Compute the model force acting on the sluice gate. (State the basic principle(s) involved.) (e) What will be the corresponding prototype discharge and force on the gate? (f) What will be the scale for the force ratio?

Gate operation may result in unsteady flow situations. If a prototype gate operation has the following characteristics: gate opening duration = 3 minutes, gate opening height = 0.5 m, what (g) gate opening duration and (h) gate opening height should be used in the model tests?

Revision exercise No. 2

An artificial concrete channel model is to be built. Laboratory facilities limit the scale ratio to 50:1 and maximum model discharge is 50 L/s. The maximum full-scale discharge is $150\,\mathrm{m^3/s}$, the cross-section of the channel is approximately rectangular (50 m bottom width) and the bed slope is 0.14 m per kilometre.

Note: The roughness height of the prototype is estimated as 3 mm while the smoothest model surface feasible has a Darcy friction factor of about $f = 0.03$.

Discharges ranging between the maximum flow rate and 10% of the maximum flow rate are to be reproduced in the model.

(1) *Undistorted* model (assume normal flow conditions)

Compute the model discharge at maximum full-scale discharge, the model flow depth, the Darcy coefficient of the model flow and the Darcy coefficient of the prototype channel.

(2) A *distorted* model is to be built

Determine the acceptable maximum (largest) value of the vertical scale ratio Z_r. Determine the acceptable minimum (smallest) value of the vertical scale ratio Z_r. For the acceptable minimum value of the vertical scale ratio Z_r, calculate the corresponding model maximum discharge.

Appendices to Part 3

A3.1 Physical modelling of movable boundary hydraulics

Introduction

Physical modelling of free-surface flows is most often performed using a Froude similitude. In the presence of sediment motion, additional parameters must be introduced.

In the following sections, we will develop further similitude criteria for a movable bed in open channel flows. These derive from the definition of the boundary shear stress:

$$\tau_o = \frac{f}{8}\rho V^2 \qquad\qquad (A3.1)$$

where f is the Darcy friction factor and V is the flow velocity.

Bed-load motion

Occurrence of bed-load motion
Experimental investigations and dimensional analysis have shown that the occurrence of bed-load motion may be predicted from the Shields diagram:

$$(\tau_*)_c = \frac{(\tau_o)_c}{\rho g(s-1)d_s} = \mathscr{F}_1\left(\rho \frac{d_s V_*}{\mu}\right) \qquad \text{Occurrence of bed-load motion} \qquad (A3.2)$$

where d_s is the sediment size, **s** is the relative density of sediment particle (i.e. $\mathbf{s} = \rho_s/\rho$) and V_* is the shear velocity (i.e. $V_* = \sqrt{\tau_o/\rho}$).

Perfect model–prototype similarity of occurrence of bed-load motion will occur if and only if:

$$((\tau_*)_c)_m = ((\tau_*)_c)_p \qquad\qquad (A3.3)$$

and

$$\left(\rho \frac{d_s V_*}{\mu}\right)_m = \left(\rho \frac{d_s V_*}{\mu}\right)_p \qquad\qquad (A3.4)$$

where the subscripts m and p refer to the model and prototype characteristics respectively.

Expressing the bed shear stress as a function of the Darcy friction factor (equation (A3.1)) and for a Froude similitude,[1] the equality of the Shields parameter in the model and prototype (equation (A3.3)) implies that the scale ratios for the sediment size $(d_s)_r$ and for the relative density$(s - 1)_r$ must satisfy:

$$(d_s)_r(s - 1)_r = L_r \qquad \text{Undistorted model} \tag{A3.5a}$$

and for a distorted model:

$$(d_s)_r(s - 1)_r = \frac{Z_r^2}{X_r} \qquad \text{Distorted model} \tag{A3.5b}$$

Equation (A3.5) implies that the scale ratio of the sediment grain size may be different from the geometric length scale ratio by changing the relative density of the model grains.

> **Note**
>
> Let us keep in mind that a Froude similitude implies that the model flow resistance will be similar to that in the prototype:
>
> $$f_r = 1$$
>
> $$f_r = \frac{Z_r}{X_r} \qquad \text{Distorted model (wide channel and flat slope)}$$
>
> If the sediment size scale ratio $(d_s)_r$ differs from the geometric length scale ratio L_r, the scale ratio of bed friction resistance might be affected.

Discussion

Note that the equality of the shear Reynolds number (i.e. (A3.4)) in the model and prototype yields:

$$(d_s)_r = \frac{\mu_r}{\rho_r} \frac{1}{\sqrt{L_r}} \tag{A3.6a}$$

$$(d_s)_r = \frac{\mu_r}{\rho_r} \frac{\sqrt{X_r}}{Z_r} \qquad \text{Distorted model} \tag{A3.6b}$$

Equation (A3.6) requires model grain sizes larger than prototype sizes (for $L_r > 1$). This is not acceptable in most cases and equations (A3.4) and (A3.6) must be ignored in practice.

Bed-load sediment transport rate

The bed load transport rate per unit width is related to the bed-load layer characteristics (see Section 10.3.2) as:

$$q_s = C_s \delta_s V_s \tag{A3.7}$$

---

[1] Assuming an identical gravity acceleration in the model and prototype.

where V_s is the average sediment velocity in the bed-load layer and C_s is the mean sediment concentration in the bed-load layer of thickness δ_s.

Using existing correlations (e.g. Nielsen 1992), the similarity condition of bed-load motion (i.e. equation (A3.3) and (A3.5)) yields:

$$(q_s)_r = \sqrt{L_r} \tag{A3.8a}$$

$$(q_s)_r = \frac{Z_r}{\sqrt{X_r}} \qquad \text{Distorted model} \tag{A3.8b}$$

Summary

Similarity of bed-load motion may be achieved if equations (A3.3) and (A3.4) are respected. In practice, only equation (A3.3) can be achieved. If equation (A3.3) (or equation (A3.5)) is satisfied, the similarity of bed-load motion inception and bed-load transport rate may be achieved.

Suspension

Occurrence of suspension

The occurrence of a suspended load is a function of the balance between the particle fall velocity and the upward turbulent flow motion. Experimental investigations showed that suspension occurs for:

$$\frac{V_*}{w_o} > 0.2 \text{ to } 2 \tag{A3.9}$$

where V_* is the shear velocity and w_o is the settling velocity. For spherical particles the terminal fall velocity equals:

$$w_o = -\sqrt{\frac{4gd_s}{3C_d}(s-1)} \qquad \text{Spherical particle} \tag{A3.10}$$

where the drag coefficient C_d is a function of the ratio $w_o d_s / \nu$.

For turbulent settling motion (i.e. $w_o d_s / \nu > 1000$), the drag coefficient is a constant. In addition, the scale ratio of the sediment suspension parameter is:

$$\left(\frac{V_*}{w_o}\right)_r = \frac{\sqrt{L_r}}{\sqrt{(d_s)_r(s-1)_s}} \qquad \text{Turbulent particle settling} \tag{A3.11a}$$

If the condition for the similarity in bed-load motion (i.e. (A3.3) and (A3.5)) is satisfied, it yields:

$$\left(\frac{V_*}{w_o}\right)_r = 1 \qquad \text{Turbulent particle settling} \tag{A3.11b}$$

In other words, if the similarity in bed-load motion is achieved, the similarity of turbulent particle settling is also satisfied (provided that the particle settling motion is turbulent in both the model and the prototype).

Suspended-sediment transport rate

The distribution of suspended matter across the flow depth may be estimated as:

$$
c_s = (C_s)_{y=y_s} \left(\frac{\dfrac{d}{y} - 1}{\dfrac{d}{y_s} - 1} \right)^{w_o/(KV_*)}
\tag{A3.12}
$$

where $(C_s)_{y=y_s}$ is the reference sediment concentration at the reference location $(y = y_s)$ and d is the flow depth.

The similarity of suspended sediment concentration is achieved if the ratio w_o/V_* is identical in the model and the prototype and if the integration parameters $(C_s)_{y=y_s}$ and y_s are appropriately scaled.

It is presently impossible to reproduce sediment suspension distributions and suspended-load transport rates in small-scale models which reproduce prototype sediment concentration distributions and transport rates.

Time scale in sediment transport

The *hydraulic time scale* characterizes the duration of an individual event (e.g. a flood event). For a Froude similitude, the scale ratio is:

$$
t_r = \sqrt{L_r}
\tag{A3.13a}
$$

$$
t_r = \frac{X_r}{\sqrt{Z_r}} \qquad \text{Distorted model}
\tag{A3.13b}
$$

In sediment transport another time scale is the *sedimentation time*. In movable-bed channels, the continuity equation for sediment is:

$$
\frac{\partial q_s}{\partial x} = -(1 - Po) \frac{\partial z_o}{\partial t} \cos\theta
\tag{A3.14}
$$

where q_s is the total sediment transport rate, Po is the sediment bed porosity, z_o is the bed elevation and θ is the longitudinal bed slope. For a flat channel, this gives the scale ratio of the sedimentation time t_s:

$$
(t_s)_r = (1 - Po)_r \frac{L_r^2}{(q_s)_r}
\tag{A3.15a}
$$

$$
(t_s)_r = (1 - Po)_r \frac{X_r Z_r}{(q_s)_r} \qquad \text{Distorted model}
\tag{A3.15b}
$$

If most sediment transport occurs by bed-load motion, the above result may be combined with equation (A3.8):

$$
(t_s)_r = (1 - Po)_r L_r^{3/2} \qquad \text{Bed load transport}
\tag{A3.16a}
$$

$$
(t_s)_r = (1 - Po)_r X_r^{3/2} \qquad \text{Bed-load transport – distorted model}
\tag{A3.16b}
$$

Comparing equations (A3.13) and (A3.15) (or (A3.16)), it appears that the sedimentation time ratio differs from the hydraulic time ratio in most cases.

A3.2 Extension of the backwater equation

Introduction

Presentation

For a rectangular channel, the continuity equation states:

$$Q = VBd \qquad (A3.17)$$

where Q is the total discharge, V is the average flow velocity, B is the channel width and d is the flow depth (measured normal to the flow direction).

At any position along a streamline, the (local) total head H is defined as:

$$H = \frac{P}{\rho g} + z + \frac{v^2}{2g} \qquad (A3.18)$$

where P is the pressure, z is the elevation (positive upwards) and v is the velocity along the streamline.

Assuming a hydrostatic pressure gradient, the average total head, as used in the energy equation, equals:

$$H = d \cos \theta + z_o + \frac{\alpha}{2g} \left(\frac{Q}{A} \right)^2 \qquad (A3.19a)$$

where d is the flow depth measured normal to the channel bottom, θ is the channel slope, z_o is the bed elevation, α is the kinetic energy correction coefficient (i.e. Coriolis coefficient), Q is the total discharge and A is the cross-sectional area. For a rectangular channel it yields:

$$H = d \cos \theta + z_o + \frac{\alpha}{2g} \left(\frac{Q}{Bd} \right)^2 \qquad \text{rectangular channel} \qquad (A3.19b)$$

Energy equation

Considering a non-uniform and steady flow, the backwater calculations are performed assuming that the flow is gradually varied and that, at a given section, the flow resistance is the same as for a uniform flow for the same depth and discharge, regardless of trends of the depth. In the s-direction, the energy equation becomes:

$$\frac{\partial H}{\partial s} = -S_f \qquad (A3.20)$$

where S_f is the friction slope, H is the mean total head and s is the coordinate along the flow direction. Using the mean specific energy this equation is transformed:

$$\frac{\partial E}{\partial s} = S_o - S_f \qquad (A3.21)$$

where S_o is the bed slope (i.e. $S_o = \sin \theta$).

Equation (A3.20) is the basic *backwater equation*, first developed in 1828 by J.B. Belanger.

> **Note**
> For uniform or non-uniform flows, the bed slope S_o and friction slope S_f are defined respectively as:
>
> $$S_f = f \frac{1}{D_H} \frac{V^2}{2g} = -\frac{\partial H}{\partial s}$$
>
> $$S_o = \sin\theta = -\frac{\partial z_o}{\partial s}$$
>
> In gradually varied flow, the friction slope and Darcy coefficient are calculated as in uniform equilibrium flow (see Chapter 4) but using the local non-uniform flow depth.

Extension of the backwater equations

Flat channel of constant width

For a flat channel of rectangular cross-section and constant width, the mean specific energy may be rewritten as a function of the Froude number and flow depth (see Chapter 3):

$$E = d(1 + \alpha \tfrac{1}{2} Fr^2) \tag{A3.22}$$

where Fr is the Froude number defined as: $Fr = q/\sqrt{gd^3}$ and q is the discharge per unit width.

The energy equation (A3.21) leads to:

$$\frac{\partial d}{\partial s}(1 - \alpha Fr^2) = S_o - S_f \tag{A3.23}$$

> **Note**
> For a flat rectangular channel, the Froude number is defined as:
>
> $$Fr = \frac{V}{\sqrt{gd}} = \frac{q}{\sqrt{gd^3}} = \frac{Q}{\sqrt{gB^2 d^3}}$$

Flat channel of non-constant width

Considering a *rectangular* flat channel, the differentiation of equation (A3.19b) with respect to the curvilinear coordinate s leads to:

$$\frac{\partial H}{\partial s} = \frac{\partial d}{\partial s} + \frac{\partial z_o}{\partial s} - \frac{\alpha}{g} \frac{Q^2}{B^2 d^2} \left(\frac{1}{B} \frac{\partial B}{\partial s} + \frac{1}{d} \frac{\partial d}{\partial s} \right) \tag{A3.24}$$

Introducing the Froude number $Fr = q/\sqrt{gd^3}$, the energy equation yields:

$$\frac{\partial d}{\partial s}(1 - \alpha Fr^2) = S_o - S_f + \alpha Fr^2 \frac{d}{B} \frac{\partial B}{\partial s} \tag{A3.25}$$

> **Note**
> The reader must understand that the above development implies that:
>
> - the variation of the Coriolis coefficient α with respect to s is neglected in the first approximation;
> - the total discharge Q is constant;
> - it is reasonable to assume $(\cos\theta \sim 1)$ for a flat channel.

Channel of constant width and non-constant slope

Considering a *rectangular* channel of constant width, the differentiation of equation (A3.19) in the s-direction is:

$$\frac{\partial H}{\partial s} = \frac{\partial d}{\partial s}\cos\theta - d\sin\theta\frac{\partial\theta}{\partial s} + \frac{\partial z_o}{\partial s} - \frac{\alpha}{g}\frac{Q^2}{B^2 d^3}\frac{\partial d}{\partial s} \qquad (A3.26)$$

The energy equation (A3.20) may be rewritten as:

$$\frac{\partial d}{\partial s}(\cos\theta - \alpha Fr^2) = S_o - S_f + d\sin\theta\frac{\partial\theta}{\partial s} \qquad (A3.27)$$

where $Fr = q/\sqrt{gd^3}$.

> **Discussion**
> Considering an inclined channel (slope θ) and assuming a hydrostatic pressure distribution, the mean specific energy equals:
>
> $$E = d\cos\theta + \frac{1}{2g}\frac{Q^2}{A^2}$$
>
> For a rectangular channel, the specific energy is minimum for:
>
> $$\frac{\partial E}{\partial d} = \cos\theta - \frac{Q^2}{gB^2 d^2} = 0$$
>
> In other words, at critical flow conditions, the Froude number equals:
>
> $$Fr = \frac{q}{\sqrt{gd^3}} = \cos\theta$$

Channel of non-constant width and non-constant slope

For a rectangular channel of non-constant width and slope, the differentiation of equation (A3.19) with respect to s leads to:

$$\frac{\partial H}{\partial s} = \frac{\partial d}{\partial s}\cos\theta - d\sin\theta\frac{\partial\theta}{\partial s} + \frac{\partial z_o}{\partial s} - \frac{\alpha}{g}\frac{Q^2}{B^2 d^3}\left(\frac{d}{B}\frac{\partial B}{\partial s} + \frac{\partial d}{\partial s}\right) \qquad (A3.28)$$

The energy equation yields to:

$$\frac{\partial d}{\partial s}(\cos\theta - \alpha Fr^2) = S_o - S_f + d\sin\theta\frac{\partial\theta}{\partial s} + \alpha Fr^2\frac{d}{B}\frac{\partial B}{\partial s} \qquad (A3.29)$$

where $Fr = q/\sqrt{gd^3}$.

General case

In the general case of a channel of non-constant width and slope, the differentiation of equation (A3.19) with respect to s leads to:

$$\frac{\partial H}{\partial s} = \frac{\partial d}{\partial s}\cos\theta - d\sin\theta\,\frac{\partial\theta}{\partial s} + \frac{\partial z_o}{\partial s} - \alpha Fr^2\frac{1}{B}\frac{\partial A}{\partial s} \qquad (A3.30)$$

where A is the cross-sectional area, B is the free-surface width and $Fr = Q/\sqrt{gA^3/B}$.

Notes

1. Both the cross-sectional area A and the free-surface width B are functions of the flow depth d. Note the relationship: $\partial A/\partial d = B$.
2. z_o and d must be taken at the same position: i.e. usually at the deepest point in the channel.
3. The definition of the Froude number satisfies minimum specific energy for a flat channel. For a steep channel, critical flow conditions are satisfied for:

$$\frac{\partial E}{\partial d} = \cos\theta - \frac{Q^2 B}{gA^3} = 0$$

Part 4 Design of Hydraulic Structures

16

Introduction to the design of hydraulic structures

16.1 Introduction

The *storage* of water is essential for providing Man with drinking water and irrigation water reserves. Storage along a natural stream is possible only if the hydrology of the catchment area is suitable. Hydrological studies provide information on the storage capability and as well as on the maximum (peak) flow in the system. Often the hydrology of a stream does not provide enough supply all year round, and an artificial water storage system (e.g. the reservoir behind a dam) must be developed.

Once a source of fresh water is available (e.g. water storage, rainfall run-off), water must be carried to the location of its use. The *conveyance* of water takes place in channels and pipes. In an open channel, an important difference is the propulsive force acting in the direction of the flow: the fluid is driven by the weight of the flowing water resolved down a slope. The course of open channels derives from the local topography and it may include drops (and cascades) and tunnels underneath embankments (e.g. roads).

16.2 Structure of Part 4

Part 4 of this book presents an introduction to the hydraulic design of structures: weirs and small dams (Chapter 17), drops and cascades (Chapter 18), and culverts (Chapter 19). The former deals with the storage of water while the latter chapters deal with the conveyance of water.

Weirs and dams are used to store water. A main hydraulic feature is the spillway system, used to release safely flood waters above the structure.

Drops and cascades may be used as an alternative to spillway structures. They are used also in waterways built in steep relief to dissipate the kinetic energy of the flow. Culverts are designed to pass water underneath embankments (e.g. railroad, highway).

16.3 Professional design approach

This material presents the application of the basic hydraulic principles to real design situations. The design approach is based on a system approach. A hydraulic structure must be analysed as part of the surrounding catchment and the hydrology plays an important role. Structural and hydraulic constraints interact, and the design of a hydraulic structure is altogether a complex exercise.

For example, the design of a culvert requires a hydrological study of a stream to estimate the maximum (design) discharge and to predict the risks of exceptional (emergency) floods. The dimensions of the culvert are based on hydraulic and structural considerations, as well as geotechnical matter as the culvert height and width affect the size and cost of the embankment. Furthermore, the impact of the culvert on the environment must be taken into account: e.g. potential flooding of the upstream plain.

Another example is the construction of a weir across a river. First, the stream hydrology and the catchment characteristics must be studied. If the catchment can provide enough water all year round (i.e. mean annual characteristics), the maximum peak inflow must be predicted (i.e. extreme events). The design of the weir is based upon structural, geotechnical and hydraulic considerations. Political matters might also affect the weir site location and the decision to build the dam. A consequent cost of the structure is the spillway, designed to pass safely the maximum peak flood. In addition, the impact of the weir on the upstream and downstream valleys must be considered: e.g. sediment trap, fish migration, downstream water quality (e.g. dissolved oxygen content), modifications of the water table and associated impacts (e.g. salinity).

The design process must be a *systems approach*. First the system must be identified. What are the design objectives? What are the constraints? What is the range of options? What is the 'best choice'? Its detailed analysis must be conducted. The engineers should ask: is this solution really satisfactory?

17

Design of weirs and spillways

17.1 Introduction

17.1.1 Definitions: dams and weirs

Dams and weirs are hydraulic structures built across a stream to facilitate the storage of water.

A *dam* is defined as a large structure built across a valley to store water in the upstream reservoir. All flows up to the Probable Maximum Flood must be confined to the designed spillway. The upstream water level should not overtop the dam wall. Dam overtopping may indeed lead to dam erosion and possibly destruction.

A conventional *weir* is a structure designed to raise the upstream water level: e.g. for feeding a diversion channel. Small flow rates are confined to a spillway channel. Larger flows are allowed to pass over the top of the full length of the weir. At the downstream end of the weir, the kinetic energy of the flow is dissipated in a dissipator structure (Fig. 17.1(a) and 17.2(a) and (b)).

Another type of weir is the *Minimum Energy Loss (MEL) weir* (Fig. 17.1(b) and 17.2(c)). MEL weirs are designed to minimize the total head loss of the overflow and hence to induce (ideally) zero afflux. MEL weirs are used in flat areas and near estuaries (see Appendix A4.2).

Practically, the differences between a small dam and a conventional weir are small, and the terms 'weir' or 'small dam' are often interchanged.

17.1.2 Overflow spillway

During large rainfall events, a large amount of water flows into the reservoir, and the reservoir level may rise above the dam crest. A *spillway* is a structure designed to 'spill' flood waters under controlled (i.e. safe) conditions. Flood waters can be discharged beneath the dam (e.g. culvert, bottom outlet), through the dam (e.g. rockfill dam) or above the dam (i.e. overflow spillway).

Most small dams are equipped with an overflow structure (called a spillway) (e.g. Fig. 17.3). An overflow spillway typically includes three sections: a crest, a chute and an energy dissipator at the downstream end. The *crest* is designed to maximize the

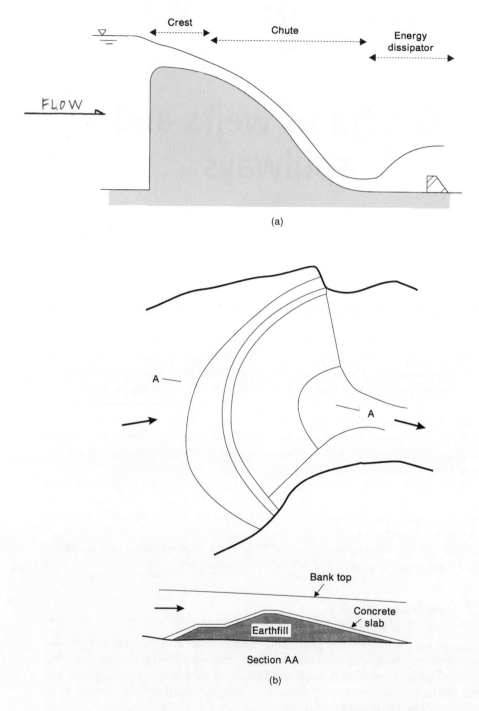

Crest

Chute

Energy
dissipator

FLOW

(a)

Bank top

Concrete
slab

Earthfill

Section AA

(b)

Fig. 17.1 Sketch of weirs. (a) Conventional weir. (b) Minimum Energy Loss (MEL) weir.

(a)

(b)

Fig. 17.2 Examples of spillway operation. (a) Diversion weir at Dalby QLD, Australia on 8 November 1997. Ogee crest followed by smooth chute and energy dissipator (note fishway next to right bank). (b) Llwyn-On dam spillway (courtesy of Mr S. Davies). Design flow conditions: 235 m^3/s, dam height: 22 m (completed in 1926), reservoir capacity: 5.5×10^6 m^3, stepped chute ($h = 0.305$ m, $W = 18.3$ m). Skimming flow on the stepped chute with stilling basin in foreground.

discharge capacity of the spillway. The *chute* is designed to pass (i.e. to carry) the flood waters above (or away from) the dam, and the *energy dissipator* is designed to dissipate (i.e. 'break down') the kinetic energy of the flow at the downstream end of the chute (Fig. 17.1(a) and 17.2).

A related type of spillway is the *drop structure*. However, as its hydraulic characteristics differ significantly from those of standard overflow weirs, it will be presented in another chapter.

(c)

Fig. 17.2 (c) Overflow above a MEL weir: Chinchilla weir at low overflow on 8 November 1997. Design flow conditions: 850 m³/s, weir height: 14 m, reservoir capacity: 9.78×10^6 m³.

Notes

1. Other types of spillways include the Morning Glory spillway (or bellmouth spillway) (Fig. 17.3). It is a vertical discharge shaft, more particularly the circular hole form of a drop inlet spillway, leading to a conduit underneath the dam (or abutment). The shape of the intake is similar to a Morning-Glory flower. It is sometimes called a Tulip intake. The Morning Glory spillway is not recommended for discharges usually greater than 80 m³/s.
2. Examples of energy dissipators include stilling basin, dissipation basin, flip bucket followed by downstream pool and plunge pool (Fig. 17.3).
3. At a Minimum Energy Loss weir, the amount of energy dissipation is always small (if the weir is properly designed) and no stilling basin is usually required. The weir spillway is curved in plan, to concentrate the energy dissipation near the channel centreline and to avoid bank erosion (Fig. 17.2(c)).
4. MEL weirs may be combined with culvert design, especially near the coastline to prevent salt intrusion into freshwater waterways without an upstream flooding effect. An example is the MEL weir built as the inlet of the Redcliffe MEL structure (Appendix A4.3).

17.1.3 Discussion

Although a spillway is designed for specific conditions (i.e. design conditions Q_{des}, H_{des}), it must operate safely and efficiently for a range of flow conditions.

Design engineers typically select the optimum spillway shape for the design flow conditions. They must then verify the safe operation of the spillway for a range of

(a)

(b)

Fig. 17.3 Examples of spillway design. (a) Overflow spillway with downstream stilling basin (Storm King dam QLD, 1954) (June 1997). (b) Overflow spillway with downstream flip bucket (Reece dam TAS, 1986) (courtesy of Hydro-Electric Commission Tasmania). Design spillway capacity: 4740 m^3/s. Overflow event: 365 m^3/s.

operating flow conditions (e.g. from $0.1Q_{des}$ to Q_{des}) and for the emergency situations (i.e. $Q > Q_{des}$).

In the following sections, we present first the crest calculations, then the chute calculations followed by the energy dissipator calculations. Later the complete design procedure is described.

(c)

(d)

Fig. 17.3 (c) Morning-Glory spillway intake (Chaffey dam NSW, 1979) (June 1997). Design spillway capacity: 800 m³/s. (d) Stepped spillway chute (Riou dam, France 1990). Step height: 0.43 m, 42 steps, chute slope: 50°.

17.2 Crest design

17.2.1 Introduction

The crest of an overflow spillway is usually designed to maximize the discharge capacity of the structure: i.e. to pass safely the design discharge at the lowest cost.

In open channels and for a given specific energy, maximum flow rate is achieved for critical flow conditions[1] (Bélanger 1828). For an ideal fluid overflowing a weir (rectangular cross-section) and assuming hydrostatic pressure distribution, the maximum discharge per unit width may be deduced from the continuity and Bernoulli equations:

$$q = \sqrt{g}(\tfrac{2}{3}(H_1 - \Delta z))^{3/2} \quad \text{Ideal fluid flow} \quad (17.1)$$

where g is the gravity acceleration, H_1 is the upstream total head and Δz is the weir height (e.g. Fig. 17.4). In practice the observed discharge differs from equation (17.1) because the pressure distribution on the crest may not be hydrostatic. Furthermore, the weir geometry, roughness and inflow conditions affect the discharge characteristics (e.g. Miller 1994). The real flow rate is expressed as:

$$q = C_D \sqrt{g}(\tfrac{2}{3}(H_1 - \Delta z))^{3/2} \quad (17.2)$$

where C_D is the discharge coefficient.

The most common types of overflow weirs are the broad-crested weir, the sharp-crested weir and the ogee-crest weir. Their respective discharge characteristics are described below.

Notes

1. Jean-Baptiste Bélanger (1789–1874) was a French professor at the Ecole Nationale Supérieure des Ponts et Chaussées (Paris). In his book (Bélanger 1828), he presented the basics of hydraulic jump calculations, backwater calculations and discharge characteristics of weirs.
2. C_D is a dimensionless coefficient. Typically $C_D = 1$ for a broad-crested weir. When $C_D > 1$, the discharge capacity of a weir is greater than that of a broad-crested weir for identical upstream head above crest ($H_1 - \Delta z$).

17.2.2 Broad-crested weir

A broad-crested weir is a flat-crested structure with a crest length large compared with the flow thickness (Fig. 17.4). The ratio of crest length to upstream head over crest must typically be greater than 1.5 to 3 (e.g. Chow 1973, Henderson 1966):

$$\frac{L_{crest}}{H_1 - \Delta z} > 1.5 \text{ to } 3$$

---

[1] Flow conditions for which the specific energy (of the mean flow) is minimum are called critical flow conditions.

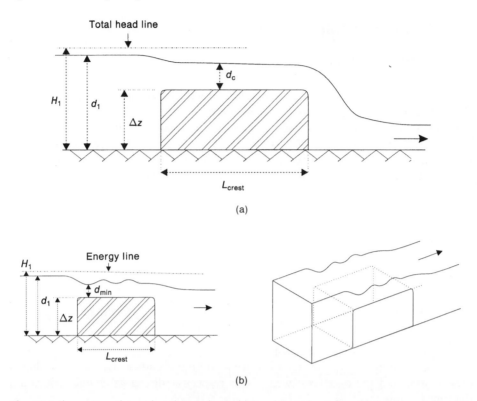

Fig. 17.4 Flow pattern above a broad-crested weir. (a) Broad-crested weir flow. (b) Undular weir flow.

When the crest is 'broad' enough for the flow streamlines to be parallel to the crest, the pressure distribution above the crest is hydrostatic and the critical flow depth is observed on the weir crest. Broad-crested weirs are sometimes used as critical depth meters (i.e. to measure stream discharges). The hydraulic characteristics of broad-crested weirs were studied during the 19th and 20th centuries. Hager and Schwalt (1994) recently presented an authoritative work on the topic.

The discharge above the weir equals:

$$q = \tfrac{2}{3}\sqrt{\tfrac{2}{3}g(H_1 - \Delta z)^3} \qquad \text{Ideal fluid flow calculations} \qquad (17.1a)$$

where H_1 is the upstream total head and Δz is the weir height above the channel bed (Fig. 17.4). Equation (17.1a) may be conveniently rewritten as:

$$q = 1.704(H_1 - \Delta z)^{3/2} \qquad \text{Ideal fluid flow calculations} \qquad (17.1b)$$

Notes

1. In a horizontal rectangular channel and assuming hydrostatic pressure distribution, the critical flow depth equals:

$$d_c = \tfrac{2}{3}E \qquad \text{Horizontal rectangular channel}$$

where E is the specific energy. The critical depth and discharge per unit width are related by:

$$d_c = \sqrt[3]{\frac{q^2}{g}} \qquad \text{Rectangular channel}$$

$$q = \sqrt{gd_c^3} \qquad \text{Rectangular channel}$$

2. At the crest of a broad-crested weir, the continuity and Bernoulli equations yield:

$$H_1 - \Delta z = \tfrac{3}{2}d_c$$

Note that equation (17.1) derives from the continuity and Bernoulli equations, hence from the above equation.

Discussion

(A) *Undular weir flow*
For low discharges (i.e. $(d_1 - \Delta z)/\Delta z \ll 1$), several researchers observed free-surface undulations above the crest of a broad-crested weir (Fig. 17.4(b)). Model studies suggest that undular weir flow occurs for:

$$\frac{q}{\sqrt{gd_{min}^3}} < 1.5 \qquad \text{undular weir flow}$$

where d_{min} is the minimum flow depth upstream of the first wave crest (Fig. 17.4). Another criterion is:

$$\frac{H_1 - \Delta z}{L_{crest}} < 0.1 \qquad \text{undular weir flow}$$

where L_{crest} is the crest length in the flow direction. The second equation is a practical criterion based on the ratio of head on crest to weir length.

In practice, design engineers should avoid flow conditions leading to undular weir flow. In the presence of free-surface undulations above the crest, the weir cannot be used as a discharge meter, and waves may propagate in the downstream channel.

(B) *Discharge coefficients*
Experimental measurements indicate that the discharge versus total head relationship departs slightly from equation (17.1) depending upon the weir geometry and flow conditions. Equation (17.1) is usually rearranged as:

$$q = C_D \tfrac{2}{3} \sqrt{\tfrac{2}{3}g(H_1 - \Delta z)^3}$$

where the discharge coefficient C_D is a function of the weir height, crest length, crest width, upstream corner shape and upstream total head (Table 17.1).

Table 17.1 Discharge coefficient for broad-crested weirs

Reference (1)	Discharge coefficient C_D (2)	Range (3)	Remarks (4)
Sharp-corner weir Hager and Schwalt (1994)	$0.85\dfrac{9}{7}\left(1 - \dfrac{2/9}{1+\left(\dfrac{H_1 - \Delta z}{L_{\text{crest}}}\right)^4}\right)$	$0.1 < \dfrac{H_1 - \Delta z}{L_{\text{crest}}} < 1.5$	Deduced from laboratory experiments.
Rounded-corner weir Bos (1976)	$\left(1 - 0.01\dfrac{L_{\text{crest}} - r}{W}\right)$ $\times \left(1 - 0.01\dfrac{L_{\text{crest}} - r}{d_1 - \Delta z}\right)$	$\dfrac{d_1 - Dz}{L_{\text{crest}}} > 0.05$ $d_1 - \Delta z > 0.06\,\text{m}$ $\dfrac{H_1 - \Delta z}{\Delta z} < 1.5$ $\Delta z > 0.15\,\text{m}$	Based upon laboratory and field tests.
Ackers *et al.* (1978)	0.95^{a}	$0.15 < \dfrac{H_1 - \Delta z}{\Delta z} < 0.6$	

Notes: [a] Re-analysis of experimental data presented by Ackers *et al.* (1978); r = curvature radius of upstream corner.

17.2.3 Sharp-crested weir

A sharp-crested weir is characterized by a thin sharp-edged crest (Fig. 17.5). In the absence of sidewall contraction, the flow is basically two-dimensional and the flow field can be solved by analytical and graphical methods: i.e. ideal fluid flow theory (e.g. Vallentine 1969, p. 79).

For an aerated nappe, the discharge per unit width is usually expressed as:

$$q = \tfrac{2}{3}C\sqrt{g(d_1 - \Delta z)^3} \qquad (17.4)$$

where d_1 is the upstream water depth, Δz is the crest height above the channel bed (Fig. 17.5) and C is a dimensionless discharge coefficient. Numerous correlations were proposed for C (Tables 17.2 and 17.3). In practice, the following expression is recommended:

$$C = 0.611 + 0.08\frac{d_1 - \Delta z}{\Delta z} \qquad (17.5)$$

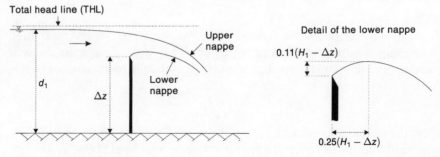

Fig. 17.5 Sharp-crested weir.

Table 17.2 Discharge coefficient for sharp-crested weirs (full-width weir in rectangular channel)

Reference (1)	Discharge coefficient C (2)	Range (3)	Remarks (4)
von Mises (1917)	$\dfrac{\pi}{\pi + 2}$	$\dfrac{d_1 - \Delta z}{\Delta z}$ very large	Ideal fluid flow calculations of orifice flow.
Henderson (1966)	$0.611 + 0.08\dfrac{d_1 - \Delta z}{\Delta z}$	$0 \le \dfrac{d_1 - \Delta z}{\Delta z} < 5$	Experimental work by Rehbock (1929).
	1.135	$\dfrac{d_1 - \Delta z}{\Delta z} = 10$	
	$1.06\left(1 + \dfrac{\Delta z}{d_1 - \Delta z}\right)^{3/2}$	$20 < \dfrac{d_1 - \Delta z}{\Delta z}$	
Bos (1976)	$0.602 + 0.075\dfrac{d_1 - \Delta z}{\Delta z}$	$d_1 - \Delta z > 0.03\,\mathrm{m}$ $\dfrac{d_1 - \Delta z}{\Delta z} < 2$ $\Delta z > 0.40\,\mathrm{m}$	Based on experiments performed at Georgia Institute of Technology.
Present study	1.0607	$\Delta z = 0$	Ideal flow at overfall.

Notes
1. Sharp-crested weirs are very accurate discharge meters. They are commonly used for small flow rates.
2. For a vertical sharp-crested weir, the lower nappe is deflected upwards immediately downstream of the sharp edge. The maximum elevation of the lower nappe location occurs at about $0.11(H_1 - \Delta z)$ above the crest level in the vertical direction and at about $0.25(H_1 - \Delta z)$ from the crest in the horizontal direction (e.g. Miller 1994).
3. At very low flow rates, $(d_1 - \Delta z)/\Delta z$ is very small and equation (17.4) tends to:

$$q \sim 1.803(d_1 - \Delta z)^{3/2} \qquad \text{very small discharge}$$

4. Nappe aeration is extremely important. If the nappe is not properly ventilated, the discharge characteristics of the weir are substantially affected, and the weir might not operate safely. Sometimes, the crest can be contracted at the sidewalls to facilitate nappe ventilation (e.g. Henderson 1966, pp. 177–178).

Table 17.3 Discharge correlations for sharp-crested weirs (full-width in rectangular channel)

Reference (1)	Discharge per unit width q (m^2/s) (2)	Comments (3)
Ackers et al. (1978)	$0.564\left(1 + 0.150\dfrac{d_1 - \Delta z}{\Delta z}\right)\sqrt{g}$ $\times (d_1 - \Delta z + 0.0001)^{3/2}$	Range of applications $d_1 - \Delta z > 0.02\,\mathrm{m}$ $\Delta z > 0.15$, $(d_1 - \Delta z)/\Delta z < 2.2$
Herschy (1995)	$1.85(d_1 - \Delta z)^{3/2}$	Approximate correlation ($\pm 3\%$): $(d_1 - \Delta z)/\Delta z < 0.5$

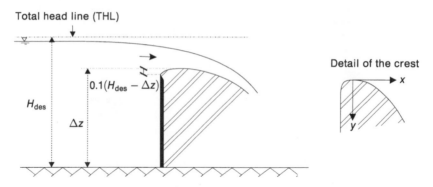

Fig. 17.6 Sketch of a nappe-shaped ogee crest.

17.2.4 Ogee crest weir

The basic shape of an ogee crest is that of the lower nappe trajectory of a sharp-crested weir flow for the design flow conditions (discharge Q_{des}, upstream head H_{des}) (Figs 17.2(a), 17.5–17.8).

Nappe-shaped overflow weir

The characteristics of a nappe-shaped overflow can be deduced from the equivalent 'sharp-crested weir' design. For the design head H_{des}, equation (17.4) leads to:

$$q_{des} = \frac{2}{3} C \sqrt{g \left(\frac{H_{des} - \Delta z}{0.89} \right)^3} \qquad (17.6)$$

where Δz is the crest elevation (Fig. 17.6). For a high weir (i.e. $(H_{des} - \Delta z)/\Delta z \ll 1$), the discharge above the crest could be deduced from equations (17.5) and (17.6):

$$q_{des} = 2.148 (H_{des} - \Delta z)^{3/2} \qquad \text{Ideal flow conditions} \qquad (17.7)$$

Discharge characteristics of an ogee crest

In practice, the discharge–head relationship differs from ideal flow conditions (equation (17.7)). For the design flow conditions, the flow rate per unit width is often presented as:

$$q_{des} = C_{des} (H_{des} - \Delta z)^{3/2} \qquad \{ \text{Design flow conditions: } q = q_{des} \} \qquad (17.8)$$

where the design discharge coefficient C_{des} is primarily a function of the ogee crest shape. It also can be deduced from model tests.

For a vertical-shaped ogee crest, typical values of the discharge coefficient are reported in Table 17.4 and Fig. 17.9(a). In Table 17.4, the broad-crested weir case corresponds to $\Delta z/(H_{des} - \Delta z) = 0$ and it is found that $(C_D)_{des} = 1$ (equation (17.1)). For a high weir (i.e. $\Delta z/(H_{des} - \Delta z) > 3$), the discharge coefficient C_{des} tends to $2.19 \, \text{m}^{1/2}/\text{s}$. Such a value differs slightly from equation (17.7) but it is more reliable because it is based upon experimental results.

(a)

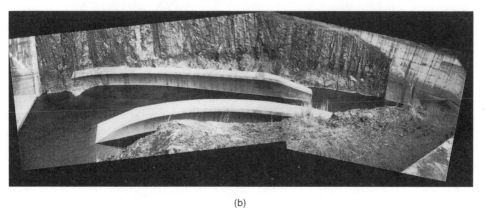

(b)

Fig. 17.7 Spillway chute of Hinze dam in September 1997 (Gold Coast QLD, 1976). Dam height: 44 m; design spillway capacity: 1700 m^3/s. (a) Ogee crest chute (view from downstream, on right bank). (b) Concrete veins at downstream end of chute, directing the supercritical chute flow into the energy dissipator (flow from left to right).

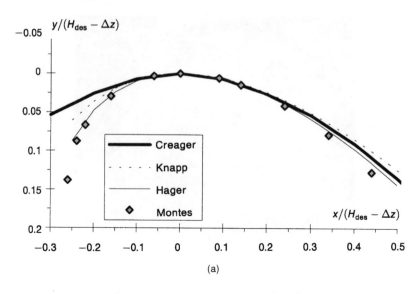

(a)

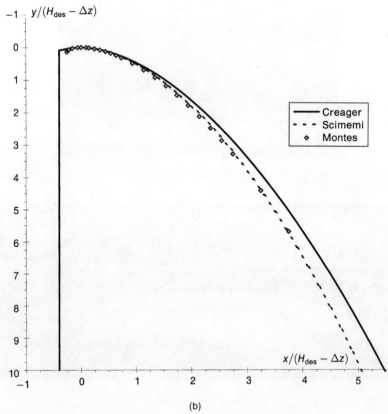

(b)

Fig. 17.8 Profiles of ogee crest. (a) Crest details. (b) Creager and Scimemi profiles.

Table 17.4 Discharge coefficient of ogee crest (vertical-faced ogee crest)

$\Delta z/(H_{des} - \Delta z)$ (1)	C_{des} (m$^{1/2}$/s) (2)	$(C_D)_{des}$ (3)	Comments (4)
0	1.7	1.0	Broad-crested weir
0.04	1.77	1.04	
0.14	1.92	1.13	
0.36	2.07	1.21	
0.50	2.10	1.23	
1.0	2.14	1.257	
1.5	2.16	1.268	
3.0	2.18	1.28	
>3.0	2.19	1.28	Large weir height

Reference: US Bureau of Reclamation (US Dept. of the Interior 1987).
Notes: $C_{des} = q_{des}/(H_{des} - \Delta z)^{3/2}$; $(C_D)_{des} = q_{des}/(\sqrt{g}(2/3)^{3/2}(H_{des} - \Delta z)^{3/2})$.

For a given crest profile, the overflow conditions may differ from the design flow conditions. The discharge versus upstream head relationship then becomes:

$$q = C(H_1 - \Delta z)^{3/2} \qquad \{\text{Non-design flow conditions: } q \neq q_{des}\} \qquad (17.9)$$

in which the discharge coefficient C differs from the design discharge coefficient C_{des}.

Generally, the relative discharge coefficient C/C_{des} is a function of the relative total head $(H_1 - \Delta z)/(H_{des} - \Delta z)$ and of the ogee crest shape (e.g. Fig. 17.9(b)).

Discussion

For $H_1 = H_{des}$ the pressure on the crest invert is atmospheric (because the shape of the invert is based on the lower nappe trajectory of the sharp-crested weir overflow).

When the upstream head H_1 is larger than the design head H_{des}, the pressures on the crest are less than atmospheric and the discharge coefficient C (equation (17.9)) is larger than the design discharge coefficient C_{des} (typically 2.19 m$^{1/2}$/s). For $H_1 < H_{des}$, the pressures on the crest are larger than atmospheric and the discharge coefficient is smaller. At the limit, the discharge coefficient tends to the value of 1.704 m$^{1/2}$/s corresponding to the broad-crested weir case (equation (17.1b)) (Table 17.5).

Standard crest shapes

With nappe-shaped overflow weirs, the pressure on the crest invert should be atmospheric at the design head. In practice, small deviations occur because of bottom friction and the developing boundary layer. Design engineers must select the shape of the ogee crest such that sub-atmospheric pressures are avoided on the crest invert: i.e. to prevent separation and cavitation-related problems. Several ogee crest profiles were developed (Table 17.6).

The most usual profiles are the WES profile and the Creager profile. The Creager design is a mathematical extension of the original data of Bazin in 1886–88 (Creager

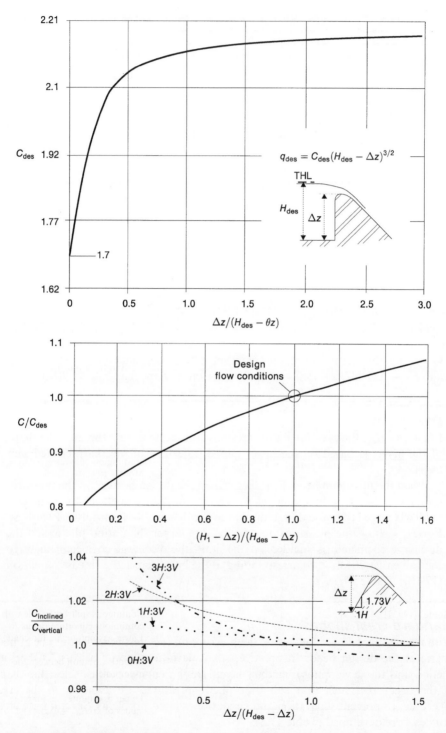

Fig. 17.9 Discharge coefficient of a USBR-profile ogee crest. Data: US Bureau of Reclamation (US Dept. of the Interior 1987).

Table 17.5 Pressures on an ogee crest invert for design and non-design flow conditions

Upstream head (1)	Pressure on crest (2)	Discharge coefficient (3)
$H_1 = H_{des}$	quasi-atmospheric	$C = C_{des}$
$H_1 > H_{des}$	less than atmospheric	$C > C_{des}$
$H_1 < H_{des}$	larger than atmospheric	$C < C_{des}$
$H_1 \ll H_{des}$	larger than atmospheric	$C \approx 1.704\,\mathrm{m}^{1/2}/\mathrm{s}$

1917). The WES-standard ogee shape is based upon detailed observations of the lower nappe of sharp-crested weir flows (Scimemi 1930) (Figs 17.5 and 17.8). Montes (1992a) stressed that the ogee crest profile must be continuous and smooth, and sudden variation of the crest curvature must be avoided to prevent unwanted aeration or cavitation. Ideally, the crest profile should start tangentially to the upstream apron, with a smooth and continuous variation of the radius of curvature.

Table 17.6 Examples of spillway profiles (vertical faced ogee crest)

Profile (1)	Equations (2)	Comments (3)
Creager (1917) profile	$Y = 0.47 \dfrac{X^{1.80}}{(H_{des} - \Delta z)^{0.80}}$	Derived from Bazin's (1888–1898) experiments.
Scimemi (1930) profile	$Y = 0.50 \dfrac{X^{1.85}}{(H_{des} - \Delta z)^{0.85}}$	Also called WES profile.
Knapp (1960)	$\dfrac{Y}{H_{des} - \Delta z} = \dfrac{X}{H_{des} - \Delta z}$ $\quad - \ln\left(1 + \dfrac{X}{0.689(H_{des} - \Delta z)}\right)$	Continuous spillway profile for crest region only (as given by Montes 1992a).
Hager (1991)	$\dfrac{Y}{H_{des} - \Delta z}$ $= 0.1360 + 0.482625\left(\dfrac{X}{H_{des} - \Delta z} + 0.2818\right)$ $\times \ln\left(1.3055\left(\dfrac{X}{H_{des} - \Delta z} + 0.2818\right)\right)$	Continuous spillway profile with continuous curvature radius: $-0.498 < \dfrac{X}{H_{des} - \Delta z} < 0.484$
Montes (1992a)	$\dfrac{R_1}{H_{des} - \Delta z} = 0.05 + 1.47\dfrac{s}{H_{des} - \Delta z}$ $\dfrac{R}{H_{des} - \Delta z} = \dfrac{R_1}{H_{des} - \Delta z}\left(1 + \left(\dfrac{R_u}{R_1}\right)^{2.625}\right)^{1/2.625}$ $\dfrac{R_u}{H_{des} - \Delta z} = 1.68\left(\dfrac{s}{H_{des} - \Delta z}\right)^{1.625}$	Continuous spillway profile with continuous curvature radius R. Lower asymptote: i.e. for small values of $s/(H_{des} - \Delta z)$. Smooth variation between the asymptotes. Upper asymptote: i.e. for large values of $s/(H_{des} - \Delta z)$.

Notes: X, Y horizontal and vertical coordinates with dam crest as origin, Y measured positive downwards (Fig. 17.8); R = radius of curvature of the crest; s = curvilinear coordinate along the crest shape.

17.3 Chute design

17.3.1 Presentation

Once the water flows past the crest, the fluid is accelerated by gravity along the chute. At the upstream end of the chute, a turbulent boundary layer is generated by bottom friction and develops in the flow direction. When the outer edge of the boundary layer reaches the free-surface, the flow becomes fully-developed (Fig. 17.10).

In the developing flow region, the boundary layer thickness δ increases with distance along the chute. Empirical correlations may be used to estimate the boundary layer growth:

$$\frac{\delta}{s} = 0.0212(\sin\theta)^{0.11}\left(\frac{s}{k_s}\right)^{-0.10} \qquad \text{Smooth concrete chute } (\theta > 30°) \qquad (17.10)$$

$$\frac{\delta}{s} = 0.06106(\sin\theta)^{0.133}\left(\frac{s}{h\cos\theta}\right)^{-0.17} \qquad \text{Stepped chute (skimming flow)} \qquad (17.11)$$

where s is the distance measured from the crest origin, k_s is the equivalent roughness height, θ is the chute slope and h is the step height. Equations (17.10) and (17.11) are semi-empirical formulations that fit well model and prototype data (Wood *et al.* 1983, Chanson 1995b). They apply to steep concrete chutes (i.e. $\theta > 30°$).

In the fully-developed flow region, the flow is gradually varied until it reaches equilibrium (i.e. normal flow conditions). The gradually-varied flow properties can be deduced by integration of the backwater equation. The normal flow conditions are deduced from the momentum principle (Appendix A4.1).

Note

On steep chutes, free-surface aeration[2] may take place downstream of the intersection of the outer edge of the developing boundary layer with the free-surface. Air entrainment can be clearly identified by the 'white water' appearance of the free-surface flow. Wood (1991) and Chanson (1997) presented comprehensive studies of free-surface aeration on smooth chutes, while Chanson (1995b) reviewed the effects of free-surface aeration on stepped channels.

The effects of free-surface aeration include flow bulking, some drag reduction effect and air–water gas transfer. The topic is still under active research (e.g. Chanson 1997).

17.3.2 Application

For steep chutes (typically $\theta \sim 45°$ to $55°$), both the flow acceleration and boundary layer development affect the flow properties significantly. The complete flow calcula-

---

[2] Air entrainment in open channels is also called free-surface aeration, self-aeration, insufflation or white waters.

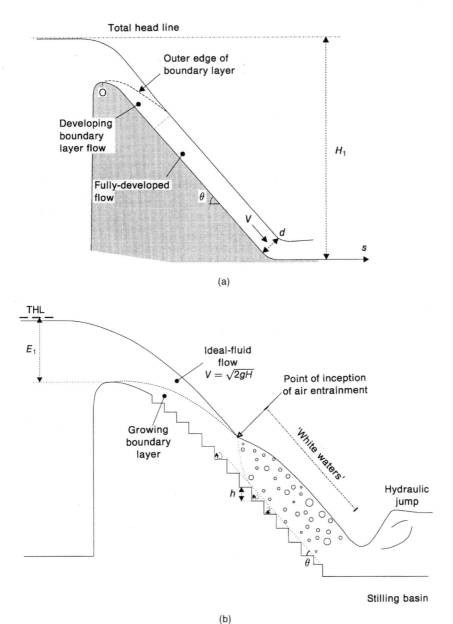

Fig. 17.10 Sketch of a steep chute. (a) Smooth chute. (b) Stepped chute.

tions can be tedious and most backwater calculations are not suitable.[3] Complete calculations of developing flow and uniform equilibrium flow (see Appendix A4.1) may be combined to provide a general trend that may be used for a preliminary

---

[3] Because backwater calculations are valid only for fully-developed flows. Furthermore most software assumes hydrostatic pressure distributions and neglects the effects of free-surface aeration.

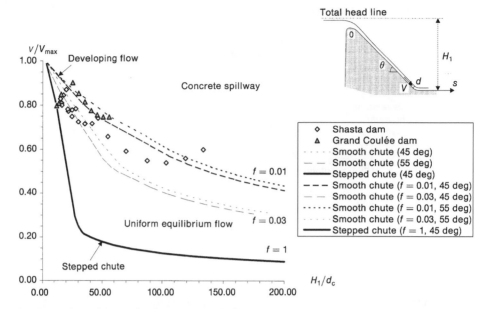

Fig. 17.11 Flow velocity at the downstream end of a steep chute.

design (Fig. 17.11). Ideally, the maximum velocity at the downstream chute end is:

$$V_{max} = \sqrt{2g(H_1 - d\cos\theta)} \qquad \text{Ideal fluid flow} \qquad (17.12)$$

where H_1 is the upstream total head and d is the downstream flow depth: i.e. $d = q/V_{max}$ (Fig. 17.10). In practice the downstream flow velocity V is smaller than the theoretical velocity V_{max} because of friction losses.

In Fig. 17.11, the mean flow velocity at the end of the chute is plotted as V/V_{max} versus H_1/d_c where V_{max} is the theoretical velocity (equation (17.12)), H_1 is the upstream total head (above spillway toe) and d_c is the critical depth. Both developing flow calculations (equations (17.10) and (17.11)) and uniform equilibrium flow calculations are shown. Fitting curves must be plotted to connect these lines (Fig. 17.11).

The semi-empirical curves (Fig. 17.11) are compared with experimental results obtained on prototype spillways (Grand Coulée dam and Shasta dam). The curves are valid for smooth and stepped spillways (concrete chutes), with slopes ranging from 45° to 55° (i.e. 1 V:1 H to 0.7 V:1 H).

17.3.3. Discussion

It is worth comparing the performances of stepped and smooth chutes. The larger mean bottom shear stress, observed with stepped chute flows, implies larger hydrodynamic loads on the steps than on a smooth invert. Stepped chutes require a reinforced stepped profile compared with a smooth chute for identical inflow conditions. On the other hand, Fig. 17.11 also shows that larger energy dissipation take place along a stepped spillway compared with a smooth chute. Hence, the size of the downstream stilling basin can be reduced with a stepped chute.

Note

Stepped chute flows are subjected to strong free-surface aeration (Chanson 1995b). As a result, flow bulking and gas transfer are enhanced compared with a smooth channel.

Discussion

The energy dissipation characteristics of stepped channels were well known to ancient engineers. During the Renaissance period, Leonardo Da Vinci realized that the flow, 'the more rapid it is, the more it wears away its channel'; if a waterfall 'is made in deep and wide steps, after the manner of stairs, the waters (...) can no longer descend with a blow of too great a force'. He illustrated his conclusion with a staircase waterfall 'down which the water falls so as not to wear away anything' (Richter 1939).

During the 19th century, stepped (or staircase) weirs and channels were quite common: e.g. in the USA, nearly one third of the masonry dams built during the 19th century were equipped with a stepped spillway. A well-known 19th century textbook stated: 'The byewash[4] will generally have to be made with a very steep mean gradient, and to avoid the excessive scour which could result if a uniform[5] channel were constructed, it is in most cases advisable to carry the byewash down by a series of steps, by which the velocity will be reduced' (Humber 1876, p. 133).

17.4 Stilling basins and energy dissipators

17.4.1 Presentation

Energy dissipators are designed to dissipate the excess in kinetic energy at the end of the chute before it re-enters the natural stream. Energy dissipation on dam spillways is achieved usually by: (1) a standard stilling basin downstream of a steep spillway in which a hydraulic jump is created to dissipate a large amount of flow energy and to convert the flow from supercritical to subcritical conditions, (2) a high velocity water jet taking off from a flip bucket and impinging into a downstream plunge pool, or (3) a plunging jet pool in which the spillway flow impinges and the kinetic energy is dissipated in turbulent recirculation (Fig. 17.12). The construction of steps on the spillway chute may also assist in energy dissipation.

The stilling basin is the common type of dissipator for weirs and small dams. Most energy is dissipated in a hydraulic jump assisted by *appurtenances* (e.g. step, baffle blocks) to increase the turbulence.

---

[4] Channel to carry waste waters: i.e. spillway.
[5] In the meaning of a uniform smooth channel bed (i.e. not stepped).

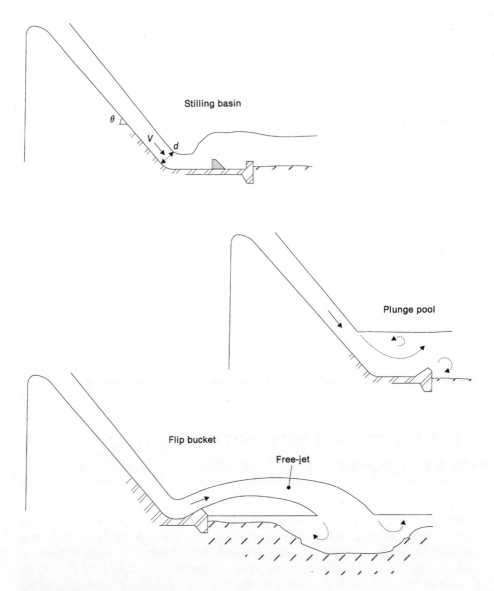

Fig. 17.12 Types of energy dissipators.

Notes
1. Other forms of energy dissipator include the drop structure and the impact-type stilling basin. The drop structure is detailed in another section because the flow pattern differs substantially from chute spillways. With impact-type dissipators, dissipation takes place by impact of the inflow on a vertical baffle (e.g. US Bureau of Reclamation 1987, p. 463).
2. On stepped chutes, the channel roughness (i.e. steps) contributes to the energy dissipation. In practice, a stilling basin is often added at the downstream end,

but its size is smaller than that required for a smooth chute with identical flow conditions (see Section 17.3.2).

17.4.2 Energy dissipation at hydraulic jumps

Introduction

A hydraulic jump is the rapid transition from a supercritical to subcritical flow. It is an extremely turbulent process, characterized by large-scale turbulence, surface waves and spray, energy dissipation and air entrainment. The large-scale turbulence region is usually called the 'roller'.

The downstream flow properties and energy loss in a hydraulic jump can be deduced from the momentum principle as a function of the upstream Froude number Fr and upstream flow depth d. For a horizontal flat rectangular channel, the downstream flow depth equals:

$$\frac{d_{conj}}{d} = \tfrac{1}{2}(\sqrt{1 + 8Fr^2} - 1) \qquad (17.13)$$

where $Fr = V/\sqrt{gd}$. The energy equation gives the head loss:

$$\frac{\Delta H}{d} = \frac{(\sqrt{1 + 8Fr^2} - 3)^3}{16(\sqrt{1 + 8Fr^2} - 1)} \qquad (17.14)$$

Notes
1. The upstream and downstream depth d and d_{conj} are referred to as conjugate or sequent depths.
2. The upstream Froude number must be greater than unity: $Fr > 1$.
3. Jean-Baptiste Bélanger (1789–1874) was the first to suggest the application of the momentum principle to the hydraulic jump flow (Bélanger 1828). The momentum equation applied across a hydraulic jump is called the Bélanger equation. Equation (17.13) is sometimes called (improperly) the Bélanger equation.

Application
In a horizontal rectangular stilling basin with baffle blocks, the inflow conditions are: $d = 0.95\,\text{m}$, $V = 16.8\,\text{m/s}$. The observed downstream flow depth is 6.1 m. The channel is 12.5 m wide. Calculate the total force exerted on the baffle blocks. In which direction does the force applied by the flow onto the blocks act?

Solution
First, we select a control volume for which the upstream and downstream cross-sections are located far enough from the roller for the velocity to be essentially horizontal and uniform. The application of the momentum equation to the control volume yields:

$$\rho q(V_{tw} - V) = \tfrac{1}{2}\rho g(d^2 - d_{tw}^2) - \frac{F_B}{B}$$

where d_{tw} and V_{tw} are the tailwater depth and velocity respectively, F_B is the force from the blocks on the control volume and B is the channel width. The total force is: $F_B = +605\,kN$. That is, the force applied by the fluid on the blocks acts in the downstream direction.

Remark

Equation (17.13) is not applicable. It is valid only for flat horizontal rectangular channels. In the present application, the baffle blocks significantly modify the jump properties.

Types of hydraulic jump

Hydraulic jump flows may exhibit different flow patterns depending upon the upstream flow conditions. They are usually classified as functions of the upstream Froude number Fr (Table 17.7). In practice, it is recommended to design energy dissipators with a steady jump type.

Table 17.8 summarizes the basic flow properties of hydraulic jump in rectangular horizontal channels.

Notes

1. The classification of hydraulic jumps (Table 17.7) must be considered as *rough* guidelines. It applies to hydraulic jumps in rectangular horizontal channels.
2. Recent investigations (e.g. Chanson and Montes 1995, Montes and Chanson 1998) showed that undular hydraulic jumps (Fawer jumps) might take place for upstream Froude numbers up to 4 depending upon the inflow conditions. The topic is still actively studied.
3. A hydraulic jump is a very unsteady flow. Experimental measurements of bottom pressure fluctuations indicated that the mean pressure is quasi-hydrostatic below the jump but large pressure fluctuations are observed (e.g. Hager 1992b).

Table 17.7 Classification of hydraulic jump in rectangular horizontal channels (Chow 1973)

Fr (1)	Definition (2)	Remarks (3)
1	Critical flow	No hydraulic jump.
1 to 1.7	Undular jump (Fawer jump)	Free-surface undulations developing downstream of jump over considerable distances. *Negligible* energy losses.
1.7 to 2.5	Weak jump	Low energy loss.
2.5 to 4.5	Oscillating jump	Wavy free surface. Production of large waves of irregular period. Unstable oscillating jump. Each irregular oscillation produces a large wave that may travel far downstream, damaging and eroding the banks. To be avoided if possible.
4.5 to 9	Steady jump	45 to 70 % of energy dissipation. Steady jump. Insensitive to downstream conditions (i.e. tailwater depth). Best economical design.
>9	Strong jump	Rough jump. Up to 85 % of energy dissipation. Risk of channel bed erosion. To be avoided.

Table 17.8 Dimensionless characteristics of hydraulic jump in horizontal rectangular channels

Fr (1)	d_{conj}/d (2)	$\Delta H/H$ (3)	L_r/d (4)
3	3.77	0.26	11.8
3.5	4.47	0.33	15.7
4.5	5.88	0.44	23.4
5.5	7.29	0.53	30.9
6.5	8.71	0.59	38.2
7.5	10.12	0.64	45.3
9	12.24	0.70	55.5
12	16.48	0.77	73.9
15	20.72	0.82	89.6

Length of the roller

The roller length of the jump may be estimated as (Hager *et al.* 1990):

$$\frac{L_r}{d} = 160 \tanh\left(\frac{Fr}{20}\right) - 12 \qquad 2 < Fr < 16 \qquad (17.15)$$

where tanh is the hyperbolic tangent function and L_r is the length of the roller. Equation (17.15) is valid for rectangular horizontal channels ($d/B < 0.1$) and it may be used to predict the length of horizontal dissipation basins.

Practically, the length of a hydraulic jump stilling basin must be greater than the roller length for all flow conditions.

17.4.3 Stilling basins

Basically, the hydraulic design of a stilling basin must ensure a safe dissipation of the flow kinetic energy, to maximize the rate of energy dissipation and to minimize the size (and cost) of the structure. In practice, energy dissipation by a hydraulic jump in a stilling basin is assisted with elements (e.g. baffle blocks, sill) placed on the stilling basin apron.

Basic shapes

The basic features of stilling basins include drop, backward-facing step (or sill), baffle block(s), sudden expansion (Fig. 17.13). Hager (1992b) reviewed the advantages of each type. A summary is given below.

Drops and backward-facing steps are simple elements used to stabilize the hydraulic jump. Drops (also called negative steps) are advised when the downstream tailwater level may vary significantly (Fig. 17.13(a)). Backward-facing steps (also called positive steps) are usually located near the toe of the jump (Fig. 17.13(b)).

Baffle blocks (or dentated sills) can be placed in one or several rows.[6] The blocks force the flow above them (as a sill) and in between them. Baffle blocks must be designed with standard shapes. They are not recommended when the inflow velocity is larger than 20 to 30 m/s because of the risks of cavitation damage.

---

[6] The single-row arrangement is comparatively more efficient than the multiple-rows geometry.

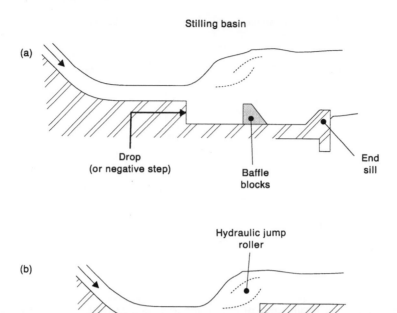

Fig. 17.13 Sketch of stilling basins.

Sudden expansion (in the stilling basin) is another technique to enhance turbulent energy dissipation and to reduce the basin length. Physical modelling is, however, strongly recommended.

Standard stilling basins

Several standardized designs of stilling basins were developed in the 1950s and 1960s (Table 17.9, Fig. 17.14). These basins were tested in models and prototypes over a considerable range of operating flow conditions. The prototype performances are well known, and they can be selected and designed without further model studies.

In practice, the following types are highly recommended:

- the USBR (US Bureau of Reclamation) Type II basin, for large structures and $Fr > 4.5$;
- the USBR Type III basin and the SAF basin, for small structures;
- the USBR Type IV basin, for oscillating jump flow conditions (Fig. 17.15).

Note

The requirements of the US Bureau of Reclamation (USBR) for its dissipators are more stringent than those of other organizations. As a result, the USBR Type III basin is sometimes too conservative and the SAF basin may be preferred for small structures.

Table 17.9 Standard types of hydraulic jump energy dissipators

Name (1)	Application (2)	Flow conditions (3)	Tailwater depth d_{tw} [a] (4)	Remarks (5)
USBR Type II	Large structures.	$Fr > 4.5$ $q < 46.5 \, \text{m}^2/\text{s}$ $H_1 < 61 \, \text{m}$ Basin length $\sim 4.4 d_{\text{conj}}$	$1.05 d_{\text{conj}}$	Two rows of blocks. The last row is combined with an inclined end sill (i.e. dentated sill). Block height = d.
USBR Type III	Small structures.	$Fr > 4.5$ $q < 18.6 \, \text{m}^2/\text{s}$ $V < 15$ to $18.3 \, \text{m/s}$ Basin length $\sim 2.8 d_{\text{conj}}$	$1.0 d_{\text{conj}}$	Two rows of blocks and an end sill. Block height = d.
USBR Type IV	For oscillating jumps.	$2.5 < Fr < 4.5$ Basin length $\sim 6 d_{\text{conj}}$	$1.1 d_{\text{conj}}$	One row of blocks and an end sill. Block height = $2d$. Wave suppressors may be added at downstream end.
SAF	Small structures.	$1.7 < Fr < 17$ Basin length $= 4.5 d_{\text{conj}} Fr^{-0.76}$	$1.0 d_{\text{conj}}$	Two rows of baffle blocks and an end sill. Block height = d.
USACE		Basin length $> 4 d_{\text{conj}}$	$1.0 d_{\text{conj}}$	Two rows of baffle blocks and end sill.

References: Chow (1973), Hager (1992b), Henderson (1966), US Bureau of Reclamation (US Dept. of the Interior 1987).
Notes: [a] Recommended tailwater depth for optimum stilling basin operation; d = upstream flow depth (inflow depth); d_{conj} = conjugate flow depth (equation (17.13)); d_{tw} = tailwater depth; Fr = inflow Froude number.

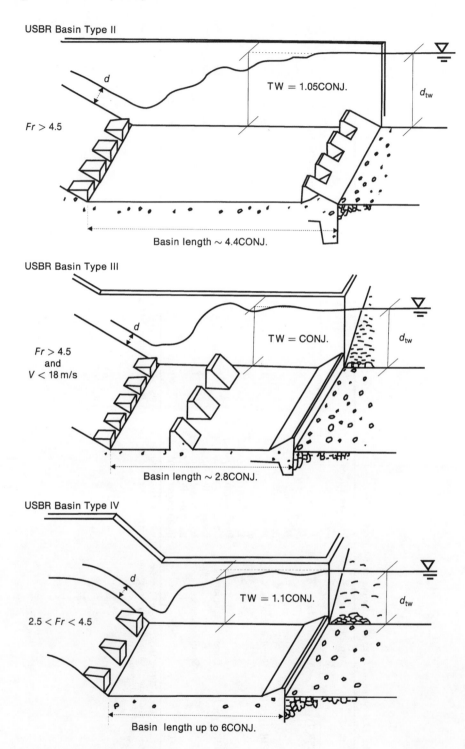

Fig. 17.14 Sketch of USBR basins.

Fig. 17.15 Prototype stilling basin: USBR basin Type 4 at Bjelke-Petersen dam, Australia on 7 November 1997. Flow direction from left to right. Design conditions: 3660 m³/s, dam height: 43 m.

17.4.4 Discussion

In practice, design engineers must ensure that a stilling basin can operate safely for a wide range of flow conditions. Damage (scouring, cavitation) to the basin and to the downstream natural bed may occur in several cases:

- the apron is too short and/or too shallow for an optimum jump location (i.e. on the apron),
- poor shapes of the blocks, sill, drop resulting in cavitation damage,
- flow conditions larger than design flow conditions,
- unusual overflow during construction periods,
- poor construction of the apron and blocks,
- seepage underneath the apron, inadequate drainage and uplift pressure build-up,
- wrong conception of the stilling basin.

17.5 Design procedure

17.5.1 Introduction

The construction of a small dam across a stream will modify both the upstream and possibly downstream flow conditions. The dam crest elevation must be selected accurately to provide the required storage of water or upstream water level rise. Furthermore, the spillway and stilling basin must operate safely for a wide range of flow rates and tailwater flow conditions.

17.5.2 Dam spillway with hydraulic jump energy dissipator

Considering an overflow spillway with hydraulic jump energy dissipation at the toe, the basic steps in the design procedure are:

Step 1. Select the crest elevation z_{crest} (bed topography, storage level).

Step 2. Choose the crest width B (site geometry, hydrology). The crest width may be smaller than the weir length (across the stream).

Step 3. Determine the design discharge Q_{des} from risk analysis and flood routing. Required information includes catchment area, average basin slope, degree of impermeability, vegetation cover, rainfall intensity, duration and inflow hydrograph. The peak spillway discharge is deduced from the combined analysis of storage capacity, and inflow and outflow hydrographs.

Step 4. Calculate the upstream head above spillway crest $(H_{des} - z_{crest})$ for the design flow rate Q_{des}:

$$\frac{Q_{des}}{B} = C_{des}(H_{des} - z_{crest})^{3/2}$$

Note that the discharge coefficient C_{des} varies with the head above crest $(H_{des} - z_{crest})$.

Step 5. Choose the chute toe elevation (i.e. apron level): $z_{apron} = z_{crest} - \Delta z$. The apron level may differ from the natural bed level (i.e. tailwater bed level).

Step 6. For the design flow conditions, calculate the flow properties d and V at the end of the chute toe (Fig. 17.10) using:

$$H_1 = H_{des} - z_{apron} \qquad \text{Upstream total head}$$

$$V_{max} = \sqrt{2g(H_1 - (q_{des}/V_{max})\cos\theta)} \qquad \text{Ideal fluid flow velocity}$$

Step 7. Calculate the conjugate depth for the hydraulic jump:

$$\frac{d_{conj}}{d} = \frac{1}{2}\sqrt{1 + 8Fr^2} - 1) \qquad \text{where } Fr = V/\sqrt{gd}$$

Step 8. Calculate the roller length L_r. The apron length must be greater than the jump length.

Step 9. Compare the *Jump Height Rating Level* (*JHRL*) and the natural downstream water level (i.e. the *Tailwater Rating Level TWRL*). If the jump height does not match the natural water level, the apron elevation, the crest width, the design discharge or the crest elevation must be altered (i.e. go back to Steps 5, 3, 2 or 1 respectively).

Practically, if the *JHRL* does not match the natural water level (*TWRL*), the hydraulic jump will not take place on the apron.

Figure 17.16 presents the basic definitions. Figure 17.17 illustrates two cases for which the stilling basin is designed to equal *TWRL* and *JHRL* at design flow conditions. In each sketch (Fig. 17.17), the real free-surface line is shown by a solid line while the *JHRL* is shown by a dashed line.

Figures 17.18 and 17.19 show prototype dissipators in operation for different flow rates, highlighting the tailwater effects on the hydraulic jump.

Notes

1. The jump height rating level (*JHRL*) is the free-surface elevation downstream of the stilling basin.

2. The tailwater rating level (*TWRL*) is the natural free-surface elevation in the downstream flood plain. The downstream channel often flows as a subcritical flow, controlled by the downstream flow conditions (i.e. discharge and downstream flood plain geometry).

Discussion: calculations of the jump height rating level (JHRL)

(A) For a horizontal apron, the *JHRL* is deduced simply from the apron elevation and the conjugate depth:

$$JHRL = z_{apron} + d_{conj}$$

(B) For an apron with an end sill or end drop, the JHRL is calculated using the Bernoulli equation:

$$d_{conj} + z_{apron} + \frac{q^2}{2g(d_{conj})^2} = (JHRL - z_{tw}) + z_{tw} + \frac{q^2}{2g(JHRL - z_{tw})^2}$$

where z_{tw} is the downstream natural bed elevation. $(z_{tw} - z_{apron})$ is the drop/sill height.

Note that the above calculation assumes that the complete jump is located upstream of the drop/sill, and that no energy loss takes place at the sill/drop.

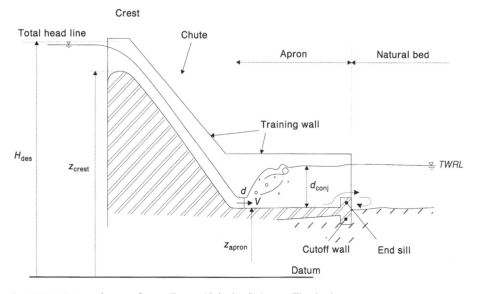

Fig. 17.16 Design of an overflow spillway with hydraulic jump stilling basin.

Matching the JHRC and the TWRC

During the design stages of a hydraulic jump energy dissipator, engineers are required to compute the *JHRL* for all flow rates. The resulting curve, called the *Jump Height Rating Curve* (*JHRC*), must be compared with the variations of natural downstream water level with discharges (i.e. *TWRC* or *Tailwater Rating Curve*).

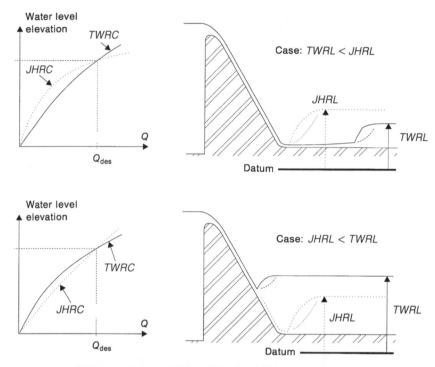

Fig. 17.17 Effect of tailwater level on the hydraulic jump location.

First, let us remember that the flow downstream of the stilling basin jump is subcritical, and hence it is controlled by the downstream flow conditions: i.e. by the tailwater flow conditions. The location of the jump is determined by the upstream and downstream flow conditions. The upstream depth is the supercritical depth at the chute toe and the downstream depth is determined by the tailwater flow conditions. The upstream and downstream depths, called conjugate depths, must further satisfy the momentum equation: e.g. equation (17.13) for a horizontal apron in the absence of baffles.

Discussion

In practice, the $TWRC$ is set by the downstream flood plain characteristics for a range of flow rates. It may also be specified. Designers must select stilling basin dimensions such that the $JHRC$ matches the $TWRC$: e.g. if the curves do not match, a sloping apron section may be used (e.g. Fig. 17.20). The height of the sloping apron, required to match the $JHRC$ and the $TWRC$ for all flows, is obtained by plotting the two curves, and measuring the greatest vertical separation of the curves. Note that if the maximum separation of the two curves occurs at very low flow rates, it is advisable to set the height of the sloping apron to the separation of the curves at a particular discharge (e.g. $0.15Q_{des}$).

When the jump height level is higher than the tailwater level, the jump may not take place on the apron but downstream of the apron. The flow above the apron becomes a jet flow and insufficient energy dissipation takes place. In such a case

(a)

(b)

Fig. 17.18 Effect of tailwater levels on the Waraba Creek weir QLD, Australia (photographs from the collection of late Professor G.R. McKay, Australia). Smooth chute followed by a stilling basin with baffle blocks. (a) Operation at very-low overflow Q_1. Flow from the right to the left. Note baffle blocks in the stilling basin. (b) Operation at low overflow Q_2 ($>Q_1$) Note the fully-developed hydraulic jump.

(i.e. $JHRL > TWRL$), the apron must be lowered or the tailwater level must be artificially raised. The apron level may be lowered (below the natural bed level) using an inclined upward apron design or a sill at the end of the apron (Fig. 17.20). The tailwater level may also be raised by providing a downstream weir at the end of the basin (and checking that the weir is never drowned).

When the jump height level is lower than the natural tailwater level, the jump may be drowned and a backwater effect take place. The jump is 'pushed' upstream onto the chute and it may become similar to a plunging jet flow. In

(c)

(d)

Fig 17.18 (c) Operation at medium flow rate Q_3 ($Q_2 < Q_3 < Q_4$). Flow from the left to the right. Note the rising tailwater (compared with Fig. 17.18(b)). (d) Operation at large discharge Q_4 ($> Q_3$). View from downstream, flow from top right to bottom left. Note the plunge pool formed in the stilling basin caused by the rising tailwater level.

this case ($JHRL < TWRL$), a downward slope or a higher apron level followed by a drop may be incorporated in the apron design (Fig. 17.20).

17.5.3 Practical considerations

For hydraulic jump energy dissipators, it is extremely important to consider the following points.

(a)

(b)

Fig. 17.19 Effect of tailwater levels on the Silverleaf weir QLD, Australia (Courtesy of Mr J. Mitchell). Timber crib stepped weir completed in 1953 (5.1 m high structure). (a) Operation at very-low flow of the stepped weir. (b) Operation at large overflow. Note the high tailwater level.

- The energy dissipator is designed for the reference flow conditions (i.e. design flow conditions).
- For overflow discharges larger than the design discharge, it may be acceptable to tolerate some erosion and damage (i.e. scouring and cavitation). *However, it is essential that the safety of the dam is ensured.*

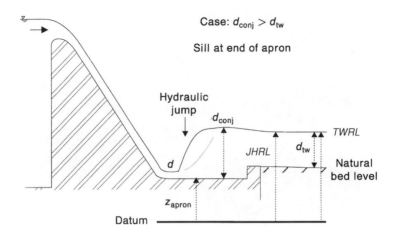

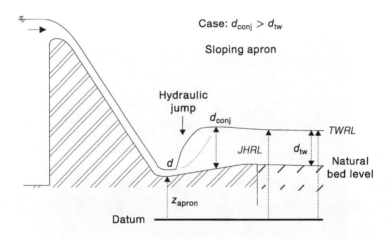

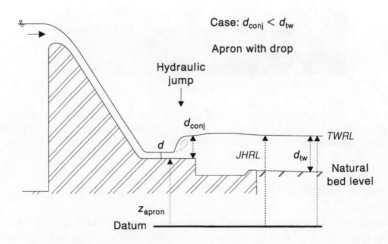

Fig. 17.20 Apron disposition to match the *JHRL* and *TWRL*.

- For discharges smaller than the design discharge, perfect performances are expected: i.e. (a) the energy dissipation must be controlled completely and it must occur in the designed dissipator, and (b) there must be no maintenance problem.

These objectives are achieved by: (1) a correct design of energy dissipation basin dimensions, (2) a correct design of sidewalls (i.e. training walls) to confine the flow to the dissipator, and (3) provision of an end sill and cut-off wall at the downstream end of the apron (Fig. 17.16). The training wall must be made high enough to allow for bulking caused by air entrainment in the chute flow and for surging in the hydraulic jump basin.

Note

Practically, it is uppermost important to remember that a hydraulic jump is associated with large bottom pressure fluctuation and high risks of scour and damage. As a result, the jump must be contained within the stilling basin for all flow conditions.

Discussion: air entrainment in stilling basin

The effects of air entrainment on hydraulic jump flow were discussed by Wood (1991). Chanson (1997) presented a comprehensive review of experimental data.

Practically, the entrained air increases the bulk of the flow, which is a design parameter that determines the height of sidewalls. Further air entrainment contributes to the air–water gas transfer of atmospheric gases such as oxygen and nitrogen. This process must be taken into account for the prediction of the downstream water quality (e.g. dissolved oxygen content).

17.6 Exercises

Numerical solutions to some of these exercises are available from the Web at www.arnoldpublishers.com/support/chanson

An energy dissipator (hydraulic jump type) is to be designed downstream of an undershoot sluice gate. The invert elevation of the transition channel between the sluice gate and the dissipator is set at 70 m R.L. The channel bed will be horizontal and concrete-lined. The gate and energy dissipator will be located in a 20 m wide rectangular channel. The design inflow conditions of the prototype dissipator are: discharge $= 220 \, \text{m}^3/\text{s}$, inflow depth $= 1.1 \, \text{m}$ (i.e. flow depth downstream of gate). Four dissipator designs will be investigated.

(i) Dissipation by hydraulic jump in the horizontal rectangular channel (70 m R.L. elevation) (no blocks).
(ii) Dissipation by hydraulic jump in the horizontal rectangular channel (70 m R.L. elevation) assisted with a single row of nine baffle blocks.
(iii) Dissipation in a standard stilling basin (e.g. USBR, SAF) set at 70 m R.L. elevation.

(iv) Dissipation in a specially-designed stilling basin. The apron elevation will be set to match the tailwater conditions.

For design (i): (a) compute the jump height rating level (*JHRL*) at design flow conditions.

For design (ii): (b) compute the force on each baffle block when the tailwater level is set at 73.1 m R.L. at design inflow conditions. (c) In what direction is the hydrodynamic force acting on the blocks: i.e. upstream or downstream?

For design (iii): (d) what standard stilling basin design would you select: USBR Type II, USBR Type III, USBR Type IV, SAF? (If you have the choice between two (or more) designs, discuss the economic advantages of each option.)

(e) Define in words and explain with sketch(es) the terms *TWRL* and *JHRL*. (Illustrate your answer with appropriate sketch(es) if necessary.)

Design a hydraulic jump stilling basin (design (iv)) to match the jump height and tailwater levels. The downstream bed level is at elevation 70 m R.L. At design discharge, the tailwater depth equals 3.1 m.

(f) Calculate the apron level in the stilling basin to match the jump height to tailwater at design flow conditions.

(g) Alternatively, if the apron level remains at the natural bed level (70 m R.L.), determine the height of a broad-crested weir necessary at the downstream end of the basin to raise artificially the tailwater level to match the *JHRL* for the design flow rate.

An overflow spillway is to be designed with an uncontrolled ogee crest followed by a stepped chute and a hydraulic jump dissipator. The width of the crest, chute and dissipation basin will be 127 m. The crest level will be at 336.3 m R.L. and the design head above crest level will be 3.1 m. The chute slope will be set at 51° and the step height will be 0.5 m. The elevation of the chute toe will be set at 318.3 m R.L. The stepped chute will be followed (without transition section) by a horizontal channel, which ends with a broad-crested weir, designed to record flow rates as well as to raise the tailwater level.

(a) Calculate the maximum discharge capacity of the spillway.
(b) Calculate the flow velocity at the toe of the chute.
(c) Calculate the residual power at the end of the chute (give the SI units). Comment.
(d) Compute the jump height rating level (*JHRL*) at design flow conditions (for a hydraulic jump dissipator).
(e) Determine the height of the broad-crested weir necessary at the downstream end of the dissipation basin to raise artificially the tailwater level to match the *JHRL* for the design flow rate.
(f) Compute the horizontal force acting on the broad-crested weir at design inflow conditions. In what direction will the hydrodynamic force be acting on the weir: i.e. upstream or downstream?
(g) If a standard stilling basin (e.g. USBR, SAF) is to be designed, what standard stilling basin design would you select: USBR Type II, USBR Type III, USBR Type IV, SAF?

Notes: In calculating the crest discharge capacity, assume that the discharge capacity of the ogee crest is 28% larger than that of a broad-crest (for the same upstream head above crest). In computing the velocity at the spillway toe, allow for energy losses by using the results presented in this book. The residual power equals $\rho g Q H_{res}$ where Q is the total discharge and H_{res} is the residual total head at chute toe taking the chute toe elevation as the datum.

An overflow spillway is to be designed with an un-gated broad-crest followed by a smooth chute and a hydraulic jump dissipator. The width of the crest, chute and dissipation basin will be 55 m. The crest level will be at 96.3 m R.L. and the design head above the crest level will be 2.4 m. The chute slope will be set at 45° and the elevation of the chute toe will be set at 78.3 m R.L. The stepped chute will be followed (without a transition section) by a horizontal channel that ends with a broad-crested weir, designed to record flow rates as well as to raise the tailwater level.

(a) Calculate the maximum discharge capacity of the spillway. (b) Calculate the flow velocity at the toe of the chute. (c) Calculate the Froude number of the flow at the end of the chute. (Comment.) (d) Compute the conjugate flow depth at design flow conditions (for a hydraulic jump dissipator).

The natural tailwater level ($TWRL$) at design flow conditions is 81.52 m R.L. (e) Determine the apron elevation to match the $JHRL$ and the $TWRL$ at design flow conditions.

(f) If a standard stilling basin (e.g. USBR, SAF) is to be designed, what standard stilling basin design would you select: USBR Type II, USBR Type III, USBR Type IV, SAF?

(Note: In computing the velocity at the prototype spillway toe, allow for energy losses.)

A weir is to have an overflow spillway with Ogee type crest and a hydraulic jump energy dissipator. Considerations of storage requirements and risk analysis applied to the 'design flood event' have set the elevation of the spillway crest 671 m R.L. and the width of the spillway crest at $B = 76$ m. The maximum flow over the spillway when the design flood is routed through the storage for these conditions is 1220 m³/s. (Note: the peak *inflow* into the reservoir is 3300 m³/s.) The spillway crest shape has been chosen so that the discharge coefficient at the maximum flow is 2.15 (SI units). For the purpose of the assignment, the discharge coefficient may be assumed to decrease linearly with discharge down to a value of 1.82 at very small discharges. The chute slope is 1 V:0.8 H (i.e. about 51.3°). The average bed level downstream of the spillway is R.L. 629.9 m and the tailwater rating curve ($TWRC$) downstream of the dam is defined as follows.

Discharge (m³/s)	TWRL (m)	Discharge (m³/s)	TWRL (m)
0	629.9	400	634.45
25	631.5	500	634.9
50	631.95	750	635.75
100	632.6	950	636.35
150	633.0	1220	637.1
250	633.65	1700	638.2

Design three options for the hydraulic jump stilling basin to dissipate the energy at the foot of the spillway as follows. (A) Apron level lowered to match the Jump Height

Rating Curve (*JHRC*) to the *TWRC*. (B) Apron level set at 632.6 m R.L. and tailwater level raised artificially with a broad-crested weir at the downstream end of the basin. (C) Apron level set at 629.9 m R.L. (i.e. average bed level) but spillway width *B* is changed to match the *JHRC* to the *TWRC*.

In each case use a sloping apron section if this will improve the efficiency and/or economics of the basin. In computing the velocity at the foot of the spillway, allow for energy losses. The design calculations are to be completed and submitted in two stages as specified below: stage 1 (calculations are to be done for 'maximum' flow *only*) and stage 2 (off-design calculations).

Stage 1: (a) Calculate apron level to match the *JHRC* to natural *TWRC* at 1220 m^3/s. (b) Calculate height of broad-crested weir to raise local *TWRC* to *JHRC* at 1220 m^3/s for an apron level at 632.6 m R.L. (c) Calculate *B* to match *JHRC* to natural *TWRC* at 1220 m^3/s for an apron level at 629.9 m R.L.

Stage 2: For both cases (A) and (B): (a) Calculate the *JHRC* for all flows up to and including 1220 m^3/s in sufficient detail to plot the curve. (b) Use the results from stage 1 to check the correctness of your calculation for 1220 m^3/s. (c) Plot the *JHRC* from (a) and the *TWRC* (natural or local, as required) on the same graph and determine the height of the sloping apron required to match *JHRC* and *TWRC* for all flows. (This is obtained from the greatest vertical separation of the two curves as plotted. If the maximum separation occurs at a flow less than 200 m^3/s, set the height of the sloping apron to the separation of the curves at 200 m^3/s.) (d) Draw to scale a dimensioned sketch of the dissipator.

Summary sheet (stage 1)

(a)	Apron level		m R.L.
(b)	Height of broad-crested weir above apron		m
	Minimum crest length		m
(c)	Spillway crest width		m
(d)	If the crest width is changed from 76 m (design (C)), and for a peak inflow into the reservoir of 3300 m³/s, will the 'maximum' spillway overflow change from 1220 m³/s?	Yes/No	
	Give reason for answer		

Summary sheet (stage 2)

(Design A)	Apron level		m R.L.
	Height of sloping apron		m
	Length of horizontal apron		m
(Design B)	Height of broad-crested weir above apron		m
	Height of sloping apron		m
	Length of horizontal apron		m

For each case (A) and (B), supply plotted curves of *JHRC* and *TWRC* (natural or raised as required). Supply dimensioned drawing of dissipators to scale.

A weir is to have an overflow spillway with Ogee type crest and a hydraulic jump energy dissipator. Considerations of storage requirements and risk analysis applied to the 'design flood event' have set the elevation of the spillway crest at 128 m R.L. and the width of the spillway crest at $B = 81$ m. The maximum flow over the spillway when the design flood is routed through the storage for these conditions is $1300\,\text{m}^3/\text{s}$. (Note: the peak *inflow* into the reservoir is $3500\,\text{m}^3/\text{s}$.) The spillway crest shape has been chosen so that the discharge coefficient at the maximum flow is 2.15 (SI units). For the purpose of the assignment, the discharge coefficient may be assumed to decrease linearly with discharge down to a value of 1.82 at very small discharges. The chute slope is 1 V : 0.8 H (i.e. about 51.3°). (Note: both stepped and smooth chute profiles will be considered.) The average bed level downstream of the spillway is R.L. 86.9 m and the tailwater rating curve (*TWRC*) downstream of the dam is defined as follows.

Discharge (m^3/s)	TWRL (m)	Discharge (m^3/s)	TWRL (m)
0	86.9	400	91.45
25	88.5	520	91.9
50	88.95	800	92.75
100	89.6	1000	93.35
150	90.0	1300	94.1
250	90.65	1800	95.2

Design three options for the hydraulic jump stilling basin to dissipate the energy at the foot of the spillway as follows. (A) Smooth concrete chute with apron level lowered to match the Jump Height Rating Curve (*JHRC*) to the *TWRC*; (B) Stepped concrete chute (step height $h = 0.15\,\text{m}$) with apron level lowered (or raised) to match the Jump Height Rating Curve (*JHRC*) to the *TWRC*; (C) Smooth concrete chute with apron level set at 86.9 m R.L. (i.e. average bed level) but spillway width B is changed to match the *JHRC* to the *TWRC*.

In each case, use a sloping apron section if this will improve the efficiency and/or economics of the basin. In computing the velocity at the foot of the spillway, allow for energy losses. The design calculations are to be completed and submitted in two stages: stage 1 for design flow conditions ($1300\,\text{m}^3/\text{s}$) and stage 2 for non-design flow conditions.

Stage 1: (a) Calculate apron level to match the *JHRC* to natural *TWRC* at $1300\,\text{m}^3/\text{s}$. (b) Calculate apron level to match the *JHRC* to natural *TWRC* at $1300\,\text{m}^3/\text{s}$ (with a stepped chute). (c) Calculate B to match *JHRC* to natural *TWRC* at $1300\,\text{m}^3/\text{s}$ for an apron level at 86.9 m R.L.

Stage 2: For both cases (A) and (B): (a) calculate the *JHRC* for all flows up to and including $1300\,\text{m}^3/\text{s}$ in sufficient detail to plot the curve. (b) Use the results from stage 1 to check the correctness of your calculation for $1300\,\text{m}^3/\text{s}$. (c) Plot the *JHRC* from (a) and the *TWRC* (natural or local, as required) on the same graph and determine the height of sloping apron required to match *JHRC* and *TWRC* for all flows (up to design flow conditions). (This is obtained from the greatest vertical separation of the two curves as plotted. If the maximum separation occurs at a flow less than

$200 \, \text{m}^3/\text{s}$, set the height of the sloping apron to the separation of the curves at $200 \, \text{m}^3/\text{s}$.) (d) Draw to scale a dimensioned sketch of the dissipator. (e) Discuss the advantages and inconvenients of each design case, and indicate your recommended design (with proper justifications).

Summary sheet (stage 1)

(a)	Apron level		m R.L.
(b)	Apron level		m R.L.
(c)	Spillway crest width		m
(d)	If the crest width is changed from 81 m, will the 'maximum' flow rate change from $1300 \, \text{m}^3/\text{s}$?	Yes/No	
	Give reason for answer		

Summary sheet (stage 2)

(Design A)	Apron level		m R.L.
	Height of sloping apron		m
	Length of horizontal apron		m
(Design B)	Apron level		m R.L.
	Height of sloping apron		m
	Length of horizontal apron		m

For each case (A) and (B), supply curves of *JHRC* and *TWRC* (natural or raised as required). Supply dimensioned drawings of dissipators to scale.

18

Design of drop structures and stepped cascades

18.1 Introduction

For low-head structures, the design of a standard weir (Chapter 17.2) may be prohibitive and uneconomical, and a drop structure may be preferred (Figs 18.1–18.3). The main features of drop structures are the free-overfall, the nappe impact and the downstream hydraulic jump. Energy dissipation takes place at the nappe impact and in the hydraulic jump.

A related form of drop structures is a stepped weir with large steps on which the flow bounces down from one step to the next one, as a succession of drop structures (Figs 18.4 and 18.5). This type of flow is called a nappe flow regime (or jet flow regime).

18.2 Drop structures

18.2.1 Introduction

The simplest case of drop structures is a vertical drop in a wide horizontal channel (Figs 18.1 and 18.2). In the following sections, we shall assume that the air cavity below the free-falling nappe is adequately ventilated.

> **Note**
> A drop structure is also called a vertical weir.

18.2.2 Free overfall

At the overfall, the flow becomes suddenly accelerated. If the upstream flow is subcritical, critical depth is expected to take place at the transition. However, a rapid pressure redistribution is also observed at the brink of the overfall, and critical

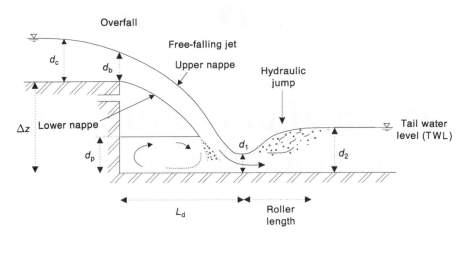

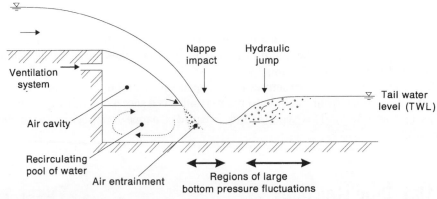

Fig. 18.1 Sketch of a drop structure.

flow depth ($d_c = \sqrt[3]{q^2/g}$) is observed, in practice, at a distance of about 3 to 4 times d_c upstream of the step edge (e.g. Henderson 1966, p. 192).

Analytical developments and experimental measurements showed that, for a horizontal overfall (wide channel), the flow depth at the brink equals:

$$d_b \approx 0.7 d_c \tag{18.1}$$

Notes

1. In a rectangular channel, the critical flow depth equals:

$$d_c = \sqrt[3]{\frac{q^2}{g}}$$

where q is the discharge per unit width and g is the gravity constant. This result is obtained for a horizontal channel with hydrostatic pressure distribution. At

Fig. 18.2 Drop structure overflow: Glenarbon weir (Texas QLD, Australia 1959) on 21 February 1998. Flow from left to right. Design conditions: 1420 m³/s, $\Delta z = 3$ m, 56 m long. Steel sheet-pile and concrete structure with a fishway in the middle.

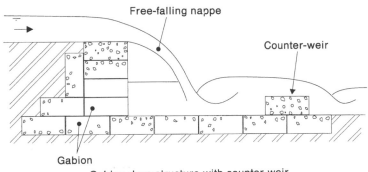

Gabion drop structure with counter-weir

Fig. 18.3. Sketch of a gabion drop structure with counterweir.

an overfall brink the pressure distribution is not hydrostatic and, as a result, $d_b \neq d_c$.

2. For a horizontal rectangular channel, a more accurate estimate of the brink depth is (Rouse 1936):

$$d_b = 0.715 d_c$$

3. Recent reviews of the overfall characteristics include Montes (1992b) and Chanson (1995b, pp. 230–236). The former reviewed the overfall properties near the edge of the step. The later describes the nappe trajectory equations and the application of the momentum principle at nappe impact.

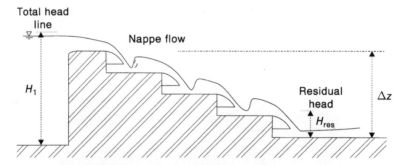

Fig. 18.4. Nappe flow regime above a stepped structure.

18.2.3 Drop impact

At the drop impact, a recirculating pool of water forms behind the overfalling jet (Fig. 18.1). This mass of water is important as it provides a pressure force parallel to the floor, which is required to change the jet momentum direction from an angle to the bottom, to parallel to the horizontal bottom.

If the downstream water level (i.e. tailwater level) corresponds to a subcritical flow, a hydraulic jump takes place immediately downstream of the nappe impact (Figs 18.1 and 18.2).

The drop length, pool height and the flow depths d_1 and d_2 (Fig. 18.1) may be estimated from empirical correlations (Rand 1955):

$$\frac{L_d}{\Delta z} = 4.30 \left(\frac{d_c}{\Delta z} \right)^{0.81} \tag{18.2}$$

$$\frac{d_p}{\Delta z} = \left(\frac{d_c}{\Delta z} \right)^{0.66} \tag{18.3}$$

$$\frac{d_1}{\Delta z} = 0.54 \left(\frac{d_c}{\Delta z} \right)^{1.275} \tag{18.4}$$

$$\frac{d_2}{\Delta z} = 1.66 \left(\frac{d_c}{\Delta z} \right)^{0.81} \tag{18.5}$$

where Δz is the drop height and $d_c = \sqrt[3]{q^2/g}$. Equations (18.2) to (18.5) were successfully verified with numerous laboratory data.

18.2.4 Design criterion

In practice, the design of the drop structure includes.

1. Select the crest elevation z_{crest} (bed topography, storage level).
2. Determine the design discharge Q_{des} from risk analysis and flood routing.
3. Choose a crest width B. The crest width may be smaller than the crest length.
4. Compute the critical depth for design flow conditions:

$$(d_c)_{des} = \sqrt[3]{\frac{Q_{des}^2}{gB^2}}$$

5. Choose the apron level (i.e. $z_{crest} - \Delta z$).
6. For the design flow conditions, compute the drop length and drop impact characteristics (equations (18.2) to (18.5)).

If the tailwater flow conditions are set, the design procedure must be iterated until the free-surface elevation downstream of the hydraulic jump (i.e. $d_2 + z_{crest} - \Delta z$) matches the tailwater rating level ($TWRL$).

---

7. Check the stability of the vertical weir (e.g. Agostini *et al.* 1987) and the risk of scouring at nappe impact.

Notes
1. In practice drop structures are efficient energy dissipators. They are typically used for drops up to 7 to 8 m and discharges up to about $10\,\text{m}^2/\text{s}$ (US Bureau of Reclamation 1987, Agostini *et al.* 1987). Note further that vertical drops can operate for a wide range of tailwater depths (usually $TWRL < z_{crest}$).
2. Large mean bottom pressures and bottom pressure fluctuations are experienced at the drop impact. The apron must be reinforced at the impingement location of the nappe and underneath the downstream hydraulic jump (Fig. 18.1).
3. A counterweir can be built downstream of the drop structure to reduce the effect of erosion (Fig. 18.3). The counterweir creates a natural stilling pool between the two structures which reduces the scouring force below the falling nappe impact.
4. An adequate nappe aeration is *essential* for a proper operation of drop structures. Bakhmeteff and Fedoroff (1943) presented experimental data showing the effects of nappe ventilation (and the lack of ventilation) on the flow patterns.

(a)

(b)

Fig. 18.5. Examples of stepped cascades. (a) Robina stepped weir No. 1 (Gold Coast QLD, 1996) in April 1997. Reno mattress weir, $h = 0.6\,$m, $l = 1.1$ to 2 m, $Q_{des} = 27\,$m^3/s. (b) Stepped diversion weir: Joe Sippel weir (Murgon QLD, 1984) on 7 November 1997. Rockfill embankment with concrete slab steps, $H = 6.5\,$m, pooled step design.

18.2.5 Discussion

The above calculations (equations (18.1) to (18.5)) were developed assuming that the flow upstream of the brink of the overfall is *subcritical*, hence critical immediately upstream of the brink.

If the upstream flow is supercritical (e.g. drop located downstream of an underflow gate), the hydraulic characteristics of supercritical nappe flows are determined by

(c)

(d)

Fig. 18.5. (c) Debris dam and sand trap system near Nanhua reservoir, Tainan County (Taiwan, Republic of China). (d) Timber crib weir: Murgon weir (Murgon QLD, 1935) on 8 November 1997.

the nappe trajectory, the jet impact on the step and the flow resistance on the step downstream of the nappe impact.

Note
Several researchers (Rouse 1943, Rajaratnam and Muralidhar 1968, Hager 1983, Marchi 1993) gave details of the brink flow characteristics and of the jet shape for

supercritical flows. For supercritical overfalls, the application of the momentum equation at the base of the overfall, using the same method as White (1943), leads to the result:

$$\frac{d_1}{d_c} = \frac{2Fr^{-2/3}}{1 + \dfrac{2}{Fr^2} + \sqrt{1 + \dfrac{2}{Fr^2}\left(1 + \dfrac{\Delta z}{d_c}Fr^{2/3}\right)}}$$

where Fr is the Froude number of the supercritical flow upstream of the overfall brink.

18.3 Nappe flow on stepped cascades

18.3.1 Presentation

When the vertical drop height exceeds 7 to 8 metres or if the site topography is not suitable for a single drop, a succession of drops (i.e. a cascade) can be envisaged. With a stepped cascade (Figs 18.4 and 18.5), low overflows result in a succession of free-falling nappes. This flow situation is defined as a *nappe flow regime*. Stepped channels with nappe flows may be analysed as a succession of drop structures.

18.3.2 Basic flow properties

Along a stepped chute with nappe flow, critical flow conditions take place next to the end of each step, followed by a free-falling nappe and jet impact on the downstream step. At each step, the jet impact is followed by a hydraulic jump, a subcritical flow region and critical flow next to the step edge.

The basic flow characteristics at each step can be deduced from equations (18.1) to (18.5) (Chanson 1995b).

Note

At large flow rates, the stepped channel flow becomes a skimming flow regime (i.e. extremely rough turbulent flow). The re-analysis of model studies (Chanson 1995b) suggests that the transition from nappe to skimming flow is a function of the step height and channel slope. It occurs for:

$$\frac{d_c}{h} > 1.057 - 0.465\frac{h}{l} \qquad \text{Skimming flow regime}$$

where d_c is the critical depth ($d_c = \sqrt[3]{q^2/g}$), h is the step height and l is the step length.

For skimming flows, chute calculations are developed as detailed in Chapter 17.2 and Appendix A4.1.

18.3.3 Energy dissipation

In a nappe flow situation, the head loss at any intermediary step equals the step height. The energy dissipation occurs by jet breakup and jet mixing, and with the formation of a hydraulic jump on the step. The total head loss along the chute ΔH equals the difference between the maximum head available H_1 and the residual head at the downstream end of the channel H_{res} (Fig. 18.4):

$$\frac{\Delta H}{H_1} = 1 - \frac{H_{res}}{H_1} \tag{18.6}$$

The residual energy is usually dissipated at the toe of the chute by a hydraulic jump in the dissipation basin. Combining equations (18.4) and (18.6), the total energy loss can be calculated as:

$$\frac{\Delta H}{H_1} = 1 - \left(\frac{0.54 \left(\dfrac{d_c}{\Delta z} \right)^{0.275} + \dfrac{3.43}{2} \left(\dfrac{d_c}{\Delta z} \right)^{-0.55}}{\dfrac{3}{2} + \dfrac{\Delta z}{d_c}} \right) \tag{18.7}$$

where Δz is the dam crest elevation above the downstream toe.

Discussion

Equation (18.7) was developed for the nappe flow regime with a fully-developed hydraulic jump. Chanson (1995b) showed that other types of nappe flow may occur: e.g. nappe flow without hydraulic jump. For such nappe flows, equation (18.7) could overestimate the rate of energy dissipation.

18.4 Exercises

A vertical drop structure is to be designed at the downstream end of a mild-slope waterway. The width of the crest and drop will be 15 m. The crest level will be at 23.5 m R.L. and the design head above crest level will be 1.9 m. The elevation of the downstream channel bed will be set at 17.2 m R.L. The impact drop will be followed (without a transition section) by a mild-slope channel. The tailwater level at design flow conditions will be set at 19.55 m R.L.

(a) Calculate the design flow rate of the drop structure. Assuming the apron level to be at the same elevation as the downstream bed level, calculate: (b) the supercritical flow depth immediately downstream of the nappe impact, (c) drop length, (d) residual power at the end of the chute (give SI units and comment), (e) jump height rating level (*JHRL*) at the design flow. (f) Determine the apron level to match the tailwater level and *JHRL* at the design flow rate.

A stepped cascade (nappe flow regime) is to be designed for 60 m³/s. The cascade is to have a broad-crest followed by five steps (1.5 m high, 2.5 m long). The crest and cascade width will be 51 m. The crest level will be at 74.0 m R.L.

At the design flow condition: (a) calculate the head above crest invert, (b) check that the flow is nappe flow, (c) calculate the residual head and the residual power to be dissipated in a downstream stilling structure. (d) If the spillway flow reaches $210\,m^3/s$, what will happen? (Discuss and comment.) (e) For the same cascade slope and width, what is the smallest step height to have nappe flow (at design flow conditions)?

19

Culvert design

19.1 Introduction

A culvert is a covered channel of relatively short length designed to pass water through an embankment (e.g. highway, railroad, dam). It is a hydraulic structure and it may carry flood waters, drainage flows, natural streams below earthfill and rockfill structures. From a hydraulic aspect, a dominant feature of a culvert is whether it runs full or not.

The design can vary from a simple geometry (i.e. box culvert) to a hydraulically-smooth shape (i.e. MEL culvert[1]) (Fig. 19.1). In this section, we will first review the design of standard culverts, then we will discuss the design of minimum energy loss culverts.

Note

A culvert below an embankment dam is usually a 'long' structure that operates full (i.e. as pipe flow). In this section, we shall focus on 'short' culverts operating primarily with free-surface flows.

19.2 Basic features of a culvert

19.2.1 Definitions

A culvert consists of three parts: the *intake* (also called inlet or fan), the *barrel* (or throat) and the *diffuser* (also called the outlet or expansion fan) (Fig. 19.2). The cross-sectional shape of the barrel may be circular (i.e. pipe), rectangular (i.e. box culvert), or multi-cell (e.g. multi-cell box culvert) (Fig. 19.2). The bottom of the

---

[1] The design of a Minimum Energy Loss (MEL) culvert is associated with the concept of constant total head. The inlet and outlet must be streamlined in such a way that significant form losses are avoided. For an introduction on Minimum Energy Loss culverts, see Apelt (1994). For a complete review of MEL waterways, see Apelt (1983).

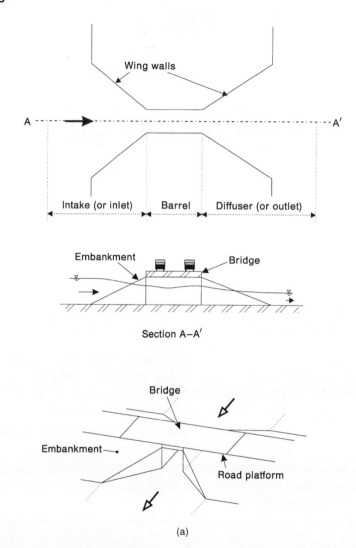

Fig. 19.1 Sketch of a culvert. (a) Box culvert.

barrel is called the *invert* while the barrel roof is called the *soffit* or *obvert*. The training walls of the inlet and outlet are called *wing walls*.

19.2.2 Ideal flow calculations

A culvert is designed to pass a specific flow rate with the associated natural flood level. Its hydraulic performances are the design discharge, the upstream total head and the maximum (acceptable) head loss ΔH. The design discharge and flood level are deduced from the hydrological investigation of the site in relation to the purpose of

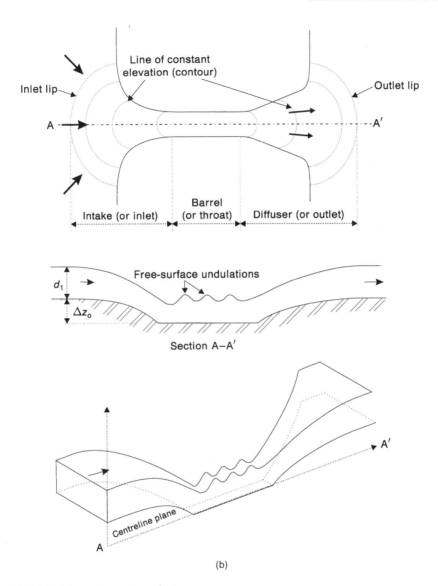

Fig. 19.1 (b) Minimum Energy Loss culvert.

the culvert. Head losses must be minimized to reduce upstream backwater effects (i.e. upstream flooding).

The hydraulic design of a culvert is basically the selection of an optimum compromise between discharge capacity and head loss or afflux, and of course construction costs. Hence (short) culverts are designed for free-surface flow with critical (flow) conditions in the barrel. Simplified hydraulic calculations are based upon the assumptions of smooth intake and diffuser, and no energy loss: i.e. the total head is the same upstream and downstream. Assuming a given upstream total head H_1, for a horizontal structure of rectangular cross-section and neglecting energy loss, the

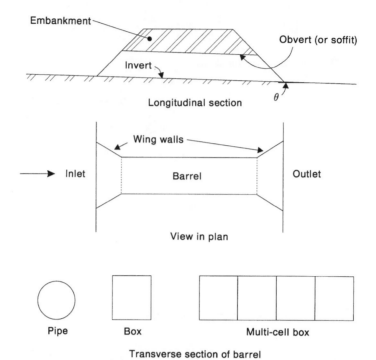

Fig. 19.2 Basic definitions.

maximum discharge per unit width is achieved for critical flow conditions[2] in the barrel:

$$q_{max} = \sqrt{g}(\tfrac{2}{3}(H_1 - z_{inlet}))^{3/2} \qquad (19.1a)$$

where z_{inlet} is the inlet bed elevation.

The minimum barrel width to achieve critical flow conditions is then:

$$B_{min} = \frac{Q_{max}}{\sqrt{g}} (\tfrac{2}{3}(H_1 - z_{inlet}))^{-3/2} \qquad (19.2a)$$

The critical depth in the barrel equals:

$$d_c = \tfrac{2}{3}(H_1 - z_{inlet}) \qquad (19.3a)$$

The barrel invert may be lowered to increase the discharge capacity or to reduce the barrel width (e.g. Fig. 19.1(b)). The above equations become:

$$q_{max} = \sqrt{g}(\tfrac{2}{3}(H_1 - z_{inlet} + \Delta z_o))^{3/2} \qquad (19.1b)$$

$$B_{min} = \frac{Q_{max}}{\sqrt{g}} (\tfrac{2}{3}(H_1 - z_{inlet} + \Delta z_o))^{-3/2} \qquad (19.2b)$$

---

[2] In open channel flows, the flow conditions, such that the specific energy is minimum, are called the critical flow conditions.

$$d_{\mathrm{c}} = \tfrac{2}{3}(H_1 - z_{\mathrm{inlet}} + \Delta z_{\mathrm{o}}) \tag{19.3b}$$

where Δz_{o} is the bed elevation difference between the inlet invert and the barrel bottom.

Notes

1. Equation (19.2) gives the minimum barrel width to obtain near-critical flow without 'choking' effects. If $B < B_{\mathrm{min}}$, choking effects will take place in the barrel (e.g. Henderson 1966, pp. 47–49).
2. The barrel width can be reduced by lowering the barrel bottom elevation, but designers must however choose an adequate barrel width to avoid the risks of culvert obstruction by debris (e.g. rocks, trees, cars).

19.2.3 Design considerations

First the function of the culvert must be chosen and the design flow conditions (e.g. Q_{des}, flood level) must be selected.

The primary design constraints in the design of a culvert are:

1. the cost must (always) be *minimum*;
2. the afflux[3] must be small and preferably minimum;
3. eventually the embankment height may be given or may be part of the design;
4. a scour protection may be considered, particularly if a hydraulic jump might take place near the culvert outlet.

19.2.4 Flow through a culvert

Most culverts are designed to operate as open channel systems, with critical flow conditions occurring in the barrel in order to maximize the discharge per unit width and to reduce the barrel cross-section (and hence its cost).

The flow upstream and downstream of the culvert is typically subcritical. As the flow approaches the culvert, the channel constriction (i.e. intake section) induces an increase in Froude number. For the design discharge, the flow becomes near-critical in the barrel. In practice, perfect-critical flow conditions in the barrel are difficult to establish: they are characterized by 'choking' effects and free-surface instabilities. Usually, the Froude number in the barrel[4] is about 0.7 to 0.9, and the discharge per unit width is nearly maximum, as shown in Table 19.1.

---

[3] The afflux is the rise of water level above the normal free-surface level on the upstream side of the culvert.
[4] If the upstream and downstream flow conditions are supercritical, it is preferable to design the barrel for Froude numbers of about 1.3 to 1.5.

Table 19.1 Flow rate in the barrel as a function of the barrel Froude number (box culvert)

Fr (1)	E/d_c (2)	q/q_{max} (3)
0.3	2.33	0.52
0.5	1.79	0.77
0.7	1.58	0.926
0.8	1.53	0.969
0.9	1.51	0.993
0.95	1.50	0.998
0.99	1.50	1.000
1	1.50	1.000
1.01	1.50	1.000
1.1	1.51	0.994
1.2	1.52	0.977
1.5	1.62	0.89

Notes: q_{max} = maximum discharge per unit width computed as equation (19.1a); E = specific energy in the barrel; Fr = Froude number in the barrel.

Comment

In mountain areas, the stream slope may be steep, and the upstream and downstream flow conditions would typically be supercritical. The culvert is designed to achieve critical conditions in the barrel.

19.2.5 Undular flow in the barrel

In the barrel, the near-critical flow[5] at design discharge is characterized by the establishment of stationary free-surface undulations (e.g. Fig. 19.1(b)). For the designers, the characteristics of the free-surface undulations are important for the sizing of the culvert height. If the waves leap on the roof, the flow might cease to behave as an open channel flow and become a pipe flow.

Henderson (1966) recommended that the ratio of upstream specific energy to barrel height should be less that 1.2 for the establishment of free-surface flow in the barrel (Table 19.2). Such a ratio gives a minimum clearance above the free-surface level in the barrel of about 20%.

Note

Chanson (1995a) observed experimentally that a 20% free-space clearance (between the mean free-surface level and the roof) is a minimum value when the barrel flow conditions are undular. He further showed that the free-surface

---

[5] Near-critical flows are defined as flow situations characterized by the occurrence of critical or nearly-critical flow conditions over a 'reasonably-long' distance and period of time.

undulation characteristics are very close to undular weir flow. Both the broad-crested weir and the culvert are designed specifically for near-critical flow above the crest and in the barrel, respectively. The similarity between different types of near-critical flows might be used by designers to gain some order of magnitude of the free-surface undulation characteristics (in the absence of additional experimental data).

Table 19.2 Flow conditions for free-surface inlet flow (standard culvert)

Reference (1)	Conditions (2)	Remarks (3)
Henderson (1966)	$(H_1 - z_{inlet}) \leq 1.2D$	Box culvert
	$(H_1 - z_{inlet}) \leq 1.25D$	Circular culvert
Hee (1969)	$(H_1 - z_{inlet}) \leq 1.2D$	
Chow (1973)	$(H_1 - z_{inlet}) \leq 1.2$ to $1.5D$	
US Bureau of Reclamation (1987)	$(H_1 - z_{inlet}) \leq 1.2D$	

Notes: D = barrel height; H_1 = total head at inlet; z_{inlet} = inlet bed elevation.

19.3 Design of standard culverts

19.3.1 Presentation

A standard culvert is designed to pass waters at a minimum cost without much consideration of the head loss. The culvert construction must be simple: e.g. circular pipes, precast concrete boxes (Fig. 19.3).

19.3.2 Operation – Flow patterns

For standard culverts, the culvert flow may exhibit various flow patterns depending upon the discharge (hence the critical depth in barrel d_c), the upstream head above the inlet invert $(H_1 - z_{inlet})$, the uniform equilibrium flow depth in the barrel d_o, the barrel invert slope θ, the tailwater depth d_{tw}, and the culvert height D.

Hee (1969) regrouped the flow patterns into two classes and eight categories altogether (Fig. 19.4, Table 19.3):

- Class I for *free-surface inlet flow* conditions, and
- Class II for *submerged entrance*.

Free-surface inlet flows [Class I] take place typically for (Table 19.2):

$$\frac{H_1 - z_{inlet}}{D} \leq 1.2 \tag{19.4}$$

In each class, the flow patterns can be sub-divided in terms of the control location: i.e. whether the hydraulic control is located at the entrance (i.e. *inlet control*) or at the outlet (i.e. *outlet control*) (Table 19.3).

(a)

(b)

Fig. 19.3 Examples of a standard culvert in Brisbane, QLD, Australia (photograph taken in September 1996). Combination of four concrete pipes and three precast concrete boxes. This culvert is not properly designed and the road is overtopped typically once or twice per year. (a) Inlet. Note the trashrack to trap debris. Note also that the barrel axis is not aligned with the creek. (b) Outlet.

Discussion

With free-surface inlet flow, critical depth is observed at the outlet when S_c is larger than S_o, where S_o is the barrel invert slope and S_c is the barrel critical slope (Fig. 19.4(a)). Note that S_o and S_c must be calculated in terms of the barrel invert slope and barrel flow conditions.

For submerged entrance cases (Fig. 19.4(b)), the flow is controlled by the outlet conditions when the barrel is full or drowned (cases 6 and 7), and by the inlet conditions when free-surface flow is observed in the barrel (cases 5 and 8).

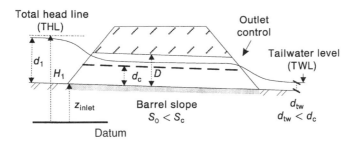

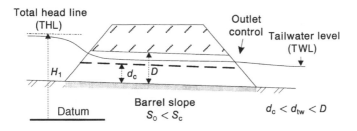

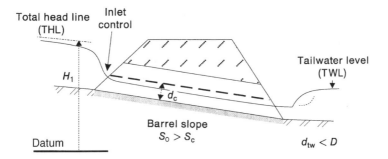

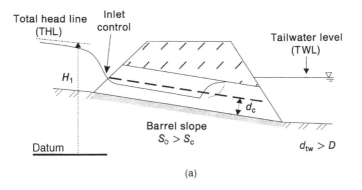

(a)

Fig. 19.4 Operation of a standard culvert. (a) Free-surface inlet flow conditions. (b) Submerged entrance.

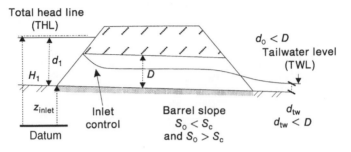

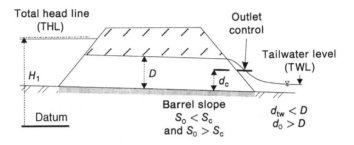

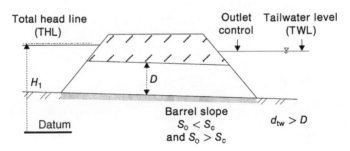

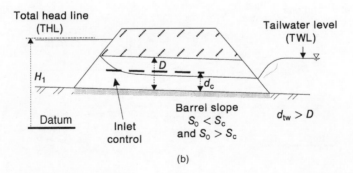

(b)

Fig. 19.4 (b) Submerged entrance.

Table 19.3 Typical conditions of standard culvert operation

Flow pattern (1)	Control location (2)	Flow conditions (3)	Remarks (4)
Class I. Free-surface inlet flow			
Case 1	Outlet control	$(H_1 - z_{inlet}) \leq 1.2D$ $d_{tw} < d_c$ $S_o < S_c$	
Case 2	Outlet control	$(H_1 - z_{inlet}) \leq 1.2D$ $d_c < d_{tw} > D$ $S_o < S_c$	
Case 3	Inlet control	$(H_1 - z_{inlet}) \leq 1.2D$ $d_{tw} < D$ $S_o \geq S_c$	Hydraulic jump takes place at outlet.
Case 4	Inlet control	$(H_1 - z_{inlet}) \leq 1.2D$ $d_{tw} < D$ $S_o \geq S_c$	Hydraulic jump takes place in the barrel.
Class II. Submerged entrance			
Case 5	Inlet control	$(H_1 - z_{inlet}) > 1.2D$ $d_{tw} < d_c$ $d_o < D$ $S_o < S_c$ or $S_o > S_c$	
Case 6	Outlet control	$(H_1 - z_{inlet}) > 1.2D$ $d_{tw} < d_c$ $d_o > D$ $S_o < S_c$ or $S_o > S_c$	Drowned barrel. Critical flow depth is observed at outlet
Case 7	Outlet control	$(H_1 - z_{inlet}) > 1.2D$ $d_{tw} > D$ $S_o < S_c$ or $S_o > S_c$	Drowned barrel. Observed usually for $d_o > D$. But might occur for $d_o < D$ if a backwater effect moves the hydraulic jump in barrel.
Case 8	Inlet control	$(H_1 - z_{inlet}) > 1.2D$ $d_{tw} > D$ $S_o < S_c$ or $S_o > S_c$	Hydraulic jump takes place at outlet. Usually observed for $d_o < d_c$. May occur for $d_o > d_c$ because of vena contracta effect at barrel intake.

Notes: $D =$ barrel height; $d_c =$ critical depth in barrel; $d_o =$ uniform equilibrium depth in barrel; $d_{tw} =$ tailwater depth; $H_1 =$ total head at inlet invert; $S_c =$ critical slope of barrel; $S_o =$ barrel invert slope; $z_{inlet} =$ inlet invert elevation. Reference: Hee (1969).

Case 7 is observed usually for $d_o > D$ where d_o is the barrel normal depth. But it might also occur for $d_o < D$ if a backwater effect (i.e. large tailwater depth) moves the hydraulic jump into the barrel. Case 8 is usually observed for $d_o < d_c$. It may also occur for $d_o > d_c$ because of the 'sluice gate' effect (i.e. vena contracta) induced by the barrel entrance.

For flat (i.e. mild) flood plains, the flow pattern is 'outlet control' if $d > d_c$ in all the waterway, as subcritical flows are controlled from downstream. Culverts flowing full are controlled by the tailwater conditions (i.e. outlet control).

Notes

1. The barrel invert slope S_o is defined as: $S_o = \sin \theta$ (Fig. 19.2). The uniform equilibrium flow conditions are related to the bed slope as:

$$S_o = \sin \theta = S_f = f \frac{V_o^2}{2g(D_H)_o}$$

where V_o and $(D_H)_o$ are the uniform equilibrium flow velocity and hydraulic diameter respectively, and f is the Darcy friction factor.

2. The channel slope, for which the uniform equilibrium flow is critical, is called the *critical slope*, denoted S_c. The critical slope satisfies:

$$S_c = \sin \theta_c = S_f = f \frac{V_o^2}{2g(D_H)_o}$$

where V_o and $(D_H)_o$ must also satisfy: $V_o = V_c$ and $(D_H)_o = (D_H)_c$. For a wide rectangular channel, the critical slope satisfies: $S_c = \sin \theta_c = f/8$.

19.3.3 Discharge characteristics

The discharge capacity of the barrel is primarily related to the flow pattern: free-surface inlet flow, submerged entrance, or drowned barrel (Fig. 19.4, Table 19.3).

When free-surface flow takes place in the barrel, the discharge is fixed only by the entry conditions (Table 19.4), whereas with drowned culverts, the discharge is determined by the culvert resistance. Table 19.4 summarizes several discharge relationships for circular and box culverts.

Nomographs are also commonly used (e.g. US Bureau of Reclamation 1987, Concrete Pipe Association of Australasia 1991). Figures 19.5 and 19.6 show nomographs to compute the characteristics of box culverts with inlet control and drowned barrel, respectively.

For short box culverts (i.e. free-surface barrel flow) in which the flow is controlled by the inlet conditions, the discharge is typically estimated as:

$$\frac{Q}{B} = C_D \frac{2}{3} \sqrt{\frac{2}{3}g} \left(H_1 - z_{inlet}\right)^{1.5} \qquad \text{Free-surface inlet flow [Class I]} \qquad (19.5)$$

$$\frac{Q}{B} = CD \sqrt{2g(H_1 - z_{inlet} - CD)}$$

$$\text{Submerged entrance and free-surface barrel flow} \qquad (19.6)$$

where B is the barrel width and D is the barrel height (Fig. 19.7). C_D equals 1 for rounded vertical inlet edges and 0.9 for a square-edged inlet. C equals 0.6 for a square-edged soffit and 0.8 for a rounded soffit (Table 19.4).

Note

Recently, computer programs have been introduced to compute the hydraulic characteristics of standard culverts. One is described in Appendix A4.4. Practically, computer programs cannot design culverts: i.e. they do not provide the optimum design. They are, however, useful tools to gain a feel for the operation of a known culvert.

Table 19.4 Discharge characteristics of standard culverts

Flow pattern (1)	Relationship (2)	Flow conditions (3)	Comments (4)
Free-surface inlet flow			
Circular culvert	$\dfrac{Q}{D} = 0.432\sqrt{g}\,\dfrac{(H_1 - z_{inlet})^{1.9}}{D^{0.4}}$	$0 < \dfrac{H_1 - z_{inlet}}{D} < 0.8$ $0.025 < \sin\theta < 0.361$	Relationship based upon laboratory experiments (Henderson 1966).
	$\dfrac{Q}{D} = 0.438\sqrt{g}(H_1 - z_{inlet})^{1.5}$	$0.8 < \dfrac{H_1 - z_{inlet}}{D} < 1.2$ $0.025 < \sin\theta < 0.361$	Relationship based upon laboratory experiments (Henderson 1966).
Box culvert	$\dfrac{Q}{B} = C_D\dfrac{2}{3}\sqrt{\dfrac{2}{3}g}$ $\times (H_1 - z_{inlet})^{1.5}$	$\dfrac{H_1 - z_{inlet}}{D} < 1.2$	Relationship based upon laboratory experiments (Henderson 1966).
	$C_D = 1$	Rounded vertical edges (radius $> 0.1B$)	
	$C_D = 0.9$	Vertical edges left square	
Submerged entrance			
Box culvert	$\dfrac{Q}{B} = CD$ $\times \sqrt{2g(H_1 - z_{inlet} - CD)}$	$\dfrac{H_1 - z_{inlet}}{D} > 1.2$	Free-surface barrel flow. Relationship based upon laboratory experiments (Henderson 1966).
	$C = 0.8$	Rounded soffit edges	
	$C = 0.6$	Square-edged soffit	
Drowned barrel			
Circular culvert	$\dfrac{Q}{D} = \dfrac{\pi}{4}D\sqrt{2g\dfrac{\Delta H}{K'}}$	K': head loss coefficient (primary and secondary losses)	Darcy equation. Pipe flow head loss calculations.
Box culvert	$\dfrac{Q}{D} = B\sqrt{2g\dfrac{\Delta H}{K'}}$	K': head loss coefficient (primary and secondary losses)	Darcy equation. Pipe flow head loss calculations.

Notes: B = barrel width; D = barrel height; H_1 = upstream total head (immediately upstream of inlet); z_{inlet} = inlet invert elevation; ΔH = head loss; θ = barrel invert slope.

19.3.4 Design procedure

The design process for standard culverts can be divided into two parts (Herr and Bossy 1965, HEC No. 5). First, a system analysis must be carried out to determine the objectives of the culvert, the design data, the constraints, etc, including the design flow Q_{des} and the design upstream total head H_{des} (basically the design upstream flood height).

In a second stage, the barrel size is selected by a test-and-trial procedure, in which both inlet-control and outlet-control calculations are performed (Figs 19.8 and 19.9). At the end:

the *optimum size* is the smallest barrel size allowing *inlet control* operation.

Calculations of the barrel size are *iterative*.

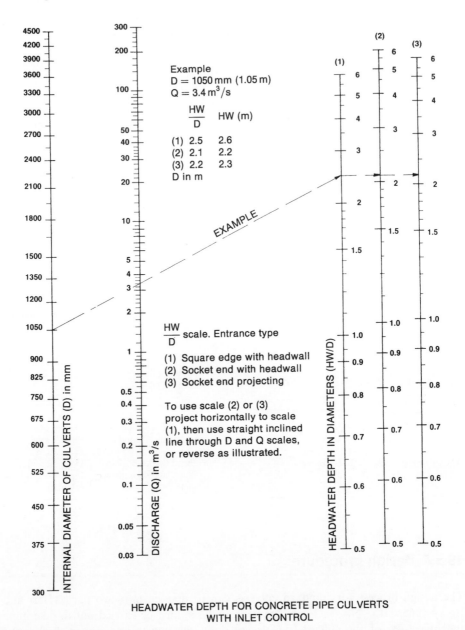

HEADWATER DEPTH FOR CONCRETE PIPE CULVERTS WITH INLET CONTROL

Fig. 19.5 Hydraulic calculations of upstream head above invert bed for box culverts with inlet control (Concrete Pipe Association of Australasia 1991, p. 39).

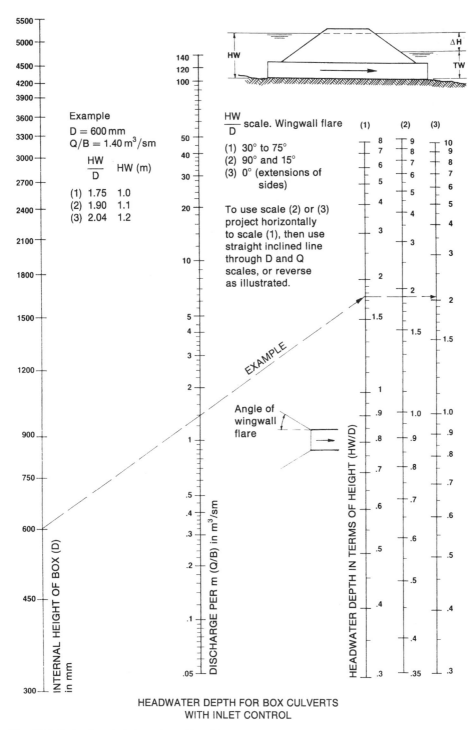

Fig. 19.6 Hydraulic calculations of total head losses for concrete box culverts flowing full (i.e. drowned) (Concrete Pipe Association of Australasia 1991, p. 41).

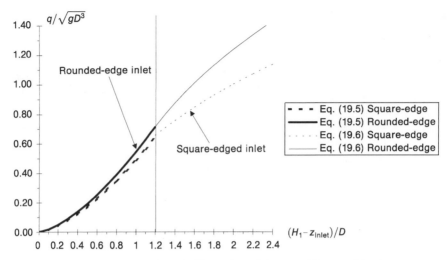

Fig. 19.7 Discharge characteristics of standard box culverts (inlet control flow conditions).

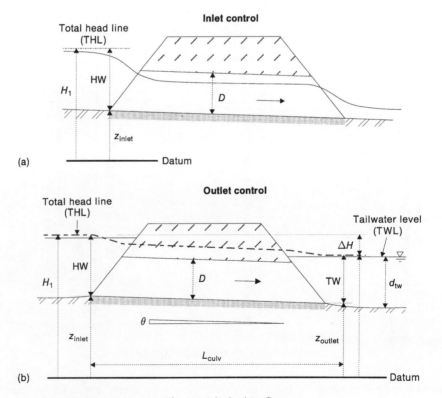

Fig. 19.8 Inlet control and outlet control for a standard culvert flow.

Fig. 19.9 Examples of box culvert operation. Box culvert model – design flow conditions: $Q_{des} = 10\,L/s$, $d_{tw} = 0.038\,m$, $S_o = 0.0035$, $B_{min} = 0.15\,m$, $D = 0.107\,m$, $L_{culv} = 0.5\,m$. (a) Design flow conditions: view from downstream (flow from top right to bottom left). $Q = 10\,L/s$, $d_{tw} = 0.038\,m$, $d_1 = 0.122\,m$. Note the hydraulic jump downstream of the culvert. (b) Non-design flow conditions: view from downstream of the drowned barrel (flow from top right to bottom left). $Q = 10\,L/s$, $d_{tw} = 0.109\,m$, $d_1 = 0.133\,m$, outlet control. (c) Non-design flow conditions: view from upstream (flow from bottom right to top left). $Q = 5\,L/s$, $d_{tw} = 0.038\,m$, $d_1 = 0.082\,m$. Inlet control and free-surface inlet flow.

Step 1. Choose the barrel dimensions.

Step 2. Assume an *inlet control* (Fig. 19.8(a)).

  Step 2.1. Calculate the upstream total head $H_1^{(ic)}$ corresponding to the design discharge Q_{des} assuming inlet control. Use the formulae given in Table 19.4 or design charts (e.g. Figs 19.5 or 19.7).

  Step 2.2. Repeat the above procedure (Step 2.1) for different barrel sizes until the upstream head $H_1^{(ic)}$ satisfies the design specifications (i.e. $H_1^{(ic)} = H_{des}$).

Step 3. Assume an outlet control (Fig. 19.8(b)).

  Step 3.1. Use design charts (e.g. Fig. 19.6) to calculate the head loss ΔH from inlet to outlet for the design discharge Q_{des}.

  Step 3.2. Calculate the upstream total head $H_1^{(oc)}$:

$$H_1^{(oc)} = H_{tw} + \Delta H$$

Step 4. Compare the inlet control and outlet control results:

$$H_{des} = H_1^{(ic)} \lessgtr H_1^{(oc)}$$

The larger value controls

When the inlet control design head H_{des} (used in Step 2.2) is larger than $H_1^{(oc)}$, inlet control operation is confirmed and the barrel size is correct. On the other hand, if $H_1^{(oc)}$ is larger than H_{des} (used in Step 2.2), outlet control takes place. Return to Step 3.1, and increase the barrel size until $H_1^{(oc)}$ satisfies the design specification H_{des} (used in Step 2.2).

Notes

1. For a wide flood plain, the tailwater head approximately equals:

$$H_{tw} \approx d_{tw} + z_{outlet}$$

   where z_{outlet} is the outlet invert elevation.

2. Note that $z_{inlet} = z_{outlet} + L_{culv} \tan \theta$, where L_{culv} is the horizontal culvert length and θ is the barrel bed slope (Fig. 19.8).

Discussion

Figure 19.9 illustrates the operation of a box culvert for design and non-design flow conditions. Figure 19.9(a) shows the design flow conditions with inlet control and free-surface flow in the barrel. Note the presence of a hydraulic jump at the outlet. In Fig. 19.9(b), the tailwater depth is larger than the design tailwater depth, leading to the 'drowning' of the barrel. Outlet control takes place and the barrel flow is similar to the flow in a rectangular pipe. Figure 19.9(c) shows the culvert operation for a discharge lower than design flow, with inlet control and free-surface inlet flow.

19.4 Design of Minimum Energy Loss culverts

19.4.1 Definition

A *Minimum Energy Loss* (MEL) culvert is a structure designed with the concept of minimum head loss. The flow in the approach channel is contracted through a streamlined inlet into the barrel where the channel width is minimum, and then it is expanded in a streamlined outlet before being finally released into the downstream natural channel. Both the inlet and outlet must be streamlined to avoid significant form losses. The barrel invert is sometimes lowered to increase the discharge capacity (equation (19.1b)). Figures 19.1(b), 19.10 to 19.12 show typical designs. Appendix A4.3 also presents illustrated examples.

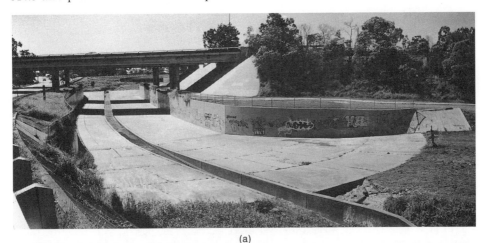

(a)

(b)

Fig. 19.10 Photographs of prototype MEL waterways and culverts (see Appendix A4.3). (a) Minimum Energy Loss waterway along Norman Creek in September 1996. Design discharge: ~200 m³/s; throat width: ~4 m. View from upstream. Note the low-flow channel, the inlet (in foreground), the road (Ridge Street) parallel to the waterway and the freeway bridge (above). (b) MEL culvert along Norman Creek, below Ridge Street in September 1996. Design discharge: 220 m³/s; barrel: 7 cells (2 m wide each). View of the outlet. Note the low-flow channel in the foreground (left).

(a)

(b)

Fig. 19.11 Photographs of the Nudgee Road bridge MEL waterway (see Appendix A4.3). Design discharge: 850 m^3/s; excavation depth: 0.76 m; natural bed slope: 0.00049; throat width: 137 m. (a) Inlet view (from left bank) on 14 September 1997. (b) Operation during a flood flow (~400 m^3/s) in the 1970s: view of the inlet fan from left bank. Note the low-turbulence of the flow (photograph from the collection of late Professor G.R. McKay, Australia).

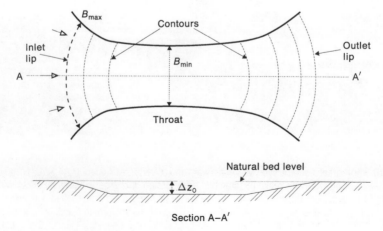

Fig. 19.12 Sketch of a Minimum Energy Loss waterway.

Apelt (1983) presented an authoritative review of the topic. In the following paragraphs, a summary of the design process is given. The same procedure applies to both MEL culverts and MEL waterways (e.g. bridge waterways).

Notes

1. MEL culverts are also called constant energy culverts, constant head culverts, minimum energy culverts.
2. MEL waterways are designed with the same principles as MEL culverts. They can be used to reduce the bridge span above the waterway, to increase the discharge capacity of bridge waterways without modification of an existing bridge or to prevent the flooding of an adjacent roadway (e.g. Fig. 19.11).
3. Compared with a standard culvert, a MEL culvert design provides less energy loss of the same discharge Q and throat width B_{min}. Alternatively the throat width can be reduced for the same discharge and head loss.

19.4.2 Basic considerations

The basic concepts of MEL culvert design are: *streamlining* and *critical flow conditions* throughout all the waterway (inlet, barrel, outlet).

Streamlining

The intake is designed with a smooth contraction into the barrel while the outlet (or diffuser) is shaped as a smooth expansion back to the natural channel.

The 'smooth' shapes should reduce the head losses (compared with a standard culvert) for the same discharge and barrel width. Practically, small head losses are achieved by streamlining the inlet and outlet forms: i.e. the flow streamlines will follow very smooth curves and no separation is observed.

Discussion: analogy between standard/MEL culverts and orifice/Venturi systems

The shape of the standard and MEL culverts can be respectively compared to a sharp-edge orifice and a Venturi-meter in a circular pipe.

At an orifice, large head losses take place in the recirculation region immediately downstream of the orifice (i.e. as for a standard culvert). In the Venturi-meter, the flow is streamlined and very small head losses are observed. From a top view, the MEL culvert sidewalls usually follow the shape of a Venturi meter (Figs 19.12, 19.13). As for a Venturi meter, the angle between straight diverging walls and the waterway centreline should be less than about 8° (for straight wing wall outlets).

Critical flow conditions

In an open channel, maximum discharge per unit width for a given specific energy is achieved at critical flow conditions (Section 19.2.2, Table 19.1).

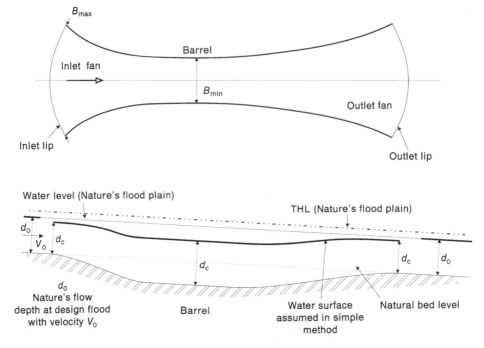

Fig. 19.13 Free-surface profile assumed in the 'simple design method'.

Minimum Energy Loss structures are designed to achieve critical flow conditions *in all the engineered waterways*: i.e. in the inlet, at the throat (or barrel) and in the outlet (Fig. 19.12). At the throat, the discharge per unit width q is maximum and it may be increased by lowering the barrel bed (Equation (19.1b)).

Notes
1. MEL structures are usually designed for critical flow conditions from the inlet lip to the outlet lip (i.e. throughout the inlet, barrel and outlet).
2. MEL waterways are not always designed for critical flow conditions. The MEL structure profile may be designed for subcritical (or supercritical) flow conditions: e.g. $Fr \approx 0.2$ to 0.5.
3. In many cases, it is assumed that the total head gradient through the MEL structure is the same as that which occurred previously in the natural channel: i.e. zero afflux.
4. Lowering the bed level at the throat gives an increase in local specific energy E and discharge per unit width q, hence the throat width may be reduced.

19.4.3 A simple design method (by C.J. Apelt)

Professor C.J. Apelt (The University of Queensland) proposed a simple method to calculate the basic characteristics of a MEL culvert. This method gives a preliminary design. Full calculations using the backwater equations are required to predict accurately the free-surface profile.

The simple method is based on the assumption that the flow is critical from the inlet to outlet lips including in the barrel.

Design process

Step 1. Decide the design discharge Q_{des} and the associated total head line (THL). The total head line is deduced assuming uniform equilibrium flow conditions in the flood plain.

(a) *Neglect energy losses*

Step 2.1. Calculate the waterway characteristics in the throat (i.e. barrel) for critical flow conditions. Compute the barrel width B_{min} and/or the barrel invert elevation z_{barrel} (neglecting energy losses).

Step 2.2. Calculate the lip width assuming critical flow conditions ($Fr = 1$) and natural bed level (i.e. $\Delta z_o = 0$). The lip width B_{max} is measured along the smooth line normal to the streamlines.

Step 3.1. Decide the shapes of the fans. Select the shape of the fan wing walls (or select the shape of the invert profile).

Step 3.2. Calculate the geometry of the fans to satisfy critical flow conditions everywhere. The contours of the fans are defined each by their width B (measured along a smooth line normal to streamlines), the corresponding bed elevation (to satisfy $Fr = 1$) and the longitudinal distance from the lip.

In the above steps, either the barrel width B_{min} is selected and the barrel invert drop Δz_o is calculated; or Δz_o is chosen and the barrel width is calculated.

(b) *Now include the energy losses*

Step 4. Adjust the bed profile of the waterway to take into account the energy losses. Calculate the total head loss available. Estimate the energy losses (friction loss and form loss) in the waterway (inlet, barrel, outlet). For a long barrel, the barrel slope is selected as the critical slope (of the barrel).

Step 5. Check the 'off-design' performances: i.e. $Q > Q_{des}$ and $Q < Q_{des}$.

Notes

1. In the simple design method, the inlet lip width and the outlet lip width are equal (Fig. 19.13).
2. The barrel width B_{min} is constant all along the barrel.
3. Another way to present the simplified approach is as follows.

First, let us assume an ideal fluid flow (i.e. no friction losses) and a horizontal bed (natural bed and barrel invert). Calculate B_{min}, B_{max}, the three-dimensional shapes of the inlet and the outlet. If the outlet shape could lead to flow separation at the expansion, increase the outlet length and repeat the outlet design procedure.

Secondly, take into account the energy losses (friction and form loss). Calculate the optimum barrel invert slope (see below), and incorporate the natural bed slope and barrel invert slope in the bottom elevation calculations (Fig. 19.13).

Comments

Figure 19.14 presents an example of MEL culvert operation at design and non-design flow conditions. In Fig. 19.14(a), design flow conditions are characterized by no afflux ($d_{tw} = d_1$) and the absence of hydraulic jump at the outlet (compared with box culvert operation, Fig. 19.9(a)). The flow in the barrel is undular and dye injection in the outlet highlights the absence of separation in the diffuser. Figure 19.14(b) shows a non-design flow ($Q < Q_{des}$). Note the subcritical flow in the barrel, the hydraulic jump in the inlet and the supercritical flow upstream (from the inlet lip).

A comparison between Figs 19.9 and 19.14 illustrates the different flow behaviour between standard-box and MEL culvert models with the same design flow conditions ($Q_{des} = 10\,\text{L/s}$, $d_{tw} = 0.038\,\text{m}$).

Discussion

- In Steps 2.1, 2.2, 3.1 and 3.2, the critical flow depth d_c satisfies:

$$q^2 = g d_c^3$$
$$d_c = \tfrac{2}{3} E$$

where E is the specific energy.

- B_{max} is deduced from the critical flow conditions at inlet lip (Step 2.2). At the inlet lip, the specific energy is known:

$$E_1 = H_1 - z_{inlet}$$

Hence the flow depth (i.e. critical depth) is known:

$$d_1 = d_c = \tfrac{2}{3} E_1$$

B_{max} is deduced from the continuity equation:

$$Q_{des} = V_c d_c B_{max}$$

- The profile (and slope) of the inlet and outlet invert is deduced from the complete calculations of (B, z, s) assuming $d = d_c$ at any position (equation (19.2)) and s is the longitudinal coordinate (in the flow direction).
- Upstream of the inlet lip, the flow depth equals the uniform equilibrium flow depth at design flood d_o. At the inlet lip, the flow depth is critical, *and* the 'simple method' assumes that the flow remains critical from the inlet lip down to the outlet lip (Fig. 19.13).
- In practice, the outlet might not operate at critical flow conditions because of the natural flood level (i.e. tailwater level) (Fig. 19.15).

  Critical depth at the exit could be achieved by installing a small 'bump' near the exit lip (e.g. Nudgee Road bridge waterway, Queensland). But form losses associated with the 'bump' are much larger than the basic exit losses. In practice, a 'bump' is not recommended. The Nudgee Road waterway was built with a 'bump' as a result of a misunderstanding between the designer and the constructor (see also Appendix A4.3).
- The barrel slope is often designed for the critical slope to account for the friction losses. However the barrel flow is not comparable to a fully-developed equilibrium flow for which the momentum, Chézy or Gauckler–Manning equations apply. In

(a)

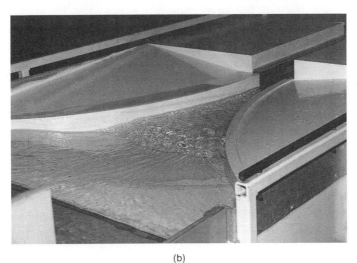

(b)

Fig. 19.14 Examples of MEL culvert operation. MEL culvert model – design flow conditions: $Q_{des} = 10$ L/s, $d_{tw} = 0.038$ m, $S_o = 0.0035$, $B_{min} = 0.10$ m, $D = 0.17$ m, $\Delta z_o = 0.124$ m, no afflux ($d_1 = d_{tw}$). (a) Design flow conditions: view from downstream (flow from top right to bottom left). $Q = 10$ L/s, $d_{tw} = 0.038$ m, $d_1 = 0.038$ m. Dye injection in the outlet highlights the streamlined flow (i.e. no separation). (b) Non-design flow conditions: view from upstream of the inlet (flow from bottom right to top right). $Q = 5$ L/s, $d_{tw} = 0.038$ m, $d_1 = 0.026$ m. Note the hydraulic jump in the inlet and subcritical flow in the barrel.

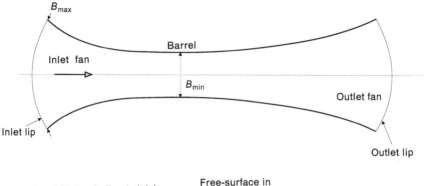

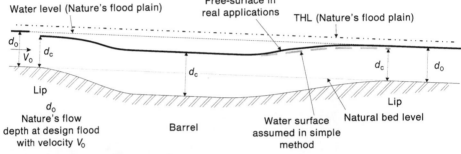

Fig. 19.15 Real free-surface profile in a MEL waterway.

practice, Apelt (1983) suggested that most barrel designs with critical slope should operate reasonably well. But he mentioned one case where the approximation did not hold: if the slope of the barrel invert is not equal to the critical slope, critical flow conditions might not be achieved in a long barrel. As a result some afflux might occur.

- The simple method calculations must be checked by physical model tests and complete backwater calculations might be conducted.

Notes

1. The lip of the inlet and outlet is the outer edge (upstream for inlet, downstream for outlet) of the inlet and outlet. The 'lip length' (or lip width) is the length of the lip curve measured along the smooth line perpendicular to the streamlines (Figs 19.12 and 19.13).
2. The inlet and outlet may be shaped as a series of curved smooth lines of constant bed elevation (i.e. contours) which are all normal to the flow streamlines. These smooth lines are equi-potential lines (cf. ideal-fluid flow theory). The streamlines and equi-potential lines (i.e. contour lines) form a flow net.
3. In most cases the barrel slope is selected as the critical slope. The critical slope is defined as the slope for which uniform equilibrium flow conditions are critical. It yields:

$$S_c = \sin \theta_c = \frac{f}{8} Fr^2 \frac{P_w}{B}$$

or

$$S_c = \sin \theta_c = n_{Manning}^2 Fr^2 \frac{g P_w^{4/3}}{A^{1/3} B}$$

where Fr, A, B, P_w are, respectively, the Froude number, cross-sectional area, width and wetted perimeter of the barrel flow. f and $n_{Manning}$ are the Darcy friction factor and the Gauckler–Manning friction coefficient respectively. *Warning*: The Gauckler–Manning equation is valid only for wide channels and it is inappropriate for most culvert barrels.

4. The barrel Froude number is usually unity at design flow conditions. In some cases, the barrel flow might be designed for a non-unity Froude number (e.g. $Fr = 0.6$ to 8) to reduce free-surface undulation amplitudes in the barrel and to reduce the outlet energy losses.

5. When the (required) barrel critical slope differs substantially from the natural bed slope, the inlet and outlet shapes must be changed in consequence. This situation requires an advanced design in contrast to the 'simple design method'. The whole waterway must be designed as a 'unity' (i.e. a complete system).

Energy losses

The energy losses in a MEL waterway include the *friction losses* in the inlet, barrel and outlet, and the *form losses* (primarily at the outlet). Figures 19.15 and 19.16 summarize the typical free-surface profile and total head line along a MEL waterway. The total head line of the natural flood plain is indicated by the dotted line in Fig. 19.16. A proper design approach must take into account:

- The total head loss available equals the natural bed drop along the culvert length plus the 'acceptable' afflux. The latter is usually zero (e.g. Fig. 19.16).
- The exit losses are always significant. As a result, the outlet flow depth is rarely equal to the critical depth.
- The exit losses are calculated from the exit velocity:

$$\Delta H_{exit} = \frac{V_{exit}^2}{2g} - \frac{V_o^2}{2g} \tag{19.7}$$

where V_o is the uniform equilibrium flow velocity of the flood plain.

When the outlet does not operate at critical flow conditions, $V_{exit} < V_c$ and the expansion ratio is smaller as $d_{exit} > d_c$. As a result, the exit losses become smaller than that calculated by the simple method (Apelt 1983).

- The calculations of the exit energy loss are very important. The exit losses must be less than the total energy loss available!
- For the design of MEL culverts the 'simple method' works well when the exit energy loss is less than one-third of the total head loss available.

Discussion: exit loss calculations

The exit head loss equals:

$$\Delta H_{exit} = \frac{V_{exit}^2}{2g} - \frac{V_o^2}{2g} \tag{19.7}$$

where V_{exit} is the mean velocity at the outlet lip and V_o is the normal velocity in the flood plain. Ideally (i.e. 'simple method' design), the exit velocity should equal:

$$V_{\text{exit}} = \frac{Q}{B_{\max}d_c}$$

where $B_{\max}$ is the outlet lip width and d_c is the critical flow depth at the outlet lip. In practice, the outlet lip depth roughly equals the normal depth d_o in the flood plain and the 'real' exit velocity is often:

$$V_{\text{exit}} = \frac{Q}{B_{\max}d_o}$$

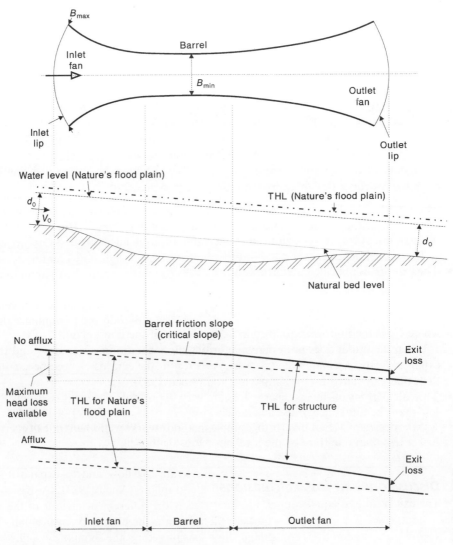

Fig. 19.16 Total head line (THL) in a MEL waterway.

19.4.4 Discussion

Challenges and practical considerations

A proper operation of MEL waterways and culverts requires a proper design. Several points are very important and must be emphasized:

- The velocity distribution must be uniform (or as uniform as possible).
- Separation of flow in the inlet and in the outlet must be avoided.
  In one case, an improper inlet design resulted in flow separation at the inlet, and four barrel cells did not operate properly, reducing the discharge capacity of the culvert by more than one third (at design flow conditions).
  Separation in the outlet is also a common design error. For straight wing outlet walls, the divergence angle between the wall and centreline should be less than about 8° (see Notes below).
- The shapes of the inlet and outlet must be within some reasonable limits.
  For the inlet, Apelt (1983) suggested that the contraction ratio should be limited to a maximum value corresponding to about $\Delta z_o/d < 4$, where Δz_o is the lowering of the bed and d is the approach flow depth.
  Further, he indicated that the 'optimum length of inlet fan is the shortest for which the flow approximates the two-dimensional (flow) condition without separation or unacceptable transverse water surface slope'. In practice, this leads to a minimum inlet length of about 0.5 times the maximum inlet width (measured along a contour line normal to the flow streamlines): $L_{inlet}/B_{max} > 0.5$.
  For the outlet: 'it is more difficult to achieve rapid rates of expansion and large expansion ratios in the outlet fan than it is to achieve rapid rates of contraction and large contraction ratios in the inlet fan' (Apelt 1983). Head losses at the outlet account for the greater proportion of the losses through the MEL structure.
- Critical flows are characterized by free-surface undulations and flow instabilities.
  In practice, free-surface undulations are observed (at design flow conditions) in reasonably long barrels (e.g. Chanson 1995a). In some cases, the wave amplitude might be as large as $0.5d_c$.
  Practically, the waterway may be designed for a barrel Froude number different from unity (e.g. $Fr = 0.6$ to 0.8) to reduce the wave amplitude and to minimize the obvert elevation.
- The head losses must be accurately predicted.

Notes

1. McKay (1978) strongly recommended limiting MEL designs to rectangular cross-section waterways. For non-rectangular waterways, the design procedure becomes far too complex and it might not be reliable.

2. *Design of the outlet fan*
   The function of the outlet fan is to decelerate the flow and to expand it laterally before rejoining the natural flood plain (Apelt 1983). A basic difference between the outlet and inlet fans is the different behaviour of the boundary layers in the two fans. In the inlet fan, the boundary layer is thin and the accelerating convergent flow ensures that the boundary layers remain thin. In contrast, in the outlet fan, the expanding flow is subjected

to an adverse (and destabilizing) pressure gradient. The boundary layer growth is more rapid. At the limit, separation might take place and the flow is no longer guided by the fan walls.

Designers could keep in mind the analogy between the MEL culvert and the Venturi meter when designing the shape of the outlet fan. Practically, the outlet is often longer than the inlet.

3. *Length of the barrel*
The selection of the barrel length depends upon the overall design (structural, geomechanics, etc.). Overall, the barrel must be as short as possible. Basically its length would be set at the top width of the embankment plus any extra length needed to fit in with the design of the slopes of the embankment.

Usually, the barrel length equals the bridge width for low embankments. For high embankments, the barrel length would be larger.

4. *Outlet divergent*
For a Venturi meter installed in a closed pipe, the optimum divergence angle is about (Levin 1968, Idelchik 1969, 1986):

$$\theta = 0.43\left(\frac{f}{\alpha_1}\frac{A_{\text{outlet}}/A_{\text{inlet}} + 1}{A_{\text{outlet}}/A_{\text{inlet}} - 1}\right)^{4/9} \tag{19.8}$$

where θ is in radians, f is the Darcy friction factor, A_{outlet} and A_{inlet} are respectively the diffuser inlet and outlet cross-sectional areas, and α_1 is an inflow velocity correction coefficient larger than or equal to 1 (Idelchik 1969, 1986). Equation (19.8) may be applied for straight outlet wingwalls of concrete rectangular MEL culverts. Assuming $\Delta z_0 = 0$, typical results are shown in Fig. 19.17, suggesting an optimum angle between 5° and 8° for straight divergent walls.

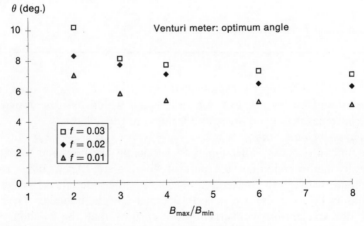

Fig. 19.17 Optimal divergence angle of Venturi meters (equation (19.8)).

In comparison with standard culverts, the design of MEL culverts must include additional factors:

- the flow velocities in the culvert are larger. Wingwalls and floors must be adequately protected;
- after a flood, ponding of water is a potential health hazard. It can be avoided by installing drains or anti-ponding pipes (e.g. Fig. 19.10, Appendix A4.3).

Design of multi-cell barrel

For construction reasons, wide rectangular culverts are often built with a number of identical rectangular cells (Fig. 19.10 and 19.18). In such cases, the flow cross-section is obstructed by the dividing walls and the *total* wall thickness between adjacent cells must be taken into account into the final design. Indeed, the minimum width of the fan is larger than the 'real' barrel width.

The late Professor Gordon McKay (The University of Queensland) proposed incorporating into the design a transition section (Fig. 19.18). The transition section is characterized by a progressive change of wall height associated with a change of bed level.

Note

It is recommended that the transition slope be less than 10% (Fig. 19.18).

Operation for non-design flow conditions

Culverts are designed for specific flow conditions (i.e. design flow conditions). However, they operate most of the time at non-design flow rates. In such cases design engineers must have some understanding of the culvert operation.

Figure 19.19 shows a typical sketch of total energy levels at the throat of a MEL waterway for non-design discharges. At design flow conditions (i.e. $Q = Q_{des}$), the barrel flow is critical, no afflux takes place and the barrel acts as a (hydraulic) control.

For $Q < Q_{des}$, the barrel flow is typically subcritical, no afflux is observed and the flow is controlled by the normal flow conditions in the flood plain (i.e. downstream flood plain). For discharges larger than the design discharge, the barrel flow becomes critical and acts as a hydraulic control. In addition, the afflux is larger than zero.

19.4.5 Benefits of MEL culverts and waterways

A main characteristic of MEL waterways is the *small head loss*. It results in a small afflux and in a small (or zero) increase of flood level. Furthermore, the *width* of the throat is *minimized*. Moreover, the culvert flow is streamlined and exhibits *low turbulence* (e.g. Fig. 19.11(c)). As a result, the erosion potential is reduced: e.g. fans can be made of earth with a grassed surface (e.g. Figs 19.11(a) and (c), see also Appendix A4.3).

For zero afflux, the size of a MEL culvert (inlet, barrel, outlet) is much smaller than that of a standard culvert with identical discharge capacity. Further Hee (1969)

Multi-cell Minimum Energy Loss culvert

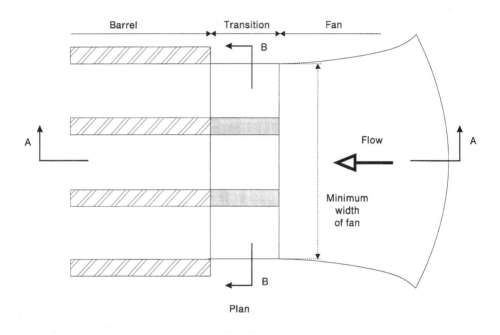

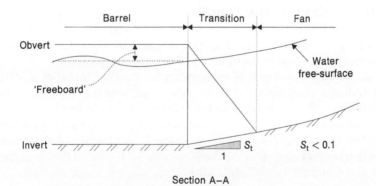

Fig. 19.18 Multi-cell Minimum Energy Loss culvert. Design of the transition section.

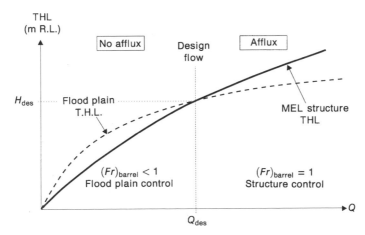

Fig. 19.19 Total energy level (TEL) in the throat of a MEL waterway for non-design flow conditions.

indicated that, for a very long culvert, the MEL culvert design tends to be more economical.

An additional consideration is the greater safety factor against flood discharges larger than the design discharge. Model and prototype observations have conclusively shown that MEL culverts can safely pass flood flows significantly larger than the design flow conditions. This is not always the case with standard culverts.

19.5 Exercises

A culvert is to be built to pass $25 \, \text{m}^3/\text{s}$ under a road embankment crossing a flood plain. The ground level is R.L. 22.000 m and the water level corresponding to this flow is expected to be R.L. 23.300 m. Both levels are at the centreline of the embankment which is 20 m wide at its base. The flood gradient is 0.004. (The culvert will be a multi-cell standard box culvert made of precast concrete boxes 1.5 m high and 1 m wide. Assume square-edge inlets.)

(a) Design a standard box culvert (with invert set at ground level) to carry the flood flow without causing any increase in flood level upstream *and operating under inlet control*. (b) If the number of boxes is reduced to five (i.e. 5-cell box culvert), (i) calculate the change in flood level if the design discharge remains unaltered assuming inlet control. (ii) Check whether inlet control is the correct assumption. (Use figure 'Flow characteristics of box culverts flowing full (Concrete Pipe Association of Australasia 1991)', assuming $k_e = 0.5$ for square-edge inlet.)

A culvert is to be built to pass $32 \, \text{m}^3/\text{s}$ under a road embankment crossing a flood plain. The ground level is R.L. 24.200 m and the water level corresponding to this flow is expected to be R.L. 25.750 m. Both levels are at the centreline of the embankment which is 20 m wide at its base. The flood gradient is 0.005 and the flood plain width is 65 m. The waterway must be designed using the principles of minimum energy loss culverts, to minimize the length of the bridge required and to carry the flood flow through the embankment without causing any increase in flood level upstream ('no afflux'). Use the 'simple method' design developed by Professor C.J. Apelt. The greatest depth of excavation allowable is 1.1 m below the natural surface. (Assume that the waterway will be concrete-lined.)

(a) For the design of MEL culverts, what are the two basic considerations?
(b) What is the inlet lip width?
(c) What is the barrel width?
(d) What is the outlet lip width?
(e) What are the exit losses?
(f) What is the maximum total head loss available? Comment.

A culvert is to be built to pass $87.0 \, m^3/s$ under a road embankment crossing a 350 m wide flood plain. The ground level is R.L. 20.000 m and the water level corresponding to this flow is expected to be 21.400 m. (Both levels are at the centreline of the embankment which is 20 m wide at its base.) The longitudinal slope of the flood plain is 0.004. The culvert is to be a multicell box structure using precast units with inside dimensions 1.8 m wide by 2.0 m high. (Assume square-edge inlets. Use $k_e = 0.5$ for square edged inlets.)

   (I) Design a minimum energy loss multicell box culvert for this situation. The greatest depth of excavation allowable is 0.900 m below the natural surface. (Use 'simple' method for design. 'no afflux' design.) Calculate: (a) inlet lip width, (b) number of cells, (c) required depth of excavation in barrel, (d) critical depth in barrel, (e) specific energy in barrel, (f) freeboard in barrel, (g) exit loss, (h) total head loss available. (Compare the results (g) and (h). Comment.)

   (II) Compare the MEL culvert with a standard multicell box culvert placed at ground level operating under inlet control for the following cases.
Design 1. The standard culvert has the same number of cells as the minimum energy loss one.

   (i) Calculate the culvert discharge capacity for the flood level to remain unaltered (i.e. no afflux). (j) Calculate the change in flood level (i.e. afflux) if the discharge remains unaltered (i.e. $87 \, m^3/s$).
Design 2. The standard culvert has sufficient cells to pass the discharge without significant afflux.

   (k) Calculate the number of cells.
Design 3. If the standard culvert has the same number of cells as the minimum energy loss one and the discharge remains unaltered, determine whether inlet control is the correct assumption.

   (l) With outlet control, calculate the upstream head above inlet invert (m). Will the box culvert operate with inlet control or outlet control?

A culvert is to be built to pass $98 \, m^3/s$ under a road embankment crossing a flood plain. The ground level is R.L. 15.000 m and the water level corresponding to the design flood is expected to be R.L. 16.700 m. Both levels are at the centreline of the embankment, which is 30 m wide at its base. The flood plain gradient is 3.5 m per km. The culvert will be a multicell standard box culvert made of precast concrete boxes 2.2 m high and 2 m wide. The total wall thickness between adjacent cells may be taken as 0.1 m. (Assume square-edge inlets.)

(a) Design a standard box culvert (with invert set at ground level) to carry the flood flow without causing any increase in flood level upstream (i.e. no afflux) *and operating under inlet control*. How many cells are required?

(b) If the number of boxes is reduced to five (i.e. 5-cell box culvert): (i) calculate the change in capacity for the flood level to remain unaltered; (ii) calculate the change in flood level if the design discharge remains unaltered assuming inlet control; (iii) check whether inlet control is the correct assumption in (ii) ($k_e = 0.5$ for square-edge inlet).

(c) Design a minimum energy loss (MEL) culvert to carry the flood with four cells using the 'simple method' design developed by Professor C.J. Apelt: (i) calculate the required depth of excavation below the natural surface; (ii) calculate the lip lengths of the inlet and of the outlet; (iii) what is the minimum inlet length that you recommend?

A road embankment is to be carried across a flood plain. You are required to design a means of carrying the flood flow (safely) below the embankment. The design data are:

Design flood flow	$98 \, \text{m}^3/\text{s}$
Natural ground level at centreline of embankment	R.L. 15.000 m
Design flood water level (at centreline of embankment)	R.L. 16.700 m
Flood plain slope	3.5 m per km
Width of embankment base	30 m

The culvert will be built as a multi-cell box culvert using precast units with inside dimensions 2 m wide by 2.2 m high. The *total* wall thickness between adjacent cells can be taken as 100 mm. An earlier study of MEL culvert calculations (for this structure) suggested using a five-cell structure (i.e. 5-cell MEL culvert). That MEL culvert was designed to carry the flood flow without causing any increase in flood level upstream (i.e. no afflux).

Compare the MEL culvert design with your design of a standard box culvert placed at ground level, operating under inlet control for the following cases.

(a) The standard culvert has the same number of cells as the minimum energy loss one. Calculate (Case 1) the change in capacity for the flood level to remain unaltered, and (Case 2) the change in flood level if the discharge remains unaltered.

(b) The standard culvert has sufficient cells to pass the discharge without significant afflux (Case 3).

Notes: Discharges through box culverts operating under inlet control can be calculated using Fig. 19.7, from Fig. 7-21 on p. 264 of Henderson (1966), or from equations (7-31) and (7-32) in Henderson (1966). Assume square-edged inlet.

(c) Repeat calculations of (a) (i.e. cases 1 and 2) by using the nomograph (Fig. 3.4) of Concrete Pipe Association of Australasia (1991).

(d) Using the nomograph (Fig. 3.6) of Concrete Pipe Association of Australasia (1991), determine whether inlet control is the correct assumption for (Case 2). Use $k_e = 0.5$ for square-edged inlets.

Summary sheet:
Flow capacity when water depth in flood plain just upstream is 1.7 m (Case 1).

Question	(a) (m³/s)	(b) (% of 98 m³/s)	(c) (m³/s)
Square-edged inlet			

Depth of water and specific energy on flood plain just upstream (Case 2).

Question	(a) Water depth (m)	(a) Specific energy (m)	(b) Water depth (m)	(b) Specific energy (m)
Square-edged inlet				

(b) Number of cells: (square-edged inlet) (Case 3)
(d) Inlet control (Case 2):   Yes/No
If No, state the new specific energy just upstream: m R.L.

A culvert is to be built to pass $32 \, \text{m}^3/\text{s}$ under a road embankment crossing a flood plain. The ground level is R.L. 4.200 m and the water level corresponding to this flow is expected to be R.L. 5.750 m. Both levels are at the centreline of the embankment which is 100 m wide at its base. The flood gradient is 0.005 and the flood plain width is 65 m. Design a means of carrying the flood flow through the embankment without causing any increase in flood level upstream ('no afflux').

(a) Design the waterway, using the principles of minimum energy loss culverts, to minimize the length of the bridge required. The greatest depth of excavation allowable is 1.1 m below the natural surface. Your design must include the details of inlet and outlet fans. Assume that the waterway will be concrete-lined.

Use the 'simple method' design developed by Professor C.J. Apelt. Estimate the exit loss. Comment.

(b) Calculate the width of waterway required to achieve the same objective of no afflux if standard box culverts are used with *rounded inverts* set at natural ground level and operating under inlet control. Select a barrel size which minimizes the construction costs (by minimizing the cross-sectional area of the barrel).

Summary sheet
(a)

Exit loss	
Total head loss available	
Minimum slope required for barrel	
Slope available	
Comment	

Specify in the table below the waterway widths and invert levels at intervals corresponding to depths of excavation equal to 0.25 × (maximum excavation).

Location	Station	Distance from embankment C.L. (+ve in d/s) (m)	Depth of excavation (m)	Ground level (m R.L.)	Waterway width (m)
Upstream lip	1				
	2				
	3				
	4				
Barrel (u/s)	5				
Barrel (midway)	6				
Barrel (d/s)	7				
	8				
	9				
	10				
Downstream lip	11				

Provide dimensioned drawing of waterway (plan and longitudinal section).

(b)

Net waterway width in box culverts (rounded inlets)	m
Recommended barrel height	m

Part 4 Revision exercises

Numerical solutions to some of these exercises are available from the Web at www.arnoldpublishers.com/support/chanson

Revision exercise No. 1

A hydraulic jump stilling basin (equipped with baffle blocks) is to be tested in a laboratory to determine the dissipation characteristics for various flow rates. The maximum prototype discharge will be 310 m^3/s and the rectangular channel is 14 m wide. Note that the channel bed will be horizontal and concrete-lined (i.e. smooth). A 42:1 scale model of the stilling basin is to be built. Discharges ranging between the maximum flow rate and 10% of the maximum flow rate are to be reproduced in the model.

(a) Determine the maximum model discharge required. (b) Determine the minimum prototype discharge for which negligible scale effects occurs in the model. (Comment on your result.)

For one particular inflow condition, the laboratory flow conditions are: upstream flow depth of 0.025 m, upstream flow velocity of 3.1 m/s, downstream flow depth of 0.193 m. (c) Compute the model force exerted on a single baffle block. State the basic principle(s) involved. (d) What is the direction of force in (c): i.e. upstream or downstream? (e) What will be the corresponding prototype force acting on a single block? (f) Compute the prototype head loss. (g) Operation of the basin may result in unsteady wave propagation downstream of the stilling basin. What will be the scale for the time ratio? (h) For a prototype design at maximum design discharge and with a prototype inflow depth of 1.7 m, what standard stilling basin design would you recommend: (i) USBR Type II, (ii) USBR Type III, (iii) USBR Type IV, (iv) SAF? (Justify your answer. If you have the choice between two (or more) designs, discuss the economical advantages of each option.)

Revision exercise No. 2 (hydraulic design of a new Gold Creek dam spillway)

The Gold Creek dam is an earthfill embankment built between 1882 and 1885. The length of the dam is 187 m and the maximum height of the embankment is at

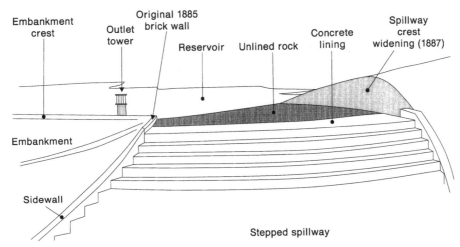

Fig. R.7 Sketch of the 1890 Gold Creek dam spillway, after an 1890s photograph.

99.1 m R.L. The reservoir storage capacity is about $1.8 \times 10^6 \, \text{m}^3$. The catchment area is $10.48 \, \text{km}^2$ of protected forest area.

An overflow spillway is located on the left abutment on rock foundation. It is uncontrolled: i.e. there is no gate or control system. The stepped chute (i.e. 'staircase wastewater course') was completed in 1890 (Fig. R.7). The crest and chute width is 55 m. The crest is nearly horizontal and 61 m long in the flow direction. It is followed by 12 identical steps. The spillway crest elevation is at 96.3 m R.L. The steps are 1.5 m high and 4 m long. The stepped chute is followed (without a transition section) by a smooth downward sloping channel ($\theta = 1.2°$) which ends with a sharp-crested weir, designed to record flow rates. The sloping channel is used as a dissipation channel. It is 55 m wide and its shape is approximately rectangular. The sharp-crested weir is rectangular without sidewall contraction (i.e. 55 m wide). It is located 25 m downstream of the chute toe and the elevation of the weir crest is set at 79.1 m R.L. (For the purpose of the assignment, the spillway crest discharge characteristics may be assumed ideal.)

The dam owner is investigating an increase of the maximum discharge capacity of the spillway. The new overflow spillway will replace the existing waste waterway structure. Two options are considered.

(i) Design of an overflow spillway with Ogee type crest and a hydraulic jump dissipator. The crest level is at the same level as the existing spillway. The chute slope is set at 45°. The elevation of the chute toe is set at 78.3 m R.L.

(ii) Design of an overflow spillway with Ogee type crest and hydraulic jump dissipator. The crest level is lowered by erasing completely one step; i.e. the new crest level is set at 94.8 m R.L. The chute slope is set at 45°. The elevation of the chute toe is set at 78.3 m R.L.

In each case, the dam wall, upstream of the spillway crest, has a 3 V : 1 H slope, and the reservoir bottom is set at 77 m R.L. In both refurbishment options, the chute is followed by a concrete apron. The tailwater level will be raised artificially with a broad-crested weir located at the downstream end of the concrete apron (i.e. geometry to be determined). In each case, use a sloping apron section if this will improve the efficiency and economics of the basin. In calculating the crest discharge capacity, assume an USBR ogee crest shape and use Fig. 17.9 for the discharge coefficient calculations. In computing the velocity at the spillway toe, allow for energy losses.

The assignment includes three parts.

1. Calculation of the maximum discharge capacity of the existing spillway $Q_{max}^{(1)}$ and estimate the hydraulic jump location at the downstream end of the spillway. Assume a skimming flow regime on the stepped chute.
2. For the new spillway design [Option (i)], compute the maximum discharge capacity $Q_{max}^{(2)}$ of the new spillway. Compare the result with the maximum discharge capacity of the present spillway.
3. For the new spillway design [Option (ii)], compute the maximum discharge capacity of the spillway, $Q_{max}^{(3)}$, design the hydraulic jump stilling basin to dissipate the energy at the foot of the spillway, and select the location, elevation and crest length of the broad-crested weir to ensure that the hydraulic jump stilling basin is always contained between the smooth chute toe and the weir for discharges ranging from $Q_{max}^{(3)}$ down to $0.1Q_{max}^{(3)}$.

Note: the maximum discharge over the crest is usually computed when the reservoir free-surface elevation reaches the dam crest elevation. For an earth embankment, it is essential that *no* overtopping (at all) occurs. For that reason, in practice, it is safer to allow a safety margin of 0.4 m (to 1 m) for wind wave effects (i.e. dam overtopping caused by wind wave action). For the assignment, use a margin of safety of 0.4 m. The calculations are to be completed and submitted in two stages as specified below.

Stage A. Calculations are to be done for the maximum flow rate *only*.

(a) Calculate the maximum discharge capacity of the present spillway and the flow velocity at the toe of the chute.
(b) Calculate the maximum discharge capacity of the refurbished spillway [Option (i)] and the flow velocity at the toe of the chute.
(c) Calculate the maximum discharge capacity of the refurbished spillway [Option (ii)] and the flow velocity at the toe of the chute.
(d) Compare the residual power at the end of the chute between the three options. In term of energy dissipation, what are the main differences between the three designs? In each case, what type of standard stilling basin would you suggest? (The residual power equals ρgQH_{res} where Q is the total discharge and H_{res} is the total head taking 78.3 m R.L. as the datum.)

Stage B.
(a) For the present spillway, calculate the hydraulic jump dissipation channel to dissipate the energy at the toe of the spillway for the *maximum flow rate* (assuming skimming flow regime) calculated in stage A. (1) Calculate the jump length. (2) Calculate the hydraulic Jump Height Rating Level (*JHRL*). (3) Calculate the Tailwater Rating Level (*TWRL*). (4) Will the hydraulic jump be contained between the chute toe and the sharp-crested weir at maximum flow rate? In the negative, what will happen? Justify and explain clearly your answer in words, and use appropriate relevant sketch(es). (5) Draw to scale the dimensioned sketch of the dissipation

channel, showing the spillway toe and the sharp-crested weir. (6) Draw the free-surface profile on the dimensioned sketch.

(b) For the spillway refurbishment [Option B], design the hydraulic jump stilling basin to dissipate the energy at the foot of the smooth chute spillway. (1) Calculate the *JHRC* for all flows ranging from $Q^{(3)}_{max}$ down to $0.1Q^{(3)}_{max}$. $Q^{(3)}_{max}$ was calculated in stage A. (2) Plot the *JHRC* (from (1)). (3) Determine the crest level of the broad-crested weir to match the *JHRL* and the *TWRL* for the maximum flow rate $Q^{(3)}_{max}$. (4) Calculate a minimum weir crest length suitable for all flow rates ranging from $Q^{(3)}_{max}$ down to $0.1Q^{(3)}_{max}$. (5) Plot the *TWRC* (created by the broad-crested weir) on the same graph as the *JHRC* for all flows ranging from $Q^{(3)}_{max}$ down to $0.1Q^{(3)}_{max}$. (6) Determine the height of the sloping apron required to match the *JHRC* to *TWRC* for all flows. (This is obtained from the greatest vertical separation of the two curves as plotted in question (5). If the maximum separation of the two curves occurs at a flow less than $0.1Q^{(3)}_{max}$, set the height of the sloping apron to the separation of the curves at $0.1Q^{(3)}_{max}$.) (7) Draw to scale the dimensioned sketch of the dissipator. (8) Draw the free-surface profile on the dimensioned sketch for $Q^{(3)}_{max}$ and for $0.3Q^{(3)}_{max}$.

Note: Use a sloping apron section if this will improve the efficiency and/or economics of the basin. You may consider an apron elevation different from the chute toe reference elevation (i.e. 78.3 m R.L.) if this is more suitable.

Summary sheet (stage A)

	Present spillway	Refurbished spillway [Option (i)]	Refurbished spillway [Option (ii)]	Units
Maximum discharge capacity				....
Flow velocity at chute toe				....
Froude number at chute toe				....
Residual power at chute toe				....
Recommended type of standard stilling basin[a]				N/A

Note: [a] USBR Types II, III or IV, or SAF.

Summary sheet (stage B)

(a)

TWRL	m R.L.
JHRL	m R.L.
Jump length	m

(b)

JHRL at maximum flow rate	m R.L.
Broad-crested weir crest elevation	m R.L.
Broad-crested weir crest length	m
Horizontal length of apron	m
Height of sloping apron	m

Supply the plotted curves of the *JHRC* and *TWRC*. Supply dimensioned drawings of the dissipation basin to scale.

Appendix. History of the dam

The Gold Creek dam was built between 1882 and 1885.[1] It is an earthfill embankment with a clay puddle corewall, designed by John Henderson. The fill material is unworked clay laid in 0.23 m layers. The length of the dam is 187 m and the maximum height of the embankment above the foundation is 26 m. The reservoir storage capacity is about $1.8 \times 10^6 \, m^3$. The catchment area is 10.48 km^2 of protected forest area. Originally the Gold Creek reservoir supplied water directly to the city of Brisbane. As the Gold Creek reservoir is located close to, and at a higher elevation than, the Enoggera dam, the reservoir was connected to the Enoggera reservoir in 1928 via a tunnel beneath the ridge separating the Enoggera creek and Gold creek basins. Nowadays the Gold Creek reservoir is no longer in use, the pipeline having been decommissioned in 1991.[2] The reservoir is kept nearly full and it is managed by Brisbane Forest Park. An outlet tower was built between 1883 and 1885 to draw water from the reservoir. The original structure in cast iron failed in 1904, following improper operation while the reservoir was empty. The structure was replaced by the present concrete structure (built in 1905).

An overflow spillway is located on the left abutment on rock foundation. It is uncontrolled: i.e. there is no gate or control system. Since the construction of the dam in 1882–1885, the spillway has been modified three times. The original spillway was an unlined rock overflow. In 1887, the spillway channel was widened to increase its capacity. In January 1887 and early in 1890, large spillway overflows occurred and the unlined-rock spillway was damaged and scoured. A staircase concrete spillway was built over the existing spillway in 1890. The steps are 1.5 m high and 4 m long. In 1975 the crest level was lowered to increase the maximum discharge capacity. Reference: Chanson and Whitmore (1996).

Revision exercise No. 3 (hydraulic design of the Nudgee Road bridge waterway)

The Nudgee Road bridge is a main arterial road in the eastern side of Brisbane (see Appendix A4.3). The bridge crosses the Kedron Brook stream downstream of the Toombul shopping-town. At that location, a man-made waterway called the Schultz canal carries the low flows. Today the Schultz canal flows east-north-east and dis-charges into the deep-water Kedron Brook flood-way, flowing parallel to and north of the Brisbane Airport into the Moreton Bay. The natural flood plain (upstream and downstream of the bridge) is 400 m wide and grass-lined (assume $k_s = 90$ mm) with a longitudinal slope of 0.49 m per km. At the bridge location, the natural ground level is at R.L. 10.0 ft.

The assignment will include three parts: the design of the old waterway, the hydraulic design of the present waterway and the upgrading of the waterway.

---

[1] It is the 14th largest dam built in Australia
[2] The tunnel was first decommissioned in 1977 but the water supply to Enoggera reservoir resumed from 1986 up to 1991.

Part 1. The old waterway (prior to 1968)
The old waterway could pass '7000 cusecs' before overtopping the road bridge.

Design flood flow	7000 ft^3/s
Natural ground level at centreline of embankment	R.L. 10.00 ft
Width of embankment base	30 ft
Road elevation	R.L. 13.00 ft
Bridge concrete slab thickness	6 in.
Flood plain slope	0.49 m per km

(a) Calculate the width of the old waterway (i.e. span of the old bridge) assuming that the waterway was a standard box culvert with an invert set at natural ground level and operating under inlet control (i) with squared-edged inlets and (ii) with rounded inlets.

(b) Use the spreadsheet you developed for the backwater calculations to calculate the flood level (i.e. free-surface elevation) for distances up to 1500 m upstream of the culvert entrance. Plot both backwater profiles and compare with the uniform equilibrium flow profile.

(c) Using the commercial software HydroCulv (see Appendix A4.4), calculate the hydraulic characteristics of the throat flow at design flood flow.

Part 2. The present Nudgee Road waterway (built in 1969)
In 1969, the new Nudgee Road bridge was hydraulically designed based upon a Minimum Energy Loss design ('no afflux'). The improvement was required by the rapid development of the Kedron Brook industrial area and the development of the Toombul shopping town upstream. The waterway was designed to pass about 849.5 m^3/s (30 000 cusec) which was the 1:30 year flood. The ground level (at the new bridge location) is R.L. 3.048 m (R.L. 10 ft) and the water level corresponding to this flow was expected to be R.L. 4.7549 m (R.L. 15.6 ft). Both levels are at the centreline of the bridge. The waterway throat is 137 m wide, 40 m long and lined with kikuyu grass (assume $k_s = 0.04$ m). (Note: the flood plain gradient is 0.49 m per km.)

Design the minimum energy loss waterway culvert for this situation. Include details of the inlet and outlet in your design. (Use the simple design method developed by Professor C.J. Apelt. Estimate exit loss. Comment.)

Part 3. Upgrading the Nudgee Road waterway (1997–1998)
The Nudgee Road waterway must be enlarged to pass 1250 m^3/s. The water level corresponding to this flow is expected to be R.L. 5.029 m (at bridge centreline, in the absence of waterway and embankment). It is planned to retain the existing road bridge and most of the road embankment by building a 0.275 m high concrete wall along the road (on the top of the bridge and embankment) to prevent overtopping. The waterway inlet, outlet and throat will be concrete-lined to prevent scouring that could result from the larger throat velocity. Two Minimum Energy Loss designs ('no afflux') are considered: (A) retain the existing bridge and excavate the throat invert; (B) extend the existing bridge and keep the throat invert elevation as in the present waterway (calculation in Part 2).

(a) Design the minimum energy loss waterway culverts. (Use the simple design method developed by Professor C.J. Apelt. Estimate exit loss. Comment.)
(b) Calculate the construction cost of both designs.
(c) For the *minimum-cost* MEL waterway design, include details of the inlet and outlet in your design.

Construction costs:

Extension of a two-lane bridge	$10 500 per metre length
Excavation cost	$9 per cubic metre
(Embankment cost	$21 per cubic metre)

Summary sheet (Part 1)

(a)	(i) Net waterway width, square edged	m
	(ii) Net waterway width, round-edged	m
(b)	Normal depth in the flood plain	m
	Increase in flood level 800 m upstream	m
(c)	Inlet Control? (Yes/No) [*HydroCulv calculations*]	m
	If No, state new Specific Energy just upstream [*HydroCulv calculations*]	
	Free-board at bridge centreline [*HydroCulv calculations*]	

Summary sheet (Part 2)

Invert level at bridge centreline	m R.L.
d_c in throat	m
Minimum slope required for throat	
Exit loss	m
Total head loss available	m
Comments	

Specify in the table below the waterway widths and invert levels at intervals in the inlet corresponding to depths of excavation equal to $0.25 \times$ (maximum excavation).

Location	Station	Distance from C.L. of bridge (+ve in d/s) (m)	Depth of excavation (m)	Invert level (m R.L.)	Waterway width (m)
Upstream lip	1				
	2				
	3				
	4				
Throat (u/s)	5				
Throat (midway)	6				
Throat (d/s)	7				
Downstream lip	8				

Supply dimensioned drawings of the waterway (plan and longitudinal section).

Summary sheet (Part 3)

	Design A	Design B	
Inlet lip length			m
Waterway width (at throat)	137.00		m
Invert level at bridge centreline			m R.L.
d_c in throat			m
E in throat			m
Minimum slope required for throat			
Exit loss			m
Total head loss available			m
Comments			
Excavation volume			m^3
Excavation cost	$	$	
Bridge enlargement cost	$	$	
Total cost	$	$	

Most economical design:
Specify in the table below the waterway widths and invert levels at intervals corresponding to depths of excavation equal to $0.25 \times$ (maximum excavation).

Location	Station	Distance from C.L. of embankment (+ve in d/s) (m)	Depth of excavation (m)	Invert level (m R.L.)	Waterway width (m)
Upstream lip	1				
	2				
	3				
	4				
Throat (u/s)	5				
Throat (midway)	6				
Throat (d/s)	7				
	8				
	9				
	10				
Downstream lip	11				

Supply dimensioned drawings of the waterway (plan and longitudinal section).

Appendices to Part 4

A4.1 Spillway chute flow calculations

Introduction

On an uncontrolled chute spillway, the flow is accelerated by the gravity force component in the flow direction. For an ideal fluid flow, the velocity at the downstream chute end can be deduced from the Bernoulli equation:

$$V_{max} = \sqrt{2g(H_1 - d\cos\theta)} \qquad \text{Ideal fluid flow} \qquad (A4.1)$$

where H_1 is the upstream total head, θ is the channel slope and d is the downstream flow depth (Fig. A4.1). In practice, however, friction losses occur and the flow velocity at the downstream end of the chute is less than the ideal-fluid velocity (called the maximum flow velocity). At the upstream end of the chute, a bottom boundary layer is generated by bottom friction and develops in the flow direction. When the outer edge of the boundary layer reaches the free-surface, the flow becomes fully-developed.

First the basic equations are reviewed for both the developing and fully-developed flows. The complete calculations are then developed and generalized.

Developing flow region

Basic equations
In the developing flow region, the flow consists of a turbulent boundary layer next to the invert and an ideal-fluid flow region above. In the ideal-fluid region, the velocity (called the free-stream velocity) may be deduced from the Bernoulli equation:

$$V_{max} = \sqrt{2g(H_1 - z_o - d\cos\theta)} \qquad \delta < y < d \qquad (A4.2)$$

where z_o is the bed elevation and d is the local flow depth.

In the boundary layer, model and prototype data indicate that the velocity distribution closely follows a power law:

$$\frac{V}{V_{max}} = \left(\frac{y}{\delta}\right)^{1/N_{bl}} \qquad 0 < y/\delta < 1 \qquad (A4.3)$$

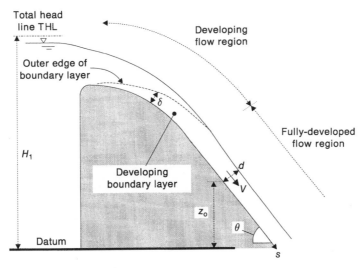

Fig. A4.1 Sketch of a chute spillway flow.

where δ is the boundary layer thickness defined as $\delta = y\,(V = 0.99\,V_{max})$ and y is the distance normal to the channel bed (Fig. A4.2). The velocity distribution exponent typically equals $N_{bl} = 6$ for smooth concrete chutes.

Combining equations (A4.2) and (A4.3), the continuity equation gives:

$$q = V_{max}\left(d - \frac{\delta}{N_{bl}+1}\right) \tag{A4.4}$$

Boundary layer growth

Several researchers investigated the boundary growth on spillway crests (Table A4.1).

Table A4.1 Turbulent boundary layer growth on chute spillways

Reference (1)	Boundary layer growth δ/s (2)	Remarks (5)
Spillway flow (smooth chute)		
Bauer (1954)	$0.024\left(\dfrac{s}{k_s}\right)^{-0.13}$	Empirical formula based upon model data.
Keller and Rastogi (1977)	Theoretical model with $k-\varepsilon$ turbulence model	WES spillway shape.
Cain and Wood (1981)	$a_1\left(\dfrac{s}{k_s}\right)^{-0.10}$	Based upon prototype data.
Wood *et al.* (1983)	$0.0212(\sin\theta)^{0.11}\left(\dfrac{s}{k_s}\right)^{-0.10}$	Semi-empirical formula. Fits prototype data well.
Spillway flow (stepped chute)		
Chanson (1995b)	$0.06106(\sin\theta)^{0.133}\left(\dfrac{s}{h\cos\theta}\right)^{-0.17}$	Skimming flow only. Empirical formula based upon model data. Includes models with various crest shapes.

Notes: k_s = equivalent roughness height; h = step height.

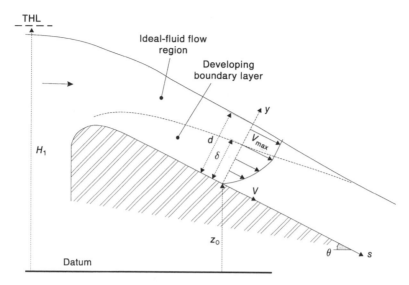

Fig. A4.2 Sketch of the developing flow region.

For smooth concrete chutes, the following formula is recommended:

$$\frac{\delta}{s} = 0.0212(\sin\theta)^{0.11}\left(\frac{s}{k_s}\right)^{-0.10} \qquad \text{Smooth concrete chute } (\theta > 30°) \qquad (A4.5)$$

where s is the distance from the crest measured along the chute invert, θ is the chute slope and k_s is the equivalent roughness height. Equation (A4.5) is a semi-empirical formula that fits well the model and prototype data (Wood *et al.* 1983). It applies to steep concrete chutes (i.e. $\theta > 30°$). For stepped chutes with skimming flow, the turbulence generated by the steps enhances the boundary layer growth. The following formula can be used to a first approximation:

$$\frac{\delta}{s} = 0.06106(\sin\theta)^{0.133}\left(\frac{s}{h\cos\theta}\right)^{-0.17} \qquad \text{Stepped chute (skimming flow)} \qquad (A4.6)$$

where h is the step height. Equation (A4.6) was checked with model and prototype data (Chanson 1995b). It applies only to skimming flow on steep chutes (i.e. $\theta > 30°$). Note that it is nearly independent of the crest shape.

Developing flow characteristics

In the developing region, the complete flow characteristics can be deduced from equations (A4.2) to (A4.5) (or equation (A4.6) for stepped chutes). The results give the flow depth, mean flow velocity, free-stream velocity, boundary layer thickness and velocity distribution at any position s measured from the crest.

Results are shown in Fig. A4.3, where the dimensionless mean velocity is plotted as a function of the dimensionless upstream head. Calculations were performed assuming an USBR ogee-shaped crest.

Figure A4.3 presents results for smooth concrete chutes and stepped chutes. The trend indicates that the mean velocity in the developing flow region departs from

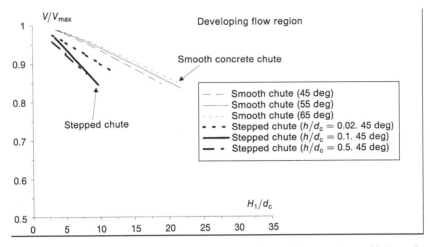

Fig. A4.3 Dimensionless mean flow velocity in the developing flow region. Note: $V = q/d$; Ogee-shaped crest (both smooth and stepped chute calculations).

the theoretical velocity with increasing upstream total head. Indeed, for a given discharge, an increase in upstream total head brings an increase in dam height, in distance s from the crest, in boundary layer thickness (equations (A4.5) and (A4.6)) and hence in the bottom friction effect.

> **Note**
> Figure A4.3 presents stepped chute calculations for the following chute geometry: smooth USBR ogee crest followed by steps, assuming $N_{bl} = 6$. If the steps start further upstream, the bottom roughness would modify the velocity profile and lower values of N_{bl} are advisable.

Fully-developed flow region

Presentation
In the fully-developed flow region, the flow is accelerated until the boundary friction counterbalances the gravity force component in the flow direction (i.e. uniform equilibrium flow or normal flow). At equilibrium, the normal conditions can be deduced from the momentum equation.

Uniform equilibrium flow
In uniform equilibrium open channel flows, the momentum equation yields (for a wide channel):

$$V_o = \sqrt{\frac{8g}{f}} \sqrt{d_o \sin \theta} \tag{A4.7a}$$

where V_o is the uniform equilibrium flow velocity, f is the Darcy friction factor, θ is the channel slope and d_o is the normal flow depth. Equation (A4.7a) is a form of the Chézy equation.

Combining with the continuity equation, it becomes:

$$V_o = \sqrt[3]{\frac{8g}{f}q\sin\theta} \qquad\qquad (A4.7b)$$

On a long chute, uniform equilibrium flow conditions are reached before the downstream end of the chute, and the downstream flow velocity V equals:

$$\frac{V}{V_{max}} = \sqrt{\frac{4\sin\theta}{f\left(\dfrac{H_1}{d_c}\sqrt[3]{\dfrac{8\sin\theta}{f}} - \cos\theta\right)}} \qquad\qquad (A4.8)$$

where V_{max} is the ideal flow velocity (equation (A4.1)) and d_c is the critical flow depth. Equation (A4.8) is valid for large values of H_1/d_c and for wide rectangular channels.

Figure A4.4 presents typical results for smooth concrete chutes ($f = 0.01$, 0.03, 0.005) and stepped chutes ($f = 1$, Chanson 1995b).

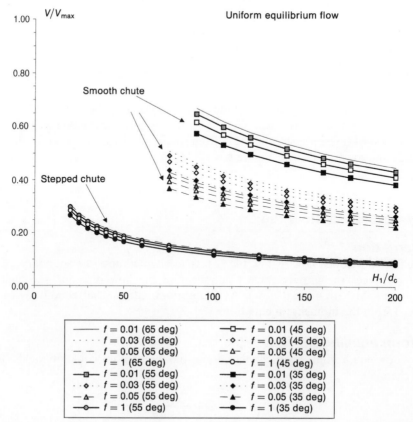

Fig. A4.4 Mean uniform equilibrium flow velocity as a function of the upstream head (equation (A4.8)).

Gradually-varied flow region

In the fully-developed flow region, the flow is gradually varied in the upstream part until becoming uniform equilibrium. In the gradually varied flow region, the flow properties can be deduced from a differential form of the energy equation:

$$\frac{\partial H}{\partial s} = -S_f \tag{A4.9a}$$

where H is the mean total energy, s is the direction along the channel bed and S_f is the friction defined as:

$$S_f = f \frac{1}{D_H} \frac{V^2}{2g}$$

D_H is the hydraulic diameter and V is the mean flow velocity.

Assuming a hydrostatic pressure distribution, the energy equation becomes:

$$\frac{\partial d}{\partial s} (\cos \theta - \alpha Fr^2) = \sin \theta - S_f \tag{A4.9b}$$

where d is the flow depth measured normal to the channel bed, α is the kinetic energy correction coefficient[1] and Fr is the Froude number. In the fully-developed flow region, the velocity distribution may be approximated by:

$$\frac{V}{V_{max}} = \left(\frac{y}{d}\right)^{1/N} \tag{A4.10}$$

A value of $N = 6$ is typical for smooth concrete chutes. Lower values are obtained for stepped chute flows (Chanson 1995b). In any case, for a power law velocity distribution, the kinetic energy coefficient equals:

$$\alpha = \frac{(N+1)^3}{N^2(N+3)} \tag{A4.11}$$

Fully-developed flow calculations are started from a position of known flow depth and flow velocity: i.e. typically at the end of the developing region when $\delta/d = 1$. Equation (A4.9) can be integrated numerically to deduce the flow depth and mean velocity at each position along the crest.

Application and practice

Presentation

For flat-slope chutes, the flow properties in the developing flow region and in the fully-developed flow region must be computed using the above development.

For steep chutes (typically $\theta \sim 45°$ to $55°$), both the acceleration and boundary layer development affect the flow properties significantly. In addition, free-surface aeration[2] may take place downstream of the intersection of the outer edge of the

---

[1] Also called the Coriolis coefficient.
[2] Air entrainment in open channels is also called free-surface aeration, self-aeration, insufflation or white waters.

developing boundary layer with the free surface. Air entrainment can be clearly identified by the 'white water' appearance of the free-surface flow (e.g. Wood 1991, Chanson 1997).

Practical considerations

On a steep channel, the complete calculations of the flow properties can be tedious. In practice, however, the combination of the flow calculations in developing flow (e.g. Fig. A4.3) and in uniform equilibrium flow (e.g. Fig. A4.4) give a general trend that may be used for a preliminary design.

Figure A4.5 combines the results obtained in Figs A4.3 and A4.4 with smooth transition curves. It provides some information on the mean flow velocity at the end of the chute as a function of the theoretical velocity V_{max} (equation (A4.1)), the upstream total head (above spillway toe) and the critical depth. In Fig. A4.5, the general trend is shown

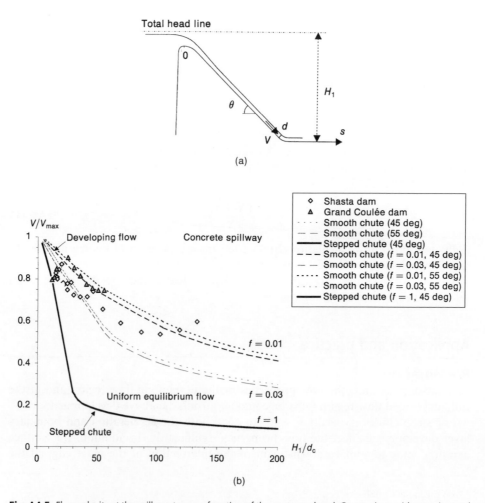

Fig. A4.5 Flow velocity at the spillway toe as a function of the upstream head. Comparison with experimental data (Bradley and Peterka 1957, paper 1403). (a) Definition sketch. (b) Results.

for smooth and stepped spillways (concrete chutes), with slopes ranging from 45° to 55° (i.e. 1 V : 1 H to 1 V : 0.7 H). Experimental results obtained on smooth-chute prototype spillways (Grand Coulée dam and Shasta dam) are also shown.

Application

Considering a 45° chute, the dam height above the spillway toe is 55 m and the design discharge per unit width is $10 \, m^2/s$. Calculate the downstream flow properties for a smooth chute and for a stepped chute (skimming flow regime). Assume a concrete chute with an USBR ogee-shaped crest.

Solution

The critical depth equals 2.17 m. The discharge versus head relationship can be estimated as:

$$q = C(H_1 - \Delta z)^{3/2}$$

with $C = 2.18 \, m^{1/2}/s$ at design discharge (the ratio of head above crest to dam height is about 0.03). At design discharge, the ratio of upstream total head to critical depth equals: $H_1/d_c = 27$.

The downstream flow velocity can be deduced from Fig. A4.5. For a smooth concrete spillway, a value of $f = 0.03$ is typical. On stepped chutes, a value of $f = 1.0$ is usual. Results are summarized below:

	V/V_{max}	V (m/s)	d (m)
Smooth chute	0.80	26.9	0.37
Stepped chute	0.35	11.8	0.85

Note that such results indicate that larger bottom friction and energy dissipation take place along a stepped spillway compared with a smooth chute. At the end of the chute, $V = 12 \, m/s$ for a stepped chute design compared with $V = 27 \, m/s$ for a smooth invert.

A4.2 Examples of Minimum Energy Loss weirs

Introduction

The concept of Minimum Energy Loss (MEL) weirs was developed by late Professor G.R. McKay. MEL weirs were designed specifically for situations where the river catchments are characterized by torrential rainfalls (during the wet summer) and by very small bed slope ($S_o \sim 0.001$). The MEL weirs were developed to pass large floods with minimum energy loss, hence with minimum upstream flooding.

At least four MEL weirs were built in Queensland (Australia): Clermont weir, Chinchilla weir,[1] Lemontree weir and a fourth structure built on the Condamine river (list in chronological order).

---

[1] Classified as a 'large dam' according to the International Commission on Large Dams.

Design technique

The purpose of a MEL weir is to minimize afflux and energy dissipation at design flow conditions (i.e. bank full), and to avoid bank erosion at the weir foot. The weir is curved in plan to converge the chute flow and the overflow chute is relatively flat. Hence, the downstream hydraulic jump is concentrated near the river centreline away from the banks and usually on (rather than downstream of) the chute toe. The inflow Froude number remains low and the rate of energy dissipation is small compared with a traditional weir.

MEL weirs are typically earthfill structures protected by concrete slabs. Construction costs are minimum. A major inconvenience of a MEL weir design is the risk of overtopping during construction (e.g. Chinchilla weir). In addition, an efficient drainage system must be installed beneath the chute slabs.

In the following sections, two practical examples are illustrated.

Hydraulic calculations

Assuming a broad-crest and no head loss at the intake (i.e. smooth approach), the discharge capacity of the weir equals:

$$Q = B_{crest} \sqrt{g} (\tfrac{2}{3}(H_1 - \Delta z))^{3/2} \qquad (A4.12)$$

where H_1 is the upstream head, Δz is the weir height above ground level and B_{crest} is the crest width (Fig. A4.6).

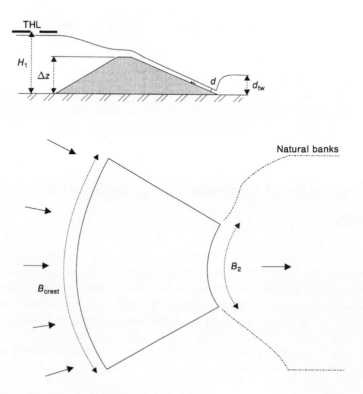

Fig. A4.6 Sketch of a Minimum Energy Loss weir.

Ideally, a MEL weir could be designed to achieve critical flow conditions at any position along the chute and, hence, prevent the occurrence of a hydraulic jump. Neglecting energy loss along the chute, the downstream width should be:

$$B_2 = B_{crest} \left(\frac{H_1 - \Delta z}{H_1}\right)^{3/2} \qquad \text{'ideal conditions'} \qquad (A4.13)$$

Practically this is not achievable because critical flow conditions are unstable and the variations of the tailwater flow conditions with discharge are always important. In practice, a weak jump takes place at the chute toe. Note that the jump occurs in an expanding channel and the downstream flow depth (d_{tw}) is fixed by the tailwater conditions.

Examples of MEL weirs

Ref. No.	MEL-WE-1
Description	MEL weir at Chinchilla (Fig. A4.7)
Location	Chinchilla QLD, Australia
River	Condamine river
Purpose	Irrigation water
Design discharge	$850\,\text{m}^3/\text{s}$
Weir characteristics	Earth embankment with concrete slabs downstream facing
	Maximum height: 14 m
	Crest length: 410 m
	Maximum reservoir capacity: $9.8 \times 10^6\,\text{m}^3$
Notes	Excellent design. Completed in 1973. Tested shortly after completion with a large overflow ($\sim 400\,\text{m}^3/\text{s}$). Listed as a large dam in ICOLD (1984). During construction, the unprotected earthfill was overtopped and damaged by a flood.
Ref. No.	MEL-WE-2
Description	MEL weir at Lemontree (Fig. A4.8)
Location	Mirmellan QLD, Australia
River	Condamine river (upstream of Chinchilla weir)
Purpose	Irrigation water
Weir characteristics	Earth embankment with concrete slabs downstream facing
Notes	Excellent design. Completed in the 1980s. Located 245 km upstream of the Chinchilla weir MEL-WE-1.
	A pumping station is located upstream of the weir. The station is equipped with three large pumps to transfer water into a (large) farm storage tank, for future water replenishment of the weir reservoir during dry periods.

(a)

(b)

Fig. A4.7 Minimum Energy Loss weir at Chinchilla on 8 November 1997 (structure Ref. No. MEL-WE-1). (a) View from the right bank. (b) View from downstream. Note the low-turbulence in the stream.

Fig. A4.8 Minimum Energy Loss weir at Lemontree on 8 November 1997 (structure Ref. No. MEL-WE-2).

A4.3 Examples of Minimum Energy Loss culverts and waterways

Introduction

In the coastal plains of Queensland (north-east of Australia), torrential rains during the wet season place a heavy demand on culverts. Furthermore, the natural slope of the flood plains is often very small ($S_o \sim 0.001$) and little fall (or head loss) is permissible in the culverts.

Professors G.R. McKay and C.J. Apelt developed and patented the design procedure of minimum energy loss waterways. These structures were designed with the concept of minimum head loss. Professor C.J. Apelt presented an authoritative review of the topic (Apelt 1983) and a documentary (Apelt 1994).

In the following sections, practical examples are described and illustrated.

Examples of MEL waterways

Ref. No.	MEL-W-1
Description	MEL waterway along Norman Creek (Fig. A4.9)
Location	Brisbane, Australia
Purpose	Passage of Norman Creek underneath the South-East Freeway and along Ridge street, Greenslopes
Design discharge	$200 \, \text{m}^3/\text{s}$ (to $250 \, \text{m}^3/\text{s}$ without flooding Ridge Street)
Geometry	Throat width: $\sim 10 \, \text{m}$
	Throat height: $\sim 4 \, \text{m}$
Notes	Excellent design
Ref. No.	MEL-W-2
Description	MEL waterway under Nudgee Road bridge (built in 1968–1969) (Fig. A4.10)

Fig. A4.9 Minimum Energy Loss waterway along Norman Creek (structure Ref. No. MEL-W-1) (September 1996). View from downstream.

(a)

(b)

Fig. A4.10 Minimum Energy Loss waterway beneath Nudgee Road bridge (structure Ref. No. MEL-W-2) (14 September 1997). (a) Detail of the throat looking downstream, Nudgee Road bridge in the foreground with the new Gateway freeway bridge over the outlet in the background. Note the cattle wandering in the waterway and the absence of lining. (b) Outlet view, looking upstream at the new freeway bridge above the outlet, the Nudgee Road bridge in background and the waterway throat underneath.

Location	Brisbane, Australia
Purpose	Crossing of Nudgee Road above Schultz canal connecting Kedron Brook to Serpentine Creek
Design discharge	$850 \, \text{m}^3/\text{s}$
Design head	Flood level: R.L. 4.755 m (at bridge centreline)
	Natural ground level: R.L. 3.048 m (at bridge centreline)
Geometry	Barrel width: $\sim$137 m (bridge length)
	Barrel length: $\sim$18.3 m
	Barrel invert drop: 0.76 m (below natural ground level)
	Culvert length: $\sim$245.3 m (including inlet and outlet, 122 m long each)
	Natural bed slope: 0.00049
Note	Excellent design. Grassed waterway. Design discharge was the 1:30 year flood observed in 1968. Tested during torrential rains in 1970, 1972 and 1974. No scour observed. Natural flood plain width: 427 m.
	Prior to 1968 the old waterway capacity was $198 \, \text{m}^3/\text{s}$ before overtopping the bridge.

Examples of MEL culverts

The information is regrouped in two paragraphs. First, efficient designs are described. Each structure has been subjected to large floods and has operated properly. Then poor (inefficient) designs are listed. Experience should be learned from each case.

Efficient designs

Ref. No.	MEL-C-1
Description	MEL culvert underneath the South-East Freeway (built in 1971) (Fig. A4.11)
Location	Brisbane, Australia
Purpose	Passage of Norman Creek underneath the South-East Freeway and Marshall Road, Holland Park West
Design discharge	$170 \, \text{m}^3/\text{s}$ (Disaster flow: $250 \, \text{m}^3/\text{s}$)
Geometry	Multi-cell box culvert (2 cells)
	Culvert length: 146 m (including inlet and outlet)
Note	Excellent design.
	Culvert located in a bend, underneath a major traffic intersection with the freeway above (the intersection).

Ref. No.	MEL-C-2
Description	MEL culvert underneath the South-East Freeway (built in 1971) (Fig. A4.12)
Location	Brisbane, Australia
Purpose	Passage of Norman Creek underneath the South-East 'Freeway and parallel to Birdwood Street, Holland Park West

Fig. A4.11 MEL culvert along Norman Creek, below the South-East Freeway (structure Ref. No. MEL-C-1) (November 1996). View of the outlet. Note the freeway embankment and freeway bridge across a traffic intersection.

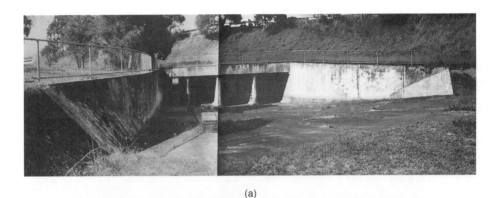

(a)

(b)

Fig. A4.12 MEL culvert along Norman Creek, below the South-East Freeway (structure Ref. No. MEL-C-2) (November 1996). (a) View of the inlet. Birdwood Street is on the left. (b) View of the outlet. Note the grassed outlet invert.

Fig. A4.13. MEL culvert along Norman Creek, below Ridge Street (structure Ref. No. MEL-C-3) (September 1996). View of the inlet.

Design discharge	$170\,\mathrm{m^3/s}$ (Disaster flow: $250\,\mathrm{m^3/s}$)
Geometry	Multiple-cell box culvert (4 cells)
	Cell width: $\sim 2\,\mathrm{m}$
	Cell height: $\sim 3\,\mathrm{m}$
	Barrel length: $\sim 110\,\mathrm{m}$
Note	Excellent design.
	Structure located immediately downstream of the MEL culvert Ref. No. MEL-C-1, both being about 1–2 km upstream of the MEL waterway Ref. No. MEL-W-1.
Ref. No.	MEL-C-3
Description	MEL culvert along Norman Creek (Fig. A4.13)
Location	Brisbane, Australia
Purpose	Passage of Norman Creek underneath Ridge Street, Greenslopes
Design discharge	$\sim 220\,\mathrm{m^3/s}$
Geometry	Multi-cell culvert: 7 cells (precast concrete boxes)
	Cell width: $\sim 2\,\mathrm{m}$
	Cell height: $\sim 3.5\,\mathrm{m}$
Note	Excellent design.
	Culvert located immediately downstream of the MEL waterway Ref. No. MEL-W-1, both structures being downstream of the MEL culvert Ref. No. MEL-C-2. Bicycle path located in one cell. Low-flow drain located in another cell.
Ref. No.	MEL-C-4
Description	MEL culvert at Wynnum underneath the Gateway Arterial (Fig. A4.14)

(a)

(b)

Fig. A4.14 MEL culvert at Wynnum underneath the Gateway Arterial on 14 September 1997 (structure Ref. No. MEL-C-4), (a) Inlet view. (b) View of the outlet. Note a burned car at the exit of one cell.

Location	Brisbane, Australia
Purpose	Crossing of the Gateway Arterial. 0.3 km South of Wynnum Road
Design discharge	$\sim$220 m^3/s
Geometry	Multi-cell box culverts (6 cells)
	Cell width: $\sim$3 m
	L_{inlet} $\sim$34 m
	Cell height: $\sim$3 m
	B_{max} $\sim$62 m
Note	Good design. But the low-flow drain does not work properly and water ponding often occurs.

Fig. A4.15 MEL culvert underneath the Gateway Arterial on 14 September 1997 (structure Ref. No. MEL-C-5). Inlet view. Note car passing along the Gateway freeway.

Ref. No.	MEL-C-5
Description	MEL culvert at Wynnum underneath the Gateway Arterial (Fig. A4.15)
Location	Brisbane, Australia
Purpose	Crossing of the Gateway Arterial; 0.6 km North of Wynnum Road
Design discharge	$\sim 100\,\mathrm{m^3/s}$
Geometry	Multi-cell box culverts (11 cells)
	Cell width: $\sim$1.8 m
	Cell height: $\sim$1.5 and 1.7 m
	L_{inlet} $\sim$60 m
	$B_{\max}$ $\sim$90 m
	L_{outlet} $\sim$75 m
Note	Good design. But the outlet system is not optimum because the outlet lip is followed by a drop of ground level.
Ref. No.	MEL-C-6
Description	MEL culvert at Redcliffe peninsula (built in 1959) (Fig. A4.16)
Location	Redcliffe QLD, Australia
Purpose	Passage of Humpy Bong Creek underneath a shopping centre parking, and prevention of salt intrusion from the sea into the creek (during dry periods)
Design discharge	$25.8\,\mathrm{m^3/s}$
Geometry	One rectangular cell
	Barrel width: 5.48 m
	Barrel height: $\sim$3.5 m
	Barrel length: 137.2 m
	Barrel invert slope: 0.0016
	Barrel invert drop: 1.16 m
	L_{inlet} = 30.5 m
	L_{outlet} = 30.5 m
	Culvert length: 198.1 m (including inlet and outlet)

(a)

(b)

Fig. A4.16 MEL culvert at Redcliffe peninsula in September 1996 (structure Ref. No. MEL-C-6). (a) View of the barrel entrance and inlet from upstream. The inlet is designed as a minimum energy loss weir to prevent salt intrusion into the creek as well as to maintain a recreational lake upstream. (b) Inlet view from the road with the upstream lake and fountains.

Note   Excellent design. Fulfils completely the design
                        expectations. The culvert is regularly used.
                        Culvert inlet combined with a minimum energy loss
                        weir to maintain a recreational lake upstream of the
                        culvert and to prevent salt water intrusion into the
                        creek (at high tide levels). The culvert discharges
                        directly into the sea.

(c)

Fig. A4.16 (c) View of the outlet with the sea in the background.

Poor designs

Ref. No.   MEL-C-X1
Description   MEL culvert along Norman Creek (Fig. A4.17)
Location   Brisbane, Australia
Purpose   Passage of Norman Creek underneath Cornwall Street, Stones
                    Corner
Design discharge $220\,\mathrm{m}^3/\mathrm{s}$ (?)
Geometry   Multi-cell culvert: 9 cells (precast concrete boxes)
                    Cell width: ~2 m
                    Cell height: ~2.8 m
                    Barrel length: 16.1 m
                    Barrel invert slope: 0.004
                    Barrel invert excavation: ~1.2 m (below natural ground level)
Note   Poor design. Cornwall Street is often overtopped during
                    torrential rains.
                    Culvert located downstream of the MEL culvert Ref. No.
                    MEL-C-3.
Discussion   The culvert is not properly aligned with the natural creek. The
                    barrel centreline is oriented almost 90° off the incoming stream
                    (see photograph). Furthermore, the inlet and outlet are both
                    extremely short.
                    During flood periods flow separation is observed at the inlet
                    (near the left bank) (Hee 1978) and four cells become ineffective
                    (i.e. recirculation region). As a result, Cornwall Street becomes
                    overtopped. Further heavy siltation is observed after the floods.
                    The real discharge capacity of the culvert is probably about
                    $120\,\mathrm{m}^3/\mathrm{s}$ (before Cornwall Street overtopping).

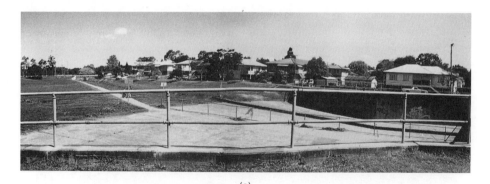

(a)

(b)

Fig. A4.17 MEL culvert along Norman Creek, below Cornwall Street in September 1996 (structure Ref. No. MEL-C-X1). (a) View of the right bank of the inlet, Norman Creek and a bicycle path. Norman Creek is on the right of the bicycle path and it connects to the low-flow channel. (b) View of the outlet (from the right bank).

A4.4 Computer calculations of standard culvert hydraulics

Introduction

Computer programs have recently been developed to calculate the hydraulic characteristics of standard culverts. Most assume a known culvert geometry and compute the free-surface elevation in the barrel, at the inlet and at the outlet (e.g. HEC-6, HydroCulv).

Fig. A4.18 Box culvert experiment. Design flow conditions: $Q_{des} = 10\,\text{L/s}$, $d_{tw} = 0.038\,\text{m}$, $S_o = 0.0035$, $B_{min} = 0.15\,\text{m}$, $D = 0.107\,\text{m}$, $L_{culv} = 0.5\,\text{m}$. Flow from bottom right to top left. View from upstream at design flow conditions: $Q = 10\,\text{L/s}$, $d_{tw} = 0.038\,\text{m}$, $d_1 = 0.122\,\text{m}$.

Discussion

Before using a computer program for hydraulic calculations, the users must know the basic principles of hydraulics. In the particular case of culvert design, they must further be familiar with the design procedure (Chapter 19) and they should have seen culverts in operation (i.e. laboratory models and prototypes).

As part of the hydraulic course at the University of Queensland, the author developed a pedagogic tool to introduce students to culvert design. It is a culvert laboratory study, which includes an audiovisual presentation, a physical model of a box culvert and numerical computations using the program HydroCulv. During the afternoon, students are first introduced to culvert design (video-presentation, Apelt 1994) before performing measurements in a standard box culvert model. The model is in perspex and the students can see the basic flow patterns in the channel, barrel, inlet and outlet (Fig. A4.18). Then the same flow conditions are input in the program HydroCulv, and comparisons between the physical model and computer program results are conducted. Later, the comparative results are discussed in front of the physical model (see below).

Audio-visual material

Apelt, C.J. (1994) '*The Minimum Energy Loss Culvert.*' Videocassette VHS, colour, Department of Civil Engineering, University of Queensland, Australia, 18 Minutes.

Physical model
The physical model characteristics are: $Q_{des} = 10\,\text{L/s}$, $B_{max} = 1\,\text{m}$, $B_{min} = 0.15\,\text{m}$, $D = 0.11\,\text{m}$, $L_{curv} = 0.5\,\text{m}$, $d_{tw} = 0.038\,\text{m}$, intake and outlet: $45°$ diffuser, $S_o = 0.0035$.

Computer program
HydroCulv Version 1.1.

Use of HydroCulv

A simple tool for culvert calculations is the program HydroCulv, designed as a Windows-based software. The program is based on solid hydraulic calculations and the outputs are easy to use. It is, furthermore, a shareware program that may be obtained from:

HydroTools Software: dwilliams@compusmart.ab.ca

Inputs
The inputs of the program are the geometry of the culvert and the hydraulic parameters. The geometric characteristics include barrel shape, height, width, length and slope. The culvert barrel is assumed prismatic and must have a constant bed slope. The hydraulic parameters include the energy loss coefficients and the boundary conditions (e.g. discharge, tailwater depth).

Outputs
The output may vary depending upon the boundary conditions (i.e. normal flow in flood plain or specified head elevation). Basically, the main result is the free-surface profile in the barrel (e.g. Fig. A4.19). In addition, the results show whether the culvert flow conditions are inlet control or outlet control. A graphical solution compares the computed free-surface profile, critical depth and normal depth in the barrel.

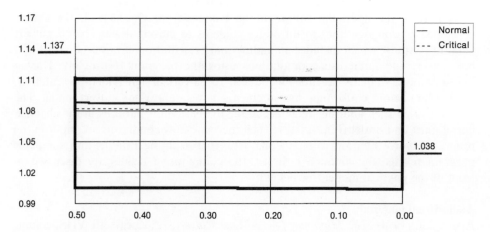

Fig. A4.19 Output of HydroCulv for the flow conditions 1: free-surface profile in the barrel.

Discussion

The calculations depend *critically* upon the energy loss coefficients entered for the calculations. Three types of loss coefficients must be input:

- roughness
  The calculations may be performed with the Gauckler–Manning coefficient 'n', the Darcy friction factor 'f' or the Keulegan friction coefficient 'k', related to the Chézy coefficient as: $C_{\text{Chézy}} = 18 \log_{10}(D_H/k) + 8.73$ (see Chapter 4).
- entrance loss coefficient
  The entrance head loss equals: $\Delta H = K_1' V^2/2$, where K_1' is the entrance loss coefficient and V is the barrel velocity. For a pipe intake, $K_1' = 0.5$ and a similar value is reasonable for a culvert intake.
- exit loss coefficient
  At the culvert outlet, the exit loss equals: $\Delta H = K_2' V^2/2$, where K_2' is the exit loss coefficient and V is the barrel velocity. $K_2' = 1$ for a pipe exit and this value might be used as an order of magnitude for the culvert outlet.

Application

Consider a simple application of the program HydroCulv. The calculations input are as follows.

HydroCulv parameter	HydroCulv input	HydroCulv units	Remarks
U/S Invert Elevation	1.004768	m R.L.	
D/S Invert Elevation	1.00301	m R.L.	
Culvert length	0.5	m	
Roughness, 'n'	0.01	SI units	n_{Manning} in s/m$^{1/3}$ for Perspex
Entrance loss coefficient	0.5	–	K_1'
Exit loss coefficient	1	–	K_2'
Culvert height	0.107	m	Barrel height D
Culvert width	0.15	m	Barrel width B_{min}
Shape	BOX		Box culvert
Boundary conditions			
Specified head elevation	Q (cms)	TW head elevation (m)	
	0.010	1.038	Conditions 1 (i.e. design flow)
	0.010	1.105	Conditions 2

Comparison of physical model–computer results

The same flow conditions (i.e. boundary conditions) were tested in the physical model. For the design flow conditions 1, the physical model has a free-surface barrel flow, a submerged entrance and inlet control. For the flow conditions 2, the flow pattern in the physical model is: free-surface barrel flow, free-surface inlet flow and inlet control.

Comparative results between the physical model and computer program are summarized below.

	Physical model tests	HydroCulv results
Flow conditions 1	Submerged entrance Free-surface barrel flow Inlet control $H_1 = 1.128\,\text{m}$	Free-surface entrance flow Free-surface barrel flow Inlet control $H_1 = 1.137\,\text{m}$
Flow conditions 2	Submerged entrance Pipe flow in barrel Outlet control $H_1 = 1.133\,\text{m}$	Free-surface entrance flow Hydraulic jump in barrel Inlet control $H_1 = 1.336\,\text{m}$

Basically, close agreement is noted at design flow conditions. But, with non-design flow conditions (e.g. conditions 2), some differences may be observed.

From the author's experience, the above example is typical of the agreement between culvert flow conditions and computer program results. In any situation for which great accuracy is required, design engineers are strongly encouraged to conduct a physical model test.

Note
The above tests were performed with the version 1.1 (32 bits), June 1997.

Problems

P1

A study of the Marib dam and its sluice system (115 BC–575 AD)

Preface

The following problem covers lecture material relevant to Basic Principles (Part I), Sediment Transport (Part II), and Physical Modelling and Numerical Modelling (Part III). The assignment was first used in 1992. Based upon real historical facts, the geometry of the hydraulic structures has been simplified to facilitate the calculations.

P1.1 Introduction

P1.1.1 Historical background

The site of the Marib dam is located on the Wadi Dhana, upstream of the ancient city of Marib in northern Yemen (Fig. P1.1). Marib was the capital over which Queen of Sheba (or Sabah) ruled around 950 BC. The Queen of Sheba is mentioned in the Old Testament for her famous visit to King Solomon in Jerusalem. The Kingdom of Sheba was prosperous and its power was based on trade, by both sea (to and from India and the Persian Gulf) and land routes (the spices roads in Arabia) (LeBaron Bowen and Albright 1958, Phillips 1955). Marib, the capital city of the kingdom, was the focal point of the trade routes, and the Sabaens built the Marib dam to irrigate the land around the city. The dam was considered one of the wonders of the ancient world and played an important role. The Kingdom of Sheba is described as a fertile land. The land was cultivated by controlling flood water in the 'wadis', and had fertile gardens and fields with fruits and spices (Eurenius 1980).

The Sabaens developed large irrigation systems in south-west Arabia using earth dams to divert flood waters into the land. Such waters contained a large amount of silt, sand and gravel sediments.

The dam was built in order to intercept the floods in the Wadi Dhana. Its purpose was to raise the level of the wadi's flow during periods of run-off, following a fall of rain in the mountains and to divert this water into the canal systems which took the waters to the city (Fig. P1.2).

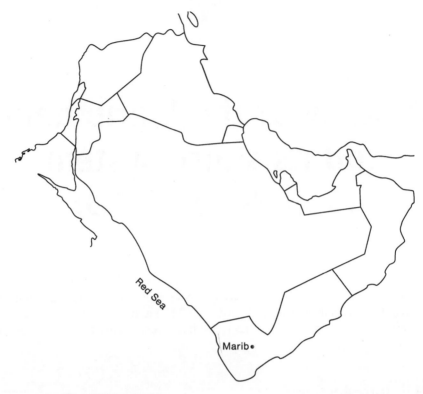

Fig. P1.1 Map of the Arabic peninsula.

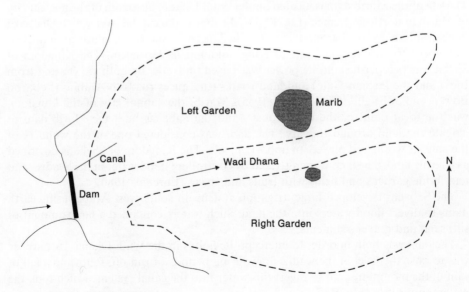

Fig. P1.2 Sketch of the Marib area.

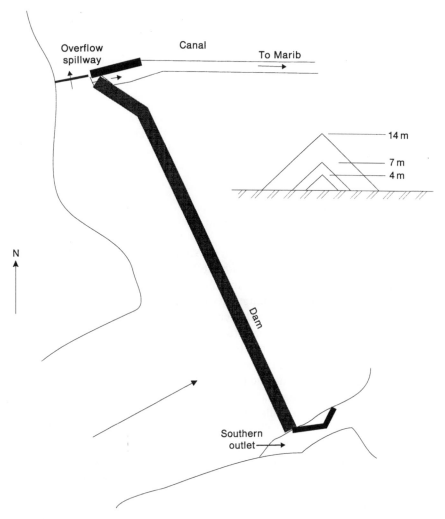

Fig. P1.3. Sketch of the successive dams on the site shown in plan and cross-section.

P1.1.2 Dam construction

The first Marib dam was built around 750 BC. The dam was a simple earth structure, 580 m in length and is estimated to have been 4 m high. It ran straight across the wadi between high rocks on the southern side to a rock shelf on the northern bank. The dam was built slightly downstream of the narrowest point in the Wadi Dhana to allow the space for a natural spillway and sluices between the northern end of the dam and the high rocky cliff to the west (Fig. P1.3).

Around 500 BC the dam was heightened. The second structure was a 7 m high earth dam. The cross-section was triangular with both faces sloping at 45° (Fig. P1.3). The water face (i.e. the upstream slope) was covered with stones set in mortar to make the dam watertight and to resist the erosive effects of waves (Smith 1971).

The final form of the Marib dam was reached after the end of the Kingdom of Sheba. From 115 BC onwards the ruling people in southern Arabia were the Himyarites and the next major reconstruction appears to be a Himyarite work (Smith 1971). This reconstruction led to a new 14 m dam with elaborate water works at both ends (Fig. P1.3).

Southern sluice system

The southern sluices system was known as 'Marbat el-Dimm'. The spillway was an overfall located about 7 m below the top of the dam, and the spillway width was 3.5 m.

Northern sluice system

The northern system of sluices included an overfall spillway and a channel outlet (Fig. P1.3).

The channel outlet was located between the spillway and the earth dam, and it contained one great wall over 140 m long, about 9 m thick and from 5 to 9 m high. This system conducted water from the northern end of the dam through a single 1000 m long and 12 m wide canal to a rectangular structure which divided the water entering into 12 different streams (LeBaron Bowen and Albright 1958).

At the dam the discharge flowing into the canal was controlled by two gates. The overfall gates were 2.5 m wide, and the maximum gate openings were 1 m and 4 m below the dam crest. In the initial section of the canal, the canal walls widened to form a basin of about 23 m wide and 65 m long (essentially the length of the northern wall) which acted as a settling basin for heavier material carried by the wadi. The 1000 m long canal had earth walls which were covered with a cemented stone lining on the inside (i.e. water side). The cross-sections of the canal and of the settling basin were approximately rectangular.

The spillway was a 40 m wide overfall with a broad unlined rock crest. The crest elevation of the overflow spillway was located 3 m below the height of the dam and the spillway was not gated. As the crest was 3 m below the crest of the dam, it is believed that the reservoir volume corresponding to the top 3 m of the dam was used to damp the flood and flood catchment.

End of the Marib dam

The dam suffered numerous breachings caused by overtopping and the maintenance works were substantial. In 575 AD, the dam was overtopped and never repaired. The final destruction of the dam was a milestone in the history of the Arabic peninsula. The fame of the Marib dam was such that its final destruction is recorded in the Koran.

'New' Marib dam

In 1986 a new 38 m high earth dam was built across the Wadi Dhana. The dam site is located 3 km upstream of the ruins of the old Marib dam (Ganchikov and Munavvarov 1991). The new dam was designed to store water for irrigating the Marib plains.

P.1.1.3 Chronological summary

Date	Comments
750 BC	Construction of the 1st Marib dam
	4 m high earth dam. Length: 580 m. Original northern spillway and canal sluices. Irrigation of the northern bank of the wadi.
500 BC	Second structure: the initial dam was heightened
	7 m high earth dam. Triangular cross-section with both faces sloping at 45°. Water face covered with stones set in mortar. Construction of southern sluices. Irrigation of both the northern and southern side.
After 115 BC	Dam reconstruction (Himyarite work)
	New 14 m high earth dam. New water works at both ends. Northern water works: raise of the spillway floor, construction of five spillway channels, two masonry sluices, masonry tank used as settling pond, and a paved channel (1000 m long) to a distribution tank.
325 AD	Final form of the northern outlet works
449 AD	Flood. Dam restoration
450 AD	Flood. Dam breach. Complete renovation
542 AD	Flood. Dam breach
543 AD	Dam restoration
575 AD	Destruction of the dam. The dam was never restored.
1984–86 AD	Construction of a new 38 m high earth dam

P.1.1.4 Bibliography

Books

LeBaron Bowen Jr, R. and Albright, F.P. (1958) *Archaeological Discoveries in South Arabia.* (The John Hopkins Press: Baltimore, USA).

Smith, N. (1971) *A History of Dams.* (The Chaucer Press: Peter Davies, London, UK).

Phillips, W. (1955) Qataban and Sheba — *Exploring Ancient Kingdoms on the Biblical Spice Routes of Arabia.* (Victor Gollancz: London, UK).

Schnitter, N.J. (1994) *A History of Dams: the Useful Pyramids.* (Balkema: Rotterdam, The Netherlands).

Journal articles

Eurenius, J. (1980) Ancient Dams of Saudi Arabia. *International Water Power and Dam Construction,* **32**(3), 21–22.

Ganchikov, V.G. and Munavvarov, Z.I. (1991) The Marib Dam (history and the present time). *Gidrotekhnicheskoe Stroitel'stvo,* No. 4, pp. 50–55 (in Russian). (Translated in *Hydrotechnical Construction,* 1991, Plenum, pp. 242–248).

Hathaway, G.A. (1958) Dams – their effect on some ancient civilizations. *Civil Engineering,* **28**(1), 58–63.

Schnitter, N.J. (1967) A short history of dam engineering. *Water Power,* **19**, 142–148.

P1.2 Hydraulics problem

Introductory note

In the problem we shall consider the final stage of the dam and its auxiliary structures (e.g. spillways, channel) with a dam height of 14 m.

P1.2.1 Study of the spillways

In the following questions, let us assume frictionless flows.

(A) Draw a sketch of the dam, both spillways and the outlet, view from the reservoir. Indicate the main dimensions on the sketch.

(B) *Southern spillway*
(B.1) We assume that the water level in the reservoir is equal to the crest of the northern spillway, and we assume that the canal gates are closed (i.e. no discharge in the canal). Using the continuity equation, what is the formula giving the total discharge Q as a function of the spillway width and the flow depth above the crest? Explain in words how this formula is obtained.
(B.2) Application: what is the maximum discharge of the southern spillway for this flow configuration?

(C) *Northern spillway*
(C.1) If the flow depth above the crest is 1 m, what is the total discharge above the spillway? Explain in words your calculations.
(C.2) What is the maximum flood discharge that the northern spillway can absorb before the dam becomes submerged? Explain in detail your calculations.

(D) We assume that the canal gates are closed (i.e. no discharge in the canal). What are the maximum discharge capacities of the southern bank spillway, the northern bank spillway and both spillways, before the dam is overtopped?

P1.2.2 Study of the channel outlet and settling basin

(A) *Basic hydraulic knowledge*
(a) What are the assumptions used to derive the Bernoulli equation? Write the Bernoulli equation.
(b) What is the definition and significance of the Coriolis coefficient? Rewrite the Bernoulli equation using the Coriolis coefficient.
(c) Assuming that the velocity distribution follows a power law:

$$\frac{V}{V_{max}} = \left(\frac{y}{d}\right)^{1/N}$$

Give the relation between the maximum velocity V_{max} and the average flow velocity. Give the expression of the Coriolis coefficient as a function of the exponent of the power law.

(B) *Change in channel width*
Consider a horizontal, rectangular channel with a change of channel width, and assume a frictionless flow.
(a) Write the continuity and Bernoulli equation for the channel.
(b) Combining the continuity equation and the Bernoulli equation, develop the differentiation of the Bernoulli equation, along a streamline in the s-direction. Introducing the Froude number, how do you rewrite the differentiation of the Bernoulli

equation in term of the Froude number? How would the flow depth vary when the channel width varies? Explain your answer in words.

(c) The upstream flow conditions are subcritical. What is the optimum width such that the flow becomes critical? Explain each step of your calculations.

(d) We consider now the flow entering into the settling basin. The upstream flow conditions are: $B = 12\,\text{m}$, $d = 1.7\,\text{m}$, $Q = 25\,\text{m}^3/\text{s}$. What are the flow conditions in the settling basin? Explain in words your calculations. Discuss the aim of the settling basin. Why did the designers adopt a settling basin at this location?

(C) *Settling basin*

We consider the same flow entering the settling basin with the following inflow conditions (upstream of the widening) $B = 12\,\text{m}$, $d = 1.7\,\text{m}$, $Q = 25\,\text{m}^3/\text{s}$. We shall assume that the flow conditions in the settling basin are those calculated above (by assuming a smooth transition between the upstream flow and the settling basin).

During a flood event, the canal inflow is usually heavily sediment-laden. Taking into account the bed roughness, we will consider the sediment motion of the inflow and of the settling basin flow.

(a) Compute the bed shear stress in the inflow channel and in the settling basin.

(b) Find out the critical particle size for bed-load motion in the inflow channel and in the settling basin. Discuss the findings.

(c) Calculate the critical particle size for bed-load motion in the inflow channel and in the settling basin.

(d) Discuss your results. For example, what is the heaviest bed-load particle size that could enter the 1000 m long canal for the above flow conditions.

P1.2.3 Study of the canal

(A) *Model study*

The King of Sheba wants a study of the 1000 m canal. He orders an (undistorted) scale model study where the length of the model is 40 m.

The prototype flow conditions to be investigated are:

$$\{1\}\ d = 1\,\text{m};\ Q = 12\,\text{m}^3/\text{s} \qquad \{2\}\ d = 3\,\text{m};\ Q = 72\,\text{m}^3/\text{s}$$

where d is the flow depth and Q the total discharge.

(a) What type of similitude would you choose? What is the geometric scale of the model? Explain and discuss your answer with words.

(b) For both prototype flow conditions, what is(are) the width(s) of the scale model(s)?

(c) For both prototype flow conditions, what are the model flow depth(s), flow velocity(ies) and total discharge(s)? Explain carefully your answer in words.

(d) Determine the minimum prototype discharge for which negligible scale effects occurs in the model.

(B) *Normal flow conditions*

The 1000 m long canal is a paved channel of 12 m width and the slope is 0.1°.

(a) What values would you choose for the (equivalent sand) roughness height and the Gauckler–Manning coefficient? Discuss and justify your choice.

(b) Considering a uniform equilibrium flow, develop the momentum equation, expressing the bed shear stress as a function of: (1) the Darcy coefficient, and (2) the Gauckler–Manning coefficient. Explain in words your development.

(c) The water discharge in the canal is $25\,\mathrm{m^3/s}$.

(c1) For the selected roughness height, what are the values of the Darcy coefficient and the normal depth computed from the Darcy coefficient? Explain and discuss your calculations.

(c2) What is the uniform flow depth using the formula of the uniform depth as a function of the Gauckler–Manning coefficient?

(c3) Discuss your results and the eventual discrepancies between the results of the above questions. Compare both the friction coefficients and the flow depth.

(C) *Backwater calculations*

We consider now the gradually varied flows of real fluid in the settling basin and in the 1000 m canal.

(a) Write the differential form of the energy equation in terms of the flow depth and local Froude number for a channel of variable width and variable slope. Comment on your result.

(b) What are all the assumptions made in order to obtain the above equation?

(c) Explain (in words) how you would integrate the resulting differential equation. Where would you start the calculations.

(d) Table P1 provides you the geometric characteristics of the 1000 m canal and the settling basin at the start of the canal. We consider a flow rate of $Q = 60\,\mathrm{m^3/s}$.

(d1) Plot the canal profile on graph paper (dimensioned sketched).

(d2) Locate the occurrence of critical flow conditions and hydraulic controls (if any).

(d2) Select the position where you will start your calculations.

(d3) Compute the flow depth, flow velocity, Darcy coefficient, Froude number and friction slope at all the positions defined in Table P1).

(d4) Plot the backwater curves (i.e. the flow depth curves) on the graph paper.

Notes

1. Question (d3) is very important for correct calculations.
2. Flow resistance calculations must be performed using the Darcy friction coefficient.
3. Assume a linear variation of the channel width B, the channel slope θ and roughness height k_s between each position (Table P1).

Table P1 Canal geometry

Point	Bed elevation (m)	s (m)	Roughness	θ (degree)	B (m)	Remarks
A	0		$k_s = 0.5\,\mathrm{mm}$	0	12	
B		30	$k_s = 0.5\,\mathrm{mm}$	0	23	Start of the settling basin
C		65	$k_s = 0.5\,\mathrm{mm}$	0	23	
D		95	$k_s = 0.5\,\mathrm{mm}$	0	23	End of the settling basin
E		120	Paved channel (*)	0.1	12	Start of the canal
F		900	Paved channel (*)	0.1		
G		901	Paved channel (*)	3.0	12	
H		1060	Paved channel (*)	3.0	12	End of the canal

Note: (*) Students are asked to use the roughness height of a paved channel previously obtained in question B and to compute the flow resistance using the Darcy equation.

P1.3 Hydrological study: flood attenuation of the Marib reservoir

The objective of this exercise is to estimate the peak flow rate (for the undammed river) of the flood that would have just caused overtopping of the 14 m high old Marib dam. To do this, the hydraulic performances of the two old spillways must first be understood (Section P1.2.1).

Information about the surface area of the old Marib reservoir is not easily available, but reasonable scaling down from the 30 km^2 lake that now serves the much larger new Marib dam, suggests that the old lake would have been about 4 km^2.

Assumptions

Assume that this area applies with the lake free surface level the same as the crest of the 40 m wide northern spillway, which is also the assumed initial condition for the approaching flood wave.

As the water level rises to overtop the old dam, assume that the lake area increases linearly to 5.5 km^2.

Assume that the time-of-concentration for this catchment is about 12 hours (note: this value is considerably less than the typical time-to-peak for the Brisbane river which has almost exactly the same catchment area, but the Marib catchment is very sparsely vegetated in relation to the local catchment).

A study of the Moeris reservoir, the Ha-Uar dam and the canal connecting the Nile river and Lake Moeris around 2900 BC to 230 BC

Preface

The following problem covers lecture material relevant to Basic Principles (Part I).

The assignment was first used in 1993. Based upon real historical facts, the geometry of the hydraulic structures has been simplified to facilitate the calculations.

P2.1 Introduction

P2.1.1 Presentation

In the history of dams and the story of civilizations, the first dams were built in Egypt and Iraq around BC 3000 where they controlled canals and irrigation works. Often, the civilizations originated in areas where irrigation was a necessity. The history of dams followed closely the rise and fall of civilizations, especially where these depended on the development of the water resources. A typical example is the Egyptian civilization. For centuries, the prosperity of Egypt relied on the annual flood of the Nile river from July to September and the irrigation systems. One of the most enormous efforts of the Egyptian Kings was the creation of the Lake Moeris in the Fayum depression and the construction of a 16 km long canal connecting the lake to the Nile. The lake was used to regulate the Nile river and as a water reservoir for irrigation purposes.

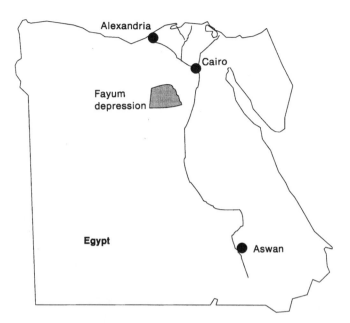

Fig. P2.1 Map of Egypt and the Nile river.

P2.1.2 The Moeris Lake[1]

The location of the Lake was the Fayum (or Fayoum) depression, located 80 km south-west of the city of Cairo (Fig. P2.1). The depression is a vast area of 1700 km^2 and the lowest point in the depression is 45 m below sea level.

Egyptian engineers connected the Nile river and the Fayum depression to lead flood flows into the depression during high floods. The connection between the river and the depression was a natural cut in the mountains. It was in existence at the time of the King Menes, founder of the 1st Egyptian dynasty (BC 2900). At that time, the Fayum depression contained only a natural lake filled from the Nile during large floods.

King Amenembat (BC 2300) widened and deepened the canal between the Nile and the Fayum depression. He converted the existing lake into an artificial reservoir (i.e. Lake Moeris) which controlled the highest flood of the Nile. The canal connecting the river to the reservoir was regulated by the Ha-Uar dam. The regulation system consisted of two earthen dams at both ends of the canal, with gates by means of which the architects regulated the rise and fall of the water.

Lake Moeris had three main purposes:

1. the control of the highest floods of the Nile during the July–September periods;
2. the regulation of the Nile river during the dry season by releasing water from the Lake; and
3. the irrigation of a large surface area around the Lake.

---

[1] There are still some arguments upon whether Lake Moeris existed or not. There are suggestions that there were in fact two lakes (e.g. Schnitter 1994, pp. 4–6).

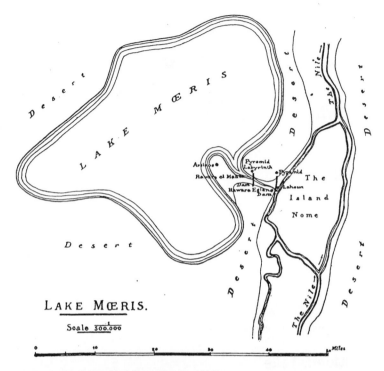

Fig. P2.2 Sketch of the Lake Moeris (after Willcocks 1919).

From its size and depth, Lake Moeris was capable of receiving the overflow of the Nile during its rising, and preventing the flooding of downstream cities like Memphis (see Fig. P2.1). When the river fell, the lake waters discharged again via the canal connecting the river to the lake, and these waters were available for irrigation.

The Ha-Uar dam had also a strategic interest. At the time when Egypt was divided in two kingdoms, Lower Egypt (i.e. Northern Egypt) and Upper Egypt (i.e. Southern Egypt), the frontier fortress of Lower Egypt was at the head of the Lake Moeris canal (Hathaway 1958). The capture of the dams controlling the canal, and the injudicious or malicious use of the reservoir could deprive a great part of the Lower Egypt (i.e. Northern Egypt) of any basin irrigation at all, for such irrigation utilized only the surface waters of the Nile flood. The importance of the fortress commanding the regulator of the canal ceased when the kingdoms were re-united.

The abandonment of the Lake Moeris was primarily caused by the fact that the Lahoun branch of the Nile (i.e. the West branch of the river, Fig. P2.2) dwindled in size and reduced the use of the reservoir. From BC 230, the canal was abandoned and the area inundated by Lake Moeris became the province of Fayoum as it is today. Nile shells can still be found in the Fayoum area near the limits of the ancient Lake Moeris (Willcocks 1919).

Geometrical characteristics

Lake Moeris was located in the Fayoum (or Fayum) depression. The geometry of the Lake was:

Lowest point	45 m below the sea level
Mean free surface level	22.5 m above sea level
Capacity	$50 \times 10^9 \, \text{m}^3$ of water (50 TL)
Surface area of the lake	$1700 \times 10^6 \, \text{m}^2$ (17 Ma)
Irrigated land surface area	$1.46 \times 10^9 \, \text{m}^2$ (14.6 Ma)

The flood regulation capacity of the Lake was $13 \times 10^9 \, \text{m}^3$ of water each year, and $3 \times 10^9 \, \text{m}^3$ extra for every year it was not used. As an example, if the reservoir was used only every two years, it was capable of taking from a flood: $13 \times 10^9 + 3 \times 10^9 = 16 \times 10^9 \, \text{m}^3$ of water.

Climatic conditions

The evaporation from the lake was about 2.5 m per annum in depth. When the lake was full, the evaporation was around $4.25 \times 10^9 \, \text{m}^3$ of water per year.

It must be noted that the artificial regime of the lake had an effect on the climate. The regulation of the Nile by the Moeris reservoir prevented or reduced stagnant and dirty waters downstream, and hence suffocating air in the cities downstream of the Ha-Uar dam (Belyakov 1991).

P2.1.3 The Ha-Uar dam

The canal connecting the Nile river to Lake Moeris was controlled by two dams at each end of the canal (Fig. P2.2). The two regulators were earthen dams.

As shown in Fig. P2.2, the Nile flowed in two channels opposite the head of the Lake Moeris canal. The upper regulator (i.e. eastern dam) consisted of the existing Lahoun[2] bank (i.e. west bank of the river), a broad spill channel, cut out of the rock to a suitable level for passing ordinary floods, and a massive earthen dam across a ravine, which was cut in dangerously high floods. The other end of the canal (i.e the western end) was a much simpler earth dam. During a flood, the dam was cut: the cutting was easy enough.

Fortresses and barracks were located on either sides of the two dams to protect them. Access to the eastern dam and to the eastern end of the canal was difficult: it was written that a fleet was essential to gain possession of the lower great dam (i.e. western dam) (Willcocks 1919).

The passing of very large floods was possible by cutting the dams. But their reconstruction after the passing of a flood entailed an expense of labour which even an Egyptian Pharaoh considered excessive!

Canal connecting the Nile and Lake Moeris

The canal connecting the river to the Fayum depression was initially a natural cut through the mountain bordering the Libyan desert (Fig. P2.2). The natural canal was 16 km long and 1.5 km wide (Garbrecht 1987b). This place is now called the Fayoum Bahr Yusuf Canal.

---

[2] Also spelled Lahun.

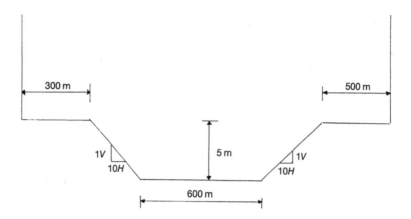

Fig. P2.3 Cross-section of the artificial canal and the natural valley (view from the Nile river).

Around BC 2300, King Amenembat started the construction of an artificial canal along the natural valley. The artificial channel was 16 km long and 5 m deep. Its shape was trapezoidal with a width 600 m at the bottom (Fig. P2.3). The slope of the banks was 1 V : 10 H to allow the use of non-cohesive rockfill and earth materials. The protection of the channel bottom and the bank slopes consisted of cut stones and cement joints. The covering blocks were placed on the bottom and on the slopes, and they were cemented together.

The average slope of the channel bed was 0.01°. The canal was inclined towards the Fayum depression.

Eastern dam

The eastern dam of the Ha-Uar dam structure blocked the valley connecting the Nile river to the Fayum depression. The axis of the valley was east–west. The dam stretched over 1550 m from the south to the north, and it consisted of three large parts.

1. In the north, the Lahoun bank blocked the natural valley over a 550 m length. The elevation of the top of the bank was 10 m above the canal bottom. The bank consisted of non-erodable material.
2. Next to the Lahoun bank, there was a broad spill channel. The channel was an ungated broad-crested weir. The weir was 400 m wide with a rectangular cross-section. The crest elevation was 3 m above the canal bottom.
3. An earthen dam blocked the southern extremity of the natural valley. The top of the dam was 9 m above the channel bottom.

A hydraulic structure was located between the broad-crested weir and the earthen dam. The structure supported a sluice gate used to release waters to the river. The gate was 5 m high and 10 m wide. The bottom of the gate opening was at the same elevation as the canal bottom.

P2.1.4 Historical events

History indicates that Joseph arrived in Egypt in the time of the Hyksos who ruled Lower Egypt (i.e. Northern Egypt) while Theban dynasties ruled Upper Egypt. It was a time of continuous wars between the two kingdoms.

One of the famines of long duration in Egypt occurred during Joseph's time and was described in the Book of Genesis. It was caused by the capture and the breaching of the Ha-Uar dam by the King of Upper Egypt (i.e. Southern Egypt). The famines were ended by the recapture and repair of the dam by the King of Lower Egypt. According to a tradition, Joseph (BC 1730) worked on the restoration of the canal and the dams. Later Jewish slaves were used for the maintenance of the works.

When the Jews fled from Egypt with Moses, stories relate that the Egyptian army was destroyed by the sea when it crossed the 'Red Sea'. In fact, the drowning of the army was probably caused by the breaching of the Ha-Uar dam.

P2.1.5 Important dates

Date	Comments
BC 2900	The connection between the Nile and the Fayum depression was in existence at the time of the King Menes, founder of the 1st Egyptian dynasty. This Egyptian King built the city of Memphis.
BC 2300	King Amenembat (12th dynasty) widened and deepened the canal between the Nile and the Fayum depression. He converted the existing lake into an artificial reservoir which controlled the highest flood of the Nile.
BC 230	Abandonment of Lake Moeris primarily due to the fact that the Lahoun branch of the Nile (see Fig. P2.2) dwindled in size and reduced the use of the reservoir.

P2.1.6 Discussion

Although there are a lot of stories about Lake Moeris, the earliest of these historians wrote his account of the lake in BC 430. There is no trace left of the western dam and only the remains of the Eastern structure may be seen (e.g. Garbrecht 1987a,b, Schnitter 1994).

In fact, there are still doubts and arguments about the existence of Lake Moeris. Schnitter (1967) mentioned the enigmatic Lake Moeris. Smith (1971) and Belyakov (1991) implied that Lake Moeris was a natural lake and not an artificial reservoir. Garbrecht (1987b) suggested that the Fayum depression was transformed by large-scale reclamation works into a fertile province around BC 1700 and was never used as an artificial reservoir. Recent studies (Schnitter 1994, Garbrecht 1996) suggested the existence of two lakes since around BC 300–250. Prior to BC 300 the water level in the Fayum depression was about 15–20 m above sea level. It dropped down to about 2 m below sea level (and later to about 36 m below sea level) as a result of land reclamation works associated with a high evaporation rate. An artificial reservoir was created around BC 300–250 to irrigate the reclaimed land. The artificial lake was in use up to the 19th century.

In any case, Willcocks (1919) and Hathaway (1958) provided a lot of evidence supporting the existence of Lake Moeris. Furthermore the constructions of the pyramids, of large temples and of a large dam (i.e. the Sadd el-Kafara dam) indicate that Egyptian engineers had the expertise and the knowledge to build large-scale civil engineering works. There is no doubt that the Egyptians were able to divert the floods of the Nile River into Lake Moeris. Furthermore, they had the technology and the engineering skills to build canals and earthen dams.

Interestingly, a new mini hydro scheme will be installed by the year 2000 (*Hydropower & Dams* 1997). The purpose of the project is the regulation of irrigation water in the depression as well as the generation of hydro-electricity using low-head Kaplan turbines.

P2.1.7 Bibliography

Books

Garbrecht, G. (1987a) *Hydraulics and Hydraulic Research: a Historical Review*. (Balkema: Rotterdam, The Netherlands).

Schnitter, N.J. (1994) *A History of Dams: the Useful Pyramids*. (Balkema: Rotterdam, The Netherlands).

Smith, N. (1971) *A History of Dams*. (The Chaucer Press: Peter Davies, London, UK).

Willcocks, W. (1919) *From the Garden of Eden to the Crossing of the Jordan*. (E & FN Spon: New York, USA).

Articles

Belyakov, A.A. (1991) Hydraulic engineering and the environment in antiquity. *Gidrotekhnicheskoe Stroitel'stvo*, No. 8, pp. 46–51 (in Russian). (Translated in *Hydrotechnical Construction*, 1992, Plenum, pp. 516–523).

Buckley, A.B. (1923) The influence of silt on the velocity of water flowing in open channels. *Minutes of the Proceedings of the Institution of Civil Engineers*, 1922–1923, **216**, Part II, 183–211. Discussion, pp. 212–298.

Garbrecht, G. (1987b) Hydrologic and hydraulic concepts in antiquity. In *Hydraulics and Hydraulic Research: a Historical Review*. (Balkema: Rotterdam, Netherlands), pp. 1–22.

Garbrecht, G. (1996) Historical water storage for irrigation in the Fayum Depression (Egypt). *Irrigation and Drainage Systems*, **10**(1), 47–76.

Hathaway, G.A. (1958) Dams — their effect on some ancient civilizations. *Civil Engineering*, **28**(1), 58–63.

Hydropower & Dams (1997) Mini hydro scheme for Egyptian oasis. *International Journal of Hydropower and Dams*, **4**, 12.

Schnitter, N.J. (1967) A short history of dam engineering. *Water Power*, April, 142–148.

P2.2   Hydraulics problem

Introductory note

We shall consider Lake Moeris and the artificial canal connecting the Nile river to the reservoir (Fig. P2.3). On Fig. P2.3, the flood plain on each side of the trapezoidal channel consisted of grass, bush and rocks. An equivalent roughness height of $k_s = 100\,\text{mm}$ could be considered, if necessary.

P2.2.1 Study of the upper regulator

(A) Draw a sketch of the eastern dam and the canal cross-section, view from the west (i.e. view from the canal). Indicate the main dimensions on the sketch. Show the north and south directions.

(B) During a flood, the discharge in the Nile river south of Ha-Uar is $8000 \, \text{m}^3/\text{s}$. At the same time, the Chief Engineer records a 1.1 m flow depth above the sill of the broad-crested weir.

In this question, you will assume that the weir crest and the transition between the weir and the channel are smooth and horizontal. For each sub-question, students are asked to explain in words each formula and assumption used. Indicate clearly each fundamental principle(s).

(B.1) What is the flow rate in the Nile river downstream of the Ha-Uar dam?

(B.2) Sketch a cross-section profile and a top view of both the broad-crested weir and the start of the channel. On the cross-section view, plot the water surface profile.

(B.3) What are the values of the specific energy: (a) upstream of the weir, (b) mid-sill, (c) downstream of the weir where the cross-section is rectangular (and 400 m wide) and (d) downstream of the weir at the start of the canal where the cross-section is trapezoidal (see Fig. P2.3)? You will assume that the channel bed elevation is the same upstream and downstream of the weir.

(B.4) Develop the dimensionless relationship between the specific energy and the flow depth (i.e. E/d_c as a function of d/d_c) for a channel of irregular cross-section. Explain clearly in words each step of your development. Deduce the expression of the dimensionless specific energy E/d_c for a rectangular channel.

(B.5) Plot the dimensionless specific energy diagram (i.e. E/d_c versus d/d_c) on graph paper. Indicate on the graph the points representing the flow conditions: (a) upstream of the weir, (b) mid-sill and (c) downstream of the weir where the cross-section is rectangular (and 400 m wide).

(B.6) Downstream of the weir, is the flow subcritical or supercritical: (a) at the end of the weir where the cross-section is rectangular (400 m width) and (b) at the entrance of the canal where the cross-section is rectangular? Justify your answers in words.

(C) The flow rate in the trapezoidal canal was initially $1500 \, \text{m}^3/\text{s}$. The initial flow depth in the canal was 2 m. A large flood arrives from the Nile river and the flow depth at the start of the canal becomes instantly 2.5 m. A surge develops and travels downstream in the canal towards Lake Moeris.

In this question, the channel will be assumed smooth and horizontal.

(C.1) Draw the appropriate sketch(s) of the travelling surge. Indicate clearly the direction of the initial flow and the direction of the surge.

(C.2) What type of surge is taking place in the channel? Can you make any appropriate assumption(s) to compute the new flow conditions. Justify your answer. If you are not able to do any calculations, go to the next question. If you can do the calculations, continue this question.

(C.3) Define a control volume across the surge front for the trapezoidal canal. Indicate the control volume on your sketch(es) (question C.1). Show on your sketch(es) (question C.1) the forces acting on the control volume. Show also your choice for the positive direction of distance and of force.

(C.4) Write the continuity and momentum equation as applied to the control volume shown on your sketch.

(C.5) Compute the surge velocity and the new flow rate.

(C.6) Neglecting flow resistance, how long would it take for the surge to reach the downstream end of the canal?

(C.7) A horseman would need 25 min to cover the distance between the two ends of the channel. Starting from the upstream end at the same time as the surge, will he reach the downstream end before the surge.

P2.2.2 Study of the channel

(A) The canal is discharging $3000\,\text{m}^3/\text{s}$.

(A.1) For a canal of irregular shape, how are the critical flow conditions defined? Explain your answer clearly. Use appropriate sketch(es) if necessary.

(A.2) For the trapezoidal artificial canal, what is the critical depth for the above discharge? Is the critical depth a function of the channel roughness and/or slope?

(B) *Uniform equilibrium flow*

(B.1) What is the definition of uniform flow? Give at least two practical examples of uniform flow situations.

(B.2) Draw a sketch of a uniform flow situation. Choose an appropriate control volume. Show on your sketch the forces acting on the control volume. Show also your choice for the positive direction of distance and of force.

(B.3) Write the momentum equation for a uniform flow in a channel of irregular cross-section.

(B.4) How do you deduce all the uniform flow conditions for a channel of irregular cross-section? Explain your answer in words. *No more* than one equation is required.

(C) The canal is discharging $3000\,\text{m}^3/\text{s}$ (as for question P.2.2.2(A)).

(C.1) What roughness height would you choose for the trapezoidal channel? Justify your answer in words.

(C.2) What is the uniform flow depth for that discharge? Is the uniform flow depth a function of the channel roughness and/or slope?

(C.3) What value of the Manning coefficient would you choose for the channel? Justify your choice. Compute the uniform flow depth using the Manning formula? Does your result differ from the result obtained in question C.2? Discuss the comparison (if any) between the results obtained in question C.2 and question C.3.

(C.4) Assuming that uniform flow is obtained in the canal, where can you control the flow in the canal (e.g. upstream, downstream)? Justify and discuss your answer.

(C.5) Give at least two examples of hydraulic controls that could be used to regulate the flow in the canal (question C.4). Discuss each example and explain clearly the difference(s) between each possibility. Sketch each example.

(D) In BC 2251, a very large flood of the Nile river discharged into the canal to Lake Moeris. The level of water in the valley connecting the river and the reservoir was 2.1 m above the bed elevation of the valley (i.e. 7.1 m above the bed elevation of the canal).

Assuming that uniform flow conditions were reached in the long canal and the valley:

(D.1) Deduce the water discharge into the reservoir.

(D.2) Provide all the flow conditions in the artificial canal (i.e. water discharge, velocity of water, flow depth, hydraulic diameter, cross-section area, wetted perimeter, friction factor).

(D.3) Provide all the flow conditions in the flood plain (i.e. water discharge, velocity of water, flow depth, hydraulic diameter, cross-section area, wetted perimeter, friction factor).

(D.4) Is the flow supercritical? Explain your answer.

Students must detail, discuss and justify every step of their method: e.g. calculation of flow resistance, choice of roughness height(s). For the trapezoidal canal, students should use the roughness height selected in question P2.2.2(C1). The roughness height of the flood plain (i.e. on each side of the artificial canal) is given at the beginning of the assignment.

P2.3 Hydrology of Egypt's Lake Moeris

Part A

(a) Using an atlas such as *The Times Atlas of the World*, find out which 2 months of the year normally yield the greatest rainfalls in the upper Nile catchment, near Khartoum (Sudan), and near Adis Abeba (Ethiopia). In this region, which direction are winds blowing *from* during those months? [This question prompts consideration of where is the likely evaporative source of moisture for this rainfall.]

(b) In the upper Nile catchment, why do winds tend to blow in that direction during that period? [This question prompts consideration of what is driving the airflow. If the wind direction does not conform to the general circulation described in class, then the driving force must be a very strong phenomenon.]

(c) Explain in a few sentences why the flood absorbtion capacity of Lake Moeris is significantly less if a major flood occurs during the previous year than if it does not, but greater still if preceded by two or more relatively dry years.

(d) What does the atlas indicate for the general order of magnitude of annual evaporation rates in the Nile region? Explain how this relates to (c) above. How can an (empty) depression form below sea-level, and remain empty?

(e) Do you think high flows in the Nile would have snow-melt as a significant contributor? Explain. In what way is soil at the bottom of Lake Moeris likely to be *infertile*?

Part B

This question addresses the potential for attenuating flood flows in the Nile river by means of the off-stream storage provided by Lake Moeris. [The previous question dealt with flood attenuation due to direct (on-stream) storage.] Just upstream of the junction between the Nile and the lake's canal, the river flow typically varies as indicated in Fig. P2.4, reaching a peak rate of just over 700 GL/day (about 8.2 ML/s) in early September. [Very wet years probably yield flows about 30% greater than this.]

For the questions below, increase the ordinates of Fig. P2.4 to somewhere between the 'typical' wet season and the 'extreme' wet season, by adding about 30%.

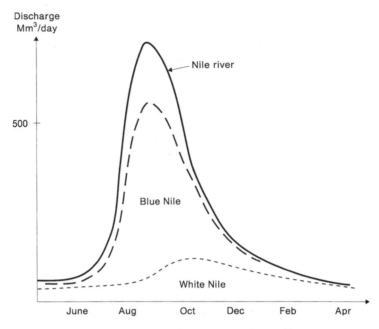

Fig. P2.4 Hydrography of the Nile River near Aswan (data: Garbrecht 1987b).

Although Part II of this assignment asks you to find the normal depth in the lake's canal at a particular flowrate, for this question you will need the whole stage–discharge curve (Q versus H) up to about 10 ML/s, which may include a portion of over-bank flow.

From Fig. P2.4, note that the flowrate in June is typically steady (at about 50 GL/day). During this period, assume that the lake's surface level is the same as the river surface, with no flow in the canal. As the wet season brings higher flows from up-river each week, assume that the river level at the junction must rise until the sum of the lake's canal flow and the downstream river flow equals the given total flowrate from upstream each week. For this exercise, assume that the canal has no dams at either end, and perform all requested calculations manually (without computer assistance). (Note: the imprecise calculations requested below are intended merely to provide a rough estimate of the attenuating effect of flood diversion to the lake. More accurate hydraulic calculations would be needed, particularly as the lake approaches 'full'.)

(a) Replot the total inflow hydrograph (appropriately rescaling the data from Fig. P2.4) dividing each month into four equal 'weeks' of 7.6 days, and showing the hydrograph in stepped form (as if each 'weekly' flow was constant).

(b) Integrate the total Nile flow volume, from 1 July to when the falling hydrograph passes half its peak value. Deduce what depth of Lake Moeris (assuming constant plan area) would be required to store *all* of this volume.

(c) Plot a stage-discharge curve for the river at (above and below) the canal junction:

$$Q(\mathrm{ML/s}) = 0.12H^2$$

[This means, for example, that when $H = 8.2\,\text{m}$, then $Q = 8.2\text{ML/s} = 8200\,\text{m}^3/\text{s}$ approximately, at which time the velocity is about $1.5\,\text{m/s}$, the sectional area is about $5500\,\text{m}^2$, and the river width is about $550\,\text{m}$.]

(d) On the plot of (c), superimpose a hypothetical [but incorrect] stage–discharge curve for the canal:

$$Q = 0.72H^{1.4}$$

[From your Section P2.2 results, plot points to show how incorrect this curve is.]

Set up a tabulation (with each row representing a 'week', starting at the end of June), showing how each 'week's' upstream inflow is divided into two components, one of which represents the diversion to Lake Moeris. A column should show the stage (H) at the canal junction. [The objective of this table is to identify how much attenuation of the upstream river flow is achieved by the diversion.] Superimpose the deduced downstream river flow on the plot of (a). Highlight the 'attenuation'.

(e) Sum the canal flow volumes to the lake each 'week' from 1 July, and deduce the new surface level of the lake at the end of each week. Stop the calculations in (d) as soon as the lake surface level matches the stage-level at the canal–river junction (i.e. the canal flow is completely 'drowned'). [From this point, continued falling river levels would result in outflows from Lake Moeris, unless these are intercepted by blocking the canal (to save the water for later irrigation purposes).]

A study of the Moche River irrigation systems (Peru AD 200–AD 1532)

Preface

The following problem covers lecture material relevant to Basic Principles (Part I) and Hydraulic Modelling (Part III). The assignment was first used in 1994. Based upon real historical facts, the geometry of the hydraulic structures has been simplified to facilitate the calculations.

P3.1 Introduction

P3.1.1 Presentation

South-American agriculture began in the mountain regions of southern Peru and northern Bolivia (Fig. P3.1). The early civilizations grew up in the semi-arid highlands and arid coastal valleys traversed by small rivers. Irrigation was the dominant agricultural technique throughout the prehistory of the north coast of Peru. The establishment of the early civilizations took place in the river valleys (e.g. Moche, Chicama rivers).

The northern coastal valleys of Peru were populated by the Mochicas around AD 200–1000 and later by the Chimus (AD 1000–1466). The Mochicas and the Chimus developed the irrigation of the area particularly on the rivers Moche and Chicama. The Moche Valley was the locus of the Mochican civilization and later held the capital of the Chimu empire, Chan-Chan.

In AD 1466, the Incas overran the Chimu empire.[1] Although the capital city of the Incas was Cuzco, in the Peruvian highlands, the Inca empire extended at its maximum from Ecuador to Chile. Inca engineers were expert in road and bridge building. Furthermore, they gained some expertise in dam construction and canal building from the Chimus. However, in AD 1535, the power of the Incas was destroyed by

---

[1] It is thought that the Incas conquered Chan-Chan after cutting its water supply system.

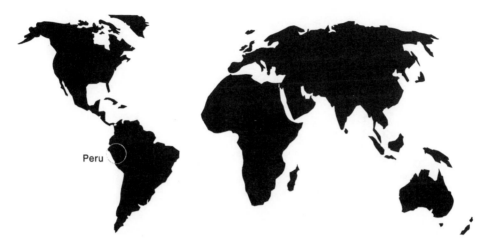

Fig. P3.1 Map of the world and location of the Moche valley.

the Spaniards. The Spanish conquistadors of Peru, led by Francisco Pizarro,[2] came seeking gold and silver.[3] Their small troupe benefitted from a civil war within the Inca empire. After the capture and killing of the Great Inca, the Incas surrendered without a fight.

Geography

Peru is located on the Pacific coast of South America. The borders of ancient Peru extended from just south of the Equator to about 20° South latitude. The countryside is dominated by the great mountain mass of the Central Andes.

The Andes, running from north-west to south-east, divide Peru into three long and narrow topographic zones: (a) the desert coast including the western slope of the mountains, (b) the highlands and (c) the tropical forests on the eastern slope of the mountains.

The Peruvian coast is one of the world's driest deserts and life is almost impossible without irrigation: e.g. the average annual rainfall is less than 5 mm per year at the estuary of the Moche river (Nials *et al.* 1979a,b). The waters along the Peruvian coast are found in small rivers that flow from the nearby Andes across the coastal desert into the Pacific and which are fed by limited seasonal rains in the high mountains.

The Mochican culture

The Mochican civilization (AD 200–1000) was located around the valley of the Moche river. It is believed that the Mochican civilization was dynamic, aggressive and well advanced in irrigation and construction. The Mochicans built large

[2] Francisco Pizarro (AD 1475–1541) was born in Trujillo (Spain). With help of his brothers Gonzalo (AD 1502–1548) and Hernando (AD 1508–1578), he conquered Peru. He was killed in Lima by the partisans of his rival Diego de Almagro (AD 1475–1538) who were led by de Almagro's son Diego (AD 1518–1542).

[3] The land of Peru was rich in gold, silver, copper, tin and other metals. South American Indians extensively used silver and gold for ornaments.

Fig. P3.2 Extent of the Chimu and Inca empires.

pyramids (e.g. Huaca del Sol, Huaca de la Luna) and some impressive irrigation systems in the Moche and Chicama valleys.

The Mochicans extensively used gold, silver and copper for ornamental purposes. Further, they excelled in the field of ceramics.

The Chimu empire

The Chimu civilization was founded in the Moche valley before it spread over the adjacent valleys (e.g. Chicama, Viru valleys). In the latter stages, the maximum extent of the empire covered around 1000 km of coastline up to the Ecuadorean border (Fig. P3.2).

The Chimu civilization was well developed and its culture was famous in South America. The Chimu engineers were also expert in road-building (e.g. inter-valley roads along the Pacific coast), irrigation canal construction (e.g. inter-valley canal) and city development (e.g. Chan-Chan).

Table P3.1 The Inca dynasty

	Name	Period	Comments
(1)	(2)	(3)	(4)
[1]	Manco Capac	AD 1200	
[2]	Sinchi Roca		
[3]	Lloque Yupanqui		
[4]	Mayta Capac		
[5]	Capac Yupanqui		
[6]	Inca Roca		
[7]	Yahuar Huacac		
[8]	Viracocha Inca		
[9]	Pachacuti Inca Yupanqui	1438–1471	Beginning of the Inca empire expansion
[10]	Topa Inca Yupanqui	1471–1493	
[11]	Huayna Capac	1493–1525	
[12]	Huascar	1525–1532	
[13]	Atahuallpa	1532–1533	Killed by the Spanish

Reference: (Mason 1957)

The capital city of the Chimu empire was Chan-Chan, located in the lower delta of the Moche valley. Chan-Chan was the largest city ever built in ancient Peru and it covered an area of about 28 km^2. Torrential rains in 1934 destroyed much of the city ruins.

The development of the Chimu empire and of its capital city Chan-Chan relied heavily upon the development of irrigation systems. The water supply of Chan-Chan was provided by large canals withdrawing waters from the Moche and Chicama rivers (see below).

The Chimu empire was defended in the south by the impressive fortress of Paramonga in the Fortaleza valley. In the north, no frontier defences have been observed and the Incas invaded the Chimu empire in AD 1466 from the north.

The Inca empire

The Incas were a small tribe in the southern highlands of Peru during the early part of the 13th century AD. Their settlements were located around Cuzco. After defeating the other tribal states of the southern highlands of Peru (around AD 1438), the Incas emerged as the strongest military power of the region. They conquered, retained and re-organized the territories of the south highlands and the nearby coastal regions.

The Inca dynasty included 13 emperors (Table P3.1). During the expansion period, Cuzco became the capital city of the Inca empire.

In the early 1460s, the Inca armies marched north conquering the highlands as far as Quito.[4] Then they invaded the Chimu empire and destroyed the Chimu armies (AD 1466). The Inca empire was extended to its ultimate limits near the end of the 15th century (Fig. P3.2). At the time of the death of Topa Inca Yupanqui (AD 1493), the Inca empire extended from northern Ecuador to central Chile, including the whole of Bolivia and Peru, a coastal distance close to 5000 km.

The Incas made their conquest through a combination of military might and diplomacy. The Incas were recognized as ferocious warriors and great engineers. Settlements were defended by hilltop fortresses (e.g. Sacsahuaman on the hill at the

---

[4] Quito is now the capital city of modern Ecuador (altitude: 2540 m).

edge of Cuzco). An impressive network of roads and bridges connected the main cities across their empire.

The last Inca emperor, Atahuallpa, was captured and executed by Pizarro in AD 1533.

P3.1.2 The irrigation system of the Moche river valley

The irrigation systems of the Mochicas and Chimus included large and long canals with flat and steep gradients, aqueducts, dams, side weirs. They were built by the Mochica civilization, later extended by the Chimus and the Incas. The two largest canals were the Vichansao canal and the Inter-valley canal (Figs P3.3 and P3.4).

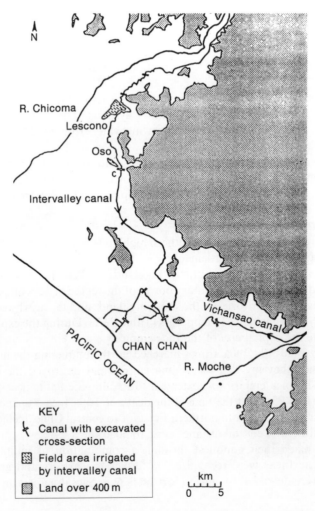

Fig. P3.3 Location of the Vichansao and Inter-valley canals (source: Farrington and Park 1978).

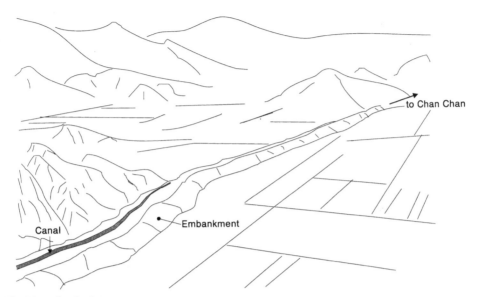

Fig. P3.4 Sketch of the inter-valley canal in the Chicama valley (Acequia de Ascope) (after photographs by Ubbelohde-Doering 1967).

Vichansao canal

The Vichansao canal was the most important prehistoric canal of the Moche valley. Its construction began around AD 0(?) to AD 250. By AD 1250, the system was over 28 km long. The Chimus subsequently added another 7 km of canal. Later the Incas extended the canal again by another 5–10 km.

The canal was 2 m wide and the height of the sidewalls was 1.5 m. Parts of the canal were made with cobble bed; other parts were made of sand and clay. The canal was fed from an artificial reservoir located 25 km upstream of the Moche river estuary. The river was dammed by a diversion weir across the river bed. The dam was made of stonework masonry and the river flow discharged over the crest. The crest of the dam was horizontal and 20 m long. The dam was 2 m high, its cross-section was trapezoidal with a vertical upstream face. The thickness of the structure was 4 m at the base and 3 m at the crest.

Inter-valley canal

When the Chan-Chan region expanded so that the Vichansao canal capacity became inadequate, the Inter-valley canal was designed to provide additional water supply. The canal was built to withdraw waters from the Chicama valley and it was later connected to the Vichansao canal (Fig. P3.3). The Chicama river is a larger stream than the Moche river and the canal enabled a more continuous water supply through-out the year, which could have been used to irrigate a second crop.

The initial canal was 79 km long. At a later stage, another portion (60 km long) was added to connect the inter-valley canal with the Vichansao canal. The canal was 7 m wide and 2 m deep.

Note that, after the Spanish conquest of Peru, the Spaniards continued to use the irrigation system of the Moche and Chicama Valleys (Kosok 1940).

P3.1.3 Important dates

Date	Comments
(1)	(2)
AD 0(?)–AD 250	First construction works of the Vichansao canal
AD 1250–AD 1462	Extension of the Vichansao canal during the Chimu empire
AD 1466	Annexation of the Chimu empire to the Inca empire
AD 1532	Landing of Pizarro in Ecuador
AD 1533	Execution of the last Inca emperor Atahuallpa

P3.1.4 Bibliography

Books

Lanning, E.P. (1967) *Peru Before the Incas*. (Prentice-Hall: Englewood Cliffs NJ, USA).

Mason, J.A. (1957) *The Ancient Civilizations of Peru*. (Penguin Books: Harmondsworth, UK).

Smith, N. (1971) *A History of Dams*. (The Chaucer Press: Peter Davies, London, UK).

Ubbelohde-Doering, H. (1967) *On the Royal Highways of the Inca Civilizations of Ancient Peru*. (Thames and Hudson: London, UK).

Articles

Cook, O.F. (1916) Staircase farms of the ancients. *National Geographic Magazine*, **29**, 474–534.

Farrington, I.S. (1980) The archaeology of irrigation canals with special reference to Peru. *World Archaeology*, **11**, 287–305.

Farrington, I.S. and Park, C.C. (1978) Hydraulic engineering and irrigation agriculture in the Moche Valley, Peru: c. A.D. 1250–1532. *Journal of Archaeological Science*, **5**, 255–268.

Kosok, P. (1940) The role of irrigation in ancient Peru. *Proceedings of the 8th American Scientific Congress*, Washington DC, USA, **2**, 168–178.

Nials, F.L., Deeds, E.E., Moseley, M.E., Pozorski, S.G., Pozorski, S.G. and Feldman, R. (1979a) El Niño: the catastrophic flooding of coastal Peru. Part I. *Field Museum of Natural History Bulletin*, **50**(7), 4–14.

Nials, F.L., Deeds, E.E., Moseley, M.E., Pozorski, S.G. and Feldman, R. (1979b) El Niño: the catastrophic flooding of coastal Peru. Part II. *Field Museum of Natural History Bulletin*, **50**(8), 4–10.

P3.2 Hydraulics problem

P3.2.1 Study of the Vichansao canal and the diversion dam (Part A)

We shall consider the Vichansao canal connecting the Moche river to the Chan-Chan region (Fig. P3.3) and the diversion dam across the Moche river. We shall assume that: (a) the Moche river discharge (upstream of the canal intake) is about $2.1\,\text{m}^3/\text{s}$, and (b) the canal operates and discharges 30% of the lowest river flow rate.

Note: later, compare the above river discharge with the results obtained in Section P3.3, question B2.

(A) *Sketch*
(A.1) Draw a sketch in elevation of the Moche river, the diversion dam and the Vichansao canal intake. Indicate the main dimensions on the sketch. Show the north and south directions.
(A.2) Draw also the canal cross-section with the main dimensions.
(A.3) On which river bank is the canal intake: (a) the right bank (i.e. NW), or (b) the left bank (i.e. SE)?

(B) *The Vichansao canal*
We consider now the waters flowing into the Vichansao canal.
(B.1) Define the concept of critical flow conditions. Explain your answer in *words* only. Do not give any equations.
(B.2) For a rectangular channel (channel width B), derive the expression of the critical flow depth. 'Derive' means that you must explain and justify your calculations before giving the expression of the critical flow depth.
(B.3) For the Vichansao canal, give the value (and units) of the critical flow depth.

(C) *The diversion dam*
In operation, the water overflows the entire crest of the diversion dam uniformly. We shall assume that the energy losses can be neglected.
(C.1) What is the flow depth above the crest of the diversion dam?
(C.2) What is the value (and units) of the specific energy: (a) on the diversion weir crest, (b) upstream of the diversion weir, and (c) downstream of the diversion weir.
(C.3) What is the free-surface elevation of the upstream reservoir above the dam crest?
(C.4) What is the flow depth downstream of the diversion dam?
(C.5) Sketch a cross-section profile of the diversion dam. On the cross-section view, plot the water surface profile.
For each sub-question, you are asked to explain in words each formula and assumption that you use. Indicate clearly each fundamental principle(s).

P3.2.2 Study of the Inter-valley canal

We shall consider now the Inter-valley canal connecting the Chicama river to the Chan-Chan region (Figs P3.3 and P3.4).

(A) *Hydraulic model study*
The Chimu emperor wants a hydraulic study of the first 79 km of the canal. He orders an (undistorted) scale model study where the length of the model equals 985 m.
   The prototype flow conditions to be investigated are:

$$\{1\} \quad d = 0.4 \, \text{m}; Q = 0.7 \, \text{m}^3/\text{s}$$

$$\{2\} \quad d = 1.5 \, \text{m}; Q = 3 \, \text{m}^3/\text{s}$$

where d is the flow depth and Q the total discharge.

(A.1) What type of similitude would you choose? What is (are) the scale(s) of the model(s)? Explain and discuss your answer *with words*.

For both prototype flow conditions (i.e. {1} and {2}), answer the following questions:

(A.2) What is (are) the width(s) of the scale model(s)?

(A.3) What are the model flow depth(s), flow velocity(ies) and total discharge(s)? Explain carefully your answer in words.

(B) *Upstream end of the canal*

The first part of the canal (i.e. first 79 km) is a paved channel (7 m width) and the bottom slope is 0.5°.

(B.1) Draw a sketch of a uniform flow situation. Choose an appropriate control volume. Show on your sketch the forces acting on the control volume. Show also your choice for the positive direction of distance and of force.

(B.2) Write the momentum equation for a uniform flow in the channel.

(B.3) Develop the formula of the uniform flow depth as a function of the Darcy friction factor. Explain in words your development.

(B.4) What values would you choose for: (a) the roughness height and (b) the Manning coefficient? Discuss and justify your choice.

(B.5) The water discharge in the canal is $2\,\text{m}^3/\text{s}$.

(B.5.1) For the selected roughness height (question B.4), what are the values of the Darcy coefficient and the normal depth computed from the Darcy coefficient? Explain and discuss your calculations.

(B.5.2) What is the uniform flow depth computed using the Manning equation and the Manning coefficient (question B.4)?

(B.5.3) Discuss your results and the (inevitable) discrepancies between the results of the above questions B.5.1 and B.5.2. Compute the values of the Chézy coefficients for each case. Compare both the Chézy coefficients *and* the flow depths.

(B.5.4) Is the uniform flow subcritical, supercritical or critical? Discuss (in words) the properties of the uniform flow regime.

For the question (C) and (D), we shall now consider the downstream end of the inter-valley canal (near the junction with the Vichansao canal). The last 3 km of the canal includes:

(A) a rectangular channel (7 m wide, 2850 m long, 3° bed slope) with concrete bottom and sidewalls,

(B) a 150 m long rectangular channel (7 m wide, 0.1° slope) with gravel bed[5] ($d_{50} = 0.2\,\text{m}$), followed by

(C) a wooden sluice gate (7 m wide) to regulate the canal.

(C) *Downstream end of the canal*

We consider that the canal discharges $1.4\,\text{m}^3/\text{s}$ of water and that the flow depth downstream of the sluice gate is 0.05 m.

(C.1) Compute the flow velocity and Froude number downstream of the sluice gate.

---

[5] For a gravel bed, it is reasonable to assume that the equivalent roughness height equals nearly the median gravel size (d_{50}).

(C.2) Calculate the flow depth, flow velocity and Froude number immediately upstream of the sluice gate. Explain clearly each step of your calculations and highlight each assumption used.

(C.3) Compute the uniform flow depth in: (a) the 2850 m long channel, and (b) the 150 m long channel (immediately upstream of the sluice gate).

(C.4) Compute the free-surface profile in both channels. Explain very clearly *in words* each step of your calculations. Discuss any assumptions used during the calculations.

(C.5) On graph paper, plot the channel bottom, the sluice gate and the free-surface profile along the 3 km long channel. Use an appropriate scale.

(D) *Gate operation*

The canal discharge remains constant (i.e. $1.4 \, \text{m}^3/\text{s}$). The initial flow conditions are the same as defined in question (C) and plotted in question (C.5). Suddenly, the wooden sluice gate is opened and the new gate opening is now 0.7 m.

(D.1) Describe and discuss *in words* the 'hydraulic event' occurring (in the 3 km long channel) immediately after the opening of the sluice gate. Use *sketches* to illustrate your discussion. Explain (in words) what type of calculations could be done. *Do not* give any equations. *No* calculation is required.

(D.2) A few hours after the gate opening, the flow characteristics become steady.

(D.2.1) Compute the new free-surface profile.

(D.2.2) On graph paper, plot the channel bottom, the sluice gate and the free-surface profile (obtained in question D.2.1) along the 3 km long channel. Use an appropriate scale.

(D.3) Compare and discuss the free-surface profiles obtained in questions (C.4), (C.5) and (D.2). What are the main differences?

P3.3 Hydrology of western Peru

P3.3.1 Background

Consult a map of Peru (on the west coast of South America) and focus on the northern coastal city of Trujillo at about 8° South latitude. Being so close to the equator, one might expect that this area would experience equatorial convective rainfall in the southern summer, but not so. Trujillo's dry (desert) climate is like that of Lima (see Table P3.2). It is very pleasant, but what about water supply! Why does it not rain there? Because the ocean's cold Humbolt current flows northward past Peru then eastward across the Pacific driven by the Walker Cell. The proximity of cold water attracts condensation, so atmospheric moisture is scarce. [About eight

Table P3.2 Average meteorological conditions in two Peruvian towns

Town	Latitude (deg. S)	Elevation (km)	Coldest temperature (Celsius)	Mon. rain (wettest) (mm/mo)	Mon. rain (driest) (mm/mo)	Ann. rain (mm/a)
Cajamarca	7.1	2.6	−4 [Dec]	115 [Feb]	5 [Aug]	720
Lima	12.1	0.1	9 [Jul]	8 [Aug]	0 [Feb]	42

times per century, a big El-Niño event kills the Walker Cell, causing warm Western Pacific water to slosh back toward Peru, cutting off (usually about Christmas) the supply of nutrients to Peru's fishing industry, and bringing Ecuador's tropical rain south to Trujillo and Lima.]

Trujillo is on the north side of the Rio Moche estuary. The catchment of this river stretches about 100 km inland to Peru's major Andes plateau at elevation 4.1 km. Its area is about 2708 km^2 (about 80 km by 30 km). The upper half of this catchment is located at elevations above 1500 m and it would receive rainfall like Cajamarca's. [Because southern Peru's mountains are higher than those in the north, winter snow plays a significant role there, but snow hardly affects the Rio Moche's hydrology.] The average annual discharge of the Rio Moche is about 9.5 m^3/s, but 71% of the annual discharge occurs between February and April and only 1.8% between July and September. The largest recorded flood occurred during the 1925 El-Niño, when the width of the river during maximum flood stage would have been about 3.5 km (about 10 km upstream of the estuary).

About 35 km up the coast from the Moche River mouth, the larger Rio Chicama also meets the sea. Its catchment area is about 4400 km^2, and its rainfall characteristics are very similar to those of the adjacent Rio Moche.

Located between the Rio Moche and the Rio Chicama, the Quebrada Rio Seco runs parallel to the Moche river, about 10 km north-west of the Rio Moche. Its catchment area is about 300 km^2. Because the drainage basin is small, there are short-term flash floods heavily loaded in sediments (from clay to boulders).

P3.3.2 Question A

In tabular form, deduce the seasonal (month-by-month) distribution of streamflows in the Moche River by applying the appropriate monthly rainfalls to the upper two-thirds of the catchment area (none to the remainder). Assume that the monthly (gross) rainfalls are distributed roughly sinusoidally between the wettest and driest values given in Table P3.2. Assume that evaporative processes remove about 1.3 mm of recent rainfall every day (i.e. 40 mm/mo). [If less than 40 mm of rain falls in any month, none of this reaches the stream.] For each month's net rainfall, assume that only 40% enters the stream directly as surface run-off in that month, the remainder infiltrates the groundwater system and appears as baseflow in seven subsequent months at rates of 22%, 14%, 9%, 6%, 4%, 3% and 2% respectively (totalling 100%). [Hint: delay converting units from mm/mo to kL/s until as late as possible.]

P3.3.3 Question B

If the people of Trujillo wanted to grow maize, they would need a steady flow of irrigation water throughout the 4 month growing season. Because the growing season includes the month of lowest river flow, the maximum area of irrigated crops that could be sustained would be governed by the lowest river flowrate (unless a storage reservoir is provided). For the case of no reservoir, calculate the maximum irrigable area (km^2), given that about 10 ML is required for each hectare (spread over 4 months), including transmission losses. Note: 1 km^2 is 100 hectares. One hectare is 100 ares. One are is a (French) garden plot ($10\,m \times 10\,m = 100\,m^2$).

P3.3.4 Question C

How big a storage volume (in litres) would be required to enable all the Rio Moche water to be used for irrigation at a steady rate, year-round? [Hint: find the mean streamflow (mm/mo), subtract this from each monthly flow to deduce which months are deficient, then sum the volumes needed to cover the deficient months (knowing that this deficiency will be replenished in the other months).

References

Ackers, P., White, W.R., Perkins, J.A., and Harrison, A.J.M. (1978) *Weirs and Flumes for Flow Measurement*. (John Wiley: Chichester, UK) 327 pages.

Agostini, R., Bizzarri, A., Masetti, M., and Papetti, A. (1987) *Flexible Gabion and Reno Mattress Structures in River and Stream Training Works. Section One: Weirs*. 2nd edn (Officine Maccaferri: Bologna, Italy).

Alembert, J. le Rond d' (1752) *Essai d'une Nouvelle Théorie de la Résistance des Fluides*. (Essay on a New Theory on the Resistance of Fluids.) (David: Paris, France).

Alexander, J., and Fielding, C. (1997) Gravel antidunes in the tropical Burdekin River, Queensland, Australia. *Sedimentology*, **44**, 327–337.

Anderson, A.G. (1942) Distribution of suspended sediment in a natural stream. *Transactions of the American Geophysics Union*, **23**(2), 678–683.

Apelt, C.J. (1983) Hydraulics of minimum energy culverts and bridge waterways. *Australian Civil Engineering Trans., I.E.Aust.*, **CE25**(2), 89–95.

Apelt, C.J. (1994) *The Minimum Energy Loss Culvert*. Videocassette VHS colour, Department of Civil Engineering, University of Queensland, Australia, 18 minutes.

Bagnold, R.A. (1956) The flow of cohesionless grains in fluids. *Philosophical Transactions of the Royal Society, London, Series A*, **249**, 235–297.

Bagnold, R.A. (1966) An approach to the sediment transport problem from general physics. *US Geological Survey*, Professional Paper 422-I, Washington DC, USA, 37 pages.

Bakhmeteff, B.A., and Matzke, A.E. (1936) The hydraulic jump in terms of dynamic similarity. *Transactions, ASCE*, **101**, 630–647. Discussion: **101**, 648–680.

Bakhmeteff, B.A., and Fedoroff, N.V. (1943) Energy loss at the base of a free overfall – discussion. *Transactions, ASCE*, **108**, 1364–1373.

LeBaron Bowen Jr, R., and Albright, F.P. (1958) *Archaelogical Discoveries in South Arabia*. (The Johns Hopkins Press: Baltimore, USA).

Bauer, W.J. (1954) Turbulent boundary layer on steep slopes. *Transactions, ASCE*, **119**, 1212–1233.

Bazin, H. (1865a) Recherches Expérimentales sur l'Ecoulement de l'Eau dans les Canaux Découverts. (Experimental Research on Water Flow in Open Channels.) *Mémoires présentés par divers savants à l'Académie des Sciences* (Paris, France), **19**, 1–494 (in French).

Bazin, H. (1865b) Recherches Expérimentales sur la Propagation des Ondes. (Experimental Research on Wave Propagation.) *Mémoires présentés par divers savants à l'Académie des Sciences* (Paris, France), **19**, 495–452 (in French).

Bazin, H. (1888–1898) Expériences Nouvelles sur l'Ecoulement par Déversoir. (Recent Experiments on the Flow of Water over Weirs.) *Mémoires et Documents, Annales des Ponts et Chaussées* (Paris, France), 1888: Sér. 6, **16**, 2nd Sem., 393–448; 1890: Sér. 6, **19**, 1st Sem.,

9–82; 1891: Sér. 7, **2**, 2nd Sem., 445–520; 1894: Sér. 7, **7**, 1st Sem., 249–357; 1896: Sér. 7, **12**, 2nd Sem., 645–731; 1898: Sér. 7, **15**, 2nd Sem., 151–264 (in French).

Belanger, J.B. (1828) *Essai sur la Solution Numérique de quelques Problèmes Relatifs au Mouvement Permanent des Eaux Courantes.* (Essay on the Numerical Solution of Some Problems relative to Steady Flow of Water.) (Carilian-Goeury: Paris, France) (in French).

Belanger, J.B. (1849) Notes sur le Cours d'Hydraulique. (Notes on the Hydraulics Subject.) *Mém. Ecole Nat. Ponts et Chaussées* (Paris, France) (in French).

Belidor, B.F. de (1737–1753) *Architecture Hydraulique.* (*Hydraulic Architecture.*) 4 Volumes (Charles-Antoine Jombert: Paris, France) (in French).

Belyakov, A.A. (1991) Hydraulic Engineering and The Environment in Antiquity. *Gidro-tekhnicheskoe Stroitel'stvo*, No. 8, 46–51 (in Russian) (Translated in *Hydrotechnical Construction*, 1992, Plenum, 516–523).

Beyer, W.H. (1982) *CRC Standard Mathematical Tables.* (CRC Press Inc: Boca Raton, Florida, USA).

Bhowmik, N.G. (1996) Impact of the 1993 floods on the Upper Mississippi and Missouri River Basins in the USA. *Water International*, **21**, 158–169.

Bos, M.G. (1976) *Discharge Measurement Structures.* Publication No. 161, Delft Hydraulic Laboratory, Delft, The Netherlands (also Publication No. 20, ILRI, Wageningen, The Netherlands).

Bos, M.G., Replogle, J.A., and Clemmens, A.J. (1991) *Flow Measuring Flumes for Open Channel Systems.* (ASAE: St. Joseph MI, USA) 321 pages.

Bossut, Abbé C. (1772) *Traité Elémentaire d'Hydrodynamique.* (*Elementary Treaty On Hydro-dynamics.*) 1st edn (Paris, France), (in French) (2nd edn: 1786, Paris, France; 3rd edn: 1796, Paris, France).

Boussinesq, J.V. (1877) Essai sur la Théorie des Eaux Courantes. (Essay on the Theory of Water Flow.) *Mémoires présentés par divers savants à l'Académie des Sciences* (Paris, France), **23**, Ser. 3, No. 1, supplément 24, 1–680 (in French).

Boussinesq, J.V. (1896) Théorie de l'Ecoulement Tourbillonnant et Tumultueux des Liquides dans les Lits Rectilignes à Grande Section (Tuyaux de Conduite et Canaux Découverts) quand cet Ecoulement s'est régularisé en un Régime Uniforme, c'est-à-dire, moyennement pareil à travers toutes les Sections Normales du Lit. (Theory of turbulent and tumultuous flow of liquids in prismatic channels of large cross-sections (pipes and open channels) when the flow is uniform, i.e. constant in average at each cross-section along the flow direction.) *Comptes Rendus des séances de l'Académie des Sciences* (Paris, France), **122**, 1290–1295 (in French).

Boys, P.F.D. du (1879) Etude du Régime et de l'Action exercée par les Eaux sur un Lit à Fond de Graviers indéfiniment affouillable. (Study of Flow Regime and Force exerted on a Gravel Bed of infinite Depth.) *Ann. Ponts et Chaussées* (Paris, France), Série 5, **19**, 141–195 (in French).

Bradley, J.N., and Peterka, A.J. (1957) The hydraulic design of stilling basins. *Journal of Hydraulic Div.*, **83**, papers 1401, 1402 and 1403. Bradley, J.N., and Peterka, A.J. (1957) The hydraulic design of stilling basins: hydraulic jumps on a horizontal apron (basin I) *Journal of Hydraulic Div.*, **83**, paper 1401, 1401-1/1401–22. Bradley, J.N., and Peterka, A.J. (1957) The hydraulic design of stilling basins: high dams, earth dams and large canal structures (basin II) *Journal of Hydraulic Div.*, **83**, paper 1402, 1402-1/1402–14. Bradley, J.N., and Peterka, A.J. (1957) The hydraulic design of stilling basins: short stilling basin for canal structures, small outlet works and small spillways (basin III). *Journal of Hydraulic Div.*, **83**, paper 1403, 1403-1/1403–22.

Bresse, J.A. (1860) *Cours de Mécanique Appliquée Professé à l'Ecole des Ponts et Chaussées.* (*Course in Applied Mechanics Lectured at the Pont-et-Chaussées Engineering School.*) (Mallet-Bachelier: Paris, France) (in French).

Buat, P.L.G. du (1779) *Principes d'Hydraulique.* (*Hydraulic Principles.*) 1st edn (Imprimerie de Monsieur: Paris, France) (in French) (2nd edn: 1786, Paris, France, 2 volumes; 3rd edn: 1816, Paris, France, 3 volumes).

Buckingham, E. (1915) Model experiments and the form of empirical equations. *Transactions, ASME*, **37**, 263–296.

Buckley, A.B. (1923) The influence of silt on the velocity of water flowing in open channels. *Minutes of the Proceedings of the Institution of Civil Engineers.*, *1922–1923*, **216**, Part I, 183–211. Discussion, 212–298.

Cain, P. (1978) Measurements within self-aerated flow on a large spillway. Ph.D. Thesis, Ref. 78-18, Department of Civil Engineering, University of Canterbury, Christchurch, New Zealand.

Cain, P., and Wood, I.R. (1981) Measurements of self-aerated flow on a spillway. *Journal of Hydraulic Div.*, **107**, HY11, 1425–1444.

Carvill, J. (1981) *Famous Names in Engineering.* (Butterworths: London, UK).

Chang, C.J. (1996) A tale of two reservoirs – Greater Taipei's water woes. *Sinorama*, **21**(12), 6–19.

Chanson, H. (1995a) *Flow characteristics of undular hydraulic jumps. Comparison with near-critical flows.* Report CH45/95 (monograph), Department of Civil Engineering, University of Queensland, Australia, June, 202 pages.

Chanson, H. (1995b) *Hydraulic Design of Stepped Cascades, Channels, Weirs and Spillways.* (Pergamon, Oxford, UK) 292 pages.

Chanson, H. (1997*)* *Air Bubble Entrainment in Free-Surface Turbulent Shear Flows.* (Academic Press, London, UK) 401 pages.

Chanson, H., and Montes, J.S. (1995) Characteristics of undular hydraulic jumps. Experimental apparatus and flow patterns. *Journal of Hydraulic Engineering*, **121**(2), 129–144. Discussion: **123**(2), 161–164.

Chanson, H., and Whitmore, R.L. (1996) The stepped spillway of the Gold Creek Dam (built in 1890). *ANCOLD Bulletin*, No. 104, 71–80.

Chanson, H., and James, P. (1998) Rapid reservoir sedimentation of four historic thin arch dams in Australia. *Journal of Performance of Constructed Facilities*, No. 2, May.

Chen, C.L. (1990) Unified theory on power laws for flow resistance. *Journal of Hydraulic Engineering*, **117**, 371–389.

Cheng, N.S. (1997) Simplified settling velocity formula for sediment particle. *Journal of Hydraulic Engineering*, **123** 149–152.

Chien, N. (1954) Meyer-Peter formula for bed-load transport and Einstein bed-load function. *Research Report No. 7*, University of California Institute of Engineering, USA.

Chiew, Y.M., and Parker, G. (1994) Incipient sediment motion on non-horizontal slopes. *Journal of Hydraulic Research*, **32**, 649–660.

Chow, V.T. (1973) *Open Channel Hydraulics.* (McGraw-Hill International: New York, USA).

Colebrook, C.F. (1939) Turbulent flow in pipes with particular reference to the transition region between the smooth and rough pipe laws. *Journal of the Institute of Civil Engineers*, 1938–1939, No. 4, pp. 133–156.

Coleman, N.L. (1970) Flume studies of the sediment transfer coefficient. *Water Res. Res.*, **6**, 801–809.

Coles, D. (1956) The law of wake in the turbulent boundary layer. *Journal of Fluid Mechanics*, **1**, 191–226.

Comolet, R. (1976) *Mécanique Expérimentale des Fluides.* (*Experimental Fluid Mechanics*) (Masson Editor: Paris, France) (in French).

Concrete Pipe Association of Australasia (1991) *Hydraulics of Precast Concrete Conduits.* 3rd edn (Jenkin Buxton Printers: Australia) 72 pages.

Cook, O.F. (1916) Staircase farms of the ancients. *National Geographic Magazine*, **29**, 474–534.

Coriolis, G.G. (1836) Sur l'établissement de la formule qui donne la figure des remous et sur la correction qu'on doit introduire pour tenir compte des différences de vitesses dans les divers points d'une même section d'un courant. (On the establishment of the formula giving the backwater curves and on the correction to be introduced to take into account the velocity differences at various points in a cross-section of a stream.) *Annales des Ponts et Chaussées*, 1st Semester, Series 1, **11**, 314–335 (in French).

Couette, M. (1890) Etude sur les frottements des liquides. (Study on the frictions of liquids.) *Ann. Chim. Phys.*, **21**, 433–510 (in French).

Creager, W.P. (1917) *Engineering of Masonry Dams.* (John Wiley & Sons: New York, USA).

Creager, W.P., Justin, J.D., and Hinds, J. (1945) *Engineering for Dams.* (John Wiley & Sons: New York, USA), 3 Volumes.

Danilevslkii, V.V. (1940) History of hydroengineering in Russia before the nineteenth century. *Gosudarstvennoe Energeticheskoe Izdatel'stvo*, Leningrad, USSR (in Russian) (English Translation: *Israel Program for Scientific Translation*, IPST No. 1896, Jerusalem, Israel, 1968, 190 pages).

Darcy, H.P.G. (1856) *Les Fontaines Publiques de la Ville de Dijon. (The Public Fountains of the City of Dijon)* (Victor Dalmont: Paris, France), 647 pages (in French).

Darcy, H.P.G. (1858) Recherches expérimentales relatives aux mouvements de l'eau dans les tuyaux. (Experimental research on the motion of water in pipes.) *Mémoires Présentés à l'Académie des Sciences de l'Institut de France*, **14**, 141 (in French).

Darcy, H.P.G., and Bazin, H. (1865) *Recherches Hydrauliques. (Hydraulic Research.)* (Imprimerie Impériales: Paris, France), Parties 1ère et 2ème (in French).[1]

Degremont (1979) *Water Treatment Handbook*, 5th edn (Halsted Press Book, John Wiley & Sons: New York, USA).

Dooge, J.C.I. (1991) The Manning formula in context. In *Channel Flow Resistance: Centennial of Manning's Formula*, B.C. Yen (ed) (Water Resources Publishers, Littleton CO, USA) pp. 136–185.

Dupuit, A.J.E. (1848) *Etudes Théoriques et Pratiques sur le Mouvement des Eaux Courantes. (Theoretical and Practical Studies On Flow of Water).* (Dunod: Paris, France) (in French).

Einstein, A. (1906) Eine Neue Bestimmung der Moleküldimensionen. *Ann. Phys.*, **19**, 289 (in German).

Einstein, A. (1911) Eine Neue Bestimmung der Moleküldimensionen. *Ann. Phys.*, **34**, 591 (in German).

Einstein, H.A. (1942) Formulas for the transportation of bed-load. *Transactions, ASCE*, **107**, 561–573.

Einstein, H.A. (1950) The bed-load function for sediment transportation in open channel flows. *US Department of Agriculture Technical Bulletin No. 1026* (Soil Conservation Service: Washington DC, USA).

Elder, J.W. (1959) The dispersion of marked fluid in turbulent shear flow. *Journal of Fluid Mechanics*, **5**, 544–560.

Engelund, F. (1966) Hydraulic resistance of alluvial streams. *Journal of Hyd. Div.*, **92**, No. HY2, 315–326.

Engelund, F., and Hansen, E. (1967) *A Monograph on Sediment Transport in Alluvial Streams.* (Teknisk Forlag: Copenhagen, Denmark).

Engelund, F., and Hansen, E. (1972) *A Monograph on Sediment Transport in Alluvial Streams.* 3rd edn (Teknisk Forlag: Copenhagen, Denmark), 62 pages.

Eurenius, J. (1980) Ancient dams of Saudi Arabia. *International Water Power and Dam Construction*, **32(3)**, 21–22.

---

[1] Work prepared and published posthumously by Bazin (1865a,b).

Falvey, H.T. (1980) *Air–Water Flow in Hydraulic Structures.* (USBR Engrg. Monograph, No. 41, Denver, Colorado, USA).

Farrington, I.S. (1980) The archaeology of irrigation canals with special reference to Peru. *World Archaeology*, **11**, 287–305.

Farrington, I.S., and Park, C.C. (1978) Hydraulic engineering and irrigation agriculture in the Moche Valley, Peru: c. A.D. 1250–1532. *Journal of Archaeological Science*, **5**, 255–268.

Fawer, C. (1937) Etude de quelques écoulements permanents à filets courbes. (Study of some steady flows with curved streamlines.) Thesis, Lausanne, Switzerland, Imprimerie La Concorde, 127 pages (in French).

Fernandez Luque, R., and Van Beek, R. (1976) Erosion and transport of bed-load sediment. *Journal of Hydraulic Research*, **14**, 127–144.

Fick, A.E. (1855) On liquid diffusion. *Philos. Mag.*, **4**(10), 30–39.

Fischer, H.B., List, E.J., Koh, R.C.Y., Imberger, J., and Brooks, N.H. (1979) *Mixing in Inland and Coastal Waters.* (Academic Press: New York, USA).

Fourier, J.B.J. (1822) *Théorie Analytique de la Chaleur.* (*Analytical Theory of Heat.*) (Didot: Paris, France) (in French).

Franc, J.P., Avellan, F., Belahadji, B., Billard, J.Y., Briancon-Marjollet, L., Frechou, D., Fruman, D.H., Karimi, A., Kueny, J.L., and Michel, J.M. (1995) *La Cavitation. Mécanismes Physiques et Aspects Industriels.* (*The Cavitation. Physical Mechanisms and Industrial Aspects.*) (Presses Universitaires de Grenoble, Collection Grenoble Sciences: France), 581 pages (in French).

Garbrecht, G. (1987a) *Hydraulics and Hydraulic Research: a Historical Review.* (Balkema: Rotterdam, The Netherlands).

Garbrecht, G. (1987b) Hydrologic and hydraulic concepts in antiquity. In *Hydraulics and Hydraulic Research: a Historical Review.* (Balkema: Rotterdam, Netherlands), pp. 1–22.

Garbrecht, G. (1996) Historical water storage for irrigation in the Fayum Depression (Egypt), *Irrigation and Drainage Systems*, **10**(1), 47–76.

Ganchikov, V.G. and Munavvarov, Z.I. (1991) The Marib Dam (history and the present time) *Gidrotekhnicheskoe Stroitel'stvo*, No. 4, 50–55 (in Russian) (Translated in *Hydrotechnical Construction*, 1991, Plenum, pp. 242–248).

Gauckler, P.G. (1867) *Etudes Théoriques et Pratiques sur l'Ecoulement et le Mouvement des Eaux.* (*Theoretical and Practical Studies of the Flow and Motion of Waters.*) (Comptes Rendues de l'Académie des Sciences: Paris, France), Tome 64, pp. 818–822 (in French).

Gibbs, R.J., Matthews, M.D., and Link, D.A. (1971) The relationship between sphere size and settling velocity. *Journal of Sedimentary Petrology*, **41**(1), 7–18.

Gilbert, G.K. (1914) *The Transport of Debris by Running Water.* (Professional Paper No. 86, US Geological Survey: Washington DC, USA).

Graf, W.H. (1971) *Hydraulics of Sediment Transport.* (McGraw-Hill: New York, USA).

Guy, H.P., Simons, D.B., and Richardson, E.V. (1966*) Summary of Alluvial Channel Data from Flume Experiments.* (Professional Paper No. 462-I, US Geological Survey: Washington DC, USA).

Hager, W.H. (1983) Hydraulics of plane free overfall. *Journal of Hydraulic Engineering*, **109**(12), 1683–1697.

Hager, W. (1991) Experiments on standard spillway flow. *Proceedings of the Institution of Civil Engineers, London, Part 2*, **91**, 399–416.

Hager, W.H. (1992a) Spillways, shockwaves and air entrainment – review and recommendations. *ICOLD Bulletin*, No. 81, 117 pages.

Hager, W.H. (1992b) *Energy Dissipators and Hydraulic Jump.* Water Science and Technology Library, Vol. 8 (Kluwer Academic: Dordrecht, Netherlands), 288 pages.

Hager, W.H., and Schwalt, M. (1994) Broad-crested weir. *Journal of Irrigation and Drainage Engineering*, **120**(1), 13–26. Discussion: **12**(2), 222–226.

Hager, W.H., Bremen, R., and Kawagoshi N. (1990) Classical hydraulic jump: length of roller. *Journal of Hydraulic Research*, **28**(5), 591–608.

Hathaway, G.A. (1958) Dams – their effect on some ancient civilizations. *Civil Engineering*, **28**(1), 58–63.

Hee, M. (1969) Hydraulics of culvert design including constant energy concept. *Proceedings of the 20th Conference of Local Authority Engineers*, Department of Local Government, Queensland, Australia, Paper 9, pp. 1–27.

Hee, M. (1978) Selected case histories. *Proceedings of the Workshop on Minimum Energy Design of Culvert and Bridge Waterways*, Australian Road Research Board, Melbourne, Australia, Session 4, Paper 1, pp. 1–11.

Hellström, B. (1941) Några Iakttagelser Över Vittring Erosion Och Slambildning i Malaya Och Australien. *Geografiska Annaler*, Stockholm, Sweden, No. 1–2, pp. 102–124 (in Swedish).

Helmholtz, H.L.F. (1868) *Über discontinuirliche Flüssigkeits-Bewegungen.* (Monatsberichte der königlich preussichen Akademie der Wissenschaft zu Berlin, pp. 215–228 (in German).

Henderson, F.M. (1966) *Open Channel Flow.* (Macmillan: New York, USA).

Herr, L. A., and Bossy, H.G. (1965) Hydraulic charts for the selection of highway culverts. *Hydraulic Engineering Circular*, US Department of Transportation, Federal Highway Administration, HEC No. 5, December.

Herschy, R. (1995) General purpose flow measurement equations for flumes and thin plate weirs. *Flow Measurement and Instrumentation*, **6**, 283–293.

Hinze, J.O. (1975) *Turbulence.* 2nd edn (McGraw-Hill: New York, USA).

Howe, J.W. (1949) *Flow Measurement.* Proceedings of the 4th Hydraulic Conference, Iowa Institute of Hydraulic Research, H. Rouse (Ed.), (John Wiley & Sons) June, pp. 177–229.

Humber, W. (1876) *Comprehensive Treatise on the Water Supply of Cities and Towns with Numerous Specifications of Existing Waterworks.* (Crosby Lockwood: London, UK).

Hydropower & Dams (1997) Mini hydro scheme for Egyptian oasis. *International Journal of Hydropower and Dams*, **4**(4), 12.

Idel'cik, I.E. (1969) *Mémento des Pertes de Charge.* (*Handbook of Hydraulic Resistance.*) (Eyrolles Editor, Collection de la direction des études et recherches d'Electricité de France: Paris, France).

Idelchik, I.E. (1986) *Handbook of Hydraulic Resistance.* 2nd revived and augmented edn. (Hemisphere: New York, USA).

International Commission on Large Dams (ICOLD) (1984) *World Register of Dams – Registre Mondial des Barrages – ICOLD.* (ICOLD: Paris, France), 753 pages.

Ippen, A.T., and Harleman, R.F. (1956) Verification of theory for oblique standing waves. *Transactions, ASCE*, **121**, 678–694.

ISO (1979) *Units of Measurements – ISO Standards Handbook 2.* (International Organization for Standardization, ISO: Switzerland).

Jevons, W.S. (1858) On clouds; their various forms, and producing causes. *Sydney Magazine of Science and Art*, **1**(8), 163–176.

Julien, P.Y. (1995) *Erosion and Sedimentation.* (Cambridge University Press: Cambridge, UK), 280 pages.

Julien, P.Y., and Raslan, Y. (1998) Upper-regime plane bed. *Journal of Hydraulic Engineering*, **124**, 1086–1096.

Kamphuis, J.W. (1974) Determination of sand roughness for fixed beds. *Journal of Hydraulic Research*, **12**(2), 193–203.

Kazemipour, A.K., and Apelt, C.J. (1983) Effects of irregularity of form on energy losses in open channel flow. *Australian Civil Engineering Transactions*, **CE25**, 294–299.

Keller, R.J., and Rastogi, A.K. (1977) Design chart for predicting critical point on spillways. *Journal of Hyd. Div.*, **103**, 1417–1429.

Kelvin, Lord (1871) The influence of wind and waves in water supposed frictionless. *London, Edinburgh and Dublin Philosophical Magazine and Journal of Science, Series 4*, **42**, 368–374.

Kennedy, J.F. (1963) The mechanics of dunes and antidunes in erodible-bed channels. *Journal of Fluid Mechanics*, **16**, 521–544 (& 2 Plates).

Keulegan, G.H. (1938) Laws of turbulent flow in open channels. *Journal of Research, National Bureau of Standards*, **21**, Paper RP1151, pp. 707–741.

Knapp, F.H. (1960) *Ausfluss, Überfall and Durchfluss im Wasserbau*. (Verlag G. Braun: Karlsruhe, Germany) (in German).

Korn, G.A., and Korn, T.M. (1961) *Mathematical Handbook for Scientist and Engineers*. (McGraw-Hill: New York, USA).

Kosok, P. (1940) The role of irrigation in ancient Peru. *Proceedings of the 8th American Scientific Congress*, Vol. 2, Washington DC, USA, pp. 168–178.

Lagrange, J.L. (1781) Mémoire sur la Théorie du Mouvement des Fluides. (Memoir on the Theory of Fluid Motion). In *Oeuvres de Lagrange* (Gauthier-Villars: Paris, France) (printed in 1882) (in French).

Lane, E.W., and Kalinske, A.A. (1941) Engineering calculations of suspended sediment. *Transactions of the American Geophysics Union*, **20**.

Lanning, E.P. (1967) *Peru Before the Incas*. (Prentice-Hall: Englewood Cliffs, NJ, USA).

Laursen, E.M. (1958) The total sediment load of streams. *Journal of Hyd. Div.*, **84**, No. HY1, Paper 1530, pp. 1–36.

Lesieur, M. (1994) *La Turbulence*. (*The Turbulence*.) (Presses Universitaires de Grenoble: Collection Grenoble Sciences: France) 262 pages (in French).

Levin, L. (1968) *Formulaire des Conduites Forcées, Oléoducs et Conduits d'Aération*. (*Handbook of Pipes, Pipelines and Ventilation Shafts*.) (Dunod: (Paris, France) (in French).

Li, D., and Hager, W.H. (1991) Correction coefficients for uniform channel flow. *Canadian Journal of Civil Engineering*, **18**, 156–158.

Liggett, J.A. (1993) Critical depth, velocity profiles and averaging. *Journal of Irrigation and Drainage Engineering*, **119**, 416–422.

Liggett, J.A. (1994) *Fluid Mechanics*. (McGraw-Hill: New York, USA).

McKay, G.R. (1978) Design principles of minimum energy waterways. *Proceedings of the Workshop on Minimum Energy Design of Culvert and Bridge Waterways*, Australian Road Research Board, Melbourne, Australia, Session 1, pp. 1–39.

McMath, R.E. (1883) Silt movement by the Mississippi. *Van Nostrand's Engineering Magazine*, pp. 32–39.

Manning, R. (1890) On the flow of water in open channels and pipes. *Institution of Civil Engineers of Ireland*.

Marchi, E. (1993) On the free-overfall. *Journal of Hydraulic Research*, **31**, 777–790.

Mariotte, E. (1686) *Traité du Mouvement des Eaux et des Autres Corps Fluides*. (*Treaty on the Motion of Waters and other Fluids*.) (Paris, France) (in French (Translated by J.T. Desaguliers, 1718) (Senex and Taylor: London, UK).

Mason, J.A. (1957) *The Ancient Civilizations of Peru*. (Penguin: Harmondsworth, UK).

Maynord, S.T. (1991) Flow resistance of riprap. *Journal of Hydraulic Engineering*. **117**, 687–696.

Meunier, M. (1995) Compte-Rendu de Recherches No. 3 BVRE de DRAIX. (Research Report No. 3 Experimental Catchment of Draix) *Etudes CEMAGREF, Equipements pour l'Eau et l'Environnement*, No. 21, 248 pages (in French).

Meyer-Peter, E. (1949) Quelques problèmes concernant le charriage des matières solides. (Some problems related to bed load transport.) *Société Hydrotechnique de France*, No. 2 (in French).

Meyer-Peter, E. (1951) Transport des matières Solides en Général et problème Spéciaux. *Bull. Génie Civil d'Hydraulique Fluviale*, Tome 5 (in French).

Meyer-Peter, E., Favre, H., and Einstein, A. (1934) Neuere Versuchsresultate über den Geschiebetrieb. *Schweiz. Bauzeitung*, **103**, No. 13 (in German).

Miller, A. (1971) *Meteorology*. 2nd edn (Charles Merrill: Colombus OH, USA), 154 pages.

Miller, D.S. (1994) Discharge characteristics. *IAHR Hydraulic Structures Design Manual No. 8, Hydraulic Design Considerations* (Balkema: Rotterdam, The Netherlands), 249 pages.

Mises, R. Von (1917) Berechnung von Ausfluss und Uberfallzahlen. *Z. ver. Deuts. Ing.*, **61**, 447 (in German).

Montes, J.S. (1992a) Curvature analysis of spillway profiles. *Proceedings of the 11th Australasian Fluid Mechanics Conference AFMC*, Vol. 2, Paper 7E-7, Hobart, Australia, pp. 941–944.

Montes, J.S. (1992b) A potential flow solution for the free overfall. *Proceedings of the Institution of Civil Engineers Water Maritime & Energy*, **96**, 259–266. Discussion: 1995, **112**, 81–87.

Montes, J.S., and Chanson, H. (1998) Characteristics of undular hydraulic jumps. Results and calculations. *Journal of Hydraulic Engineering*, **124**, 192–205.

Moody, L.F. (1944) Friction factors for pipe flow. *Transactions, ASME*, **66**, 671–684.

Morelli, C. (1971) *The International Gravity Standardization Net 1971 (I.G.S.N.71)* (Bureau Central de l'Association Internationale de Géodésie: Paris, France).

Navier, M. (1823) *Mémoire sur les Lois du Mouvement des Fluides. (Memoirs on the Laws of Fluid Motion.)* (Mém. Acad. Des Sciences: Paris, France), Vol. 6, pp. 389–416.

Nials, F.L., Deeds, E.E., Moseley, M.E., Pozorski, S.G., Pozorski, S.G., and Feldman, R. (1979a) El Niño: the Catastrophic flooding of coastal Peru. Part I. *Field Museum of Natural History Bulletin*, **50(7)**, 4–14.

Nials, F.L., Deeds, E.E., Moseley, M.E., Pozorski, S.G., Pozorski, S.G., and Feldman, R. (1979b) El Niño: the catastrophic flooding of coastal Peru. Part II. *Field Museum of Natural History Bulletin*, **50**(8), 4–10.

Nielsen, P. (1992) Coastal bottom boundary layers and sediment transport. *Advanced Series on Ocean Engineering*, Vol. 4 (World Scientific: Singapore).

Nielsen, P. (1993) Turbulence effects on the settling of suspended particles. *Journal of Sedimentary Petrology*, **63**, 835–838.

Nikuradse, J. (1932) *Gesetzmässigkeit der turbulenten Strömung in glatten Rohren. (Laws of Turbulent Pipe Flow in Smooth Pipes.)* (VDI-Forschungsheft, No. 356) (in German) (Translated in NACA TT F-10, 359).

Nikuradse, J. (1933) *Strömungsgesetze in rauhen Rohren. (Laws of Turbulent Pipe Flow in Rough Pipes.)* (VDI-Forschungsheft, No. 361) (in German) (Translated in NACA Tech. Memo. No. 1292, 1950).

Phillips, W. (1955) *Qataban and Sheba – Exploring Ancient Kingdoms on the Biblical Spice Routes of Arabia.* (Victor Gollancz: London, UK).

Pitlick, J. (1992) Flow resistance under conditions of intense gravel transport. *Water Res. Res.*, **28**, 891–903.

Poiseuille, J.L.M. (1839) Sur le mouvement des liquides dans le tube de très petit diamètre. (On the movement of liquids in the pipe of very small diameter.) *Comptes Rendues de l'Académie des Sciences de Paris*, **9**, 487 (in French).

Prandtl, L. (1904) Über Flussigkeitsbewegung bei sehr kleiner Reibung. (On fluid motion with very small friction.) *Verh. III Intl. Math. Kongr.*, Heidelberg, Germany (in German) (Also NACA Tech. Memo. No. 452, 1928).

Prandtl, L. (1925) Über die ausgebildete Turbulenz. (On fully developed turbulence.) *Z.A.M.M.*, **5**, 136–139 (in German).

Rajaratnam, N. (1967) Hydraulic jumps. *Advances in Hydroscience*, Vol. 4, V.T. Chow (ed) (Academic Press: New York, USA), pp. 197–280.

Rajaratnam, N., and Muralidhar, D. (1968) Characteristics of the rectangular free overfall. *Journal of Hydraulic Research*, **6**, 233–258.

Rand, W. (1955) Flow geometry at straight drop spillways. *Proceedings, ASCE*, **81**, No. 791, pp. 1–13.

Raudkivi, A.J. (1990) *Loose Boundary Hydraulics*. 3rd edn (Pergamon Press: Oxford, UK).

Raudkivi, A.J., and Callander, R.A. (1976) *Analysis of Groundwater Flow.* (Edward Arnold: London, UK).

Rayleigh, Lord (1883) Investigation on the character of the equilibrium of an incompressible heavy fluid of variable density. *Proceedings of the London Mathematical Society,* **14,** 170–177.

Rehbock, T. (1929) The River Hydraulic Laboratory of the Technical University of Karlsruhe. In *Hydraulic Laboratory Practice* (ASME, New York, USA), 111–242.

Reynolds, O. (1883) An experimental investigation of the circumstances which determine whether the motion of water shall be direct or sinuous, and the laws of resistance in parallel channels. *Philosopical Transactions of the Royal Society London,* **174,** 935–982.

Richter, J.P. (1939) *The Literary Works of Leonardo da Vinci.* 2nd edn (Oxford University Press: London, UK) 2 volumes.

Rijn, L.C. van (1984a) Sediment transport, Part I: bed load transport. *Journal of Hydraulic Engineering,* **110,** 1431–1456.

Rijn, L.C. van (1984b) Sediment transport, Part II: suspended load transport. *Journal of Hydraulic Engineering,* **110,** 1613–1641.

Rijn, L.C. van (1984c) Sediment transport, Part III: Bed forms and alluvial roughness. *Journal of Hydraulic Engineering,* **110,** 1733–1754.

Rijn, L.C. van (1993) *Principles of Sediment Transport in Rivers, Estuaries and Coastal Seas.* (Aqua: Amsterdam, The Netherlands).

Rouse, H. (1936) Discharge characteristics of the free overfall. *Civil Engineering,* **6,** 257.

Rouse, H. (1937) Modern conceptions of the mechanics of turbulence. *Transactions, ASCE,* **102,** 463–543.

Rouse, H. (1938) *Fluid Mechanics for Hydraulic Engineers.* (McGraw-Hill: New York, USA) (also Dover: New York, USA, 1961, 422 pages).

Rouse, H. (1943) Energy loss at the base of a free overfall – Discussion. *Transactions, ASCE,* **108,** 1383–1387.

Rouse, H. (1946) *Elementary Mechanics of Fluids.* (John Wiley & Sons: New York, USA), 376 pages.

Sarrau (1884) *Cours de Mécanique.* (*Lecture Notes in Mechanics.*) (Ecole Polytechnique: Paris, France) (in French).

Schetz, J.A. (1993) *Boundary Layer Analysis.* (Prentice Hall: Englewood Cliffs, USA).

Schlichting, H. (1979) *Boundary Layer Theory.* 7th edn (McGraw-Hill: New York, USA).

Schnitter, N.J. (1967) A short history of dam engineering. *Water Power,* **19,** 142–148.

Schnitter, N.J. (1994) *A History of Dams: the Useful Pyramids.* (Balkema: Rotterdam, The Netherlands).

Schoklitsch, A. (1914) *Über Schleppkraft un Geschiebebewegung.* (Engelmann: Leipzige, Germany) (in German).

Schoklitsch, A. (1930) *Handbuch des Wasserbaues.* (*Handbook of Hydraulic Structures.*) (Springer: Vienna, Austria).

Schoklitsch, A. (1950) *Handbuch des Wasserbaues.* (*Handbook of Hydraulic Structures.*) 2nd edn (Springer: Vienna, Austria).

Scimemi, E. (1930) *Sulla Forma delle Vene Tracimanti.* (*The Form of Flow Over Weirs.*) (L'Energia Elettrica: Milano), Vol. 7, No. 4, pp. 293–305 (in Italian).

Shields, A. (1936) *Anwendung der Aehnlichkeitsmechanik und der Turbulenz Forschung auf die Geschiebebewegung.* (Mitt. der Preussische Versuchanstalt für Wasserbau und Schiffbau: Berlin, Germany), No. 26.

Smith, N. (1971) *A History of Dams.* (The Chaucer Press, Peter Davies: London, UK).

Spiegel, M.R. (1968) *Mathematical Handbook of Formulas and Tables.* (McGraw-Hill, New York, USA).

Stokes, G. (1845) *Transactions of the Cambridge Philosophical Society,* **8.**

Stokes, G. (1851) On the effect of internal friction of fluids on the motion of pendulums. *Transactions of the Cambridge Philosophical Society*, **9**, Part II, 8–106.

Straub, L.G. (1935) Missouri River Report. *House Document 238* (US Government Printing Office: Washington DC, USA).

Streeter, V.L., and Wylie, E.B. (1981) *Fluid Mechanics*. 1st SI Metric Edition, (McGraw-Hill: Singapore).

Strickler, A. (1923) *Beiträge zur Frage der Geschwindigkeitsformel und der Rauhligkeitszahlen für Ströme, Kanäle und Geschlossene Leitungen. (Contributions to the Question of a velocity Formula and Roughness data for Streams, Channels and Closed Pipelines.)* Vol. 16 (Mitt. des Eidgenössischen Amtes für Wasserwirtschaft, Bern, Switzerland) (in German) (Translation T-10, W.M. Keck, Laboratory of Hydraulics and Water Resources, California Institute of Technology, USA, 1981).

Sumer, B.M., Kozakiewicz, A., Fredsøe, J., and Deigaard, R. (1996) Velocity and concentration profiles in sheet-flow layer on movable bed. *Journal of Hydraulic Engineering*, **122**, 549–558.

Sutherland, W. (1893) The viscosity of gases and molecular forces. *Phil. Mag.*, Series 5, 507–531.

Swanson, W.M. (1961) The magnus effect: a summary of investigations to date. *Journal of Basic Engineering, Transactions ASME, Series D*, **83**, 461–470.

Takahashi, T. (1991) *Debris Flow*. IAHR Monograph (Balkema: Rotterdam, The Netherlands).

Thompson, P.A. (1972) *Compressible Fluid Dynamics*. (McGraw-Hill: New York, USA) 665 pages.

Tison, L.J. (1949) Origine des ondes de sable (ripple-marks) et des bancs de sable sous l'action des courants. (Origin of sand waves (ripple marks) and dunes under the action of the flow) *Proceedings of the 3rd Meeting of the International Association for Hydraulic Structures Research*, Grenoble, France, 5–7 September, Paper II-13, 1–15 (in French).

Tricker, R.A.R. (1965) *Bores, Breakers, Waves and Wakes*. (American Elsevier: New York, USA).

Ubbelohde-Doering, H. (1967) *On the Royal Highways of the Inca Civilizations of Ancient Peru.* (Thames and Hudson: London, UK).

US Department of the Interior (1987) *Design of Small Dams*. 3rd edn (Bureau of Reclamation: Denver CO, USA).

Vallentine, H.R. (1969) *Applied Hydrodynamics*. SI edition. (Butterworths: London, UK)

Vanoni, V.A. (1946) Transportation of suspended sediment in water. *Transactions, ASCE*, **111**, 67–133.

Wan, Z., and Wang, Z. (1994) *Hyperconcentrated Flow*. IAHR Monograph, (Balkema: Rotterdam, The Netherlands), 290 pages.

Wasson, R.J., and Galloway, R.W. (1986) Sediment yield in the barrier range before and after European settlement. *Australian Rangeland Journal*, **2**(2) 79–90.

White, M.P. (1943) Energy loss at the base of a free overfall – discussion. *Transactions, ASCE*, **108**, 1361–1364.

Willcocks, W. (1919) *From the Garden of Eden to the Crossing of the Jordan*. (E.&F.N. Spon: New York, USA).

Wisner, P. (1965) Sur le rôle du critère de Froude dans l'étude de l'entraînement de l'air par les courants à grande vitesse. (On the role of the Froude criterion for the study of air entrainment in high velocity flows.) *Proceedings of the 11th IAHR Congress*, Leningrad, USSR, Paper 1.15 (in French).

Wood, I.R. (1991) Air entrainment in free-surface flows. *IAHR Hydraulic Structures Design Manual No. 4, Hydraulic Design Considerations* (Balkema: Rotterdam, The Netherlands), 149 pages.

Wood, I.R., Ackers, P., and Loveless, J. (1983) General method for critical point on spillways. *Journal of Hydraulic Engineering, ASCE*, **109**, 308–312.

Yalin, M.S. (1964) Geometrical properties of sand waves. *Journal of Hyd. Div.*, **90**, No. HY5, 105–119.

Yalin, M.S., and Karahan, E. (1979) Inception of sediment transport. *Journal of Hyd. Div.*, **105**, No. HY11, 1433–1443.

Yen, B.C. (1991a) Hydraulic resistance in open channels. In *Channel Flow Resistance: Centennial of Manning's Formula*, B.C. Yen (Ed), (Water Resources: Littleton CO, USA) pp. 1–135.

Yen, B.C. (1991b) *Channel Flow Resistance: Centennial of Manning's Formula.* (Water Resources: Littleton CO, USA), 453 pages.

Notes

Idel'cik or Idelchik refer to the same Russian author.

Darcy and Bazin (1865) was published posthumously by Bazin (1865a,b).

Additional bibliography

The following includes several materials of pedagogical value. They may assist the reader (student or lecturer) in gaining a good feel for open channel hydraulics and to visualize practical applications of the lecture material.

Bibliography: history of hydraulics

Carvill, J. (1981) *Famous Names in Engineering*. (Butterworths: London, UK).

Galbrecht, G. (1987) *Hydraulics and Hydraulic Research: a Historical Review*. (Balkema: Rotterdam, The Netherlands).

Rouse, H., and Ince, S. (1957) *History of Hydraulics*. (Iowa Institute of Hydraulic Research: Iowa City, USA), 269 pages.

Schnitter, N.J. (1994) *A History of Dams: the Useful Pyramids*. (Balkema: Rotterdam, The Netherlands).

Smith, N. (1971) *A History of Dams*. (The Chaucer Press, Peter Davies: London, UK).

Bibliography: Audio-visual material

Apelt, C.J. (1994) *The Minimum Energy Loss Culvert*. Video cassette VHS, colour, Department of Civil Engineering, University of Queensland, Australia, 18 Minutes.
Comments: Utilized to reduce flooding of stormwater plains drains, the benefits of minimum energy loss culverts, designed by Gordon McKay and Colin Apelt, are illustrated by comparison with the flow capacity of the standard culvert. *It is a very good teaching tool* to introduce the concept of specific energy and the application to culvert design.

Lerner, B. (1994) *After the Flood*. Video cassette VHS, colour, SBS – The Cutting Edge, 48 Minutes.
Comments: In order to control flooding of the river Brahmaputra, Bangladesh, water engineers propose to change the width and course of the river. Along the Mississippi, USA, similar water engineering is the alleged cause of the Mississippi's flooding in 1993. Archival film helps to illustrate some of the problems to be overcome. produced by Bettina Lerner. *Very good documentary* dealing with practical applications of open channel hydraulics, sediment transport, catchment hydrology and environmental impact of hydraulic structures.

Mississippi Floods 1993. Video cassette VHS, colour, Australian Channel News, 4 Minutes.
Comments: News footage of the Mississippi flood in 1993. Footage from Australian News
Channels 7, 9, 10, SBS.

St Anthony's Falls Hydraulic Laboratory (1947) *Some Phenomena of Open Channel Flow.* Video
cassette NTSC, B&W, SAF Hydraulic Laboratory, Minneapolis MN, USA, 33 Minutes.
Comments: In this programme, open channel flow lecture material is demonstrated. It looks
at: supercritical and subcritical flow, hydraulic jumps, hydraulic drops, specific energy curve,
pressure momentum curve, critical depth, travel of surface waves in channels flowing at
critical, subcritical and supercritical velocities, uphill flow, abrupt gate closure, movable
bed channels, and more.

US Bureau of Reclamation (1988?) *Challenge at Glen Canyon Dam.* Video cassette VHS,
colour, US Department of Interior, Denver, Colorado, USA, 27 Minutes.
Comments: This programme is divided into two parts. The first part examines flood waters of
the Colorado River system. The second part describes the damage caused to the Glen Canyon
dam spillways following the excessive amount of water that flowed into Lake Powell due to
heavy snow falls late in the season. The programme then goes on to examine the method
used to repair the damage after the flood has passed. It is a *superb educational movie* for
both civil, environmental and hydraulic engineering students. It is quite entertaining.

US National Committee for Fluid Mechanics (1967) *The Hydraulic Surge Wave.* Video cassette
VHS, B&W, Education Development Center, USA, 4 Minutes.
Comments: Film of experiments illustrating the hydraulic surge wave, the hydraulic jump
and the analogy between hydraulic jump and surge.

Suggestion/correction form

Thank you for your comments and suggestions regarding the book. They will be most helpful to improve the book in the near future. If you find a mistake or an error, please record the error(s) on this page and send it to the author:

Dr H. Chanson

Department of Civil Engineering, The University of Queensland, Brisbane QLD 4072, Australia

Fax: (61 7) 33 65 45 99

Email: h.chanson@mailbox.uq.edu.au

Suggestion/correction form

Contact

Name	
Address	
Tel.	
Fax	
E-mail	

Description of the suggestion, correction, error

Page number	
Line number	
Figure number	
Equation number	

Proposed correction

Author index

Ackers, P. 322, 323, 411, 412
Agostini, R. 359
Albright, F.P. 437, 440
Alembert, Jean Le Rond D' xvi
Alexander, J. 161, 238
Anderson, A.G. 215
Apelt, C.J. xxvii, 235, 236, 365, 389, 391,
    393, 421, 431, 481
Avellan, F. xx

Bagnold, R.A. 191, 200
Bakhmeteff, B.A. xxv, 359
Bauer, W.J. 411
Bazin, H. xxi, xxiv, 82, 265, 329
Belahadji, B. xx
Bélanger, J.B. xvii, xxv, 37, 57, 108, 265,
    285, 319, 335
Belidor, B.F. De xviii, xxii
Belyakov, A.A. 449, 451
Beyer, W.H. 131
Bhowmik, N.G. 150
Billard, J.Y. xx
Bizzarri, A. 359
Bos, M.G. 44, 322, 323
Bossut, Abbé C. xviii
Bossy, H.G. 377
Boussinesq, J.V. xviii, xxvii, 28
Boys, P.F.D. Du xviii, 196, 198
Bradley, J.N. 416
Bremen, R. 64, 337
Bresse, J.A. xix, 110, 265
Briancon-Marjollet, L. xx
Brooks, N.H. 215
Buat, P.L.G. Du xix, xxv
Buckingham, E. 265
Buckley, A.B. 452

Cain, P. 411
Callander, R.A. 163
Carvill, J. 18, 481
Chang, C.J. 155
Chanson, H. 7, 61, 64, 216, 238, 249, 270,
    330, 333, 336, 349, 358, 362, 363, 370,
    393, 406, 411, 412, 414, 415
Chen, C.L. 27, 85, 178, 218, 220
Cheng, N.S. 169
Chien, N. 199
Chiew, Y.M. 187
Chow, V.T. 63, 64, 85, 109, 287, 319, 336,
    339
Clemmens, A.J. 44
Colebrook, C.F. 75, 77, 260
Coleman, N.L. 215
Coles, D. 177
Comolet, R. 75, 80, 82
Concrete Pipe Association Of Australasia
    376, 378, 379, 397, 399
Cook, O.F. 464
Coriolis, G.G. xxi, 29
Couette, M. xxi
Creager, W.P. xxi, 56, 327, 329

Danilevslkii, V.V. xxiv
Darcy, H.P.G. xxi, xxiv, 82
Deeds, E.E. 459
Degrémont, 129
Deigaard, R. 191, 235
Dooge, J.C.I. 83
Dupuit, A.J.E. 265

Einstein, A. 222
Einstein, H.A. 198, 199, 226
Elder, J.W. 215

Engelund, F. 169, 199, 226, 235, 237, 238, 239, 240, 241, 242, 244, 246, 247
Eurenius, J. 437

Falvey, H.T. 61
Farrington, I.S. 462, 464
Favre, H. 199, 204
Fawer, C. xxxiii, 63
Fedoroff, N.V. 359
Feldman, R. 459
Fernandez Luque, R. 201
Fick, A.E. xxiii
Fielding, C. 161, 238
Fischer, H.B. 215
Fourier, J.B.J. xxiii
Franc, J.P. xx
Frechou, D xx
Fredsøe, J. 191, 235
Fruman, D.H. xx

Galbrecht, G. 451
Galloway, R.W. 248
Ganchikov, V.G. 440
Garbrecht, G. 18, 449, 481
Gauckler, P.G. xxvi, 288
Gibbs, R.J. 169, 174
Gilbert, G.K. 199
Graf, W.H. 159, 171, 173, 196, 199, 204, 215, 217, 222, 226
Guy, H.P. 226, 238, 240

Hager, W. 320, 322, 329, 337, 339, 361
Hager, W.H. xxxi, 29, 57, 64, 65
Hansen, E. 169, 199, 226, 238, 239, 240, 241, 242, 244, 246, 247
Harleman, R.F. xxxi
Harrison, A.J.M. 322, 323
Hathaway, G.A. 448, 452
Hee, M. 371, 375, 395
Hellström, B. 249
Helmholtz, H.L.F. xxvi
Henderson, F.M. 24, 42, 44, 46, 65, 71, 72, 73, 74, 75, 77, 81, 101, 113, 115, 159, 259, 264, 269, 272, 275, 290, 294, 319, 323, 339, 356, 369, 370, 377, 399
Herr, L.A. 377
Herschy, R. 323
Hinds, J. 56
Hinze, J.O. 190
Howe, J.W. xxviii
Humber, W. 333

Hydropower & Dams 452

Idelchik, I.E. 77, 85, 260, 394
Imberger, J. 215
Ince, S. 481
International Commission On Large Dams 152, 417
Ippen, A.T. xxxi
ISO 129

James, P. 249
Jevons, W.S. xxv
Julien, P.Y. 161, 165, 173, 191, 213, 217, 223, 224
Justin, J.D. 56

Kalinske, A.A. 215
Kamphuis, J.W. 235
Karahan, E. 184
Karimi, A. xx
Kawagoshi, N. 64, 337
Kazemipour, A.K. 235, 236
Keller, R.J. 411
Kelvin, Lord xxvi
Kennedy, J.F. 161, 238
Keulegan, G.H. 82
Knapp, F.H. 329
Koh, R.C.Y. 215
Korn, G.A. 131
Korn, T.M. 131
Kosok, P. 463
Kozakiewicz, A. 191, 235
Kueny, J.L. xx

Lagrange, J.L. xxxiv
Lane, E.W. 215
Lanning, E.P. 464
Laursen, E.M. 161, 238
LeBaron Bowen Jr, R. 437, 440
Lerner, B. 481
Lesieur, M. xxxiii
Levin, L. 394
Li, Damei 29
Liggett, J.A. 18, 29, 41, 42, 43, 72, 74, 258
Link, D.A. 169, 174
List, E.J. 215
Loveless, J. 411, 412

Manning, R. xxvi, 83, 288
Marchi, E. 361
Mariotte, E. xxvii

Masetti, M. 359
Mason, J.A. 461, 464
Matthews, M.D. 169, 174
Matzke, A.E. xxv
Maynord, S.T. 235
McKay, G.R. 393
McMath, R.E. 149
Meunier, M. 224
Meyer-Peter, E. 198, 199, 204
Michel, J.M. xx
Miller, A. 127, 319, 323
Miller, D.S. 319, 323
Mises, R. Von 323
Montes, J.S. 329, 336, 358
Moody, L.F. 75, 76
Morelli, C. 125
Moseley, M.E. 459
Munavvarov, Z.I. 440
Muralidhar, D. 361

Navier, M. xxvii
Nials, F.L. 459
Nielsen, P. 171, 198, 201, 204, 220, 303
Nikuradse, J. xxviii

Papetti, A. 359
Park, C.C. 462, 464
Parker, G. 187
Perkins, J.A. 322, 323
Peterka, A.J. 416
Phillips, W. 437
Pitlick, J. 208
Poiseuille, J.L.M. xxviii
Pozorski, S.G. 459
Prandtl, L. xxvii

Rajaratnam, N. 61, 361
Rand, W. 358
Raslan, Y. 161
Rastogi, A.K. 411
Raudkivi, A.J. 163, 191
Rayleigh, Lord xxv
Rehbock, T. 323
Replogle, J.A. 44
Reynolds, O. xxiv, 265
Richardson, E.V. 226, 238, 240
Richter, J.P. 333
Rijn, L.C. Van 171, 172, 183, 186, 187, 191,
        196, 201, 204, 206, 208, 215, 218, 219,
        226, 235, 237, 238, 244
Rouse, H. 25, 35, 44, 56, 213, 358, 361, 481

Sarrau 265
Schetz, J.A. 177
Schlichting, H. xxii, 75, 77, 177, 178, 186,
        190, 272
Schnitter, N.J. xxiv, xxvii, 447, 451, 481
Schoklitsch, A. 198
Schwalt, M. 320, 322
Scimemi, E. 329
Shields, A. 183, 198
Simons, D.B. 226, 238, 240
Smith, N. xxxiv, 439, 440, 451, 481
Spiegel, M.R. 125
St Anthony's Falls Hydraulic Laboratory
        482
Stokes, G. 168, 169
Straub, L.G. xxxii, 198
Streeter, V.L. 10, 28, 42, 71, 72, 77, 81, 101,
        126, 186
Strickler, A. 84
Sumer, B.M. 191, 235
Sutherland, W. 128
Swanson, W.M. xxvi

Takahashi, T. 224
Thompson, P.A. 43
Tison, L.J. 238
Tricker, R.A.R. 67

Ubbelohde-Doering, H. 463, 464
US Bureau of Reclamation 339, 359, 376, 482
US Department of the Interior 328, 482
US National Committee for Fluid
        Mechanics 482

Vallentine, H.R. 322
Van Beek, R. 201
Vanoni, V.A. 213

Wan, Zhaohui 221, 222, 223, 224
Wang, Zhaoyin 221, 222, 223, 224
Wasson, R.J. 248
White, M.P. 362
White, W.R. 322, 323
Whitmore, R.L. 406
Willcocks, W. 448, 449, 452
Wisner, P. 61
Wood, I.R. 7, 61, 270, 330, 349, 411, 412
Wylie, E.B. 10, 28, 42, 71, 72, 77, 81, 101,
        126, 186

Yalin, M.S. 184, 238
Yen, B.C. 85

Subject index

Page numbers in **bold** indicate particularly important references

Abutment xvi
Académie des Sciences de Paris xvi
Accretion xvi, **229–33**
Adiabatic xvi, 126
Advection xvi, 210–12
Advective diffusion, sediment suspension 210–16
Adverse slope 109–10, 188
Aeration device (or aerator) xvi
Afflux xvi, **369**, 386, 395
Aggradation xvi, **229–33**
Air xvi
Air bubble 4
Air concentration xvi
Air entrainment 7, 57, 61, 267, 273, 335, **349**
Air flow 184, 186, 267, 272
Air, physical properties **125–8**
Alembert (d') xvi, 181
Alternate depth(s) xvi, **35–6**, **143–5**
Analogy, gas flow and free-surface flow 42–**4**, 267, 272
Analytical model xvi
Angle of repose **171–2**, **183–7**
Antidunes **157–61**, 238, 239, Plate 9
Apelt, C.J. xvii, 383, 386–92, 421
Apron xvii, **341–3**, 348
Aqueduct xvii, Plate 5
Arch dam xvii, 150–2, 248–51, Plate 6
Arched dam xvii
Archimedes xvii
Ariostotle xvii
Armouring xvii, 187
Assyria xvii
Avogadro number xvii, xviii

Backwater xvii, **106–13**, 257, 284–97
Backwater calculation xvii, **106–8**, 108–13, **284–97**, 305–8
Backwater equation **107**, **284–5**, 305–8
Baffle blocks 333, 335–6, **337–8**, 340–1
Bagnold, R.A. xvii, 191, 200
Bakhmeteff, B.A. xvii
Barrage xvii
Barré de Saint-Venant, A.J.C. xvii, 14
Barrel xvii, **365–6**, 367–97
Bazin, H.E. xvii, xxi, 82, 265, 327–9
Bed armouring 187
Bed form(s) xvii, **157–61**, 231–3, 233–40, Plates 8 and 9
Bed load xviii, 157, **175–88**, **195–208**, 301–3
Bed load, formulae for **196–202**, **237**, 302–3
Bed slope 79–80, 90–1, 99–100, 274
Bélanger, J.B.C. xviii, 57, 108, 285, 319, 335
Bélanger equation xviii, **57**, 290, 297, 335
Belidor, B.F. de xviii
Bernoulli, D. xviii, 18
Bernoulli equation **17–18**, **21–48**
Bessel, F.W. xviii
Bidone, G. xviii
Biesel, F. xviii
Bingham plastic fluids **222–4**
Blasius, H. xviii, 77
Boltzmann, L.E. xviii
Boltzmann constant xviii
Borda, J.C. de xviii, 236
Borda–Carnot formula **236**
Borda mouthpiece xviii
Bore xviii, **67**
    see also Surge

Bossut, C. xviii
Bottom outlet xviii
Boundary layer xviii, **176–8**, 186, 330–1,
    393–4, 410–13
Boussinesq, J.V. xviii, 28
Boussinesq coefficient xviii, 17, **28–9**, 54
Boussinesq–Favre wave xviii
Boys, P.F. du xviii, **196–8**
Braccio xviii
Braided channel xix
Bresse, J.A.C. xix, xxii, 265, 110
Brink depth at a free overfall 289–91, **355–8**
Broad-crested weir xix, **319–22**
Buat, P.L.G. du xix
Buckingham pi-theorem 265
Buoyancy xix, 165, 167–8
Buttress dam xix
Byewash xix, 333

Candela xix
Carnot, L.N.M. xix
Carnot, S. xix, 236
Cartesian co-ordinate xix
Cascade xix, 355, **362–3**
Cascade design **362–3**
Cataract xix, 7
Catena d'Acqua xix
Cauchy, A.L. de xix, 266
Cauchy number xliii
Cavitation xx, 327, 337
Celsius, A. xx
Celsius degree (or degree centigrade) xx,
    130
Centrifugal pressure 23–5
Chadar xx
Chan Chan 458–67
Channel cross-section, most efficient shape
    100–1
Chézy, A. xx, 81, 288
Chézy coefficient xx, **80–2**, 83–5, **286–9**
Chézy equation **80–2**, **286–8**
Chimu xx, 458–64
Choke xx, 46
Choking flow xx, 46
Chord length xx
Chute blocks, *see* Baffle blocks
Chute-pool, bed form 157–9
Clausius, R.J.E. xx
Clay xx, 162, 173, 187, 194
Clepsydra xx
Cofferdam xx

Cohesive sediment xx, 161, 187
Colbert, J.B. xx
Colebrook–White formula **75–7**, 85, 234,
    260
Compressibility 11, 42–4, 126
Computer program 257, 284, 293–4, 297,
    430–4
Conjugate depth(s) xxi, **55**, 335, 442
Conservation of energy, *see* Energy equation
Conservation of mass, *see* Continuity
    equation
Conservation of momentum, *see* Momentum
    equation
Constriction, *see* Contraction in channel
    width
Continuity equation **10–11**, 16, 26, 53,
    230–3
Continuity equation, for sediment **230–3**
Continuity, application to open channels 11,
    16, 26, 53, 257–8
Contraction in channel width 45–6
Control(s) xxi, **44–6**, 111–12, 289–92
Control section xxi, **44–5**
Control surface xxi
Control volume xxi
Control, effect on backwater profiles **45**,
    111–12, **289–92**
Coriolis, G.G. xxi, 29
Coriolis coefficient xxi, **28–9**, 73, 258, 285,
    287, 305–8
Couette, M. xxi
Couette flow xxi
Craya, A. xxi
Creager profile xxi, 326–9
Crest of spillway xxi, 313–15, **319–29**
Crib xxi
Crib dam xxi, 347, 361
Critical depth xxi, **32–3**, 38, 47–8
Critical flow and control **44–5**
Critical flow conditions xxi, **31–8**, **41–8**,
    108–9, 319, 355–8, 368–9, 385–6
Critical flow conditions, definition **31–2**
Critical flow conditions, occurrence **44–5**
Critical slope **100**, 390–1
Critical-depth meter **44**
Culvert(s) xxi, 311–12, 365, 421
Culvert design **366–97**, 421–34
Culverts, minimum energy loss 365, **382–97**,
    421–30
Culverts, standard box 365, **371–82**, 430–4
Cyclopean dam xxi

Danel, P. xxi
Darcy, H.P.G. xxi, 73
Darcy equation **73**, 85
Darcy law xxi, 163
Darcy–Weisbach friction factor xxi, **73–8**,
    85–7, 234–5, 259–60
Debris xxii, 192
Debris flow 192
Degradation xxii, **229–33**
Density 3, 125–6, 161–4, 222
Depth, alternate, *see* Alternate depth
Depth, conjugate, *see* Conjugate depth
Depth, critical, *see* Critical depth
Depth, normal, *see* Normal depth
Depth, sequent, *see* Sequent depth
Depth, uniform equilibrium, *see* Uniform
    equilibrium flow depth
Descartes, R. xxii
Diffusion xxii, 210–16
Diffusion coefficient xxii, 210–16
Diffusivity xxii, 210–16
Dimensional analysis xxii, **264–70**
Discharge 7, 11
Discharge calculations 35–8, 319–29, 356,
    368, 376
Discharge coefficient, for broad-crested
    weirs 321–2
Discharge coefficient, for ogee crest weirs
    324–9
Discharge coefficient, for sharp-crested weirs
    322–3
Discharge, sediment 195–245
Distorted model study **274–5**, 276–7, 278–9
Diversion channel xxii
Diversion dam xxii
Drag coefficient **74**, 271
Drag coefficient, particle settling **165–9**
Drag reduction xxii, 108
Drag, form **234–7**, 271
Drag, total resistance 234, 241
Drainage layer xxii
Drop xxii, 311
Drop structure(s) xxii, 311, 315, **355–62**
Droplet xxii
Du Boys (or Duboys), *see* Boys, P.F.D. du
Du Boys equation, *see* Boys, P.F.D. du
Du Buat (or Dubuat), *see* Buat, P.L.G. du
Dune bed form resistance 234–7
Dune(s) **157–61**, 231–3, 234–40, Plate 8
Dupuit, A.J.E.J. xxii, 265
Dynamic similarity 262, 266–7

Earth dam xxii
Ecole Nationale Supérieure des Ponts et
    Chaussées xxii
Eddy viscosity xxii, 212
Einstein bed-load formula 198–9
Embankment xxii
Energy dissipator(s) 64–5, 315–16, 333–41,
    355–63
Energy equation **18–20**, **57–60**, **107–8**, 258,
    284–5, 335
Energy loss **57–60**, 335, 363, 391
Energy loss, bed forms 234–7
Energy, specific, *see* Specific energy
Entrainment of air, *see* Air entrainment
Entrainment of sediment 180–4
Erosion, channel bed **229–33**
Escalier d'Eau, *see* Water staircase
Euler, L. xxii, 14
Euler equation 14
Extrados xxii

Face xxii
Fall velocity 165–71
Favre, H. xxiii
Fawer jump, *see* Undular hydraulic jump
Fayum depression 446–52
Fick, A.E. xxiii, 210–12
Fixed bed hydraulics 3–113
Fixed-bed channel xxiii
Flash flood xxiii
Flashboard xxiii
Flashy xxiii
Flat bed (bed form) **157–63**
Flettner, A. xxiii
Flip bucket xxiii, 24, 286
Flood plain, calculations 85–7, **90–1**
Fog xxiii
Forchheimer, P. xxiii
Fortier, A. xxiii
Fourier, J.B.J. xxiii
Free overfall **355–8**
Free-surface xxiii
Free-surface aeration xxiii
    see also Air entrainment
French revolution (Révolution Française)
    xxiii
Friction factor **73–91**, 259–60
Friction factor, Chézy, *see* Chézy
    coefficient
Friction factor, Darcy–Weisbach, *see*
    Darcy–Weisbach friction factor

Friction factor, Gauckler–Manning, *see* Gauckler–Manning coefficient
Friction factor, Strickler, *see* Strickler coefficient
Friction slope **80**, 99, 107–8, 285–9
Frontinus, S.J. xxiii
Froude, W. xxiii, 265, 273
Froude number xxiv, **39–46**, 265, 272–9

G.K. formula xxiv
Gabion xxiv
Gabion dam xxiv
Gas transfer xxiv
Gas-flow analogy 31–44
Gate xxiv
    see also Radial gate; Sluice gate
Gauckler, P.G. xxiv, 82, 288
Gauckler–Manning coefficient **82–3**, 84, 276, 286–9
Gauckler–Manning equation 82–3, 286–8
Gay-Lussac, J.L. xxiv
Ghaznavid xxiv
Gold Creek dam 402–6
Gradually-varied flow xxiv, 106, 285
Gradually-varied flow, calculations 105–13, 285–96, 305–8
Graphical method, discharge 35–7, 378–9
Graphical method, specific energy 31–6
Gravity, effects 7–8, 40, 79–80, 272–5
Gravity acceleration 124–5
Gravity dam xxiv
Grille d'eau xxiv

Hasmonean xxiv
Ha-Uar dam 446–57
Head loss **73–4**, 236, 258–9, 335, 363, 382, 391–2
Head loss coefficient 73, 236, 377, 433
Head loss, in contraction 433
Head loss, in expansion 236, 391–2, 433
Helmholtz, H.L. von xxiv
Hennin, G.W. xxiv
Hero of Alexandria xxiv
Himyarite xxiv, 440
Hohokams xxv
Hokusai, Katsushita xxv
Huang Chun-Pi xxv
Hydraulic diameter xxv, 47, **73–4**
Hydraulic fill dam xxv

Hydraulic jump xxv, 16–17, 38, **56–7**, 68–9, 105, 289–90
Hydraulic jump, at drop structure 358
Hydraulic jump, basic equations **56–67**
Hydraulic jump, classification **63–4**, 335
Hydraulic jump, energy loss 57–60, 335
Hydraulic jump, length 60, 64, 337
Hydraulic jump, numerical modelling 289–90, 297
Hydraulic jump, stilling basin 66–7, 333–41
Hydraulic jump, undular 58, 63–4
Hydraulic radius 74
Hydrostatic pressure distribution **4–6**, 22–5
Hyperconcentrated flow xxv, 192–4, 221–4

Ideal fluid xxv, 17–18, 21
Idle discharge xxv
Inca xxv, 458–64
Inflow xxv
Inlet xxv, 365–8
Intake xxv, 365–8
Interface xxv
International system of units, *see* Système international d'unités
Intrados xxv
Invert xxv, 327, 366–8
Inviscid flow xxv
Irregular channel(s) 85–91
Irrotational flow xxv

JHRC xxv, 342–8
JHRL xxv, 342–8
Jet d'eau xxv
Jevons, W.S. xxv

Karman, T. von xxvi
Karman constant (or von Karman constant) xxvi, 27, **177–8**, 212–13, 215, 217–18
Kelvin (Lord) xxvi
Kelvin degree 130
Kelvin–Helmholtz instability xxvi
Keulegan, H.H. xxvi, 82, 433
Kinematic similarity 262
Kinetic energy correction coefficient, *see* Coriolis coefficient

Lagrange, J.L. xxvi, 446–57
Lake Moeris 446–57
Laminar flow xxvi, **74**, 77, 166, 168–9

Laplace, P.S. xxvi
LDA velocimeter xxvi
Left abutment xxvi
Left bank (left wall) xxvi
Leonardo da Vinci xxvi, 333
Lift on sediment particle 180–1
Lining xxvi
Longitudinal free-surface profile(s) 105–15, 284–97
Lumber xxvi

Mach, E. xxvi, 265
Mach number, *see* Sarrau–Mach number
Magnus, H.G. xxvi
Magnus effect xxvi
Manning, R. xxvi, 83, 288
Manning coefficient, *see* Gauckler–Manning coefficient
Manning equation, *see* Gauckler–Manning equation
Marib dam 437–45
Mariotte, E. xxvii
Masonry dam xxvii
Mathematical functions 131–42
McKay, G.M. xxvi, 393, 394, 417, 421
Meandering channel xxvii
MEL culvert, *see* Minimum Energy Loss culvert
Metric system, *see* Système métrique
Meyer-Peter bed-load formula 196–9
Mild slope **99–100**
Minimum energy loss culvert(s) xxvii, 365–6, **382–97**, 421–30
Minimum energy loss weir(s) 313–14, 417–21
Mississippi river 149–50, 274
Mixing length xxvii
Moche river 458–69
Mochica xxvii, 458–64
Model(s) 257, **262–308**
Model(s), computer **284–97**, 305–8, 430–4
Model(s), physical **262–79**, 301–4
Mole xxvii
Momentum correction coefficient, *see* Boussinesq coefficient
Momentum equation 9, **11–18**, **52–91**, 258, 335–6
Momentum exchange coefficient xxvii
    see also Eddy viscosity
Momentum flux 15–16, 54, 59
Momentum, definition **15**, **59**

Momentum, function 55
Momentum, rate of change **15**, **53**
Moody diagram **75–7**, 85, 271
Moor xvii
Morning-Glory spillway xxvii, 316, 318
Motion equation 9, **11–15**
Mount St Helens 206–8
Movable boundary hydraulics **149–253**, 275, 301–4
Mud xxvii, 150, 192
Mughal (or Mughul or Mogul or Moghul) xxvii

Nabataean xxvii
Nappe flow xxvii, 355–63
Natural rivers 7, 149–50, 274–5
Navier, L.M.H. xxvii, 14
Navier–Stokes equation xxvii, **11–15**, 17
Negative surge xxviii, **71–2**
Newton, I. xxviii, 4
Newton's law of motion **9–10**
Newton's law of viscosity **3–4**, 222–3
Nielsen's bed-load formula 198, 200–2
Nikuradse, J. xxviii, 77
Nile river xix, 7, 149, 446–52
Non uniform equilibrium flow xxviii, **105–13**, 149, 284–97
Non-Newtonian fluid **4**, 222–4
Non-rectangular channel sections **46–8**, 85–6, 90–1, 100–1
Nonuniform flow **105–13**
Normal depth xxviii, **99–100**, 101–5
Nudgee road waterway 384, 406–9, 421–3

Obvert xxviii, 366–8
One-dimensional flow xxviii
One-dimensional model xxviii, 284
Orifice flow xviii, 385
Outflow xxviii
Outlet xxviii, 365–8

Particle fall velocity, *see* Fall velocity
Pascal (units) xxviii, 125, 129–30
Pascal, B. xxviii, 4–6
Pascal's law 5
Pelton turbine (or wheel) xxviii
Pervious zone xxviii
Piezometric head 21–2
Pitot, H. xxviii, 25
Pitot tube xxviii, **25–6**
Pitting xxviii

Plato xxviii
Plunging jet xxviii, 333–4
Poiseuille, J.L.M. xxviii
Poiseuille flow xxviii
Poisson, S.D. xxviii, 14
Positive surge xxix, **67–72**
Potential flow xxix
Prandtl, L. xxix, 25
Pressure **4–6**
Pressure distribution, hydrostatic **4–6, 22–3**
Pressure distribution, non hydrostatic **22–5**
Prony, G.C.F.M.R. de xxix
Prototype 149–55, 258, 261–79

Quasi-steady-flow analogy 67–92

Radial gate xxix
Rankine, W.J.M. xxix
Rapidly varied flow xxix, 56–72, **105–7**,
    285–6, 293
Rayleigh xxix
Rectangular channel sections 21–46, 52–84
Reech, F. xxix, 265, 273
Rehbock, T. xxix, 323
Renaissance xxix
Reservoir sedimentation **150–5**, 248–53
Reynolds, O. xxix, 265
Reynolds number xxix, **74–5, 265–72**
Rheology xxix, 192, 222–3
Riblet xxix
Richelieu xxix
Riemann, B.G.F. xxix
Right abutment xxix
Right bank (right wall) xxix
Ripples **157–61**, 240
Riquet, P.P. xxix
Rockfill xxix
Rockfill dam xxx
Roll wave xxx
Roller xxx, 56–61, 333–8
Roller Compacted Concrete (RCC) xxx
Roughness 75–89, 233–8
Roughness, bed forms 233, **235–8**
Roughness, equivalent roughness height
    75–7, 235
Roughness, skin friction 75–7, 233–5
Rouse, H. xxx, 213
Rouse number **213**

SAF xxx, 338–9
Sabaen xxx

Saint-Venant, *see* Barré de Saint-Venant,
    A.J.C.
Saltation xxx, 195–6
Sarrau xxx, 265
Sarrau–Mach number xxx, 42–3, 265
Scalar xxx
Scale effect xxx, **267–70**, 274–7
Scale in model studies 262–4
Scale in model studies, distortion 274–5
Scale in model studies, sediment transport
    275, 301–4
Scour xxx, **229–33**
Sediment xxx, 157
Sediment concentration 199–201, 210–17
Sediment load xxx, 157
Sediment transport xxx, 157, 157–240,
    301–4
Sediment transport capacity xxx, 225–40,
    301–4
Sediment yield xxx
Seepage xxx
Sennacherib (or Akkadian Sin-Akhkheeriba)
    xxx
Separation xxx, 181, 393–4
Separation point xxxi
Sequent depth(s) xxi, xxxi, 55, 335, 342
    see also Conjugate depth(s)
Settling motion **165–71**
Settling velocity **165–71**, 172–3
Settling, individual particle 165–9
Sewage xxi
Sewer xxxi
Shear Reynolds number **74–5**
Shear stress **74–5**, 79–80, 234–8
Shear stress, in natural river 233–8
Shear velocity **74–5**, 176–9, 182–6
Shields diagram **183–6**
Shields parameter **183–6**
Shock waves xxxi
Side-channel spillway xxxi
Silt 162, 173, 187, 194
Siltation, individual particle, *see* Settling
    motion
Siltation, reservoir, *see* Reservoir
    sedimentation
Similarity, dynamic **261–4**, 266–7
Similarity, geometric **261–4**
Similarity, kinematic **261–4**
Similarity, scale effects **267–70**
Similitude xxxi, 40–1, **261–79**
Siphon xxxi

Siphon-spillway xxxi
Ski jump 24, 286
    see also Flip bucket
Skimming flow xxxi, 315, 330–1, 362,
    411–12, 417
Slope xxxi
Sluice gate xxxi, 33–5
Small dams, hydraulic design 313–63
Soffit xxxi, 366, 368
Specific energy xxxi, **21–46**
Spillway xxxi, 313–63
Spillway design 313–49
Spillway overflow 313–63
Spillway overflow, chute **330–3**, 410–17
Spillway overflow, crest **319–29**
Spillway overflow, energy dissipation
    333–41
Splitter xxxi
Spray xxxi
Stage-discharge curve xxxi
Stagnation point xxxi
Staircase xxxi
Stall xxxi
Standing waves (bed form) **157–61**, 238
Static fluid **4–6**
Steady flow xxxii
Steep slope **99–100**
Stepped spillway 315, 318, 330–3, 355–63,
    410–17
Stepped spillway, nappe flow 355–63
Stepped spillway, skimming flow 315, 330–1,
    362, 411–12, 417
Step-pool, bed form 157–9
Stilling basin(s) xxxii, 64–5, **333–41**
Stokes, G.G. xxxii, 14
Stokes' law 166, **168–9**
Stop-logs xxxii
Storm water xxxii
Storm waterway xxxii
Straub, L.G. xxxii
Stream function xxxii
Stream tube xxxii
Streamline xxxii, 385
Streamline maps xxxii
Subcritical flow xxxii, **32–5**, **45**, 56–61,
    158–9, 333–6, 344
Submergence, culvert barrel 373–7
Submergence, culvert intake 371–7
Subsonic flow xxxii, 42–4
Supercritical flow xxxii, **32–5**, **45**, 56–61,
    158–9, 330, 333–6

Supersonic flow xxxii, 42–4
Surface tension xxxii, **4**, 125
Surfactant (or surface active agent) xxxii
Surge(s) xxxii, **67–72**
Surge wave xxxii, 67
Surges, negative, *see* Negative surge
Surges, positive, *see* Positive surge
Suspended load xxxiii, 157, **210–21**
Suspension 190–4, 210–24
Système international d'unités xxxiii, **128–30**
Système métrique xxxiii, 128–30

TWRC xxxiii, 342–8
TWRL xxxiii, 342–8, 382
Tailwater depth xxxiii, 342–8
Tailwater level xxxiii, 342–8, 382
Tainter gate xxxiii
Time scale, hydraulic flow 275, 304
Time scale, sediment transport 275, 304
Total head xxxiii, **22–5**, 30, 57, 284, 316–43,
    363, 366–9
Training wall xxxiii, 343, 349
Trapezoidal channel sections 47, 101
Trashrack xxxiii
Turbulence xxxiii, 171, 210–16
Turbulent flow xxxiii, 74–8, 166–9, 176–8,
    181
Turriano, J. xxxiii
Two-dimensional flow xxxiii

USACE xxxiv, 294, 339
USBR xxxiv, 328, 334, 338–41
Ukiyo-e xxxiii
Undular hydraulic jump xxxiii, 58, 63–4
Undular surge xxxiii, 69
Uniform equilibrium flow xxxiii, **79–91**,
    98–105, 108, 176–80, 289–90
Uniform equilibrium flow, depth 79–80,
    98–105
    see also Normal depth
Uniform equilibrium flow, velocity **79–84**,
    98–105, 175
Universal gas constant (also called molar
    gas constant or perfect gas constant)
    xxxiv
Unsteady flow xxxiv, 67–92
Uplift xxxiv, 63–4, 341
Upstream flow conditions xxxiv

VNIIG xxxiv
VOC xxxiv

Validation xxxiv, 257, 275, **284**, 284–9, 296–7
Vauban, S. xxxiv
Velocity coefficients 28–9
Velocity distribution 27–9, 176–80, 410–12
Velocity potential xxxiv
Vena contracta xxxiv
Venturi meter xxxiv, 385
Villareal de Berriz, B. xxxiv
Viscosity xxxiv, **3–4**
Vitruvius xxxiv
Von Karman constant, *see* Karman constant

WES xxxv, 294, 329
WES ogee crest profile xxxv, **326–9**
Wadi xxxiv
Wake region xxxiv, 181
Warrie xxxiv
Waste waterway xxxiv
Wasteweir xxxiv
Water xxxiv, 4, 7, 9
Water, fluid properties 125

Water clock xxxv
Water staircase (or 'Escalier d'Eau') xxxv
Waterfall xxxv
Water-mill xxxv
Weak jump xxxv, 63, 336
Weber, M. xxxv, 265
Weber number xxxv, 265
Weir(s) xxxv, 313–63
Weirs, broad-crested 38–9, **319–22**
Weirs, hydraulic design **313–63**
Weirs, sharp-crested **322–4**
Weisbach, J. xxxv, 69
Wen, Cheng-Ming xxxv
Wetted perimeter xxxv, 47, 73, 101
Wetted surface xxxv
White water sports xxxv
White waters xxxv
Wing wall xxxv, 366–8
Wood, I.R. xxxv

Yellow river 150, 192–4, 221–4
Yunca xxxv

The Italian American Experience

ITALIANS IN THE UNITED STATES

A REPOSITORY OF RARE TRACTS
AND MISCELLANEA

ARNO PRESS

A New York Times Company

New York — 1975

Reprint Edition 1975 by Arno Press Inc.

Copyright © 1975 by Arno Press Inc.

The Italian American Experience
ISBN for complete set: 0-405-06390-3
See last pages of this volume for titles.

Publisher's Note: The selections in this
anthology were reprinted from the only
available copies.

Manufactured in the United States of America

———————◆———————

Library of Congress Cataloging in Publication Data
Main entry under title:

Italians in the United States.

   (The Italian American experience)
   Reprints of pamphlets previously issued by various
publishers.
   Four of the 16 texts in Italian.
   CONTENTS: Italy America Society. Constitution,
adopted March 30, 1920. 1920.--An open letter from a
member of the Circolo Italo-Americano di Boston. 1908.
--Italy America Society. List of members, March 10,
1920. 1920. [etc.]
   1. Italian Americans--History--Sources. 2. Ameri-
canization--History--Sources. I. Series.
E184.I8I84   917.3'06'51   74-17934
ISBN 0-405-06406-3

CONTENTS

Italy America Society
CONSTITUTION ADOPTED MARCH 30, 1920. New York, [1920]

AN OPEN LETTER FROM A MEMBER OF THE CIRCOLO ITALO-AMERICANO DI BOSTON. [Boston, 1908]

Italy America Society
LIST OF MEMBERS: MARCH 10, 1920. New York, [1920]

Italy America Society
SECRETARY'S ANNUAL REPORT. APRIL 2, 1920. New York, [1920]

Frisco Lines
COLONIA ITALIANA: Tontitown, Arkansas, Stati Uniti D'America (Italian Colony: Tontitown, Arkansas, United States of America) Italian Text. St. Louis, Missouri, [1899]

Wright, Frederick H.
THE ITALIANS IN AMERICA. New York, 1913

Carr, John Foster
IMMIGRANT AND LIBRARY: Italian Helps, with Lists of
Selected Books. New York, 1914

Sheridan, Frank J., et al.
GL'ITALIANI NEGLI STATI UNITI (Italians in the United
States) Italian Text. Rome, 1909

Lodge C. Colombo, I. V. No. 414. Order Sons of Italy
DIRECTORY: FIFTH GRAND MASQUERADE BALL.
[Rutland, Vermont, 1921]

ITALY AMERICA SOCIETY.
New York, [1924]

Griel, Cecile L.
I PROBLEMI DELLA MADRE IN UN PAESE NUOVO
(The Problem of Women in a New Country) Italian Text.
New York, [1919]

Oppenheimer, Francis J.
THE TRUTH ABOUT THE BLACK HAND. New York,
[1909]

CONSIGLI AGL' IMMIGRANTI (Advisements to
Immigrants) Italian Text. [1904]

Adams, Joseph H.
IN THE ITALIAN QUARTER OF NEW YORK. New York,
[1903]

Italy America Society
ITALY AMERICA DAY CELEBRATION IN RECOGNI-
TION OF THE FIFTH ANNIVERSARY OF ITALY'S
ENTRANCE INTO THE WAR. [New York], 1920

Kansas City & Memphis Railway Company
TONTITOWN, ARK.: The World's Ideal Vineyard. Rogers,
Arkansas. [1901]

ITALY AMERICA SOCIETY

▽

CONSTITUTION

adopted March 30, 1920

ITALY AMERICA SOCIETY

23 WEST 43rd STREET

NEW YORK

▽

"*. . . To create and maintain between the United States and Italy an international friendship based upon mutual understanding of their national ideals and aspirations and of the contributions of each to progress in science, art and literature, and upon co-operative effort to develop international trade.*"

CONSTITUTION

adopted March 30, 1920

▽

ARTICLE I

Name

The name of the Society shall be the Italy America Society.

ARTICLE II

Purpose

The purpose of the Italy America Society shall be to create and maintain between the United States and Italy an international friendship based upon mutual understanding of their national ideals and aspirations and of the contributions of each to progress in science, art and literature, and upon co-operative effort to develop international trade.

ARTICLE III

Officers

1. The officers of the Society shall be a President; two Honorary Presi-

dents, who shall be the Italian Ambassador to the United States and the American Ambassador to Italy; a Vice-President; one or more Honorary Vice-Presidents, among whom shall be included men who have filled the office of American Ambassador to Italy; a Secretary, and a Treasurer.

2. The said officers shall be elected annually by the membership of the Society.

ARTICLE IV

Members

1. The members of the Society shall be such persons, firms, or organized groups as shall be elected to membership by the Executive Committee upon recommendation of the Membership Committee.

2. Membership in the Society shall be divided into the following classes:

(a) Life Membership;

(b) Annual Membership;

(c) Commercial Membership, which shall be open exclu-

4

sively to business concerns, Chambers of Commerce, or other organized groups.

3. Life Members shall pay a membership fee of $100.

4. Annual Members shall pay membership dues of $10 per year, payable on the date of election to membership in the Society and yearly thereafter upon the anniversary of such date.

5. Commercial Members shall pay membership dues of $100 per year, payable on the date of election to membership in the Society and yearly thereafter upon the anniversary of such date.

6. In addition to the classes of membership already specified, Honorary Members may from time to time be elected by the Executive Committee upon recommendation of the Membership Committee. Such Honorary Members shall have no vote at the Annual Meeting of the Society and shall not be subject to payment of dues, but shall be subject to all other rules and regulations of the Society.

ARTICLE V
Election to Membership

1. Candidates for election to membership shall be proposed by one member and seconded by another. The candidate's name in full, his business, and his address shall be entered on a form provided for the purpose, and shall be endorsed by the proposer and seconder, or by their authority.

2. All proposals for membership shall be submitted to the Membership Committee, whose duty it shall be to recommend to the Executive Committee those whom it deems suitable for membership.

3. Election shall be by vote of the Executive Committee. One black ball in five shall exclude. If twice rejected, a candidate cannot again be put up for election without the unanimous consent of the Executive Committee.

ARTICLE VI
Cancellation of Membership

1. If a member's annual dues shall be three months in arrears, he shall cease to be a member of the Society.

2. The Executive Committee shall have power to reinstate a member upon a satisfactory explanation of failure of due payment, provided that he shall have paid the full amount of his indebtedness at the date of such reinstatement, including all dues which would have accrued if he had remained continuously a member of the Society.

ARTICLE VII

Trustees

1. There shall be a Board of Trustees, which shall act in an advisory capacity to the officers and Executive Committee.

2. The Trustees shall be elected annually by the membership of the Society.

3. The Trustees shall elect a President of the Board from their own number, who shall be empowered to call meetings of the Board at such times as he may deem necessary, or upon the request of three or more members of the Board.

ARTICLE VIII

Executive Committee

1. There shall be an Executive Committee which shall be charged with the duty of directing the activities of the Society and enforcing the provisions of the Constitution. The Executive Committee shall be composed of the President, Vice-President, Secretary, Treasurer, the President of the Board of Trustees, and the Chairman of the Advisory Committee, ex-officio, and seven other members who shall be elected by the membership of the Society at its Annual Meeting.

2. The Executive Committee shall elect its Chairman from its own membership.

3. The Committee shall have power to fill vacancies in its own body until the next Annual Meeting after the occurrence of the vacancy.

4. The Committee shall make such rules and regulations as it may from time to time think necessary for the management and well-being of the Society.

5. Meetings of the Executive Committee shall be held at the call of the Chairman of the Committee.

6. Five members of the Committee shall constitute a quorum.

ARTICLE IX
Advisory Committee

1. The Executive Committee shall appoint an Advisory Committee composed of fifteen or more members of Italian extraction, the functions of which shall be to advise with the officers and Executive Committee in all matters relating to the program and policies of the Society.

2. The Advisory Committee shall have power to appoint such sub-committees from its own membership as it may from time to time think advisable, and to delegate to such sub-committees any part of the functions and powers appertaining to the Advisory Committee as a whole.

ARTICLE X
Committees

1. The Executive Committee shall appoint the following Standing Com-

mittees from the membership of the Society:

(a) A Committee on Membership

(b) A Committee on Ways and Means.

The Executive Committee may also appoint such other Standing or Special Committees from the membership of the Society as it may from time to time deem necessary to carry out the program and activities which the Executive Committee determines upon, and shall assign appropriate powers to such committees and prescribe their duties.

2. Such Standing or Special Committees shall hold office during the pleasure of the Executive Committee, to which they shall be responsible, and to which they shall report at such times as the Executive Committee may direct.

3. The Executive Committee shall have power to appoint sub-committees for any purpose or object from its own membership and to delegate to such sub-committees authority to carry out the functions for which

such sub-committees were designated.

ARTICLE XI

Executives

1. The Executive Committee shall have power to appoint a Manager, who shall have charge of the Society's office and employees, and shall have supervision over all activities of the Society under the direction of the Executive Committee. The Manager shall hold office during the pleasure of the Executive Committee, to which he shall at all times be responsible.

2. The Manager shall be ex-officio a member of all Standing or Special Committees, but shall not be a member of the Executive Committee.

3. The Executive Committee shall appoint an Assistant Treasurer, who sha" be empowered to draw checks against the Society's funds upon vouchers approved in writing by the Treasurer or Manager of the Society or the Chairman of the Executive Committee.

11

ARTICLE XII

Annual Meeting

The Annual Meeting of the Society shall be held in March of each year, on such day and at such place as the Executive Committee may direct, for the election of officers, members of the Executive Committee, and Trustees, and for the transaction of such other business as may properly be brought before such meeting. Two weeks' notice of such meeting shall be given by the Secretary in writing to every member of the Society.

ARTICLE XIII

Fiscal Year

The fiscal year shall begin on March first.

ARTICLE XIV

Amendments

The Executive Committee shall have power to amend the Constitution by a majority vote at any regular meeting of the Committee or at a special meeting called for such purpose, provided .nat notice of the proposed amendment shall have been sent in writing to each member of the Committee at least two weeks before the date of such meeting, and provided further that no such amendment shall be effective beyond the next Annual Meeting if disapproved by the membership of the Society by a majority vote at such Annual Meeting.

AN OPEN LETTER FROM A MEMBER OF THE CIRCOLO-AMERICANO DI BOSTON

AN OPEN LETTER FROM A MEMBER OF THE CIRCOLO ITALO-AMERICANO DI BOSTON.

Dear Fellow Members of the Circolo Italo-Americano:

It is now almost ten years since we formed into an organization, and determined that, by binding ourselves to a common purpose, we would become a factor for strength and righteousness in the construction of our community life.

We have come to the winter which, in its passing, will leave behind a full decade of effort — to the winter which is the threshold of the decade to come. And so it seems the part of wisdom to review our life up to this point, that the ten years past may serve as power and light for years to come.

The first startling mark of our success as an organization is that we are Alive! How many clubs and societies have been started during our lifetime, only to fall in the conflict of internal dissension or to sink in the torpor of inertia! We have had no deadly periods of dormancy — we are alive with the virility of nine full years' worth of action and progress.

Our constitution says our object shall be to promote social and educational activity, and to

1

bring about a friendly relationship between Italians and Americans. Have we done what we started out to do? Surely there is but one answer.

Socially, we remember first our Friday evening meetings at Denison House — that House which is truly a "Casa del Popolo," with which we affiliate ourselves, and whose good works are so numerous that a special article is needed to tell of them. These are for entertainment and recreation, but into them from year to year we have also brought educational features. Among these we recall with pleasure the papers read about the author's home city or town in Italy. Why not have more of these informal talks when some Southern member wanders in fields of memory and calls to his Northern brother? We like our informal music, our singing together, our occasional little plays, our chatting while having a sociable bit of refreshment.

We recall vividly our spring and summer festas, when friends of Italy or friends of members have extended to us the open hand of hospitality and called us to the country. Indelibly graven, beautiful pictures come before our eyes as we mention Wellesley, Walpole, Weston, Waltham. Will you ever forget? Or the receptions which have been given in our honor! Away from the city we think of Dana Hall. In Boston we recall first of all the reception tendered to us by our president, Professor Vida Dutton Scudder of Wellesley College. Then others, extended also

2

with exquisite courtesy, come before our minds, bringing happy recollections of music and gaiety and glad hospitality.

Nor have we remained passive recipients in our social life. You remember the reception the Circolo tendered to the Baron and Baroness Franchetti, of Rome, and the hospitality we offered them and other Italian and American friends at the Denison House? And how pleasantly we American members of the Circolo think back upon the entertainment given us by the Italian members one winter, in the Twentieth Century Club Rooms.

We will not try to enumerate one by one all the social occasions in which we have met one another, and made new acquaintances also outside the membership of our Circolo. Nor can we tabulate those many mutually helpful relationships which have grown through the touch established by means of the Circolo. Especially I think of those developed between Italians and Americans, as surely opportunity for a point of contact between two races is but rarely offered save in the field of low politics, and there it is usually to the disadvantage of the newest comer.

Turning now to give a look back over our educational work. The public Sunday lecture-concerts are our first thought. These are indeed a valuable part of Circolo activity; they constitute the only Italian forum in the whole city. Though they are not financed by the Circolo funds, they must legitimately be numbered among its good

3

works. They were inaugurated by the Circolo; they have been sustained by its good will. Many of the orators have been Circolo members; Circolo members also many of the musicians. It spreads the news of them; it attends them; it suggests and counsels for them. From the first one they have been distinctly under the auspices of the Circolo, and we, as its members, may well feel pride in their continued success. On the Sunday platform we have received Boston's noted citizens; they have felt it worth while to address us Professors of Harvard University and of Boston University, of Tufts College and of Wellesley College, have spoken for us.

Edward Everett Hale has counselled us. We have had the high honor to pay tribute from our platform to Julia Ward Howe and to Charles Eliot Norton. Guglielmo Ferrero has lectured for us. Italy and America have given of their best, and shall we not be proud? Through our conferences we have touched the spirit of the great liberators of Italy and America; we have considered social and industrial problems; we have applauded addresses ranging through Art, Literature, Civics, Hygiene, History, Scientific Advance.

We must remember, too, all the publications issued under the auspices of the Circolo, — the "Vade Mecum," a book made up of leaflets, published first separately, written by prominent citizens, in regard to civic life, its duties, privileges, etc.; the "Decalogo Civile"; the "All' Aria

4

Aperta." Requests came for copies of these from the country over, and so other far-away organizations have followed in our footsteps.

You know the Civic Reader used in the evening schools? This materialized as the result of the Circolo's "Vade Mecum." From time to time the sub-groups have met to make special study of English or politics or civics, or for the lighter but still useful purpose of working up some play to present for the edification of ourselves and our guests.

We want to remember, too, our large and well-attended exhibition of Italian Arts, to which Ex-Governor Curtis Guild gave his endorsement, not only by his presence, but by an address to a large audience of enthusiastic Italians and Americans. This, again, though not entirely financed from our treasury, sprang from our initiative. We were its center and its spirit. Our vice-president was Italian chairman and one of our members American chairman. "Viva la Fratellanza" was our slogan, stamped in gold letters on the white badges of honor, and, as the guest of honor, she whom the Circolo termed the Great World Patriot, Julia Ward Howe, granted by her presence the inspiration and incentive of one who had worked a life long toward the brotherhood we sought.

Twice during these years the work we are doing, or with which we have been closely associated, has been represented in great Congresses in Italy. At one of these we were fortunate in

5

having as our representative the Honorable Baron Franchetti. At the other we received honorable mention, and our president still holds the medal sent us in award.

Are we then fulfilling our object? Yes; we have done more than fulfill the letter. We have entered into the spirit of the purpose of those, Italians and Americans together, who formed the club — a spirit which can hardly be interpreted in the formal words of a constitution; a more intangible, inarticulate something, which is better expressed through the symbol we adopted some years ago, with the permission of its master sculptor, George Gray Barnard. You remember? It was always printed on the front page of our Bolletinos. In one of these our president, Miss Scudder, wrote of the original as follows: "It represents the struggle of two comrades to reach each other through a great mass of crude rock which separates them. Perhaps the rock symbolizes race; perhaps language; perhaps religion. It stands for any prejudice or constraint that divides man from his brother and hinders fellowship. And the two comrades, who are straining so hard toward an embrace, may represent any people who are trying to pass through such barriers, and achieve that fellowship which is Heaven. Already, looking close, one can see that their finger tips are touching. The earnestness of their attitude shows clearly that they will succeed in time, will stand face to face, with no obstacle between, and reach the brotherly embrace they crave."

As a club, we mean to ally ourselves in spirit to those causes which are working toward that fellowship which is Heaven. As individuals many members have worked in practice toward that end. Some have taught in children's classes; some have given their services as interpreters for relief-giving agencies; some have served as friendly visitors in homes of the distressed; some have stood in readiness to speak for the language-tied patient in the hospital. In many ways, and with willing spirit, members of the Circolo offer such service as shall make our common life sweeter and lighter.

As a member of the Circolo since its first days, with faith in its purpose and belief in its achievement, I send out to my fellow members this incomplete, sketchy review, knowing that it is good to stop, in the haste and press of things, and consider our reason for being; to call ourselves to account; to see if our past warrants our future.

In Union is Strength; so let us with renewed consciousness stand as "The Circolo Italo-Americano" for the Big Things. There are some things upon which we all can stand united — about which there can be no dissent. We are with the Campaign against Tuberculosis; we are with the Campaign against White Slavery; we are with the Campaign against Child Labor; we are with these and many others. And, as individuals, let

7

us try unceasingly to increase the usefulness of
that organization, of which we are members, by
trying to understand the life about us; by want-
ing to strengthen it where it is weak; by offering
our services freely where humanity calls them.

Viva la Fratellanza, and Viva il Circolo Italo-
Americano, to help toward it!

MEMBER.

ITALY AMERICA SOCIETY

▽

LIST OF MEMBERS
March 10, 1920

ITALY AMERICA SOCIETY

23 WEST 43rd STREET

NEW YORK

▽

"... To create and maintain between the United States and Italy an international friendship based upon mutual understanding of their national ideals and aspirations and of the contributions of each to progress in science, art and literature, and upon co-operative effort to develop international trade."

LIST OF MEMBERS
March 10, 1920

▽

HONORARY MEMBER
Howells, William Dean

LIFE MEMBERS
Alexander, Charles B.
Aspegren, John

Bache, Jules S.
Baker, George F.
Baker, George F., Jr.
Barney, Hiram
Bedford, A. C.
Benedict, William Leonard
Bennett, Mrs. John Hudson
Bibb, William Garrett
Brady, Nicholas F.
Brady, Mrs. Nicholas F.
Brown, Franklin Q.
Byrne, Mrs. James

Cannon, Henry White
Cochran, Alexander Smith
Coe, William R.

3

Coffin, Charles A.
Costantini, David A.

D'Antoni, S.
De Vecchi, Paolo
Di Giorgio, Joseph
Di Giorgio, Salvatore
Dinsmore, Mrs. W. B.
Dodge, Cleveland H.
Duryea, Mrs. Harmanus B.

Fahnestock, Mrs. Gibson
Fontana, A. G.
Francolini, Joseph N.
Freeman, Charles D.

Gayley, Mrs. Gardiner
Gerli, Emanuel
Gerli, Mrs. Emanuel
Gerli, Joseph
Giannini, A. P.
Goelet, Mrs. Ogden
Goelet, Mrs. Robert
Guggenheim, William
Guggenheim, Mrs. William
Gurnee, Augustus Coe

Hemphill, Alexander J.
Hughes, Charles E.
Huntington, Archer M.
Hurlbut, Margaret Crane

Jennings, Annie Burr

4

Kirby, Thomas E.

Lamont, Thomas W.
Lane, Mrs. James Warren

Mackay, Clarence H.
Mc Gregor, Robert
Mc Millin, Emerson
Mellon, Edward P.
Mercadante, Joseph
Morgan, J. P.
Morgan, William Fellowes

O'Neill, Thomas
Osborn, William Church
Osborn, Mrs. William Church

Pedrazzini, Guido
Perera, Lionello
Perkins, George W.
Piva, Celestino
Portfolio, A.

Rice, Alexander Hamilton
Riso, Osvaldo
Rumsey, Mrs. Charles

Sabin, Charles H.
Satterlee, Mrs. Herbert L.
Sheldon, Edward W.
Solari, Luigi
Stone, Charles A.

Todd, Henry Alfred

Traverso, Mrs. Estelle MacMillin
Tritonj, Romolo
Vanderbilt, Mrs. Cornelius
Villa, Alfonso P.
Villa, Silvio
Virden, Charles E.

Waterman, Frank D.
Wickersham, George W.

ANNUAL MEMBERS

Adams, Edward D.
Ahlstrom, C. F.
Aldrich, Chester H.
Alexander, Pugliese Dott
Allen, Frederic W.
Arbib-Costa, Alfonso
Atkinson, Henry A.

Bache, Mrs. Jules S.
Bacon, Mrs. Francis McN.
Baisley, Charles T.
Baldwin, Le Roy W.
Baldwin, William Delavan
Barlow, Peter Townsend
Bartlett, E. O.
Bava, Felice
Baylies, Edmund L.
Beck, James M.
Belmont, August
Benington, Arthur
Berardo, Luigi

6

Berizzi, Stefano
Bertelli, Riccardo
Bettini, Gianni
Biasutti, Gaetano
Blackiston, H. C.
Bliss, Cornelius N., Jr.
Blumenthal, George
Booth, Willis H.
Borg, Sidney C.
Bradley, Mrs. Charles Burnet
Braman, Chester A.
Breton, Albert
Brown, James
Bruce, J. C.
Buckner, Thomas A.
Bullard, Mrs. Arthur
Burchell, Henry J., Jr.
Burleigh, George W.
Burton, Theodore E.
Butler, Nicholas Murray
Byrne, James

Caglieri, Guido E.
Caldwell, Rebecca A.
Camera, A. U. N.
Campanella, Pasquale
Campora, Vincenzo
Cantu, D. G.
Caproni, Gianni
Carse, Henry R.
Caruso, Enrico

7

Case, James F.
Cerruti, Camillo
Chanler, Winthrop
Chanler, Mrs. Winthrop
Chapman, John Jay
Child, Richard Washburn
Childs, William Hamlin
Clarke, Lewis L.
Collins, Joseph
Conte, Arminio
Conway, William P.
Corlett, W. W.
Corliss, Charles A.
Cornwell, William C.
Corti, Louis F.
Cosenza, Mario E.
Cotillo, Salvatore A.
Cravath, Paul D.
Criscuolo, Luigi
Crise, Stewart S.
Curtis, F. Kingsbury
Curtis, William E.
Cushman, Mrs. James Stewart
Cutler, Otis H.
Cutting, R. Fulton

Dahl, Gerhard M.
Darlington, James Henry
Davison, Henry P.
Day, Joseph P.
de Biasi, Agostino

8

Delano, William Adams
de Lima, E. A.
De Luca, Vincent
de Schauensee, Baroness
de Vita, Salvatore
de Wolfe, Elsie
Dillon, Clarence
di Radione, Conte Dino Spetia
Ditson, Mrs. Charles H.
Dodge, Mrs. Arthur M.
Downer, Charles A.
Drake, Mrs. John Adams
Duff, Mrs. Sarah Robinson
Duggan, Stephen P.
Dunbar, Douglas L.

Edgar, Mrs. Newbold LeRoy
Elkus, Abram I.
Ellis, George Adams
Englar, D. Roger
Erit, John B.

Fabbri, Ernesto G.
Fabbri, Mrs. Ernesto G.
Fagnani, Charles P.
Fanciulli, Romolo A.
Fanoni, Antonio
Finley, John H.
Finocchiaro, Francesco Paolo
Freschi, John J.
Frissell, Algernon S.
Frugone, Frank L.

9

Fuller, Samuel L.

Gaudiosi, Pasquale
Gidoni, Domenico
Gilbert, Cass
Granata, Genserico
Greene, Edward A.
Greenough, William
Guardabassi, Francesco M.
Guthrie, William D.

Hammond, John Henry
Hannigan, George A.
Harris, Hayden B.
Harrison, John P. S.
Hauser, M. H.
Heaton, John L.
Hereford, W. R.
Hine, Francis L.
Holt, Constance
Holt, George C.
Holt, Mrs. George C.
Holt, Hamilton
Holt, Mrs. Hamilton
Holt, Henry C.
Holt, Sylvia
House, Mrs. E. M.
Howe, Edward Leavitt
Howland, Harold
Howland, Karl V. S.
Hoyt, Colgate
Hurrey, Charles D.

Jaccaci, Mrs. Mabel Thayer
James, Arthur Curtiss
James, Mrs. Walter B.
Johnson, Robert Underwood
Johnston, John S.
Joline, Mrs. Adrian Hoffman
Jones, Raymond E.

Kahn, Otto H.
Kelsey, Clarence H.
Kennedy, George P.
Kent, Fred I.
Kingsley, Darwin P.
Kirby, Mrs. Gustavus T.
Knapp, Joseph P.
Knoke, Herman
Koyl, George S.

Ladenburg, Mrs. Adolf
La Farge, C. Grant
Lamond, Felix
Lancashire, Mrs. J. Henry
Lathrop, Alanson P.
Lawrence, Margarita W.
Leeds, Mrs. Warner M.
Léon, Maurice
Lerro, Luigi
Lewis, John J.
Lewisohn, Adolph
Lowell, Guy
Lynch, Frederick

11

Mc Clellan, George B.
Mc Clellan, Mrs. George B.
McDonnell, Peter
McFadden, George H.
Macdonough, Joseph M.
Macfarland, Charles S.
Marbury, Elisabeth
Marcone, Paul P.
Margarella, Pasquale
Markoe, Francis H.
Marling, Alfred E.
Marshall, H. Snowden
Maxwell, George
Mead, William Rutherford
Metz, Sigmund
Miele, Stefano
Miller, Mrs. Benjamin
Mitchell, C. E.
Monte-Sano, Vincent
Moree, Edward A.
Moulton, Mrs. Arthur J.
Moulton, May Taylor
Murphy, Frank J.
Muschenheim, Mrs. Fred A.

Nardi, Pasquale
Nicoll, Mrs. Benjamin
Nollen, John S.
North, Frank Mason
Norton, Charles D.

Oldrini, Alexander

12

Olcott, Eben E.
Olmsted, Frederick Law
Osborn, Henry Fairfield

Parker, Alton B.
Peabody, George Foster
Pendleton, Francis Key
Perera, Guido
Perkins, Robert P.
Pinto, Angelo A.
Platt, Charles A.
Plimpton, George A.
Polk, Beach
Poole, Robert A.
Porter, William H.
Potter, Julian W.
Potter, Mark W.
Prentice, Bernon S.
Previtali, Giuseppe
Previtali, Mrs. Giuseppe

Quagliotti, Gerard
Quattrone, F.

Racca, Vittorio
Ricci, Louis D.
Riley, Oscar E.
Riotti, Eugene A.
Robinson, Mrs. Douglas
Robinson, Edward
Ronconi, Romeo R.
Rossano, Augusto Thomas
Rounds, Arthur C.

Rousseau, Theodore
Ruggieri, Carlo Ciulli
Russell, Lindsay

Sachs, Samuel
Salt, Albert L.
Sampson, Charles E.
Sapelli, Alessandro
Savidge, Eugene Coleman
Savini, Carlo
Scaramelli, Louis J.
Scheftel, Edwin K.
Schiff, Jacob H.
Schiff, Mrs. Mortimer L.
Schmid, John F.
Schurman, George W.
Scoville, Herbert
Scribner, Charles
Sedgwick, Henry Dwight
Serafini, Amedeo
Shepard, Finley J.
Silvi, Mariano
Simondetti, E. T.
Simpson, Ernest L.
Sisca, Marziale
Sisson, F. H.
Sleicher, John A.
Smith, Irwin
Smith, Munroe
Smith, Robert A. C.
Snyder, Valentine P.

Spalding, Albert
Stanchfield, John B.
Steele, Charles
Stella, Antonio
Stella, Mrs. Antonio
Stern, Albert
Stetson, Eugene W.
Stetson, Francis Lynde
Stone, Galen L.
Storey, John De Raimes
Strauss, Frederick
Strong, Jane S.
Sumner, Malcolm
Sutphen, Henry R.
Swope, Gerard

Taylor, George C.
Thaw, Alexander Blair
Thayer, Eugene V. R.
Thomas, Eugene P.
Thompson, William Boyce
Thorne, Samuel, Jr.
Thornton, Lucile
Tolentino, Raoul
Tomlinson, Roy E.
Trowbridge, S. Breck P.
Truda, Dominic A.

Vanderbilt, Cornelius
Vick, Walker W.
Villa, Mrs. Alfonso P.
Vitale, Ferruccio

Vitelli, G. B.

Warren, Whitney
Webb, F. Egerton
Webb, Mrs. F. Egerton
Wexler, Sol.
White, J. G.
Wiley, Louis
Williams, Edgar I.
Wing, Daniel G.
Wykes, Hunter

Yaselli, E. Paul

Zaccaria, Marchese Guiscardo
Zazzali, Charles P.

COMMERCIAL MEMBERS

American Italian Commercial Corp.
Atlas Finishing Company
Banca Commerciale Italiana
Cravath & Henderson
General Silk Importing Company
Gio. Ansaldo & Company
Hotel Astor
Independent Corporation
Italian Discount & Trust Company
Kidder, Peabody & Company
Transatlantica Italiana
United States Transport Company
A. P. Villa & Bros., Inc.
Warren & Wetmore
Woodward, Wight & Company

▽

ITALY AMERICA SOCIETY

▽

SECRETARY'S ANNUAL REPORT
April 2, 1920

ITALY AMERICA SOCIETY

23 WEST 43rd STREET
NEW YORK

▽

"... To create and maintain between the United States and Italy an international friendship based upon mutual understanding of their national ideals and aspirations and of the contributions of each to progress in science, art and literature, and upon co-operative effort to develop international trade."

SECRETARY'S ANNUAL REPORT

Presented at the Annual Meeting
April 2, 1920

▽

Mr. Chairman and Gentlemen:

THE year through which the Italy America Society has passed since the last Annual Meeting must be characterized broadly as a year of transition and preparation rather than as a year of achievement. In that interval the Society has brought its war-time program to a close, has formulated its peace-time program, has expanded and strengthened its organization, and today looks forward to a period of greater effectiveness than ever before.

At the date of its last Annual Meeting, the Society had a total membership of 297, made up of 65 life members and 232 resident or non-resident members paying annual membership dues. Today it has a membership of 400, made up of 80 life members, 305 annual members, and 15 business firms enrolled in the recently constituted class of commercial members. In addition, the Chairman of the Committee on Membership informs me that he has in hand today the applications of 37 candidates for personal membership and of 14 candidates for commercial membership.

The Banquet to Baron Avezzana

THE most conspicuous of the Society's activities during the past year was the banquet held at the Hotel Astor on February 24th in honor of His Excellency the Royal Italian Ambassador and Baroness Romano Avezzana, at which a brilliant assemblage of more than nine hundred persons was in attendance. Addresses were made by Judge Charles E. Hughes, who presided; Robert Underwood Johnson, American Ambassador to Italy; F. H. La Guardia, acting mayor of New York City; General Charles P. Summerall of the United States Army; C. Grant La Farge of the American Academy in Rome, and Baron Avezzana.

The Ambassador has especially requested that his appreciation of the Society's tribute be expressed to the members of the Society. His telegram, addressed to Judge Hughes, reads as follows:

"On the occasion of the next meeting of the Italy America Society, which I hear is taking place to-morrow, I wish to express to you and to ask of you to convey to the members of the Italy America my appreciation and gratitude for their efforts in organizing such a magnificent manifestation of sympathy and friendship towards my country and my person as the one which was constituted by the dinner of February 24. No greater proof could have been given by the Italy America of their efficiency in promoting cordial understanding between the two countries.

Romano Avezzana."

Program of Activities

AFTER numerous conferences, the Executive Committee and Board of Trustees have definitely formulated and authorized the Society's program of activities for the ensuing year. The program may be divided roughly into efforts along the following lines:

Commercial and economic
Educational
Travel
Literary
Artistic
Social

The work along commercial and economic lines began with the publication of the first three issues of a *Commercial and Economic* series of Trade Bulletins. Two special monographs are definitely projected, the one on the past and future development of Italian commerce on the high seas, and the other on the development of hydraulic power. The series will be extended to include careful studies of Italian activity in commercial, industrial and economic fields.

The President of the Society has been authorized to appoint a special Committee on Economic Relations to study questions of trade, finance, the tariff, immigration, and all other economic matters of mutual interest to Italy and the United States, and he now has the personnel of the Committee under advisement.

Educational

In its efforts along educational lines, the Society is fortunate in having established an entente cordiale with the Institute of International Education, of which Prof. Stephen P. Duggan is Director, and in having secured Prof. Duggan as Chairman of its Committee on Education. The Committee's program includes exchange professorships and scholarships, the exchange of books and current periodicals, etc. The Committee on Education will have its headquarters at 419 West 117th Street, New York City.

7

The Society has taken preliminary
steps toward the organization of a

Travel
Bureau

Travel Bureau to en-
courage and facili-
tate travel to and
in Italy. If the pres-
ent plans materialize, the Travel
Bureau will aim to serve as a clearing
house of information, where a pros-
pective traveller can secure data on
steamship lines, hotels, railroads,
automobile tours, etc. The Bureau
also plans to furnish letters of introduction to members of the Society
who propose to visit Italy.

In addition to the News Bulletin,
the Trade Bulletin, and the other

Publications

commercial and eco-
nomic booklets that
have already been
mentioned, the Society's Committee
on Publications has formulated plans
for several general or special publications to be issued during the current year, the principal of which are:

A booklet containing the
speeches delivered at the
banquet of February 24th
(now on the press) ;

8

A pamphlet reprint of the list of members of the Society (now in the printer's hands);

A pamphlet reprint of the Revised Constitution (now in the printer's hands);

The Year Book of the Society;

An American edition of Premier Nitti's recent book, "The New Italy".

The Committee on Publications also plans to issue from time to time a series of special monographs dealing with Italian achievement or development in social, intellectual, scientific or artistic life, material for the first of which is already being assembled.

The Society's work along artistic lines has been entrusted to a Committee on the Arts, *The Arts* which already has d e f i n i t e achievements to report. Through the initiative of the Chairman of the Commit-

9

tee, the International Exposition of Arts at Venice has taken the unprecedented action of accepting for exhibition a group of paintings by living American artists, collected by Mrs. Harry Payne Whitney.

The Committee's plans include a reciprocal exhibit of the works of Italian artists, to be held here next season under the auspices of the Society, as well as an exhibition of industrial arts.

The Committee on the Arts has taken the initiative in organizing a nation-wide recognition of the 400th anniversary of Raphael's death, which falls on April 6th of this year.

Social

As for its social activities, the Society hopes that it may continue to act as host to many of the distinguished Italians who visit this country, bringing them into contact with the leaders of American thought, and helping to secure a wide distribution for the messages which they bring to America.

Administrative Arrangements

THE Society has, of course, been forced to enlarge its organization very greatly to enable it to cope with its enlarged program of activities. In addition to the special committees already mentioned, Committees on Membership and on Ways and Means have been appointed. A strong Advisory Committee, with its own sub-committees, is working with the Executive Committee. The office personnel of the Society has been increased, and Mr. Irwin Smith has been brought in as Manager.

On the 15th of the present month, the Society will remove from its present offices to its new headquarters at 23 West 43rd Street, New York City.

Honors Received

THE Society has been greatly honored in the fact that His Majesty the King of Italy has conferred decorations upon several of its officers in recognition of their work as officials of the Italy America Society.

11

The Degree of Grand Cross Knight of the Crown of Italy has been conferred upon Charles E. Hughes, President of the Society. The Decoration of Officer of the Crown of Italy has been conferred upon Hamilton Holt, Lindsay Russell, and John J. Freschi. The decoration of Officer of the Saints Maurizio and Lazarro has been awarded to William Fellowes Morgan, Chairman of the Executive Committee. The decoration of Knight of the Crown of Italy has been conferred upon Theodore Rousseau, Douglas L. Dunbar, and Francis H. Markoe.

The Society is also honored in the appointment of one of its Trustees, Robert Underwood Johnson, to the high post of American Ambassador to Italy.

Respectfully submitted:

Theodore Rousseau

Secretary

COLONIA ITALIANA

TONTITOWN
ARKANSAS

STATI UNATI D'AMERICA

La Storia della Colonia Italiana a Tontitown, Arkansas.

U NO DEI più meravigliosi progressi di una parte del paese, da apparente sterilità in una bella, fiorente e vantaggiosa produttività che sia mai stata conosciuta in questi paesi è quella della Colonia Italiana—Tontitown—fondata tredici anni fa, a sei miglia all' ovest di Springdale, Arkansas, una piccola città sulla linea della Ferrovia del Frisco. Tredici anni fa, i dintorni del sito attuale della colonia erano sterili, incolti colli dell' Arkansas come abbiamo già detto. Oggi è un meraviglioso sviluppo di perita e ben diretta agricoltura. In primavera è un giardino di rose imbalsamato dalla fragranza delle piante che germogliano, viti ed alberi che promettono un' abbondante raccolta; odorosi in autunno per la fragranza dell' uva, un buon campo di raccolti per grano turco e frumento, un vero frutteto, che produce le migliori mele, pere e pesche, prugne ed altri frutti, un giardinaggio di grande dimensioni, che coltivano per i mercati i più scelti legumi, fragole, more, lamponi ed altri piccoli frutti.

La colonia di Tontitown ha questo ed anche di più. L' entusiasmo non trova parole sufficenti per farne una buona descrizione, giacchè ha mostrato uno sviluppo tale che è difficile

1

Bestiame al Pascolo Negli Ozarks.

descrivere in spazio limitato da un articolo.
Il medesimo terreno sul quale i poveri contadini americani inesperti, diventavano sempre più poveri ogni anno rende ora questi
coloni dal di la dell' oceano, uomini donne e
fanciulli agiati, comodi e felici.

I dettagli dello sviluppo sembrano quasi
una favola quando si leggono. Non sono una
storia di grande prosperità dal principio, ma
sebbene una storia di grandi stenti di uno
stuolo di italiani, senza cognizione dell' America, degli Americani o della lingua di America,
contro tremende difficoltà, facendo spropositi
quando non era necessario di farli, ma sempre
sotto la ferma, perseverante ed intelligente
direzione—quella del Padre Bandini, un sacer-
dote cattolico romano, che stava coi coloni fin
dal principio della loro dura lotta per la loro
pace presente e la loro prosperità.

Venti anni fa il Padre Bandini lasciò la
bella Italia e la sua famiglia e gli amici per
venire in America a studiare la condizione

2

Avanti e dopo

degli emigranti italiani in questo paese. Egli
trovò che la loro condizione in New York City
era assai deplorevole, e si mise a cercare una
maniera per porvi argine, e dare un sollievo ai
suoi infelici connazionali.

Il suo primo passo fu di organizzare la
Società di San Raffaele per la protezione degli
emigranti italiani, una società esistente tut-
tora, ed è di sommo vantaggio agli immigrati.
Lo scopo di questa società è quello di prestare
aiuto spirituale e materiale all' emigrante
italiano. Gli ufficiali della Società vanno incontro a tutti i vapori che conducono emigranti
italiani a bordo in Ellis Island. Essi dirigono
lo straniero alla sua destinazione, cercano i
suoi amici e li avvisano del suo arrivo, o
cercano impiego per quelli senza amici. Fanno

L' Allevamento di Pollame è Molto Lucroso in una Fattoria Degli Ozarks.

le dovute investigazioni nei casi in cui certi non ottengono il permesso di entrare in paese, e fanno si, che questi siano dovutamente fatti ritornare alla loro vecchia patria.

Quantunque questo fosse un passo sulla buona via, il Padre Bandini presto capì che questa non era ancora una soluzione al gran problema di prender cura dell' emigrante, o di sollevarne la posizione in una maniera pratica. Egli presto comprese che molto pochi di essi sapevano in che parte dell' America dovevano andare, o che sorta di lavoro o che professione avevano intenzione di seguire. Dopo minuziose ricerche egli trovò che novanta per cento di essi erano esperti orticoltori ed agricoltori. Per molti di loro l' America voleva dire semplicemente la città di Nuova York. Molti erano buoni coltivatori e avrebbero fatto bene nei paesi agricoli ma

4

Vigneti a Tontitown

perchè arrivarono senza denari, o perche avevano gia speso il poco di denaro che avevano, o in alberghi, o camminando per le strade senza lavoro, o l' avevano dato a qualche agente del lavoro, che aveva mancato alla promessa di procurar loro impiego, essi erano abbandonati nelle strade di New York senza tetto, pane o lavoro.

Il Padre Bandini capi subito l' importanza di far andare questa gente alla campagna. Egli ben realizzò che era la miglior cosa per gli emigranti, spiritualmente, moralmente, fisicamente e finanzialmente, e la miglior cosa per la prosperità della loro patria adottiva, dove il terreno aveva sommo bisogno di chi lo coltivasse con perizia. Egli comprese che la colonizzazione era la sola soluzione del prob-

5

Meloni Degli Ozarks.

lema. Molte compagnie di terreni, corporazioni ed altri interessi finanzieri offrirono proposte, ma furono fatte solamente nell' interesse dell' instituzione e non dell' emigrato. Il Padre Bandini rifiutò di accettarle essendo il suo solo pensiero l' emigrante ed il suo benessere.

Durante l' anno 1895, Il Signor Augustine Corbin, un ben noto capitalista, coll' aiuto del Principe Ruspoli, allora Sindaco di Roma stabilì una colonia italiana a Sunny Side nella parte del Sud Est dell' Arkansas. Il Principe Ruspoli doveva mandare per cinque anni consecutivi 100 famiglie all' anno a Sunny Side, per la via di New Orleans, ed alla sua volta il Signor Corbin doveva somministrare a ciascuna famiglia 20 acri di terreno, una casa, una stalla e tutti gli atrezzi agricoli che erano necessari per coltivare il terreno. Per questo dovevano pagare $100 all' acre per il terreno al 10/100 di interesse dentro i venti anni.

6

Chiesa e casa del Parroco di Tontitown

Le prime 100 famiglie arrivarono il 4 Dicembre, 1895, ed il secondo gruppo, pure di 100 famiglie, il 5 di Gennaio, 1897. Siccome i coloni erano del tutto sprovvisti di fondi, solo potevano ottenere le loro provviste dalla bottega della compagnia, il quale importo veniva aggiunto al loro debito.

A richiesta del Sig. Corbin, e mosso dal suo proprio interesse nel benessere degli emigranti, il Padre Bandini visitò la colonia a Sunny Side. Egli fece minuziose ricerche sulle condizioni e trovò che sarebbe quasi impossibile per i coloni di adempire le condizioni dei contratti, e suggerì al Sig. Corbin che queste venissero modificate. Questo gli fu concesso. Egli insistette pure che i coloni avessero una chiesa ed una scuola loro propria per il

7

Raccolta di Fragole.

loro bene spirituale e per l' educazione dei loro
figli, perchè la gente fosse più contenta. Queste
furono provviste, ed il Signor Corbin sollecitò
il Padre Bandini a prendersi carico della
colonia, ed egli vi acconsentı.

La colonia sembrava essere in un certo
stato di prosperità e di 'felicità, per far onore
al suo nome. Ma nè il Padre Bandini nè i suoi
uomini sapevano che i terreni erano infetti
dalla malaria. Molti dei suoi uomini contras-
sero la febbre, si ammalarono e morirono, e
Padre Bandini supplicò urgentemente che si
facesse qualche cosa per rimediare a questo
stato di cose.

Il Signor Corbin desiderando di assicurare
il successo della colonia, ed essendo interessato altamente nel benessere dei coloni si pose
immediatamente a cercare di rimediare alle

8

Ricevimento al Barone des Planches
Ambasciatore Reale Italiano.

condizioni esistenti. Egli mandò pompe a vapore per fornire l' acqua dal fiume Mississippi: ordinò che fossero scavati dei pozzi artesiani, e che i terreni fossero asciugati dalle paludi: ma aventi che si potesse nemmeno incominciare questi lavori, il Signor Corbin morì. Disgraziatamente gli eredi non dividevano il suo entusiasmo nella colonia, nè il medesimo interesse nell' opera, ed i piani non furono mai posti in effetto.

Molti dei coloni morirono, altri erano ammalati, indeboliti dalle febbri e incapaci di lavorare, e non potevano pagare le loro terre, Essi si scoraggiarono e cominciarono a sbandarsi, alcuni cercando lavoro nelle mine, altri

Un Orto di Meli Nell' Arkansas.

andando nelle città. Quelli che potevano ritornarono in Italia. Quaranta delle famiglie si portarono nella Montagne degli Ozark nel Missouri a Knobview, una città sulla Ferrovia Frisco 95 miglia da St. Louis. Qui, essi godono buona salute, prosperano e sono contenti e felici.

Il Padre Bandini intanto fece ricerche sul clima ed altre condizioni nella parte del Nord ovest dell' Arkansas sugli altipiani. Egli trovò il clima salubre a 1500 piedi al di sopra del livello del mare, le stagioni erano moderate e lunghe. Egli fece esaminare il terreno e trovò che era produttivo se fosse ben coltivato, e ben lavorato. Egli indusse venti sei famiglie ad andare con lui in quelle parti.

Egli comprò 300 acri di terra nella contea di Washington, sei miglia all' est di Springdale, una cittadella sulla Ferrovia del Frisco, facendo un primo pagamento di $1,500, garantendo ogni cambiale sulla ipoteca del terreno

Interno della Chiesa Cattolica

per il resto; egli comperò altri 500 acri di
terreno. Nel mese di Marzo di quell' anno,
Padre Bandini e i suoi connazionali presero
possesso del terreno. Vi era una casa di quattro stanze, una capanna di tronchi d' albero, una
stalla di tronchi d' albero, ed un granaio di
tronchi d' albero sui primi 300 acri. In questi,
il Padre Bandini e la sua gente abitarono per i
primi sei mesi. Divisero la terra in piccoli tratti,
dai 5 ai 20 acri, facendo tali pagamenti dai
$10.00 ai $25.00 come meglio potevano, e dando
loro cambiali per il resto.

Egli diede il nome alla colonia di Tontitown
in onore di Enrico Tonti, un Italiano che era
luogotenente del General LaSalle nei primi
tempi, e che fondò una piccola fortezza militare

Mele Dell' Arkansas.

ed una colonia vicino al Fiume Arkansas, che è ora chiamato Arkansas Post, e diede il nome di Arkansas a tutto l' intero territorio.

Quella gente cominciò ad incoraggiarsi. Respiravano aria sana della montagna, il sole di primavera mandava i suoi caldi raggi sulla terra che rispondeva alle sue carezze, e gli alberi e gli arbusti germogliavano colla promessa di una raccolta abbondante. Parte del terreno era coperto di legname, ed alcuni incominciarono a fabbricare capanne di tronchi, secondo la foggia del contadino dell' Arkansas, ed alcuni fabbricarono case di mota. Essi coltivarono pazientemente la terra, piantando tali raccolte in primo duogo che dassero loro il pane. Poscia piantarono viti, alberi fruttiferi e piccoli frutti.

In inverno per guadagnare denaro per poter tirar innanzi, gli uomini ed i ragazzi che erano

12

Scuola di Tontitown

grandi abbastanza si misero al lavoro nelle
mine di carbone nell' Oklahoma, lasciando le
donne ed i bambini sotto la custodia del Padre
Bandini per prender cura delle loro case.

Non subito ottennero successo. Essi sop-
portarono molte privazioni, ed il loro lavoro
fu lungo e arduo. Vi erano molte difficoltà
da superare. Essi non ricevettero nessun incoraggiamento nè aiuto dalla gente di fuori.
I filantropi e quelli che si interessano nel benessere pubblico del paese non vedevano di buon
occhio quel movimento. Essi riguardavano l'
esperimento del Padre Bandini poco favorevol-
mente, e predissero che sarebbe un insucesso.
Erano stranieri in un suolo straniero, ma
Padre Bandini aveva fiducia, e diciotto delle
venti sei famiglie avevano fede in lui. Erano

13

in buona salute e forti, i loro figli stavano bene, i loro visi davano tutti gli indizi di buona salute, gli occhi erano vivi e sani, erano felici e contenti di continuare a lavorare.

Durante il secondo anno, quando che le condizioni erano più scoraggianti, il Padre Bandini cominciò a fabbricare il fondamento della Chiesa. La loro religione non era stata nè dimenticata, nè negletta, era stato il loro conforto nelle ore di dure prove, un rozzo altare era stato eretto in una capanna di tronchi d' albero, e là andavano ad ascoltare la messa e a far le loro divozioni. Sapendo che la Chiesa è il simbolo della patria—materialmente parlando come pure spiritualmente—per gli Italiani, egli decise di dimostrar loro che questa era la loro patria e la loro casa. La capanna di tronchi non si raccomandava agli animi della sua gente, come il campanile e la campana. Questo li renderebbe più contenti, infonderebbe loro la speranza, vorrebbe dire permanenza il quel luogo.

Durante il primo anno il Padre Bandini insegnò ai ragazzi fra molti altri doveri che gli erano ricaduti sulle spalle. Il primo anno apersero una scuola, completando un edificio che era stato incominciato dal distretto. Era suo desiderio che i ragazzi della colonia fossero educati per meglio conoscere i costumi della loro patria adottiva, e la gente di quel paese. Molti dei ragazzi americani che stavano vicini, presto s' inscrissero e frequentarono questa scuola. La Chiesa fu trasportata dalla capanna di tronchi d' albero all' edifizio della scuola dove si facevano le funzioni finchè fosse finita la Chiesa.

Banda dei giovani di Padre Bandini suonando
pel Barone des Planches

Non vi era nessuna parte della loro vita
che Padre Bandini lasciasse inerte. Egli li
consigliò spiritualmente e finanzialmente e
procurò loro onesti divertimenti. Egli organ-
izzò e diresse una banda; egli scrisse comedie
sulle loro dure prove e stenti, insegnò ai
ragazzi e alle fanciulle a recitare, cosi incorag-
giando uomini e donne a ridere dei loro passati
stenti.

La fortuna arrideva al loro assiduo lavoro
e al loro grande coraggio ed essi prospera-
vano. Fra alcuni anni, tutto il terreno fu
pagato e ne comperarono dell' altro. Le viti
e gli alberi fruttiferi fiorivano e produeevano
un' abbondante raccolta. Il grano ed altre
raccolte crescevano in abbondanza. I fanciulli

15

arrivavano ad essere uomini e donne, sani di corpo e di mente.

Ben presto tutto il terreno fu pagato, delle case comode furono fabbricate, e vennero comperati altri acri, fino che ora la colonia possiede 4,760 acri di buon terreno produttivo. Dalla piccola comitiva di meno di cinquanta persone, ve ne sono ora 700, tutti felici e prosperi. Dal principio la colonia consisteva solamente di italiani, ma ora vi sono alcuni tedeschi, francesi, irlandesi e americani. Tutte famiglie attive oneste e che osservano le leggi ben viste e ben accolte senza badare alla nazionalità.

Parlando dei 300 primi acri che i coloni comperarono dall' agricoltore dell' Arkansas, questi aveva detto; Mio padre mi diede il terreno, ed è cosi sterile che non ho mai potuto riuscire a guadagnarmi la vita per la mia piccola famiglia, e sono sicuro che anche gli Italiani non vi riusciranno. Non potranno mai arrivare a pagarlo, ed io riceverò di ritorno il mio terreno ed avrò pure il loro denaro dato in acconto. Oggi vi sono dodici famiglie—grandi famiglie —che abitano su quei 300 acri comodamente e che fanno denari, hanno tutti delle belle case dalle sei alle dodici camere, circondate da bellissime spianate d' erba. Oltre queste dodici famiglie, la Chiesa, la casa del Padre Bandini, l' Accademia di Santa Maria, la prima scuola e il primo ufficio postale sono collocati su questo terreno. La colpa non era del terreno, ma del metodo di lavorarlo.

Sempre col pensiero della condizione dei suoi connazionali nella città di New York, Il Padre Bandini, fino dal principio aveva fatto

16

*Il Barone des Planches Regio Ambasciatore
Italiano in cammino per Tontitown,
via Frisco Lines.*

dei piani per somministrare impiego proficuo
e piacevole, che non fosse il lavoro dei campi,
per i giovani e le giovanette, credendo che se
fossero impiegati cosi, avrebbero meno tentazione di lasciare la campagna. Capi pure che
se l' agricoltore sviluppava i suoi prodotti in
maniera da produrre il prodotto finito della
sua campagna, invece che di vendere i primi prodotti, il guadagno sarebbe più grande, e vi
sarebbe meno materiale sprecato e meno danno
al terreno. Con questo scopo in vista, la gente
cominciò a produrre il vino, Evaporizzatori di
frutta, fabbriche di sidro e di aceto, fabbriche di
formaggi, e di burro, fabbriche di mattoni, e
forni per la calcina. Queste industrie sono

17

divenute importanti fattori nello sviluppo e nel progresso della colonia.

Il clima mite, le lunghe stagioni, gli inverni senza neve, l' aria pura e acqua abbondante rendono Tontitown e le sue vicinanze in condizioni ideali per le frutta, i legumi e per la coltivazione dei grani, e per l' allevamento del pollame. Una nuova ferrovia, la Arkansas, Oklahoma & Western Railway ormai completata, percorrerà in mezzo a questa prospera colonia. La strada sarà in operazione avanti il tempo delle fragole nella primavera del 1911. Con questi mezzi di trasporto alla mano, le industrie e la prosperità della colonia aumenterà rapidamente. L' unico inconveniente è' sempre stato il lungo trasporto alla stazione che era costoso in tempo ed in denaro. La nuova ferrovia è fiancheggiata da ambe parti da vigne e da frutteti per due miglia.

I coloni hanno impiegato i metodi dell' Europa nel coltivare i loro campi, quello della coltivazione intensa, successive raccolte e dell' ingrassamento dei terreni. Il Padre Bandini ha studiato attentamente il terreno, le condizioni del clima e le relazioni delle Stazioni Agricole degli Stati Uniti, e ha diretto la sua gente secondo le medesime facendo loro delle conferenze e somministrando loro stampati su questi soggetti nella loro propria lingua. Da principio i coloni per mancanza di mezzi usarono gli istrumenti più primitivi—la forca, il badile e il rastrello, ma oggi essi usano gli attrezzi rurali i più perfezionati e le macchine agricole le più moderne.

Viene dato molta attenzione alla grandezza ed alla cura del campo. Il Padre Bandini mi

18

disse: Uno dei piu grandi pericoli che minacciano un agricoltore è che possa diventare povero di terreni, cioè, che possieda più terreno di quello che egli possa coltivare con vantaggio. Quaranta acri e tutto quello che un uomo con famiglia può lavorare vantaggiosamente. Con venti acri egli può mantenere una famiglia comodamente e risparmiare un po' di denaro. Con quaranta acri può diventare un uomo agiato, se è assiduo el lavoro e si prende buona cura del suo terreno. Deve vedere che il terreno gli dia buoni prodotti pel denaro investito. Quando egli ha esausto la forza di produzione dal terreno colle raccolte, deve alimentare il terreno, se vuole nuove ed abbondanti raccolti.

I coloni di Tontitown seguono questi insegnamenti. Essi arricchiscono il terreno colla rotazione delle raccolte e colla fertilizzazione. Colle lunghe stagioni buone, possono far crescere due raccolte dei medesimi legumi nella stessa stagione per mezzo dell' intensa coltivazione e della fertilizzazione. Cosi, per esempio, essi piantano cipolle primaticcie della primavera per il mercato nel terreno fra giovani peschi. Dopo che le cipolle si sono vendute al mercato, piantano fagiuoli verdi sul medesimo terreno. Quando queste sono state vendute, piantano raccolti che producono nitrogeno —come il piselli. I piselli sono usati come foraggio per il bestiame, cosi il pisello serve come un ingrasso direttamente ed indirettamente e la produttività del suolo è aumentata.

Ogni contadino nella colonia ha una vigna dai 4 ai 12 acri che può portargli un introito dai $500 ai $600 per acre di viti di quattro anni.

Coll' uva si fa il vino. I coloni producono vini nostrani che sono eguali a quelli importati. I mercanti sanno che questo vino trova molto favore per il consumo corrente ed è venduto a buoni prezzi.

I loro frutteti ora sono all' età di produrre sono e buone sorgenti di introito. Impaccano con eleganza le loro migliori mele per la spedizione, e le vendono a buoni prezzi. Le seconde sono poste in latte, evaporate, o che se ne fa sidro o aceto, così portanto buoni guadagni, meglio ancora che se fossero stati venduti direttamente allo speditore. Furono formate due compagnie anonime fra i coloni, appena che i primi alberi fruttiferi cominciarono a dare i loro frutti, e si stabilirono due fabbriche per l' evaporazione elettrica dei frutti, ed una fabbrica per il sidro e per l' aceto, dando così impiego alle donne ed ai ragazzi. Le pesche scelte sono spedite, e le seconde messe in latte, ed evaporate.

Hanno tre cascine per la crema, allo scopo di far il miglior uso del latte. Una di queste cascine è dedicata alla confezione del burro, e le altre per la manifattura di formaggi, un' arte nella quale il campagnuolo straniero è veramente maestro. I fatto che questi coloni manifatturano una qualità di formaggio, superiore e dello stesso sapore dell' importato è riconosciuto dai rivenditori e dai sensali del paese, e vi è una gran domanda di questo prodotto.

Hanno una fabbrica di scope, dove la ginestra coltivata dai campagnuoli di Tontitown, è usata per far scope, che sono vendute ai coloni dei paesi vicini. Hanno pure fabbri ferraii e calzolai.

Tra gli uomini della colonia sono esperti lavoratori, in lavori da muratori, carpentieri ed altri mestieri ed arti.

Oltre al convertire i prodotti della campagna in prodotti finiti per scopi commerciali, i coloni convertono ed utilizzano le risorse naturali in prodotti commerciali per il loro proprio uso.

Così, per esempio, le pietre dei campi sono usate per fabbricare le cinte, almeno come fondamenti di cinte. La pietra calcarea ed il granito vengono scavate ed usate per i fondamenti delle case, molte volte fabbricati intieri sono fatti di questi materiali. Tre fratelli sono ora molto occupati nel fabbricare una casa di pietra di tre piani, ed il materiale usato è stato preso dal loro proprio terreno. Le muraglie sono di uno spessore di un piede e mezzo, ed in bellezza di linee, ed in proporzione architettonica è uguale a qualunque fabbricato di città. Gli uomini lavorano dietro alla fabbrica a tempo perduto, che rimane dopo la dei loro raccolti. Il primo fabbricato tredici anni fa era una casa di fango di una sola camera. Ogni agricoltore ha ora una bella casa comoda e ben costruita, molti di loro hanno case grandi ed eleganti con bellissime spianate. Vi sono tre fabbriati di pietra e quattro di mattoni nella colonia; gli altri sono nitide costruzioni in legno, moderne e piacevoli alla vista per la loro struttura. Tutti hanno pozzi e cisterne; si scava ordinariamente il pozzo prima, por si costruisce la casa, buone stalle rimesse piene di grano e di buon bestiame, cavalli, muli ed altri animali.

Il Padre Bandini fece esaminare una roccia e trovò che conteneva il novanta nove per cento di pura calce. Si formò una compagnia di azionisti fra i coloni e questa collina fu comperata dal proprietario americano che lo credeva una perdita di denaro giacchè non era altro per lui che un mucchio di pietre inutile. Fu convertit dai coloni in un industria col costruire una fornace di calcina e vendendo la calcina ai paesi vicini, e facendone spedizioni, cosi pure somministrandola agli agricoltóri del vicinato, dando cosi impiego alla gente. Circa 1200 barrili di calce si possono produrre e spedive da questa fornace ogni settimana. Quando la strada ferrata sarà finita, questo sarà un affare assai proficuo.

Cosi pure venne esaminata l' argilla, e si trovò che si prestava per una buona qualità di mattone. Fu subito eretta una fornace di mattoni per fornirli per le costruzioni. I mattoni sono di buona qualità ed i coloni ora vendono i mattoni ai contrattori del paese.

Queste industrie sono ancora nella lora infanzia, ma hanno prodotto meravigliosi risultati. Si sono fatti dei piani e si organizzarono delle compagnie di azionisti collo scopo di erigere piu grandi fabbriche e stabilimenti per aumentare la produzione appena che sarà completa la ferrovia.

Il valore del terreno dei dintorni ha molto aumentato durante gli ultimi anni, a cagione del meraviglioso sviluppo del paese dovuto ai coloni. Il terreno che poteva esser comperato tredici anni fa per $15.00 all' acre, e questo si considerava un alto prezzo, ora porta dai $50.00 ai $150.00 all' acre, e per 12 miglia intorno alla colonia è difficile trovare terreno a meno di $25.00 all' acre.

Si è costituito una cittadina nella colonia. Si trova nel centro di 150 miglia quadrate di ricco territorio, ed è già un prospero villaggio. Un hotel moderno di venti stanze è stato fabbricato, provvisto di luce elettrica, misure sanitarie bagni di acqua calda e fredda, e tutte le convenienze moderne. Costa un poco più di $8,000.

Vi sono pure tre botteghe, un ufficio postale, una fabbrica di scope, un ufficio per l' acquisto e vendite dei terreni, una sala, la prima scuola, circa venti fabbricati moderni, la chiesa che costa $4,000, la casa parrochiale che costa $1,000 e l' Accademia di Santa Maria che costa $8,000.

L' Accademia di Santa Maria è una grande struttura di legno, fornita con tutte le convenienze moderne, per il riscaldamento e por la distribuzione dell' acqua. Contiene cinque grandi sale, una camera pel giardino d' infanzia, ed appartamenti per abitazione per quattro Suore della Misericordia e le due maestre secolari. Vi sono ora cento e trenta ragazzi che vanno all' Accademia. Il corso inchiude dal primo dipartimento fino al grado di Università, e musica e belle arti. Si farà presto qualche aggiunta al fabbricato della scuola, giacchè non vi è più molto posto, e il Padre Bandini ha ricevuto molte richieste da tutte le parti del Sud Ovest, da genitori che desiderano di mandare i loro fanciulli all' Accademia di Santa Maria.

Dei ragazzi delle prime famiglie che si stabilirono a Tontitown, nove delle ragazze ora sono belle giovani—e maestre, avendo ottenuto certificati di primo o di secondo grado dallo Stato, e sono insegnanti. La prima maestra della scuola di Tontitown, Miss Bernardetta Brady, è professora nell' istituto Normale dello Stato. Due signorine sono stenografe competenti, un giovanotto ha il posto di agente dell' express sul Frisco ed un altro occupa un posto di responsabilità come esperto contabile, e due sono esperti stenografi.

Tre signorine, ora maestre, che hanno perduto il loro padre durante il primo anno della colonia, hanno appena finito di costruire e fornire di mobiglia una casetta moderna e regalata alla loro madre, oltre di cio hanno finito di pagare il terreno che il loro padre comperò avanti la sua morte. Questo lo hanno affittato con profitto.

Oltre ad avere una comoda casa, i coloni di Tontitown hanno pagato le loro terre, e bene provviste, le loro vigne ed orti aumentano in area ogni anno. Essi hanno molte vigne che portano frutti, e i seguenti alberi fruttiferi;

1,000,000	piante di fragole.		
200,000	viti di uva.		
6,000	MELI.	500	Ciliegi.
4,000	Peschi.	500	Peri.

La loro vita sociale non è stata trascurata. Hanno società dramatiche, che danno recite, società corali, circoli letterarii, circoli sociali, le loro feste di giardino, i loro balli ed altri divertimenti.

Lo scopo della lor vita sociale si è di tenere i genitori interessati negli affari dei giovani, e di tener vivo lo spirito e l' amore della famiglia. I giovani sono cosi ben vestiti, come la gente da bene nelle città. Sono bene educati, e si tengono al corrente dei tempi, prendono interesse negli eventi, che corrono, compiti, raffinati, e colti, e sopra tutto sono sani, fisicamente e moralmente, felici e contenti.

Per biglietti di ferrovia, carte geografiche, ed ampie informazioni circa a un viaggio al Texas, rivolgersi quanto prima al Signor A. Hilton, Agente Generale pei passeggieri, Frisco Lines, Frisco Building, St. Louis, Missouri, U. S. A.

A435

The
Italians in America

By
FREDERICK H. WRIGHT

Superintendent
Italian Missions of the
Methodist Episcopal
Church

NEW YORK
MISSIONARY EDUCATION MOVEMENT
OF THE UNITED STATES AND CANADA
1913

General Outline

PAGE

I. Italian Democracy......... 5

II. Taxes......... ....··· 8

III. Italians in the United States 10

IV. Sobriety............................... 10

V. Poverty...... 12

VI. Crime.............. 13

1. *Jealous Guarding of Home Life*............... 13

2. *Statistics Not a Fair Index* 14

3. *Laxity of City Officials* 16

4. *Inherited Hatred of Governmental Restrictions.*... 16

VII. Occupations...... 18

VIII. Italian Renaissance........... 20

IX. Religious Work............. 22

X. Social Opportunity... .... ·····················24

The Italians in America

THE Italian has an intense love for the land of his birth. He thinks Italy is the most beautiful country in the world, and there are many who agree with him. "Come le piace l'Italia?"—How do you like Italy?—is one of the first questions asked of an American by an Italian when he discovers that you have lived in Italy. The American asks a similar question of a visitor to this land. "How do you like our institutions?" The Englishman or German rarely asks such a question. His national pride would not so degrade itself. He thinks his country is all right, and your estimate of it is a matter of positive indifference to him. Perhaps in this attitude of the Italian may be found the secret of his unconscious affiliation with American ideas.

I. Italian Democracy

There is much in common between the Italian and the American. A rugged democracy characterizes this dweller in his sunny southland. He believes in the divine rights of man, and a man-made nobility has small value in his eyes. The Englishman loves a lord, but the Italian idolizes a man of noble character whether he be aristocrat or plebian. Victor Emanuel II, Cavour, Mazzini, Garibaldi, are names to conjure with, and while the two first-named were aristocrats by birth, the favorite cognomen of the King was "Re Gallantuomo," which indicated his bluff democratic spirit, and in diplomacy Count Cavour was the plain man of the people. A title of nobility was offered Garibaldi by a grateful king and nation, but like the great commoner, Gladstone, he chose to preserve his identity, and his name is honored more in Italy than any other earthly name. There is not a city, town, or hamlet in that fair land where the name of this great man is not perpetuated in some way or other. He is the Washington-Lincoln of Italy.

This democratic instinct has molded the life of the people, strengthened the family tie, and developed an almost patriarchal life. I remember a home in Sicily which I have often visited, in which there lived together in perfect harmony a widowed mother with four sons and their wives, two unmarried daughters, besides the smaller

6

children of the family—I think there were sixteen in all. A large round dining-table accommodated more than a dozen at one sitting, and the affectionate relations were delightful. The dreadful earthquake at Messina swept away the entire family, except one. That family represented a type of Italian life beautiful to behold. Anyone familiar with the life of the Italian in America will know that the same spirit characterizes them in the land of their adoption. Not only the immediate relatives are cared for, but uncles, aunts, cousins, nephews, nieces, brothers and sisters-in-law, fathers and mothers-in-law are all united under one roof. This means a generous hospitality, suggestive of our own southland. It also means a strong family affection. An Italian father may punish his child, even severely, but no father loves his children more than the Italian.

The feudal system, although abolished by law, is still practically in vogue. Most of the land is the property of large land-holders. and absentee landlordism, as in Ireland, is the curse of Italy. The Italian nobleman usually farms out his land to a " piccolo proprietario " for a certain sum, and then lives in Palermo, Naples, Rome, Florence, or some other city, on his income. The lesser landlord then sublets to the " contadino " or peasant, and, of course, expects the lion's share of the proceeds. The peonage system, also, is in practical operation, and many a poor peasant is hopelessly in debt to his landlord.

II. Taxes

Then there are the taxes, and it is generally conceded that Italy is the most heavily taxed nation in Europe. A recent writer in a prominent review insists that the national budget is unjust. The poorer classes of Italy are burdened with fifty per cent. of the national tribute. The majority of luxuries escape taxation, while the essentials—corn, salt, petroleum, and like products—are exorbitantly taxed. The "Lotto," or national gambling association—a heritage from papal rule—nets about twenty-seven millions of lire ($5,400,000) yearly to the state, depleting the public pocket to the extent of nearly seventy million lire ($14,000,000), coming principally from the small wage-earner, and the laboring classes. The duties on salt benefit the Treasury from fifty-four to fifty-nine million lire, or $10,800,000. Two hundred and twenty pounds of salt, which costs the state about thirty-two cents, is sold to the people for $8.00. Petroleum, which costs the government $3.50, is sold at $13.00.

The operative forces of the Italian struggle for national independence and constitutional liberty were essentially democratic, as we have already indicated, and the national sentiments and institutions are nominally so to-day. Yet we find not only successive governments but even the local administrations of communes and provinces following in practise a course diametrically opposite. For example, in the South, the saddle-horse and the four-in-hand of the rich aristocrat pays no tax, because, forsooth, such luxuries cost money but bring in no income. On the other hand, the donkey of the contadino or peasant, which carries his produce to market or draws his antiquated plow, being considered an implement of labor and consequently a source of income, must pay the tax. This is obviously unjust, and the effect upon the Italian laboring man, especially in the South, has been to cause him to seek other lands for a living.

The enormous emigration to America is the direct result of the deplorable economic conditions of Italy, otherwise very few would leave their fatherland. " L'Italia e' bella ma povera,"—Italy is beautiful but poor,—the immigrant declares with tears in his eyes, and it is easy to see that they would stay there if they could ; indeed, many return, after a few years' sojourn in this country, to spend the rest of their days in "la bell' Italia." Rarely, if ever, are they weaned from their old love.

III. Italians in the United States

It is a very difficult matter to compute the number of Italians who come to the United States. There is an incessant coming and going, but a recent official item from Rome estimates that fully 3,000,000 are still in the United States, and possibly 600,000 are in Greater New York. The rest are distributed throughout the New England States, Pennsylvania, New York, Maryland, Delaware, New Jersey, and California, though it would be hard to find a State in the Union where the ubiquitous Italian cannot be found. Like his Irish prototype of years ago, whose place he has in a large measure taken, he is everywhere. As market gardener and farmer he is making vast strides in the central States, and future years will see greater developments than ever in this direction.

IV. Sobriety

The moral, social and religious condition of the Italians in this country is an interesting study. In this brief outline there is not space enough to enter into detail. But, in general, we may unhesitatingly declare that the social life is exceptionally good, at least, when the immigrant arrives in this country. His sobriety is worthy of note. The *Strassburger Post* has recently devoted space to a comparative statistical study of the amount of liquor consumed by the inhabitants of various European states. The Dane leads the continent, his average being 104 quarts of beer, very little wine, but 24 quarts of brandy each year. The Italian, on the other hand, is the most temperate: he drinks the least beer, a mere two quarts, and the least alcohol, one and a quarter quarts. His wine consumption is 98 quarts. *No nation in the world has such a record for temperate drinking.* Time and again, during our six years' residence in Italy, we remarked on the decent aspect of the Italian laboring man on holidays. It is a delight to see him coming home after a day's outing, with his wife and family,—a jolly, laughing, good-natured, sober crowd,—a striking contrast to his American compeer. In the United States, wine is expensive, and he resorts to beer and whisky. It is sad to note that Italians are developing a taste for strong drink. He drinks beer because he sees his American fellow workingman doing the same. He is

11

very impressionable, and becomes a part of all he meets. He comes in contact with the poorer class of Americans, and patterns after them. Yet withal, his record for sobriety is noteworthy, even in America.

V. Poverty

The Italian is industrious. It is a very rare thing to see an Italian pauper in America; they are too busy to beg. Those who have been in Italy know that one of the blights on that fair land is the trail of the professional beggar. He is a common nuisance, and tries sorely the patience of the tourist by his persistency. The American sculptor, Story, tells of Beppo, the king of the beggars in Rome, who became a very rich man. The very genius of the Roman religion encourages begging. Believing in salvation through works, and that he who "giveth to the poor lendeth to the Lord," and heaps up merit to himself, the Italian is sharp enough to see that there is a decided business advantage in making his appeal for charity in order to help to save the soul of the charitably disposed; hence the beggar is much in evidence to assist men and women to protect their souls. But this type of Italian is not in evidence in America, thus prov-

12

ing that this undesirable class does not and cannot emigrate Jacob Riis is authority for the declaration that among the street beggars of New York City, the Irish leads with 15 per cent., the native Americans follow with 12, the Germans with 8; while the Italian shows but 2 per cent. We are told further that "in the almhouses of New York the Italian occupies the enviable position of having the smallest representation, Ireland having 1,617 persons and Italy but 19, while the figures for the entire United States are equally favorable." When it is remembered that the Italian population of the metropolis is more than double that of the Irish, the figures are striking. Perhaps I ought to say that this comparison is made not for the purpose of reflecting on the Irishman, who has made his record, but for the purpose of placing the Italian in the right light.

VI. Crime

1. *Jealous Guarding of Home Life.* The home life of the Italian is jealously guarded, and no nation has a higher record of social virtue. Their chief crimes are the result of jealousy,—a fear that their home life has been invaded. After all, it is a proof of the intense love which dominates their lives, for jealousy, at the last analysis, is in-

tensified love, or love gone mad. Their criminal record is not so bad as it appears. The newspapers seem to take particular delight in exploiting their deeds, yet Dr. S. J. Barrows, the expert criminologist states: "There are vile men in every nationality, and it does not appear by any substantial evidence that the Italian race is peculiarly burdened, though it has been unwarrantably reproached through ignorance." In August, 1911, in Wilmington, Delaware, with a total population of 77,000, there were 350 of all nationalities who had been sentenced to the workhouse, and yet with an Italian colony of 6,000, only 12 were inmates of that institution, whereas their proportionate number on the basis of their population would be 27. This is no exception to the rule. The Black Hand gang and the blackmail scares have created a prejudice in the minds of Americans who have not studied the actual relation of things.

2. *Statistics Not a Fair Index*. Professor E. A. Steiner, known as the immigrant's friend, makes the significant statement which follows, based upon his wide experience: "I have visited nearly all the penitentiaries in the eastern and western States; not to ask how many foreigners there are in jail, but to ask why and how they were convicted, what their present behavior

is ; to look the men and women squarely in the face and to converse with them. Let me say here again, emphatically, that the statistics are misleading, and that in spite of the large number of Italians in prison, there are by far *fewer* criminals among them than the statistics *indicate*. In a large number of cases, the crimes for which the Italian suffers, have grown out of local usage in his old home. None the less are they justly punished here, lest they be permitted to perpetuate themselves in the new home. Most of the Italians in prison have used the stiletto and the pistol too freely, just as they used them at home when jealousy made them mad, or when they were in pursuit of vengeance for real or fancied wrongs. There are not a few real criminals who have used the weapon for gain, but in the majority of cases the stabbing or shooting was an affair of honor with those concerned, and even the aggrieved parties preferred to suffer in silence and die, bequeathing their grudge to the next generation, rather than bring the affair before a sordid court. Testimony in such cases is very hard to get, and I have seen many a wounded Italian bite his lips, inwardly groaning, and suffering in silence, unwilling to let strange ears hear the proud secret of which he was the keeper and the victim. Italian burglars have not reached proficiency enough to have a place in the 'Hall of Infamy,' and bank robbers and ' hold-up' men need not yet fear serious competition from that source. The prisons contain many Italians who transgressed out of ignorance as well as from passion ; numbers suffer because they do not know the language of the court, and did not have enough ' coin of the realm.' "

3. *Laxity of City Officials.* An unfortunate reason for the apparent prevalence of crime is too often found in the laxity of the city officials. So long as the Italian kills his own kind, the American representatives of justice take little time and trouble to sift out the crimes, and accordingly many go off scot free, and crime is thus encouraged. Not so when he kills an American : then he is hunted to the death and glaring headlines tell of the awful crime.

4. *Inherited Hatred of Governmental Restriction.* The Italian has an inherited hatred of all governmental restriction. For generations he has been the slave of the conquering nations, which in turn dominated his fair peninsula, and he has ever been a rebel. This explains his attitude to all authority, and accounts somewhat for his criminal record. Lillian Betts, herself a Roman Catholic, blames the ignorance and

bigotry of the immigrant priests who have set themselves against American influence, and knowing, as she does, these Italians both sympathetically and critically, her words have considerable weight. She says: '' In New York, the streets the Italians live in are the most neglected, the able head of this department claiming that cleanliness is impossible where the Italian lives. The truth is that preparation for cleanliness in our foreign colonies is wholly inadequate. The police despise the Italian except for his voting power. He feels the contempt, but with the wisdom of his race he keeps his crimes foreign and defies this department more successfully than the public generally knows. He is a peaceable citizen in spite of the peculiar race crimes which startle the public. The criminals are as one to a thousand of these people. On Sundays watch these colonies. The streets are literally crowded from house-line to house-line, as far as the eye can see, but not a policeman in sight, nor occasion for one. Laughter, song, discussion, exchange of epithet, but no disturbance. They mind their own business as no other nation, and carry it to the point of crime when they protect their own criminals. Like every other human being in God's beautiful world, they have the vices of their virtues.

It is for us to learn the last to prevent the first."

VII. Occupations

The Italian in this country belongs to the laboring class, and chiefly to unskilled labor. Nevertheless, he is moving up, and in New York alone there are fifteen hundred or more lawyers, five hundred physicians, besides a growing number of merchants, bankers, and business men. The young Italians are flocking to our higher schools of learning, and are making an enviable record. Last spring (1911) a new system was established in Columbia University, of graduation with honor. The first and only man to graduate under the new rule was an Italian. He stood an oral examination before the professors of the various departments, in which he took honors. Yet with all these good indications, there is no use disguising the fact that the crowded condition of our little Italies is a menace to our nation. Sociologists have done a great deal to ameliorate the awful conditions which obtain, but much more will have to be done to avert a dire calamity. Socialism of a revolutionary type —suggestive of violence, and something more than a suggestion—is rampant among the Italians. It is an inheritance from the

18

homeland. The spirit of discontent, nurtured by the unfortunate economic conditions there, has a freer scope in this land of liberty, and unless some high moral and religious restraint is brought to bear upon it, the results may be disastrous both for the Italians and for this country.

A great many have the impression that the Italians who come to this country are chiefly of the riffraff element, but this is very far from the truth. They are the brawn and muscle and undeveloped brain of Italy. The proof of the first is seen in the mass of unskilled labor which lays our suburban tracks, digs our ditches, tunnels our subways, and builds our railroads. What would we do without them? The proof of the second is seen in the wonderful progress they make in our public schools. The invariable testimony of teachers is to the alertness of mind of children of the first generation. I remember visiting an Italian Sunday-school in New York City a short time ago. The secretary, a young lady, was born in Southern Italy. I thought she was an American, she had so absorbed her environment. In conversation with her I found that she was studying typewriting and stenography, that she had made good progress, and I have since learned that she occupies a very responsible position in the city. She was the daughter of a New York City street-sweeper and her parents could neither read nor write. America had produced that change in a life by the larger opportunity, and all in the first generation. In a certain city of the East, settlement work was taken up by a Christian church. Italian boys of the immigrant class frequented the settlement. One of the boys is now judge of a juvenile court, another is an instructor of Italian and Latin in one of the well-known universities, another is a successful physician, while still another is a lawyer of prominence and a graduate of Harvard. In commercial life they are forging their way to the front, and in the professions they have already won their place, thus demonstrating their virile qualities. The story of the decadence of the Italian people is the story of the rise of the papacy, and with it the suppression of free thought.

VIII. Italian Renaissance

In the eleventh century Sicily was the center of culture and refinement, then came the fall of Constantinople and the subsequent development of the so-called Holy Roman Empire with its domineering hierarchy. All tendencies to mental and religious liberty were checked with an iron hand, and though there were many who resisted even to death the encroachments of a tyrannical ecclesiasticism, little by little the papal Church throttled free speech and free thinking, and developed in its own territory a nation of ignoramuses, until in 1870, when the record of illiteracy was fully eighty-five per cent. The Italian Renaissance saved Europe, but Italy itself was cursed by the dominance of a medieval priestcraft and remained in abject poverty and despair. Symonds has well said : " The history of the Italian Renaissance is not the history of arts, or of sciences, or of literature, or even of nations. It is the history of the attainment of self-conscious freedom by the human spirit manifested in the European races." The three R's of the Middle Ages were *Renaissance*, —the new birth, the era of Medicean influence and Florentine elegance ; *Reformation*, — the struggle of Luther and Zwingli against scholasticism and papal supremacy; *Revolution*,—the culmination of the irresistible forces of Renaissance and Reformation in the European life and thought. Italian Renaissance, Lutheran Reformation, and French Revolution, form the trinity of influences begun through a domineering yet decadent ecclesiasticism.

The Italian of the twentieth century—the

immigrant to the new world—is the product of an effete hierarchy. A child of the country which saved Europe and then the world, he himself remained a victim of a world-bound papacy, and as he touches our American environment so many are ready to declare him a degenerate son of a degenerate people, but the thoughtful observer will look into the philosophy of history and draw a conclusion juster and truer, namely: that the modern Italian is the undeveloped product of a degenerate Church. Breaking away from the traditions of the past, this same Italian breathes the free air of a Protestant environment in the new world, and rises to his old-time power and influence.

IX. Religious Work

Something must be done to meet the immediate needs of the Italian immigrant. American Protestants for a long time did nothing, believing that every Italian was a Roman Catholic, but after a careful investigation it was found that a vast number had no affiliation with the mother Church except by tradition and baptism, and that if nothing was done for them by Protestantism, they would come under no religious influence whatsoever.

The socialistic element of Southern Italy is decidedly anticlerical. It knew nothing but the Roman type of Christianity, and the revolt from that was spontaneous and positive ; hence it became antichristian. It was the swinging of the pendulum. Contact with American life has brought larger views on religious thought, and these erstwhile antichristian Italians cannot resist the force of Protestant ideas. Protestantism has made its greatest inroads in the socialist camp, and thousands are now rejoicing in the light of the gospel of Christ, who, without such influences, would have been completely separated from the Christian church.

We have no quarrel with the Roman Church ; we are simply trying to reach the vast majority of Italians who for one reason or another have lost faith in her. This is not the place to discuss the reasons for this lack of faith ; the fact is undeniable. Apart from the socialistic element, which is aggressively antichristian, the greater number of Italians in this country are absolutely indifferent regarding Christian faith and life. The almost tragic relation of the state to the Church is lamentable The fiftieth anniversary of the unity of Italy was celebrated in 1911 with great pomp in Rome and Turin, but not a prayer was publicly offered up for the blessing of the God of nations. This total disregard of Christian teaching augurs ill for a people who in their natural tendencies are religious. The emotional dominates their lives, and yet the cold, bare facts reveal a complete absence of interest in everything which is Christian.

The Italians in America, under the present environment, are fast becoming an ungodly people, and this is our danger and theirs. The various denominations are active in establishing churches in their midst, and the response they have received indicates that these happy-hearted children of Italy are not impervious to the teachings of Christ.

X. Social Opportunity

Evening classes in English and instruction in United States Government, together with the presentation of the New Testament type of Christianity, can be made splendid channels of propaganda among the Italian immigrants. Naturally, they are suspicious of unselfish effort in their religious behalf·; they have been so used to ulterior motives that they stand on guard, but as they begin to see the disinterested spirit which prompts evangelical Christians to point them to the Lamb who taketh away the sin of the world, their hearts become responsive to the truth, and their simple, childklike, yet withal strong and stalwart natures, assert themselves and prove the power of their intellect and the warmth of their hearts.

Americans must break away from the silly prejudices of the past and that unreasoning ignorance which destroys Christian bonds and creates a chasm between races. By a mutual sympathy and Christlike spirit, we, as Protestant Christians can do mighty things for these lovable people, whose forbears, long years ago, gave our fathers the gospel, and who now themselves need the same old gospel of mercy, compassion, and love.

Italy has played her part well in the history of the world. Notwithstanding all the suppression of free thought, she has produced some of the greatest representatives in literature, science, art, and religion. Dante, Petrarch, and Tasso sing as sweetly as any poets. She has produced a Mondino, the father of modern anatomy, a Falloppio who preceded Harvey. Galileo studied the heavens and turned astrology into astronomy. The barometer, the timepiece, the improvement of the mariners' compass are the gifts of Italian genius. Galvani gave us the first principles of electrical science, followed by Volta and Meucci with the

25

telegraph and telephone, while Marconi crowned the whole with his wireless messages. Michelangelo, Raphael, Andrea del Sarto, Perugino, Carlo Dolci, Giulio Romano, Da Vinci, Titian, and Fra Angelico peopled the galleries of the world with matchless art. Bellini, Verdi, and Donizetti discoursed to us sweet music, while Paganini and his associates reproduced in instrument and song the masters of Italian symphony. Her roll of statesmen and generals is no mean record. From Julius Cæsar down to Garibaldi and Napoleon Bonaparte she has shown masterful generalship, and from Lorenzo to Cavour she has produced men of state and giant nation-builders. In the religious world she has developed her resourceful and intrepid missionaries who went forward and won our Anglo-Saxon fathers to the Christian faith, and even Ireland pays its tribute to Italian evangelism in the person of her patron saint Patrick, himself the son of Italian parents, who was Christian without being Romanist and Catholic without being Protestant, and who in his "Confessions" insists on justification by faith only.

This is the story of Italy. What a debt we owe her ! As her sons and daughters of yeoman blood, illiterate yet virile, undeveloped yet brainy, crowd upon our shores, let us give them our heartiest greetings and an unstinted welcome, and by the grace of God they will do for our America what they have done for their glorious Italy.

IMMIGRANT AND LIBRARY:
Italian Helps

WITH LISTS OF SELECTED BOOKS

By
JOHN FOSTER CARR

Author of

"A Guide to the United States for the Italian Immigrant"

IMMIGRANT EDUCATION SOCIETY
241 FIFTH AVENUE
NEW YORK
1914

Index

Page.

The Library and the Immigrant 7
The Librarian and the Italian 12
"The Lists that Follow" 14
Biography 19
Travel, Description, History 23
In the United States 27
Italy 30
Books of Literature and Education 38
Fiction—Italian 43
Fiction from Other Languages 49
Books for Children 53
Poetry 57
Drama 65
Music 68
Art, and the Sciences and their Application 71
Other Useful Books 75
Books of Reference 81
Periodicals and Newspapers 84
Library Rules and Helps in Italian 89

Pronouncing Italian

It is very difficult to pronounce Italian perfectly—very easy to pronounce it so that you can be understood. The spoken word faithfully follows the spelling. There are no mute vowels, and no silent consonants except *h,* which is never pronounced. Every syllable begins, if possible, with a consonant. Double consonants are usually divided.

The letter *a* is always pronounced as in *far;* *e* sometimes as in *there,* sometimes like *a* in *gate; i* always has the sound of *i* in *machine; o* is pronounced as in *rose*—shorter before two consonants; *u* always as *oo* in *boot.*

Italians pronounce *c* before *e* and *i* like *ch* in *church*—otherwise as in English; *ch* has the sound of *k* before *e* and *i.* And so *g* is pronounced as in *gentle* before *e* or *i*—otherwise as in English, and *gh* is always hard; *gn* sounds like *ni* in *union;* *gl* like *lli* in *million.* The *r* is rolled—heavily rolled by some Italians—and *t* is pronounced as in *return,* never as in *nation.*

The Library and the Immigrant

The library was long a sort of institutional Lord Bacon. All learning was its province. Now its province has become all life, and it is already the greatest of our popular universities. It is ever seeking larger powers of usefulness, and striking is its development along simpler and humbler lines.

In the new duties that immigration has brought, it is unquestionably meeting the greatest educational problem yet unattempted in this country. The Census Bureau states the size and significance of that problem when it announces that there are four million foreign-born white men of voting age in the United States, who are not citizens, and two million men and women either foreign-born, or wholly or partly of foreign parentage, who are illiterates.

It is of the very first national importance that the foreign-born who are to remain among us, should be made an effective part of our democracy,—a vital part of our own people. But how shall they gain quick interest in our collective life, our citizenship, our government? How shall they be given our English and a knowledge of American conditions that will make their daily living easier, improve their working skill and wages, and reduce by one half, as English does, their liability to industrial accident?

When Firmin Roz, keenest of French publicists, wrote the other year of the marvels of the United States, he put in the very forefront of his preface the most astounding thing he had found: "There," he said, "the aged and out-worn races of the world repair, cast aside old age like a garment, and renew their youth in American life." I believe that if you see the working life of our foreign-born at its normal, all of these toilers seem in marvellously rapid process of Americanization. But they are often in close touch with the worst and not the best side of our civilization; they too often lose the restraints and ideals of the old world and find nothing to replace them.

In this great work of education, the library has a far greater opportunity than the school. Friendly and helpful, its aid is more inviting and less formal. It makes less strenuous demand upon the attention of a man who is often very tired after a long day's work. It welcomes those who think themselves too old for school. It is open throughout the year, where the night school at most is open only seven months of the year. It can furnish papers and books in the immigrant's own language and thus provide a familiar and homely air. A common meeting ground with Americans, it gives him a sense of joint right and ownership with us in the best things of our country, and this with no suggestion of patronizing interest. Best of all, I think, it can put him in effective touch with American democracy, and American ideals; and so, better than any other agency, destroy the impression of heartless commercialism, that many of our immigrants, in their colonies, continually assert is the main characteristic of our civilization.

Work for our immigrants is not wholly a new thing in American libraries. It dates back many years; but it is new in the extent of its present enterprise and interest. Its progress has never been without opposition. Many have insisted that the immigrant should have no books in his own tongue. Many have wished him to forget everything he was or thought before coming to America, and they have been jealous of foreign languages, insisting on English.

We sometimes forget that no naturalized citizen can ever be a good American unless he has first been a good Italian or German or Greek,—unless he has the reverent instinct of loyalty to the land of his birth. If he is to be a good American, we must give him some sufficient reason for respecting and loving our land. And how better than through the library can this country of ours be made alluring, accepted in love? Alluring certainly is the library's invitation to personal progress and self-betterment, and in its friendly rooms are an American environment and the atmosphere of our spoken English.

Even though the foreigner may never become a citizen, but remain only as a visitor of seasonal labor, he may still have in the library some helpful experience of American good-fellowship, pass profitable hours in at least one place that is filled with respect for the land of his birth, and this to the definite advantage of our state. The wider use of the library by the foreign-born directly stimulates among us an increasing and truer knowledge of other lands, and a more generous understanding of their national ideals.

It is the unvarying experience of librarians that every attempt made in opening the libraries to our recent immigrants has had large and unexpected success. Providence reports that the hunger for books among the foreign-born is keen and universal. Boston, welcoming the unskilled laborer as well as the cultured student of the classics, has made striking progress in these new efforts the last three or four years, and incidentally has discovered—eloquent testimony to the ambition in the homes of these workers—that the "children of foreign-born parents read a better class of books than their American

—8—

brothers and sisters." A Brooklyn branch lets it be known that men coming from work with their dinner pails are welcome. And at once the library reaches a point and has success of service before unknown.

The result of broad and aggressive work in the New York Public Library has had an instant return. During 1913, the last report records, the circulation of Italian books increased by nearly ten thousand—a remarkable growth when a moment's calculation shows you that it amounts to nearly 27 per cent, falling less than 4 per cent behind the Yiddish, read by the most eager frequenters of our libraries.

And here another significant matter may be learned, useful for quoting to those who think the dominance of our English threatened by the foreign languages. In this same report the large total is set down of the circulation of German books—by far the largest circulation for books in foreign tongues. Yet, figuring again, it appears that for all the new inducements and attractions of the library, the annual gain had barely passed one half of one per cent.

The community life of our foreign colonies rapidly passes. Its picturesqueness and foreign customs vanish, its theatres and festivals. Even its music dies. And in spite of every effort its speech is lost.

The generation of the great mass of our German immigrants is, of course, rapidly passing—so rapidly that by the last census, in spite of an immigration of seven hundred thousand for the decade, our total German-born population decreased by over three hundred thousand. This goes far to explain a stationary circulation. But it is also clear that these same people, the most literate, and the most tenacious of their national culture of all our earlier immigrants, have come so far into the practice of the English language, forgetting their own, that further increase of German readers in our libraries is hardly to be looked for.

It is important for the immigrant to learn English more rapidly, and the library can greatly help in this. It is also important that the knowledge of foreign languages should be seriously cultivated among us. It could now easily be made a national accomplishment as it is in many countries of the Continent. Here again the library should greatly help.

But such results as those attained in New York only come as the consequence of hard and earnest work. There are difficulties a plenty in the way. Our foreign-born working men and women oftentimes know nothing of the existence of the library. Or they have a strange fear to enter, and need much persuasion before they can believe that they will be welcome visitors in such splendid buildings. Often, too, they seem to fear that the library may be connected with a church that is trying to proselytize them, or that some advantage may be taken of them. They need to learn that the library, like the school, is non-sectarian and non-political; that it is the property of the public, and that full privilege of it belongs to every man and woman and reading child. For this reason their priests and rabbis make the librarians' most helpful friends.

And once the immigrant workman is persuaded to enter the library, he needs immediate personal attention. He needs to have the different rooms of the library explained, the few simple rules given him to read in his own language. Index cards are impossible to him. The open shelf is generally almost useless. He knows little or nothing of the proper use of books; often he has never even handled one. He requires the librarian's aid in the mysteries of selecting and registering books. In short, he requires much painstaking individual help.

But how bring the immigrant to the library? In a number of places, very ambitiously, lists were made, classified by nationalities, of all the foreign-born families living within the radius served by the library; and to each family an attractive postal card notice was sent. But in many of our cities such work would be an almost impossible task. In such cases, and generally, very effective publicity has been found in the distribution of cards and leaflets bearing lists of appealing books. These have been sent to the multitude of national societies and clubs of various kinds that exist, as well as to drug, stationery and grocery stores, to the rooms of trade unions and to factories. Many librarians are regularly sending boxes of books to such very practical distributing centres. And public schools, night schools, parochial schools are being pressed more and more widely into the service, and the teachers' help very effectively claimed.

In some of the New York branches rooms have been assigned for the use of literary and historical societies, and here meetings with music have been held for the discussion of literature, history, folk lore and social questions. By one admirable and popular plan a special visit is invited of a group of men and women of the same nationality. The librarian receives them and one of their own countrymen explains in their native tongue the privileges of the library. Most of our foreign friends are used to being read to, and an adaptation of the story hour has brought excellent results. It has proved fruitful in the independent and more careful reading of books, and has sometimes directly opened the way to the formation of library clubs.

In New York, also, lessons in English have been given, the library itself often supplying the text-books needed. This has promptly caused a greater demand for simple books in English. Librarians report that every effort such as these described not only increases membership, and revives the use of cards that have fallen into disuse, but gives a profitable opportunity for intensive study of the neighborhood.

Successful experiments of great variety have been made in providing evening entertainments organized directly by the library. These have included simple lectures, often illustrated by the stereopticon; addresses by men, often leading men, of different nationalities to those of their own speech; musical entertainments, vocal and instrumental; dramatic recitations, with national music on the phonograph; exhibitions of photographs of Italian art and lace. As many children are too young to leave alone, there is a suggestive instance at the Mt. Vernon Library that invites parents to bring their little ones to the children's room for separate entertainment.

Emphatically it is a work that is fast growing, spreading usefully over the country. To develop it efficiently within the borders of the state, Massachusetts through its Free Public Library Commission is carefully organizing effort, learning the location of the foreign colonies, their nationalities, and library facilities. The active interest of the leaders of the various groups has been secured; and with the help of a traveling secretary specially provided by the new law to take up this educational work, the results achieved within a single year have been so very promising that it is hoped that these efforts may be greatly extended. And where one state has so practically led the way, others must soon follow.

All this reveals the broad field of service now opening to our libraries. We are apt to forget that a man becomes an American, that his blood becomes American, when the judge signs his second citizenship paper. Whether he becomes a good American or a bad American depends in some measure upon ourselves. The great virtues and ideals that we are fond of thinking characteristically our own are often equally the national ideals of other lands. *Italianità* and Americanism are hard to distinguish in a moral definition. And if we find in America some special glory and leading, even some tang of the air, that no other land could give, we may be sure that our nation, for all the races of our origin, will never become great on its cosmopolitan plan, unless we respect and nourish the culture and all the precious heritage of the centuries, developed by other countries at such heavy sacrifice, and brought us, however humbly and indirectly, by the millions of our immigrants.

The Librarian and the Italian

Beyond all doubt it means trouble and work to start an Italian department in a library; but it is trouble and work that always, so it seems, bring delighted satisfaction to the librarian. The Italian readers who come to our libraries are hard-working folk, who were but yesterday, the majority of them, shy and simple peasants. They are extremely grateful for everything done for them, and scrupulously follow the rules, once they are understood. They are full of respect, and so honest that the Boston Public Library reports the disappearance of only three Italian books in 1913.

It is the common habit of these Italians of ours to live in closer human fellowship than ever we do. Their gracious manners expect cordial ways and smiles and kindly acts; and for these they make polite and generous return. Following the children, they may call you "Teacher," but easy is the transition from teacher to friend.

The Italian, it is true, usually has a background of reserve. He is patiently submissive to authority of every sort, yet he is hurt by brusqueness and, perhaps more than any other of our immigrants, resents unofficial intrusion. "Sir," was the complaint of an Italian workingman, "these investigators are as smoke in our eyes." But welcome to them is such friendship as the librarian can give, neither curious nor officious, but full of good will and usefulness. And they are so responsive to tactful suggestion that its practical helpfulness is soon proved.

A gift for remembering faces and names is a wonderful aid in capturing their friendship. Learn something of the history of Italy and the things that Italians are proud of. Find out from what province each man comes. Learn a few words of Italian and ask their help in learning more. One librarian of my acquaintance has found memorable pleasure and great profit for her work in a summer's trip to Italy.

It is no difficult matter to make a beginning with these good people. Simple books, like the story of Silvio Pellico's imprisonment, or one of Vallardi's little manuals, or some carefully chosen volume from the children's shelf, sent home from time to time by a small boy or girl, may finally bring you a man who thinks he is too old to learn any English, and yet has secretly yearned to have some book in his own tongue.

Take the new-comer to the Italian books, even though he can only be shown by gestures, for certainly the Italian, quickest of all, understands gesture. You will find it a language of large possibilities. You may not know a single word of Italian, and yet simply by gestures be able to explain quite satisfactorily the Italian resources of the library as he needs to know them. And incidentally you will discover his tastes, and be able to offer him some specific book, perhaps opened at an interesting chapter.

At the time of your new reader's enrollment, be sure impressively to call his attention to the rules, which should be printed in Italian. And, a thing of great importance, to avoid otherwise inevitable misunderstandings, point out for his special reading the rule about fines, and the one restricting the number of novels that may be taken out—if you have such a rule. And as many are careless in the handling of books,— none will damage them wilfully—it would be well if every volume on the Italian shelves carried such a paster of advice as that given in the last chapter of this book.

Our Italian readers are fond of the heroic, the dramatic, the ideal and the noble, and they like the picturesque. Poetry, drama, fiction, biography, history, travel are popular with them and in about the order named. But you will meet with many surprises. Many a rough looking laborer you will find who will revel in the classics. It was a workman, grimy from the shops, who returned Hamerton's "Intellectual Life" to a librarian in a little Massachusetts town, with: "That's what I call a good book!"

In the New York Public Library Italians make constant demand for simple books of agriculture. And though nothing may rival in enthralling power "I Reali di Francia," it is surprising how often an immigrant with no discoverable schooling will be a keen judge of the practical value of a book. You will find them grow enthusiastic about the library. None are inspired so easily as they by the artful words: "The more you use these books and magazines, the more of them we shall buy."

For a last word: Cultivate in the Italian a love for the land of his birth. Give him books about Italy and the Italian things he craves. So, he will come to find America a friendly home, worthy of love and better knowledge. In the personal touch of friendship is all the magic of dealing with these simple and lovable people.

"The Lists That Follow"

If the Italian people formerly read little, it was because the stress of life made books and reading luxuries that few could afford; and the books urged were forbidding, learned, so burdened with rigorous logic and philosophical generalities that they were called *"noiosi"* by the common people who ventured to open them,—and utterly wearisome they mostly were.

But during the last ten years, following the rapid development of national wealth, life on easier terms has brought greater leisure, and with the growth of the schools, and the fight on illiteracy in the South, the habit of reading is fast becoming national, and popular and excellent books in great variety have appeared. Often sold for a few *centesimi,* they have had the great merit of being simple, informing, and at the same time readable. And strangely enough, among the most attractive and popular of them have been school books, until now in every land the special abomination of the adult. The great success of their first experiments has led to more ambitious attempts by the publishers, and new demands have come for new books of the most varied sort.

Unfortunately, for our purpose, as will be seen, there are few books for the immigrant dealing helpfully with American life. One will hunt in vain for accurate and readable books of American history; for simple and practical books on American business, industries and agriculture, and their opportunities; or for American biographies, books of description, travel, popular customs, sports. Grammars and manuals for learning English, with the promising title *"metodo accelerato,"* are many and often expensive, but they are of little use to the people for whom they are mainly intended.

The excellent model gift libraries so generously distributed by the *Società Nazionale Dante Alighieri* have formed the basis of the lists that follow. Many helpful suggestions were found in the admirable *Bibliotechine* of the *Federazione Italiana delle Biblioteche Popolari.* as well as in the *Catalogo,* and its supplement, of the *Commissione Permanente per le Librerie dei Marinai.* In our own country, the *Catalogo dei Libri Italiani* issued by the New York Public Library has been of the greatest

—14—

service. The lists of the Public Libraries of Providence and Springfield, Mass., have been most carefully and profitably read. Most helpful and suggestive has been that excellent little pamphlet, "Aids in Library Work with Foreigners" by Miss Marguerite Reid and Mr. John G. Moulton, first issued by the Massachusetts Library Club, and now published by the American Library Association. To these should be added the latest catalogues of the Italian publishers, with their very attractive and important announcements.

In preparing these lists, special attention has been given to the needs of readers of limited educational experience, and a considerable number of elementary books have been included. But the uniform experience of librarians is that the reading of our foreign born is steadily progressive; and in the great mass of our immigrants there are many with excellent educations, who have bravely been willing to start life anew in a new land, sometimes taking rough manual work and showing no outward mark of that education which they really have. And so these lists have been formed with a definite idea of progression. For this reason they should also be of service as a reading guide for Americans, who, with varying mastery of the language, are coming to take a deeper interest in Italian literature and the development of Italian national life.

For our selection, the first purpose has been to seek books of fine human quality, of great and attractive simplicity. Some famous names have been omitted for a variety of sufficient reasons. It has been the purpose to include no expensive books, no difficult books of science or of research, of narrow or purely scholarly interest, no unwholesome or trivial books, no books representing the aesthetic or decadent schools, no English books—for though the number of these is restricted, many useful lists already exist. No religious books will be found here, except "The Little Flowers of St. Francis" and "Pilgrim's Progress."

The descriptive notes make no ambitious attempt to give a complete critical estimate of authors, but merely in an elemental way to be suggestive and informing. As these lists are not exclusively for the use of librarians, no knowledge whatever of Italian books or literature is presumed. It is hoped that no apology is needed for a somewhat unvarying note of praise—a deliberate attempt has been made to select books that are worthy of praise, the best that are available of those now in print.

In the description of each book here listed, the author's name comes first, except in the case of works that are anonymous. Then follow the title of the book—translated into English when necessary; the name, in Italian, of the place of publication; the name of the publisher; the date of printing of the edition chosen; a brief bibliographical description, and the price.

Where the date of an author is not given, it means that he is either a contemporary, or has recently been writing. The many asterisks do not indicate books that are equally important or desirable or simple. They are merely a note of attention, suggesting a first choice, subject to conditions that must vary with each library.

As these lists have come for a final revision, they have made upon me a double impression deeply marked. They seem to mirror the sentiment, the life, the whole progress of modern Italy. And most impressive of all, in almost every section is the inspiring patriotic note. The poets' dream of a united Italy, constant throughout the centuries from the time of Dante, was at last given substance through the heroic work of the Revolution. And that great struggle of the Risorgimento—the long and bitter years stretching from 1814 and the Congress of Vienna to the September day in 1870, when the armies of Victor Emanuel entered Rome,—forms a basis upon which nearly everything that is worthy in the life and in the literature of the new Italy has since been built. We Americans would do well to know more of the great men of that time and their great deeds, that did in fact accomplish the "resurrection" of Italy. In Italian patriotism we should find refreshment for our own.

Ordering Books

With most of the larger Italian publishers it is possible to correspond in English, and orders may be sent them directly. But generally, to avoid mistakes and misunderstandings, unless correspondence can be in Italian, it is strongly advised that all purchases be made through a responsible American agent. Brentano's, 229 Fifth Avenue, New York, and G. E. Stechert & Company, 151 West 25th Street, New York, are well-known importers and carry a considerable stock of Italian books. Books that are not on their shelves, they will import promptly direct from the publishers. Such orders require from six to eight weeks. The prices are invariably given in Italian money, the lira. Its American equivalent will vary from 21c to 25c, depending to some extent upon the cost of the book, and the amount of the purchase. Books may be had by post in from three to four weeks at an additional cost of about 10%.

Unless otherwise stated, the books listed are in paper. Books so quoted can only be had in this way, practically unbound, and usually need immediate and substantial binding. Many of the cheaper books issued by Italian and other continental publishers are so poorly bound in paper, and so poorly stitched, that they can hardly be cut and opened for the reading of a single chapter without falling apart. This matters little in a land where bookbinding is cheap, and either economy or taste requires books in paper, but it often means serious difficulties for the innocent American purchaser.

With the exception of a few manuals, every book included in these lists has been seen and examined. Particular effort has been made to select editions of moderate price, and there has been equal effort to choose those well printed on good paper. But in many cases the only form in which a book may be had, and this is particularly true of translations, is in a very cheap edition.

It is believed that every volume described in these lists is now in print. A very large proportion of the titles—nearly 70%—are of books that have been printed within the last five years. They are all to be found in current catalogues, with the exception of a few, and for these, pains have been taken to make certain that they are held in ample stock.

It is the experience of librarians that it is necessary to purchase in greater number duplicates of books that are popular with these new Americans, than of popular books for any other class of readers. Their disappointment is keen if they fail to secure a book on which they have set their hearts. A second failure often results in abandonment of all further effort.

Biography

The great popularity of books of biography among Italian readers is frequently a surprise to our librarians. But there is no mystery in the preference. It is a choice that is chiefly due to the thrilling heroic and dramatic appeal in the life stories that have been written of many men, who were great patriots, who sacrificed and suffered much, and won liberty so splendidly for the oppressed and divided land of their birth.

* ALFIERI, VITTORIO. See also "Drama." 1749-1803. *Vita e rime scelte.* Roma: Albrighi, Segati e C. 1902. 5" x 7½". pp. 240. L 1.80.

The autobiography and selected poems of the great poet, who Carducci said, "created the Revolution in Italy." The "Life" of this fierce, impetuous and forthright man to whom liberty was a religion makes thrilling and inspiring reading—thrilling for its adventures, inspiring for its story of long struggle and patience and work. It is besides a strongly personal book. He pictures himself: "A resolute soul, most obstinate and unconquered; a heart filled to overflowing with love of every sort; and with that strange mixture of love and all its furies, I had in me a profound and most ferocious rage and horror for tyranny in every form."

BACCINI, IDA. See also "Books for Children." *La mia vita—Ricordi autobiografici.* Roma: Albrighi, Segati e C. 1904. 5" x 7½". pp. 297. L 3.

The life story of a very popular writer—a school teacher of Florence, whose writings, largely for children, are marked by great simplicity and gentleness of spirit. Widely honored and loved.

* BELLIO, V. *Cristoforo Colombo.* Milano: Ulrico Hoepli. 1892. 4¼" x 6". pp. 159. L 1.50. Illustrated and bound.

A popular story of Columbus, interesting and written with special appeal to Italians.

* BELTRAMI, LUCA. *Alessandro Manzoni.* Milano: Ulrico Hoepli. 1898. 4¼" x 6". pp. 191. L 1.50.

A charming little book, crowded with personal detail, and in spite of its few pages, with illustrations. Manzoni was a great patriot as well as a great novelist and poet. His were the words, famous in Italian history: "We shall not be free, if we are not united."

CAPPELLETTI, LICURGO. *Napoleone I.* Milano: Ulrico Hoepli. 1908. 4" x 6". pp. 272. L 2.50.
Admirable little biography, well illustrated. •

* CELLINI, BENVENUTO. 1500-1571. *La vita scritta da lui medesimo.* Edited by Gaetano Guasti. Firenze: G. C. Sansoni. 1908. 5¼" x 7¾". pp. 205. L 1.50.
One of the famous autobiographies of the world and one of Italy's classics—the swashbuckling story of the great 16th century artist. Expurgated edition.

* D'AZEGLIO, MASSIMO. See also "Fiction—Italian." 1798-1866. *I miei ricordi.* Edited by Prof. A. Pippi. Firenzi: G. Barbèra. 1910. 5" x 7½". pp. 361. L 2.
The very popular and inspiring autobiography of "the spotless knight without fear"—il cavaliere senza macchia e senza paura—of the heroic struggle for the regeneration of Italy. Painter, novelist, soldier, statesman, prime minister.

* FRANKLIN, BENJAMIN. *La vita, scritta da se medesimo.* Translated by Pietro Rotondi. Firenze: G. Barbèra. 1912. 5" x 7". pp. 296. L 2.
Several editions of Franklin's autobiography have been printed in Italy, and have been widely read.

GOTTI, AURELIO. See also "Italy." *Italiani del secolo XIX.* Cittá di Castello: S. Lapi. 1911. 6½" x 10". pp. 417. L 3.
A selection of biographical sketches of twenty important men of Italy of the 19th century.
I due primi re d'Italia. Roma: Albrighi, Segati e C. 1912. 5½" x 8". pp. 288. L 2.
The story, simply told, of the lives of the first two kings of Italy.

* KING, BOLTON. *Mazzini.* Firenze: G. Barbèra. 1905. 5" x 7". pp. 400. L 4.
The most popular volume of this admirable series. A very inspiring story of "the apostle of liberty," who, preaching and organizing, dedicated his whole life to Italy and its cause of freedom and union.

* MARIO, JESSIE WHITE. *La vita di Garibaldi.* Milano:
Fratelli Treves. 1902. 5" x 7½". pp. 285, 303. 2 vols. L.1
each.
The popular biography of Garibaldi, the great hero of the
Revolution.

* MORANDI, LUIGI. See also "Literature and Education."
Come fu educato Vittorio Emanuele III. Roma: G. B.
Paravia e C. 1905. 5½" x 8". pp. 141. Illustrated. L 1.50.
Tells in a very entertaining way of the education and
training of the present King of Italy—of the Spartan dis-
cipline to which he was subjected as a boy, and the great
diversity and shrewd practical usefulness of his studies.

* PELLICO, SILVIO. 1788-1854. *Le mie prigioni.* Milano:
Bietti. 1911. 4½" x 7½". pp. 253. L 1.
A famous popular book of Italy. This story of the prison
life of Pellico had an important part in the great struggle for liberty. "It was a real battle against Austria." For
its political effect it has been compared to "Uncle Tom's
Cabin." Celebrated for its beauty and mildness of spirit,
its simplicity of language.

* PLUTARCH. *Vite degli uomini illustri.* Milano: Sonzogno.
1913. 5" x 7½". pp. 400. L 1.50.
Includes the most famous "Lives."

* RAMBALDI, PIER LIBERALE. *Amerigo Vespucci.*
Firenze: G. Barbèra. 1898. 5" x 7". pp. 229. L 2.
Italian readers in our libraries are keenly interested in
the stories of the two great Italian discoverers. Amerigo
Vespucci, no less than Columbus, appeals to their nation-
al pride.

RICCI, C. *Michelangelo.* Firenze: G. Barbèra. 1904. 5" x 7".
pp. 208. L 2.
An excellent short life, clearly and simply written, with
abundant human detail.

* SCHIATTAREGIA, PROF. BENEDETTO. *Vita di Vittorio Emanuele II.* Roma: Albrighi, Segati e C. 1896.
5" x 7½". pp. 103. Illustrated. L 1.
A very simple story of the life and times of Victor Emanuel II. A popular account of the patriotic and heroic
aspect of the long struggle for the liberty and union of
Italy. It includes many of the most famous pieces of
patriotic poetry of the days of Victor Emanuel.

VASARI, GIORGIO. 1511-1574. *Narrazioni scelte dalle vite.* Firenze: G. Barbèra. 1905. 4¾" x 7½". pp. 315. L 2.

For ordinary library purposes, probably the best selection of Vasari's "Lives of the most Excellent Painters, Sculptors and Architects"—contains nearly all the famous passages.

VILLARI, PASQUALE. 1827—. *Storia di Girolamo Savonarola.* Firenze: Successori Le Monnier. 1898. 6" x 8½". 2 vols. pp. 532 & CLXVIII, 261 & CCCLVI. L 6. each.

The life of Savonarola is considered the best work of Italy's foremost living historian, who in learning and critical power is frequently compared by Italian scholars to Macaulay.

* ZANICHELLI, DOMENICO: *Cavour.* Firenze: G. Barbèra. 1905. 5" x 7". pp. 427. L 4.

Admirable life within small limits of the wise statesman and prime minister, whose political skill had so great a part in the creation of the new Italy. Cavour, like King Victor Emanuel, is called the "Father of his Country."

Travel—Description—History

* ABRUZZI, DUCA DEGLI. *La "Stella Polare" nel mare artico.* 1899-1900. Milano: Ulrico Hoepli. 1912. 6½" x 9½". pp. 518. Profusely illustrated—2 maps—bound in cloth. L 8.50.

A graphic account of the voyage of the "Polar Star" (1899-1900) and the Arctic explorations of the Duke of the Abruzzi. Very popular.

BARETTI, GIUSEPPE. 1719-1789. *Lettere famigliari.* Edited by Gioachino Brognoligo. Roma: Albrighi, Segati e C. 1906. 5" x 7½". pp. 194. L 2.

Fascinating letters of an 18th century traveller in Spain and Portugal. Humorous, rapid, keen-eyed observer of men and national customs, picturesque, farcical. Baretti was a friend of Samuel Johnson.

* BARZINI, LUIGI. *La metà del mondo vista da un automobile.* (The Half of the World as Seen from an Automobile). Milano: Ulrico Hoepli. 1910. 6½" x 9½". pp. 523. Bound and illustrated, with maps L 9.50.

From Pekin to Paris in 60 days in an automobile. Delightful book of travel—vivid and brilliant pages—by one of the ablest of living journalists. Extraordinary photographic power.

BEVIONE, GIUSEPPE. See also "Italy." *L'Inghilterra d'oggi.* (The England of To-day). Torino: Fratelli Bocca. 1910. 5" x 7½". pp. 439. L 5.

Highly instructive and interesting book, brilliant and sympathetic, written by one of the most serious of the younger newspaper men of Italy, who was for several years London correspondent of "La Stampa" of Turin: English Life—The Theatre—Sport—Journalism—A study of the new Imperialism.

* *L'Argentina.* Torino: Fratelli Bocca. 1911. 5" x 7½". pp. 239. L 3.50.

Careful journalistic studies of social, economic, and political conditions in the Argentine Republic, of great interest to Italians, because of the large Italian population in that country. Almost everyone has friends there, particularly the northern Italian. This book pays special attention to this phase of interest.

* BIASIOLI, U. *Piccola storia del popolo argentino.* Milano: Antonio Vallardi. 1910. 4¾" x 7½". pp. 149. Illustrated. L .60.

This admirable little book gives a brief history of the Argentine, some account of its original peoples, of its present population, of its geography, climate, institutions, customs, industry and commerce, with a very useful chapter on the Italian immigrant in the Republic.

BORGHESE, G. A. *La nuova Germania.* Torino: Fratelli Bocca. 1909. 5" x 7½". pp. 495. L 5.

This is not merely a serious and exact study of the life of the new Germany; but in telling "how the Germans live" it becomes a picturesque book of travel, rapid, photographic, crowded with illustrative anecdotes.

* BRAGAGNOLO, G. See also "Music." *Storia di Francia.* Milano: Ulrico Hoepli. 1905. 4" x 6". pp. 424. L 3.

A rapid and concise account of the history of France "from the most remote times" to our own days.

* *Storia d'Inghilterra.* Milano: Ulrico Hoepli. 1906. 4" x 6". pp. 367. L 3.

Like the author's history of France, this little history of England is notable for rapid narration that sacrifices nothing to interest. It carries the story from Roman days almost to the present.

* CASTELLINI, GUALTIERO. *I popoli balcanici nell'anno della guerra.* Milano: Fratelli Treves. 1913. 5" x 7½". pp. 211. Illustrated. L 3.50.

A political study of the Balkan peoples, made during the year of the war, from the Italian point of view. Filled with human interest and well illustrated. The author effectively answers the question: "What are these people?"

* COLOMBO, E. *Argentina.* Milano: Ulrico Hoepli. 1909. 4¼" x 6". pp. 330. L 3.50.

A historical account of the Republic of Argentina that includes much detail on present economic conditions, statistics, notes of geography, etc.

COCCHIA, E. *Il Giappone vittorioso.* (Victorious Japan). Milano: Ulrico Hoepli. 1913. 5" x 7½". pp. 408. L 5.50.

Tells the story of the rapid evolution of the Japanese spirit, its historical causes, and its consequences, political

— 24 —

and social. The sub-title "The Rome of the Far East" completes the description of an extremely readable and instructive book.

* DE AMICIS. EDMONDO, See also "Literature and Education", "Fiction—Italian", "Books for Children", and "Poetry." 1846-1908.

The first popular writer of Italy. An unusually long list of his books is recommended. His simple and direct style, friendly spirit, imbued with love for humanity, his healthy sentiment and sympathy, passionately Italian, make him a particularly effective "public educator," as he has been called, for young men and workmen to whom so large a part of his work was consecrated. His stories are delightful, filled with genial humor, hopefulness, vivacity. The vivid pages of his books of travel read like romances.

These six charming books of travel and description devoted to Constantinople, London, Morocco, Holland, Paris, and "Upon the Ocean" form a remarkably interesting and instructive series.

Costantinopoli. Milano: Fratelli Treves. 1905. 5" x 7½". pp. 579. L 6.50.

Ricordi di Londra. (London). Milano: Fratelli Treves. 1901. 6½" x 10". pp. 111. Illustrated. L 1.50.

Olanda. (Holland). Milano: Fratelli Treves. 1909. 5" x 7½". pp. 477. L 4.

Marocco. Milano: Fratelli Treves. 1907. 5" x 7½". pp. 485. L 5.

Parigi. (Paris). Milano: Fratelli Treves. 1913. 5" x 7½". pp. 331. L 1.

Sull'Oceano. Milano: Fratelli Treves. 1910. 5" x 7½". pp. 423. L 5.

Brilliant sketches and stories of life at sea. The tropic seas. Emigrants embarking and at sea. The voyage to South America.

GREEN, J. R. *Breve storia del popolo inglese, dalle origini ai tempi nostri.* (Short History of the English People). Translated by Sofia Santarelli. Firenze: G. Barbèra. 1884. 5" x 7½". pp. 912. L 6.

* MARCO POLO. 1254-1323. See "Books for Children." *I viaggi.* (The Voyages).

* MONACHESI, PROF. G. *Piccola storia del popolo brasiliano.* Milano: Antonio Vallardi. 1913. 4¾″ x 7½″. pp. 150. Illustrated. L.60.

This little story of the Brazilian people, besides history gives a short, but very interesting account of the geography, ethnography, products, industry, commerce, and public institutions of Brazil. Italian immigration to Brazil has fallen off lately, but that country still has a large Italian population and this little book accordingly devotes two chapters to the subject, detailing something of Italian progress, and not forgetting to give some very necessary advice to the immigrant.

SERAO, MATILDE. See also "Fiction—Italian." *Nel paese di Gesù (Ricordi di un viaggio in Palestina).* Napoli: Francesco Perrella. 1910. 5½″ x 7½″. pp. 366. L 3.

Brilliant and rather mystical record of a journey in Palestine by the ablest Italian woman prose writer of our day.

SOLERIO, G. P. *La rivoluzione francese.* Milano: Ulrico Hoepli. 1914. 4″ x 6″. pp. 176. L 2.

Excellent little book, giving, within narrow limits, an adequate account not only of the events of the French Revolution, but also of its causes and effects.

In the United States

The list of helpful and informing books written in Italian about the United States and life in America is an exceedingly short and unsatisfactory one. Several good books are out of print, and with a few exceptions those that are available are either inadequate in one way or another, or inaccurate, or wholly out of date. But with our present heavy Italian immigration, our rapidly developing commerce with Italy, and the consequent steady increase of vital mutual interests, the time must soon come when it will be important for Italy to have adequate knowledge of every side of our national life. And better books describing America should be a consequence of that need.

* ARBIB-COSTA, ALFONSO. *Lezioni graduate di lingua inglese.* New York: Italian Book Co. 1906. 5½″ x 7½″. pp. 286. $1.00.
Not a simple book on learning English for a man with little or no educational experience and without a teacher. But still almost certainly the best book of the sort whether for library or school use. In some libraries in which it has been placed, it has proved so serviceable that it is in constant use.

* BOTTA, CARLO. See also "Italy." 1766-1837. *Storia della guerra dell'independenza degli Stati Uniti d'America.* Torino: Unione Tipografico—Editrice Torinese. 1859. 4½″ x 7″. 3 vols. pp. 283, 291, 275. L 4.15 for the set.
This history of the American Revolution, first published over a hundred years ago, is the most important book ever written about America by an Italian. It is also Botta's greatest work—the result of love and long labor. His sincerity and nobility, his faith in justice and democracy, his clear picturing of the purposes of our Revolution made this history an inspiration to the ardent young men of his day, who were to make ready Italian liberty. And through the years it has been the best interpretation of America to his countrymen. In many ways it is antiquated; its style is difficult, yet today it is still being urged for educational use in Italy on account of its power "in forming the hearts and souls

of youth." A recent reprint from the plates of an old edition.

CARNEGIE, ANDREW. *Nel regno degli affari.* (The Empire of Business). Translated by Giulio De Rossi. Firenze: G. Barbèra. 1912. 5" x 7". pp. 260. L 2.

* CARR, JOHN FOSTER. *Guida degli Stati Uniti per l'immigrante italiano.* New York: Immigrant Education Society. 1913. 5" x 7¼". pp. 79. Illustrated, with map, 20c. postpaid. Aims to give the immigrant the practical information he needs about life in the United States. Separate chapters tell him where to go for work, how to travel, how to learn English; about the schools and libraries and other educational advantages, about the opportunities of agriculture, and the successes made by Italians in farming. Other chapters tell of the geography and climate of the United States, of the federal and state governments, of citizenship and the qualifications for it. There are, too, chapters on the laws of the United States, on health, savings banks, postal rates, our money, and weights and measures, with tables of information about the States and a chapter of special advice.

FERRERO, GUGLIELMO. See also "Italy." *Fra i due mondi.* Milano: Fratelli Treves. 1913. 5" x 7½". pp. 430. L 5.

The author warns the reader that this book is not a novel, nor a book of travel, nor a drama, nor a treatise of philosophy and sociology. But the reader will find it something of all of these, though the historian of Rome has ably fulfilled his intention of writing a record of his trip to America that would be a comparison between the old world and the new. And a brilliant comparison he has made it.

* GIANI, RODOLFO. *Storia degli Stati Uniti d'America.* Milano: Paolo Carrara. 1902. 5" x 8". pp. 285. L 2.

On the whole, a poor and inaccurate history of the United States, based in great part on the work of Romussi, written nearly forty years ago. But excellent in parts, and for all its shortcomings, the only work of the sort available in Italian.

MONDAINI. GENNARO. *Le origini degli Stati Uniti d'America.* Milano: Ulrico Hoepli. 1904. 5" x 7½". pp. 459. L 6.50.

Deals in very careful and scholarly way with the origins of the Republic. Does not go beyond colonial days.

* PECORINI, ALBERTO. *Gli americani nella vita moderna osservati da un italiano.* Milano: Fratelli Treves. 1909. 5" x 7½". pp. 448. L 5.

The best modern book about the United States written by an Italian, a treasury of accurate information about American life. Not altogether a simple book, but able, interesting, at times brilliant, written with an obvious friendly intention—usually successful—to be fair. Opens with a good chapter of history. Contains a very clear analysis of our government and politics. Separate chapters are devoted to woman, religion, journalism, capital and labor, industry and commerce, the railroads, education, the army and navy, Indians, negroes, American cities, art, literature, music and immigration.

ROMUSSI, C. *Storia degli Stati Uniti d'America.* Milano: Sonzogno n. d. 4¼" x 6½". pp. 63. L .20.

This little pamphlet in the "Biblioteca del Popolo" series only carries the History of the United States to the Philadelphia Exposition of 1876.

ROOSEVELT, THEODORE. *Vigor di vita.* (The Strenuous Life). Translated by Hilda di Malgrà. Milano: Fratelli Treves. 1905. 5" x 7½". pp. 305. L 3.

* ROSSI, ADOLFO. *Un italiano in America.* Milano: Fratelli Treves. 1912. 5" x 7½". pp. 325. L 1.

The experiences of an Italian journalist travelling as an immigrant workman in the U. S. 1879-1881. The first and most popular book of the well-known war correspondent and member of the Royal Italian Emigration Commission. He is now Minister to Uruguay.

STRAFFORELLO, GUSTAVO. *Letteratura americana.* Milano: Ulrico Hoepli. 1898. 4" x 6". pp. 158. L 1.50.

Again a book, "excellent in parts," but wholly out of date, even in its revised edition.

Italy

History—Description—War with Turkey—Nationalism

With the rapid economic development of Italy during the last ten years, there has been also a marked growth of national consciousness and of patriotic interest in all things Italian. Many exceedingly interesting and well-written books, accurate and yet filled with characteristic patriotic fervor, have appeared, covering almost every side of national life. The old provincial distinctions and jealousies are rapidly passing away, and among Italians themselves there is seen the growth of a deep interest in every detail of the progress of the nation, not merely politically and industrially, but also ideally,—an intense feeling that Italy has a definite national mission in the world; and of this, particularly on the humanitarian and cultural sides, eloquent testimony will be found in many places in these pages. For this section the usual alphabetical arrangement of the list has been almost wholly discarded for a grouping of books of related interest.

History

FERRERO, GUGLIELMO. *Grandezza e decadenza di Roma.* (The Greatness and Decline of Rome). Milano: Fratelli Treves. 1908-1910. 5" x 7½". 5 vols. pp. 526, 562, 599, 379, 423. L 22. for the set.

The original edition of Ferrero's great work. Italians are taking an increasing interest in the story of ancient Rome. Its history and traditions they are coming to feel are a part of their own—an inspiration of definite political importance these last years since the beginning of the war with Turkey.

* GALANTI—ZIPPEL—RAULICH.

Manuale di storia ⎰ *del medio evo.* 476-1313.
moderna d'Europa e specialmente d'Italia. 1313-1748.
contemporanea d'Europa e specialmente d'Italia. 1750 *ai nostri giorni.*

Torino: G. B. Paravia e C. 1909-1910. 5½" x 8½". 3 vols. pp. 472, 357, 433. L 11. for the set.

These three volumes grouped together give in considerable detail the history of Europe, and especially of **Italy** from 476 A.D. to the death of King Humbert in 1900. Each volume has been prepared by a specialist. All three are clearly written with admirable perspective. Maps accompany each volume. Written for the use of secondary schools.

* ORSI, PIETRO. See also "Books for Children." *Breve storia d'Italia.* Milano: Ulrico Hoepli. 1911. 4" x 6". pp. 285. L 2.50.

A brief history of Italy from prehistoric times to the earthquake of 1908.
* *Italia moderna.* Milano: Ulrico Hoepli. 1914. 5" x 7½". pp. 535. Illustrated, with three maps. L 7.50.

The history of Italy from 1750 to the close of 1913. The best and the most popular book of the sort available.

* BOTTA, CARLO. 1766-1837. *Narrazioni di storia patria.* Edited by Prof. G. Finzi. Roma: Albrighi, Segati e C. 1897. 5" x 7½". pp. 249. L 1.75.

Contains 20 selections of the most famous passages of this famous historian, covering the story of Italy from the days of Alexander de'Medici to the coronation of Napoleon.

* BERTOLINI, L. *Storia del risorgimento italiano.* Milano: Ulrico Hoepli. 1905. 4" x 6". pp. 208. L 1.50.

Admirable short account—perhaps the best within so few pages—of the whole struggle that ended in the creation of modern Italy—1814-1870.

* LA VITA ITALIANA NEL RISORGIMENTO. 1815-1861. Firenze: R. Bemporad e Figlio. 1911. 4¾" x 7". 12 vols. About pp. 175 each. L 1. each.

This little series covers in a very popular way the entire story of the struggle for the union of Italy, in the several points or history, letters, arts and sciences. It has been carefully prepared by the co-operation of some of the ablest living writers of Italy

CARDUCCI, GIOSUÈ. See also "Literature and Education" and "Poetry." 1836-1907. *Letture del risorgimento italiano.* 1749-1870. Bologna: Nicola Zanichelli. 1912. 5" x 7½". pp. 534. L 3.

Remarkable selection of interesting readings from original sources, showing the history and development of the ideas and of the literature of the Risorgimento—the rise. development and union of modern Italy.

* CAVALLOTTI, FELICE. See also "Drama." 1842-1898.
Martirologio italiano. Milano: Sonzogno. 1898. 4½" x 6½".
pp. 106. L .30.

Cavallotti, killed in a duel in 1898, was a strange combination of idealist and rebel. He wrote many beautiful poems filled with patriotic ardor. This selected volume is of prose and is devoted to a few of the early martyrs of the great struggle for the independence and union of Italy.

* ERRERA (R) and TRENTO (I). *Italia.* Milano: Giacomo Agnelli. 1912. 5½" x 8". pp. 562. L 3.

A patriotic reader carrying the motto: "To increase love through knowledge." It might carry the sub-title "Italy in Song and Story," for it tells of Italy's wonders and beauties, as described in famous passages of literature—the heroic things of the present as well as the past—the homely and lovely things, and not forgotten, an appealing note of human brotherhood. Side by side with the great authors of the past are the most brilliant writers of to-day. A model of a patriotic reader.

GOTTI, AURELIO. *Quadri e ritratti.* Roma: Albrighi, Segati e C. 1910. 5½" x 8". pp. 378. L 2.50.

An exceedingly readable story of the creation of modern Italy, told chiefly in vivid sketches of the lives of its great men.

DE GUBERNATIS, ANGELO. 1840-1913. *L'Italia.* Roma: Albrighi, Segati e C. 1911. 5½" x 8". pp. 337. L 2.

A clear and simply written book about Italy by a well-known historian and educator. The first part of the book is devoted to an account of the physical characteristics of Italy and local history of cities and provinces. About a hundred pages are given to the history of Italy, from the founding of Rome to the death of Victor Emanuel and a hundred pages are given to sketches of the lives of great Italians.

VALLARDI, ANTONIO. *Mezzo secolo di vita italiana.* Milano: Antonio Vallardi. 1911. 12" x 15". pp. 215. L 5.

Twenty-six popular, well informed articles, by specially qualified writers, abundantly illustrated and covering every phase of the development of Italy during the last fifty years: Politics, army, navy, industry, art, science, literature, journalism, religion, schools, finance.

SANTORO, CAV. AVV. MICHELE. *L'Italia nei suoi progressi economici dal* 1860 *al* 1910. Roma: Tipografia Popolare. 1911. 7" x 10". pp. 527. L 10.

A careful account of the economic progress of Italy made during a half century—in considerable part a detailed statistical study. Extremely interesting is the comparison between economic conditions in the old Italy and the new.

MOSSO, ANGELO. See also "Art and the Sciences." 1846-1910. *Vita moderna degli italiani.* Milano: Fratelli Treves. 1906. 5" x 7½". pp. 430. L 4.

A series of sympathetic studies by a distinguished scientist, dealing chiefly with emigration, and the economic and social problems affecting the peasantry and working people of Italy. A clear headed and charming book with a purpose: that the reader may know more accurately something of these problems that so vitally affect Italian national life, and that he may learn "to love the poor."

NITTI, FRANCESCO S. *L'Italia all'alba del secolo XX.* Torino: Società Tipografico-Editrice Nazionale. 1901. 5½" x 9". pp. 215. L 2.50.

"Italy at the Dawn of the Twentieth Century." A collection of addresses to the young men of Italy by one of the ablest economists and constructive Italian statesmen of our time. Together they form a clear and purposeful statement of national resources, of national problems of population, political divisions and industry, and they have made a profound impression in Italy, inspiring for the practical progress of the nation. Signor Nitti was Minister of Agriculture, Industry and Commerce during the last Ministry. His chief interests for many years have been the industrial development of his country—particularly southern Italy—the development of its water power, and emigration and education.

NOVIKOW, GIACOMO. *La missione dell'Italia.* Milano: Fratelli Treves. 1903. 5" x 7½". pp. 339. L 3.

This book by the great Russian pacifist is celebrated in Italy. Written at a time of almost national pessimism, it furnished a scientific and practical analysis of economic, political and intellectual conditions, and became a true prophecy of the rapid advance of Italy during these last years. It showed Italy's capacity for leadership among Latins, based on the remarkable Italian progress in the Argentine, and it held that the union of Italy, not based on the subjugation of any part of the nation, is the type that can best be followed in a peaceful federation of Europe.

— 33 —

Description

* ABBA, GIUSEPPE CESARE. See also "Books for Children."
Le Alpi nostre e
  il Veneto montano.
  il Monferrato.
  il Piemonte.
  la Lombardia montana tra la Sesia e l'Adda.
  la Lombardia montana, Adda-Mincio.
Bergamo: Istituto Italiano d'Arti Grafiche. 1901. 5" x 7½".
5 vols. pp. 176,172, 178, 170, 176. Well illustrated, with maps.
L .60 each.

The Italian Alps, in their different divisions. First part general and same in all books. Second devoted to the section named in title. School books that make excellent reading. Topics, simply and very interestingly treated, are: geology, glaciers, crevasses, mountains, valleys, rivers, lakes, clouds, fauna, flora, tunnels, agriculture, hospices, famous passes, peoples, dialects and languages, history, cities, towns, traditions, famous men, art, poetry.

BACCELLI, ALFREDO. *Vette e ghiacci.* Roma: Albrighi, Segati e C. 1901. 5" x 7½". pp. 216. Illustrated L 2.50.
Italy is so largely divided between mountain and shore, Italians are always interested in mountains. This title might be translated "mountain summits and fields of ice." It is almost entirely devoted to the Italian Alps, a healthy, enthusiastic outdoor book, filled with beauty of thought—not without humor.

REY, G. *Alpinismo acrobatico.* Torino: S. Lattes e C. 1914. 6" x 9". pp. 313. L 6.
A thrilling book of mountain climbing, beautifully illustrated.

* VECCHI, A. V. See also "Books for Children."
L'Italia marinara e il lido della patria.
  Liguria e Toscana.
  Calabria, Ionica, Puglie e Marche.
  Cilento, Calabria, Tirrena e Sicilia.
  Lazio, Campania e Sardegna.
  Romagna e Veneto.
Bergamo: Istituto Italiano d'Arti Grafiche. 1901. 5" x 7½".
5 vols.
pp. 147, 151, 155, 152, 150, well illustrated. L .60 each.

Italy and the Sea. School books of splendid interest, exceedingly simple and readable. All Italians are interested in the sea as in the mountains. The first part of each volume is the same and is given to the general topics indicated by the series. The second and third parts deal with the particular section of the Italian sea coast suggested by the title. Each book has a clear map. Topics are of a great variety of interest: The geography, history and antiquities of the Italian coast, its towns and cities, industries and commerce, fisheries; the navy and merchant marine; salt making, coral, light houses and life saving, legends and descriptions, emigration; the physical, moral and economic influence of the sea.

* STOPPANI, ANTONIO. 1824-1891. *Il bel paese.* Milano: L. F. Cogliati. 1908. 5" x 7". pp. 663. L 2.

Stoppani was geologist, philosopher, priest. A famed and instructive book. Talks upon the natural beauties, the geology and physical geography of Italy. Very popular among young people. This is the 75th "economic edition."

* CITTÀ (LE) D'ITALIA ILLUSTRATE. (The Cities of Italy Illustrated). Milano: Sonzogno. 1908-1914. 8" x 11". 4 vols. pp. 96 to 104 each. L 1.25 each. Already published: Roma, Milano, Venezia and Torino.

Give in brief and popular form the history of the great cities of Italy, including admirable descriptions of important monuments, good accounts for each city of its industry and commerce, education, charities, public service activities, population, traveller's guide. Profusely and carefully illustrated.

* MOSCHINO, ETTORE. *La bella Napoli.* Milano: Fratelli Treves. 1911. Folio. L 3.50.

A supplement to "L'Illustrazione" devoted to a description of Naples—the old city and the new, its monuments and life, beautifully illustrated.

CHIESI, GUSTAVO. *La Sicilia illustrata nella storia, nell'arte, nei paesi.* Milano: Sonzogno. 1892. 8½" x 12". pp. 720. L 9.

A large and profusely illustrated book, still very satisfactory in spite of the date of its publication.

SAN GIULIANO, ANTONIO DI. *Condizioni presenti della Sicilia.* Milano: Fratelli Treves. 1896. 5" x 7½". pp. 225. L 1.

Written nearly twenty years ago by the Marquis di San Giuliano, the brilliant Minister of Foreign Affairs of Italy's new cabinet. This book has maintained its importance not merely for historical reasons, but because in spite of the progress and changes of recent years, in its most significant parts it still remains the best study of many economic and social conditions that seem permanent in Sicilian life.

PITRÈ, GIOVANNI. *Usi popolari.* Catania: Cav. N. Giannotta. 1912. 5" x 7½". pp. 250. L 1.

Very popular in Sicily. Deals with Sicilian folk lore and customs. Very well known, not only in Sicily, but also in Germany.

MARTINI, FERDINANDO. See also "Drama." *Nell'Affrica italiana.* Milano: Fratelli Treves. 1895. 5" x 7½". pp. 357. L 2.

Succinct and interesting record of things seen in Eritrea, Italy's first African colony, on the Red Sea. It includes not only description of places and peoples—their character and customs—but a very careful study of the economic possibilities of the country. Martini was Governor-General of Eritrea for many years.

War With Turkey—Nationalism

BEVIONE, GIUSEPPE. *Come siamo andati a Tripoli.* Torino: Fratelli Bocca. 1912. 5" x 7½". pp. 425. L 5.

Collection of letters and dispatches to "La Stampa" of Turin, written from Tripoli in the spring and autumn of 1911 (2 Apl. to 4 Nov.) Complete story of the conquest by an eye witness.

CORRADINI, ENRICO. Ablest and best known writer of Nationalism—in Italy these last years a movement of reaction against socialism and internationalist anti-patriotic ideals. He has been its precursor and apostle, calling for a greater faith in Italy and a more virile political policy.

La conquista di Tripoli. 5" x 7½". pp. 233. L 3.50.
Sopra le vie del nuovo impero. 5" x 7½". pp. 242. L 3.50.
Milano: Fratelli Treves. 1912.

The first volume describes the capture of the city of Tripoli in a series of letters that were written on the scene at the time—6th October to 12th December, 1911. The second with the capture of the islands of the Aegean Sea and the problems created by the result of the African war, their moral value and effect on national character.

Both books are remarkable for their brilliant, vigorous, picturesque and eloquent writing.

La patria lontana. See "Fiction—Italian." Milano: Fratelli Treves. 1910. 5" x 7½". pp. 258. L 3.50.

The most popular of nationalist novels, inspired by lofty patriotism.

SIGHELE, SCIPIO. *Pagine nazionaliste.* 5" x 7½". pp. 244 L 3.50.

Il nazionalismo e i partiti politici. 5" x 7½". pp. 259. L 3.50 Milano: Fratelli Treves. 1910 & 1911.

Books of Nationalist faith and propaganda, by the President of the first Nationalist Congress. Give the history of the movement and the development of Nationalist ideas with their application to the foreign and domestic politics of Italy.

Books of Literature and Education

* DE AMICIS, EDMONDO. 1846-1908. *L'Idioma gentile*. Milano: Fratelli Treves. 1910. 5" x 7½". pp. 440. L 3.50.

A simple, *readable*, witty, and to Italians and some others, an inspiring book of language. It drew from a Minister of Public Instruction a special letter to the heads of the secondary schools of Italy advising its diligent use. A book of patriotism, well loved for the love there is in it of native land and language. One of De Amicis' later books, but many times reprinted.

* FERRARI, VITTORIO. *Letteratura italiana, moderna e contemporanea*. 1748-1911. Milano: Ulrico Hoepli. 1911. 4¼" x 6". pp. 340. L 3.

A convenient little manual of modern Italian literature. By far the best of its kind.

* PIZZI, ITALO. *Storia della letteratura italiana*. Torino: Hans Rinck. 1912. 5½" x 8". pp. 413. L 3.

One of the best brief accounts of Italian literature from its beginnings to the date of publication.

CARACCIOLO, FRANCESCO. *Antologia italiana di prosa e poesia*. Torino: G. B. Paravia e C. 1902. 5" x 8". pp. 389. L 3.

An exceedingly interesting and useful anthology of Italian prose and verse, arranged in strictly chronological order. Has a very attractive popular character.

MESTICA, ENRICO. *Antologia letteraria*. Torino: G. B. Paravia e C. 1905. 5" x 8". pp. 672. L 3.

An anthology of prose and verse, very useful as a companion volume to the preceding book of Caracciolo's. The selections of famous passages of the most diverse sorts are here arranged entirely according to literary classifications.

* MORANDI, LUIGI. *Letture educative*. Città di Castello: S. Lapi. 1912. 5" x 7½". pp. 350. L 2.
Prose e poesie italiane. Città di Castello: S. Lapi. 1912. 5" x 7½". pp. 828. L 3.50.

Two famous school readers. When they first appeared, the objection was made that they were so interesting

that the children would read them out of school, and so
find nothing of fresh interest in them in the class room.
"I hope so," said the author, "and I hope that their
fathers and mothers will read them, too." And so it
proved. "Teaching good and useful things" as well as
reading, the first of these books consists of selections
from the best of the world's literature, Franklin, Cervantes, Sterne, Heine, with the greatest Italian authors of all
time, and journalists of today. The second is an anthology that mirrors the life of modern Italy. Both have a
wealth of lively anecdotes; they contain a large amount
of practical information about history, geography, social
and political life, hygiene, agriculture, industry and literature. With a very inspiring note of hero-worship and
patriotism, their importance from the humanitarian side
brought them special approval from the Peace Congress
of Berne.

* CARDUCCI, GIOSUÈ, & BRILLI, UGO. *Letture italiane.*
Bologna: Nicola Zanichelli. 1898. 4¾" x 7½". pp. 759. L 3.
An anthology of prose and verse of very unusual selections from the great masters of Italian style, chosen by
the poet and critic with his characteristic enthusiasm and
broad knowledge, in order "to accustom young people to
the reading, to the understanding and to the love of the
great writers, who are classics because they were first of
all Italians."

CARDUCCI, GIOSUÈ. 1836-1907. *Prose.* Bologna: Nicola
Zanichelli. 1911. 5½" x 7¾". India paper. pp. 1486. Bound.
L 10.
Comprehensive collection of his prose writings—those the
author himself judged significant "in literary or political
history." In prose Carducci is considered the ablest literary critic of modern Italy—polished, nervous, clear—as
in his poetry, classic in form, master of language. Powerful controversialist, using always irony and humor.

CARLYLE, THOMAS. *Gli eroi.* (Heroes and Hero
Worship). Translated by Maria Pezzè Pascolato, Firenze: G.
Barbèra. 1912. 5" x 7". pp. 327. L 2.50.

D'ANNUNZIO, GABRIELE. See also "Poetry." *Prose
scelte.* Milano: Fratelli Treves. 1909. 5" x 7½". pp. 399.
L 4.
Some 50 pages consist of selections from essays and addresses. The remainder of the volume is given to selections from his novels. Characteristic and popular pas-

sages showing every side of this remarkable stylist at his best, including admirable examples of his great de scriptive power.

* DE AMICIS, EDMONDO. 1846-1908. *Letture scelte.* Milano: Fratelli Treves. 1911. 5" x 7½". pp. 340. L 2.

Selection of his most popular prose—widely read stories and description.

Ricordi d'infanzia e di scuola. Milano: Fratelli Treves. 1910. 5" x 7½". pp. 444. L 4.

Chiefly recollections and stories of childhood and youth. Contains a number of his famous short stories.

Fra scuola e casa. Milano: Fratelli Treves. 1912. 5" x 7½". pp. 437. L 4.

Like "Ricordi d'infanzia," a book of sketches and stories.

La carrozza di tutti. Milano: Fratelli Treves. 1911. 5" x 7½". pp. 476. L 4.

"Not exactly a novel—a series of pictures, of observations, of studies of modern life and all the questions of the day, but pleasant enough to include stories of Anatole France and tales of Mark Twain."

ESOPO. *Favole.* (Fables). Milano: Sonzogno. 1910. 4½" x 6½". pp. 95. L .30.

FIORETTI DI SAN FRANCESCO E IL CANTICO DEL SOLE. Milano: Ulrico Hoepli. 1907. 5" x 7½". pp. 335. L 1.50.

"The Little Flowers of St. Francis," a collection of popular and monastic legends, very simply and devoutly describing the life of peace and love and compassion toward all God's creatures, lived by the great saint of the middle ages.

GIACOSA, GIUSEPPE. See also "Drama." *Conferenze e discorsi.* Milano: L. F. Cogliati. 1909. 5" x 7½". pp. 291. L 3.25.

A book of lectures and addresses by the brilliant Piedmontese dramatist.

KIDD, BENJAMIN. *L'evoluzione sociale.* Firenze: G. Barbèra. 1904. 5" x 7". pp. 320. L 3.

A complete translation of Kidd's most important book, which is still much read in Europe.

LEOPARDI, GIACOMO. See also "Poetry." 1798-1837. *Le prose morali.* Edited by Ildebrando della Giovanna. Firenze: G. C. Sansoni. 1912. 5¼" x 7¾". pp. 409. L 2.50.

Leopardi was great as a prose writer as well as a poet. This volume includes his best dialogues and the "Thoughts." Here, as always, he is "the great solitary of pessimism." His writing, particularly in the "Thoughts", is clear, incisive, effective, often profound. He covers a wide range of subjects in a vigorous, and frequently stimulating way.

* LESSONA, MICHELE. 1823-1894. *Volere è potere.* Firenze: G. Barbèra. 1910. 5" x 7½". pp. 496. L 3.

"Will is power" is a well known book of high educational value, written for young men. Lessona was prominent as a scientist, was the translator of Darwin, and, as Carducci said, "learned in many arts." He had broad and generous and noble views of life,—ideals that he made most effective in this best of all his books, whose lively pages never seem to lose their inspiration for the young men of Italy.

MACHIAVELLI, NICCOLO. *Il principe ed altri scritti politici.* Edited by Francesco Costèro. Milano: Sonzogno. 1905. 5" x 7½". pp. 336. L 1.50.

This excellent little edition contains not only "The Prince"—always interesting and always eagerly discussed—but the important Dialogue of the Art of War, and other significant selections from his political writings. Machiavelli stands high among the Italian classical writers. His knowledge of the science of government was unrivalled. Keenly and profoundly he knew human nature. Because of his clearness, brevity, precision, limpid reasoning, he is still used as a model in the schools. And Machiavelli also has a popular quality of high patriotic appeal.

MAETERLINCK, MAURIZIO. *La saggezza e il destino.* Translated by Enrico Malvani. Torino: Fratelli Bocca. 1910. 5" x 8". pp. 302. L 3.50.

The translation of one of the most popular of Maeterlinck's books of essays: "Wisdom and Destiny."

* MAZZINI, GIUSEPPE. 1805-1872. *Doveri dell'uomo.* Roma: Commissione Editrice degli Scritti di Giuseppe Mazzini, via Torino, 122. 1905. 4¾" x 7". pp. 96. L .15.

Selected by the Italian Government to be widely used in the public schools of Italy as best representing the thought of Mazzini, the great "world apostle of liberty," bearing on the formation of character in the young. On broad and religious lines.

SETTEMBRINI, LUIGI. 1812-1876. *Pagine scelte.* Selected by Francesco Torraca. Roma: Albrighi, Segati e C. 1913. 5″ x 7½″. pp. 175. L 1.80.

Admirable example of books of selections used in the schools of Italy. This from the works of a great patriot and literary character of the Risorgimento is made up of descriptions, stories, letters, historical sketches, recollections, in a style that is simple, clear, rapid, colloquial.

SPENCER, HERBERT. *Educazione intellettuale, morale e fisica.* Translated by Sofia Fortini-Santarelli. Firenze: G. Barbèra. 1910. 5″ x 7½″. pp. 220. L 1.30.

TEDESCO, PROF. LUIGI. *Il mare: Antologia di prose e poesie.* 2 vols. (I & III). Savona: D. Bertolotto e C. 1896. 5½″ x 8″. pp. 546, 432. L 2.25 and L 2. respectively.

Written for the young men of Italy: Attractive to people like the Italians, that in many provinces are almost wholly men of the sea. It seems as if he had searched out everything about the sea in all modern literature that is beautiful either in form or thought; for this anthology includes not merely descriptions and stories, heroic deeds, strange sights, but biographies of great sailors and marine inventors, thoughts and proverbs of the sea, famous voyages. For those who love "the traditions of the sea, that inspire a holy and lofty ideal for the new destinies that the sea reserves for our Italy."

FICTION
Italian

The novels given in the following list include a selection of the famous books of Italian fiction—among them famous patriotic books that are always read and always popular. To these have been added a certain number of the simplest and most popular works of more recent fiction.

* BARRILI, ANTON GIULIO. 1836-1908.
Versatile and popular writer of many novels—pleasant and simple love stories, attractively written. Widely read, and, in Italy, especially recommended to young ladies.
L'olmo e l'edera. Milano: Fratelli Treves. 1910. 5" x 7½". pp. 328. L 1.
Cuor di ferro e cuor d'oro. Milano: Fratelli Treves. 1910. 2 vols. 5" x 7½". pp. 286, 274. L 1. each.
One of the most widely read of Barrili's novels.

CANTÙ, CESARE. 1805-1895. *Margherita Pusterla.*
Firenze: A. Salani. 1908. 5" x 7½". pp. 393. L 1.
A famous historical novel of the days of the Viscontis, written by the historian, Cantù, in prison. Exciting tale of conspiracies and misfortunes, but full of lofty and noble sentiments. Almost a classic.

CAPUANA, LUIGI. See "Books for Children." *Passa l'amore.* Milano: Fratelli Treves. 1913. 5" x 7½". pp. 343. L 3.50.
Thirteen delightful short stories. Capuana has been called the de Maupassant of Italy. By many he is considered the best short story writer of Italy. He is a Sicilian, but his stories are popular with all Italians.

CARCANO, GIULIO. 1812-1884. *Angiola Maria.* Milano: Cesare Cioffi. n. d. 5" x 7¼". pp. 319. L 4.
Very popular historical novel of the old fashioned sentimental sort. There is something of Scott in it, something of Manzoni, of whom Carcano was the friend, and devoted follower.

* CASTELNUOVO, ENRICO. Very popular writer of many novels, which are simple and easy in form, with well drawn characters taken from daily life. For the purity of his work, Castelnuovo is compared to Barrili, and he is, as they say in Italy, most "simpatico." His two most popular books, perhaps, are:

Dal primo piano alla soffita. Milano: Fratelli Treves. 1912. 5" x 7½": pp. 316. L 2.
L'onorevole Paolo Leonforte. Milano: Fratelli Treves. 1913. 5" x 7½". pp. 352. L 2.

* CORRADINI, ENRICO. *La patria lontana.* See "Italy—War with Turkey—Nationalism."

* D'AZEGLIO, MASSIMO. 1798-1866. *Ettore Fieramosca.* Firenze: Successori Le Monnier. 1895. 4¾" x 7". pp. 255. Bound L 1.

D'Azeglio's first novel—a historical romance, written with the patriotic purpose, as he said, "to put fire into the souls of Italians." Instantly popular, "it made hearts beat with a new love of country."

Niccolò De'Lapi. Firenze: Successori Le Monnier. 1909. 4¾" x 7". pp. 594. Bound L 1.75.

D'Azeglio's second novel, also historical and patriotic,—"to stir to flames the fires of patriotism."

* DE AMICIS, EDMONDO. 1846-1908. *La vita militare.* Milano: Fratelli Treves. 1910. 5" x 7½". pp. 453. L 1. Better paper L 4.

Stories and sketches of military life—in large part recollections of his service as officer in the army. With "Cuore" (See "Books for Children") most famous of all his books. Contains pages widely known and loved.

DELEDDA, GRAZIA. *Anime oneste.* Milano: Fratelli Treves. 1910. 3½" x 5½". pp. 273. L 3.

The first novel of this talented Sardinian authoress, written in the purest Tuscan; dealing almost entirely with the primitive people of Sardinia, and abounding in local color. One writer says that this has "all the charm of an unpremeditated autobiography."

Cenere. Milano: Fratelli Treves. 1910. 5" x 7½". pp. 322. L 3.50.

One of the best of her novels, and characteristic of the many of them that are inclined to be sad.

I giuochi della vita (and other stories). Milano: Fratelli Treves. 1911. 5" x 7½". pp. 321. L 3.50.

* FARINA, SALVATORE. *Tesoro di donnina.* Torino: Società Tipografico—Editrice Nazionale. 1907. 5" x 7½". pp. 380. L 3.50.

One of the best known novels of this popular writer— admired for his simple and natural plots and his pleasant and moving stories, not without humor. "Wholesome reading" is the comment of Ferrari.

Amore bendato. Milano: Galli. 1895. 5" x 7½". pp. 207. L 2.50.

Il signor Io. Torino: Società Tipografico—Editrice Nazionale. 1909. 5" x 7½". pp. 189. L 1.50.

Pe'belli occhi della gloria. Torino: Società Tipografico—Editrice Nazionale. 1906. 5" x 7½". pp. 294. L 3.50.

* FOGAZZARO, ANTONIO. 1842-1911. *Piccolo mondo antico.* Milano: Baldini e Castoldi. 1911. 5" x 7½". pp. 470. L 5.

By many considered the first of recent Italian writers of fiction. An idealist, "noble and dignified," says Pizzi, "even in his comic passages." This and "Piccolo mondo moderno" are his best and most popular novels. Lively and patriotic stories.

Piccolo mondo moderno. Milano: Ulrico Hoepli. 1912. 5" x 7½". pp. 461. L 5.

FUCINI, RENATO ("NERI TANFUCIO"). See also "Poetry." *Le veglie di Neri—Paesi e figure della campagna toscana.* Milano: Ulrico Hoepli. 1905. 5" x 7½". pp. 239. Illustrated. L 2.50.

Book of delightful short stories of the Tuscan countryside—filled with local color, and the clear air of the hills—witty and gay. A book of good fellowship, marked above all by Fucini's happy way of becoming democratically a countryman himself.

GIOVAGNOLI, RAFFAELLO. *Spartaco.* Milano: Paolo Carrara. 1889. 5" x 7½". 2 vols. pp. 427, 412. L 2.50 each.

Of the school of Manzoni. This is the best of his series dealing with subjects of ancient Rome.

* GROSSI, TOMMASO. 1791-1853. *Marco Visconti.* Firenze: Successori Le Monnier. 1911. 4¾" x 7". pp. 415. Bound. L 1.75.

A story of the 14th century. A sentimental melancholy romance—the author a close friend of Manzoni. Always popular.

* GUERRAZZI, F. D. 1804-1873. *Assedio di Firenze*. Milano:
Bietti. n. d. 4" x 6". 2 vols. pp. 275, 291. L 1.60 for the two.

The most fiery and warlike of the many popular novelists
who powerfully helped the union of Italy. He wrote this
particular book, he said, because he was unable to fight
a battle.

* MANZONI, ALESSANDRO. See also "Poetry." 1785-1873.
I promessi sposi. Milano: Ulrico Hoepli. 1911. 5" x 7½".
pp. 575. L 1.

This simple, thrilling love story of peasants is the greatest
novel of Italy. After its publication in 1825, when Manzoni told Sir Walter Scott that he had had his inspiration
from him, Scott gracefully replied: "In that case I consider 'I promessi sposi' my best work." And Goethe de-
clared his belief that no greater heights could be reached
in fiction. It was the first readable novel printed in Italy,
and its humor and simplicity made wide appeal. Its simplicity indeed gave literature to the common people, and
so far created a popular language for the new Italy, that
it has been said that no writer since Dante has rendered
greater services to the Italian tongue.

MISASI, N. *Racconti calabresi*. Napoli: Salvatore Romano. 1905. 5" x 7½". pp. 200. L 1.50.

A collection of stories of Calabria, very vividly recalling
the days of the brigands.

* NIEVO, IPPOLITO. 1832-1861. *Le confessioni d'un ottua-
genario*. Milano: Fratelli Treves. 1908. 5" x 7½". 3 vols.
pp. 314, 303, 339. L 1. each.

This political romance written by a young soldier of
Garibaldi's is coming into new popularity in Italy.

NOVELLINO, IL. Roma: Albrighi, Segati e C. 1911. 5" x
7½". pp. 135, L 1.

A quaint book of short medieval tales, written in the
thirteenth century. For its simple and interesting character and beautiful style always widely read. Italian
critics are fond of calling it "a golden book."

* PALADINI DI FRANCIA, STORIA DEI. Piacenza: Pon-
tremolese. 1910. 5" x 7½", abridged. pp. 351. L 3.50.

Extremely popular with the common people. The romances of chivalry, which form the basis of most mari-
onette plays in Italy, are usually either taken from this
work, or from the book which follows: "I reali di Francia."

* REALI, (I) DI FRANCIA (ANDREA DI JACOPO DA BARBERINO). Firenze: A. Salani. 1908. 5" x 7½". pp. 548. Illustrated L 3.

Elaborate prose romance of chivalry, of a heroic sort, written in the fifteenth century. Symonds says it has never ceased to be the most widely popular of all books written in Italian.

* ROVETTA, GEROLAMO. See also "Drama." 1854-1910. One of the most popular of Italian novelists, both idealist and realist. His novels are clever in plot, swiftly moving, filled with color and gaiety. The two chosen are among his best.

Il processo Montegù. Milano: Fratelli Treves. 1897. 5" x 7½". pp. 306. L 1.

Sott'acqua. Milano: Fratelli Treves. 1883. 5" x 7½". pp. 273. L 3.50.

* RUFFINI, GIOVANNI. 1807-1881. *Lorenzo Benoni.* Milano. Luigi Trevisini. n. d. 5½" x 8½". pp. 495. L 3.

A political romance originally written in English—largely autobiographical—which was powerful in winning English sympathy for the cause of Italian unity. Ruffini was a friend and fellow exile of Mazzini, who appears in the book in the character of Fantasio.

Dottor Antonio. Firenze: A. Salani. 1911. 5" x 7½". pp. 272. L 1.

Originally written in English like "Lorenzo Benoni," with which it must be read. Records Ruffini's story, the story of his brothers, the tragedy of his family and his country, of the group of young liberals who surrounded Mazzini— the picture of the conspiracies for liberty in Italy from their beginning to 1848.

* SALGARI, EMILIO. 1863-1911. See "Books for Children."
The thrilling books of Salgari are almost as popular among grown-ups, as among boys.

SERAO, MATILDE. See also "Travel." *Paese di cuccagna.* Napoli: Francesco Perrella. 1910. 5" x 7½". pp. 479. L 4.
Her best book. A true and imaginative picture of Neapolitan life, social classes, customs. Fanciful, vivacious, showing keen powers of observation.

VERGA, GIOVANNI.
A powerful realist, describes the peasant life of Sicily, with much local color. He is fond of depicting the sombre and violent side of Sicilian life.

Storia di una capinera. Milano: Fratelli Treves. 1907. 3½" x 5½". pp. 250. L 3.

Cavalleria rusticana. Milano: Fratelli Treves. 1912. 5" x 7½". pp. 267. L 3.

Basis of the well known opera by Mascagni. Contains several other popular short stories.

Mastro don Gesualdo. Milano: Fratelli Treves. 1911. 5" x 7½". pp. 333. L 3.50.

Fiction Translated Into Italian From Other Languages

Even our Italian immigrants who have little or no education usually have a vague consciousness of the glories of Italian literature, and you will occasionally find a man who is quite illiterate, reciting some bit of Tasso or Dante. But there are still large numbers of them who know nothing of their own literature and develop here a habit of reading. Their children, ardent little Americans through the power of our public schools, often lead them to select books that are translations of our own classics,—sometimes school classics, sometimes very simple children's books. In this way "Robinson Crusoe" and "Ivanhoe" have achieved a very marked popularity in a number of our libraries. This suggests that a helpful introduction to things American may·often be had through American novels and tales, many of which have now been translated into Italian. The best of these have been chosen, and grouped with the best of the world's fiction.

BELLAMY, E. *Nell'anno 2000*. (Looking Backward). Milano: Fratelli Treves. 1910. 5″ x 7½″. pp. 308. L 1.

BUNYAN, JOHN. *Il pellegrinaggio del cristiano*. (Pilgrim's Progress). Translated by Stanislao Bianciardi. Firenze: Tipografia Claudiana. 1904. 5″ x 7½″. pp. 374. L 1.50.

* CERVANTES, MICHELE—DI SAAVEDRA. *Don Chisciotte della Mancia*. (Don Quixote). Milano: Sonzogno. 1911. 4½″ x 7″. pp. 395. L 1.50.

* CLEMENS, S. L. (MARK TWAIN). See "Books for Children." *Le avventure di Tom Sawyer*. Firenze: R. Bemporad e Figlio, 1911. 6″ x 8½″. pp. 133. Illustrated. L .95.
Racconti umoristici. Translated by Livia Bruni. Torino: S. Lattes e C. 1906. 5″ x 7½″. pp. 180. Illustrated. L 2.50.
A selection of well known stories illustrating the author's characteristic humor.

* COOPER, J. F. *La spia*. (The Spy). Milano: Cesare Cioffi. n. d. 5″ x 7½″. pp. 319. L 2.

* CRAWFORD, F. M. *Saracinesca.* Milano: Fratelli Treves. 1898. 2 vols. 5" x 7½". pp. 299, 283. L 1. each.

"Saracinesca," and its sequels, "Sant'Ilario" and "Don Orsino," have proved very popular in Italy, as stories of Italian life. They cover the period 1865-1887. One of the characters in the story is Cardinal Antonelli, Secretary of State to Pius IX.

Sant'Ilario. Milano: Fratelli Treves. 1910. 2 vols. 5" x 7½". pp. 284, 274. L 1. each.

Don Orsino. Milano: Fratelli Treves. 1910. 2 vols. 5" x 7½". pp. 294, 324. L 1. each.

Corleone. Milano: Fratelli Treves. 1900. 2 vols. 5" x 7½". pp. 308, 329. L 1. each.

"Corleone" is enjoying a popularity in Italy second only to the "Saracinesca" series.

DAUDET, ALPHONSE. *Tartarino di Tarascona.* Firenze: A. Salani. 1912. 5" x 7½". pp. 230. L.75.

Tartarino sulle Alpi. Firenze: A. Salani. 1904. 5" x 7½". pp. 223. L.75.

* DEFOE, DANIEL. *Robinson Crusoè.* See "Books for Children." In this version also popular with adults.

DICKENS, CHARLES. See "Books for Children." *Cantico di natale.* (Christmas Carol). Milano: Ulrico Hoepli. 1888. 3½" x 5". pp. 237. L 3.

Memorie di Davide Copperfield. Milano: Fratelli Treves. 1910. 2 vols. 5" x 7½". pp. 357, 356. L 1. each.

* DUMAS, ALEXANDER. *Monte Cristo.* Firenze: A. Salani. 1909. 5" x 7½". pp. 1076, Illustrated. L 4.

I tre moschettieri. (Three Musketeers). Firenze: A. Salani. 1900. 2 vols. 5" x 7½". pp. 265, 248. L 1. each.

Venti anni dopo. (Twenty Years After). Milano: Bietti. 1907. 5" x 7½". pp. 351. L 2.50.

ELIOT, GEORGE. *Romola.* Milano: Fratelli Treves. 1906. 8½" x 12". pp. 135. L 5.

This inconveniently shaped volume is the only form in which Romola may now be had in Italian.

FRANCE, ANATOLE. *Il delitto di Silvestro Bonnard.* (The Crime of Sylvestre Bonnard). Milano: Fratelli Treves. 1910. 5" x 7½". pp. 299. L 1.

HUGO, VICTOR. *Nostra Donna di Parigi.* (Notre Dame de Paris). Firenze: A. Salani. 1903. 5" x 7½". pp. 431. Illustrated. L 1.50.

* *I miserabili.* (Les Miserables). Milano: Bietti. 1914. 6½" x 9½". pp. 659. Illustrated. L 3.50.

LOTI, PIERRE. *Pescatori d'Islanda.* (An Iceland Fisherman). Firenze: A. Salani. 1900. 5" x 7½". pp. 253. L 1.

* LYTTON, BULWER. *Ultimi giorni di Pompei.* (Last Days of Pompeii). Milano: Baldini e Castoldi. 1911. 5" x 7½". pp. 278. L 1.50.

* MILLE E UNA NOTTE. (Arabian Nights). Translated by Armando Dominicis. Firenze: A. Salani. 1908. 5" x 7". pp. 1018. L 4.

* POE, EDGAR ALLEN. *Racconti straordinari.* Translated by G. A. Santini. Firenze: R. Bemporad e Figlio. 1911. 6" x 8½". pp. 127. Illustrated. L .95.

Nuovi racconti straordinari Firenze: R. Bemporad e Figlio. 1909. 6" x 8½". pp. 135. Illustrated. L .95.

These two volumes include most of Poe's famous stories.

* SCOTT, SIR WALTER. *Lucia di Lammermoor.* (The Bride of Lammermoor). Firenze: A. Salani. 1909. 5" x 7½". pp. 377. L 1.20.

This is popular among Italian readers, because it was taken as the basis of the story in Donizetti's opera.

Ivanhoe. Milano: Fratelli Treves. 1910. 7" x 10½". pp. 677. L 5.

SIENKIEWICZ, H. *Quo vadis.* Milano: Fratelli Treves. 1910. 5" x 7½". pp. 379. L 1.

STERNE, LAWRENCE. *Viaggio sentimentale.* (Sentimental Journey). Milano: Sonzogno. 1910. 4½" x 6½". pp. 124. L .30.

STEVENSON, R. L. *Rapito.* (Kidnapped). Milano: Fratelli Treves. 1910. 5" x 7½". pp. 294. L 1.

* STOWE, HARRIET B. *La capanna dello zio Tom.* (Uncle Tom's Cabin). Firenze: R. Bemporad e Figlio. 1911. 6" x 8½". pp. 256. Illustrated. L .95.

* SUTTNER, BARONESS VON. *Abbasso le armi!* (Lay Down Your Arms). Milano: Fratelli Treves. 1910. 2 vols. 5" x 7½". pp. 282, 298. L 1. each.

A translation of Baroness von Suttner's moving novel that in Italian, as in so many other languages, has had a profound effect in the cause of international peace.

— 51 —

* SWIFT, JONATHAN. *Viaggi di Gulliver.* (Gulliver's Travels). Milano: Fratelli Treves. 1896. 5" x 7½". pp. 285. Illustrated. L 1.50.

THACKERAY, WILLIAM M. *La fiera della vanità.* (Vanity Fair). Milano: Fratelli Treves. 1910. 3 vols. 5" x 7½". pp. 324, 335, 299. L 2. each.

TOLSTOI, L. *Anna Karenine.* Milano: Fratelli Treves. 1904. 2 vols. 5" x 7½". pp. 319, 321. L 1. each.
Guerra e pace. (War and Peace). Milano: Fratelli Treves. 1910. 4 vols. 5" x 7½". pp. 323, 323, 295, 306. L 1. each.

TURGHENIEFF, I. *Vergini terre.* (Virgin Soil). Milano: Fratelli Treves. 1902. 5" x 7½". pp. 340. L 1.

* VERNE, J. *Cinque settimane in pallone.* (Five Weeks in a Balloon). Milano: Bietti. 1907. 5" x 7½". pp. 254. L 1.50.
Viaggio al centro della terra. (A Journey to the Center of the Earth). Milano: Bietti. 1912. 5" x 7½". pp. 254. L 1.50.
20,000 leghe sotto i mari. (Twenty Thousand Leagues under the Sea). Milano: Paolo Carrara. 1909. 5" x 7½". pp. 634. Illustrated. L 5.
Il giro del mondo in ottanta giorni. (Around the World in Eighty Days). Milano: Fratelli Treves. 1910. 5" x 7½". pp. 315. L 1.

VOLTAIRE, F. *Candido.* (Candide). Milano: Sonzogno. 1909. 4½" x 6½". pp. 108. L .30.

* WALLACE, LEW. *Ben Hur.* Milano: Baldini e Castoldi. 1902. 5" x 7½". pp. 483. L 4.

WELLS, H. G. *Nei giorni della cometa.* (In the Days of the Comet). Milano: Fratelli Treves. 1906. 5" x 7½". pp. 353. L 3.
La guerra nell'aria. (The War in the Air). Milano: Fratelli Treves. 1909. 5" x 7½". pp. 424. L 3.
Novelle straordinarie. Milano: Fratelli Treves. 1905. 6½" x 10". pp. 213. L 3.
A collection of strange stories.

Books for Children

Italian children in this country learn English so very quickly that the use of children's books in Italian is somewhat limited. But still they will often be the best introduction to the library, for those children who have learned to read before coming to America. Such books will also serve very valuably to help these little folk retain a knowledge of their own language. This is not only important for the practical use of knowing Italian as well as English. It is even more important because it will foster in them a respect and love for the land of their race, which they so promptly lose in their speedy Americanization. To preserve this respect and love will help bridge the gulf that quickly divides parents and children among our immigrants.

Italians have an actual cult for children, and this is well shown in the great number of admirable books, especially written for them. Among these there are many that combine definite educational value with stories of thrilling interest. And recently many translations have been made of children's books from other languages. A selection of all of these has been included in our list and here no asterisks are needed.

ABBA, CESARE. *Storia dei Mille.* Firenze: R. Bemporad e Figlio. 1904. 6½″ x 10″. pp. 212. Illustrated. L 3. A cheaper edition at L 2. is now available.

A stirring story of the Revolution in Italy, written for children.

ALCOTT, LOUISA M. *Piccole donne.* (Little Women). Lanciano: R. Carabba. 1914. 5″ x 7½″. 2 vols. pp. 341, 379. L 2. each.

ANDERSEN, H. C. *Novelle.* (Fairy Tales). Translated by Giuseppe Fanciullo. Firenze: R. Bemporad e Figlio. 1911. 2 vols. 6″ x 8½″. pp. 121, 125. Illustrated. L .95 each.

BACCINI, IDA. *Memorie di un pulcino.* Firenze: R. Bemporad e Figlio. 1911. 5″ x 7½″. pp. 115. L 1.

The story of a chicken is the most widely popular book in Italy for small children. See also "Biography."

Cristoforo Colombo. Torino: G. B. Paravia e C. 1909. 5″ x 7½″. Bound. pp. 47. L 1.

The story of Columbus written for children.

BARRIE, J. M. *Peter Pan.* Translated by F. C. Ageno. Firenze: R. Bemporad e Figlio. 1913. 7" x 9½". pp. 142. Illustrated and bound. L 6.

CAPUANA, LUIGI. *C'era una volta.* (Once upon a Time). Firenze: R. Bemporad e Figlio. 1910. 5" x 7½. pp. 315. L 2.50.

Capuana is very popular not only as the author of charming short stories for grown ups, but as a writer for children and especially for boys.

Scurpiddu. Torino: G. B. Paravia e C. 1907. 6½" x 9½" pp. 172. Illustrated. L 3.

Book for boys, recommended by the Italian Minister of Public Instruction—also recommended by the delight with which the boys themselves read about the boy Scurpiddu.

CLEMENS, S. L. (MARK TWAIN). *Le avventure di Tom Sawyer.* Firenze: R. Bemporad e Figlio. 1911. 6" x 8½". pp. 133. Illustrated. L .95.

"COLLODI, C." (LORENZINI, C.) *Avventure di Pinocchio.* Firenze: R. Bemporad e Figlio. 1907. 5" x 7½". pp. 300. L 2.50.

A famous book for small boys.

Racconti delle fate. Firenze: R. Bemporad e Figlio. 1909. 5" x 7½". pp. 267. L 2.

Translation of a collection of fairy tales: Little Red Riding Hood, Sleeping Beauty, Cinderella, Hop o' My Thumb, Beauty and the Beast, etc.

"CORDELIA." (VIRGINIA TEDESCHI). *Piccoli eroi.* Milano: Fratelli Treves. 1911. 5" x 7½". pp. 290. L 2.

The 55th edition of this extremely popular book for children (9-14 years), teaching them life through the lessons of life itself. Romance of science and modern industry, humble heroes, unknown sacrifices, inspiring readings—open air fun.

DE AMICIS, EDMONDO. 1846-1908. *Cuore.* Milano: Fratelli Treves. 1912. 5" x 7½". pp. 340. L 2.

One of the most famous of all books for boys. Translated into nearly every modern language. The edition here listed is the 618th..

DEFOE, DANIEL. *Robinson Crusoè.* Firenze: A. Salani. 1907. 5" x 7½". pp. 545. L 2.

DICKENS, CHARLES. *Cantico di natale.* (Christmas Carol). Milano: Ulrico Hoepli. 1888. 3½" x 5". pp. 237. L 3.

GRIMM (BROTHERS). *Novelle.* Translated by B. Vettori. Firenze: R. Bemporad e Figlio. 1911. 6" x 8½". pp. 128. Illustrated. L .95.

A selection from the old familiar fairy tales.

KIPLING, RUDYARD. *Il libro delle bestie.* (Just So Stories). Firenze: R. Bemporad e Figlio. 1913. 7" x 10". pp. 165. Well illustrated. L 3.50.

MARCO POLO. 1254-1323. *I viaggi.* (The Voyages). Milano: Sonzogno. 1906. 4½" x 7". pp. 148. L .30.

In this edition the language has been carefully modernized.

ORSI, PIETRO. *Come fu fatta l'Italia.* (Might be translated: How Italy Became a Nation). Torino: Società Tipografico—Editrice Nazionale. 1914. 4¾" x 7¼". pp. 233, Illustrated. Bound. L 3.

This story of the great struggle for the union of Italy is so popularly written that it should prove very interesting to children.

SALGARI, EMILIO. 1863-1911. *La scimitarra di Budda.* (The Scimitar of Buddha). Milano: Fratelli Treves. 1909. 6½" x 10". pp. 251. Illustrated. L 3.

Salgari is the Jules Verne of Italy. His stories are usually of wild adventure and imagined wonders of science applied to life. Extremely popular and this is one of the most read of all his many books. It was first published in a paper for children.

La città dell'oro. (The City of Gold). Milano: Fratelli Treves. 1898. 6½" x 10". pp. 365. Illustrated. L 3.

Another of Salgari's early stories—widely read.

Il re dell'aria. Firenze: R. Bemporad e Figlio. 1907. 6½" x 10". pp. 324. Illustrated. L 3.50.

In this "The King of the Air" they say that Salgari invented the first dirigible of fiction.

SETON, ERNEST THOMPSON. *Animali eroi.* Translated by Laura Torretta. Milano: L. F. Cogliati. 1910. 6" x 8". pp. 346. L 5.50.

Well printed on good paper with the original illustrations. A collection of nine of the best of Seton's stories of animals.

SOCCI, ETTORE. *Umili eroi della patria e dell'umanità.* Milano: Libreria Editrice Nazionale. 1903. 5" x 7½". pp. 232. L 2.

This book of "humble heroes of our country and of humanity" is chiefly concerned with the desperate struggle of the Revolution in Italy. A simple and beautiful book—few books so well known popularly; not written to glorify war but to glorify the nobility of sacrifice, the giving of life by the humble for a noble end—not stories of the aristocracy of heroism. Highly educational these tales—from that of Goretti, the clown, to Federigo Comandini, the peasant.

"VAMBA." (LUIGI BERTELLI). *Ciondolino.* Firenze: R. Bemporad e Figlio. 1910. 5" x 7½". pp. 228. L 2.50.

In "Ciondolino" a boy becomes an ant, and learns the habits and customs of ant life. The author is a naturalist, humorist, journalist. This little book has been highly praised by scientists, and boys like it so well that it has gone through many editions.

VECCHI, A. V. *Racconti di mare e di guerra.* Firenze: R. Bemporad e Figlio. 1903. 5" x 7½". pp. 276. L 2.

These stories largely of the sea, were written with the object of interesting children, through dramatic tales, in science and nature, and of inspiring them with interest in animals and affection for them.

VIDOTTO, GIACOMO. *Garibaldi.* Roma: Albrighi, Segati e C. 1899. 4¾" x 7¼". pp. 110. L .75.

The life story of Garibaldi told for children.

Poetry

Italians have not lost, as we have, the habit of reading poetry and loving it. They have a national fondness for verse, shared by cultured and uncultured alike. In its humblest forms, simple songs and refrains of a hundred kinds, elemental poetry is linked with music. A Roman gardener of my acquaintance finds his chief delight in composing sonorous octaves. And in line with Homeric tradition is the Calabrian discovery of a friend. "My uncle is a poet," proudly proclaimed the waiter. "You must let me see one of his books," said my friend. "But he cannot write and so they have never been printed." Declamation revealed an undoubted epic!

Owing to the great number and unusual excellence of the Italian poets, the selection of a necessarily short list presents unusual difficulties—particularly in dealing with the poets of the Risorgimento. Many important, many popular names have been omitted in order that this selection might not lose a certain essential proportion. But if Berchet and Cavallotti, Monti and Prati, and many others fail of a special title, they will all be found in their best work in the fine anthology of Barbiera.

In any list of poetry such as this, popularity must be a determining thing. And so I have included a selection of popular poets of dialect. Of these, particularly living writers, Italy has many, for every part of the country has its own poet in dialect, who sometimes in comic and satirical vein, sometimes seriously and nobly, mirrors the simple life of his province. Work and love, death and adventure are his theme. His characters are peasant and priest, petty provincial officials, the hundred classes of townsfolk, the conscript boys doing service in army and navy. The dialect of one province is often almost unintelligible beyond its narrow boundaries, yet very remarkably, in one way or another, nearly all of these poets give evidence of the traditional and fundamental Italian unity that has at last made one people of the men of all provinces.

The Italian is very fond of reading poetry aloud and those who cannot read, as well as many who can, listen delightedly. For this custom—a classical inheritance—and for many readers, the simpler and more dramatic poetry should be chosen. But dialect poets like Pascarella and Martoglio, or a stern sentimentalist like Rapisardi, or Leopardi's "Ginestra" may make more appeal than Tasso. I have found Dante many times in the hands of workingmen. Some will be fond of De Amicis, and Ada Negri—some thrilled with the patriotic "Garibaldean Rhapsodies" of Marradi.

ALEARDI, ALEARDO. 1814-1878. *Canti.* Firenze: G. Barbèra. 1905. 5" x 7½". pp. 499. L 4.

This poet appeals to many for his marked moral and religious qualities. His harmonious and dignified verses often strike a national note. But on the whole a sentimental and melancholy bard.

* ALFIERI, VITTORIO. See "Biography." *Vite e rime scelte.* See also "Drama."

ALIGHIERI, DANTE. 1265-1321.
In the case of Dante alone has it seemed necessary to describe a choice of editions to meet different needs. His greatness as a world classic calls for no note in this book. But for Italians it is not merely that he was their first and greatest writer—in a literary sense almost the creator of their language. Throughout the centuries he has represented the best in character and ideals to which they have looked. And national love and reverence have given him the popular name of "Padre Dante."

Tutte le opere di Dante Alighieri. Edited by Dr. E. Moore. Oxford: Oxford University Press. 1904. 5½" x 8". pp. 490. $2.25.

A complete edition of all the works of Dante in one volume. Handy, scholarly. Without notes except index of proper names and notable things mentioned.

La divina commedia. Edited, with notes by G. A. Scartazzini. Milano: Ulrico Hoepli. 1911. 5" x 7½". pp. 1047 & 124. L 6.

Contains full notes of the famous Dante scholar and a *rimario*—an index of verses arranged according to their rhymes. A mine of learning in the literature of Dante and his commentators—abridged for school use from greater work. On this work Scartazzini lavished the love and labor of a life time. In one form or another it has been the standard and most generally used edition for more than a generation.

* *La divina commedia.* Edited, with notes by Francesco Torraca. Roma: Albrighi, Segati e C. 1908. 5" x 7½". pp. 952. L 4.50.

Very popular in Italy for schools and general reading. Torraca's chief care is to make plain Dante's meaning and to "collect and explain" the secrets of his art. Torraca is a brilliant and accomplished man of letters and his notes are full of literary and human interest. The best edition for general use, where notes are needed.

* *La divina commedia.* Illustrated by Gustave Dorè, with the notes of E. Camerini. Milano: Sonsogno. 1911. 10" x 14". pp. 688. L 10.

Dorè's illustrations make this a very eagerly chosen book in libraries.

* ARIOSTO, LUDOVICO. 1474-1533. *L'Orlando furioso.* Edited by Augusto Romizi. Roma: Albrighi, Segati e C. 1912. 5½" x 8". pp. 542. L 3.50.

The great poet of Italy who has been most widely and delightedly read by people of every class. The "Orlando Furioso" is, so some believe, the human comedy grotesquely staged in the world of chivalry. The stories of knights and ladies, the wars of Christians and Saracens, bloody catastrophes, delicate sentimentalities, incantations, visits with Dante in the Inferno, with St. John to the moon, all verging from tragic to comic, from majestic to simple, make such alluring reading that Baretti said the wonderful pleasure of it ought to be allowed only as a prize and recompense to those who render their country a great service.

* BARBIERA, RAFFAELLO. *I poeti italiani del secolo XIX.* Milano: Fratelli Treves. 1913. 5" x 7½". pp. 1400. Bound and illustrated with portraits. L 10.

Unquestionably the best anthology of the Italian poets of the Nineteenth Century. It contains an excellent introduction, biographical sketches, notes and full indices. Every school is well represented: Classicism, Romanticism, Realism, Symbolism. There is a particularly good selection of the poets of the Risorgimento. For a number of well known poets nearly all their best work is included and there is a notable selection of important verse by little known poets not to be found in any other anthology.

* CARDUCCI, GIOSUÈ. 1836-1907. *Poesie.* 1850-1900. Bologna: Nicola Zanichelli. 1911. 5½" x 7¾". pp. 1075. Bound. India paper. L 10.

Foremost poet of modern Italy. Deliberately classic in style and form. In spirit modern and national. Beauty, dignity, vigor characterize his work. He was powerful and fearless in support of his ideals. His verse is often difficult. Immensely respected.

— 59 —

D'ANNUNZIO, GABRIELE. Of the poetical works of this leader of the Italian aesthetic school, three books have been chosen, beginning with the volume that contains the splendid "Naval Odes." In these are found beautiful and noble ideas, and lofty sentiments inspired by the national glories of Italy, joined with rare beauty of form. These are nearly free from the obscurities, affectations and other objections that may be urged against his later work, and prevent his books from becoming broadly popular.

Poema paradisiaco; Odi navali. Milano: Fratelli Treves. 1913. 3¾" x 5¾". pp. 226. L 4.

Canzone di Garibaldi. Milano: Fratelli Treves. 1909. 7½" x 11½". pp. 64. L 1.50.

In morte di Giuseppe Verdi. Milano: Fratelli Treves. 1913. 7½" x 11½". pp. 28. L 1.

* DE AMICIS, EDMONDO. 1846-1908. *Poesie.* Milano: Fratelli Treves. 1907. 4" x 6". pp. 268. L 4.

Very popular among all classes of Italians. "To my mother" is widely known and loved. Besides much that is tender and lovely this little book contains many examples of delightfully humorous verse.

* DI GIACOMO, SALVATORE. *Poesie.* Napoli: Riccardo Ricciardi. 1909. 5" x 7½". pp. 442. L 4.

The best of the Neapolitan poets of dialect. His poetry is of high lyrical quality. His homely verses are free from every trace of vulgarity, and are liked for their "sweet melancholy."

* FOSCOLO, UGO. 1779-1827. *Scelta di poesie e di prose.* Edited by Dr. Pio Spagnotti. Milano: Ulrico Hoepli. 1901. 4½" x 7". pp. 352. L 2.50.

Many selections have been made of the work of this early poet of the Risorgimento, but this in several respects is the best of all. Foscolo was a brilliant, erratic, romantic genius, a revolutionary, a wanderer. He has been likened to Byron. He had so great an effect in helping the cause of liberty in Italy, that Mazzini said: "Without him, we perhaps should not have been what we are." This volume, of course, contains the famous "Carme dei Sepolcri."

* FUCINI, RENATO. ("Neri Tanfucio.") See also "Fiction— Italian." *Le poesie.* Firenze: R. Bemporad e Figlio. n. d. 4" x 6". pp. 367. L 2.50.

Writes the gayest and most spontaneous of popular Tuscan poetry. His verses are filled with humor and life.

and seem caught from the lips of the people. Yet his art is exquisite and restrained.

GIUSTI, GIUSEPPE. 1809-1850. *Poesie complete*. Firenze: A. Salani. 1909. 5" x 7½". pp. 431. L 1.50.
Giusti was above all a satirical poet—the most read of his time, and still has a very considerable popular appeal. He attacked the shameful, the vicious, the ridiculous things in the life of his day. His work became national and powerfully he attacked the foreign oppressor. Yet in many of his greatest poems his interest is not limited to Italy, but is broadly humanitarian. He uses irony and satire with extraordinary vigor and dramatic power, and pathos and delicate fancy abound in his pages.

* LEOPARDI, GIACOMO. 1798-1837. *I canti*. Edited by G. Tambara. Milano: Antonio Vallardi. 1912. 5" x 8". pp. 339. L 2.50.
The great poet of pessimism. Of first importance in literature because he gave new form and new force to Italy's poetry. His work, classic in style, is still vigorous and beautiful. His love of his country was ardent. Graf said: "There have been greater poets, but none is his equal." This selection includes the famous "Ginestra"—"all thunder and lightning and funereal light"—that may very likely prove to be what it is often called, "immortal."

MANZONI, ALESSANDRO. See also "Fiction—Italian." 1785-1873. *Le tragedie, gli inni sacri, le odi*. Milano: Ulrico Hoepli. 1907. 5" x 7½". pp. 400. L 2.50.
The poetry of Manzoni is distinguished by lofty moral and religious value, by great dignity and idealism. When he treats of national themes, it is with deep patriotic fervor.

* MARRADI. *Rapsodie garibaldine*. Firenze: G. Barbèra. 1902. 5" x 8". pp. 128. L 2.50.
A Tuscan poet of the country side, deals often, as in this book, with stirring national themes. Widely popular.

* MARTOGLIO, NINO. *Centona*. Catania: Giannotta. 1913. 5" x 7½". pp. 350. L 3.
The dialect verse of a very popular Sicilian poet. Abounds in local color with many touches of quaint humor.

* MELI, GIOVANNI. 1740-1815. *Le bucoliche*. Milano: Sonzogno. 1903. 4½" x 6½". pp. 173. L .60.
Greatest Sicilian poet in the vernacular. His poetry is

Arcadian, delicate and beautiful; contains little local color, but is very popular. This edition has on opposite pages an Italian translation in verse.

* NEGRI, ADA. *Maternità.* Milano: Fratelli Treves. 1912. 4" x 5½". pp. 285. L 4.
Considered the first of living poets among Italian women. The devoted and affectionate partisan of the working classes and of the rights of the humble. Rather sombre but very popular.

* OXFORD BOOK OF ITALIAN VERSE. Edited by St. John Lucas. Oxford: Oxford University Press. 1910. 4½". x 6¾". pp. 576. $2.
Selections chiefly from the classical poets, from the 13th to the 19th century. A delightful anthology.

* PARINI, GIUSEPPE. 1729-1799. *Le poesie scelte.* Edited by Michele Scherillo. Milano: Ulrico Hoepli. 1906. 5" x 7½". pp. 378. L 2.50.
Excellent representative selection from the poetry of the great poet priest of Milan—for generations used for the instruction and inspiration of Italian youth. His poems of classic beauty nearly always deal with the useful and practical, and in "Il Giorno" his grave irony became a powerful popular demand for human equality and for justice against the privileges of the great and nobly born.

* PASCARELLA, CESARE. *Sonetti.* Torino: Società Tipografico—Editrice Nazionale. 1911. 6½" x 9". pp. 180. L 4.

First of all living Italian poets of dialect. Writes in the dialect of Rome. Extremely popular. Combines the grotesque and comic with much dignity and nobility. Considered one of Italy's greatest living writers. The Discovery of America (Scoperta dell'America) in this volume is famous and is often chosen for recitation.

* PASCOLI, GIOVANNI. · 1855-1912. *Limpido rivo—Prose e poesie.* Bologna: Nicola Zanichelli. 1912. 5" x 7½". pp. 247. L 3.
One of the best of the poets of modern Italy. Noted for his smooth and exquisite verse, his descriptive powers and his serene and noble views of life. This volume contains a selection of his best known verse—highly polished minute descriptions of country life to which he principally owes his fame, together with well chosen examples of his prose.

* PETRARCA, FRANCESCO. 1304-1374. *Il canzoniere*. With the notes of Giuseppe Rigutini. Edited by Michele Scherillo. Milano: Ulrico Hoepli. 1910. 5" x 7½". pp. 474. L 3.50.

Scholarly edition, with a wealth of interesting and human notes. The introduction gives an excellent account of the place of Petrarch in Italian literature and Italian life: Not only the exquisite poet of love, the last of the troubadours—the last great writer of the Middle Ages; but also the first of the humanists, reviving the study of the classics of Greece and Rome—first of moderns, philosopher, courtier, antiquary, constant traveller, free of mysticism, with all the restlessness and curiosity of the modern world. He was also as they are saying these late days, "a true Italian," for he tried to reconcile the jealous discords with which the Italy of that day was torn; he had the dream of continuing the glories of ancient Rome, a dream that is now a power and inspiration in the progress of Italy.

* RAPISARDI, MARIO. 1844-1913. *Poesie religiose*. Milano: Sonzogno. 1908. 4½" x 6½". pp. 91. L .30.

Rapisardi, the literary antagonist of Carducci, is greatly admired, particularly in Sicily. In his native province of Catania even the peasants call him "Il gran padre." His dignity sometimes becomes pompous; he is often frankly a pessimist. But he has great imaginative and descriptive power, and high and rigorous purposes. These poems are called religious—so is the reason given—"because they represent the triumph of reason, of love, of sacrifice—the protest of humanity in behalf of truth and the humble." Garibaldi spoke of Rapisardi's poetry, as "a great work of moral emancipation heroically begun."

La palingenesi. Sesto S. Giovanni: Madella. 1912. 5" x 7½". pp. 239. L 2.

Perhaps the best example of Rapisardi's longer symbolical poems. This deals with the hope of civil and religious reform, bringing peace to the earth. On its first printing, Victor Hugo wrote the poet: "You hold in your hands two torches: the torch of poetry and the torch of truth—the great Italian heart beats everywhere in your generous book."

* TASSO, TORQUATO. 1493-1569. *La Gerusalemme liberata*. Edited by Riccardo Cornali. Roma: Albrighi, Segati e C. 1901. 5" x 7½". pp. 348. L 2.

Says Prof. Cornali: "The minds of young people are fires to be lighted, not vases to fill." And so with human and interesting notes he has prepared perhaps the very best edition for general reading of this great epic—a story of the Crusades, and of the rescue of the tomb of Christ from the infidels. The "Gerusalemme" has always been popular not only for its thrilling accounts of adventure of its knightly champions, its duels and thundering battles, and sorcery, but also for the very affecting love stories that are a prominent part of its plot. Yet it is dignified, noble, religious. Anciently it used to be sung through the streets of the Italian cities; gondoliers sang it on the canals of Venice within the memory of men still living; its beautiful and sonorous verses are well known and loved now, and in Italy one often hears its favorite passages declaimed.

* TRILUSSA (CARLO ALBERTO SALUSTRI). *Sonetti romaneschi.* Roma: Enrico Voghera. 1909. 6½" x 9½". pp. 201. L 4.

Very popular satirical poet; writes facile, musical, and often humorous verse in the dialect of Rome.

POETRY TRANSLATED FROM THE ENGLISH.

* LONGFELLOW, H. W. *Miles Standese.* Translated by Giacomo Zanella. Milano: Ulrico Hoepli. 1884. 3½" x 5". pp. 180. L 3.

* *Evangelina.* Translated by Giacomo Zanella. Milano: Ulrico Hoepli. 1883. 3½" x 5". pp. 172. L 3.

MILTON, JOHN *Paradiso perduto.* Translated by Lazzaro Papi. Milano: Sonzogno. n. d. Folio pp. 296. L 4. An inconvenient size, but the only form in which "Paradise Lost" is now available in Italian.

WHITMAN, WALT. *Canti scelti.* Translation and introduction by Luigi Gamberale. Milano: Sonzogno. 1908. 4½" x 6¾". 2 vols. pp. 104, 128. L .30 each.

Admirable selection of the poems of Whitman, covering every phase of the poet's work.

Drama

The reading of plays strongly appeals to the dramatic instincts of the Italian, and is today, as it has always been, popular with him. In fact, it is the frequent experience of librarians that their Italian readers are often more apt to choose a book of poetry or drama than of fiction.

* ALFIERI, VITTORIO. 1749-1803. See also "Biography." *Tragedie scelte.* Firenze: G. C. Sansoni. 1912. 5¼" x 7¾". pp. 397. L 2.50.

A group of the most famous plays of the "Father of Italian Tragedy." The Italian Revolution was the work of thinkers and poets, and in this Alfieri in a large sense led the way. Pizzi says: "His was the first voice to cry 'liberty,' after a silence of centuries·" With the broken, rapid, breathless dialogue, with the precipitate rush of action, love of liberty and his country fills all his tragedies.

* BENELLI, SEM. *La cena delle beffe.* Milano: Fratelli Treves. 1910. 5½" x 7½". pp. 152. L 3.

Brilliant historical tragi-comedy—the one real success of the Italian stage of the last few years. Thoroughly Tuscan.

* BRACCO, ROBERTO. *Teatro (Vol. V.)* Napoli: Remo Sandron. 1911. 5" x 7½". pp. 338. L 3.

This vol. contains *Maternità—Il frutto acerbo.*

Neapolitan dramatist—versatile genius, keen observer, realist—sometimes wild, noisy—melancholy in his love songs. Popularity seems sometimes to reach point of fascination.

CAVALLOTTI, FELICE. 1842-1898. *Il cantico dei cantici.* Milano: Carlo Barbini. 1909. 4½" x 6½". pp. 69. L 1.20.

In the drama, even more than in his verse, Cavallotti won popularity. The play selected is a graceful and witty domestic idyl.

* GIACOSA, GIUSEPPE. 1847-1906. Considered the first of the Italian dramatists of our time. Able, vivacious and graceful writer, with lofty moral standards. The play that gives title to the first book selected, and "Come le foglie" illustrate the two radically different kinds of his work—one the sprightly historical comedy—the other, modern, with the problems of our day·

Una partita a scacchi—Trionfo d'amore—Intermezzi e scene. Milano: Fratelli Treves. 1908. 5" x 7½". pp. 263. L 3.

Come le foglie. (19th Edition). Milano: Fratelli Treves. 1907. 5" x 7½". pp. 275. L 4.

* GOLDONI, CARLO. 1707-1793. *Commedie scelte.* Introduction by Raffaello Nocchi. Firenze: Successori Le Monnier. 1910. 4¾" x 7". pp. 487. Bound. L 1.75.

A good selection of the most popular plays of the great founder of the Italian drama. Gayly satirizing the follies and extravagances of Venetian life, "painting nature without spoiling it," as he said, Goldoni became for Italy what Moliere was for France. He is still unequalled as a writer of comedy and a number of his plays, notably "La Locandiera" included in this volume, are frequently acted with unfailing success. "La Locandiera" is one of the favorite plays, and one of the favorite parts, of Eleanora Duse, the greatest living Italian actress.

MANZONI, ALESSANDRO. 1785-1873. See "Poetry."

* MARTINI, FERDINANDO. *Chi sa il gioco non l'insegni.* Firenze: R. Bemporad e Figlio. 1906. 5" x 7½". pp. 245. L 3.

A book of delightful one act comedies, by the gifted Minister of Finance in Italy's new cabinet—the most distinguished literary man in a cabinet of literary men. These plays can always be relied upon to draw crowded houses. They are called "Proverbi" because each one takes its title from some well known proverb, and wittily points the moral of the bit of popular wisdom selected.

* MOLIÈRE, G. B. *Commedie scelte.* Translated by Alcibiade Moretti. Milano: Fratelli Treves. 1912. 5" x 7½". 2 vols. pp. 339, 345. L 1. each.

Excellent selection—excellent translation.

* NICCOLINI, GIAMBATTISTA. 1782-1861. *Arnaldo da Brescia.* Milano: Sonzogno. 1910. 4½" x 6¾". pp. 161. L .30.

The most important drama, artistically and politically, of the great tragic poet of the Italian Revolution. In this play, as in his "Sforza," love of his own country is so joined with hatred of Austria that the representation of his plays is still forbidden in Trieste.

* ROVETTA, GEROLAMO. 1852-1911. *Romanticismo.* Milano: Baldini e Castoldi. 1911. 5" x 7½". pp. 254. L 3.50.

Rovetta, a popular novelist and dramatist, lively, satirical, with a keen eye for the comic. In this play, his best, he leaves the intrigues of contemporary society and deals with a patriotic, historical subject.

SHAKESPEARE, WILLIAM. The Plays. Translated by Diego Angeli. Each play in a separate volume, well printed on excellent paper. Milano: Fratelli Treves. 1911-1914. 5½" x 8". pp. 175 to 250. L 3. each.

This translation, now in course of publication, by Signor Angeli, the well known novelist, art critic and satirist of the "Giornale d'Italia," is an exceedingly good one—by common consent the best in Italian. The plays that have already appeared are La Tempesta, Giulio Cesare, Macbeth, Amleto, Come vi pare (As You Like It), La bisbetica domata (Taming of the Shrew), Antonio e Cleopatra, Otello, La notte dell'epifania (Twelfth Night), Il sogno di una notte di mezza estate (Midsummer Night's Dream). Several other.of the plays by the same translator are in preparation, and will soon be published. The Italian readers in our libraries are, of course, apt to be
* especially interested in Othello, Julius Caesar, Romeo and Juliet and the Merchant of Venice. The last two of these are not yet published in this series, and may be had in inferior editions·

Tragedie scelte: *Otello, Macbeth, Mercante di Venezia.* About pp. 300. 1 vol. L 1.50.

Giulietta e Romeo. L .30.

Milano: Sonzogno. 1913 catalogue.

Teatro completo di Shakspeare. The complete dramatic works. Translated by Carlo Rusconi. Torino: Unione Tipografico—Editrice Torinese. 1859. 7 vols. 5" x 7½". About pp. 400 each. L 8.50 for the set.

One of the old editions, unsatisfactory in many respects, but still readable and serviceable for ordinary use. It is newly printed from old plates that are in fair condition. The paper is passable.

Music

Among Italian workingmen you will often find not only a passion for music, but a very astonishing knowledge of it. I have a memory of my first summer night in Venice when there was wondrous singing of *Rigoletto* and *Don Giovanni* across the Piazza, while the bells of the Campanile marked the hour of two. I thought that they were certainly artists returning from the opera. But they were not. They were night workers of the first shift, who had been coaling ships in the harbor. A surprising experience, but soon matched after you have made friends with Italian workingmen.

In Mount Vernon, New York, a city of 31,000 inhabitants, the librarian has found that opera librettos have been very eagerly read—and this to such an extent, that the library now has more than 200 of them on its shelves, each stitched at the library in a heavy red paper cover. Pains have been taken to select those editions giving English on one side, and Italian on the other. In many cases it has been possible to choose librettos which contain excerpts of the most famous music in the opera score.

LIBRETTOS.

G. Ricordi & Company, 14 East 43rd Street, New York, the American Branch of the great Italian firm of music publishers, issue many librettos in Italian, a few in English and Italian. 6¼"x9¼" Prices vary from L .50 to L 2. No music included in these. Almost any special needs concerning Italian music can be met by Ricordi. Catalogue on request.

Charles H. Ditson & Co., 8 East 34th Street, New York, publish 47 librettos in English and Italian, all with selections of popular airs from the vocal score. 6¾" x 9½". 25c each. Catalogue on request.

G. Schirmer, 3 East 43rd Street, New York, publishes 7 librettos in Italian and English, 2 sizes, 6"x8½" and 7"x10½", 25c and 35c. These include the newest operas by Wolf-Ferrari, very popular with Italians. No music included. Catalogue on request.

COLLECTIONS OF SONGS.

Anthology of Italian Song, 17th and 18th centuries. New York: Schirmer, 1898, 2 vols. 7½"x10¾" pp. 144, 145.

Paper $1.00 .each, cloth $2.00. The words of the songs are in Italian and English, biographical notes of the composers in English only; music, with piano accompaniment.

The Prima Donna's Album. Edited by Josiah Pittmann, New York: Schirmer, 1898, 7" x 10½". pp. 328. Paper $1.50, cloth $2.50. Forty-one famous arias and cavatinas, representing all of the leading Italian composers, as well as Mozart, Weber, and Meyerbeer, whose operatic writings are largely in the Italian style; text in Italian and English, music with piano accompaniment.

Operatic Anthology. Celebrated arias selected from the works of old and modern composers. Edited by Max Spicker, New York: Schirmer, 1903; 5 vols., arranged by voices: soprano, alto, tenor, baritone, bass, pp. 285, 263, 234, 261, 254, respectively. 7¾" x 11", paper $1.50 each, cloth $2.50. The words of the songs are in their original language, and in English; arias in the "Prima Donna's Album" are not repeated here. The six volumes give an admirable summary of the best in the operatic literature of all time.

Neapolitan Songs. Edited by Eduardo Marzo, New York: Schirmer, 1905. 7¾" x 11", pp. 80 $1.00. 19 Neapolitan folk and popular songs, words in Italian and English, music with piano accompaniment.

Echoes of Naples. Edited by Mario Favilli, New York: Ditson. 1909, 9¼"x12", pp. 87, $1.25. 30 Neapolitan songs, words in Italian and English.

FOUR BOOKS.

* BRAGAGNOLO, (G.) e BETTAZZI (E.) *La vita di Giuseppe Verdi.* Milano: G. Ricordi e C. 1905. 5" x 7½". pp. 350 L 2.
Well illustrated—Given prize at national memorial ceremonies in Milan in honor of Verdi.

CHECCHI, E. *La vita di Rossini.* Firenze: G. Barbèra. 1898. 5" x 7". pp. 183. L 2.
A short and simple biography.

* MAGRINI, G. *Manuale di musica.* (Theory & Practice for Families and Schools). Milano: Ulrico Hoepli. n. d. 4½" x 6". pp. 414. L 4.

* UNTERSTEINER, ALFREDO. *Storia della musica.* Milano: Ulrico Hoepli. 1910. 4¼" x 6". pp. 423. L 4.
Admirable little summary of the history of music.

THE PHONOGRAPH.

Librarians are more and more taking up the question of combining entertainment with instruction. The phonograph offers great possibilities, especially for work with immigrants, and is comparatively inexpensive.

For $50 can be bought a moderate sized machine, sufficient to give enjoyable music in a hall seating 300 people. The most popular foreign records are double faced, 10" discs, costing 75c. Records of the more famous singers vary in price from $2 to $7. These records average 3 to 4 minutes in length, and with one winding, 3-10" or 2-12" records may be played. Such machines do not easily get out of order, and if care is taken that they are not scratched, the records are practically indestructible.

Italian lists of records in the catalogues offer a rather remarkable selection of dance and opera music, folk songs, national airs, and some speaking records in English and Italian. The opera records give in great variety, not only the characteristic work of the most famous Italian composers, but also a very generous representation of the more recent Italian school that has proved so popular. With regard to the popular songs and speaking records in Italian, a word of caution is necessary: They should be selected by some dependable adviser, preferably an Italian, as it has been found that records containing some very objectionable material are on sale.

Art, and the Sciences and Their Applications

* CHERUBINI, EUGENIO. *Storia dell'arte.* (The History of Art). Firenze: R. Bemporad e Figlio. 1909. 5" x 7½" pp. 231. Illustrated. L 2.50.

Devoted largely to Italian art. Though written for children, this book would be read with pleasure by many adults. Simple and very attractive in manner and form.

LIPPARINI, G. *Storia dell'arte.* Firenze: G. Barbèra. 1909. 5" x 7½". pp. 448. Illustrated. L 4.

This is a slightly more advanced book than Cherubini's, is profusely illustrated, and is also devoted largely to Italian art.

* SERRA, L. *Storia dell'arte italiana.* Milano: Francesco Vallardi. n. d. 5¾" x 8¾". pp. 558. Bound. 525 Illustrations. L 8.

The history of Italian art exclusively. The book selected by the national society of the Dante Alighieri as a part of its gift libraries.

* GALILEI, GALILEO. 1564-1642. *Prose.* Edited by Augusto Conti. Firenze: G. Barbèra. 1908. 5" x 7½". pp. 276. L 1.30.

The dialogues of Galileo are considered one of the great glories of Italian literature. This selection prepared for school use is highly thought of for its practical educational value, apart from its importance in a literary way.

* STRAFFORELLO, G. *La scienza ricreativa.* Torino: Fratelli Bocca. 1900. 4¾" x 8". pp. 223. L 3.

A delightful book of popular readings in science. In short chapters—never more than six or seven pages in length—such subjects are discussed as the story of the telegraph, sleep, sugar, photography, hunger, thirst, the coal supply of the world, the spectroscope, our teeth, tobacco —its use and abuse.

* MACH, DR. ERNESTO. *Letture scientifiche popolari.* Translated by A. Bongioanni Torino: Fratelli Bocca. 1900. 4¾" x 8". pp. 259. L 3.50.

Interesting book of science very popularly written. Chapters on the form of liquids, the explanation of harmony, the velocity of light, why a man has two eyes, symmetry, the conservation of energy, the part that chance has in inventions and discoveries.

POKORNY-FISCHER. *Storia illustrata del regno minerale.* Translated by G. Piolti and L. Colomba. Torino: Ermanno Loescher. 1907. 6" x 9". pp. 1,76. Illustrated. L 2.50.

This illustrated "history of the mineral kingdom" is clearly and interestingly written and is about of high school standard.

CARUEL, TEODORO. *Storia illustrata del regno vegetale.* Edited by Oreste Mattirolo. Torino: Ermanno Loescher. 1913. 6" x 9". pp. 331. L 3.25.

"Illustrated history of the vegetable kingdom"—companion volume to foregoing.

CAVANNA, GUELFO. *Zoologia.* Firenze: G. C. Sansoni. 1909. 5½" x 8". 2 vols. pp. 223, 213. Illustrated. L 3. for the two.

Excellent book of zoology, well illustrated, and rather simpler than the two preceding books.

GARBASSO, ANTONIO. *I progressi recenti della fisica.* Roma: Albrighi, Segati e C. 1911. 6" x 9". Illustrated. pp. 300. L 4.

A collection of lectures by well known Italian scientists. Not an elementary book. Treats of such subjects as electricity and matter, the electric spark, spectrum analysis, submarines, dirigibles, etc.

MACH, DR. ERNESTO. *I principii della meccanica.* Translated by D. Gambioli. Roma: Albrighi, Segati e C. 1909. 5½" x 8". pp. 547. Illustrated. L 6.

Story of the history and development of the principles of mechanics, written with the useful, the interesting, the significant always in view. Very clearly and simply written, but still requires some education in the elements of algebra and geometry.

VEROI, ING. GOMBERTO. *L'abbici dell'elettrotecnica—Libro per gli operai*. The A B C of Electro-technics. Roma: Albrighi, Segati e C. 1911. 6" x 9". pp. 291. Profusely illustrated. L 4.50.

Covers rapidly in a popular but effective way the principal applications of electricity to the uses of our civilization, from the magnet to the wireless telegraph. Included are trolleys, dynamos, lighting—from arc light to mercury vapor—and motors. The sub-title, "A book for workingmen," needs the explanation that this volume is intended for the higher technical workingmen, who in Italy in increasing numbers are going to evening continuation schools. It is not as simple a book as the manuals noted in the following list, but requires a certain knowledge of algebra and geometry to be read with advantage.

* FAUSTINI, A. *Orrori e meraviglie dell'universo*. Roma: Albrighi, Segati e C. 1912. 5½" x 8". pp. 302. Well illustrated. L 3.

"To create interest in the infinite problems of the universe, to excite curiosity and inspire to more profound study." A wonder-book of nature, popularizing science by telling dramatically its story of power and mystery. Subjects: The Heavens, the Earth, the Sea, Man.

* PIPERNO, DOTT. ARRIGO. *Salute*. Roma: Albrighi, Segati e C. 1907. 5½" x 8". pp. 221. L 2.

"Health," a school book, but like so many other Italian school books, alluringly readable. A book of literature as well as of hygiene. Covers whole life of our body, its care and perils. Wholesome, simple, not without humor.

SALVADORI, DR. ROBERTO. *Elementi di chimica*. Firenze: Successori Le Monnier. 1912. 5½" x 8". pp. 516. L 5.

A simple and fairly comprehensive book, intended for the standard of the secondary school.

* MOSSO, ANGELO. 1846-1910. *Fatica*. Milano: Fratelli Treves. 1911. 5" x 7½". pp. 351. L 3.50.

A popular book, dealing in a very practical way with the causes and effects of "Fatigue," from social and individual points of view.

A favorite book of Verdi's. Contains account of author, a distinguished scientist, and his work.

* GOURAUD, DR. F. X. *Che bisogna mangiare?* Translated by Dr. A. Cutolo, Municipal Chemist of Naples. Napoli: Società Commerciale Libraria. 1911. 5" x 7½". pp. 364. L 3.

A popular and "rational" discussion, admirably arranged,

of food values and digestive processes. Excellent chapters on pure food, canned meats, alcoholic drinks, vegetarianism, "education of the table."

GIACOMELLI, ANTONIETTA. *Il gran nemico.* Milano: Rivista "Contro L'Alcoolismo." 1912. 5" x 7". pp. 59. L .20.
The "Great Enemy" Alcohol—a booklet widely sold and distributed by the principal Italian temperance society.

* "DONNA CLARA." *Dalla cucina al salotto.* Torino: S Lattes e C. 1909. 5" x 7½". pp. 364. L 3.
A little encyclopedia for housewife and mother. A chatty and sensible book crowded with useful notes. Chapters on the kitchen and the preparation of food, with several hundred recipes. Also chapters on the care of the body and clothes, the care and ventilation of the house; social suggestions; flowers in the house, first aid, the medicine closet, family finances, and an admirable, well illustrated chapter on the baby. All written from the view point of simple science.

Other Useful Books
The Popular " Manuals "

As part of an important movement in Italy to popularize knowledge, several series of useful and simple books have been issued by different publishers. They have been of so dependable and practical a character that they have been widely used in Italy, and are often found on the shelves of Italian book-stores in the United States. But often these books "in stock" are so wholly out of date that a special note of warning is needed. The usefulness of such manuals depends in great measure upon their frequent revision. Those that are least popular are least frequently reprinted. It is therefore all the more necessary to secure the latest edition, which in every case should be ordered specifically, either directly of the publisher, or through a responsible agent.

MANUALI HOEPLI. (**The Hoepli Manuals**) : Ulrico Hoep of Milano, has been publishing for many years a series of manuals of wide educational scope, treating in a popular form the various divisions of letters, arts, the sciences and industry. Nearly 1300 of these have now been published. They are the work of thoroughly qualified specialists. They are fully illustrated and at every reprinting are carefully corrected and brought up to date. The books are all of the same size 4"x6"; are substantially bound in cloth; but vary, as will be seen, in number of pages and price. Catalogue may be had on request of the publisher. The following **are** some useful titles:

AGRICOLTORE, IL LIBRO DELL'. By A. Bruttini. pp. 446. 303 illustrations. L 3.50.
Covers nearly the whole field of farming in a very simple and practical way. In spite of differences in climate and soil, this book would be a great practical help as an introduction to farming in this country; for the greater part of the book is concerned with agriculture in Northern Italy.

ARITMETICA E GEOMETRIA DELL'OPERAIO. By E. Giorli. pp. 220. L 2.
This is a simple book of elementary mathematics especially prepared for the working man.

ASTRONOMIA. The translation of the well known little book by Lockyer, revised. pp. 255. L 1.50.

BOTANICA. Translation of Hooker's primer, carefully revised to 1910. pp. 144. L 1.50.

CHAUFFEUR. By Pedretti. pp. 902. 881 illustrations. L 6.50.
This is the text book of the principal schools for chauffeurs and of the aviation schools in Italy. It has been found very widely useful, even in this country, and a new edition has just appeared.

CHIMICA, (Chemistry). pp. 231. L 1.50.
A new book by E. Ricci, based upon Roscoe.

DISEGNO. By C. Boito. 5th edition. pp. 206. L 2.
The principles of design.

ELETTRICITÀ. By G. Marchi—in press.
This is an elementary book, dealing with the principles of electricity.

EVOLUZIONE. By C. Fenizia. pp. 389. L 3.
This manual gives succinctly the history and development of the theory of evolution, with a bibliography covering the ground in fuller detail.

FABBRO-FERRAIO. By G. Belluomini. pp. 242. 233 illustrations. L 2.50.
This is a practical handbook for the blacksmith, containing much elemental, as well as much advanced, information useful to him, including elements of mathematics, principles of measurement, tempering and working of the metal, etc.

FALEGNAME. By I. Andreani. pp. 295. 264 illustrations. L 3.
The carpenter's handbook, giving the detail of the work of his trade, from the first steps in handling wood and making the simplest joints, to the advanced work of the master mechanic.

FISICA. By O. Murani. pp. 710. 407 illustrations. L 4.
Handbook of physics—the ninth edition—carefully revised and brought up to date.

FOTOGRAFIA. By L. Sassi. pp. 205. Numerous illustrations. L 2.
The first steps in photography.

FRUTTICOLTURA By D. Tamaro. pp. 232. 113 illustrations. L 2.50.

This is the sixth edition, revised and enlarged, and gives much information that, in spite of differences in climate and soil, would still be very useful to the Italian farmer or fruit-grower in this country.

GEOGRAFIA. By 'G. Grove. Translated by G. Galletti. pp. 160. L 1.50.

GEOLOGIA. This is Geikie's book, translated, rewritten and revised by A. Stoppani and G. Mercalli. pp. 180. Many illustrations. L 1.50.

GIARDINIERE. By A. Pucci. 2 vols. L 3.50 each.

The first volume has to do with the garden and flower culture, and takes up such details as the elements of botany, planning of the garden, preparation of the soil, selection and planting of the seed, care of young plants, acclimation, the diseases of plants, animals and insects harmful to plants, the vegetable garden, selling. The second volume deals entirely with ornamental plants.

LAVORI FEMMINILI. By Teresita and Flora Oddone. pp. 543. 822 illustrations, with 48 plates. L 5.50.

This book is entirely devoted to embroidery, lace-making, crocheting and knitting, and various forms of fancy work.

MECCANICA. By R. Stawell Ball, translated by J. Benetti. 5th edition, revised. pp. 198. Numerous illustrations. L 1.50.

MICROSCOPIO. By C. Acqua. pp. 230. L 2.

An elementary handbook for the use of the microscope.

MINERALOGIA GENERALE. By L. Bombicci. 3d edition, revised. pp. 220. With numerous illustrations, and colored plates. L 1.50.

A general introductory book on mineralogy.

MURATORE, IL. By I. Andreani. pp. 290. 235 illustrations. L 3.

A practical handbook for the mason.

NAVIGAZIONE AEREA. By A. De-Maria. pp. 338. 103 illustrations. L 3.50.

A new edition of this book is now in press.

OPERAIO, MANUALE DELL'. pp. 272. L 2.

A book of useful information for the workman. By workman in this book is meant chiefly the metal worker. The book consists largely of practical suggestions with regard to the mixing of alloys and the preparation of solders, casting and working metals, together with recipes for the preparation of varnish, chapters on the transmission of power, and the building of pumps, with a variety of useful tables.

PANE, IL. By G. Ercolani. pp. 261. Numerous illustrations and tables. L 3.
This book goes in considerable detail into the grinding of the grain, the qualities of flour, the preparation of yeast and the entire operation of making bread.

PARRUCCHIERE, MANUALE DEL. By A. Liberati. pp. 219. Numerous illustrations. L 2.50.
A practical barber's book.

PASTICCIERE E CONFETTIERE MODERNO. By G. Ciocca. pp. 274. 300 illustrations, 36 colored tables. L 8.50.
The modern pastry cook and confectioner's handbook, a collection of many recipes and practical directions.

POLLICOLTURA. By G. Trevisani. pp. 224. With numerous illustrations. L 2.50.
This manual has to do not merely with the ordinary barnyard fowls, but includes also doves, pigeons and pheasants.

SARTO TAGLIATORE ITALIANO. By G. Peterlongo. pp. 232. With 47 tables. L 3.50.
A theoretical and practical manual for the cutter of men's clothes.

TELEGRAFO SENZA FILI E ONDE HERTZIANE. By O. Murani. pp. 397. With numerous cuts. L 4.50.
This is the second edition of a very popular little book on wireless telegraphy and Hertzian waves.

VETERINARIO. By C. Roux and V. Lari. pp. 306. With illustrations. L 3.50.
The veterinary's handbook, filled with useful information of many kinds and numerous recipes.

VITICOLTURA. By O. Ottavi. pp. 232. Illustrated. L 2.
This is the sixth edition, revised and enlarged, of this very popular book on vine-growing.

VALLARDI'S BIBLIOTECA POPOLARE DI COLTURA.
The publishing house of Antonio Vallardi, of Milano, has recently begun the publication of a series of popular Manuals known as the "Biblioteca Popolare di Coltura."

Most of them are practical in purpose and extremely simple in form. They are well illustrated, have a strong paper binding,—size 4¾" x 7¼"—contain from 128 to 144 pages, and are sold at the uniform price of L .60 each. Each book has been prepared by a thoroughly competent writer, and all those examined—twelve have been carefully examined—have been brightly and interestingly written. The following titles are selected from the thirty-five books that have been published.

No. 1. PALLONI DIRIGIBILI—Dirigible Balloons.

No. 3. POLLI E POLLAI—An excellent little book on chickens and chicken houses.

No. 4. LA LOCOMOTIVA—The history, development and gradual perfection of the locomotive.

No. 6. LA CERAMICA NELLA STORIA, NELL'ARTE E NELL'INDUSTRIA—Ceramics, in history, art, and industry.

No. 8. DINAMO E MOTORI—Dynamos and motors, their history, theory, construction and working.

No. 11. L'AEROPLANO—The history and theory of the aeroplane.

No. 12. CONCIMI E CONCIMAZIONI—Fertilizers.

Nos. 13 & 14. L'AUTOMOBILE—The history and theory of the automobile, its motors, etc.

No. 15. LA NAVE E LA NAVIGAZIONE.
An extremely interesting and informing little book, telling of the development of the ship, from the earliest times in its simplest forms, to the giant ocean steamers of today. Admirably illustrated.

No. 17. IL MONDO POLARE—The Polar world.

No. 18. LA CARTA—The history, manufacture and special products of paper.

No. 19. I RAGGI RÖNTGEN—The Röntgen rays and their most important applications.

No. 20. NOZIONI DI FRUTTICOLTURA—Hints on fruit growing.

No. 21. MICROBII—A very useful and interesting little book on popular hygiene: The germ theory of disease, and methods of disinfection and prevention.

No. 23. I CIELI—Popular introduction to astronomy.
One of the most successful and interesting books of the series. It opens with a little history of astronomy. There are excellent chapters on the sun, the earth, the measure of time, the problem of origins.

No. 24. GLI ALIMENTI E LE LORO FALSIFICA-
ZIONI—Foods and their adulteration.

No. 27. IL RICAMO NELLA STORIA E NELL'ARTE
—Embroidery in history and art.

No. 28. GLI ARABI NELLA STORIA E NELLA
CIVILTÀ—The Arabs in history and in civilization.

No. 29. IL CEMENTO E LE SUE APPLICAZIONI—
Cement and its use.

No. 31. IL VINO—Wine.

No. 34. LA SALUTE DELL'OPERAIO—The health of
the workman. An excellent little book, especially recommended.

No. 35. FERRO, ACCIAIO E LORO LAVORAZIONE
—Iron and steel and how they are worked.

Books of Reference

ATLAS.

MARINELLI, PROF. OLINTO. *Atlante scolastico di geografia moderna.* Milano: Antonio Vallardi. 1912. 2 parts. 14" x 12". 20, 18 maps. L 3.50 each.

A very simple, clear and practical atlas. Of special interest and use to the Italian for the importance given to Italy and Central Europe, the United States, Brazil and the Argentine.

DICTIONARIES.

PETROCCHI, P. *Dizionario universale della lingua italiana.* Milano: Fratelli Treves. 1910. 7" x 10½". 2 vols. pp. 1286, 1288. L 25 for the two.

The popular unabridged Petrocchi. By many considered the best for practical use of the many good Italian dictionaries.

Piccolo dizionario universale. Milano: Antonio Vallardi. 1906. 4½" x 6½". pp. 1159. L 4.50.

Excellent Italian dictionary in small compass, includes useful little encyclopedia of arts, sciences, biography, history, mythology, geography.

MILLHOUSE (JOHN) and BRACCIFORTI (FERDINANDO). *English-Italian and Italian-English.* New York: D. Appleton & Co. 1912. 2 vols. 5" x 8". pp. 741, 854. $5.50 for the two.

The well known and long the standard dictionary in both languages, simple in arrangement and use.

TAUCHNITZ POCKET DICTIONARY. *English-Italian and Italian-English.* By J. E. Wessely, revised and rewritten by G. Rigutini and G. Payn. New York: Lemcke & Buechner. 1912. 1 vol. Bound. 4¾" x 6½". pp. 226, 199. $.75.

Clearly printed on good paper. Proper names and irregular verbs both English and Italian. Strongly bound. For its size and cost the best little dictionary available.

EDGREN (DR. HJALMAR), BICO (GIUSEPPE) & GERIG (JOHN L.). *Italian and English Dictionary.* Italian-English and English-Italian. New York: Henry Holt & Co. 1902. 6" x 8½". 1 vol. pp. 576, 452. $3.00.

A serviceable dictionary in both languages. Regarded as in many respects—scholarly respects—an advance over all similar dictionar'es, but it is not simple, and is emphatically a book for the student.

CARENA-SERGENT-GORINI. *Nuovo vocabolario di arti e mestieri.* Milano: Francesco Pagnoni. n. d. 4¾" x 7". pp. 394. L 3.

A useful little dict'onary of arts and trades, arranged under subject headings.

Nuovo vocabolario domestico. Milano: Francesco Pagnoni. n. d. 4¾" x 7". pp. 413. L 3.

A dictionary of the household and of daily life, arranged under subject head:ngs.

ENCYCLOPEDIA.

GAROLLO, G. *Piccola enciclopedia Hoepli.* A-D. (To be completed in three volumes). Milano: Ulrico Hoepli. 1913. 4¾" x 7". pp. 1522. L 12.50 for first vol.

Very comprehensive little encyclopedia, and the only one moderate in price and recent. Articles are short, but accurate and unusually informing. The first volume is ready; the second is in press; and it is expected that the third will be delivered this year.

YEAR BOOKS.

ALMANACCO ITALIANO. Firenze: R. Bemporad e Figlio. 1914. 5" x 7½". pp. 1000. L 3.50.

Annual handbook of stat'stics and facts of the world's progress, with special reference to Italy. To be compared with our popular almanacs of reference. But abundantly illustrated; and is as well a kind of readable and popular encyclopedia. Contains interesting chapters on the Italian Government and the colonies of Italy, model workingmen's dwell:ngs in Italy, on "Europe in Figures," on astronomy, art, industry, aeroplanes, sport, hygiene.

ALMANACCO DELLO SPORT. 1914. Firenze: R. Bemporad e Figlio. 1914 5" x 7¼". pp. 350. L 1.50.

In addition to topics of sport in which an American would be interested, the Italian sportsman's almanac pays special attention to wrestl:ng, swimming, running, mountain climbing, shooting and fencing. There are also good articles on the relations between sport and literature, sport and art, and sport and hygiene.

ANNUARIO DELL'ITALIA ALL'ESTERO E DELLE SUE COLONIE. Issued by the Istituto Coloniale Italiano. Roma: Aristide Staderini. 1911. Bound in boards. 7½" x 10". pp. 759. L 5.

A. treasury of facts and statistics, regarding Italian colonies and Italians and Italian interests beyond the seas.

LA NOSTRA FLOTTA MILITARE. Torino: S. Lattes e C. 1913. 6" x 9". pp. 207. L 4. Well illustrated, with maps.

Our Italian immigrants are deeply interested in Italy's navy, and this attractive book gives abundant detail of its development and the present condition of every arm of the service, with many pictures and detail drawings of every kind of war vessel.

Periodicals and Newspapers

With the rapid industrial progress of Italy during the last ten years, journalism has been making rapid strides. The newspapers are far more widely read. They have grown larger; they are fuller of news, and more interestingly and popularly written. And in certain respects, the collection and treatment of the news, the larger and more striking headlines, the breaking up of the formal columns, newspaper standards have been set that might be called characteristically American. The signed article is a very distinctive feature of the Italian press. Many eminent authors, university professors, scientists and statesmen write for the daily papers of Italy over their own signatures. Notable in this way is the important article or interview on burning questions of the hour, given by long distance telephone.

The list selected is a very short one, and of necessity has been confined to the smallest possible number of newspapers that would represent in a summary way the most important sections of Italy and the very best of Italian journalism.

A number of considerations bear upon the choice of an Italian paper for a library. It is not always a question of the selection of the best, or one politically acceptable, or one attractive for a popular staff of writers. The choice will often be dictated by learning from what part of Italy the library's readers chiefly come. The Sicilian, or the Neapolitan, will naturally prefer to read a paper that will give him the news from his home province than one giving much space to local items from a wholly different section of Italy.

Italian journalism has made rapid progress in the United States during the last five years. More than 200 Italian papers and periodicals of various kinds are now being printed in this country in Italian. Their quality is continually improving, and they are being more and more widely read. Some of them reprint the most important articles published in the papers of Italy. Some have a cooperative arrangement with the large papers of Italy, and are able to publish simultaneously news of special interest to Italians, and cablegrams from the Argentine Republic and Brazil. On the whole, the Italian papers in the United States are of local importance, and choice among them is apt to be rather imperatively dictated by the Italian readers of each library.

With the development of the new national habit of reading, the Italian magazines and reviews have rapidly improved in quality and interest; and they have very rapidly increased in number. Their development has been along the lines of both French and American periodicals, and in the list here printed, the most popular, representative and useful have been chosen.

Periodicals

LA NUOVA ANTOLOGIA. Illustrated Monthly. Piazza di Spagna. Roma. L 46 yearly.

The monthly magazine of greatest distinction and importance published in Italy. It suggests a combination of the North American Review, the Atlantic and the World's Work. It deals with politics, science, art, poetry, literature. Many reviews. Prints fiction. Live and open-minded. Prof. Rava, the new Minister of Finance, is a distinguished member of its staff.

RASSEGNA CONTEMPORANEA. Fortnightly. C. A. Bontempelli, Corso Umberto I, 160, Roma. L 36 yearly.

Ably edited review. Scope: Literature, politics, history, economics, sociology, fiction. To be compared in certain respects to some of the English reviews. Has a marked scholarly flavor. Open to the new currents of thought.

RIVISTA POPOLARE. Fortnightly. Corso Vittorio Emanuele, 115, Napoli. L 8 yearly.

A serious and extremely well informed and interesting review of politics, letters and social science. Edited by the distinguished economist and sociologist and uncompromising republican, Prof. Napoleone Colajanni. Contains an excellent department of Review of Reviews, with reprints of important articles from the best newspapers and magazines of other nations. Pays much attention to questions of international politics and the social progress of the world.

MINERVA. Fortnightly. Unione Tipografico-Editrice Torinese, via Cicerone, 56, Roma. L 14 yearly.

The Italian Review of Reviews. Not merely in plan, but actually, international. Gives more space to foreign publications than it does to those of Italy. Follows closely all important questions and has the art of making clippings interesting.

L'ILLUSTRAZIONE ITALIANA. Illustrated Weekly. Fratelli Treves. Milano. L 48 yearly.

Gives a graphic weekly story of the world's progress, paying special attention to the life and important events of

Italy. It gives considerable space to literary, scientific, artistic matters, and important discoveries. Beautifully illustrated. The most popular of all Italian periodicals. In the hotels, restaurants and "caffès" of Italy you always have to wait until somebody else finishes with the "Illustrazione."

LA LETTURA. Illustrated Monthly. via Solferino, 8, Milano. L 8 yearly.

A cheap monthly magazine with illustrated cover. Contains many short stories and continued novels. also short and interesting articles: travel, politics, biographical sketches, art, etc. Very popular, and clean.

LA VITA ITALIANA ALL'ESTERO. Monthly. via Due Macelli, 9, Roma. L 12 yearly.

Devoted to the interests of Italy and Italians abroad. Emigration is the first of these interests, but foreign politics and the colonies receive much attention. Well written, with distinguished list of contributors.

EMPORIUM. Illustrated Monthly. Istituto Italiano d'Arti Grafiche, Bergamo. L 13 yearly.

Monthly review of art, also including in a minor way literature and science. Well and profusely illustrated. Though devoted chiefly to Italian art, it gives much space to the art of other countries. Well informed.

LO SPORT ILLUSTRATO. Illustrated Fortnightly. Corso di Porta Nuova, 19, Milano. L 15 yearly.

Well printed and well illustrated record of every sporting interest. Water sports, wrestling, motoring, the aeroplane, running and jumping, tennis fill most of the space. A new recent interest, now exciting wider and wider attention, is foot-ball.

Newspapers

IL CORRIERE DELLA SERA. Daily. via Solferino, 28, Milano. L 36 yearly. With "La Lettura," (See "Periodicals.") L 43. yearly. With La Lettura & La Domenica del Corriere (a colored Sunday supplement—*popular*). L 50 yearly.

Commonly considered the best newspaper of Italy and with high rank among the newspapers of the world. Honest, clean,—"a family paper"—thoroughly abreast of the world's progress. Has excellent foreign news service. Printed in Milan, but read throughout Italy and abroad on account of its quality. Articles by prominent statesmen frequently appear in its columns. Minister of State Luzzatti, Minister of Education Daneo and Minister of Agriculture Cavasola have written for it for years. The vivid articles —see "Travel"—on Mexico by Luigi Barzini, one of its travelling correspondents, famous throughout Italy, were reprinted day by day in the London Telegraph and the New York World.

LA STAMPA. Daily. via Davide Bertolotti, 3, Torino. L 35.50 yearly.

Stands high even compared with the best papers of the continent. It also has excellent foreign news service, and an able staff of writers. Giuseppe Bevione—see "Travel" —has contributed several important series of articles to its columns, which in matters of foreign policy affecting the Argentine Republic, Tripoli and Asia Minor have powerfully influenced public opinion, and government action. The "Stampa" is often compared to the "Corriere." Though serious in character, it is, however, more brightly written.

LA TRIBUNA. Daily. via Milano, 67, Roma. L 34 yearly.

The official organ of the successive governments of Italy. An admirably written, progressive, well informed paper, which has maintained its high character unchanged for many years. Like nearly all continental papers, and many of our own, it runs a continued novel, and prints many clever sketches. It has an able staff of writers, among them "Rastignac," the pen name of Vincenzo Morello, one of the best known editorial writers of Italy.

IL GIORNALE D'ITALIA. Daily. Palazzo Sciarra, Roma. L 34 yearly.

Ably represents one of the wings of the liberal party. Honest, clean, well written. Prime Minister Salandra and Marquis San Giuliano, now Minister of Foreign Affairs, have been for a number of years two of its principal writers. So, too, Diego Angeli, well known in Italy as novelist, art critic, satirist and Translator of Shakespeare. A number of well known Nationalist writers, such as Federzoni, Bellonci and Maraviglia contribute frequently to its columns.

IL MATTINO. Daily. Galleria Umberto I, Napoli. L 33 yearly.

The best paper of Naples, ably edited by the well known journalist Enrico Scarfoglio. The Mattino has made rapid progress during the last few years, enlarging its scope and gaining in popularity and influence.

IL GIORNALE DI SICILIA. Daily. Piazza Stazione Centrale, Palermo. Sicily. L 55. yearly.

Commonly considered the best of the papers of Sicily. Like the other important papers of Italy, it devotes much space to foreign news. But it gives more attention than the others mentioned to local matters, specializing on Sicilian interests.

Library Notices, Rules and Friendly Helps in Italian

The greatest possible care should be taken to see that all notices in Italian are correct in every detail of wording and spelling.

The following notice in Italian, now revised, with its familiar appealing "thou's", has been pasted on the covers of all Italian books in the Library at Mount Vernon, N. Y. It has proved very helpful in inducing readers to take better care of the books:

Amico Lettore!
Tratta questo libro come tratteresti un caro amico.
Non spiegazzarlo; *non sporcarlo;* non stracciarlo; non segnarlo con la matita o con la penna; e non portare le dita alla bocca per voltarne le pagine.
Pensa che esso deve anche servire ad altri tuoi compatrioti.
Se lo stracci o *lo sporchi,* dai un cattivo esempio, e impedisci che altri italiani ne traggano vantaggio.
Rispetta questo libro per il buon nome e per il vantaggio degli italiani!

Friend Reader!
Treat this book as thou wouldst a dear friend.
Do not rumple it; *do not soil it;* do not tear it; do not mark it with a pencil or with a pen; do not moisten your fingers to turn its pages.
Think that it must also serve others who are thy compatriots.
To tear it or *to soil it* would set a bad example, and prevent other Italians getting benefit from it.
Respect this book for the good name and for the advantage of Italians.

For those who abuse books the plan has been formed, also at Mount Vernon, of showing a copy of a new book that had been borrowed only once and then returned in so bad a condition that it could not again be sent out, and comparing this with a copy of Dante that was printed in Venice in 1529, whose pages are as clean, and in many cases almost as white, as when it left the press nearly 400 years ago.

A SELECTION OF NOTICES AND RULES, REVISED, THAT ARE BEING USED IN THE PUBLIC LIBRARIES OF NEW YORK CITY, PROVIDENCE AND SPRINGFIELD, MASS.

USO DELLA BIBLIOTECA.

La Biblioteca è gratuita per tutti.

I libri sono nella Biblioteca perchè voi possiate leggerli o studiarli o portarli a casa, se avete la tessera.

CHI PUÒ PORTARE LIBRI A CASA?

Voi lo potete. I vostri bambini lo possono.
Chiunque in città lo può.

OR

Il lettore deve far firmare la carta da una persona che lo conosca, e il cui nome sia registrato nel Directory.

OR

Qualsiasi persona che risieda nella città, dando opportune referenze, può ottenere il permesso di portare libri a casa.

COME I LIBRI SI POSSONO PRENDERE.

La tessera è data *gratis*. La prima volta che voi venite, se non potete parlare inglese, portate con voi un fanciullo o qualche persona che vi possa far da interprete. E' necessario dare le seguenti indicazioni: Nome e cognome, residenza, occupazione, luogo dove si ha la occupazione.

Noi vi faremo vedere dove sono i libri italiani.
Quando avete trovato il libro che desiderate, date la

THE USE OF THE LIBRARY.

The Library is free for all.

The books are in the Library in order that you may read them, or study them, or take them home, if you have the card.

WHO MAY TAKE BOOKS HOME?

You may.
Your children may.
Anyone in the city may.

OR

The reader must have the required form signed by some one who knows him, and whose name is in the Directory.

OR

Anyone who lives in the city, giving the necessary references, may obtain the loan of books to take home.

HOW BOOKS MAY BE TAKEN OUT.

The card is given without charge. The first time you come, if you cannot speak English, bring with you a boy, or some person who may act as interpreter for you. It is necessary to give the following information: Your full name, residence, occupation, and the address of the place where you work.

We will show you where the Italian books are.
When you have found the book that you wish, give

vostra tessera e il libro che avete scelto alla bibliotecaria. Ella ne prenderà nota e vi consegnerà il libro da portare a casa.

Se i libri desiderati non si trovano in Biblioteca, si farà di tutto per ottenerli.

Il libro che sia fuori quando voi lo domandate, vi verrà riserbato dalla bibliotecaria, se voi gliene fate richiesta.

La sezione pei ragazzi e ragazze è ben provvista di libri istruttivi e dilettevoli, e ha una sala dove si possono comodamente studiare le lezioni di scuola.

Perchè non mandate i vostri figlioli alla Biblioteca, dove possono ottenere aiuto per il loro lavoro di scuola?

VOLETE I M P A R A R E L'INGLESE?

FREQUENTATE LA SCUOLA G R A T U I T A ALLA B I B L I O T E C A PUBBLICA.

LE LEZIONI COMINCIANO ....

ALLE ORE ....P. M.

PER ALTRE INFORMAZIONI D O M A N D A R E ALLA BIBLIOTECARIA.

IL REGOLAMENTO.

Gli adulti possono prendere non più di quattro (due) volumi alla volta—di cui uno soltanto romanzo e una rivista; i ragazzi non più di due volumi alla volta. I libri devono essere restituiti dopo non più di due settimane—e alcuni dopo non più di una settimana sola.

Le riviste del mese in corso e del mese precedente non

your card and the book that you have chosen to the Librarian. She will make a note of it, and will give you the book to take home.

If the books that you wished are not in the Library, every effort will be made to secure them for you. When a book that you have asked for is out, it will be reserved for you by the Librarian, if you ask her to do so.

The Department for boys and girls is well provided with instructive and entertaining books, and it has a room where they may very comfortably study their lessons for school.

Why do you not send your children to the Library, where they may obtain help for their school work?

DO YOU WISH TO LEARN ENGLISH?

COME TO THE FREE SCHOOL AT THE PUBLIC LIBRARY.

THE LESSONS COMMENCE ....

AT .... P. M.

FOR OTHER INFORMATION ASK THE LIBRARIAN.,

RULES.

Adults may take out no more than four (two) books at one time — of which one alone may be a novel and one a magazine; children not more than two books at a time. The books must be returned within two weeks—and some of them within one week.

Magazines of the current month and of the preceding

possono ottenersi in prestito per più di tre giorni.
La multa di un soldo è imposta per ogni giorno di ritardo nel restituire i libri, e la Biblioteca ha il diritto di mandare a ritirarli a spese del richiedente, che non potrà ottenere altri libri, finchè non avrà pagato tutte le spese.

Ordinariamente il prestito dei libri potrà essere rinnovato per altre due settimane, facendone domanda.
La Biblioteca sta aperta, per il prestito e la restituzione dei libri, dalle 9 a. m. alle 9. p. m. di ogni giorno, eccetto la domenica.
I richiedenti che trovano nei libri segni di matita, pagine lacere o mancanti, debbono farlo notare alla bibliotecaria.

month may not be borrowed for more than three days. A fine of one cent is imposed for each day of delay in returning the books, and the Library has the right to send to get them at the expense of the one who has borrowed them, and he will not be able to obtain other books, until he has paid all the expenses.
Ordinarily the loan of books may be renewed for an additional two weeks, by making request.
The Library is open for the loan and return of books from 9 A. M. until 9. P. M. every day, except Sunday.

Those who take out books and find in them pencil marks, pages torn or missing, should call them to the attention of the Librarian.

FRIENDLY HELPS IN ITALIAN.

With Apologies to the Berkshire Athenaeum, Pittsfield, Mass.

Buon giorno. *Good morning; good day*—(Greeting used until late afternoon).
Buona sera. *Good evening*—(Greeting used late afternoon and evening).
A rivederci. *Good-bye*—until we meet again.
Ci farete sempre piacere. *It will always give us pleasure to see you.*
Parlate inglese? Leggete l'inglese? *Do you speak English? Do you read English?*
Leggete l'italiano? Scrivete l'italiano? *Do you read Italian? Do you write Italian?*
Come vi chiamate? *What is your name?*
Leggete questa carta e firmate il vostro nome qui.
Read this paper and sign your name here.
Scrivete anche il vostro indirizzo e il nome della persona presso cui lavorate.
Write also your address, and the name of the man you work for.
Voi dovete darci il nome di una persona che noi conosciamo.

che sia disposta a firmare questa carta ed essere risponsabile per voi.
You must give us the name of some one that we know, who is willing to sign this paper and be responsible for you.
Troverete il regolamento di questa Biblioteca in questa tessera.
You will find the rules of the Library on this card.
I libri italiani sono qui, tutti insieme.
Our Italian books are all here together.
Venite qui e scegliete quel che vi piace.
Come here and choose what you wish.
Potete prendere un romanzo e un altro libro in una sola volta
You may take a novel and one other book at a time.
Vostra moglie e i vostri figli possono prendere libri se vogliono.
Your wife and children may take books if they wish.
Ma bisogna che anche loro abbiano delle tessere.
But they too must have cards.

Biblioteca della RIVISTA POPOLARE
XXX.

SHERIDAN - BERNARDY - MEADE - COLAJANNI

Gl' Italiani negli Stati Uniti

Lavoro, salari, risparmi;

agglomeramento nelle città;

vita rurale;

laboriosità, delinquenza ecc.

ROMA NAPOLI
Presso la RIVISTA POPOLARE
1909

PREFAZIONE

Il pregiudizio della razza, *quantunque ridotto al nulla dalla storia e dalla statistica, dalla scienza in genere, impera sovrano tra le moltitudini anglo-sassoni; tra le pretese* razze *superiori. Questo pregiudizio ha fatto condannare gli italiani, specie del Sud, nel Nord-America come elementi* undesirables, *come elementi non assimilabili dalla civiltà nella Repubblica . delle stelle. Contro questo pregiudizio rimane, io credo, inconfutato il mio libro* Latini e anglo-sassoni. *Una delle maggiori riviste Nord-Americane,* The Nation, *non potendo confutare una sola linea di quel libro, pur occupandosene benevolmente, lo annunziò con un titolo offensivo:* Il **verme** latino si agita.

La distruzione di questo pregiudizio non è semplicemente una questione di dignità nazionale; non ha soltanto un interesse teorico ed umanitario ; ma ha una importanza pratica crescente. Circa un milione e mezzo di italiani vivono , o permanentemente o temporaneamente, negli Stati Uniti e vi sono mal visti, disprezzati, maltrattati.

Tale numero nell'avvenire sembra destinato ad accrescersi anzichè a diminuire. Ora è utile dal punto di vista morale, politico ed economico che i nostri concittadini vi siano meglio conosciuti per esservi meglio trattati.

Per riuscire meglio nello scopo, però, si deve cominciare dal dare agli stessi italiani la coscienza di quello che sono realmente, senza esagerare le tinte da una parte, senza attenuarle dall'altra.

La percezione esatta dalle verità soltanto può indurre alla correzione dei difetti e alla valutazione giusta dei pregi.

Se a questa impresa mi fossi accinto da solo, le mie parole, le mie difese, avrebbero potuto sembrare suggerite da male inteso amor di patria, da chauvinisme. Perciò la vera documentazione l'ho fatta colla riproduzione, di poco abbreviata, di alcune pubblicazioni di scrittori nord-americani, che sottraendosi all' influenza dell'ambiente, hanno giudicato gli italiani, che vivono nella Repubblica delle Stelle colla più scrupolosa imparzialità.

Sono numerosi gli articoli delle riviste Nord-Americane, e gli scrittori che mettono le cose a posto e che agli italiani danno ciò che è dovuto; alcuni occupano dei posti eminenti nel mondo ufficiale: tale ad esempio l'Austin, direttore del Board of trade, che nella North American Review (aprile 1904: Is the new immigration dangerous to the Country?) esaminò in base ai documenti ufficiali se la nuova immigrazione — di cui è magna pars, quella italiana — sia pericolosa per la repubblica e rispose negativamente.

Non riproduco tale eccellente articolo perchè si riferisce ai dati del Censimento del 1890; ed ora si posseggono quelli più recenti e più completi del 1900; ma ho creduto opportuno riprodurre qui: 1.° un lavoro esau.

riente di Frank J. Sheridan, che espone esattamente le condizioni dei lavoratori italiani negli Stati Uniti; 2.° un discorso della signorina Amy A. Bernardy pronunziato a New York, nel quale sinteticamente e lucidamente sono spiegate le ragioni per le quali gli italiani si agglomerano nelle città americane; 3.° una monografia di Emily Fogg Meade che dimostra come gli italiani si possano addire alla vita agricola ed americanizzarsi; 4.° infine, quasi, come riepilogo, un paragrafo del mio Manuale di Demografia (2ª Ed.), nel quale, in base ai documenti ed ai giudizii dei Nord-Americani, esamino una ad una le accuse che oltre l'Atlantico si formulano contro gli undesirables e principalmente contro gl'italiani.

Raggiungerò lo scopo, che mi propongo ? Non lo so; ne dubito anche, perchè se gl' italiani, che vivono in patria leggono poco, leggono ancora meno quelli che valicano l'Oceano. Comunque, io ho fatto il dover mio tentando di dare agli Italiani, che vivono negli Stati Uniti la coscienza di quello che valgono e spronandoli a correggere i difetti più o meno gravi, che si prestano all'opera di denigrazione, che si ripercuote sulla madre patria.

DOTT. NAPOLEONE COLAJANNI

Napoli, Agosto 1909.

Frank J. Sheridan

SALARI, ORE DI LAVORO, CONSUMI E RISPARMI
degli Italiani negli Stati Uniti

Salari, ore di lavoro, consumi e risparmi degli Italiani negli Stati Uniti [1]

Gl' immigrati squalificati (unskilled) negli Stati Uniti. — In questo studio ci occupiamo dei lavotori *unskilled* immigrati di razza italiana, slava ed ungherese per mostrare come essi si siano adattati alla vita industriale degli Stati Uniti. Quantunque il titolo si riferisca agli immigrati di tali razze, seguendo la classificazione del *Bureau of Immigration and naturalization,* pure in tali razze sono compresi altri immigrati che sono della stessa condizione, *unskilled*, ma non delle razze nominate; così, ad es. tra gli Slavi sono compresi gli Ebrei dell' Europa orientale, i Rumeni ecc. che non sono di razza Slava.

Stranieri che si trovavano negli Stati Uniti sino a Giugno 1906

	Totale degli stranieri	Italiani, Slavi e Ungheresi	Percentuale sul totale degli stranieri
Sino al 1º Giugno 1900	10,356,644	1,885,896	18,21 %
Arrivati da Giugno 1900 a 30 Giugno 1906	4,933,741	3,176,291	64.37 »

(1) Frank I. Sheridan : *Italian, Slavic and Hungarian unskilled immigrantlaborers in the United States.* (Dal *Bulletin of the bureau of Labor* di Washington n. 72). Di questo lavoro accuratissimo presento tradotto i punti principali, che riguardano il lavoro, i guadagni e la vita degli Italiani.

N. Colajanni

Professione degli immigranti. — Nell'anno fiscale terminato a 30 giugno 1906 gl'immigranti secondo la loro condizione si dividevano così:

	Italiani	Slavi e Ungheresi	Italiani Slavi e Ungheresi	Di ogni nazionalità
Lavoratori *unskilled*	184,832	247,320	432,152	598,731
Senza occupazione (comprendente donne e fanciulli)	59,729	129,320	189,049	285,490
Lavoratori *skilled*	37,561	69,695	107,256	177,122
Professionisti e miscellanea	4,692	6,829	11,521	39,422
Totale	286,814	453,164	739,978	1,100,735

	Percentuale ital. nel totale	Percent. Slavi e Ungheresi	Percent. Italiani Slavi ecc. nel totale generale
Lavoratori *unskilled*	30,87 %	41,31 %	72,18 %
Senza occupazione (comprendente donne e fanciulli)	20,92 »	45,30 »	66,22 »
Lavoratori *skilled*	21,21 »	39,35 »	60,56 »
Professionisti e miscellanea	11,90 »	17,32 »	29,22 »
Totale	26,06 »	41,17 »	67,23 »

Dei 432,152 immigrati *unskilled* Italiani, Slavi e Ungheresi 268,281 o 48,2 % prima di andare in America erano agricoltori.

Nella immigrazione dal giugno 1900 a giugno 1906 gli agricoltori (*Farmers* e contadini: giornalieri *hand farmers*) furono 559,598; di cui *farmers* 20,891 e contadini 538,707.

La quota degli immigranti agricoltori si comprenderà mettendola in rapporto col numero totale degli agricoltori delle rispettive nazionalità alla data dell'ultimo censimento.

In Italia nel 1901 gli agricoltori erano 9,611,001 (1); e degli 82,115 immigranti agricoltori italiani nel 1906 il 90,2 % erano del mezzogiorno e addetti alla cerealicoltura in grande maggioranza; nel 1904 avevano prodotto il 70 % del vino e il 99 % dell'olio di oliva.

In Austria (1900) 13,709,204 persone o il 52,4 % del totale erano addetti all'agricoltura; in Ungheria (1900) 13,175,083 o il 68,4 % pure dediti all'agricoltura compresa la pastorizia ecc.

I salari di questi lavoratori nella loro patria sono bassissimi. Essi emigrano: 1º per estremo bisogno; 2º per isfuggire agli obblighi di leva e ad altri obblighi; 3º per divenire indipendenti.

Al loro arrivo negli Stati Uniti questi *unskilled* raramente si danno ai lavori agricoli. Gl'Italiani preferiscono darsi alle costruzioni ferroviarie, alle escavazioni, ai lavori delle officine industriali ecc.; gli Slavi e gli Ungheresi si danno ai lavori, nei quali c'è bisogno di forza fisica e vi sono più alti salari.

Si avverta che gli agricoltori italiani in patria non amano la terra, vivono in grossi centri e percorrono molti chilometri giornalmente per andare sul luogo del lavoro.

♣

Le crisi e la loro influenza sull' emigrazione. — Sulla immigrazione esercitano una grande influenza le crisi degli Stati Uniti. Questa influenza è chiarissima sulla immigrazione polacca, come risulta da queste cifre:

---

(1) Il Sheridan erroneamente afferma che in grandissima parte gli agricoltori erano nel Sud.

N. Colajanni

Immigrati polacchi

1891 :	27,497.	1896 :	691.
1892 :	40,536.	1897 :	4,165.
1893 :	16,374.	1898 :	4,726.
1894 :	1,941.	1905 :	47,224.
1905 :	790.		

L'anno 1892 fu anno di grande prosperità negli Stati Uniti ; perciò fu seguito da una considerevole immigrazione ; col 1893 cominciò la crisi e dopo quell'anno l'immigrazione decrebbe. Riprese col 1905, anno di prosperità (1)

✤

Come sono distribuiti gli immigranti unskilled. — Nel 1906 (e negli altri anni la distribuzione differì poco) gl'immigranti prevalentemente si portarono in sette Stati dell'Unione nelle proporzioni seguenti

	Immigranti di ogni nazionalità	Italiani	Slavi e Ungheresi
Massachussetts	73,863	18,089	20,107
Connecticut	27,942	10,144	11,539
New Iersey	58,415	16,199	27,368
New York	374,708	130,103	147,790
Pennsilvania	198,681	54,405	110,717
Ohio	47,397	6,718	28,639
Illinois	86,539	14,102	44,554
Totale	867,545	249,760	390,714

Perciò in questi sette Stati andò l'87,08 % degli Italiani; l'86,22 % degli Slavi ed Ungheresi; il 62,94 %° delle altre nazionalità ; il 78,82 % della immigra - zione totale.

Questa distribuzione fu razionale : corrisponde alla ricerca di lavoro, ai salari ed alla prosperità degli Stati che ricevettero gl'immigranti. Infatti nel 1900 il 37 % della popolazione totale, che ap-

(1) La crisi del 1907 arrestò di nuovo l'immigrazione nel 1908

N. Colajanni

parteneva a questi sette Stati produsse il 61 % di tutta la produzione manifatturiera e mineraria im-- piegandovi 3,529,168 individui o il 59,85 % dei salariati dell'intera Unione impiegati in tale produzione. La produzione agricola di tali sette Stati fu di dollari 1,170,114,388 o il 25 % della produzione totale. Nessuna meraviglia, adunque, se nel 1900 essi attrassero il 73,2 % degli Immigranti Italiani, Slavi e Ungheresi e ne attrassero l'86,55 nel 1906.

Nel 1900 nel totale di 313,484 immigranti Italiani, Slavi e Ungheresi in Pensilvania: 63,586 vivevano in Filadelfia e 26,893 in Pittsburg; il resto, cioè, il 71 % era distribuito nelle piccole città e nei villaggi. In Filadelfia c'erano 28,951 Russi, in maggior parte Ebrei.

Nelle miniere di antracite della Pensilvania nel 1905 ci furono 3975 Italiani e 36,046 Slavi o Ungheresi; rispettivamente al totale degli operai che vi erano impiegati i primi erano il 4,3 % e gli altri il 39,0 %. Negli altri Stati gl'immigranti erano occupati in varie industrie.

♣

I salari degli unskilled italiani, slavi e ungheresi.— I salari degli operai *unskilled* variano tra le tre nazionalità a seconda che lavorano negli Stati del Nord o del Sud.

Il salario medio di 36,176 lavoratori *unskilled* chiamati da New-York tanto al Nord, quanto al Sud fu il seguente: per 16,328 Italiani dollari 1,46 al giorno; per 3,099, Slavi e Ungheresi dollari 1,46; per 16,749 delle altre nazionalità dollari 1,41.

Però se la media del salario negli Stati del Nord e del Sud risulta uguale, ci sono differenze sulle proporzioni dei salari bassi ed alti. Per una giornata di 10 ore gl'*Italiani* ebbero diversi salari con queste percentuali:

Negli Stati del Nord : salari di Doll   1,25   il   5,93 %

           »   »   1,36   »   29,87   »

           »   »   1,50   »   56,53   »

           »   »   1,61   »   4,71   »

           »   »   1,76   »   2,96   »

           Media   1,46   100

Negli Stati del Sud : Salari di Doll. 1,22   il   10,40 %

           »   »   1,37   »   11,09   »

           »   »   1,50   »   74,70   »

           »   »   1,60   »   1,63   »

           »   »   1,75   »   2,18   »

           Media   1,46   100

Corrisponde la ordinata massima pel salario che si avvicina di più a quello medio di doll. 1,50 al Nord e al Sud; nel Nord, però, sono meno numerosi i salari minimi e molto più numerosi nel Sud-America; il contrario pei salari alti.

La condizione degli Italiani rispetto agli altri risulta da questi confronti :

Salari degli Slavi ed Ungheresi negli Stati del Nord

per una giornata di 10 ore Doll. 1,25   il   4,77 %

           »   1,37   »   14,69   »

           »   1,50   »   55,46   »

           »   1,63   »   5,54   »

           »   1,78   »   19,54   »

           Media   1 46   »   100

Negli Stati del Sud : Salari di Doll. 1,21   il   19,40 %

           »   »   1,34   »   32,35   »

           »   »   1,50   »   30,63   »

           »   »   1,60   »   14,56   »

           »   »   1,79   »   3,06   »

           Media   1,41   100

**Salari degli " unskilled „ di altre nazionalità
per la giornata di 10 ore**

Negli Stati del Nord: Salari di Doll. 1,25 il 5,23 %

»	»	1,37	»	34,19	»
»	»	1,50	»	37,28	»
»	»	1,62	»	12,45	»
»	»	1,85	»	10,45	»

Media   1,50   100

Negli Stati del Sud: Salari di Doll. 1,24 il 46,87 %

»	»	1,36	»	5,63	»
»	»	1,50	»	47,01	»
»	»	1,78	»	0,49	»

Media   1,37   100

Come si vede le altre nazionalità hanno più alti salari nel Nord,. assai più bassi nel Sud; perciò la media generale risulta inferiore a quella degli Italiani, degli Slavi e degli Ungheresi.

Sono interessanti le notizie sui salari degli operai *skilled* (qualificati) e variano nel Nord e nel Sud. Per 439 Italiani negli Stati del Nord colla giornata di lavoro di 8 a 10 ore ci sono salari per ora: *minimi* di 10 cent. di dollaro per gli aiutanti carpentieri; *massimi* di 50 cent. pei tagliatori di legna, fabbricanti di tegole, di mattoni, muratori; di 52 $^{1}/_{2}$ per gli intonacotori (*plasterer*). Nel Sud la giornata di lavoro per gli *skilled* italiani è di 9 ore di solito; il salario non discende a 10 centesimi per ora; ma non sale nemmeno a 50; oscilla tra 20 a 33 cent. per ora. Non risultano superiori i salari degli *skilled* delle altre nazionalità. I salari vanno, adunque, da un *minimum* di L. 5 al giorno ad un *maximum* di L. 27.

Nelle escavazioni per i fondamenti di costruzioni, nei tunnels, ecc. in New-York nel 1906 per una giornata di lavoro da 9 a 10 ore gli Italiani ricevevano da un minimo di dollari 1,75 ad un massimo di 3 dollari; gl' Irlandesi per una giornata di 8 ore negli stessi lavori ricevevano da 3 a 5 dollari. Nelle miniere di carbon fossile della Pensilvania nel 1905 i salari per Italiani, Slavi e Ungheresi variavano da doll. 2,36, a dollari 2,57 per giorno. I salari crebbero dal 1903 in seguito alla applicazione della scala mobile: *sliding scale*.

❧

Ciò che spendono gli Italiani. Il loro tenore di vita. — La vita degli Italiani è stata studiata in vari gruppi. Un gruppo di 135 uomini, quasi tutti Italiani, e tra i quali 16 erano meccanici, in un cantiere di lavoro indipendente, si forniva di viveri dal Commissario e viveva in una barracca (*shanty*) di legno.

Molti lavoratori stanno all'impiedi mentre mangiano, quantunque non manchino rozze tavole e sedie. Gli articoli di vestiario non sono forniti dal Commissariato, ma vengono comprati nei vicini villaggi. Le qualità e varietà dei cibi dei meccanici sono superiori a quelle degli operai braccianti. I braccianti di ordinario si astengono dal comprare vini, liquori (eccettuata la birra), sigari dal Commissario; in maggioranza consumano un mezzo litro di birra al giorno. La minima spesa per cibi e tabacco fu di L. 30 al mese; la più alta di L. 55; la media di L. 45. Venti braccianti, i più siciliani, spendono L. 40 al mese. I meccanici spendono in più da L. 25 a L. 50 al mese.

I prodotti di maggior consumo comprati dal Commissariato costano:

Pane	*libbra* L. 0,10	Pomodoro (conserva		
Maccher. indig.	» » 0,30	mezza scatola)	L.	0,40
» importati »	» 0,50	Patate	*libbra* »	0,80
Condimento(me-		Cipolle	» »	0,25
dia di varie		Caffè abbrustolito »	»	1,00
qualità)	» » 1,20	The	» »	1,80

Formaggio indig. *libbra* L.	1,60	Zucchero	*libbra* L.	0,30
» romano » »	1,80	Birra (bottiglia)	»	0,25
Carne di bue » »	0,75	Vino di California		
Sardine (scatola) »	0,25	(quarto di bottiglia) »		1,25
Salmone » »	0,55	Vino italiano »	»	2,00
Lardo » »	0,80	Tabacco (pacchetto)	»	0,25
Latte condensato scat. »	0,50	Sigarette »	»	0,25

In un giorno da 38 lavoratori furono consumati trentatrè pani di 1 libbra e mezza per uno, 21 libbre di maccheroni indigeni, 7 lib. e mezza di condimento uso Bologna, 5 lib. di formaggio indigeno, 11 scatole di sardine, 4 casse di legumi in conserva, 13 casse di pomidoro in conserva, 17 libbre di patate, 12 di cipolle, 3/4 di lardo, 7 e 1/4 di *fatback* (specie di cotenna lardacea), 1/4 di caffè, 5 scatole di carne in conserva, 27 bottiglie di birra, 13 bicchieri di vino di California, 6 bottiglie di Soda water, 3 porzioni di olive di 25 centesimi ciascuna, 10 pacchetti di tabacco.

La spesa per alimentazione in questi cantieri indipendenti è migliore che in quelli a *padrone system* delle costruzioni ferroviarie per la maggiore varietà dei cibi e per l'uso della birra e del vino.

Vicino a Pittsburg (1906) c'è una compagnia impegnata in una grande costruzione, che impiega 1200 operai, dei quali 600 italiani. Uno dei socii della Compagnia è italiano. Un Commissario vi è stabilito per la vendita dei cibi per gl'Italiani. Vi sono stati costruiti alloggi in ferro e in legno e vi si paga da L. 22,50 a L. 25 per settimana per alloggio e pensione. Vi è stata istituita per gl'Italiani una scuola per imparare a leggere e scrivere inglese. I salari variano da L. 7,50 a L. 8,75 per giornata di dieci ore; la spesa di alimentazione per gl'Italiani è di L. 42,80 a L. 50 al mese.

I pasti degli italiani si compongono così : *colazione*: pane e caffè, *desinare a mezzogiorno*: pane, formaggio o sardine o condimenti, the o caffè; *cena* : maccheroni (cotti separatamente), piccola fetta di carne fresca, patate, cavoli, o legumi ed una bottiglia di birra.

I lavoratori Italiani sono stati incoraggiati a consumare più della loro abitudine, perchè negli Stati Uniti nel lavoro si richiede un maggiore sforzo fisico, che in Italia; dove nell'Agricoltura sono bassi i salari e pel clima mite il lavoro non esige un grande sforzo. Il consumo di cibi vi è deficiente e scarsissimo l'uso della carne. Per quanto oggi sia migliorata l'alimentazione dei lavoratosi Italiani, gl'intraprenditori ritengono ch'essa sia deficiente negli Stati Uniti. Questo è il parere unanime.

Pure negli Stati Uniti gl'Italiani consumano una quantità e varietà di cibi maggiore che in Italia, benchè i loro consumi siano sempre inferiori a quelli degli altri immigranti europei e dei nord-americani.

Un esempio della loro deficiente alimentazione è il seguente. A New York City nell'està del 1905 una intrapresa impiegò 500 Italiani, la maggior parte di Sicilia. Il salario era di L. 8,75 al giorno. Più di 400 tra loro non spesero al di là di L. 20 per mese per cibi. Essi mangiavano solo pane e cipolle e 2 o 3 scatole di sardine per settimana. Non usavano maccheroni, formaggio, condimento o altri cibi. Il lavoro era duro; molti ammalarono e divennero inabili al lavoro.

Nelle costruzioni ferroviarie e in altre intraprese si è osservato che gl'Italiani si mantengono sempre isolati e bisogna costruire speciali barracche per loro e il Commissariato deve provvedersi dei cibi da loro a preferenza consumati; perciò si sottraggono a tutte le influenze americanizzanti e non imparano nè la lingua, nè i costumi degli anglo-americani. I padroni li incoraggiano in tali metodi e costumi per potere continuare a sfruttarli meglio.

Gl'Italiani preferiscono in generale i prodotti importati dall'Italia, anche pagandoli più cari. Ma quando non possono trovarli si abituano facilmente ai prodotti americani e ci si trovano bene. L'abbandono del vino e la sostituzione della birra è la migliore illustrazione di ciò.

La spesa media degli Italiani per cibi in molti cantieri ferroviari e in altre intraprese nel Nord e nel Sud è la seguente:

Venticinque pani L. 10; trenta libbre di maccheroni L. 10,50; condimento, sardine e formaggio L. 8,75; lardo L. 1,50; in tutto L. 29,50. Aggiungendovi L. 15 per birra, sigari o tabacco, L. 5 per alloggio la spesa totale della vita risulta di L. 49,50 al mese.

In tali cantieri per ogni lavoratore di diverse nazionalità la spesa per alimentazione e alloggio era di L. 90 al mese; per gl'Italiani di L. 34,50; differenza L. 55,50.

La spesa degli Italiani varia di poco in più o in meno nei diversi cantieri dei vari Stati dell'Unione.

I pasti degli operai delle altre nazionalità contengono cibi più vari e più copiosi. Essi preparano i pasti in comune e per loro le compagnie preparano carri e barracche più comode; spendono di più per l'alloggio.

◆

Ciò che guadagnano, consumano e risparmiano i lavoratori Italiani. Dalla cortesia delle Direzioni di tre Compagnie ferroviarie negli Stati di New York, Pennsilvania e New Iersey abbiamo avuto le notizie sui salari, consumi e risparmi degli Italiani impiegati nelle costruzioni e riparazioni nel 1905 e 1906.

Nel 1905 in un mese 44 gruppi (*gangs*) comprendenti 679 uomini lavorarono 178,147 ore e guadagnarono L. 117,077. Per alimentazione, alloggio, vestiario, sapone, tabacco ecc. spesero L. 24,445. Risparmiarono L. 92,632. Nel 1906, 89 gruppi (*gangs*) comprendenti 1530 uomini in un mese lavorarono 404,699 ore; guadagnarono L. 283,621; spesero L. 51,952; risparmiarono L. 231,669.

L'entrata media di ogni uomo in un mese nel

1905 fu di L. 172,45 ; il costo della vita di L. 36,00; il risparmio di L. 136,45. Nel 1906: entrata media Lire 185,35; costo della vita Lire 34,95; risparmio L. 150,40.

Numerose altre osservazioni in diversi altri cantieri e in diversi Stati dell'Unione danno risultati analoghi.

◆

Comparazioni tra Italiani e lavoratori di altre nazionalità nelle entrate e pel costo della vita. Tedeschi, Svedesi, Norvegiani, Slavi, Ungheresi e Inglesi accettano la pensione ordinaria, che viene loro fornita dalle imprese nei cantieri di lavoro pe L. 70 80 o 90 al mese. Gli Ungheresi e gli Slavi, il cui lavoro è di carattere permanente, formano gruppi di 25 o 30 uomini; si procurano una casa o una baracca e vivono in forma cooperativa. Essi stipendiano una donna, di ordinario la moglie di uno di loro, a cui pagano L. 5 al mese, che prepara loro i pasti e accudisce all'alloggio. Essi non comprano dal magazzino del Commissario, ma comprano tutto nei villaggi vicini o in altri magazzini.

Alla fine del mese ciascuno paga una rata uguale della spesa. Gli usi dei Polacchi sono uguali.

La carne fresca e salata fa parte dei loro pasti ordinari. Ecco come sono composti i pasti di un gruppo di Ungheresi e Slavi in Hansford, Pa (1906) che spendevano L. 60 al mese : *Colazione*, pane e caffè ; *lunch* (seconda colazione) : quattro o cinque sandwichs con carne, *desinare;* zuppa, carne bollita e arrostita — mezza libbra o tre quarti di libbra per ciascuno — vegetali e caffè; *cena*, più variata.

Le famiglie slave e ungheresi quando il lavoro è stabile in un punto, mettono casa ed hanno quasi tutti—circa il 75 %—dei pensionisti per diminuire la spesa propria.

Gli anglo-sassoni di ordinario non ne hanno perchè vogliono la casa consacrata alla sola famiglia.

Il costo comparativo della vita pei soli cibi approssimativamente è il seguente:

Italiani. Media mensile del costo presso
   il Commissario delle ferrovie . L. 25,65 a L. 27,40
»   Media mensile del costo presso
   il Commissario per contratto . » 29,50 » » 35,00
»   Media mensile del costo in can
   tieri eccezionali. . . . . . » 42,50 » » 50,00
Slavi e Ungheresi. Minima spesa col si
   stema cooperativo . » 42,15 » » 55,00
»   Massima spesa a pen-
   sione (vitto e allog.) » 60,00 » » 90,00
Altre nazionalità e Indigeni. Al cantiere
   o in pensione. . . » 80,00 » » 90,00

Il carico massimo per l'alloggio degli Italiani è di L. 5.

◆

Entrata, spesa, risparmio per sei mesi di lavoro in un cantiere a contratto.

Nazionalità	Entrata	Spesa vitto e alloggio		Risparmio	
		Ammontare	percent. dell'entrata	Ammontare	percent. dell'entrata
Italiani	L. 1170	240	20,51 %	930	79,49 %
Slavi e Ungheresi	» 1170	360	30,77 »	810	69,23 »
Altre nazionalità	» 1170	540	46,15 »	630	53,85 »

◆

Somme mandate dagli Italiani e dagli altri lavoratori dagli Stati Uniti in patria. Dai dettagli precedenti risulta chiaro che a salario uguale gl'Italiani risparmiano più di tutti i lavoratori appartenenti alle altre nazionalità. Questa indicazione logica trova la sua conferma piena ed intera nei dati statistici.

Dal 1900 al 1906 è stato calcolato che dagli Stati Uniti furono mandate all'estero 514,568,230 lettere e ne furono ricevute 282,857,558. Nello stesso periodo dall'ufficio postale di New York furono mandati all'estero 12,304,485 vaglia postali per un valore di 239,367,047,55 dollari, dei quali: 4,263,633 vaglia per dollari 119.757,895,86 o il 50,03 % del totale della somma furono mandati in Italia, in Ungheria e nei paesi Slavi.

La parte rispettiva di ogni invio risulta da questi dati:

Vaglia postali internazionali emessi negli Stati Uniti e spediti in Italia e nei paesi Slavi ed Austro-Ungarici:

VAGLIA SPEDITI

Anno	In Italia, Austria, Ungheria e Russia		Negli altri paesi	
	Numero	Ammontare (1)	Numero	Ammontare
1900	163,691 L.	19,226,255	840,416 L.	59,285,880
1901	263,599 »	31,259,435	901,039 »	63,714,725
1902	395,641 »	52,332,235	1,015,904 »	72,997,185
1903	594,565 »	86,928,795	1,144,121 »	84,412,270
1904	700,611 »	94,989,825	1,204,861 »	88,614,675
1905	966,718 »	130,070,110	1,356,859 »	100,838,100
1906	1,179,805 »	183,992,805	1,577,652 »	128,183,905
Totale	4,264,633 L.	598,789,475	8.039,852 L.	595,045,755
Media di ogni vaglia	L. 140		L. 70	

Anno	Ammontare vaglia spediti in Italia	Numero immigranti italiani	Ammontare vaglia spediti in Austria Ungheria e Russia	Numero immigranti austro-ungarici e russi
1900	L. 6,810,830	484,207(2)L.	12,215,420	1,401,689
1901	» 9,528,055	135,996 »	21,731,375	198.647
1902	» 18,038,975	178,375 »	34,283,255	279,336
1903	» 38,646,285	230,622 »	48,282,505	342,104
1904	» 43,90?,225	193,296 »	51,088,55?	322,297
1905	» 55,462,230	221,479 »	74,607,875	460,590
1906	» 81,495,670	273,120 »	102,797,335	480,803
Totale	L. 253,583,340	1,717,095 L.	345,206,135	3,485,466

(1) Ogni dollaro è calcolato L. 5 italiane; ma vale L. 5,18.

(2) In questo numero sono compresi gli Italiani, che si tro-

Perciò nell'intero settennio ogni immigrante italiano ha mandato in vaglia postali L. 147,70 in Italia, ogni Austro-ungarico e Russo L. 99,05 nel rispettivo paese.

Ammontare medio di ogni vaglia postale mandato dagli Stati Uniti : in Italia L. 202,55 ; Ungheria 176,05 ; Austria 144,000 ; Russia 95,95 ; Gran Brettagna 66,40 ; Germania 74,80 ; Svezia 103,00; Norvegia 114,45 ; Grecia 222,60. Da questo confronto risulta che i soli Greci mandano in patria più degli Italiani.

L'importo totale dei vaglia postali spediti all'Estero nel 1906 fu di lire 312,176,710 ; di cui il 58,9 % fu spedito in Italia e nei paesi Slavi. Durante tutti i sette anni—1900-906—il 43,3 % dell'intero ammontare dei vaglia postali spediti in Austria, Ungheria, Russia e Italia fu mandato nella sola Italia, mentre i suoi immigrati non costituiscono che il 33 %.

E' probabile che gl'invii di denaro nei paesi Slavi siano inferiori a quelli fatti in Italia, perchè gli Slavi emigrano di più in una alle loro famiglie.

Gl'invii per mezzo dei vaglia postali, però non sono che la parte minore. Una buona parte del denaro è mandata per mezzo dei banchieri italiani. Come indicazione della importanza di queste spedizioni riferiamo ciò che scriveva da Milano il Console Dunning in luglio 1906 « Più di 3,000,000 di lire all'anno sono mandati annualmente dagli Stati Uniti in moneta americana dagli Italiani emigrati temporaneamente ad una singola banca di Napoli ».

Probabilmente le maggiori somme sono portate personalmente dagli emigranti, che ritornano. Come venne dimostrato il 95 % degli Italiani risparmiano lire 125 a lire 150 al mese sui loro salari: per otto

---

vavano già negli Stati Uniti ; così pure nel numero degli immigranti Austro Ungarici Russi del 1900 sono compresi quelli che ci vivevano alla stessa data.

mesi dell'anno, adunque sono circa lire 1000. Si calcola che il 40 % degli Italiani ritornano al principio dell'inverno. Essi ritengono più economico pagare le spese di viaggio e ritornare in Italia anzicchè rimanere disoccupati nell'inverno in una strada soprapopolata di New York.

Ed ora a completare questa parte dello studio si aggiungono le spese cui gl'immigrati vanno incontro per andare negli Stati Uniti.

Le spese sono le seguenti: 1° Spese di preparazione in patria pel viaggio; 2° Spese di trasporto dal domicilio al porto di mare; 3° Tassa pagata al governo proprio per ogni emigrante; 4° Tassa pagata al governo degli Stati Uniti; 5° Spesa pel viaggio dall'Europa agli Stati Uniti; 6° Spesa all'agenzia di collocamento per assicurare il lavoro al porto di arrivo; 7° Spesa di trasporto dal porto di arrivo negli Stati Uniti al luogo del lavoro; 8° Costo della vita dal porto di arrivo al luogo del lavoro.

Omettendo il calcolo della spesa per gli articoli 1, 2, 6 e 8 si calcola approssimativamente che la spesa dei 470,534 braccianti, che si portarono nei Sette Stati nominati in principio — Massachussetts Connecticut, New Iersey, New York, Pennsilvania, Ohio e Illinois — e che sono quelli che assorbono la maggior parte della immigrazione nel 1906 fu di lire 95,266,620; per tutti i 640,474 Italiani, Slavi e Ungheresi immigrati in tali sette Stati la spesa fu calcolata a lire 129,311,585.

◆

Da chi furono collocati i lavoratori Italiani ecc.— Le agenzie di collocamento distribuiscono piccola parte della immigrazione. Le agenzie di New York City annualmente non collocano al lavoro che circa 70,000 persone; piccolissimo è il numero delle persone collocate dalle agenzie di Chicago nell'Illinois. Così nelle altre grandi città. Ciò logicamente perchè tale modo di collocamento è il più dispen-

dioso. E' più efficace e più economico il collocamento per mezzo degli amici e dei parenti e delle
società di navigazione.

Vi sono Società esclusivamente di beneficenza che
s'incaricano del collocamento. La più importante
è il *Labor Information Office for Italians* istituito
in marzo 1906, che non ha carattere politico o re-
ligioso. Dall' aprile all' ottobre di tale anno dette
33,058 informazioni per richieste di lavoro da ogni
parte e collocò 3.705 Italiani di cui circa la metà
nella Città e Stato di New York (1783).

La *immigrant's Home and Free Employment office
of the Hungarian Relief Society* in un anno sino a
novembre 1906 collocò 1,407 Ungheresi. L'*Austrian
Society of New York* fu istituita nel 1906 e in sei
mesi sino a tutto ottobre collocò 514 persone. La
St. Ioseph's Home e la *Polish Immigrant Society
of New York* nel 1906 sino a 31 dicembre collo-
carono 13,250 Polacchi e Lituani ; dei quali 2,150
donne.

Vi sono varie altre società di beneficenza che si
occupano del collocamento.

Il Padrone system (1).—S'intende per *Padrone system* il sistema degli speculatori, accaparratori della
forza di lavoro, che cedono a condizioni vantaggiose
per loro tale forza a coloro che la richiedono e che
negli Stati Uniti, come da pertutto dove c'è qualche
cosa di simile (*gang system* in Inghilterra, il sistema dei *caporali* nell'agro romano ecc.), ha dato
luogo ad un esoso sfruttamento dei lavoratori.

L'atto del Congresso del 4 luglio 1864 si crede

---

(1) Sul *Padrone system* si trovano notizie interessanti in
una monografia di Korèn: *The padrone system and padrone
bank*. (Bulletin of the Bureau of Labor N.º 9, Washington).
Il ladrocinio dei così detti *banchieri* italiani vi è illustrato, E
da allora ad oggi, non ostante l'azione del Banco di Napoli;
i furti dei *banchieri* a danno dei nostri concittadini sono aumentati. *N. Colajanni*

abbia dato la prima origine al *Padrone system*. Era intitolato: *atto per incoraggiare l' immigrazione* e stabiliva che l' immigrante in patria impegnava il proprio lavoro per un termine non superiore a 12 mesi e pagava le spese per la immigrazione ch' erano state anticipate dallo speculatore, che acquistava un diritto privilegiato sul salario degli immigranti.

Nel 1893 appena cominciata la crisi industriale il Ministro del Tesoro il 13 giugno nominò una Commissione onde studiare le conseguenze di tale legge. Il 7 ottobre 1895 la Commissione sul *Padrone system* riferì che questo era ritenuto responsabile della depressione per la concorrenza nel lavoro, per l'arruolamento di lavoratori a bassi salari per mezzo di agenti che erano stati mandati in Italia, e che gl' Italiani sarebbero stati le prime vittime del sistema. La Commissione ritenne ch'erano necessarie più ampie indagini per giudicare. Ma l'opinione pubblica giudicò contro la legge in modo definitivo.

La legge del 4 luglio 1864 fu abolita nel 1868 e sino dal 1882 non ci fu negli Stati Uniti alcuna legge che regolasse l' immigrazione; nel 1885 fu proibito il contratto di lavoro.

Le statistiche dell'immigrazione intanto provano che non fu la legge del 1864 a provocare l' immigrazione dei lavoratori *unskilled*. Eccone la prova:

	Dal 1864 al 1869	Dal 1869 al 1875
Totale immigranti . .	1,289,323	6,527,845
Immigranti italiani. .	5,740 o 0,45 %	169,649 o 2,60 %
Id. austriaci e ungher.	1,686 » 0,13 »	227,483 » 3,48 »
Id. russi	2,595 » 0,19 »	135,545 » 2,16 »

Basta confrontare le cifre della immigrazione col regime del 1864 e colle leggi posteriori e con quelle dal 1900 al 1906, sempre più restrittive, per comprendere come sia stata scarsa l'efficacia legislativa nel favorire o contrariare l'immigrazione.

Ma gl'inconvenienti del *padrone system* non sono

meno veri. La Relazione del 1895 li enumera e
mostra come e quanto i *padroni* guadagnassero sugli
immigranti, specialmente italiani, nell'arrivo e nella
partenza. Cinquanta *Padroni* avevano ciascuno 500
o 600 lavoratori alla propria dipendenza sui cui
salari prelevavano il 10 o il 15 °/₀ al giorno per
aver loro procurato un collocamento.

Nel solo anno 1894 da una dozzina di *banchieri* o
padroni di New York, Boston e Newark furono
guadagnati non meno di L. 500,000. La relazione
constata come una circolare dello Stato di Wi-
sconsin rivolta ai *padroni* per la ricerca dei lavoratori autorizzasse il sospetto che i pubblici poteri
riconoscessero il sistema; enumera le truffe com-
messe dai *padroni* che esigevano compensi (*fees*)
per occupazioni che nemmeno procuravano; e che
gl'Italiani, specialmente i meridionali o siciliani,
erano le vittime più numerose a causa della loro
grande ignoranza, della mancanza di conoscenza
della lingua, di relazioni ecc. ed anche pel timore
provato dai *cafoni*, come li chiama la relazione,
della vendetta dei *bosses*, dei *padroni*, se si sottrae-
vano a loro. La *bossatura*, cioè il prezzo pagato al
padrone per avergli procurato lavoro, secondo le
stagioni e i luoghi, varia da L. 5 a L. 50 per operaio. Il *boss* specula in tutti i modi e nel campo
del lavoro stabilisce il *diritto di Madonna*, il *diritto
di lampa*, la contribuzione per la *Santa Vergine* ecc.
Fornisce alloggio e vitto a condizioni usuraie. I
peggiori *bosses* si sono avuti in Pensilvania e nella
Virginia occidentale. I *padroni italiani* si dividono
in tre classi: 1° Piccoli *bosses*, i più numerosi, che
speculano e commettono ogni sorta di frodi nel
procurare lavori fantastici (*odd jobs*) a singoli individui o piccoli gruppi; 2° *bosses*, che forniscono
lavoratori in numero considerevole agli intraprenditori; 3° *bosses* che sono regolarmente iscritti come
intraprenditori. Questi ultimi sono pochissimi —
una dozzina a New York, quattro a Filadelfia, tre
o quattro a Boston — e trattano umanamente i la-

voratori. I peggiori sono i primi e si dice che ce
ne siano 2000 in New York e dintorni.

Tutti gli abusi denunziati nel 1897 sono conti-
nuati negli anni successivi; e furono innumerevoli
e incredibili.

Il 27 aprile 1904 lo Stato di New York gromulgò
una legge, modificata in aprile 1906. Colla legge fu
istituito un Commissariato adibito esclusivamente
(*Commissary system*) ad impedire gli abusi del *Pa
drone system* che non può più essere esercitato
senza regolare licenza L'attività del Commissariato
risulta da questo dato. Dal 1º Maggio 1905 al 1º
Maggio 1906 il Commissariato iniziò 812 procedure
per restituzione di paghe dei padroni; furono re-
stituite L. 10,650; furono istituiti 47 procedimenti
innanzi alla Corte criminale; si ottennero 15 con
danne; furono revocate 13 licenze e rigettate 22.

Gli abusi furono molto limitati e migliorata la
condizione dei lavoratori. Una delle innovazioni
essenziali è stata quella di fare conoscere ad ogni
lavoratore le condizioni che devono essere rispettate
nei contratti di lavoro dai *boxes* e dagli intrapen-
ditori. Ad ogni lavoratore si da copia in Inglese e
in Italiano di tali condizioni, che sottoscrivono liberamente nell'accettare il contratto di lavoro. Il
Commissario stabilisce i prezzi dei generi di consumo che si danno agli operai sul cantiere del lavoro e le condizioni dell'abitazione. I prezzi del
1906, a 100 miglia da New York, in generale sono
inferiori del 20 o 25 % a quelli adoperati col *padrone system* in New York nel 1897.

Ma per quanto le leggi del 1904 e 1906 sul *Commissary System* abbiano migliorato e frenato il *Padrone system* quelle leggi devono essere abolite per
le ragioni seguenti.

Le leggi del 1904 e 1906 provvedono specificata-
mente a regolare l'italiano *padrone system* ed impedire che gl'Italiani siano truffati. Così per la protezione di questi immigranti, questa specie timidis-
sima di viaggiatori, — *this very timid kind of tra-*

veller — furono create una serie di agenzie munite
diogni sorta di poteri per render loro dei servizi
in ogni occasione ; e quando essi cominciano a lavorare in America, essi per un eccesso di attenzione
e di protezione sono resi più impotenti e più timidi
di quello che erano quando vi arrivarono.

In vista del fatto che così il lavoratore italiano
non é messo a contatto colle influenze americanizzanti, si è inclini a pensare che esso diverrebbe un
uomo più preveggente e meglio adatto ad ottem-
perare ai doveri e alle condizioni della cittadinanza
americana se fosse messo un poco nelle condizioni
(*landed*) dello spirito del pioniere, che si assicura
da sè l'occupazione senza altro estraneo aiuto, se
non quello che si accorda agli altri immigranti,
mescolandosi cogli altri uomini ed affrontando tutte
le comuni contingenze.·

Questo fu il metodo, mercè il quale furono as-
similati tutti i *timidi* e *ignoranti* emigranti delle
altre nazioni.

Sino a tanto che il *Commisary sistem*, buono o
cattivo, sarà in vigore, l' Italiano rimarrà straniero
in terra straniera (1).

---

(1) Per parte nostra la proposta del Sheridan la troviamo
improntata alla brutalità liberistica nord-americana che liberamente ammette ogni iniquo sfruttamento.

Egli ha proclamato buoni i risultati delle leggi del 1904-906
pessimi quelli del 1864-1885. Perchè non continuare nell'uso
delle buone ?   *N. Colajanni*

Amy A. Bernardy

PERCHÈ GLI ITALIANI SI ADDENSANO
NELLE CITTÀ AMERICANE

Perchè g' Italiani si addensano nelle città americane [1]

La distribuzione proporzionale degli Italiani negli Stati Uniti è dimostrata dalle statistiche. So benissimo che coloro che vi hanno interesse sanno cercarle e cercarle da sè nel *Census* o nelle altre fonti; e quelli che non vi hanno un interesse scientifico preferiscono farne a meno. Quindi le ometterò, ma desidero far notare almeno questo, che il 72 per cento degli Italiani immigrati agli Stati Uniti si accumula nella North Atlantic Division, che la North Central ne ha l'11,4 per cento e che la South Atlantic finora non sembra aver attratto che meno di una trentesima parte del numero ascritto alla North Atlantic, poco più di un cinquantesimo del numero totale. In altri termini ancora, il 62,4 per cento degli Italiani nel 1900 gravitava in 160 città, e la sproporzione è venuta crescendo continuamente.

---

[1] Questo scritto della signorina Amy A. Bernardy è la riproduzione di un discorso pronunziato all' *Exhibit on Congestion of population* in New York in Maggio 1908. (Dal *Bollettino dell'Emigrazione* 1908 N. 17). Vi si dimostra che se gl' Italiani si addensano nelle grandi città americane ciò avviene per ragioni naturali senza colpa degli Italiani e con grave responsabilità dei governanti americani, che nulla hanno fatto e fanno per ovviare al male.   *N. Colajanni*

Tremenda sproporzione, quando si pensi alla condizione delle città del Nord e specialmente delle colonie italiane in queste città, e alla vasta distesa di terre che soprattutto nel Sud attendono ancora per produrre l'opera fecondatrice dell'uomo; al numero d'Italiani che cercano lavoro qui e alla quantità di lavoro che cerca braccia altrove! E quando si osserva che per ora fra questi due estremi non c'è comunicazione o trasfusione, sembra imperiosa la necessità di provvedere in qualche modo.

E' chiaro che l'immigrante che arriva a New-York, Boston o Philadelphia trova facile lo stabilirsi nella grande città. E' assai poco arduo il trovare occupazione nella città porto di mare, dove ci è lavoro per tutti e ne avanza tanto che l'immigrante risparmia tempo e fatica accettandolo lì, su due piedi, qualunque esso sia.

E' naturale che si fermi lì, perchè, inoltrandosi nel paese per lui sconosciuto, troppo spesso riscontrerà che la sua ignoranza della lingua, le condizione d'isolamento, l'ambiente, tutto in genere si manifesta contrario a qualsiasi tentativo di irradiazione. Finchè è coi suoi amici, parenti, compaesani, si sente a posto. Ed è naturale, date le condizioni in cui gli verrà fatto di trovarsi fuori della città affollata che non si senta a posto se non quando è nel *congested district* cittadino. A noi tocca ora fare in modo che fuori non si senta sperduto.

Non è, come troppo comunemente si crede, non è che l'immigrante si voglia fermare nel *congested district* perchè è sporco, perchè vi costa poco l'alloggio, perchè è malsano e antigienico o per tutte quelle altre immaginarie ragioni che altri, specialmente gli osservatori inesperti e superficiali, adducono a suo nome o emettono come opinione propria. Si ferma lì perchè lì trova lavoro, richiesta e paga per il suo lavoro, perchè quello che gli offrono è il lavoro che gli occorre, da eseguirsi a condizioni che gli convengono, in mezzo ad un ambiente familiare, l'ambiente paesano.

Finchè voi gli offrite condizioni migliori di guadagno nella città, non possiamo aspettarci di trovarlo pronto a faticare e soffrire fuori della città, in omaggio alle teorie della sociologia e magari ai desiderii di questo ottimo Comitato e autorevolissimo Congresso.

L'immigrante non conosce le teorie sociologiche o i principii dell'igiene, conosce la pratica della vita dal punto unilaterale ma ultra importante del guadagno, e prima di tutto cerca di accomodarsi dove la sociologia o chi per lei o contro di lei gli darà da mangiare, vestire e guadagnare per sè e la famiglia. E' vero che ci rimette un tanto di salute e che la razza degenera, ma egli non ha l'idea del valore etico e civile della salute e dell'integrità della razza.

Non accusatelo di favorire la congestione: la congestione è favorita dalle nostre condizioni industriali, dalla nostra connivenza ad un ordine di delitti sociali che non tentiamo nemmeno di condannare, nonchè combattere. Noi siamo l'elemento consapevole: l'immigrante è inconsciente. I doveri sono nostri. Provvedere alla distribuzione tocca a noi: non dobbiamo aspettarcela da lui. Si è detto e strepitato tanto intorno all'analfabetismo dell'immigrante, e ora siamo noi, noi che sappiamo di letteratura e di matematica e *de quibusdam aliis* che abbiamo il coraggio di domandare all'analfabeta e di attendere da lui la soluzione di questi gravi problemi? A noi tocca fare, se qualcosa si possa fare, come è chiaro che fare si deve.

L'immigrante è designato a lavorare di braccia ed è pronto a farlo, non ad occuparsi della cosa pubblica.

Nè per risolvere il problema vogliamo l'esclusione. Distribuzione é necessaria, non esclusione. Se mettete la gente fuori del campo, non avrete battaglia, ma nemmeno vittoria. Mettetela sul campo ma datele buoni condottieri e dalla battaglia emergerà il trionfo.

Per assicurarci questo trionfo dobbiamo dunque mettere l'immigrante in condizioni tali che favoriscano l'evoluzione delle sue buone qualità, e per far ciò dobbiamo studiare sui luoghi dove queste buone qualità hanno avuto occasione di dimostrarsi e di esemplificarsi. Attraverso tutto il paese, dappertutto dove ci sono delle *Piccole Italie*, fuori dei terribili *congested districts*, troverete del buono.

Troverete che la criminalità tanto deplorata non esiste nè meno sporadicamente dove la *Piccola Italia* è davvero tale, dove alla gente è possibile vivere in condizioni decorose, se pure modestissime.

Che cosa sa l'immigrante delle vere condizioni della vita civile in questo paese? Egli è messo a marcire nel distretto più miserabile per condizioni, ambiente e affinità; i primi cittadini americani che incontra sono *policemen* e *salonisti* (tenitori di *bars* e spacci di liquori). Ciò che egli vede e che tutti si fanno un dovere d'imprimergli bene in mente è la violazione, la corruzione o l'applicazione della legge per proprio conto; della stessa legge non arriva mai a vedere l'ordine, la maestà, la bellezza Molte volte non sa che questa legge esista finchè non si trova condannato per averle contravvenuto. Coi cittadini del buon governo, colle file di coloro che in politica combattono la buona battaglia, l'immigrante non viene a contatto mai.

Fortunatamente queste condizioni accennano a cambiare in alcune delle minori città, dove il piccolo commercio e la specializzazione dei mestieri rendono più facile l'accesso alla vita cittadina.

Voi ricordate come, nel libro famoso ancora nei nostri giovanissimi anni, *Topsy* era « sbocciata » in questo modo. Ebbene, allo stesso modo mentre ci sono state delle *Piccole Italie* premeditate e premeditatamente stabilite, ci sono anche delle *Piccole Italie* che si trovano sbocciate senza saper come, e fanno ottima riuscita. Andate a vedere quegli immigranti rurali e li troverete al lavoro sopra piccole e sopra grandi *farms*, uniti nell'onesta opera quotidiana, troverete che sono buoni Italiani, e buon[i] Italiani e buoni Americani allo stesso tempo, che le preziose qualità ereditarie dell'italiano e il senso dell'ordine pubblico americano hanno in loro fruttificato. Ricordo di aver parlato in città e villaggi con diversi cittadini e magistrati indigeni, e tutti hanno avuto parole di lode per il *setter* italiano. Dal Michigan, all'Alabama e dal New Jersey all'Arkansas troveremo sui nostri passi una vera fioritura di coloniette italiane notevoli per le loro buone attività; altre ne troviamo nell'estremo Ovest; altre le abbiamo qui alle porte della città.

Piantano, seminano, trafficano, vendemmiano, queste coloniette permanenti, mentre le masse dei lavoratori della pala e del piccone, peregrinando per tutte le regioni del paese, fanno un altro lavoro non meno necessario e fecondo.

Nel Sud, ove la terra spopolata chiede braccia e può quindi grandemente contribuire all'opera di sfollamento da noi auspicata, ci troviamo tuttavia di fronte a gravissimi problemi. Il Sud ha bisogno del lavoro italiano, questo è un fatto. Dovunque l'italiano ha trovato da fare, ha dimostrato che meritava la buona occasione, ha dimostratto nelle *farms* che è competente in materia agricola, e che il suo lavoro è così infinitamente superiore a quello del negro che ogni paragone sarebbe assurdo. Ma appunto per questo deve essere trattato da *uomo* e non dev'essere nella mente di un *foreman* bestiale qualcosa d'intermedio fra l'uomo bianco e l'uomo nero come qualche volta di lui si pensa.

A parte questo, l'italiano nei distretti agricoli del Sud se la cava e se la caverebbe bene, ha buone *chances*, e trova condizioni spesso simili a quelle che ha lasciato a casa. Ma come provvedere a che l'italiano appena arrivato sappia di queste condizioni e come eccitare in lui il desiderio di sperimentarle, quando l'ambiente che lo circonda e crea le prime impressioni, è così essenzialmente diverso ? Ecco uno dei più serii problemi nella vostra impresa.

Nell'Ovest le condizioni del lavoro sono un poco diverse. In teoria parrebbe senza paragone migliore il Sud. I problemi dell'Ovest sono più vasti: l'agricoltura diventa quasi meccanica ed industriale, e perciò meno adatta alle abitudini dell'italiano che ha la tradizione della coltura intensiva e non ha l'abitudine delle macchine e delle imprese agricole industriali su larga scala.

Ma, a parte il lavoro agricolo, nell'Ovest c'è posto e necessità di ogni specie di mano d'opera; e la mano d'opera si paga bene, perchè l'Ovest ha il denaro, lo spazio, il bisogno dell'attività su vasta scala ed è pronto a pagarla. Vi accennavo che nel Sud si coltivano fragole. Nell'Ovest si costruiscono ferrovie. Ora, senza fragole eventualmente si può andare avanti, senza ferrovie no. E la massa italiana che può coltivar fragole nel Sud può anche costruir ferrovie nell'Ovest. E allora fa ciò che facevano una volta in Europa le legioni romane, organizza la rete stradale della nazione. Ricordiamolo, qualche volta, a maggior gloria del troppo maltrattato immigrante.

Maltrattato, ho detto. E qualche volta, anzi, ingiustamente offeso anche dalle più rugiadose buone intenzioni. Io conosco delle signore, mondanissime o intellettualissime o tutt'e due assieme, che s'interessano al problema sociale press'a poco così : « Che bel tempo che fa stamani, vero ? Oh ! vediamo di metterci a far qualcosa. Cosa si potrebbe fare ? Andiamo a incivilire l'immigrante ».—Ora, non di questo ha bisogno l'immigrante. Che abbia bisogno di aiuto, nessuno lo nega, ma dev'essere aiuto da uomo a uomo, non da protettore a pezzente. Deve essere lo stesso aiuto che non offenderebbe noi stessi, se fossimo nella necessità di riceverlo. Poichè io non vedo per che ragione l'immigrante non deve ricevere lo stesso trattamento, se corre gli stessi rischi, e maggiori, degli altri cittadini. Qui non è questione d'influenza o di posizione sociale: è questione della comune umanità di tutti. Ciascuno di noi può far molto per l'immigrante, e deve farlo, non perchè questi sia un italiano o un irlandese o uno svedese o uno slavo, ma perchè è un uomo in un paese in cui si garantisce la più ampia affermazione dei diritti dell'uomo.

Ora, perchè accusare l'immigrante italiano di favorire ostinatamente la congestione ? Domandiamoci piuttosto che cosa abbiamo fatto noi per favorire lo sfollamento. Che cosa hanno fatto il Sud e l'Ovest per attirare questo immigrante ? Lo hanno trattato uniformemente, fin dal giorno del suo arrivo, come se fosse fatto di pasta diversa dal resto di noi, come una cosa, come una macchina — anzi nemmeno come una macchina, perchè una macchina costa denaro, e un guasto al macchinario è una perdita di capitale. Ma se qui si massacra o si danneggia un italiano, che importa ? Io domando per quale ragione deve il *dago* (1) lasciarsi fare a pezzi nelle miniere o andarsi a far azzoppare nell'Ohio in omaggio allo sfollamento, quando fuori dell'ambiente cittadino non ha salvaguardia nè garanzia ?

Ho notato recentemente su vari piroscafi in viaggio di ritorno in Italia, le vittime degli infortuni sul lavoro, i martiri dell'augurato sfollamento: gente che si era ridotta inabile per la vita fra la Pennsylvania e il Western New York, il New-Hampshire e il Sud.

Sfolliamo pure, dobbiamo sfollare ; ma provvediamo a che l'immigrante disperso non sia solo come un cane quando dopo una disgrazia ha bisogno di protezione e d'aiuto. Si rimproverano all'Italia i denari che gli emigrati rimandano in patria. Ebbene, a questo prezzo io vi dico che sono

---

(1) L'Italiano dai nord-americani viene chiamato *dago*, com'è chiamato *gringo*, sempre in segno di disprezzo nell'Argentina.

N. *Colajanni*

troppo cari. Se queste sono le condizioni del lavoro
fuori delle città, e l'Italia deve ritrovarsi i suoi
uomini ridotti così quando sfuggono alla lenta
morte della *congestion*; molto meglio sarebbe po-
terli tenere a casa fin dal principio. L'Italia fa il
possibile per aiutarli e proteggerli. Vediamo di
fare il possibile anche qui. Vediamo di aiutarli a
distribuirli in modo che, se vanno nel Sud, non
cadano negli orrori del *peonage*, o l'infortunio sul
lavoro non ce li rimandi in patria miserabili e
mutilati, e che, se vanno nell'Ovest, vi trovino non
solo compaesani d'Italia, ma concittadini americani.
Concittadini, non protettori rugiadosi o dispregia-
tori insolenti. Io non vi chiedo per loro un *benvenuto* sentimentale, ma un'accoglienza onesta. Fate
che si accorgano di essere in mezzo ad una cittadinanza che, se non altro, non è loro ostile e non
li disprezza. Essi faranno presto a trovare la loro
via e a diramarsi, se ne date loro convenienti occasioni ed eque garanzie. Se non possiamo fidarci
dei mezzi privati o locali, intervenga il Congresso
federale colla sua autorità e sotto le garanzie della
Costituzione, ad aiutare, ad imporre se sia necessario, lo sfollamento. L'Italiano, normalmente, è
un elemento buono e sano e desiderabile da tutti
i punti di vista, e miglior prova fa e farebbe quando
si trova fuori del luridume della congestione; in
cui, del resto, non ha poi un pazzo desiderio di
crogiolarsi. Il fatto è che lì c'è una richiesta costante per il suo lavoro, ed egli preferisce stare
dove ha la certezza del lavoro e del guadagno maggiore ed immediato, tanto più che, e voi lo sapete,
l'Italiano rifugge dal pauperismo e dalla *bread line*.
Distribuite il lavoro, e la distribuzione dell'immigrato sarà senz'altro un fatto compiuto.

Emily Fogg Meade

GL'ITALIANI NELL'AGRICOLTURA

Gl'Italiani nell' Agricoltura [1]

La presente agitazione per ulteriori restrizioni alla immigrazione è specialmente rivolta contro i Greci, i Polacchi, gli Austro-Ungarici e gli Italiani del mezzogiorno e della Sicilia che, dal 1870 ad oggi non hanno fatto altro che aumentare sempre di numero. Gli immigranti nel 1905 furono 1,026,499, e gli Italiani in numero di 221,479 vi tenevano il primo posto. Il generale Francis A. Walker fu uno dei primi oppositori alla ammissione degli emigranti di questa nazionalità. Egli diceva : « Essi non hanno nessuno di quegli istinti e tendenze ereditarie che li rendono facilmente comparabili alla emigrazione del passato. Essi sono i vinti di razze vinte, la debolezza peggiore nella lotta per la esistenza ». Recenti articoli di riviste e giornali accennano al pericolo della ammissione di questa gente. Si afferma che essi sono malnutriti, male sviluppati, spesso malati, inabili, analfabeti, mancanti del senso

---

(1) Emily Fogg Meade : *The Italian on the land: a study in immigration.* (Dal *Bulletin of the Bureau of labor* di Washington N. 70). Lo scritto del Fogg Meade prova che gli Italiani sanno adattarsi alla vita rurale e possono *americaniʒʒarsi.* La traduzione letterale dello studio sarebbe: *Gl'Italiani sulla terra.*

N. Colajanni

della responsabilità, con un acuto senso della loro
inferiorità e l'assenza assoluta di abilità per trarre
partito dalle circostanze ; non sono *desiderabili* perchè il tenore della loro vita è basso e perchè non
cercano di migliorarlo quando le loro condizioni
sono prospere ; e che dopo tutto essi sono sempre
un probabile carico per il paese come inquilini di
ospedali, di manicomi e di ricoveri di mendicità.
Di più si afferma che non sono i forti ed indipendenti che vengono, ma i deboli e gli incapaci fra i
quali hanno effetto i miraggi presentati dalle compagnie di navigazione. Questa gente si affolla nelle
nostre grandi città complicando i problemi delle
autorità municipali.

Nei primi tempi il lavoro di sterratore — il duro
lavoro spiacevole agli Americani — era fatto dai
Tedeschi. Man mano che i Tedeschi hanno elevato il tenore della loro vita, ad essi si sono sosti-
tuitigli Irlandesi, poi a loro successero i Polacchi,
e gli Ungheresi ed ora finalmente gli Italiani. Gli
Italiani sono stati specialmente stigmatizzati per il
genere del loro lavoro. Gli americani li considerano come forestieri sudici e mingherlini, che suonano l'organetto, tengono banchi di frutta, spazzano
le strade, o lavorano nelle mine, nei tunnels, nelle
vie ferrate, o in opere di muratura. I giornali son
pieni di luridi racconti di risse, nei quali lo stiletto
è messo in evidenza, di codarde pugnalature nella
schiena, di bande organizzate di criminali italiani;
ed è frequente la manifestazione del dubbio se gli
Italiani potranno mai elevarsi sulla scala sociale
come gli immigranti delle altre nazionalità. Al tempo
stesso l'offerta di lavori pesanti e duri aumenta, e
gli italiani affluiscono in gran numero e si fermano
nelle città porto di mare, dove essi sono frequentemente senza lavoro durante l'inverno. Soltanto in
questi ultimi anni c'è stato qualcuno che ha detto
una buona parola in loro favore. I punti principali
dei differenti articoli a proposito della immigrazione

Italiana sono stati recentemente riassunti in: « *Italiani in America* » (1).

Gli Italiani del mezzogiorno costituiscono una classe di contadini che vivono in estrema miseria, e danno la maggior parte di ciò che producono alle tasse, a l' affitto e alle decime. Essi sono forti agricoltori, i cui sforzi sono magramente ricompensati ; essi considerano l'America come la terra del dollaro, e nel loro desiderio di far denaro, ipotecano la loro piccola proprietà, il loro bestiame ed i loro strumenti ad altissimo interesse per fare il denaro sufficiente a mandare il capo della famiglia od un figlio a New York.

Questa gente non viene da fattorie isolate, ma da grossi villaggi affollati ; essi sono naturalmente socievoli e ricercano quelli che parlano la loro lingua. Essi ignorano le condizioni delle nostre fattorie; la pratica del lavoro mal retribuito in Italia li ha disgustati dal lavoro dei campi, e non posseggono denaro per affittare terre nel mezzogiorno, o nell'occidente degli Stati Uniti. A New York e nelle altre grandi città essi trovano lavoro e amici; la loro abilità nell'agricoltura non ha occasione di manifestarsi, ed essi sono calcolati come la più bassa classe dei manovali.

Gli Americani generalmente, non coscienti delle recondite cause della affluenza degli Itali ni nelle grandi città, li credono inadatti ai lavori agricoli. Nel 1896 il Governo fece dimandare ai governatori di diversi Stati di esprimere i loro desideri a proposito della immigrazione. Due soli Stati desiderarono gli Italiani. I vaiii commissari d' emigrazione dei diversi Stati del Sud mostrano anche ora un simile

---

(1) Eliot Lord, John I. D. Frennor, e Samuel I. Barrows: *The Italian in America* (New York, Buck et Company, 160 Fifth Avenue , 1905). E' la più calorosa difesa che sia stata fatta dagli Italiani da Nord-Americani. Me ne gioverò spesso nel mio studio sugli *Undesirables*.

N. Colajanni

pregiudizio. Ma c'è una chiara evidenza che gli Italiani nelle nostre grandi città, malgrado la pericolosa influenza dei quartieri operai, progrediscono socialmente e si americanizzano. Nondimeno la minaccia dell'affollamento, il continuo deperimento fisico di questa gente nelle città, ed il bisogno di sviluppare grandi aree nelle terre del Sud sembra essere una ragione per cercare di dirigere questa immigrazione, verso tali parti del paese.

Come un contributo del crescente movimento per attrarre l'immigrazione alla terra, un dettagliato studio delle condizioni economiche sociali e morali degli Italiani di una tipica colonia rurale, è stato fatto, mostrante come gli Italiani meridionali, — la più bassa classe degli emigranti — può progredire in mezzo ad una comunità Americana agricola; e per rispondere alla domanda: Può l'immigrante Italiano diventare un buon Americano? La città di Hammonton, nella Atlantic County (New Jersey) è stata scelta per questo studio, perchè l'aumento della popolazione Italiana in questo punto è stato naturale; non è stato stimolato nè aiutato in alcun modo da Americani, e l'emigrante ha dovuto contare soltanto su le proprie risorse, ed è stato lasciato libero di seguire la sua inclinazione. Altri significanti caratteri della colonia Hammonton sono i seguenti: 1) La immigrazione non è stata stimolata che dagli agricoltori italiani unicamente; 2) i più vi risiedono permanentemente; 3) l'immigrazione annua degli Italiani dalla città ad Hammonton, durante la raccolta, mette gli Italiani di Hammonton in stretti rapporti con i loro compaesani e diffonde la conoscenza della colonia tra gli Italiani in Filadelfia e in New York; e 4) gli agricoltori Italiani, meno poche eccezioni, vengono dal mezzogiorno d'Italia e dalla Sicilia. Hammonton, in brevi parole, è una delle poche colonie Italiane negli Stati Uniti dove una popolazione Americana ed una Italiana sono cresciute insieme, mosse dal medesimo impulso di andare alla città, seguenti le

stesse occupazioni, e viventi fianco a fianco in rapporti di buon vicinato. In Hammonton, si sono trovati i resultati d' un contatto ventenne d'una popolazione Americana tipica con la più bassa classe
d' immigranti Siciliani. E' una prova certa che ciò
che gli Italiani hanno potuto fare in Hammonton
essi possano farlo dovunque esistono circostanze
analoghe.

L' avvento degli Italiani

E' ora più di trenta anni che gli Italiani arrivarono per la prima volta ad Hammonton, ed è passato a bastanza tempo perchè una seconda generazione sia cresciuta a dimostrare che specie di cittadino Americano, puo essere formato da un Italiano nato
ed allevato nel paese e in rapporti con Americani,
come vicini, a scuola, o negli affari. I pionieri Italiani vennero nella New Jersey meridionale per la
stessa ragione, che vi vennero i coloni di New York
e della New England.

Essi cercavano di fissarsi in luoghi non troppo
lontani dalla costa, dove il clima fosse congenere,
il suolo a poco prezzo. Il New Jersey meridionale
era un nuovo territorio. Prima del 1850 le sterili
pinete si stendevano su vasto territorio, e veramente
erano sterili dal punto di vista del pastore o del-
l' agricoltore. Il clima e la foresta, nondimeno, attrassero alcuni coloni prima del 1860, quando il suolo
fu, per la prima volta, messo in vendita. La guerra
civile acuiva la dimanda di frutta e di vegetali, per
le quali culture il suolo sabbioso è specialmente
adatto, e dopo il 1865 l'apertura di mercati all'in-
grosso nelle grandi città fece della coltivazione delle
frutta, una industria rimunerativa.

Malgrado ciò lo sviluppo del New Jersey meridionale è stato lento assai, perchè molto lavoro e
grandi spese sono state necessarie per adattare il
suolo, il cui carattere era scoraggiante ; e gli immigranti attrattivi per coltivarlo avrebbero preferita
la più fertile terra dell' Ovest. Se non fosseri stati

gli Italiani, le vicinanze di Hammonton sarebbero probabilmente ancora un deserto. Gli Italiani raccolsero le coccole per il mercato ; diboscarono il suolo e lo prepararono per la coltivazione; i risparmi del loro lavoro sono stati investiti nelle terre o nelle ferrovie, ed essi hanno acquistato le tenute dai possessori i cui figli sono andati alla città o nel Far West (1). Ciò che gli Svedesi ed i Tedeschi hanno fatto dell' Illinois, del Jowa, del Minnesota, questi agricoltori Italiani hanno fatto di Hammonton: uno dei più attrattivi punti nella regione dei paesi del New Jersey meridionale.

Per dire come sorse questa comunità Italiana basta fare la breve biografia di alcuni dei suoi pionieri.

Probabilmente i primi Italiani che si fermarono ad Hammonton furono la famiglia Lo Grasso, dei musicanti, che scelsero abitazione in un punto ora tutto abitato da Italiani. Poco tempo dopo il Lo Grasso fu raggiunto da un certo Matteo Calabrese, nativo di Palermo che comprò della terra, sposò una tedesca e riuscì ad avere una bene avviata fattoria. Egli era un massone ed un ex-soldato della Grande Armata, ed aveva rapporti di amicizia con alcuni Americani. Poco di poi Isacco Nicolai, un toscano, e Michele Rubertone di Pastena vicino a Napoli, vennero a stabilirsi a Hammonton. Nicolai era un uomo di grande intelligenza; egli aveva conosciuto a San Francesco il magistrato Byrnes che lo indusse ad andare ad Hammonton, dove il figlio di Nicolai si naturalizzò. Rubertone lasciò Pastena nel 1877 vagò un pò per l' Europa suonando il violino, in compagnia dei suoi quattro figliuoli, poi arrivato a New York vi conobbe il Nicolai che lo consigliò di venire ad Hammonton. Quì il Rubertone acquistò 16 ettari di terra, vi si stabilì allargando poco a poco il suo possesso a 78 ettari. Otto ettari furono poi acquistati da un Giovanni Berri che aveva sposato

---

(1) Si chiamano così gli Stati del Sud-Pacifico.

un'Italiana, ma egli stette poco tempo in Hammonton, suo figlio invece sposò un'americana e accumulò una grossa proprietà.

Fino all'arrivo dei fratelli Campanella, dopo il 1866, c'erano pochi italiani in Hammonton. I Campanella nativi di Gesso (Messina) prosperarono, acquistarono della terra, sposarono delle americane, chiamarono quà dalla Sicilia di dove erano nativi, i loro parenti ed amici e furono i veri conduttori di una emigrazione diretta in Hammonton.

Pietro Raneri, uno zio dei Campanella, divenne fattore dello scrittore di cose orientali Scott, e questi quando partiva, lasciava la direzione della sua casa al Raneri, che era, egli diceva, il più onesto degli uomini. Col Raneri si stabilì ad Hammonton anche Antonio Cappelli, un cugino dei Campanella. Questi primi venuti della emigrazione Italiana erano chiamati dagli abitanti, *piccoli negri*, e non erano considerati italiani.

Il capo della emigrazione Napoletana, i veri Italiani com'essi dicevano, fu Domenico Tonsola, e veniva da Casalnuovo, vicino a Napoli. Dopo pochi anni di permanenza comprò della terra ed è, ora, proprietario di 120 ettari di terreno ben coltivato. Anche quelli che lo seguirono subito dopo Biazza Crescenzo da Casalvelino, Angelo Foglieti di Monteroduni, che arrivò a Filadelfia con cinquanta soldi in tasca, ed altri chiamati in Hammonton da loro prosperarono, acquistarono della terra, aprirono alberghi, botteghe e fattorie. Il numero maggiore degli Italiani arrivò ad Hammonton fra il 1885 e il 1890.

Da principio gli Italiani erano molto poveri e si ritenevano autori di piccoli furti di cavoli e di patate, ma poco a poco, le condizioni generali migliorando, i furti cessarono, e, tutto considerato, si è trovato che gli Italiani sono atti a formare dei buonissimi cittadini, e la loro presenza è stata un'ostacolo all'aumento della popolazione negra. Gli Americani valutarono presto l'abilità dell'Italiano come contadino, ed in molti casi questi si è rivelato ottimo agricoltore. Non solo egli ha bonificato molto suolo selvaggio per gli americani, ma è riuscito ad acquistare e bonificare molta terra per se. Gli Italiani ridussero a terreno coltivabile la grande pineta a sud della città, e vi vennero e vi lavorarono quando il problema della vita nella pineta era serio assai.

Dal 1880 al 1895 molti dei vecchi residenti si ritirarono o morirono, e le loro terre e fattorie abbandonate dai loro figlioli furono comperate da Italiani. Alcuni esempi illustreranno questa affermazione. 1) Un residente nel 1885 vendette la sua proprietà di 24 ettari, una delle migliori di Hammonton, ad uno dei primi arrivati fra gli Italiani per 5000 dollari contanti, e comprò una casa in città e 120 ettari di terreno incolto a 3 miglia dal paese. Dopo ch' egli la rese coltivabile la rivendette ad Italiani, fra i quali uno che aveva lavora·o sotto la sua dipendenza. 2) Un Inglese nel 1857, coltivò in Hammonton 8 ettari di terra. Poco prima della sua morte la rivendette, a contanti, ad un Italiano per 1600 dollari. 3) Nel 1897 un Italiano comprò 8 ettari di terra per 1000 dollari da uno che vi era venuto nel 1865 e aveva bonificato il suolo; ma non ne aveva ricavato un buon risultato. La proprietà vale ora 2500 dollari ed il compratore Italiano vi ha aggiunto 8 ettari. 4) Quarantadue anni fa un tratto di 28 ettari fu comprato per circa 550 dollari e poco di poi rivenduto ad un Americano per 3200, quando il suolo era in parte diboscato. Questa terra fu rivenduta di nuovo, ed è, ora, tutta nelle mani d' Italiani. 5) I figli di uno dei primi venuti lo abbandonarono, e la sua terra, quantunque ben lavorata, era in cattive condizioni ed improduttiva. Un povero Italiano la comprò per 1400 dollari, a credito. Lavorò penosamente, la liberò delle erbe parassite e la pagò col primo provento delle sue raccolte.

Questi esempi ci danno il carattere della immigrazione Italiana in Hammonton. Non c' è stato nessun incoraggiamento da parte degli Americani. Gli amici hanno aiutato gli amici, un Italiano ne ha assistito un altro, ed ogni immigrante ha saputo, arrivando in Hammonton, che c' era posto anche per lui. E' così che si sono formate anche molte città, e che le comunità rurali americane hanno attirato l' immigrazione. E' molto difficile però stimolare la immigrazione in massa degli Italiani alle fattorie del Sud e dell' Occidente perchè bisogna prima che là, come altrove, vi si sieno già stabilite famiglie d' Italiani.

Hammonton
sede d'una tipica comunità Italiana

Secondo il censimento del 1885 Hammonton aveva 2525 abitanti saliti a 3838 nel 1890. Nel 1895 questa cifra cadde a 3428; ma il censimento del 1905 segnò l'aumento a 4,334.
La seguente tavola mostra la percentuale delle nazionalità in Hammonton.

Numero e percentuale della popolazione di Hammonton secondo la nazionalità nel 1905

Natività	Numero	Percentuale
Americani	2875	66.4
Inglesi	66	1.5
Irlandesi	36	0.8
Tedeschi	62	1.4
Italiani	1223	28.2
Diversi	72	1.7
Totale	4334	100.0

Inclusi nei 2875 nati in America vi sono i figli d'Italiani nati negli Stati Uniti e viventi ad Hammonton.
Il censimento delle scuole dà 677 ragazzi frequentanti la scuola; alcuni di essi sono Italiani ma questo calcolo è incompleto. E' probabile che vi sieno 700 persone di parentela Italiana in Ham-

monton. Il risultato di questo censimento è interessante poichè dimostra non solo la larga proporzione della popolazione Italiana, ma altresì la preponderanza degli Italiani su la mistura delle altre nazionalità che si trovano nella regione. Quando si parla degli Italiani nel New-Jersey generalmente si cita Vineland perchè fu fondato da un Italiano senza alcun aiuto o iniziativa estranea. Nell'ultimo censimento a Vineland si trovarono 475 persone di nazionalità Italiana.

« Il sud del New-Jersey è una formazione geologica recente. Il paese è piano, bene irrigato naturalmente, coperto da foreste che danno, dopo il diboscamento, un terreno fertilissimo. Il diboscamento puro costa caro—quantunque sia redditizio— e il costo del diboscamento va dai 10 ai 35 dollari e in certi casi ai 65. Non è però così fertile tutto, anzi dal sud di Manchester a Mullica River il paese è assolutamente desolato ».

« Se si eccettuano gli orti lungo la strada provinciale non più del 2 % del suolo in questa regione è soggetto a coltivazione. Quà e là si trovano traccie di fabbriche abbandonate, villaggi deserti, e i resti di vetrerie nella boscaglia. Un grave silenzio pervade dovunque». Questa descrizione è data da un geologo incaricato dallo Stato di un rapporto nel 1888.

La seguente è data del paese prima che gli Italiani venissero a redimerlo. « Il suolo, varia di elevazione, è ghiaioso, sabbioso, e le parti basse ricche di fango. Quest'ultimo è fertile essendo di origine alluvionale. Quà e là spazi di suolo fertile si scuoprono fra le sabbie e le ghiaie. Le pinete occupano il suolo più povero. Per renderlo produttivo bisogna fertilizzarlo molto. Gli incendi di foreste sono frequenti e dannosissimi. Nondimeno con molta cura questo terreno può essere reso utilissimo ». E lo è difatti, ora.

I vegetali ed i frutti, lamponi, fragole, pesche, mele, uva, poponi vi crescono meravigliosamente; il New Jersey è rinomato per le sue ottime patate, e per la prodigiosa loro quantità. Generalmente se ne ritirano da 200 a 300 *bushels* per *acre* e qualche volta anche 450; ed hanno il valore doppio di quelle del Delaware e della Virginia.

Il clima è buonissimo, ed ora grazie alla coltivazione non vi sono più paludi, e la malaria vi è ignota.

Le foreste e la salubrità del clima sono due delle attrazioni del paese. Due colonie tedesche furono fondate nel paese; una a Folsom a circa 3 miglia da Hammonton e l'altra a Egg Harbor, quando i tumulti contro gli stranieri obbligarono gli immigrati di Filadelfia a cercare dimora altrove.

Forte impulso al popolamento del paese fu dato nel 1856 quando fu intrapresa la costruzione della ferrovia *Camden and Atlantic.* Allora circa 1857 coloni Inglesi, Tedeschi e Irlandesi arrivarono, e si diedero alla coltivazione del grano, di altri cereali e dei vegetali; ma con poco successo.

La tavola seguente mostra la diminuzione della produzione nei due generi principali.

Produzione del frumento e del granturco
nelle contee Cumberland e Atlantic nel 1879 e 1889

Anni	Contea Atlantic		Contea Cumberland	
	Frumento (bushels)	Granturco (bushels)	Frumento (bushels)	Granturco (bushels)
1879	98173	10519	602546	157952
1889	63970	1152	491590	117037

L'inizio della guerra civile portò lo scoraggiamento nella colonia di Hammonton, ma fu la guerra altresì che provocò la grande domanda di frutta che ne rese possibile lo sviluppo.

Dopo la guerra, molti soldati che avevano avuto agio di osservare le possibilità che offriva Ham-

monton alla coltivazione dei frutti tornarono a
stabilirvisi. Nel 1866 c'erano 1422 abitanti con 2031
acri in coltivazione, dei quali: 304 a fragole, 212
a more, 40 a bacchi di mortella, con 53000 viti, 23907
peri, 829 susini, 53767 lamponi.

In un solo giorno del 1865, 70000 *quarts* di fragole furono venduti sul mercato Le noci furono
il frutto che rese di più e larghi tratti di terreno
furono dedicati esclusivamente alla coltivazione
delle medesime.

Si trovarono, grazie ai nuovi tracciati ferroviarii,
nuovi sbocchi al commercio della frutta in New York
e Filadelfia ed anche in Boston, Providence e Pittsburg. Questo fu il periodo in cui gli Italiani arrivarono numerosissimi sia come raccoglitori, sia
come coltivatori permanenti.

Così lo Stato del New Jersey si è accresciuto di
una ricchissima regione che pareva soltanto atta
allo sfruttamento del legname delle foreste. L'ispe-
zione geologica del 1888 riconobbe che circa 20000
persone s'erano stabilite nel paese e il governatore
nel 1890 si è congratulato dello sviluppo che ha preso.
La infusione di emigrati vigorosi, Inglesi, Tedeschi,
Irlandesi ha rafforzato il carattere degli abitanti e
ha costituito in Hammonton una popolazione altamente stimata. Nei passati venti anni lo Stato ha
spinto alla colonizzazione facendo notare la fertilità
del suolo, la vicinanza dei mercati etc. Una trasformazione agricola si rese necessaria e ai cereali
si sostituirono ortaggi e frutta. Questa trasformazione: 1° esige lavoro più individuale ed assiduo;
2° rende possibili le piccole fattorie; richiede più
conoscenze scientifiche nel coltivatore; ma la concorrenza del sud soffocò la produzione dei cereali
nel paese; il lavoro divenne meno vantaggioso, ed
una crisi ne risultò in seguito alla quale molti
tratti di terreno furono abbandonati, parecchi coltivatori rovinati — e fra questi alcuni Italiani — e
la produzione diminuì rapidamente.

L'*Annuario* del New Jersey per il 1901 dà le seguenti cifre di suolo attualmente coltivato (1).

27000 acri a more
900 » a fragole
800 » a lamponi
400 » uva
200 » pere
300 » bacche di mortella.

In più si raccolgono 3000000 *quarts* di frutta che crescono selvaggie nella foresta.

Ora però si coltivano molto le fragole le quali fanno una forte concorrenza a quelle del sud. Però i frutti sono colpiti dalla malattia e le condizioni diventano più svantaggiose. Alcuni americani hanno fallito il loro scopo, altri continuano con i vecchi sistemi, sperando in una felice soluzione del problema. Gli Italiani hanno mostrato di essere buoni lavoratori, e fin ora con buoni successi in generale; rimane da vedere se essi svilupperanno l'intelligenza atta a fronteggiare le nuove condizioni

Gli Italiani quali coltivatori di frutta e fittavoli

L'estensione media delle fattorie in Hammonton è di 20 acri: quella delle fattorie Italiane è di 14,6.

Un uomo che lavori solo non può coltivare più di 20 a 30 acri di questo suolo. Nel 1905, 288 Italiani possedevano 4,226 acri di terreno, divisi in queste proporzioni:

88 poderi di uno e due acri
38 » da due a cinque acri
39 » da cinque a dieci acri
50 » da dieci a venti acri
54 » da venti a cinquanta acri
16 » da cinquanta a cento
3 » da cento a centocinquanta.

_____

(1) L'*acre* equivale a poco più di 0,40 *ettaro*.

La Rivista

Gli Italiani usano affittare 5 acri di terra che lavorano in aggiunta ai loro lavori ordinari.

Coltivano in questi loro piccoli possessi fragole, lamponi, patate dolci e patate bianche, e uva; ma la fragola è una delle coltivazioni da loro prediletta. Non si curano molto della cultura scientifica, e pochi fra loro possiedono verzieri.

Coltivano i legumi per il loro uso privato, e alcuni mandano al mercato pomodori, patate, ed altro per mezzo dei loro ragazzi. Essi hanno introdotto la cultura di parecchie verdure Italiane, e crescono con successo cavoli, peperoni, prezzemolo, sedani e asparagi. Tosto che un italiano lo può acquista un cavallo per portare al mercato le sue derrate. Le vacche non sono utilizzate. Le capre danno sovente l'unico latte possibile. Quasi ogni italiano possiede un porco ed i polli per il proprio uso, e ne ha molta cura. Una caratteristica dei coltivatori italiani è la economia, i primitivi metodi di coltivazione e la iniziativa. Il loro lavoro di zappa ha bonificato il suolo, e le loro donne, in questo, sono state utilissime.

Il loro lavoro di mondatura accuratissimo, ha salvato intiere piantagioni dalla distruzione. Sono attaccatissimi ai vecchi sistemi e pregiudizi, perfino nella scelta del tempo per piantare.

La economia degli italiani è notevole. Raccolgono le foglie per le vie per nutrire i loro animali. Sono riluttanti a potare i loro alberi, e nei loro orti ogni pollice di terreno è utilizzato; anche lo spazio che gli americani usano destinare ai fiori od al gioco. Questo sistema degli Italiani è utilissimo per fertilizzare questo terreno che ne ha tanto bisogno. Avendo veduto che il concime artificiale dà ottimi risultati ne sono divenuti avidi compratori, ma non lo usano intelligentemente. Alcuni di essi hanno assistito alle lezioni di agricoltura delle cattedre ambulanti, ed hanno capito la utilità della rotazione delle coltivazioni e la praticano, e si può dire che un coltivatore italiano può stare a fronte d'un tedesco vantaggiosamente.

La coltivazione delle bacche è assai facile, e gli Italiani sono impiegati anche dai loro connazionali alla raccolta. Essi, ingaggiati da un padrone, lavorano a cottimo; vengono anche da altre città con le donne ed i ragazzi e guadagnano da 1 dollaro a 1 dollaro e mezzo al giorno. Essi le impacchettano nelle cassette e le caricano su i treni che le trasportano ai varii mercati.

Le spese del colono italiano sono ridotte al minimo. Il lavoro di tutta la famiglia che aiuta il capo e opera ai campi insieme a lui, rende possibile ad una famiglia italiana di vivere là dove una famiglia americana morirebbe di fame; una famiglia di 8 persone vive comodamente sul prodotto di circa 3 a 5 acri di terreno. Gli italiani dicono che il loro successo è dovuto alla persistenza ed anche alla parsimonia dei loro esperimenti e della loro pratica. Il suolo del New Jersey è troppo povero per potervi tentare esperimenti costosi.

Il successo degli Italiani è dimostrato in questa lettera del Prof. Iohn B. Smith: « Ho veduto con molto interesse il progresso dei coltivatori italiani. Originariamente essi vennero portati nel New Jersey da Filadelfia unicamente come raccoglitori di bacche. Gradualmente alcuni di essi si stabilirono e acquistarono piccoli lotti di terreno. I loro metodi di coltivazione erano primitivi. Non avevano macchine, avevano pochi strumenti e tutto era da loro fatto alla mano. Erano sporchi nei loro costumi, e nelle loro famiglie; non si elevavano quanto il loro vicinato; ma riuscivano a possedere altrettanta moneta, se non più, perchè spendevano meno. Da allora le cose sono mutate. Il loro modo di vivere è migliorato; ed altresì il loro sistema di coltivazione. Gli italiani non usano la macchina tanto quanto gli americani; ma lavorano di più il loro raccolto. Il resultato è che guadagnano di più spendendo meno. I migliori fra i coltivatori italiani non valgono i migliori americani; ma in generale valgono quanto la generalità di questi. Essi capiscono i danni prodotti dagli insetti e quando si è riuscito a persuadere uno di essi dei rimedi da adottare, tutti gli altri lo seguono ».

E' notevole anche uno dei danni della loro economia. Essi coltivano e producono frutta tanto belle quanto quelle degli Americani, ma siccome le impaccano in brutte casse e senza cura, il loro prodotto è meno apprezzato sul mercato: ma siccome alcuni italiani si sono accorti di questo danno, si può sperare che presto tutti gli altri vi porranno rimedio.

Alcuni esempi sul profitto netto del raccolto possono essere dati con interesse. 1) Un italiano nato in America, nel 1903 ricavò 500 dollari su circa due acri di lamponi: 2) Un giovine italiano ricavò 160 dollari da circa 1/4 di acre. 3) Un altro colono ricavò 180 dollari da un acre di fragole. 4) 5 acri di fragole produssero 200 dollari.

Ancora è utile vedere i profitti che gli italiani sanno ricavare. 1) Un'italiano della prima generazione stimò 75 dollari annui all'acre il prodotto dei suoi dieci acri di suolo coltivato a bacche e patate. 2) Un'altro della seconda generazione, con 8 acri di bacche ed una di viti ricavò 700 dollari all'anno. 3) Un giovane italiano con circa 12 acri ricavò 400 dollari all'anno, dopo pagatine 50 per fertilizzare, e da 75 a 125 per la raccolta. 4) Uno dei primi venuti si contentava di 1000 dollari su 150 acri. 5) Un italiano, dopo quindici anni di residenza, che non parlava inglese, e aveva il suolo coltivato a mais e ribes si diceva guadagnasse da 3000 dollari a 4000 dollari; nel 1903 si dice guadagnasse 4570 dollari netti su i suoi 50 acri di lamponi. 6) Un giovine italiano che lavorò per un americano due consecutive stagioni di sei mesi, e insieme a suo padre, col denaro messo da parte lavorando undici anni prima presso un'altro, comprò un podere per 2000 dollari contanti ed una ipoteca di 500. In una stagione (1906) pagò l'intera somma e gli avanzarono 800 dollari. 7) Un altro da barbiere fattosi agricoltore comprò un podere di 70 acri : ha ora una bottega ed è agente di una compagnia di concime artificiale.

La tavola che riportiamo alla pagina seguente mostra il numero degli italiani ed americani della Unione di coltivatori di frutta classificati secondo le loro entrate lorde per vari anni dal 1889 al 1897.

I coltivatori di frutta di Hammonton sono stati sempre soggetti allo sfruttamento delle compagnie ferroviarie, e per difendersi essi hanno dovuto ricorrere alla associazione cooperativa, e farsi essi stessi i trasportatori dei loro prodotti.

Una delle loro prime società non ebbe successo; la più riuscita è quella che ha per titolo *Shipper's Union*, di Hammonton composta di Italiani e Americani: anzi nel 1900 su sessantacinque membri ventitre erano italiani. E dopo questo il Club degli agricoltori di Elm. Il Club di questa città è composto unicamente di agricoltori ed è riuscito con la indefessa opera a fare onore ai suoi membri e a fare il vantaggio dei loro propri mercati; ha magazzini di deposito, mezzi di trasporto e risparmia il 3 % delle commissioni, permettendo ai suoi membri dei dividendi del 2 1/2 per cento su un totale di dollari 61,018,75 nel 1902, salito a dollari 76,511,00, nel 1905.

Si è detto che fra gli italiani primi venuti pochi amavano il lavoro quantunque fossero buoni lavoratori. Questo può anche esser vero; ma l'agricoltura nel sud del New Jersey non è assai rinumerativa per permettere di vincere la concorrenza senza molto lavoro. E' venuto il tempo in cui l'importanza di sforzi intelligenti in questo genere di agricoltura è diventato evidente. La seconda generazione di italiani si è sviluppata con idee americane: ed ora essi migliorano i loro metodi per dare sicurezza che nel prossimo futuro essi sapranno

NUMERO DEI MEMBRI CON UNA ENTRATA IN DOLLARI

Anno	Nazionalità	Sotto 50	da 50 a 100	da 100 a 200	da 200 a 300	da 300 a 400	da 400 a 500	da 500 a 600	da 600 a 700	da 700 a 800	da 800 a 900	da 900 a 1000	da 1000 a 2000	Sopra 2000
1889	Italiani......	4	2	3	5	5	5	3	–	2	2	1	7	7
»	Americani...	7	12	28	23	15	12	12	8	5	8	4	25	15
1897	Italiani......	4	2	14	13	9	6	4	5	3	4	2	12	—
»	Americani...	6	10	9	5	5	4	7	6	2	3	2	10	3

progredire in conformità ai cambiamenti che hanno
avuto luogo nella agricoltura del New Jersey.

Gli Italiani nell' industria

Hammonton ha la fortuna di combinare la industria all' agricoltura. Alla seconda generazione
Italiana, come all'indigeno americano il lavoro a-
gricolo è tedios; gl' Italiani dunque anelano l'ecci-
tamento febbrile dell'industria. In città vi sono due
fabbriche di scarpe; due vetrerie, un lanificio, una
fabbrica di calze, a Winslaw Junction una mattonaia, e una segheria etc.

Nel 1906 queste varie industrie davano lavoro a
403 uomini e 113 ragazze, e di questi: 211 uomini
e 64 ragazze erano Italiani.

La tavola a pag. 62 mostra i loro guadagni.

La vetreria fu aperta nel 1899 con impiegati
Americani, e nel 1903, 20 Italiani e 30 Americani vi
lavoravano. Dopo lo sciopero del 1900 il numero
degli Italiani, specialmente politori — lavoro che
gli Americani non amano fare — aumentò; nel 1905
c'erano su 46 maschi e 6 femine 25 Italiani maschi
e 2 femine. E così pure per le altre fabbriche dove
il numero degli Italiani è sempre rilevante.

Uno dei caratteri degli Italiani è di essere poco
stabili al lavoro. Molti di essi lavorano anche la terra
e qualche volta, nel tempo della raccolta, le fabbriche hanno dovuto soffrire per la loro irregola-
rità, e tanto che spesso durante la raccolta delle
more la fabbrica di calze ha dovuto esser chiusa.
Ma poco a poco si sono familiarizzati con le esi
genze del lavoro ed il desiderio di guadagnare
denaro ha fatto sì ch'essi hanno finito per essere
puntuali, sempre, e tanto nella fabbrica di calze,
come alla calzoleria si sono abituati alle macchine.
La resistenza degli Italiani è superiore a quella degli Americani, ma mettono più tempo ad abituarsi
ai meccanismi ed i loro progressi sono lenti.

Contrariamente agli Americani le difficoltà non

li scoraggiano, e s'ingegnano pazientemente a superare gli ostacoli.

Industrie	Numero degli impiegati		Guadagni approssimativi degli Italiani
	Totale	Italiani	
Mattonai . . .	250 masc.	150 masc.	da 1,25 dollaro a 1,40 al giorno
Fabb. di scarpe	71 maschi	30 maschi	da 5,50 dollari a 10 per settimana
	50 femm.	10 femm.	da 1 dollaro a 4, 2 per settimana
Vetreria . . .	69 maschi	30 maschi	da 3 dollari a 6 per settimana
	8 femmine	4 femmine	da 4 dollari a 6 per settimana
Fabb. di calze .	55 femm.	50 femm.	da 4 dollari a 6 per settimana
Segheria . . .	13 maschi	1 maschio	— —
Ferrovia . . .	—	10 a 15 m.	da dollari 1,25 a 1,30 al giorno
Magazzini. . .	6 maschi	4 maschi	da dollari 5,50 a 10 per settimana
	8 femmine	1 femmina	
Mercato della carne . .	4 maschi	1 maschio	
Merceria . . .	2 femmine	1 femmina	
Mag. alimentare.	5 maschi	4 maschi	

Il pregiudizio del lavoro a minor mercato era contro di loro. Alla mattonaia gli Italiani sono pagati 1 dollaro al giorno. Nel 1894 scioperarono per

essere pagati quanto gli Americani cioè 1 dollaro
e 20. Gli Italiani guadagnano ora da 1 dollaro e
25 a 1 e 40. Alla calzoleria i salari vanno da dollari 5,50 a 18 la settimana, gli Italiani pigliano le
paghe di dollari 5,50; alcuni di loro da 9 a 10 dol-
ari. E così in altre industrie.

Nella prima generazione di Italiani in Hammonton c'erano un fabbro, alcuni falegnami, un ri-
quadratore e due fornai: nella giovine vi è un ope-
ratore telegrafista, un sellaio, ed altri.

Ve ne sono anche impiegati in lavori speciali
per i quali guadagnano da 1 dollaro e 15 a 1,30 al
giorno. Lavorano anche come gazisti a 1 dollaro
e 50 al giorno. Siccome il lavoro di piombaio e
fontaniere cresce ogni anno, gli Italiani trovano
facile impiego nelle opere di terrazziere e mano-
vale.

Sono numerosi nelle intraprese pei pubblici ser-
viz (acqua, gas). In alcuni casi essi vogliono intra-
prendere qualunque specie di lavoro benchè non
sempre l'efficienza corrisponda alla buona volontà.

Alcuni hanno botteghe o sono appaltatori. Uno
dei principali appaltatori della città è un figlio di
Italiani nato in America. Fra i meridionali vi sono
un ghiacciajolo e i Fratelli Turchi sono i principali
droghieri di Elm. Molte delle botteghe tenute da Italiani sono piccole e frequentate solo da Italiani.

Siccome molti Italiani fanno acquisti in botteghe
americane è un vantaggio per i proprietari avere
fra i loro garzoni anche degli Italiani. Il magazzino
generale della città impiega una donna Italiana e
4 uomini più un'altro, pure Italiano, come stalliere.
La donna è molto stimata e prende una buona
paga. Questo magazzino impiega giovani Italiani
perchè sono superiori in cortesia e capacità ai più
degli impiegati Americani. Raramente le ragazze
italiane servono in case Americane, e ciò soltanto
per un accomodamento temporaneo, perchè i genitori non amano averle fuori di casa specialmente
la sera.

Alcune Italiane sono abilissime lavandaie e stiratrici. Anche nei lavori casalinghi, nel cucire di bianco e in corredi le donne italiane sono molto apprezzate. Prendono lavori di cucito e ricamo e lavorano nelle loro case. A riportare il lavoro sono assai inesatte, specialmente per le ore fissate, ma è accaduto più di una volta che hanno scrupolosamente riportato del denaro pagato loro in più.

D'inverno lavorano più che d'estate perché in questa stagione si occupano anche del lavoro nei campi e della raccolta. Anche il lavoro di pantalonaia per le Filippine, il West India e l'Australia si fa molto ad Hammonton, e quando piove, e se il lavoro è urgente le donne eseguiscono le loro commissioni anche durante l'estate e riescono a guadagnare da 1 a 8 dollari la settimana.

Molte ditte Americane che prima si servivano esclusivamente di Americani ora ricorrono agli Italiani in ogni specie di lavori.

L' acquisto della proprietà

Per mezzo degli accertamenti del fisco é facile vedere la proporzione di proprietari fra Americani e Italiani. Secondo il più recente (1906-07) gli Italiani pagano dollari 4493,67 cioè il 17,70 % del totale delle tasse in 25407 dollari e 79 cent. Secondo l'accertamento del 1906-07 il valore di 26421 acri di terreno era di dollari 1,149,021, e c'erano 323 Italiani possessori di 4846 acri per doll. 176,575.

Ci sono nel libro delle tasse 1370 nomi; di questi circa un terzo, 448 sono Italiani, dei quali 96 pagano solo la tassa di famiglia. Quarantotto Italiani pagano la tassa sugli immobili, ma non pagano la vera tassa su la proprietà.

La tassa sugli immobili pagata dagli Italiani rappresenta il 18/4 per 100 sull'imponibile e il 15,4 per 100 sul valore accertato. Il grande valore delle proprietà americane è dovuto in parte alle migliori proprietà. Le case Italiane ordinariamente, sono gravate per 200 dollari.

La proprietà immobiliare degli Italiani, secondo l'accertamento del 1906-1907 su la proprietà fondiaria va, da un acre a ottanta acri — 82 Italiani, possiedono un acre o meno per uno, 13 possiedono da 50 a 80 acri per uno — del valore da un minimo di 25 dollari ad un massimo di 2300.

Nella tassa *per capita* 183 Italiani pagano ciascuno 25 dollari, il minimo; 1 paga il massimo cioè 575 dollari, ripartiti in 400 per fabbrica e 175 per tassa individuale. Le alte quote pagate da Italiani sono rappresentate da un minimo di dollari 25,84 pagati su un valore di 1350 dollari ed un massimo di 96,22 su un valore di dollari 5175.

Nel suo N.º del 18 ottobre 1902 un giornale del New-Iersey stampava : « Ogni settimana o quasi una casa è costruita fra Rosedale e Winslow dagli Italiani ».

Questo rappresenta la tendenza italiana ad acquistare un acre di terreno e costruirvi una casa. Sul finire del 1905, 43 case erano state erette in Hammonton, 20 delle quali da Italiani.

La tassa personale grava su le mobilia della casa, i cavalli, carri, magazzini ecc. In questo gli Italiani sono gravati più degli Americani, 8,4 % della somma pesa su soli 304 Italiani. L'Americano ordinario con una comoda casa è tassato per 50 dollari, l'Italiano per la sua misera casuccia per 25 e se ha un cavallo, per 50. E questa è la dimostrazione di una ingiustizia fatta agli Italiani.

La città di Hammonton è stata costruita dalle sue cooperative di costruzione e dalle associazioni di prestiti, e gli Italiani hanno largamente usato di queste due forme di aiuto per costruirsi le case. Vi sono due associazioni: l'Associazione di costruzioni e prestiti di Hammonton, organizzata nel 1871, e la Associazione operaia di costruzioni e prestiti. La prima nel 1904 aveva 460 azionisti, settantanove dei quali erano Italiani. Il totale dei prestiti fu di dol-

lari 182,156 dei quali 56,600 rappresentavano la parte
di Italiani. La seconda aveva nel 1904, 553 azionisti dei quali 123 Italiani. Le donne hanno diritto
a sottoscrizione in comune con i loro mariti.

Gli Italiani usano costruire case di due stanze di
16 piedi per 16 piedi e costano circa 350 dollari.
Le case più costose con maggior numero di stanze
e portico e veranda sono costruite dalla associazione col metodo della ipoteca. Sovente in queste
compre su contratto gli Italiani pagano somme in-
debite. La ricchezza degli Italiani di Hammonton
può essere calcolata da queste cifre.

Un italiano possiede sei stanze 14 acri di terra valu-
tata 10,000 dollari; un altro possiede una fattoria ed un
commercio che gli rappresentano 15,000 dollari; un
terzo ha accumulato una proprietà dai 20,000 ai 30,000
dollari. Il risparmio è la caratteristica degli Italiani

Sul primo tempo molti di loro portavano seco il
loro denaro, o lo tenevano nelle loro case. Poi lo
depositarono presso fidati amici Italiani o Americani, ma dacchè c'è la Banca essi se ne servono
per il deposito.

Alcune cifre saranno utili. La Banca del Popolo
che è la istituzione di risparmio, e paga il 6 %
sul capitale, fu aperta nel 1887. Nel settembre del
1890 c'erano 450 depositi per 87080 dollari. Di tali
depositi tre erano di Italiani e sommavano a meno
di 500 dollari.

Nel 1904 su 260,779 dollari in deposito 56,614 o
il 21 % erano di Italiani; e su 88768 dollari in
conto risparmio paganti il 3 %, 26231 cioè il
29,6 % erano di Italiani.

Gli Italiani fanno facilmente buoni guadagni,
e ciò unito alla loro sobrietà, ed al ricavato delle
loro terre permette loro l'accumulazione di discreti
risparmi. Abbiamo detto delle altre paghe nel lavoro agricolo, diremo ora che per dieci ore di lavoro le paghe oscillano da dollari 1.25 a 1.50 per gli
uomini e per le donne da 50 a 75 cent.

Molto denaro è mandato in Italia per aiutare le

famiglie, chiamare gli amici, o sotto forma di risparmio. E' difficite stabilire la somma precisa: il solo Ufficio postale di Hammonton nel 1903 mandò in Italia per 408 mandati ammontanti alla somma di dollari 8774.39. Mandò pure 519 lettere raccomandate molte delle quali dovevano certamente contenere denaro. L'Italiano è raramente a carico della città. Nel 1900 la spesa per i poveri fu di soli dollari 400; nel 1901 di dollari 582.40.

Nel 1901 fu notata la morte di un solo italiano che figurava tra i poveri.

E' evidente che la proprietà posseduta dagli Italiani in Hammonton rappresenta i resultati di un lavoro duro e ostinato. Gli Italiani arrivati con quasi niente, hanno migliorato le loro condizioni e mettendo in cultura nuove terre hanno accresciuto la capacità produttiva e consuntiva della comunità.

Il Regime di vita

Per apprezzare i cambiamenti di vita di questa gente è opportuno conoscere la vita che fanno nei loro villaggi di origine, ove vivono accatastati in piccole case insieme al bestiame. I mercati per i loro prodotti sono poveri ed i contadini non guadagnano che magri salari. Quando lavorano per altri sono pagati 20 e 30 cent. al giorno; e pagano su ciò tasse molto alte. Il costo della vita è basso e si può vivere con cinque soldi al giorno. Mangiano polenta, maccheroni, verdura e frutta e quasi punto carne. Il loro vestiario è poverissimo.

La misera condizione di questi Italiani meridionali che vengono ad Hammonton è notevole fra i raccoglitori di frutta: per i quali gli alloggi provveduti sono miserissimi senza che essi se ne lagnino: i proprietari cercano di giustificare queste condizioni dicendo che questa gente non si cura di aver meglio, nè saprebbe del resto mantenerlo. Le donne portano vecchie sottane, giacchette e grembiali, con fazzoletti vistosi in testa. Gli uomini vecchi abiti di confezione; ed i bambini una sola veste malandata, e spesso vanno nudi. I calabresi vivono in grande parte di pane secco che infilano su un bastone e vi pompano sopra l'acqua prima di mangiarlo. I Siciliani vivono un po' meglio; in questi ultimi anni il loro modo di vita e di apparenza è migliorato un po'. Fra i residenti Italiani vi sono varii gradi di sviluppo. I nuovi venuti sono sozzi. Essi non hanno alcuna idea della cura fisica dei bambini, e tengono le loro case piene di cani e galline. D'altra parte le case di quelli della seconda generazione sono pulite Ogni Italiano desidera avere la sua casa e questo sembra essere uno dei caratteri della razza. E per avere una casa e costruirla lavorano tutti e tutti partecipano alla costruzione spendendovi tutto il tempo che possono risparmiare sul lavoro, specialmente la domenica. Le vicinanze delle loro case non sono bene mantenute. Intorno essi buttano tutti i rifiuti, tutti i rigetti: non tengono, come gli Americani fiori su i loro davanzali e le loro verande. Ma gli Italiani della seconda generazione sono già diversi e si vedono dei segni incoraggianti di miglioramento nella cura delle loro case.

Ci sono comparativamente poche piccole case in Hammonton, e gli Italiani se le sono costruite da se su tre tipi: 1) Casa di due stanze; 2) Casa di quattro stanze; 3) Modificazioni del secondo tipo. La casa di due stanze è di 16 piedi per 16, la seconda è di 32 per 16 e questa qualche volta serve a due famiglie, ed ha due porte d'ingresso. Ultimamente hanno migliorato questo tipo aggiungendovi balconi e portico. Alcune di queste case sono state costruite nel 1905.

Le case sono di legno, e, benchè piccole, solide e ben costruite. Ogni casa ha una tettoia in fondo al giardino, sotto la quale d'estate fanno la cucina. Hanno anche il forno per fare il pane. Le case sono intonacate e al muro fornite di chiodi che fanno l'ufficio di attaccapanni.

I mobili sono ridotti al minimo strettamente necessario. Alcuni oggetti da cucina, poche sedie e tavole ordinarie, e casse e bauli.

Le case di altri che son quà da più lungo tempo sono ammobiliate meglio: ma in tutte il letto si di stingue per la pulizia. Quantunque la casa sia sporca il letto è pulito, e ornato da coperte a maglia e ricami su le fodere e i bordi delle lenzuola: ricami fatti a mano dalle donne. Ci sono qualche volta degli scrittoi, ma la toeletta manca sempre.

Le zanzariere, e le tendi bianche alla finestra accennano il progresso degli abitanti di una casa: Un salotto e varie camere costituiscono un altro passo avanti nel progresso: e le case della seconda generazione sono fornite di tavola da pranzo, tappeti, tende, buone sedie e sovente anche del piano-forte. Quando l'Italiano ha danaro da spendere imita volentieri, nell'arredamento della casa, gli Americani. Siccome in Hammonton è stata messa l'acqua potabile anche il bagno è diventato familiare ai giovani italiani, ed il telefono è stato rapidamente messo in uso da loro e fra gli abbonati di Hammonton, vi sono 29 italiani.

Come abbiamo detto essi migliorano rapidamente il loro sistema di vita; basta che le possibilità di migliorarlo si mostrino loro favorevoli. In un caso una famiglia visse per anni in un domicilio provvisorio; nel 1905 questa famiglia comprò una proprietà in città e vi costruì una bella casa e vi aggiunse una bottega. Un'albergatore Italiano ha migliorato una vecchia proprietà, facendone una piacevole costruzione ed aggiungendovi la serra.

Un altro venuto qua con niente, dopo pochi anni comprò una proprietà per 1500 dollari, e quantunque fosse improduttiva egli ha saputo farla redditizia e si è acquistato ora un podere di 20 acri, e così di seguito per molti altri.

Gli italiani che vengono in Hammonton vivono

e mangiano come i raccoglitori di bacche, sul principio, con l'aggiunta di carne e uova. Si procurano
l'olio d'oliva, la salsa di pomodoro e i maccheroni
dall'Italia, dai loro orti ricavano la verdura e le
frutta, ed il loro vitto in Hammonton è superiore
a ciò che mangiano gli Italiani in altre città americane. Le donne italiane s'intendono pochissimo
di cucina, e se ne occupano poco.

Quanto al vestiario ne abbiamo parlato; è opportuno dire che gli italiani della seconda generazione hanno abbandonata l'idea di mandare nudi o
quasi nudi i loro bambini e d'inverno li coprono e li
vestono di abiti caldi. Alcune donne italiane posseggono macchine da cucire e se ne servono abil
mente. Hanno imparato dalle americane l'uso della
cassa e del prezzo fisso nelle spese. La vita fuori di
casa è molto praticata per necessità dagli italiani.
Le donne ed i loro ragazzi lavorano nel giardino,
e molti lavori della casa sono fatti in giardino.

Ora i giovani prendono presto le abitudini americane, quantunque abbiano l'idea che « costano
troppo ».

Il regime di vita ha grande importanza rispetto
alla mortalità. E' noto che molti degli Italiani che
vengono in America muoiono per malattia od in-
infortuni. La febbre palustre, la malaria, il tifo
sono malattie comuni fra gli Italiani. Ma la malaria ed il tifo non vi sono nella regione dei pini
e Hammonton sotto questo rispetto è preziosa. Il
suo clima è anche notevolmente indicato contro la
tubercolosi. Le statistiche della mortalità l'accennano piccola fra gli adulti, ma assai forte fra i
bambini di meno di 10 anni, quantunque la percentuale sia minore che nelle città italiane e nelle
grandi città americane. I ragazzi nati in Hammonton sono notevoli por la floridezza, la forza e la
salute.

Uno degli argomenti contro la immigrazione italiana è stato che essi avendo vissuto tanto tempo
in un bassissimo regime di vita, sono inadatti a

migliorarsi: ebbene la colonia italiana di Hammonton dimostra che essi, quando ne hanno l'.oppor-
tunità, rapidamente migliorano le loro condizioni:
prendono altre abitudini, e si conformano presto
agli usi del loro vicinato americano.

Le relazioni sociali

La vita di .famiglia degli Italiani, come quella
degli Ebrei può offrire agli Americani molti soggetti di emulazione· I legami familiari sono molto
stretti fra loro ed i piaceri della vita sono goduti
da tutti della famiglia insieme. Aver bimbi è considerato da loro una fortuna, una sfortuna non
averne e li amano moltissimo. Il matrimonio è da
loro considerato seriamente; le ragazze sono molto
sorvegliate e tenuta in grande pregio è la castità delle
donne. La gelosia degli uomini, e la chiesa impe-
discono molti. falli alle donne italiane. Le ragazze
godono pochissima libertà, e per andare ai balli
devono ottenere il permesso dai genitori.

E' uso fra gli italiani di fare i matrimoni secondo
le età. Il padre conclude per il matrimonio anche
contro la volontà delle figlie, e spesso mandano a
cercare i mariti in Italia. Le statistiche del matrimonio sono interessanti. Frequentemente i giovani
italiani sposano donne di altra nazionalità e più
spesso donne americane e i risultati di questi ma
trimoni sono soddisfacentissimi. Dal 1876 al 1901
ci sono stati 672 matrimoni dei quali 254, il 30 %
erano di italiani, e da allora il numero è aumentato.

Dall' 89 al 901 essi costituirono il 49 % dei matrimoni, mentre nel 1904, 1905, 1906 essi hanno
raggiunto la metà del totale.

La statistica mostra che in 109 matrimoni sul
totale di 250 le donne erano al di sotto dei 20
anni. C'era una sposa di 13 anni, due di 14, otto
di 15, diciotto di 16, quindici di 17, trentasei di
18 e ventinove di 19. La differenza normale fra la
età degli uomini e quella delle donne è mostrata

da alcune combinazioni: 27 e 37, 26 e 50; 21 e 31; 22 e 33.

Gli italiani hanno la teoria di tenere soggette le donne e castigarle, e spesso dicono di voler piuttosto tornare in Italia che sottomettersi alla legge che impedisce loro di battere le donne.

Le relazioni fra loro sono fraterne, e spesso vanno dall'una all'altra casa nelle sere d'inverno per sentire raccontare storie, ballare, fumare e intrattenersi. Sovente si vede la domenica le famiglie, andarsene sui carri a visitare amici e parenti che vivono lungi dalla città. Anche le visite a Filadelfia, ad amici e parenti sono frequenti.

Generalmente gli italiani sono cattolici. Ma in occasione di un matrimonio una famiglia si fece protestante. C'è fra essi un Battista, e la Chiesa Presbiteriana mantiene una missione dove un frate predica al popolo. Questa missione ha avuto buoni resultati. Ha una bella chiesa. Nel 1897 c'erano 31 membri con 63 bambini alla scuola domenicale e nel 1905 i membri erano cresciuti fin a 50.

La prosperità della Chiesa Cattolica di San Giuseppe è dovuta all' interessamento della colonia Irlandese cattolica, gl' Italiani erano troppo poveri per potere essere dei contribuenti efficaci, e per un periodo di tempo il rev. Gagley dovette fare in casa sua il servizio divino. La Chiesa ha, ora, un prete italiano e un tedesco, ma gli Italiani non sono tanto devoti quanto gli Irlandesi, e le loro idee in fatto di religione sono più da superstiziosi che da credenti.

La festa religiosa è il 16 luglio e in Hammonton è celebrata con più solennità che altrove. Dapprincipio le processioni erano proibite, ora sono permesse e gli Italiani spendono molto per questa festa perchè vogliono vestir bene, e rinnovano in questa occasione i loro abiti.

Usano anche far pagare i loro conti per il 15 luglio perchè intendono spendere liberamente il 16 che è la festa della Madonna del Monte Carmelo e per questa occasione fanno i fuochi artificiali, le luminarie. Nel 1904 spesero per questa festa 400 dollari e 600 nel 1905.

Anche i matrimoni sono speciali occasioni di grande festa fra Italiani e li praticano come si usa in Italia nelle campagne meridionali con spari di mortaretti e getto di confetti.

La principale società Italiana è la Società di Beneficenza per la quale la sottoscrizione è di 50 cents il mese e il sussidio di 6 dollari alla settimana. La società ha il suo medico.

La società per aumentare i suoi fondi dà delle feste alle quali intervengono numerosi anche i non Italiani. Vi sono anche molti Circoli e società di trattenimento, ma pochi Italiani vi sono ascritti, e questi tutti della seconda generazione.

Nelle loro relazioni con gli Americani, gli Italiani sono buoni vicini, e volentieri fanno regali, che consistono generalmente in mazzi di fiori da loro accomodati con molta arte, e prestano volentieri il loro aiuto in ogni occasione.

Gli Italiani sono essenzialmente un popolo sociale e cordiale, e la naturale cortesia del vecchio popolo è piena di attrazioni ed essi la manifestano molto facilmente. Il carattere pacifico di queste popolazioni è caratterizzato dal fatto che gli Americani non ne hanno paura. Essi non molestano gli Americani, non cercano aver con loro litigi o ragioni di offesa. Ciò che in loro offende gli Americani è la loro facilità nel mentire e nel rubare — vizi di popolo da lunghi secoli povero ed oppresso— ma essi abbandonano questi vizi non appena si trovano a contatto di libertà e di popoli superiori. In generale l'opinione su la loro onestà è buona. In Hammonton i botteganti dichiarano che gli Italiani pagano i loro debiti meglio degli Americani, e spesso nel pagarli si regolano con molta giustizia, pagando primi quelli che più li aiutarono ed ebbero in loro maggiore fiducia. Quando altri di altre nazionalità chiedono sussidi gli Italiani si accomodano a tirare avanti con i soli loro propri mezzi.

Rispettano molto la legge ma sono molto litigiosi, e spendono parecchio denaro nelle cause legali.

Del resto é bene, che essi imparino a sostituire i procedimenti legali alla violenza diretta nel raddrizzare i torti. Gli Italiani quistionano e si accapigliano facilmente tra loro. Gli Americani hanno molta paura del loro coltello o del loro stiletto. Le loro risse sono determinate per lo più da quistioni di donne o di bettola, quantunque essi non siano molto bevitori. Essi sono primitivi nella collera. Sono frequenti le piccole vendette per ogni sorta di motivi. La delinquenza occasionale non é rara; si sono verificati alcuni brutali omicidi e talora per denaro. Nelle risse spesso vengono adoperati i *revolvers*. Si disse che gli Italiani non sono ubbriaconi; e se tali rarissimamente provocano disordini pubblici. Tutto ciò che si riferisce alla loro delinquenza riguarda gl'immigrati arrivati da recente. I giovani della seconda generazione sono quieti e si comportano bene; si danno allo *sport* e spesso consacrano alla musica le ore di ozio.

Hammonton non é un sobborgo ma gli Italiani non pigliano una parte attiva nella politica; sono lavoratori ed uomini d'affari e si occupano esclusivamente dei loro interessi. Pigliano parte, ed anche piccola, soltanto alle elezioni.

Nel 1904 nella prima circoscrizione elettorale ci erano 394 votanti non italiani e 63 italiani, nella seconda 361 non italiani: un totale di 948 voti dei quali soli 193, il 20 per 100 italiani. I figli d'italiani nati ad Hammonton sono attualmente 1232 ma ce ne sono pochissimi naturalizzati. Essi prendono poco interesse negli affari politici della città, malgrado il loro numero rilevante o quantunque essi spesso prendono parte ai meetings, per protestare contro noie o ingiustizie a loro fatte.

Gli italiani di Hammonton si rivelano dunque come una popolazione socievole, con gusti semplici e naturali, con grande amore alla famiglia ed ai loro bambini. Sono ignoranti, primitivi, infantili, ma questi loro difetti si correggono facilmente con l'educazione ed al contatto dei costumi americani. La loro gentilezza e cortesia, il loro amore alla vita libera ed ai semplici piaceri sono contributi alla vita americana. Essi si dimostrano capaci di sviluppare al contatto degli Americani le loro migliori qualità e prendere una parte normale nella vita della comunità

La seconda generazione

Nelle pagine presenti si è spesso parlato del rapido miglioramento degli Italiani. Questo miglioramento può essere facilmente constatato ma non sottoposto a rilievi statistici. I migliori esempi ci sono offerti dalle scuole per i bambini.

Il problema della numerosa immigrazione nelle grandi città è diventato assai grave per noi e gli Italiani ne sono stati i più colpiti. Gli Italiani in quasi tutte le grandi città americane sono alloggiati male, in pessimi quartieri e luride case; spesso abbandonate da Americani come a Nuova Orleans, su i confini della città come a Chicago: la sovrapopolazione e l'agglomeramento generano malattie, vizi e delitti. Il loro vitto consiste sovente di sole frutta o di rifiuti di mercato. Questo genere di vita è dannoso specialmente ai bambini ed alle donne.

Le donne, abituate alla vita all'aria aperta nel loro paese sono facilmente colpite dalla tubercolosi e dalle altre malattie che dipendono dalla miseria, dal cattivo vitto e dalla mancanza d'aria e di luce. Gli effetti morali di queste condizioni sono pessimi. I bambini non hanno altro svago che la strada e ne corrono tutti i pericoli finchè hanno l'età di entrare nelle fabbriche, dove sono sottoposti allo sfruttamento più esoso. La dura vita della città costringe i genitori a toglierli presto dalla scuola per mandarli al lavoro, per conseguenza essi non hanno nessun modo per svilupparsi fisicamente e moralmente, o per compararsi agli americani ed emularli. Sono gli abitanti della « *piccola Italia* » e perdono i benefici della loro patria. Gli sforzi fatti dagli Americani e da loro stessi per migliorarsi sono neutralizzati dagli effetti delle orribili abitazioni.

In Hammonton le cose si svolgono in modo molto diverso. I bimbi Italiani trovano qui: 1) un ambiente campagnolo; 2) il contatto con gli Americani; 3) le scuole pubbliche. In Hammonton gli Italiani possono praticare la loro vita all' aria aperta, non c' è il contrasto brutale fra la vita di campagna e di città, per conseguenza i loro ragazzi sono robusti e la loro salute buona perchè possono vivere liberamente, ed abitano in case decenti, aerate e pulite. Il vitto è fornito loro, in maggior parte, dagli orti; e quando lavorano, il lavoro sfibrante a Hammonton è impossibile. Siccome le abitudini viziose non possono intervenire a rovinarli, mantenuti come sono sotto lo sguardo vigile dei genitori, essi sviluppano tutte le buone qualità della natura Italiana, e possono acquistare anche le virtù degli Americani, posti come sono a continuo contatto con loro. Le scuole pubbliche sono uno dei vantaggi di Hammonton e grazie a disposizioni legislative del New Jersey e ad aiuti di Atlantic City, i bimbi Italiani non appaiono più come un peso, ma bensì come una speranza della città. Gli Italiani pagano poche tasse ed hanno il vantaggio di una scuola centrale che va dal giardino infantile alla scuola secondaria. Vi sono poi sette scuole suburbane, ed i maestri ne sono abili ed i sorveglianti ed il Comitato sono i migliori uomini della città.

Nel 1903 fu fatto il censimento delle scuole e risultò che su 400 famiglie 242 erano di Italiani. Sul totale di 1340 bambini 1294 erano in età di andare a scuola e di questi 645 erano maschi e 649 femmine. In totale 299 maschi e 318 femmine non erano Italiani; 346 ragazzi e 331 femine erano Italiani.

E' notevole la superiorità dei figli nelle famiglie italiane. La media è di 2,29 per i non Italiani e 2,71 per gli Italiani. La più numerosa famiglia italiana ha 10 figli: la più numerosa non italiana ne ha 8. Di famiglie di 6 bambini e più ve ne sono 28 italiane e 15 non italiane; di 5 figli 25 italiane 15 non italiane; di 4:36 italiane e 20 non italiane; di 3:51 italiane e 62 non italiane; di 2:54 italiane e 60 non italiane.

Pochi Italiani seguono i corsi superiori, generalmente essi abbandonano la scuola dopo la quinta classe. Il numero proporzionale degli Italiani nelle scuole del distretto è del 62,2 a 100 %: e dell'85,3 % in media nelle sette scuole. Più gli Italiani vivono nella comunità e tanto più apprezzano il valore della educazione. Tempo indietro un'Italiano non educava le figlie, considerando che esse col matrimonio abbandonavano la famiglia: attualmente alcuni maschi e femmine seguono le scuole superiori, quantunque si trovino ancora delle bambine di 10 a 12 anni che non sono mai state a scuola. Sarà bene che la legge provveda seriamente al rimedio.

Per comparare lo sviluppo fisico degli Italiani a quello degli Americani è opportuno citare alcune cifre. Nel 1905, 591 ragazzi dai 4 ai 14 anni di età furono esaminati. 188 maschi e 163 femine erano Italiani; si trovò che le bimbe italiane di 5 anni erano alte pollici 41.4 le americane 42.5; i ragazzi italiani 41.6 gli americani 41.8; i ragazzi italiani pesavano libbre 41.7 e gli Americani 40.5 In peso gli Italiani sono superiori, in altezza inferiori. Con l'andare degli anni le caratteristiche della razza si affermano. A 14 anni un ragazzo americano è circa 3 pollici più alto di un italiano, e pesa 6 libbre di più.

Una misurazione comparativa fra i ragazzi italiani nati a Torino e quelli nati a New York city

dimostrò la medesima superiorità di peso e statura
dei nati in America. Per esempio, un ragazzo di
Torino a 6 anni d'età misura pollici 40.7 e una
bimba 40.2 e pesano rispettivamente libbre 36.8 e
36.2. Un ragazzo Italiano di New York city misura
se maschio pollici 42.4 se femina 41.4 e pesano
rispettivamente 44.2 e 40.9.

Statura comparata in media fra ragazzi
Italiani di Torino, New-York e Hammonton

	Media della statura in pollici					
Località	Ragazzi di età di:			Ragazze di età di:		
	5 anni	7 ar.ni	10 anni	5 anni	7 anni	10 anni
Torino . . .	38,2	44,3	49,8	38,0	43,0	50,1
New-York . .	39,6	44,6	48,2	39,0	44,2	49,2
Hammonton .	41,6	45,5	51,2	43,1	46,8	52,2

Peso comparato in media fra ragazzi
Italiani di Torino, New-York e Hammonton

	Media del peso in libbre					
Località	Ragazzi dell'età di:			Femine dell'età di:		
	5 anni	8 anni	10 anni	5 anni	8 anni	10 anni
Torino . . .	33,5	45,6	54,7	33,1	41,9	54,5
New-York . .	38,9	50,1	57,2	37,2	48,4	61,1
Hammonton .	41,7	51,0	59,5	40,9	49,1	62,3

Queste cifre dimostrano che i ragazzi Italiani
migliorano fisicamente agli Stati Uniti; migliora-

mento che si accentua anche più in un ambiente rurale.

Altre differenze sono rivelate all'esame fisico. Un agente di assicurazioni trovò che l'apparenza dei ragazzi Italiani di New York non era bella. Alcuni erano anemici, con grandi occhiaie, e pallidi, parecchi rachitici e questi erano il 9.8 % su 604 ragazzi esaminati: invece secondo il rapporto dello stesso agente medico i ragazzi di Hammonton presentarono per l'assicurazione rischi molto minori. La rachitide è una malattia ignota in Hammonton ed essi furono trovati in massima buoni per la società assicuratrice mentre quelli di New York furono giudicati invarie occasioni: 44 cattivi rischi su 100; 20 su 94; 64 su 200.

La storia recente degli Stati Uniti ci ha dimostrato quale grande vantaggio sia per la popolazione l'afflusso di giovani maschi e femine dalla campagna nella città ; e come ragazzi che entrano dalla campagna nella vita della città sieno molto più atti alla lotta per la vita che i nati nella città stessa. Alcune statistiche mostrano come i giovani venuti dalla campagna possano accedere facilmente a professioni che richiedono un certo grado di intelligenza e di capacità.

I lavori scelti da Italiani rivelano nei nuovi nati di Hammonton la tendenza ad una futura importanza dell'elemento italiano immigrante. Una grande maggioranza di essi vivono in vicinanza della campagna e ne godono i benefici ed i loro figli, la terza generazione, cresceranno in simile ambiente.

La soluzione del problema della assimilazione della immigrazione Italiana sta, probabilmente, nel trovar loro il modo di stabilirsi in località ove essi possano svilupparsi e non soffrire per le abitudini o il clima. E' dunque opportuno che gli emigranti non si fermino a New York appena arrivati.

Tre cose sono necessarie per arrivare ad una buona distribuzione degli emigranti Italiani : 1°) un piano bene organizzato; 2°) aiuto finanziario ; 3°) la convinzione che gli Italiani sono un'acquisto *desiderabile*. Le varie organizzazioni create per aiutare gli emigranti dovranno cooperare alla realizzazione di queste tre cose. Quando il bisogno di lavoro si è fatto sentire nel Sud ed in varie altre località, gli Italiani sono stati considerati recentemente come i possibili sostituti degli Europei del Nord. Se questo studio avrà servito a mettere in luce le belle qualità di questo popolo, mostrandolo volenteroso, industrioso, progressivo lo scopo che lo scrittore si è proposto sarà stato raggiunto.

Prof. Napoleone Colajanni

Deputato al Parlamento

I NON DESIDERABILI
(THE UNDESIRABLES)

I NON DESIDERABILI [1]
(The undesirables)

Gli scritti del Sheridan, dell'Amy Bernardy e del Meade, pubblicati precedentemente sulla condizione degli Italiani negli Stati Uniti hanno già fatto conoscere, che essi nella Grande repubblica vi sono assai malvisti e che sul loro conto corrono pregiudizi numerosi e sono accreditate delle esagerazioni pericolose sui loro reali difetti.

Negli Stati Uniti vivono circa un milione e mezzo d'Italiani e l'emigrazione ha già ripreso la sua via verso la Repubblica della Stelle.

Senza discutere per ora sulla convenienza o meno della emigrazione, che costituisce un gravissimo problema per l'Italia, è certo che interessa moralmente, politicamente ed economicamente a tutti

---

[1] Questo scritto, con lievi rimaneggiamenti e qualche aggiunzione, fa parte del capitolo sulla *Immigrazione* della 2ª Ed. del *Manuale di Demografia*. Sotto molti aspetti questa 2ª edizione è un'opera nuova, di cui ho cercato mutare il carattere. Il *Manuale di Demografia* che vedrà la luce tra poco, (Editore Luigi Pierro, Napoli) più che un libro scolastico spero che riuscirà un'opera di cultura generale, cui potranno attingere utilmente gli studiosi di politica positiva e di scienze sociali.

gl' Italiani, che i propri concittadini vengano giudicati e trattati come meritano e che sul conto loro non corrano pregiudizi e giudizi ingiusti.

Scrittori eminenti degli stessi Stati Uniti — Austin, Meade, Sheridan, Steiner, Commons, Lord, Trenor, Barrows e tanti altri — hanno preso le loro difese, e calorosamente.

A me sembra, però non solo opportuno, ma utile e doveroso vagliare una ad una le accuse, che si formulano contro di essi; dimostrarne l'insussistenza, combattere la esagerazione e ristabilire la verità. Se e quando c'è del vero su tali accuse bisogna lealmente riconoscerlo e fare opera, ciascuno nei limiti della sua possibile sfera di azione, affinché gl'Italiani vogliano e sappiano correggersi, nello interesse proprio e in quello della loro patria di origine.

Per questo motivo, a complemento di ciò che sinora ho pubblicato mi occupo adesso degli *undesirables*, dei *non desiderabili*, come gl'Italiani e gli immigrati di altre nazionalità che li rassomigliano, vengono chiamati dai Nord-Americani.

L'immigrazione ha creato dei problemi etnicosociali in talune colonie inglesi e in alcuni Stati dell'America, problemi, che allarmano, che si discutono vivamente, che provocano alcuni provvedimenti della politica dell'immigrazione.

Il *pericolo giallo*, cioè la immigrazione dei Cinesi e dei Giapponesi preoccupa l'Australia e ispira la sua legislazione; lo stesso pericolo si è aggiunto al pericolo nero nell'Africa Australe e vi è causa di preoccupazione e di perturbamenti.

Nel Brasile si è voluto creare un *pericolo tedesco*; ma è poca cosa se non del tutto insussistente.

Nell' Argentina e in altre repubbliche del Sud e del Centro America, non è stata ancora segnalato alcun pericolo, per quanto i *gringos* — come spesso in segno di disprezzo si chiamano gli Italiani — non poche volte e in parecchi punti siano accolti e trattati con poca simpatia.

Dove il problema, pel volume della immigrazione e per la sua composizione etnica e sociale, ha assunto grandi proporzioni è negli Stati Uniti. Ivi costituisce il problema degli *undesirables* — dei non desiderabili — che i più benevoli chiamano *obiectionables*. Ha importanza in genere ; ne ha una speciale per gli Italiani, che rappresentano l'elemento se non più numeroso, certo più avversato, degli *undesirables*.

Chi sono e quanti sono gli *undesirables* ?

Gli *undesirables* negli Stati Uniti sono i Cinesi i Giapponesi, gli Slavi, gli Ungheresi e sopratutto gli Italiani. Come e quanto lo siano questi ultimi si può argomentarlo da questo dato. Nel paese del *Negro-problema* e dove il *linciaggio* ha preso quasi il carattere di istituzione riconosciuta dai pubblici poteri, *la presenza degli Italiani* , secondo Nestler-Tricoche, *ha cagionato una specie di reazione in favore dei negri* (1).

L'allarme è stato suscitato sopratutto dalla proporzione degli immigrati e dal mutamento nella loro composizione etnica.

Gl' immigrati, rispetto alla popolazione totale se si fosse fatto il censimento nel 1906 , secondo il Sheridan (pag. 404) in base alla popolazione di 84,000,000 si sarebbero elevati al 17,6 %. Indubbiamente la proporzione sarebbe forte. Si riduce a

---

(1) R. Gonnard: *L'emigration européenne anno XIX siede* (A. Colin. Paris. 1906 pag. 214.

meschina cosa se il rapporto si fa colla densità. Su
di una superficie 9,420,670 chil. quadrati con l'eccedenza degli immigrati sugli emigrati ridotta a
2,456,870 rappresenterebbe un aumento di densità
di 0,26 per chilometro quadrato. E la densità nel
territorio proprio della repubblica nel 1906 in base
alla popolazione che comprendeva già gl' immigrati
non era che di 8,9 ! Sulla base della densità italiana — 116 ab. per chil. quad. — gli Stati Uniti
potrebbego avere una popolazione di circa *un miliardo* e 100 milioni.

Ma il pericolo, che sembra in questa guisa tanto
piccolo da farlo considerare insussistente, è maggiore pel fatto che gl' immigrati non si distribuiscono su tutto il territorio della repubblica ma con
grande prevalenza negli Stati e nelle grandi città
della regione Nord–Atlantica.

Sono maggiori le preoccupazioni dei Nord americani per il mutamento nella composizione etnica
e nazionale degli immigrati che costituiscono la
cosidetta *nuova immigrazione*.

Le proporzioni si sono mutate così :

Immigrati			
Canadesi, Britannici, Scandinavi, Tedeschi	Irlandesi	Austro-Ungar., Italiani, Russi, Polacchi	Altri paesi
1821-50 1,037,364 42,3°[o	1,038,824 42,4°[o	5,924 0,02°[o	374,703 15,2°[o
1871-80 1,892,510 67,3 „	430,871 15,5 „	180,982 6,4 „	301,828 10,7 „
'91-900 1,149,747 31,1 „	390,170 10,5 „	1,846,616 50,0 „	301,072 8,1 „

Gli elementi più *desiderati* — Britannici ecc. —
in cinquantanni quindi sono discesi dal 43,3 °/₀
al 31,1 °/₀ ; invece gli *undesirables* sono cresciuti
da 0,02 °/₀ a 50 °/₀ !

Il maggiore aumento di questi ultimi è avvenuto
a spese degli Irlandesi, che non sono stati mai

considerati alla stregua dei Britannici, Tedeschi e Scandinavi; ma non si ritengono completamente *undesirables*.

La situazione è peggiorata, dal punto di vista nord-americano, negli ultimi anni. Sopra un totale di 4,933,741 immigrati dal 1900 al 30 giugno 1906 erano *undesirables* — cioè Italiani, Slavi e Austro-Ungheresi — ben 3,176,291; cioè: il 64,3 %! Se essi sono realmente *undesirables* l'allarme dei Nord-americani è pienamente giustificato.

Perchè sono considerati *undesirables* gl'immigrati delle nazionalità cennate?

A parte i Cinesi e i Giapponesi, pei quali, c'è il vero antagonismo di razza, spiegabile anche colla differenza del colore, dalle dichiarazioni del generale Walker, di Shattuc, di Lodge, di Prescot F. Hall e di cento altri (1) gli Slavi austriaci e russi, gli Ungheresi e sopratutto gl'Italiani del Mezzogiorno sono *undesirables* per parecchi motivi. Si dice: sono poveri e vivono a carico dello Stato e dei corpi locali; sono analfabeti; sono *unskilled*, sudici e con un basso tenore di vita (*standard of life*); sono troppo fecondi e determinano un peggioramento della composizione della popolazione col loro più rapido moltiplicarsi; rifuggono dai lavori agricoli utili alla repubblica; si accentrano nelle grandi città creando o aggravando un pericolo sociale; sono uccelli di passaggio e non fissandosi negli Stati Uniti ne sottraggono ingenti capitali; hanno un'alta criminalità; fanno concorrenza nel lavoro a gli in-

---

(1) Il PHILIPPOVICH non direttamente interessato li considerò come *undesirables* sin dal 1892 (Auswanderung und Anwanderungs politik in Deutschland. Leipzig. Dunker et Humblot, 1892); anche il Commons non è entusiasta della *nuova immigrazione* (*Races-and Immigrants in America*. (New-York. Macmillam C. 1907).

digeni; non si assimilano e non si *americanizzano*.

Ce n'è abbastanza per giustificare qualunque allarme e qualunque avversione per un popolo , che si crede immune da quelle stigmate (1).

Esaminiamo , rapidamente , una ad una queste accuse.

1.° *Sono poveri e vivono a carico della collettività.* Nessuno può negare che gli *undesirables* siano poveri; se tali non fossero non emigrerebbero, andando incontro ad ogni sorta di disagi materiali e morali tormentosissimi. Ma che le nazionalità *undesirables* abbiano la stessa inclinazione a vivere di accattonaggio ed a spese del paese d'immigrazione non è vero; non lo è sopratutto per gl'Italiani.

La loro difesa , in base ai dati del censimento del 1890, è stata fatta, senza possibilità di risposta, da Austin, Capo dell'Ufficio di Statistica del Dipartimento federale del Commercio e lavoro di Washington in un onesto articolo (2).

Il censimento del 1900 e le successive pubblicazioni ufficiali hanno luminosamente provato che sotto questo aspetto gl'Italiani dovrebbero essere i più *desiderabili*. A 31 Dicembre 1903 negli Stati Uniti c'erano ricoverati (3).

---

(1) I Nord-Americani si preoccupano specialmente degli *un desirables* italiani, perciò molti sono venuti a studiare le condizioni dell'emigrazione in Italia. Nello scorso anno venne anche una Commissione ufficiale composta dei Senatori Dillingham e Latimer e dei deputati Howel, Bennet e Burdett.

(2) *Is the new immigration dangerous to the Country ?* Nella *North American Review*, Aprile 1904.

(3) *Special reports. Paupers in Almhouses. 1904* (Washington 1906 pag. 19 e 99); *The insane and feble minded. 1904* (Washington 1906 pag. 23 e 100).

	Asili dei poveri		Manicomi	Su 100 pazzi	Istituti pei deboli di mente (1)
Irlandesi . . .	14,923	46,4 %	13,664	29,0	7,0
Tedeschi . . .	7,477	23,3 »	12,644	26,9	26,4
Inglesi	2,871	8,7 »	3,311	7,8	9.7
Canadiani . . .	1,744	4,8 »	3,049	6,5	14,4
Scandinavi . .	1,573	4,9 »	5,409	15,7	11,7
Scozzesi . . .	788	2,5 »	785	1,7	1,8
Pugheria Boemi.	314	1,6 »	1,054	2.2	3,5
Russi e Polacchi	490	1,5 »	2,064	4,4	12,2
Italiani. . . .	331	1,0 »	1,084	2,4	3,7

A New York, dove Italiani e Irlandesi si pareggiano, nel 1904 furono ricoverati nell'Istituto di Blackwells 1564 Irlandesi e soli 16 Italiani (2).

Le osservazioni anteriori di Iacob Riis; dello *State Board of Charities* di New York nel 1901; la relazione di Iohn Keller Presidente del Dipartimento della Pubblica carità sui *Robleurs of the Almshouse*, in New York City; la relazione della *Associated Charities* di Boston nel 1902 assegnano una percentuale incalcolabile di poveri e di assistiti italiani. Gl'Italiani bisognosi si aiutano tra loro (3).

Ciò che si è osservato a New York nella più grande città e nell'ambiente più sfavorevole per gli Italiani, viene riconfermato pei piccoli centri rurali in una pubblicazione ufficiale (4).

Gl' Italiani inoltre non danno vagabondi, non

---

(1) Su 100 ricoverati bianchi (*The insane* ecc. pag. 261).

(2) Preziosi : *Il problema dell'Italia d' oggi* (Palermo R. Sandron 1907 pag. 78).

(3) I. Barrow nell'opera : *The Italian in America* , pagine 191 a 198.

(4) *The Italian on the land* di Emily Fogg Meade. Nel *Bulletin of the Bureau of Labor* di Washington n. 70. Riferendomi a questa pubblicazione citerò solo il nome dell'autore. Anche Steiner (*On The Trail of the Immigrants* Fleminges New-York e Londre) nel paragone tra Italiani e Irlandesi riconosce la enorme superiorità dei primi.

danno mendicanti, come hanno riconosciuto e Forbes, capo del *Mendicanus Departement of the Charity Organization Society* di New York e Petrosino (1).

Sotto questo aspetto, adunque, gl' Italiani sono *desiderabilissimi*.

✿

2.° *Sono analfabeti.* Che l' analfabetismo sia un male nessuno lo pone in dubbio; che possa nuocere allo sviluppo economico dell' Unione americana quello degli immigrati lo negano gli stessi Nord–Americani. Johnston nella *North American Review* ricordò che erano analfabeti i fondatori della Repubblica.

Ed Eliot Lord ironicamente osserva: Se l'America doveva essere riservata alle persone istruite, Colombo doveva essere avvertito di non scoprirla con un equipaggio di analfabeti (*The Italian in America* pag. 183).

E' vergognoso l'analfabetismo degli Italiani; ma costituisce una colpa dell' Italia, non un pericolo per gli Stati Uniti, che di quell' analfabetismo si giovano.

Ma l'accusa di analfabetismo non è che un pretesto per combattere l' immigrazione; l' ho dimostrato in *Latini* e *anglo-sassoni*; l' ha dichiarato esplicitamente il Ward dell' università Harward di Boston. Meglio ancora; l'analfabetismo tra gli *undesirables* della seconda generazione diminuisce più rapidamente che tra gl'indigeni. Lo dimostrò l'Austin in base ai dati del censimento del 1890; e venne riconfermato dal censimento del 1900. (*Latini* ecc. pag. 397). E gl' Italiani fra tutti brillano per la intelligenza.

---

(1) Francis I. Oppenheimer : *The Truth about tuc Black Hand* nel *Denver republican*, Gennaio 17 1909.

Lawrence Franklin ha notato che gl' insegnanti
americani sono assai contenti della intelligenza e
dell'amore allo studio dei fanciulli italiani, che accanto agli Ebrei sono i migliori scolari. Riis aggiunge che gli effetti della istruzione dei fanciulli
italiani si ripercuotono sui loro genitori e sulle loro
famiglie. (Eliot Lord : *The Italian in America*, pagine 240 e 241). Ai loro accusatori toglierebbero
anche questo pretesto di avversione se avessero
maggiore cura di secondare le associazioni locali,
che si sforzano di attrarre i loro figli nella scuola.

Del resto anche da questa accusa li difende Kate
Holliday Claghorn, *assistant Registrar of Records of
the Tenement House Department* di New York. « Non
è vero , essa dice , che gl' Italiani non mandino i
figli a scuola per risparmiare qualche spesa. Anche
i più poveri, e con gravi sacrifizi , spesso provvedono all'istruzione dei loro figli ». (*Eliot Lord* pagina 245).

♣

3.° *Sono unskilled*. La grande massa d'immigranti
— e non tra i soli *undesirables* — essendo di braccianti e di agricoltori, che nella Repubblica non si
danno al lavoro dei campi , necessariamente viene
compresa nella categoria degli *unskilled*. Il Sheridan,
in proposito, dà esatti dettagli sul loro numero e
sui lavori, cui si addicono.

Ma essi sono per lo appunto gli elementi di cui
hanno bisogno e di cui vanno in cerca gli Stati
Uniti; dove il lavoro rozzo, senza capacità, il lavoro
row material, è sfuggito dai nativi. Era prima asse-
gnato agli Irlandesi ed ora agli Italiani, agli Slavi,
agli Ungheresi.

Senza il lavoro degli *unskilled* sarebbe arrestato
lo sviluppo degli Stati Uniti ; regredirebbe sicura-
mente la loro vita economica e sociale.

❧

4.° *Il basso tenore di vita.* E' innegabile e deriva dalle condizioni economiche e intellettuali degli immigrati *undesirables.* Essi sono usi a consumi limitati in patria e sentono meno degli altri il bisogno di elevarli negli Stati Uniti. Nella spesa sono limitati altresì dal desiderio di risparmiare per raggranellare il gruzzolo mensile, che mandano a casa o quello, che raccolgono per portarlo essi stessi al ritorno in patria.

Tra gli *undesirables* innegabilmente il più basso tenore di vita lo hanno gl'Italiani: per la casa, pei vestiti, per la nutrizione spendono meno degli Slavi e degli Ungheresi. Perciò risparmiano di più.

Questo basso tenore di vita porta seco fatalmente una certo grado di sporcizia che li rende invisi; la nutrizione deficiente rispetto alla intensità del lavoro e alle esigenze del clima, ne fiacca la fibra, li espone alla degenerazione biologica, che si traduce in aumento di tubercolosi, di nevrastenia, di psicopatie. In conseguenza li rende anche meno apprezzati come lavoratori, pel rendimento minore del loro lavoro.

Tutto ciò, però, non interessa direttamente la Repubblica; il lato, che può interessare è quello delle condizioni antigieniche delle loro abitazioni. Su questo, si osservi, che, pel sudiciume in ispecie, non c'è nulla che costituisca un carattere di razza. Non sono pochi gli stessi americani che riconoscono che gl'Italiani quando possono tengono le case meglio degli Irlandesi e degli Ebrei e di altri *undesirables* (*Preziosi* pag. 79, *Schultze*). Lo avevo constatato in *Latini e Anglo-sassoni* sulla scorta di diversi scrittori; lo confermano ora il Sheridan e il Meade per gl'Italiani di Hammonton e di altri piccoli centri agricoli.

Gl' Italiani, pel tenore di vita, non devono essere giudicati solo da quelli degli *slums* e delle *Little Italies;* poichè dove e quando le circostanze lo consentono il loro *standard of life* si eleva. Comunque, gl' Italiani, come osserva Steiner (pag. 275 e 277) non sono degenerati, non sono esauriti; e sopratutto, lo si è visto, meno di tutti gli altri immigrati *undesirables* cadono a carico dello Stato e dei corpi locali (1).

Inoltre deve avvertirsi che gl'Italiani, anche quelli, che vivono nei peggiori quartieri di New York, mostrano una continua tendenza a migliorare la loro condizione; oggi le loro abitazioni sono in uno stato migliore di quello in cui lo trovarono quando per la prima volta le occuparono. Gl' Ispettori del *Tenement House Department* di New York City dichiarano, che le loro abitazioni sono migliori e più pulite di quelle dgli Ebrei e degli Irlandesi. Il malfamato Mulberry Street è stato da loro redento negli ultimi dieci anni ; così altri quartieri sono stati da loro migliorati moralmente e materialmente. Si deve agli Italiani l'elevazione del valore di molte case. Sono migliorate tutte le bitazioni tra Macdougal Str. e West Broad-way, della Tird Str. Minetta Lane, da Hancock Str. ecc. dopo che sono andati via i Francesi e i Negri e sono state occupate dagli Italiani. (*Eliot Lord* pag. 74-76).

♣

5.° *Si moltiplicano rapidamente.* Il Gonnard (p. 217)

---

(1) Pei consumi, pel tenore di vita degli Italiani e degli altri *undesirables* si riscontrino le notizie precise che dà Sheridan e per gli anni anteriori le seguenti pubblicazioni dell' Ufficio del lavoro di Washington : *The slums of Baltimora , Chicago , New York and Philadelphia.* (Washington 1894); *The Italians in Chicago.* (Washington 1897).

sulla fede del Mosso afferma che a New York
l'eccedenza delle nascite sulle morti tra gli Italiani
arriverebbe alle proporzioni inverosimili del 62 per
mille. Ciò preoccupa gli americani, che scorgereb-
bero in questa straordinaria fecondità il pericolo
di vedere rapidamente moltiplicare gl'inferiori.

A ragione si considera inverosimile questa eccedenza che suppone una natalità di 80 °/₀₀ almeno;
e si può essere sicuri ch'è errata, come sono errati
tanti dati statistici che non riguardano la fisiologia
in tutte le pubblicazioni del Mosso. Se fosse vera
gli americani non avrebbero, che a provvedere al-
l'allevamento dei nati. Il problema è d'indole generale e preoccupa demografi e antropologi che
rimangono impressionati dalla decrescente natalità
delle classi superiori e dall'alta natalità di quelle
inferiori. Il fenomeno non data da oggi; e il progresso sociale non è stato arrestato. Nel paese in
cui si teme il *suicidio della razza* e dove Roosevelt
tuona contro la scarsissima natalità, messa in evidenza nei suoi vari aspetti da Commons (p. 198 a
203), si dovrebbe rimanere assai contenti della
fecondità degli Italiani.

✚

*f) Sono uccelli di passaggio e portano via dal-
l'America ingenti capitali.* Mentre sono scarsi coloro
che si preoccupano della rapida moltiplicazione degli Italiani — e non dev'essere minore quella degli
altri *undesirables*, perchè in Europa la natalità degli
Slavi e degli Ungheresi è superiore a quella degli
Italiani — sono più numerosi, invece, quelli che in
senso contrario deplorano che essi siano degli *uccelli di passaggio*, che non prendono stabile dimora
in America; lo deplorano sopratutto perchè sottrag-

gono annualmente ingenti somme alla Repubbblica
e le portano in Europa.

Gl' Italiani sulle loro entrate risparmiano più degli altri come si può scorgere dallo studio del Sheridan; in conseguenza per questo titolo sono rite-
nuti più *undesirables* degli altri.

Tutto questo è vero. Molti Americani, però non
considerano che negl' immigrati trovano uomini
produttivi già belli e pronti a carico dei quali non
c' è costo di allevamento sino all'età, in cui diven-
gono produttivi.

Ciò che opportunamente ha ricordato ai proprii
concittadini *Eliot Lord*, che ha fatto proprio il calcolo di Frederick Knapp, commissario di emigrazione per lo stato di New York. Secondo il Knapp
il valore economico di ogni immigrante maschio
sopra i 20 anni è di 1,125 dollari; cioè di L. 5,625
(*The Italian in America*, pag. 12).

Dimenticano sopratutto gli Americani che gli *undesirables* portano seco o mandano in Europa il 10 %
della ricchezza prodotta e lasciano il 90 % in A-
merica.

Austin aveva già notato il fatto che la produzione
maggiore e la ricchezza maggiore tra gli Stati della
Unione si riscontravano per lo appunto in quelli
che accolgono il maggior numero degli immigrati
unskilled (Colajanni : *Latini* ecc. *pag. 398*), Sheridan
si associa alle stesse considerazioni e Commons
insiste nel dimostrare che la curva della prosperità
della Repubblica segue quella della immigrazione
(*pag. 119, 121, 124, 158* ece.).

Gli *undesirables*, quindi, sono creditori e non de-
bitori verso la repubblica , e da ogni critica sotto
questo aspetto brillantemente li difende lo Steiner
(*pag. 319* e *320*).

In ogni modo gli stessi Commissari dell' Emi-

grazione avvertono che aumentano le proporzioni
delle donne e dei fanciulli tra gl'immigrati italiani;
che aumenta il numero di quelli, che prendono
moglie in America, vi acquistano proprietà e si
naturalizzano : tutti indizi che gli *uccelli di passaggio*
prendono dimora stabile nella repubblica. (*Eliot
Lord* pagg. 9, 10 e 11) (1).

✣

g) L'*addensamento nelle città.* Si rimprovera agli
undesirables l'addensarsi nelle città, aggravando i
mali dell'urbanismo, e il rifuggire dai lavori agricoli. Ciò si è rimproverato specialmente agli Italiani. E' assai forte il sovraffollamento in alcuni
quartieri da loro abitati. (*Eliot Lord* pagg. 70 e 71).

Se si guarda al fatto in sè il rimprovero è me-
ritato, poichè è realmente enorme il numero degli
unskilled, degli *undesirables*, che si fermano nelle
città, specialmente nei porti di sbarco; a New-York
sopratutto.

Ma questo fenomeno è inevitabile, è il prodotto
di una legge economica. Il Sheridan giustamente ha
osservato che gl'immigrati si distribuiscono precisamente a seconda della richiesta del lavoro e del
livello dei salari.

Possono essi trasportarsi dove non vengono ri-
chiesti e vengono pagati meno? Inoltre, rimanendo
nelle città dove trovano nuclei preesistenti di connazionali, essi come osserva lo Steiner (*pag. 263*),
si salvano dal ridicolo e dalla crudeltà, con cui in
generale gli Americani trattano tutti gli stranieri,
che non hanno i loro costumi e non parlano la
loro lingua.

---

(1) Pel numero dei proprietari si riscontrino le mie opore :
Latini e anglo-sassoni o il mio *Manuale di Demografia* (*Le
colonie senza bandiera*), nonchè i dati che riporta Eliot Lord
(*The Italian ecc.*, pagg. 78 82).

Ed Eliot Lord ha deplorato l'offesa che si arreca alle sensibilità degli Italiani mettendoli in ridicolo, perchè non parlano bene l' inglese. (*The Italian in America*, pag. 66).

Gli Italiani, gli *undesirables*, gli *unskilled* arrivano negli Stati Uniti senza conoscere la lingua, i costumi, le condizioni e i luoghi del lavoro. Orbene non sarebbe interesse degli stessi governi degli Stati e del Governo Federale, impedire l' aggravarsi dell' urbanismo e dell'addensamento favorendo la razionale distribuzione degli immigrati? Ma in quest'opera di assistenza e di prevenzione il governo federale e quello degli Stati spiegano un'azione deficientissima. Ciò venne esaurientemente dimostrato nella inaugurazione della *Exihibit on Congestion of population* in New York (', a 23 marzo 1908), ed eloquentemente dalla signorina Amy A. Bernardy, che dimostrò come e perchè gli italiani si fermino a New York, vadano ad abitare i *districts congested* e formino le *Littles Italies*.

Eliot Lord in risposta all'accusa che il *New York Fvening Post* ha rivolto agli Italiani di avere trascurato il lavoro e le occupazioni agricole onestamente, e prima della Bernardy, aveva osservato: « Invece di « far carico agli Italiani di trascurare i grandi paesi « rurali del Sud e dell'Ovest, non sarebbe più cor- « retto di ricercare perchè essi li hanno trascurati? « Che cosa hanno fatto gli Stati del Sud e del- « l'Ove, eccettuato la Califormia e la Luigiana, per « attrarre o promuovere l'immigrazione e la colo- « nizzazione italiana? Che cosa hanno fatto i pro- « prietari e i *farmers* per offrire lavoro a condi- « zioni accettabili agli Italiani? Le fattorie dell'oc- « cidente furono occupate dagli Immigranti del- « l' Europa, prima che l'immigrazione fosse consi- « derevole. I tempi dell'occupazione della terra sotto

« la legge dell' *homestead* sono passati, e gli occu-
« patori attuali a preferenza chiamano immigranti
« della loro propria nazionalità. Essi ignoravano il
« valore del lavoro italiano; ed ora lo giudicano in
« base ai pregiudizi, che corrono ed ai giudizi er-
« ronei che sul loro conto si emettono nelle città ».

« I pregiudizi contro gli Italiani sono divisi an-
« che dal mondo ufficiale. Così Gordon nella sua
« recente relazione come Presidente della *Georgia*
« *Industrial Association* ha rammentato, che una
« legge della Carolina del Sud restringe l'*immigra-*
« *tion* ai cittadini bianchi nativi di America, agli
« Irlandesi, Scozzesi, Svizzeri, Francesi e a tutti gli
« stranieri di origine Sassone ». (*Te Italian in A-*
merica pagg. 114-117).

Che rifuggano dal lavoro agricolo per indole, per
determinazione volontaria non è vero. Il Meade ha
già dimostrato come gl'italiani, i maggiori accusati,
riescano bene nei lavori agricoli ; il Gonnard (pagina 215) descrive i *settlements* prosperi d'Italiani;
agli italiani si devono i meravigliosi progressi agricoli della California ; sono numerose le *farms* di
Italiani nel Texas. Eliot Lord ne enumera moltissime e tutte prospere in vari Stati del Sud e del-
l'Ovest e ricorda che si deve quasi dapertutto agli
Italiani la coltivazione degli alberi da frutta e degli
ortaggi attorno alle grandi città, (pagg. 85, 118
a 153).

Sarebbero assai più numerose le *farms* condotte
da Italiani se essi meglio fossero protetti con particolarità negli Stati del Sud e del Centro, contro la
crudeltà e le preporenze dei *bosses*; contro le violenze
collettive delle società segrete criminose, come i *Ca-*
valieri della notte; contro gli scellerati linciaggi.

Agli Americani per tutti gl'inconvenienti, che
sorgono dal contributo che danno gl'immigrati

undesirables in ciò che di più pericoloso presenta
l' urbanismo si può ben dire: *imputetur vobis*. Roo-
sewelt, per lo appunto ha detto : « ogni miserabile
« *tenement*, la cui esistenza è tollerata da una città,
« si vendica su di essa diventando un semenzaio
« di malattia e di pauperismo. Tende ad abbassare
« continuamente il livello della nostra vita civica
« e sociale : per cui la presente agitazione per il
« miglioramento delle *tenement houses* è uno sforzo,
« che mira a tagliare alla radice le malattie , che
« corrodono l'organismo sociale e politico ».

♣

h) Concorrenza del lavoro. Molti dei motivi,
che si adducono per giustificare l' avversione verso
gli *undesirables* non sono che pretesti bugiardi, come
quello dell'analfabetismo, ed un poco quello della
delinquenza. Questi pretesti servono a mascherare
il motivo reale e principale: la concorrenza nel lavoro che fanno gli *unskilled*, gli *undesirables* agli operai indigeni, deprimendone i salari e anche il tenore
di vita. Commons tratteggia rapidamente e bene
questo motivo di avversione (*pag. 125, 148 a 151*).

Questo motivo determinò anche l'avversione dei
lavoratori francesi, svizzeri e tedeschi contro gli
immigranti italiani e provocò le *cacce* agli italiani,
che assunsero in certi momenti parvenza di antagonismo nazionale, specialmente in Francia dopo
il 1881 (1).

Allora definii gli emigranti italiani i *Cinesi d'Eu-*

---

(1) Determinai il vero carattere di questi fenomeni all' indomani della occupazione di Tunisi e dei fatti di Marsiglia
nel 1881 in un giornale politico di Torino; all' indomani di
Aygues Mortes (1893) lo illustrai meglio in un articolo della
Rivista Popolare che fu pubblicato a parte : *Una quistione
ardente* (*La concorrenza del lavoro*) Roma 1893.

ropa. D'allora in poi, in Europa, la concorrenza del lavoro degli italiani è diminuita, per opera specialmente dei socialisti, che li hanno indotti a far parte dei sindacati locali ed a richiedere gli stessi salari dei lavoratori indigeni. Ma in Australia, al Capo, negli Stati Uniti; la concorrenza dura e perciò si temono e si avversano gli *unskilled* e si considerano come *undesirables*.

La realtà sull'avversione di certe popolazioni contro gli immigrati fu denunziata cinquant'anni or sono da Duval; il quale notava che in Australia, come in California fosse antica verso i Cinesi e che la si volle giustificare colla loro immoralità sessuale derivante dal celibato forzato. Ma era una ipocrisia. La razza cinese vi era perseguitata dalle leggi, eco delle passioni popolari, perchè essa è una concorrente che lavora a migliore mercato della razza europea (1).

Duval ingenuamente scriveva nel 1862: «L'uguaglianza dei trattamenti, che la giustizia non ha ottenuto, la politica la reclamerà senza dubbio. Dopo che, gli inglesi avranno forzato coll'appoggio dei francesi, i porti della Cina, per ottenere l'esecuzione sincera di un trattato che loro apre liberamente l'Impero, come mantenere contro i Cinesi, nei possedimenti inglesi, i rigori fiscali che violano apertamente le condizioni del trattato? » (*pag. 387*).

Se ritornasse in vita Duval apprenderebbe che la mostruosità politica e morale non è scomparsa; ma si è allargata l'avversione dai Cinesi ai Giapponesi, dai gialli ai bianchi (Italiani, Austro-Ungarici, in America ed anche Anglo-Sassoni in Australia).

I lavoratori nord-americani ed australiani, che

---

(1) Duval: *Histoire de l'emigration européenne* (Paris, Guillaumin et C.ie, 1862 pag. 309).

temono la concorrenza del lavoro degli *undesirables*
pei loro salari più bassi hanno ragione se tengono
conto del fatto in sè ed isolato dallo insieme delle
sue conseguenze; ma hanno torto se tengono conto
di due circostanze: 1° dello enorme incremento della
ricchezza, che produce il lavoro di questi *undesirables*
e che si ripercuote su tutta la vita economica della
popolazione totale elevandone lo *standard of life;*
2° del fatto che gli *undesirables* hanno sostituito
gl'Irlandesi in tutti quei lavori, ai quali non si vogliono più addire i nord-americani. Ciò può umi-
liare Italiani, Austro–Ungarici, Russi; ma non nuocere ai lavoratori indigeni degli Stati Uniti.

♣

i) Gli undesirables non si americanizzano. Il
sogno, il desiderio ardente e legittimo degli americani è quello di veder fondere nel seno della loro
società gl'immigrati. Roosevelt, il loro uomo rap-
presentativo, sopra tutto desidera e vuole che « l'immigrato impari a parlare, a pensare, ed agire come
un cittadino degli Stati Uniti ».

Ma gli *undesirables*, osservano gli avversari del-
l'immigrazione nuova, non si assimilano, non si
americanizzano, mentre il processo di fusione colla
immigrazione antica era facile e rapido; d'onde il crescente pericolo per gli Stati Uniti di una degenerazione profonda ; pericolo denunziato tra i primi
dal generale Francis Walker e che ora è stato preso
a bandiera di combattimento contro la nuova immigrazione da molti scrittori e da alcune associazioni (1).

---

(1) Tutti gli argomenti degli avversari della nuova immi
grazione sono riassunti nel libro di PRESCOTT F. HALL: *Im
migration and its effects upon the United States* (New York
1906). E' sorta una *Immigration restriction League*, che

Coloro che vogliono darsi le arie più scientifiche in questa propaganda contro la nuova immigrazione invocano, al solito, la differenza psicologia tra gli Anglo-sassoni e i Latini, gli Slavi e i Magiari. Contro questi vieti infondati pregiudizi sta il mio libro: *Latini e anglo-sassoni*, che non ha ricevuto risposta. Che i Tedeschi, gl'Inglesi, gl'Irlandesi, gli Scandinavi si assimilino e si americanizzino più facilmente e più rapidamente è fuori di ogni dubbio. Non c'è da sorprendersene: la lingua, la religione, i costumi, la professione, la condizione sociale di costoro li avvicinano agli elementi analoghi dei nord-americani, assai di più che non quelli degli *undesirables*. Ma questi dovunque si fissano, gradatamente si americanizzano coi contatti sociali, per imitazione, coll'azione delle scuole, per l'opera della *Children's Aid Society* e di altre analoghe società. Se resistono gli adulti — e così dev'essere perchè non si può rinnovare la coscienza già formata, nè mutarsene la psicologia con un colpo di bacchetta magica — si trasformano rapidamente i fanciulli e alla seconda generazione si può ritenere che l'americanizzazione è completa. Questo l'avviso dl Mayo-Smith, di Ionsthon, di Steiner, di Commons, di Meade. E il Meade, come si è visto, chiuse il suo studio affermando che esso avrebbe raggiunto lo scopo se fosse riuscito a dimostrare che gl'Italiani sono *desiderabili* e *americanizzabili*.

Gli *undesirables* in generale e gli Italiani in ispecie, adunque, non sono un pericolo per gli Stati Uniti, ma sono una sorgente di ricchezza.

Che non siano un pericolo lo dimostrò Austin,

pubblica numerosi opuscoli per una feroce propaganda contro gl'*Italiani* e contro gli altri *undesirables*.

In senso opposto si sono costituite: *The National liberal immigration league e The new immigrants protection league.*

uno dei più eminenti statistici nord americani e lo
confermano oggi quanti studiano l'opera da loro
compiuta (1); che siano anzi, *desiderabili*, se ne
sono maggiormente accorti gli americani ora che
in seguito alla crisi del 1907 hanno constatato i
danni e le conseguenze sinistre dell'esodo degli
immigrati nuovi, che ad essa seguì nell'ultimo trimestre del 1907 e per tutto il 1908 (2).

✤

Le accuse false, o esagerate o reali, che vengono
formulate negli Stati Uniti contro gli *undesirables*
in genere e contro gl'Italiani in ispecie hanno determinato tutta la legislazione americana sulla *immigrazione ;* ma questa non può essere giudicata
convenientemente ed onestamente se non se ne
conoscono alcuni precedenti ed alcune particolarità.
Perciò dal *Manuale di demografia* (2ª Edizione) riproduco quella parte di tale legislazione, che riguarda gli *undesirables* europei, di cui sono *magna
pars* gl'Italiani (3).

Nella grande Repubblica delle stelle benchè sia
tanto vasta la superficie e tanto debole la densità,
da tempo antico si cominciò a protestare contro
l'immigrazione. A Filadelfia nel secolo XVIII si

---

(1) O. P. AUSTIN: *Is the new immigration dangerous to the
country ?* Analogamente pensano STEINER , COMMONS , LORD,
TRENNOR e BARROWS (*The Italians in America*).

(2) N. BEHAR, presidente della *National liberal immigration*
si è mostrato preoccupatissimo di questo esodo. Della impor
tante quistione delle criminalità degli Italiani in America mi
occupo con particolarità nello studio seguente.

(3) Come la *politica della emigrazione* e quella della *immigrazione* si svolgano in senso inverso nei paesi di emigra
zione e in quelli d'immigrazione a seconda che prevalgono le
classi lavoratrici o quelle capitalistiche ho dimostrato nei re
lativi capitoli del *Manuale di demografia* (paragrafi 108 e 114)

lamentavano della indigenza degli immigrati e il
governo temendo di vedere germanizzare il paese
impose una tassa di 4 scellini su ogni immigrante.
Ma non fu esatta per lungo tempo. Una legge severa
nel 1798 fu promulgata sulla naturalizzazione. Prima
del 1860 inoltre, più volte fu ventilato il progetto
di proibire l'acquisto delle terre agli immigrati.
(*Duval* pag. 48 e 196). Questi precedenti sono
interessantissimi perchè dimostrano che vennero
considerati come *undesirables* anche i Tedeschi e
gl'Inglesi, che costituivano la massa degli immigrati
nel secolo XVIII e nella prima metà del secolo XIX.

Quando cominciò ad accentuarsi la nuova immigrazione e acquistarono maggiore potenza i lavoratori o le loro associazioni — furono attivissimi
in questo senso i *Cavalieri del lavoro* e il loro
Presidente Powderley — le proteste si fecero più
vive. Nel 1890 non meno di 12 progetti di legge
furono presentati al Congresso degli Stati Uniti
contro l'immigrazione degli *undesirables* (*Philippovich*, pag. IX).

La legge del 3 marzo 1903 intese porre dei freni
all'immigrazione degli stranieri; con quella del
1° Febbraio del 1907 non solo si elevò da 2 a 4
dollari la tassa che deve pagare ogni immigrato
all'entrata nel territorio della repubblica, ma furono
meglio precisate le categorie delle persone alle
quali era inibito lo sbarco per ragioni fisiche, men-
tali, economiche e morali (cioè agli idioti, imbe-
cilli, deboli di mente, pazzi, epilettici, persone
affette da tubercolosi, o malattia ributtante, tigna ec.
da tracoma; agli indigenti, ai poligami, agli anarchici, ai fanciulli sotto i 16 anni non accompagnati
dai genitori, alle prostitute ed ai *souteneurs*, agli
operai sotto contratto di lavoro ecc.). Altri freni
furono posti facendo obbligo ai vapori che traspor-

tano emigranti di rispettare molte regole igieniche,
alcune delle quali capricciose; e col 1° gennaio 1909
la capacità di trasporto dei vapori sarà ridotta in
media di un quarto per assicurare agli emigranti,
uno spazio considerevole.

Ora si è anche deciso — e non del tutto legal-
mente — che non potranno sbarcare gl'immigrati i
quali non posseggano almeno 25 dollari.

Ma tutti questi freni, come osservò il Sheridan,
non valsero ad arrestare l'immigrazione. I respinti,
tra più di un milione e 100 mila immigranti nel
1901, non furono che 13,064 cioè appena l'1 %!
Perciò da coloro che volevano proporzionare i mezzi
allo scopo si pensò ad un freno assai più poderoso
di quelli sopraccennati: a respingere cioè gli analfabeti. Così ebbe origine il *Lod;e bill* (1), cui fu
apposto il *veto* dal Presidente Cleveland; così ri-
nacque la stessa proposta col *Satthuc bill* nel 1906.
E solo con una legge che proibisca l'entrata agli
analfabeti si potrebbe riuscire a ridurre al 50 %
circa gl'immigrati *undesirables*. Gl'italiani, si sa,
sarebbero i più colpiti.

Ogni misura ed ogni sentimento di giustizia e
di rispetto al criterio delle reciprocità, com'è facile
immaginare, sono stati messi da parte nella lotta
contro gli *undesirables* asiatici — i gialli della Cina
e del Giappone. Sono noti gli allarmi pel pericolo
di guerra tra Giappone e Stati Uniti pel trattamento
che lo Stato di California ha fatto ai Giapponesi —
d'onde il viaggio della grande flotta americana nel

---

(1) Fu presentato la prima volta nel 1895. Dopo il *veto*
presidenziale fu votato di nuovo dalla Camera dei Deputati;
ma la sessione si chiuse e il Senato non potè votarlo. Nel
1898 lo votò il Senato; ma la guerra di Spagna impedì che
lo votasse la Camera dei Deputati (*Commons*, pag. 234).

Pacifico, quasi a far comprendere ai vincitori di Tsoushima, che i Nord-Americani sono un osso più duro dei Russi; e sono noti del pari gli sforzi lodevolissimi di Roosevelt per indurre la legislatura di California e i cittadini di S. Francisco a più onesti consigli.

Il vero significato dell'avversione e della legislazione Nord-americana contro gli *undesirables* e contro gli Italiani, infine, viene dato dall'avversione e dalla legislazione ancora più rigorosa che dal 1890 in poi si è cominciata a praticare contro gli *undesirables* nell'Australia e che ha raggiunto il suo punto culminante coll' *Immigration restriction act* del 1901. Negli Stati Uniti gli *undesirables* oltre i gialli, sono i Russi, gli Austro-Ungheresi, gl'Italiani; ma in Australia, queste nazionalità, sono rappresentate sinora da quantità incalcolabili e gli *undesirables* sono i Tedeschi e i Britannici che possono fare concorrenza ai lavoratori della *Commouwealth* australiana. Le *razze superiori* e quelle che si considerano come *inferiori*, adunque, vengono trattate alla stessa stregua!

♣

Ho esposto tutto ciò ch'era doveroso esporre a difesa degli *undesirables* e degli Italiani immigrati negli Stati Uniti, ma non vorrei essere scambiato per uno *chauvin*.

Perciò, quantunque io già abbia esplicitamente riconosciuto ciò che c'è di biasimevole nei nostri concittadini, mi pare utile e doveroso d'insistere su questo tasto affinchè la conoscenza esatta e la coscienza piena della realtà, induca tutti e ciascuno degli Italiani della madre patria e delle colonie, a provvedere, ciascuno nella misura delle proprie forze e delle proprie condizioni a correggere ed a modi-

ficare le condizioni morali e materiali della nostra
colonie, delle nostre *Piccole Italie*, come vengono
chiamate e segnalate all' altrui disprezzo nel Nord-
America.

Su questo riguardo credo di non poter far meglio
che associarmi a ciò che ha scrittto di recente la
signorina Amy A. Bernardy (1). « Il cosidetto problema italiano negli Stati Uniti, essa dice sta tutto
nel problema di adattamento alla civiltà industriale
della massa atavicamente destinata alla vita agricola
della massa più rurale e rudimentale che abbia l'Italia, alla vita più accanitamente industriale e mec-
canizzata che abbia il mondo ».

« Dal punto di vista americano, la *Piccola Italia*,
anche come definizione, ha in sè la sua condanna.

« La *Piccola Italia* è qualche cosa che fa parte
per sè stessa, che non subisce le leggi della con-
venzionalità comune, che si rinchiude e si accentra
nel proprio cerchio, che per indifferenza o per timi-
dità o per cocciutaggine o per superbia o per in-
capacità di fare altrimenti (e nella esclusione delle
nostre *Piccole Italie* dal mondo americano circostante
c' è un po' di tutti questi elementi) non dimostra
il desiderio di fare come gli altri fanno; la deferenza
ai sistemi prevalenti, la pronta volontà dell'assimi-
liazione, è senz'altro, e anche fuori dei termini del-
l' *immigration problem* una cosa che indispone l'ame-
ricanità. Uno straniero poi che viene in America,
diverso di lingua, di volto, di costumi, di pensiero
di religione, e via dicendo, ha già senz'altro un titolo
sufficiente al biasimo e al sospetto della comunità.

« Aggiungete poi, com' è il caso per l'immigrazione italiana, che accrescano questo biasimo e que-

---

(1) *Vita Italiana agli Stati Uniti. Piccola Italia.* (Nella
rivista: *L' Italia all'estero*, 20 gennaio 1909).

sto sospetto fondamentale i vari elementi a cui abbiamo accennato altra volta, di analfabetismo, di rassegnata adattabilità, di *low standard*, di criminalità passionale, ecc., e si vedrà chiaro, evidente e doloroso, quale sia il complesso di cause della riprovazione americana così laconicamente espressa nell'appellativo : *Piccola Italia,* più dolorosa ancora della verità, perchè materiata di inutile ingiustizia e di dannoso equivoco, la leggenda che finora ha circondato come un cerchio di ferro e segnato come d'un marchio d'infamia e che appena ora accenna timidamente a squarciarsi qua e là — la leggenda, dico, che ha circondato la *Piccola Italia americana* ».

« Così anche, una siepe di rovi preclude l'ingresso a un *hortus conclusus* dove si crede non vegetare altro che rospi ed erbacce, mentre poi nel fatto qualche altra cosa c'è: un fiore che attende la sua primavera ».

« Se non vogliamo veder il male che c'è, (come troppo spesso d'altra parte non vogliamo poi destare in noi le buone energie del nazionalismo, dell'individualismo civile e pratico, dell'attività alacre e sana) possiamo ficcar la testa sotto l'ala compiacente della rettorica e del facile scetticismo e pretendere che il male non esiste perchè non abbiamo il coraggio di guardarlo in faccia, e che il bene sta tutto in casa d'altri perchè preferiamo ammirarlo là anzichè che destarlo e praticarlo e riconoscerlo in casa nostra quando s'è visto che una tale attività e un tale riconoscimento portano seco l'obbligo di una certa qual responsabilità personale e civile. Ma la rettorica alla quale, sia in fatto di male che di bene, di constatazioni pessimiste che di xenofilie inerti, degne solo di anime tributarie (e tanto più vergognose per le giovani energie d'Italia quanto più si affermano e si scatenano nel più vasto mondo

gli attivi e fattivi orgogli nazionalisti e imperiali-
sti degli altri), la rettorica alla quale già troppo
ignavamente e a lungo indulgemmo e indulgiamo,
(per il maledetto vizio che abbiamo di ammirare
altrui senza agire per noi, e di conservare, mentre
siamo così pronti alle scialbe adorazioni dall'estero
una fallace suscettibilità d'amor proprio interno che
ci fa confondere le oneste confessioni e dichiarazioni dei nostri guai col delitto di lesa patria e di
alto tradimento), questa rettorica troppo spesso vede
roseo là dove più si addensa il nero, e ombre dove
una qualche luce di speranza pur brilla. Quindi è
che se il giudizio americano intorno alla colonia è
per molte parti e per molte ragioni incerto e malforme, per molte altre ragioni e parti è incerto e
malforme anche, sulla colonia, il giudizio dell'Italia.
Quello che anche e specialmente in fatto di « problema italiano » dobbiamo perseguire e volere, è
una austera coscienza da parte nostra del nostro
danno e della nostra vergogna dov' è, per correg-
ger l' una e respinger l'altro se si può, o almeno
perchè in noi se non altro non si perpetuino er-
rori, illusioni ed incoscienze che dalla leggerezza
han fatto presto a trasmodare nella colpa; da parte
altrui un giusto apprezzamento dei nostri meriti,
specialmente in materia d' immigrazione poichè anche nella più triste colonia c' è tale e tanta virtù
d'onestà, d'energia, diciamo pure d'eroismo, da re-
dimere in larga parte il danno e la vergogna, mentre che il danno e la vergogna dura ».

« Ma per ottener questo, bisogna prima abbat-
tere la siepe. Diciamolo francamente, la realtà è
triste. Ed è ben poca nella vita coloniale in sè e
per sè la luce ideale, se pur ve n' è alcuna apparente, che questa triste realtà redima e conforti. Il
valore della silenziosa tenacia e della irriducibile co-

stanza con cui la miseraglia amorfa e ancora onesta prosegue la sua via e che finirà, io credo, col
convincere anche l'inesorabile America, noi lo vediamo quando siamo riusciti ad isolarlo idealmente
dalle altre manifestazioni della vita e della psiche
coloniale nella sua troppo spesso dubitosa esplica-
zione ».

« Il disprezzo che per troppi americani nel nord
identifica l'Italiano nella miglior ipotesi col tipo
del cittadino *undesirable*, sporco, straccione e ma-
scalzone, nell'ipotesi meno benevola del criminale
della peggiore specie che unisce la mala fede allo
impulso della brutale malvagità; quello che per troppi
americani del sud accomuna ancora l'italiano al
negro, e non dico altro, non è pur troppo senza
una qualche ombra di fondamento, se non nella
piantagione e nella *farm*, certo nei bassi fondi e
qualche volta anche nei livelli, diremo un poco più
alti, della città. Non sempre si disprezza l'italiano
per preconcetto o per pregiudizio, ma perchè molti
Italiani, purtroppo, hanno associato sè e l'Italia
nella mente e nella opinione pubblica con le forme
più abbiette della extra-legalità e della corruttela
pubblica e privata ».

« Tanto vero, che si cercherebbe forse invano
una forma parlamentare per descrivere adeguata-
mente certi elementi che negli Stati Uniti forse
anche più spudoratamente che in altri paesi d'immigrazione infamano colonie in cui sono pure tante
buone energie e tante oneste e semplici coscienze
e, sia pure, incoscienze, se non ci avesse pensato
il padre Dante con un verso famoso, quello che
finisce: — « ... e simile lordura.

« Onde, quando il cittadino estraneo agli intrighi
e agli imbrogli e peggio, protesta, e trascende alle
enormità e alle ingiustizie dei giudizi complessivi

e inesorabili, e a tutta l'Italia piccola e grande fa
colpa delle colpe di quei tali, bisogna, è vero ,
rinfacciargli prima i mascalzoni e i mestatori indigeni e dirgli poi che impari a distinguere fra italiani e italiani, ma la reazione e la ribellione non
possono essere nè così sincere, nè così violente, nè
così persuasive, nè così sicure, perchè un fondo di
triste verità nella condanna, c'è ».

E la Bernardy conchiude:

« Dappertutto, Dio ci aiuti, si fa male e bene ;
dappertutto si erra, si folleggia; dappertutto c'è da
perdonare e da essere perdonati. E quelli che rimproverano alla *piccola Italia* il suo voto venduto ,
son quelli stessi che sarebbero infelicissimi se non
potessero comprarlo; quelli che sparlano della sua
criminalità, della sua malafede finanziaria , del suo
low standard e via dicendo, hanno al loro attivo le
cronache giudiziarie più brutali, la stampa più gialla,
la corruzione politica e civica più spudorata, i fal-
limenti più colossali che siano al mondo: sono essi
stessi — coscientemente o incoscientemente a noi
non tocca indagare ora — gli ospiti e i fautori di
questa gravissima condizione di cose , sicchè ogni
discussione dei relativi meriti e demeriti, ogni rin-
facciare di mutue colpe e agitare di reciproci di-
spregi sarebbe, a ogni modo, vertenza di galeotto
a marinaro.

« Ma ogni più grigia nube è foderata di sole; e
come gli americani si risentono quando si consta-
tano in loro quelle cose, così han diritto la *piccola*
e la *grande Italia* sotto la grigia nube alla loro
parte di sole. Ascoltiamo, ascoltiamo piuttosto ,
tutti, americani e italiani, prosseni e meteci, au-
toctoni di due generazioni e immigrati di ieri, che
tanta storia di dolore vediamo e viviamo quotidia-
namente nella colonia cittadina, la voce del lavoro

che attraverso tutte queste tristezze e tutte queste
brutture, accanita, alacre, pugnace, agglomerando
legname, aprendo canali, scavando fossi, collocando
rotaie, innalzando argini, gettando ponti, costruendo
dighe, preparando la via ai carri elettrici, alle automobili, alle locomotive, ai convogli , ai fili del
telegrafo e a quelli del telefono, popolando campi
e miniere, coltivando orti e verzieri , scavando le
fondamenta di case che stanno per diventare nuclei di città, attraverso tutta la grande America
compie la forza tenace della *Piccola Italia*. Della
Piccola Italia più vera e migliore, che è fondamental-
mente un buon elemento di rappresentanza per
l'Italia, di acquisto per gli Stati Uniti. E come tale
meriterebbe dall'una e dall'altra parte la migliore
considerazione ed il più largo aiuto, se non che le
molteplici circostanze di fatto e di dritto troppo
spesso infaustamente convergenti a proteggere gli
elementi men degni e a giustificarne facilmente
l'azione deleteria e corrosiva troppo spesso le tol-
gono anche quel poco usbergo che per noi si potrebbe offrirle o agli altri richiedere per lei ».

L'opera di risanamento delle nostre *Piccole Italie,*
per parte mia conchiudo, pel decoro e pel bene della
Grande Italia , non si può compiere in un giorno,
nè da un solo individuo geniale , nè da masse di
oscuri, nè da forti associazioni, nè dagli enti pubblici isolati; ma tutte le forze devono convergere.
La parte più difficile sta certamente nella trasformazione morale degli individui, degli elementi, che
compongono le *Piccole Italie*; e dev'essere compiuta
tanto in patria — colla opportuna selezione degli
emigranti; col somministrare loro delle istruzioni
sulle leggi, sui costumi, sulle condizioni tutte del
paese dove immigrano; col mostrar loro gli enormi
benefici morali e materiali, che ritrarrebbero dalla

ιɔro trasformazione ecc. ecc. — quanto nelle colonie. E qui occorrerebbero propagandisti intelligenti e volenterosi, che potrebbero essere forniti e pagati, assai proficuamente per la causa della Italianità, dalla *Dante Alighieri*; propagandisti che dovrebbero essere scelti con cura tra gli italiani delle regioni degli emigrati : Siciliani pei Siciliani, Calabresi pei Calabresi ecc. — visto che oltre l' Atlantico i nostri concittadini si aggruppano per regioni e si mostrano assai diffidenti verso i connazionali di altre regioni. Di più e molto dovrebbe fare lo Stato per mezzo dei Consoli e degli agenti suoi, mentre disgraziatamente sinora per la deficienza qualitativa e quantitativa dei suoi rappresentanti ha fatto ben poco e spesso ha fatto male.

L'Italia che non possiede territorii, ma che ha numerose e vigorose le cosidette *colonie senza bandiera* trarrebbe giovamento e lustro maggiore se ad esse consacrasse quei milioni sciupati nell'Eritrea e nel Benadir.

Ma temo pur troppo che il mio voto rimarrà insoddisfatto. I nostri diplomatici e i nostri generali, se esso venisse accolto, non potrebbero raccogliere gli allori... che tutti conoscono!

Prof. Napoleone Colajanni

Deputato al Parlemento

LA CRIMINALITÀ DEGLI ITALIANI
negli Stati Uniti

LA CRIMINALITA DEGLI ITALIANI

I.

L'avversione vivissima che si prova e si manifesta negli Stati Uniti contro gli *undesirables* in genere e particolarmente contro gl'Italiani deriva sopratutto dalla loro criminalità.

Questo motivo è sincero e reale nella parte più eletta degli Americani, che combattono la cosidetta *nuova immigrazione;* serve di pretesto per nasconderne altri — ad esempio: il timore nella concorrenza del lavoro — a non pochi, che assumono a buon mercato l'aria di difensori della moralità della grande repubblica.

L'assassinio feroce perpetrato a Palermo in persona di Petrosino, il coraggioso luogotenente della polizia di New-York venuto in Italia per studiarvi la *mafia* e la *camorra* e le sue classi delinquenti, ha dato a tale avversione delle proporzioni fantastiche, inverosimili.

Il *New York World* è arrivato a pubblicare un suo telegramma da Roma nel quale si afferma che un Principe romano — di quelli agli ordini del Papa, l'eterno nemico dell'Italia? — gli abbia dichiarato: « Tutta la polizia napoletana e siciliana è « affiliata sia con la *camorra* che con la *mafia*. La « premura con cui la polizia di Palermo si è impa- « dronita dei documenti segreti di Petrosino detta-

« glianti i risultati delle sue investigazioni è molto
« significativa e lamentevole. I delinquenti di qui e
« di America segnati sui libri di Petrosino adesso
« son salvi. Sono stati bene avvisati ».

« Un altro punto interessante è che il lavoro per-
« sonale eseguito da Petrosino in Sicilia, era consi-
« derato dalla polizia siciliana come un'intrusione ed
« una tacita sfiducia del suo lavoro. Essa aveva ra-
« gione di temere che Petrosino avesse potuto di-
« sotterrare i suoi misfatti ».

« Crescono i sospetti che Petrosino sia stato una
« vittima dei suoi confratelli siciliani, o che pure
« sia stato denunziato da essi alla *mafia* ».

E un altro giornale di New York: *The Evening
Sun* (9 Aprile 1909) invita il governo italiano ad
abbandonare le procedure legali per combattere la
mafia « Non si può trovare una via — un uomo —
« un Cromwell se è necessario — per distruggere
« la peste, la cangrena nel Regno d' Italia? Il go-
« verno di Washington non vorrà reclamare affinchè
« si provveda e presto? Il sangue di Giuseppe Pe-
« trosino deve gridare vendetta inascolta:o? » L'avversione si è talmente concentrata contro la patria
degli uccisori di Petrosino, che ogni grande reato
commesso a New York si attribuisce ai Siciliani,
anche quando ne sono conosciuti gli autori che Siciliani non sono; esagerazione di un pregiudizio
che ha spinto un magistrato nord-americano, l' avv.
Francis Corrao — di origine Siciliana — a protestare pubblicamente nell'*Aurora* di Brooklyn (N.º del
21 Agosto 1909) (1).

Se con questa accusa ci troviamo nel campo della

---

(1) Dall'egregio avv. Guglielmo Di Palma, direttore del *Labor information office for Italians* apprendo, mentre correggo
le bozze di stampa, che nelle richieste di mano d'opera degli
intraprenditori americani dopo l'assassinio di Petrosino, si è
acuita la esclusione dei meridionali in genere e dei siciliani in
ispecie. Nell'analago *Labor* ecc. nord-americano si è verificato
lo stesso fenomeno.

esagerazio.ie calunniosa, altre, però, ne vengono formulate, suffragate da cifre e sostenute da persone, che hanno speciale competenza nella materia, che impongono un diligente esame, per dare loro le giuste proporzioni ed assegnare agli Italiani la dovuta responsabilità; ma non oltre di quella ch'è loro dovuta.

Il grido di allarme più documentato contro l'incremento della delinquenza grave e il numero degli omicidi messo in rapporto coll'aumento della immigrazione italiana è stato dato da Shipley ed anche da Bingham, ex commissario capo della polizia di New York.

Lo Shipley con copia di statistiche ha cercato dimostrare che in tutte le città americane — specialmente a Newark, New Haven, Buffalo, Rochester, Siracusa, Filadelfia ecc. e sopratutto a New-York — a misura ch'è aumentata l immigrazione degli *undesirables* è aumentata la delinquenza e con particolarità l'omicidio. E' questo pure l'avviso del Mc Clure.

Bingham, del pari, riconosciuta la grande delin quenza degli Ebrei, specie contro la proprietà, afferma che quella degli italiani sta alla prima come 50 sta a 20, mentre il numero degli Italiani sta a quello degli Ebrei come 1 a 2. I malfattori, i *banditi* italiani, gli afflliati alla mafia ed alla camorra, egli dice, costituiscono la più grande minaccia contro la legge e contro l'ordine (1).

I dati statistici che questi accusatori citano sono esatti? Quanto c'è di vero in tali accuse? Questa è l'indagine imparziale che mi propongo.

---

(1) Maynard Shipley: *The effects of immigration on homicide in American cities.* (In *The popular science monthly*. Agosto 1906 pag. 162 e 163); A. Bingham: *Foreign criminals in New York.* (In: *The north american review* Settembre 1908 pag. 385); S. Mc Clure: The *Tammanyzing of a Civilization* (In: *Mc Clure's Magazine* Novembre 1909).

II.

Ogni difesa, che si volesse tentare del buon nome italiano fallirebbe allo scopo se fosse fatta a base di sentimentalismo. I fatti esclusivamente devono costituirne la base; ed essi devono, se favorevoli, ridurre alle giuste proporzioni le accuse esagerate, addurre le attenuanti, se ce ne sono; esporre, infine, la causa della reale criminalità italiana in America, per correre ai provvedimenti opportuni. In questo esame una osservazione preliminare s'impone: quale fede meritano le statistiche nord-americane ? rispecchiano esse esattamente il fenomeno della criminalità? Negli Stati Uniti non si pubblicano statistiche penali comparabili a quelle degli Stati Europei. Il governo federale pubblicava pel passato soltanto i dati sul numero dei detenuti nelle prigioni federali e dei singoli Stati quali risultavano dai censimenti decennali; ora si fa qualche altra pubblicazione senza attendere un nuovo censimento. I singoli Stati provvedono poi ad altre pubblicazioni senza uniformità di criteri e di metodo e che a giudicarne da quelle, che sono riuscito a procurarmi, principalmente sullo Stato di New York, sono molto difettose ed incomplete e niente affatto comparabili colle nostre.

Alcuni degli inconvenienti gravi di queste rilevazioni statistiche sono stati segnalati da chi non può essere sospettato di voler denigrare il proprio paese: da Iohn Koren, ch'è l'autore e il relatore delle investigazioni dell'ultimo censimento sui detenuti(1).

Il Koren nella sua relazione sui *Prisoners and juvenile delinquents in Institutions 1904* avverte:

---

(1) Lettera di North, direttore generale dei lavori del censimento ad Oscar Straus segretario di Stato pel *Trade and Commerce* del 10 agosto 1907 nel volume : *Prisoners and juvenile delinquents in institutions 1904.* Washington 1907. Il Koren ha seguito i metodi e il piano di P. Roland Falkner.

« Deve ritenersi bene che la statistica delle prigioni per quanto intelligentemente fatta non può servire se non superficialmente come un mezzo di valutare il movimento del delitto, perchè essa non tiene conto di parecchi i quali, benchè condannati sfuggono alla prigione col pagamento di una multa, colla sospensione della sentenza o con altre forme di attenuazione. Le persone condannate alla prigione forse formano la più larga parte dell'elemento che entra nel movimento generale della criminalità quantunque, ciò non sia chiaramente provato » (pag. 12 e 13).

Un gravissimo incoveniente, rilevato dallo stesso Koren e sul quale ritornerò in appresso è il seguente: i dati statistici sulla popolazione totale si riferiscono al 1900; quelli sui detenuti e sugli arrestati al 1904.

Altro ancora più grave: gli Stati e le Città dell'unione americana non procedono con eguali criteri nella classificazione dei reati e negli arresti. Si arrestano per omicidio, ad esempio, con estrema facilità in alcuni Stati e in alcune città, mentre si tratta di reati molto meno gravi; e il contrasto risulta evidente quando si fa il paragone tra il numero degli arresti e il numero dei morti per omicidio pei pochi Stati, che raccolgono e pubblicano la statistica delle cause di morte (1).

Della incertezza e della inesattezza di queste statistiche americane si può avere un' idea appena approssimativa dall' errore commesso da un distinto magistrato americano, il Cleland, che assegnò a Chicago 17 assassinati per 100,000 abitanti mentre lo Shipley, che inclina tanto ad esagerare riduce la cifra nientemeno a 7,30 (2).

_____

(1) Lo stesso Shipley avverte in una *nota*, che la statistica degli arresti per omicidio non risponde bene alle esigenze dello studio, ma che se ne serve in mancanza di altre statistiche (pag. 164).

(2) Questa osservazione la tolgo da un altro articolo di

Premesse queste riserve vediamo quale risulta la criminalità nei singoli Stati e nelle principali città degli Stati Uniti secondo la nazionalità dei detenuti.

III.

Una prima dimostrazione della maggiore delinquenza degli Italiani la si è voluta fare stabilendo il rapporto tra il numero dei detenuti in ciascuno degli Stati dell'Unione e il numero degli Italiani che vi risiedono; la seconda più determinata si fonda sulle percentuali di detenuti che danno gli immigrati negli Stati Uniti secondo la loro nazionalità; la terza infine, é ancora più particolereggiata e concerne la delinquenza omicida delle principali città degli Stati Uniti messa in rapporto col numero degli Italiani che vi dimorano.

1° *Numero dei detenuti nel 1890 e 1904 e d'Italiani residenti nei singoli Stati secondo il censimento del 1900.* — Questa comparazione è la meno conclusiva, ma serve sempre a dimostrare non essere vero: che i detenuti e quindi i reati sono più numerosi dove sono più numerosi gl'Italiani, e che i reati aumentano come aumenta l'immigrazione italiana.

Non potendo riportare le proporzioni degli Italiani — che darò nell'occuparmi dell'omicidio in ispecie — e dei detenuti per i singoli Stati, porrò il paragone solo per le singole grandi divisioni; riporterò le cifre dei singoli Stati della *North Atlantic Division*, perchè questi Stati dei 484,207 Italiani — nati in Italia da distinguere dagli Italiani nati in America — nel 1900 ne ospitavano ben 382,065 cioè oltre il 78 per 100:

Shipley sui *delitti di sangue a Chicago ed a New York* nella stessa *Popular Science Monthly* di Agosto 1908 riprodotto nella Rivista *Minerva* del 27 Settembre 1908.

	Percentuale italiani nel 1900	Detenuti per 100,000 ab. nel 1890	nel 1904
Stati Uniti	**0,6**	**131,5**	**100,6**
North Atlantic. Div.	1,6	162,4	121,6
South » »	0,1	128,8	100,5
North Central »	0,2	88,8	75,2
South » »	0,1	146,6	95,7
Western »	1,0	222,1	169,4
Stati della North Atlantic. Div.			
Maine	0,2	77,4	70,6
New Hampshire	0,2	85,3	97,7
Vermont	0,6	60,2	78,7
Massachussetts	1,0	233,5	187,2
Rhode Island	2,1	162,1	130,6
Connecticut	2,1	137,5	115,4
New York	2,5	191,2	126,7
New-Iersey	2,2	169,9	131,9
Pennsilvania	1,0	123,4	92,3

Se nella *North Atlantic Division* il numero dei detenuti è più elevato che nella media degli Stati Uniti, il fenomeno non deve attribuirsi al maggior numero degli Italiani, ma al fatto che la *North Atlantic Div.* contiene più popolazione urbana e più grandi città del resto della Unione.

I detenuti sono più numerosi, e in modo sensibile nella *Western Division*; ed ivi sono assai meno numerosi gl'Italiani. Nella California, dove raggiungono la percentuale di 1,5 il numero dei detenuti da 281,2 per 100,000 ab. nel 1890 discese a 210,2 nell'anno 1904.

Similmente nella *South Central Div.* sono numerosi gl'Italiani nella Louisiana con 1,2 per 100, ma i detenuti da 143,8 nel 1890 discesero a 112,8 nell'anno 1904.

La percentuale degli Italiani anche dove sono più numerosi, comé negli Stati di New Jersey, Connecticut, Rhode Island, Massachussetts e Pensilvania è tanto esigua che non potrebbe influire sul numero dei detenuti; tanto più che la loro delinquenza è di natura tale, che in cifre assolute dà il minimo contingente della popolazione carceraria.

Tra il 1890 e il 1904 gl'Italiani degli Stati Uniti aumentarono e forse si quadruplicarono — erano appena 0,29 % nel 1890 e passarono a 0,6 % nel 1900; ma si sa che il numero degli Italiani aumentò assai più rapidamente dopo il 1900 — e il numero dei detenuti intanto diminuì in tutta l'Unione e nelle singole divisioni, anche negli Stati che contengono il maggior numero d'Italiani.

Su questa diminuizione di detenuti tra le due date, però, il relatore Koren fa le seguenti giuste riserve: « In giugno 1904 il numero dei detenuti esistenti nelle varie prigioni (dello Stato, degli Stati, delle Contee, nei riformatori, nelle prigioni comunali ec.) fu di 81,772 o di 100,6 per 100,000 abitanti. Nel 1890, 1° giugno, fu di 82,329 o di 131,5 per 100,000, abitanti. Ma le due cifre non sono comparabili perchè nel 1904 parecchi gruppi non sono compresi nel numero dei detenuti mentre lo erano stati nel 1890. In tutto nel 1890 i detenuti che non furono compresi nel 1904 furono 15,526. La proporzione del 1890 deducendo questa cifra si ridurrebbe a 106,7 per 100,000 abitanti. La diminuizione nel 1904 non indica una diminuita tendenza al delitto, ma semplicemente un decremento nel numero dei casi in cui i delinquenti furono condannati alla prigione (*Prisoners* ecc. pag. 13 e 14).

2° *La delinquenza secondo le varie nazionalità.* — Il precedente rapporto è troppo generico ed indeterminato; ed ha quindi scarsa importanza, per quanto esso risulti piuttosto favorevole agli italiani. Ne ha una molto maggiore la distinzione della delinquenza secondo le varie nazionalità.

I detenuti sono divisi prima in due grandi categorie secondo, che appartengono ai *major offenders* o ai *minor offenders*; divisione che corrisponderebbe forse a quella antica del codice italiano in *crimini* e *delitti*. Nei *minor offenders* sono anche compresi i detenuti per *contravvenzioni* puniti col carcere (ubbriachezza, vagabondaggio ecc.).

Nella speciale delinquenza distinta per nazionalità il *Report* di Koren somministra due generi di dati: 1° quello dei detenuti, che si trovavano nelle prigioni degli Stati Uniti a 3ɔ giugio 1904; 2° l'altro dei detenuti per reati commessi durante il 1904.

E' evidente che questo seeondo dato avrebbe una importanza molto maggiore se il numero dei detenuti fosse proporzionato a quello del rispettivo numero di abitanti secondo le nazionalità.

Il numero assoluto dei detenuti stranieri a 30 giugno 1904 secondo le nazionalità principali era il seguente:

MAJOR OFFENDERS (1)

	Omicidio	Assault	Robbery	Rape	Burglary	Larceny	Fraude
Austriaci	26	42	7	9	28	59	1
Canadiani	22	36	25	15	118	161	6
Francesi	6	6	1	2	8	22	—
Inglesi	9	19	15	5	85	116	9
Irlandesi	16	52	20	6	89	112	2
Italiani	96	175	31	26	52	107	7
Polacchi	13	45	3	4	19	58	2
Russi	10	23	10	8	52	107	3
Svedesi	13	4	4	3	21	32	2
Tedeschi	33	66	22	12	148	212	19
Ungheresi	7	13	—	3	6	16	—

MINOR OFFENDERS (2)

	Vagabondaggio	Ubbriach.	Assaults	Robbery	Burglary	Larceny	Fraude	Approp.
Austriaci	163	81	95	1	5	100	16	6
Canadiani	546	1478	142	2	20	376	29	8
Francesi	75	50	13	1	4	34	3	—
Inglesi	852	1110	99	1	13	221	25	12
Irlandesi	2888	6100	337	5	28	592	24	12
Italiani	221	113	344	3	22	195	109	5
Polacchi	154	166	132	7	11	162	16	2
Russi	156	103	124	—	12	162	15	9
Svedesi	321	220	47	—	9	100	8	1
Tedeschi	1232	658	238	2	37	438	37	26
Ungheresi	30	56	66	1	2	40	8	2

(1) *Prisoners* ec. pag. 158-159. (2) *Prisoners* ec. pag. 156. 157. *L'assault* corrisponderebbe alle lesioni, alle aggressioni ed all'omicidio mancato o tentato. *Robbery, Burglary, Larceny, Fraude* sono forme varie di reati contro la proprietà e

Il *Report* divide tutta la delinquenza in tre principali gruppi : *reati contro la società* (tutti reati contro lo Stato, contro i costumi e contro l'ordine della famiglia ecc.), *contro le persone e contro le proprietà*.

Su 100 detenuti secondo questa divisione ce ne erano per reati :

	Contro la società	Contro le persone	Contro le proprietà
Austriaci	11,4	39,8	47,9
Canadiani	14,9	20,4	63,4
Inglesi	14,4	15,3	69,7
Irlandesi	26,9	22,2	49,3
Italiani	12,6	57,1	30,1
Polacchi	15,1	35,5	48,9
Russi	12,3	19,7	66,5
Svedesi	14,0	24,0	62,0
Tedeschi	11,4	39,8	47,9

Scendendo a qualche dettaglio tra gl'Italiani detenuti fra i *major offenders* su 100 ce n'erano per reati (1).

Contro la Società				12,6
Contro le persone	omicidio	16,2	
»	»	assaults	29,5	
»	»	robbery	5,2	
»	»	rape	4,4	
»	»	altri	1,9	57,1
Contro le proprietà	incendio	0,3	
»	»	burglary	8,8	
»	»	larceny	18,8	
»	»	Forgery	1,9	
»	»	Fraude	1,2	30,1
Diversi				0,2
				100,0

corrisponderebbero con maggiore esattezza a *rapina, a furto con scasso e qualificato, furto semplice e appropriazione indebita e frode*. Il *rape* tradotto letteralmente sarebbe *ratto*, che può essere reato in sè contro la persona o mezzo per compiere un reato contro la proprietà.

(1) *Prisoners* ecc. pag. 45.

Tra i reati che hanno importanza o perchè sono molto numerosi, o perchè gravi, in ordine decrescente le nazionalità su 100 detenuti ne avevano, sempre a 30 gingno 1904:

Tra i *minor offenders*: *Per vagabondaggio* : tedeschi 36,6; Inglesi 29,5; Scozzesi 29.0, Svedesi 23,7; Irlandesi 23,6; Austriaci 20,5; Polacchi 19,5; Canadesi 17,8; Russi 16,4; Italiani 14,3 (il minimum).

Per ubbriache₂₂a e disordini : Irlandesi 63,9 ; Canadesi 54,0; Scozzesi 54,0; Svedesi 51,0; Inglesi 50,1; Austriaci 35,5 ; Tedeschi 35,4, Russi 33,5; Polacchi 29,0; italiani 28,0 (il minimum).

In ordine decrescente su 100 detenuti tra i *major offenders* nelle principali categorie ce n'erano 1° *Contropersone. Omicidio* : Italiani 16,2; Svedesi 13,0; Austriaci 13,0; Polacchi, 7,0; Tedeschi 5,0; Russi 3,7; Irlandesi 3,6: Scozzesi 3,0; Inglesi 2,8. *Assaults* : Italiani 29,5; Polacchi 24,2; Austriaci 19,9; Scozzesi 12,0; Irlandesi 11,8; Tedeschi 9,9 ; Russi 8,6; Canadesi 7,3; Inglesi 5,8; Svedesi 4,0. 2° *Contro la proprietà. Larceny :* Russi 39,8 ; Inglesi 35,5; Canadiani 32,5, Svedesi 32,0; Tedeschi 31,9 ; Polacchi 31,2; Austriaci 28,0; Scozzesi 27,0; Irlandesi 25,3; Italiani 18,0. *Burglary*: Inglesi 26,0; Canadiani 23,8; Scozzesi 23,0; Tedeschi 22,3; Svedesi 21,0; Irlandesi 20,1; Russi 19.3; Austriaci 13,3; Polacchi 10.2; Italiani 8,8 (*Prisoners* ecc. pàg. 46).

Nella proporzione dei detenuti a seconda della natura dei reati è evidente dunque il triste primato degli italiani nei reati contro le persone e la loro eccellente posizione in quelli contro la proprietà.

Vediamo ora, queli sono le proporzioni secondo le nazionalità dei detenuti (1) per reati commessi durante il 1904. Tra i *major offenders* le cifre assolute sono le seguenti: *Totale detenuti stranieri*

---

(1) La relazione del censimento parla di *prisoners enumerated* o *convicted* a 30 giugno 1904; e di *committed* durante l'anno 1904.

4131; tra questi: Austriaci 211; Francesi 69; Inglesi
327; Irlandesi 422; Italiani 594; Polacchi 186; Russi
269; Svedesi 100; Tedeschi 665; Ungheresi 64 (*Prisoners* ecc. p. 154),

I detenuti stranieri come *minor offenders* per reati
commessi nel 1904 erano 30,962 ; tra i quali : Austriaci 795; Francesi 294; Inglesi 2886; Irlandesi 12,258;
Italiani 1549; Polacchi 852, Russi 933; Svedesi 930;
Tedeschi 3668; Ungheresi 360 (*Prisoners* ecc. p. 155).

Gli italiani arrestati come *major offenders* lo furono in gran parte nella *North American division*
con 503 sopra il totale di 594 ; e di questi 255,
nel solo Stato di New York ; cioè rispettivamente
l' 86,3 % e il 42,9 %, mentre secondo il censimento
del 1900 nella *North Atlantic div.* c' era poco più
del 78 % di tutti gl'Italiani, che erano nell'Unione.
Questa maggiore criminalità degli italiani nella
North Atlantic div. non è che apparente , perchè
precisamente negli Stati che ne fanno parte dopo
il 30 giugno 1900 é enormemente aumentato il
numero degli immigrati italiani come si vedrà in
appresso.

Anche tra i *minor offenders* per reati commessi
durante il 1904 il posto degli italiani è eccellente.

Queste cifre e queste graduatorie non danno un
concetto perfettamente esatto della criminalità secondo la nazionalità.

Qualche cosa di più si apprende dai dati del
Report di Koren. Riproduco quelli per i *major offenders* (*Prisoners*, ecc. p. 44).

	Percent. tra gli stranieri a 30 Giugno 1900 (1)	Percent. tra detenuti à 30 Giugno 1904
Austriaci	2,7	5,1
Canadiani	11,4	12,0
Francesi	1,0	1,5
Inglesi	9,0	7,9
Irlandesi	15,6	10,7
Italiani	4,7	14,4

(1) Queste sono le percentuali su *100 stranieri* e non nella
popolazione totale.

Polacchi	3,7	4,5
Russi	4,1	6,5
Svedesi	5,5	2,4
Tedeschi	25,8	16,1
Ungheresi	1,4	1,5

Il rapporto più favorevole tra popolazione e detenuti *major offenders* è quello dei tedeschi — delinquenza minore di un terzo alla popolazione ; il peggiore quello degli italiani che hanno una delinquenza tripla e degli austriaci che la presentano doppia.

Ma gl'Italiani non danno lo stesso contributo ai reati commessi nel 1904 in tutte le Divisioni degli Stati Uniti come non lo danno gl Irlandesi, coi quali spesso i primi sono paragonati come si può rilevare da questo specchietto (*Prisoners* ecc., pagina 42 e 43).

REATI COMMESSI 1904.

	Popolazione	ITALIANI Delinquenza totale	Major offenders	Minor offenders
Stati Uniti	4,7	6,1	14,4	5,0
North Atlantic Div.	7,4	7,1	21,8	5,6
Soulh Atlantic »	4,9	9,7	—	9,4
North Central »	1,3	2,2	4,2	1,8
South Central »	7,3	4,0	4,7	2,8
Western division	4,8	6,0	5,6	9,9
		IRLANDESI		
Stati Uniti	15,6	36,2	10,7	39,6
North Atlantic Div.	23,4	41,9	12,5	44,9
South Atlantic »	16,9	22,4	—	24,4
North Central »	8,4	21,3	8,6	23,5
South Central »	8,9	7,9	2,4	16,5
Western division	9,9	26,1	9,1	29,9

Ci sono notizie più recenti sulla delinquenza degli italiani e degli altri immigrati negli Stati Uniti e si rilevano dal numero dei detenuti alla fine dell'anno fiscale (giugno) 1908. Si ritrovano nell' ultimo *Report of the Commissioner of immigration for year ended iune 1908* (Washington 1908); ma non ci fanno conoscere la natura dei reati commessi o attribuiti agli immigrati e danno soltanto la distinzione tra *minor* e *major offenders*. Mancano le proporzioni col totale e colle rispettive popolazioni e queste non sono divise sècondo le nazionalità, ma secondo le razze o i gruppi etnici. Così, ad esempio, sono enumerati: Boemi, Moravi, Slavoni, Dalmati, Bosniaci, Slovachi che dovrebbero essere compresi quasi tutti tra gli Austriaci; gli Ebrei possono essere Russi, Polacchi, Austriaci, Rumeni; i Ruteni possono appartenere all' Austria o alla Russia; i Croati all'Ungheria; i Finnici alla Russia ecc; Svedesi, Norvegiani e Danesi sono compresi sotto la denominazione di Scandinavi. Gli stessi Italiani poi sono divisi in Italiani del nord e Italiani del sud.

Comunque mi servo di questi ultimi dati relativi alle nazionalità ben distinte per confrontare il numero assoluto del 1904 con questo del 1908 (*Report* pag. 100 e 101):

	Detenuti a 30 Giugno 1908	
	Major offenders	Minor offenders
Francesi	341	328
Inglesi	679	469
Irlandesi	395	1108
Italiani	2077	1037

I risultati della comparazione sono paradossali; dapoicchè i quattro gruppi d'immigrati nel 1908 darebbero una diminuizione colossale, inverosimile tra i *minor offenders* per i Francesi, gl'Inglesi e gli Irlandesi ed un leggerissimo aumento per gli Italiani.

Evidentemente nella cifra non saranno compresi i detenuti per ubbriacchezza e vagabondaggio che nel 1904 erano 1962 per gl'Inglesi e 8988 per gl'Irlandesi. Rimasero quasi identiche le cifre dei *minor offenders* italiani, perchè essi anche nel 1904 davano uno scarsisissimo contributo, 334, a tali reati.

Tra i *major offenders* l'aumento è enorme tra i Francesi: 69 nel 1904 e 342 nel 1908 e trà gl'Italiani: 594 nel primo anno e 2077 nell'ultimo; è assai forte tra gl'Inglesi che passano da 258 a 679; è minimo tra gl'Irlandesi che da 297 salgono a 395.

Nel 1908 sul totale di 15,323 stranieri detenuti

gl' Italiani con 3,114 rappresentavano il 20 %; pero mentre tutti gli stranieri davano il 53,2 % alla delinquenza grave (*major offenders*) e il 46,7 alla lieve (*minor offenders*) gl' italiani ne davano rispettivamente 66,6 e 34,4 %. Questa preponderanza nella delinquenza grave si ripercuote nella durata della detenzione. Il *Report* citato (pag. 102) infatti dà il *probabile periodo di detenzione* ch'è il seguente:

	Sotto 2 anni	Sopra 2 anni	A vita
Stranieri	59,6 %	35,6 %	4,7 %
Italiani	48,9 »	47,0 »	3,9 »

Riflettendo all'alto contributo che gl'italiani danno all'omicidio sorprende che la percentuale dei condannati a vita sia tra i nostri concittadini inferiore a quella degli stranieri in genere. Forse perchè sono frequenti le esecuzioni capitali ?

Nel 1908 la distribuzione geografica dei detenuti su per giù corrisponde a quella del 1904. Infatti nella *North Atlantic Division* ce n' erano. (*Report* pag. 105):

	Major offenders	Minor offenders
Italiani	76,4 %	86,5 %
Irlandesi	49,6 »	88,4 »
Inglesi	42,2 »	70,1 »

E' evidente che nella delinquenza lieve la *North Atlantic Division* offre per così dire, un terreno più fertile. La manifestazione statistica, però, può essere l'effetto di una legislazione più rigorosa verso certe contravvenzioni; di una polizia meglio organizzata e più vigile; e sopratutto della più alta percentuale di popolazione urbana, che favorisce tanto la piccola delinquenza.

Sin qui mi sono valso delle proporzioni che ha dato il Koren nel suo *Special Report* sui *Prisoners* ec. Ma queste proporzioni servono soltanto a farci comprendere il posto che ogni nazionalità occupa nella delinquenza tra gli stranieri; non ci danno la delinquenza propria di ciascuna nazionalità. Perciò pei *major offenders* soltanto rifaccio le proporzioni per 100,000 abitanti in base agli stranieri delle principali nazionalità immigrati secondo il censimento del 1900 (*Census Report*. Vol. 1° Parte 1ª *Population* pag. CLXXI).

DETENUTI MAIOR OFFENDERS PER 100,000 ABITANTI a 30 GIUGNO 1904

	Omicidio	Assault	Robbery	Rape	Burglary	Larceny	Fraude
Austriaci	9,5	15,5	2,5	3,2	10,1	21,3	0,3
Francesi	5,7	5,7	0,9	1,8	7,6	21,0	—
Inglesi	1,0	2,2	1,7	0,6	10,0	13,7	1,0
Irlandesi	0,9	3,2	1,2	0,3	5,7	7,5	0,1
Italiani	19,8	36,1	6,4	5,3	10,7	22,0	1,4
Polacchi	3,3	11,7	0,7	1,0	4,9	15,1	0,5
Russi	2,3	5,1	2,3	1,8	12,2	25,2	0,7
Svedesi	2,2	0,6	0,7	0,5	3,6	5,5	0,3
Tedeschi	1,2	2,4	0,8	0,4	5,5	7,0	0,7
Ungheresi	4,8	8,9	—	2,0	4,0	10,9	—

Il triste primato degli Italiani appare evidente; solo nella *burglary* e nella *larceny* sono superati dai Russi. Se poi si riflette che i detenuti censiti a 30 giugno 1904 rappresentano l'accumulo degli anni precedenti e che le condanne più lunghe devono essere quelle per l'omicidio, per l'*assault* ec. comparirà minore la relativa delinquenza delle emigrazioni più antiche (Tedeschi, Inglesi, Irlandesi, Svedesi) e molto più grave quella degli emigrati che accorrono negli Stati Uniti da tempo più recente; cioè degli Italiani, Russi, Austriaci ecc.

Ma nel paragone colla delinquenza degli italiani in Italia questa degli italiani negli Stati Uniti risulta discretissima e, tranne che per l'omicidio, veramente tenue. Infatti i condannati in una delle migliori regioni italiane e in una delle peggiori nel 1900 furono per 100,000 ab.:

	Piemonte	Sicilia
Omicidio di ogni specie	3,07	11,53
Lesioni	54,56	165,98
Furti di ogni specie	72,70	243,80
Truffe, rapine, ecc.	14,67	47,42

Meglio rispecchiata viene la criminalità dagli arresti avvenuti per reati commessi nel solo 1904, per quanto ci sia l'inconveniente, su cui ritornerò,

di un rapporto tra la popolazione censita nel 1900
e i reati commessi nel 1904.

SU 100,000 INDIVIDUI DI CIASCUNA NAZIONALITÀ

	Minor offenders	Major offenders
Austriaci	287,7	76,3
Francesi	281,7	61,3
Inglesi	342,7	38,8
Irlandesi	756,7	27,2
Italiani	319,9	122,6
Polacchi	222,1	48,4
Russi	224,7	63,4
Svedesi	164,0	17,4
Tedeschi	137,0	24,9
Ungheresi	246,8	43,8

Quest'altro confronto conferma i primi e assegna
il primato nella criminalità agli Italiani; i quali
solo tra i *minor offenders* sono superati dagli Inglesi e dagli Irlandesi.

La proporzione tra reati e numero d'individui
appartenenti a ciascuna nazionalità migliora di
molto la posizione degli Svedesi, che nell'omicidio, ad
esempio, venivano immediatamente dopo gl'Italiani
nella percentuale tra detenuti di ciascuna nazionalità mentre vengono dopo gl'Italiani, gli Austriaci,
i Francesi, gli Ungheresi, i Polacchi e i Russi. Solamente gl'Irlandesi, gl'Inglesi e i Tedeschi danno
un minor numero di omicidi degli Svedesi. Tra
gli arrestati durante il 1904 tra i *major offenders*
mentre gl'Italiani occupano il posto peggiore se-
guiti a grande distanza da Austriaci e Russi, gli
Svedesi occupano il migliore.

La condizione degl'Italiani apparirà peggiore vol-
gendo lo sguardo alla delinquenza dei minorenni.

Il *Report* del Koren non somministra i dati per
nazionalità censiti a 30 giugno 1904; invece li dà
per i minorenni detenuti per reati commessi durante il 1904, distinguendo i minorenni nati all'e-
stero da quelli con genitori stranieri, e che possono essere nati negli Stati Uniti.

Riproduco testualmente le due Tavole XV e XVI da pag. 238 e 239 del *Report* sui *Prisoners* ecc.

TAVOLA XV. — *Distribuzione per paese di nascita dei minorenni delinquenti nati all' estero e detenuti per reati commessi* (juvenile delinquents committed) *durante il 1904.*

PAESI DI ORIGINE	STATI UNITI		NORTH ATLANTICO DIV.		NORTH CENTRAL DIVIS.		Nelle altre div.
	Cifre assolute	Percentuale	Cifre assolute	Percentuale	Cifre assolute	Percentuale	Cifre assolute
Totale	1125	100	846	100	211	100	68
Austria	48	4,3	40	4,7	8	3,8	—
Canadà	112	10,0	77	9,1	25	11,8	10
Inghilterra Galles	78	6,9	53	6,3	17	8,1	8
Germania	115	10,2	50	5,9	61	28,9	4
Ungheria	18	1,6	13	1,5	5	2,4	—
Irlanda	40	3,6	31	3,7	4	1,9	5
Italia	317	28,2	287	33,9	21	10,0	9
Polonia	55	4,9	32	3,8	21	10,0	2
Russia	211	18,8	188	22,2	16	7,6	7
Scandinavia	24	2,1	5	0,6	18	8,5	2
Altri paesi	107	9,5	70	8,3	15	7,1	21

TAVOLA XVI. — *Distribuzione, secondo la nazionalità dei genitori, dei minorenni delin-quenti. Reati commessi durante il 1904.*

Paese di origine del padre	STATI UNITI		NORTH ATLANT. DIVISION		NORTH ATLANT DIVISION		NORTH CENTRAL DIVISION		SOUTH CENTRAL DIVISION		WESTERN DIV.	
	Cifre assol.	Percentuale	Cifre assol.	Percentuale	Cifre assol.	Percentuale	Cifre assol.	Percentuale	Cifre assol.	Percentuale	Cifre assol.	Percentuale
Totale	**4,652**	**100**	**3,043**	**100**	**131**	**100**	**1,203**	**100**	**51**		**233**	**100**
Austria	128	2,8	89	2,9	1	0,8	31	2,6	4		3	1,3
Canadà	321	6,9	213	7,0	—	—	90	7,5	—		18	8,1
Inghil. Galles.	265	5,7	142	4,7	11	8,3	79	6,6	2		31	13,9
Francia	67	1,4	32	1,1	2	1,5	26	2,2	1		6	2,7
Germania	903	19,4	377	12,4	39	29 5	442	36,7	7		38	17,0
Ungheria	44	0,9	33	1,1	—	—	10	0,8	—		1	0,4
Irlanda	954	20,5	731	24,0	38	28,8	143	11,9	5		37	16,6
Italia	763	16,4	676	22,2	14	10,6	44	3,7	14		15	6,7
Polonia	190	4,1	93	3,1	7	5,3	88	7,3	—		2	0,9
Russia	515	11,1	440	14,5	14	10,6	46	3,8	8		7	3,1
Scandinavia	166	3,6	38	1,2	1	0,8	105	8,7	—		22	9,9
Scozia	85	1,8	47	1,5	2	1,5	24	2,0	3		9	4,0
Altri paesi	251	5,4	132	4,3	3	2,3	75	6,2	7		34	15,2

Proporzionato il numero dei detenuti italiani minorenni nati in Italia col numero degli Italiani censiti negli Stati Uniti e nati del pari in Italia i primi sarebbero 65,46 per 100,000 abitanti; la proporzione sarebbe più che doppia tra i minorenni italiani nati negli Stati Uniti.

Dalla Tavola a pag. 282 si rileva la natura dei reati commessi dai minorenni. Tra i 763 nati da genitori italiani : 427 avevano commesso *reati contro la Società* (ubbriachezza, vagabondaggio, incorregibilità ecc.); 24 *contro le persone* (assaults , robbery ecc.) e 257 *contro la proprietà*; 55 reati vari. Tra i minorenni italiani di origine , adunque , la criminalità contro le persone è relativamente tenue, mentre, come si è visto è altissima tra gli adulti.

Ora dai prospetti risulta enorme la proporzione dei minorenni nati negli Stati Uniti da genitori Italiani e superiore di gran lunga a quelli nati in Italia. La proporzione dei minorenni di nostra nazionalità nati in Italia è tanto più notevole, in quanto si sa che la proporzione degli emigranti minorenni è minima rispetto a quella degli Inglesi e dei Tedeschi.

Potrebbe sembrare considerevole la proporzione dei minorenni nati negli Stati Uniti da genitori Tedeschi e Irlandesi; ma si rifletta che mentre i censiti nati in Italia nel 1900 erano 484,207; invece i Tedeschi erano 2,666,990 e gl'Irlandesi 1,618,567. Le proporzioni degli Irlandesi appaiono peggiori tenendo conto della loro scarsa natalità. I Russi danno un forte contingente.

In quanto alla distribuzione gl' Italiani minorenni prevalgono nella *North Atlantic Division ;* i Tedeschi nelle *North Central* (1).

-----

(1) Guglielmo Di Palma, chiese informazioni precise sui minorenni e sulle minorenni detenuti o ricoverate nelle Istituzioni dello Stato di New York ed ebbe le seguenti risposte: Joseph F. Scott Sopraintendete del *New York State Reformatory di Elmira* il 7 aprile gli scrisse : « Nella popolazio-

Il relatore Koren, che si mostra sempre imparziale nei suoi apprezzamenti constata a proposito
dei minorenni delinquenti: « è evidente che gli
« stranieri di recente immigrazione hanno contri-
« buito sproporzionatamente e in maggior numero
« alla delinquenza dei minorenni, mentre le emi-
« grazioni antiche, come i Tedeschi, Irlandesi e
« Inglesi danno una minore proporzione (pag. 258).
Nel *Report of immigration for year 1908* (pag. 48)
si avverte che nel 1900 51.9 % degli stranieri erano
vissuti per 15 anni e più negli Stati Uniti; ma tra
i condannati *Major offenders* — nel 1904 quelli con
più di 15 anni di soggiorno furono 36.2 % ; tra
minor offenders 52.6; nel totale 50.7.

E' evidente la benefica influenza del lungo soggiorno negli Stati Uniti.

Si può indurre dal precedente confronto che sui
minorenni italiani nati in America sia mancata
tale benefica azione ? Non essendo nota la cifra assoluta nella popolazione totale dei minorenni Italiani nati in patria e di quelli nati al di là del-
l'Atlantico sarebbe azzardata qualunque induzione.
Dal confronto, però tra i condannati in Italia e
gl'Italiani condannati o arrestati negli Stati Uniti
tale benefica influenza è innegabile tra gli adulti.

IV.

3° *La partecipazione deli Italiani all' omicidio.*
Dalla esposizione imparziale fatta sinora risulta
evidente che nella grave delinquenza gl'Italiani

« ne totale di 1500 detenuti (*inmates*) in questo momento
« vi sono 138 nati in Italia 44 nati in America da genitori
« italiani ». Miss Alice E. Curtin soprintendente della *Western
House of refuge for Women* in Albion (Stato di New York)
in data 30 marzo 1908 gli annunziava che nello Istituto c'era
una sola ragazza Italiana. La seconda cifra conferma la scarsissima partecipazione delle donne italiane al delitto, come
alla prostituzione, negli Stati Uniti.

negli Stati Uniti occupano il primo posto. Il reato,
che maggiormente allarma gli Americani e provoca
la loro avversione contro i nostri concittadini è,
però, l'omicidio.

La realtà intorno a questo punto è triste e dolorosa; ma questa realtà si è talmente esagerata da
farne una paurosa leggenda, che si riassume in
tutto ciò che si scrive sulla *mano nera*, cui oramai
si addossano tutti i più tenebrosi delitti, che si
commettono negli Stati Uniti, cui si attribuisce
una terribile organizzazione, da farne uno Stato
entro lo Stato, che altro scopo non si propone se
non l'assassinio e il più profondo sovvertimento
sociale.

La grande responsabilità degli Italiani nell'omicidio da recente venne affermata, come si è visto,
da Shipley; ma era stata anche accettata prima da
uno statistico Italiano eminente, il Bosco, in base
al Censimento dei detenuti del 1890.

Shipley stabilì questa graduatoria: per 100,000
abitanti: omicidi tra i Messicani 121; tra i Cinesi 65;
tra gl'Italiani 50,2; tra gl'Inglesi e Tedeschi 4; tra
gl' Irlandesi 3 (*The popular science monthly* pagina 162-163). Ma Bosco aveva già rincarata la dose
e aveva ammesso che gl' Italiani negli Stati Uniti
commettevano 58,1 omicidi per 100,000 abitanti (1).

La graduatoria *chauviniste* dello Shipley stabilisce
poi che i negri commettono 100 volte di più omicidii, che i bianchi indigeni e solo tre volte di più
che i bianchi stranieri. Per gl' Italiani in ispecie
egli scrive: « Che vi sia un pericolo nel grande
« aumento della immigrazione del mezzogiorno d'I-
« talia è dimostrato dal fatto che per ogni 100,000
« Italiani negli Stati Uniti, 50,2 sono carcerati
« (*held*) per omicidio. Qui non c' è un pregiudizio

---

(1) *L' omicidio negli Stati Uniti*. Roma, 1897, pag. 53 e
54. Accettai anche io, sulla meritata e grande autorità, di
cui godeva il compianto statistico questa proporzione nell'articolo: *L'omicidio in Italia* (Nella *Rivista penale*, genn. 1901).

« di razza, come si può scorgere dal diagramma ».
E il diagramma di pag. 163, infatti, dà tale cifra
di detenuti per omicidio tra gl'Italiani.

Il numero dei detenuti per un dato reato, come
avvertii, non può fare argomentare che in ogni
anno si commettano tanti reati quanti sono i detenuti; pure i detenuti per omicidio a 30 giugno
1904 non erano che 19,8 per 100,000 abitanti! Il
dato dello Shipley, dunque, è esagerato fino alla
diaffmazione; serve, però, a stabilire qual' è la tendenza criminosa di un popolo rispetto agli altri popo-
ri. Ma lo Shipley non si è contentato di argomentare
contro gl'Italiani e contro la nuova immigrazione
in genere (Italiani, Austriaci, Ungheresi, Russi, Polacchi ecc.) dal numero dei detenuti, ma dalla progressione degli *arresti* per omicidio che si è veri-
ficata nelle 161 città, che hanno più di 25,000 abitanti e dove a preferenza si fermano gl'immigrati,
per conchiuderne che l'omicidio aumenta perchè vi
aumenta la nuova immigrazione.

Ha inoltre dimenticato che la delinquenza delle
città è quasi sempre più alta di quella delle popo
lazioni rurali e quindi non può assumersi come
indice della delinquenza totale.

Egli presenta un prospetto (pag. 165) che riproduco per le città che indicherebbero progressione,
o che si prestano a speciali considerazioni:

ARRESTI PER OMICIDIO PER 100,000 ABITANTI

Città	1880	1890	1900	Media annuale Periodo
San Iosè	15,91	55,37 (?)		
Lexington	12,00	23,18		40,07—1901-904
San Francisco	13,25	19,39	25,09	19,69—1899 903
Louisville	16,16		17,09	17,41—1900 905
Charlestown	1.00		9,00	16,12—1901·904
New York	3,06	6,73	13,12	13,23—1898-903
St. Louis		7,00		11,30—1904 905
Chicago	2,79	4,45		9,30—1904
Newark	1,46	1,10	6,90	6,19—1899 904
Baltimore		6,90	8,21	7,74—1898 903
Bridgeport	3,61	8,18		

ARRESTI PER OMICIDIO PER 100,000 ABITANTI

Città	1880	1890	1900	Media annuale Periodo
Hobocken	6,45	6,94		
Philadelphia		0,76	2,74	4,93—1899 904
Patterson	3,91	1,27		4,51—1901 904
Buffalo	2,60	3,52		2,93—1902-904
New Haven	1,59	2,46		4,16—1901-904
Boston	1,93	1,33		1,98—1904 905
Millvankce	0,86	0,97	1,07	1,77—1898 904

A questo quadro lo Shipley aggiunge speciali osservazioni; alcune delle quali meritano ricordo. In S. Francisco dal 1889 al 1893 e nel 1902 e 1903 furono uccise 199 persone; delle quali 111 o il 55,6 % da stranieri, mentre questi nella popolazione totale non rappresentano che il 34,1 %. Tra gli stranieri i mongoliani (Cinesi e Giapponesi) contribuirono col 28, 6 % mentre nella popolazione essi non sono che il 6,0 %. La statistica della mortalità per San Francisco dava dal 1872 al 1897, 469 omicidi commessi da Cinesi, con una media annua di omicidi di 18,76 per 100,000. D'onde si può conchiudere che quasi tutti gli omicidi di S. Francisco si devono attribuire ai Cinesi.

Un'altra tavola (p. 167) dimostrerebbe che dal 1874 al 1904 in S. Francisco non ci sarebbe stato aumento negli arresti per assassinio (*murder*) consumato e per assassinio tentato ; tutto l' aumento si sarebbe verificato nell'omicidio (*Manslaughter*) ; da 3,27 a 11,55 per 100,000 ab.

Ma mentre nella statistica dell' assassinio c' è la distinzione tra *consumato* e *tentato* manca per l'omicidio. Gli assassini *tentati* nel 1901 furono 31,68 per 100,000 ; i *consumati* appena 8,05. Il non aumentato numero di assassini si spiega col fatto che nel 1850-74 gli abitanti erano in massima parte immigrati stranieri; oggi la grande maggioranza è d'indigeni , perchè l' immigrazione nuova si ferma quasi tutta negli Stati dell' Est. La stessa spiegazione vale per Los Angeles dove c' è stata forte diminuizione di omicidi tra il 1890 e 1901-904: da 13,15 a 2,17 (pag. 167).

In Newark il peggioramento enorme, da 1,10 a 6,19 tra il 1890 e il 1899.904 sarebbe avvenuto per l'enorme aumento della immigrazione italiana, polacca, russa, austro-ungarica e balcanica. Così pure in New York, Buffalo, Filadelfia, Millvankee. Nel Rhode Island gl'Italiani imprigionati per omicidio sono 11,58, mentre nella popolazione totale rappresentano il 2,09 %; nel Massachussets gl'italiani sono l'1 % della popolazione totale, ma tra i detenuti per omicidio sono il 26,1 %. A Kansas City durante due anni terminati a 31 ottobre 1905 ci furono 17,64 omicidi consumati — non arresti — per 100,000 ab.; ma più di quattro quinti furono commessi da negri e da immigrati *undesirables* — non vi sono, però italiani; negri ed immigrati *undesirables* non arrivano al 7,5 % della popolazione. Invece a Rochester, Minneapolis, Syracuse, ecc. vi è stata diminuizione di omicidii; ma prevalgono i nativi o gl'immigrati tedeschi, scandinavi ecc. (pag. 168 e 172).

Negli Stati del Colorado e Nevada gli omicidi sono numerosi, perchè la popolazione è incolta, troppo sparsa; mentre in Pensilvania, nel West Virginia quantunque vi sia numerosa la popolazione delle miniere, gli omicidi sono poco numerosi per la maggiore cultura e per la maggiore densità. Però se l'insieme del West Virginia dà questo risultato, scendendo a più esatte distinzioni tra Contee minerarie e Contee agricole si trova che nelle prime i detenuti sono 20 volte più numerosi che nelle seconde (pag. 172 a 174).

Attenendomi per ora ai soli dati statistici forniti dallo Shipley si può osservare che non ostante l'enorme aumento della nuova immigrazione *undesirables* in molte città di quelle di cui riporta le statistiche, tra il 1880 o il 1890 e il 1900 o epoca più recente c'è stata più o meno sensibile diminuizione nello omicidio; così a Kansas City, Omaha, Cleveland,

Des Moines, Iersey City, Washington, Los Angeles, Hartford, Rochester, Providence, Cincinnati. Non annovero tra queste New-Orleans che nel 1880 dava 9,5 omicidi e 1,30 nel 1890 perchè certamente si tratta di un errore di stampa. A Providence il numero degli stranieri è forte: 38,1 % della popolazione totale e tra gli stranieri il 10 % sono italiani; pure gli omicidi tra il 1890 e 1901-904 discesero da 2,17 a 1,70. A Denver (Colorado), invece benchè molto alta la proporzione degli omicidi per tutto lo Stato—nientemeno che 52,5 per 100,000 abitanti (?!) — la proporzione scende ad 8,21 : cifra sempre altissima, quantunque vi siano scarsissimi gli *undesirables*. La ingarbugliata dimostrazione che dà lo Shipley del rapporto tra popolazione nativa e immigrati per Cincinnati e Cleveland non toglie che in entrambi le città sia in diminuzione l'omicidio; a Cleveland discende da 13,01 nel 1890 a 9,56 nel 1903-904; a Cincinnati da 6,66 nel 1880 a 4,01 nel 1890, a 4,29 nel 1900, a 6,60 nel 1898-1904. Ciò che egli dice sugli arresti nel 1904 deporrebbe in favore degli immigrati, gli immigrati formano l'81,39 per 100 della popolazione e tra gli arrestati non figurano che pel 64,04 %. Confessa che prevalgono gli arresti tra i nativi; ma aggiunge che il fenomeno si deve ai Negri che sono il 4,4 % della popolazione e dettero il 9,60 per 100 di arrestati (p. 170).

Ai dati e alle conclusioni, cui viene lo Shipley mi permetterò una critica a base di statistiche ufficiali, che dimostrano assai incerto e contraddittorio, il rapporto tra omicidio, Italiani in ispecie e *undesirables* in genere. Aggiungo che le cifre degli arresti per omicidio verranno rettificati colle statistiche dei morti per omicidio secondo le statistiche delle *cause di morte*.

Comincio dal confrontare la percentuale degli Italiani (censimento 1900) nelle grandi divisioni e la proporzione dei detenuti per omicidio a 30 giugno 1904 e di quelli arrestati durante il 1904; ma

siccome in talune divisioni se sono scarsissimi gli
Italiani sono molto numerosi i Negri ai quali si
attribuisce la maggiore delinquenza omicida così
darò anche la percentuale dei Negri.

Il maggior numero degli italiani si trovava col
1,6 per 100 nella *North Atlantic Division*, minimo
quello dei negri : 1,9 per 100 ; il minimo numero
degli Italiani invece lo dava la *South Central* e la
South Atlantic division rispettivamente con 0,1 e 0,1
Italiani e 35,8 e 30,5 Negri.

Ora i detenuti per omicidio a 30 giugno erano
5,6 e quelli per reati commessi durante il 1904, 1,4
per 100,000 abitanti nella *North Atlantic* mentre
salivano a 21,3 e 4,1 nella *South Attantic* e a 24,5
e 6,2 nella *South Central*. E' evidente la influenza
dei Negri ; non quella degli Italiani. Nella *North
Central* del pari gl'Italiani sono molto meno numerosi che nella *North Atlantic* perché non arrivano a
0,2 e i Negri sono del pari poco numerosi : 2,1 per
100; ma i detenuti per omicidio sono più numerosi:
alle due date rispettivamente 8,4 e 1,9 per 100,000
abitanti. Ciò che sta contro l'influenza degli Italiani
ed attenua di molto quella dei Negri.

Esaminando i singoli Stati, che compongono ciascuna divisione si perviene, alle stesse conclusioni.

Il massimo numero d'Italiani: 2,5 per 100 si trovava
nello Stato di New York e i detenuti a 30 giugno
e quelli arrestati per omicidio commesso durante
il 1904 erano alle due date 6,1 e 1,2 ; il minimo
numero d' Italiani ed anche di Negri 0,2 e 0,3 era
nello Stato del Maine che aveva detenuti 6,2; e 0,3;
nel Connecticut : Italiani 2,1 e Negri 1,8: detenuti
a 30 giugno 1904 7,6 e per omicidio commesso durante il 1904 1,3. Lo Stato di Rhode Island collo
stesso numero d'italiani 2,1 e un maggior numero
di Negri 2,2 presentava un minor numero di detenuti a 30 giugno 5,2 e per omicidio commesso durante il 1904 1,3 del Connecticut; e cioè smentiva

sempre il rapporto tanto per gl'Italiani quanto pei Negri.

In tutti gli Stati della *South Atlantic division* è evidentissima la nessuna influenza degli Italiani e quella massima dei Negri. La percentuale di questi ultimi sale a 58,4; 46,7; e 43,7 nella South Carolina, nella Georgia e nella Florida e i detenuti per omicidio a 30 giugno salgono alla cifra enorme di 24,0, 33,3 e 52,0 per 100,000 abitanti; quelli per omicidio commesso durante il 1904 rispettivamente a 4,4; 6,2 e 7,0. Soltanto il West Virginia, la regione mineraria, contraddice: con una minima percentale d'Italiani: 0,3 ed una piccola: 4,5 di Negri ha una elevatissima cifra di detenuti a 30 giugno: 25,3 e per omicidio commesso durante il 1904: 5,4.

Nella *North Central div.* in tutti gli Stati il numero degli Italiani è incalcolabile; nè è molto elevato quello dei Negri, che raggiungono solo il 5,2 nel Missouri e nel South Dacota; quasi in tutti i dodici Stati che la compongono il numero dei detenuti a 30 Giugno e di quelli per omicidio commesso durante il 1904 è assai più elevato che nella *North Atlantic division*. Merita un cenno speciale lo Stato di Kansas: gli italiani sono 0,6 per 100 e i negri appena 3,7; i detenuti a 30 Giugno salgono alla cifra altissima di 14,6 e quelli per omicidio commesso durante il 1904 a 7,2 per 100.000.

Quest'ultima elevatissima cifra dice che la spiegazione data dal Koren per l'alta proporzione dei detenuti censiti a 30 Giugno, derivante, cioè, dalla esistenza nello Stato di una prigione federale, non regge.

Tra gli Stati della *South Central division* dapertutto si rileva l'influenza maggiore o minore dei Negri, e quella incalcolabile degli Italiani, che soltanto nella Lousiana raggiungono 1,2 per 100; ma ivi i Negri sono 57,2; i detenuti per omicidio a 30 Giugno salgono a 36,3 e quelli per omicidio commesso durante il 1904 a 10,3 per 100,000 ab.

Nella *Western division* sono più incerti i rapporti

tra detenuti a 30 Giugno, detenuti per omicidio commesso durante il 1904, italiani e negri. Ad esem pio nel Wyoming con 0,8 italiani e 3,8 negri i detenuti a 30 Giugno arrivano a 39,1 e gli altri a 3,8.

L'influenza degli italiani si potrebbe ammettere nel Nevada: Italiani 3,0 %; negri 16,4 %; detenuti a 30 Giugno 49,6; detenuti per omicidio commesso durante il 1904, 14,2 per 100,000 ab. (1).

Ai dati forniti dallo Shipley per le città faccio seguire l'esame di altri dati ufficiali che correggono o contraddicono i precedenti. Comincio da quelli che risultano dalla *Statistics of cities* del 1902 e 1905 e li pongo in ordine decrescente della percentuale della popolazione italiana nel 1900:

Città	Percentuale Italiani Censimento 1900 (2)	Arresti per omicidio 1902 (3)	per 100,000 ab. 1905 (4)
New Haven	4,8	4,4	10,9
Waterbury	4,3	6,4	9,7
New York	4,2	17,5	15,8
Hobocken	3,9	3,9	15,1
Utica	3,6	1,7	3,5
Providence	3,5	2,2	2,5
Newark	3,4	10,5	9,5
Youngstown	2,9	2,2	6,6
Boston	2,4	5,5	2,6
San Francisco	2,1	15,7	16,6

(1) Il Mc Clure nel citato articolo (*The Tammany^ing* ecc.) accetta una statistica dell'omicidio della *Tribune* di Chicago, secondo la quale sarebbero aumentati terribilmente gli omicidi dal 1881 in poi, perchè negli Stati Uniti aumentò parellela- mente la *nuova immigrazione*. Ma gli stessi dati statistici della *Tribune* dimostrano e lo stesso Mc Clure confessa che gli omicidi cominciano a diminuire dal 1895-96 in poi. Ed è questo precisamente il periodo in cui si accentua fortemente la *nuova immigrazione* !

(2) *Census 1900*. Parte 1ª Vol. 1° *Population* (p. 176ª — 179ª) e Elliot Lord (pag. 8-9).

(3) *Statistics of cities* (Bulletin of the Department of Labor. Settembre 1902 pag. 914).

(4) *Statistics of cities having over 25,000. Special Reports Census* (Washington 1907 pag. 318-323).

Città	Percent. Italiani Censimento 1900	Arresti per omicidio 1902	per 100,000 ab, 1905
New Orleans	2,0	15,6	26,1
Bridgeport	2,0	2,6	5,6
Trenton	1,8	13,7	6,4
Jersey City	1,8	6,0	5,1
Pittsburg	1,7	7,4	—
Buffalo	1,6	5,1	2,3
Filadelfia	1,3	4,4	4,5
Scranton	1,2	1,9	6,7
Syracuse	1,1	0,5	4,5
Chicago	0,9	2,0 (?)	16,6
Cleveland	0,8	7,9	7,3
Rochester	0,7	—	2,1
Kansas City	0,6	12,2	2,3
Worcester	0,5	4,9	3,5
Allegheny	0,5	3,7	7,5
Baltimora	0,4	4,0	7,0
S. Louis	0,3	10,3	2,0
S. Paul	0,3	0,5	7,8
Washington	0,3	4,1	—
Cambridge	0,3	1,0	—
Detroit	0,3	1,6	2,7
Cincinnati	0,2	11,4	1,3
Millwankee	0,2	1,6	2,8
Fall River	0,2	—	0,9
Louisville	0,1	20,4	18,8
Minneapolis	0,1	0,9	1,7
Lowell	0,06	1,0	1,0

I confronti diligenti mostrano che si possono spiegare colla presenza degli Italiani le alte proporzioni degli arresti per omicidio nel 1902 a New York, Newark, San Francisco, New Orleans e in minor grado in poche altre; ma Hoboken, Providence, Youngstown, Boston, Bridgeport contradicono perchè pur avendo percentuale elevata d'Italiani l'hanno minima di omicidii—o almeno molto inferiore alle altre città con più alto quoziente di omicidi e minore percentuale di Italiani. A Trenton, Chicago, Kansas City, S. Louis, Cincinnati e Louisville gli italiani con una tenuissima percentuale non possono esercitare influenza sul quoziente altissimo degli arresti per omicidio. Per alcune città come Trenton, Chicago, Kansas City, Cincinnati non si può nemmeno dire che gli Italiani nell'azione criminogena vengono sostituiti dai Negri, perchè anche questi vi sono in piccola proporzione.

In quanto all'aumento degli arresti per omicidio tra il 1902 e il 1905: tra 32 città, che offrono i dati per la comparazione, in venti ci fu aumento; ma certo in molte non determinato dall'aumento della immigrazione italiana che vi era tenuissima nel primo anno come a Kansas City, Cincinnati ecc. Gli aumenti di New Haven, Waterbury, Hobocken, Youngstown notevoli e dove gl'Italiani erano già numerosi si possono spiegare colla loro aumentata immigrazione; ma viceversa hanno un valore più alto le diminuizione di New York, di Newark, di Boston, di Iersey City, di Buffalo dove pure erano numerosi.

I fatti, adunque, scrupolosamente esposti, escludono lo stretto rapporto tra percentuale d'Italiani ed omicidio, desunto dal numero degli arresti — criterio assai fallace—e della correlazione di sviluppo tra i due fenomeni (1).

Un criterio alquanto più esatto vien fornito dal numero dei *morti per omicidio*, che si rileva dalle *statistiche delle cause di morte*. Questa statistica non c'è per tutti gli Stati dell'Unone.

Per gli Stati, che la danno, riporto i dati dei morti per omicidio messi al confronto colle percentuali d'Italiani e di Negri nella popolazione totale (1).

---

(1) Il Barrow dimostra che in alcune grandi città che nel 1900 avevano numerosi Italiani, come New York, Boston, Providence, Paterson e nel Massacchusetts, nello stesso anno gli Italiani arrestati furono in proporzione inferiore agli Italiani liberi (*Pauperism, Disease and Crime* nel volume : *The Italian in America*. New York, Buck et Company, 1905 pagina 210 a 213).

(1) *Mortality 1905* (Washington 1907 pag. 194 e 195), *Mortality 1906* (Washington 1908 pag. 224 e 225). La media è quella del quinquennio 1901-1906 per molti Stati. Pel

Stati	Percentuale Italiani	Percentuale Negri	Morti p. omic. p. 100,000 ab Città	Distretti rurali
Maine	0,2	0,3	1,7	0,7
New Hampshìre	0,2	0,2	1,1	1,3
Vermont	0,6	0,3	2,0	0,6
Massachussetts	1,0	1,3	1,3	0,5
Rhode Island	2,1	2,2	3,0	2,1
Connecticut	2,1	1,8	0,7	0,3
New-York	2,5	1,5	2,2	0,5
New Iersey	2,2	3,8	1,2	0,7
Pennsilvania	1,0	2,5	5.8	4,8
Maryland	0,2	19,8	4,6	2,2
Indiana	0,05	2,3	2,7	1,1
Michigan	0,2	0,9	1,04	1,0
South Dacota	0,08	5,2	—	2,2
Colorado	1,2	2,0	12,0	15,6
California	1,5	5,5	11,3	8,3

Questi dati mostrano la nessuna influenza degli Italiani tanto più che la manifestazione statistica dei morti per omicidio si riferisce agli anni posteriori al censimento del 1900, quando, cioè, la immigrazione italiana vi aveva preso le maggiori proporzioni; quella dei Negri, appare evidente nel Maryland. Sono enormi le proporzioni del Colorao e della California e superano le cifre delle peggiori provincie italiane, tanto da dubitare della esattezza dei dati. In California dallo Shipley sappiamo che il fenomeno puo spiegarsi colla presenza dei gialli.

Connecticut è degli anni 1903 e 905 pei distretti rurali; pel Vermont nelle città degli anni 1904-905; per la California, Pennsilvania, South Dacota e Colorado pel solo 1906; per il Maine, Michigan, New Hampshire e Rhode Island pel 1902 1906. Il Maryland ci dà una distinzione preziosa, quella dei morti per omicidio tra bianchi e negri nel 1906.

	Città	Distretti rurali
Bianchi	2,7	1,6
Negri	14,8	4,5
Media generale	4,6	2,2

V.

La città che maggiormente ha richiamato l'attenzione degli studiosi per la sua prevalente importanza economica, politica, sociale ed etnica—su tutte è la metropoli: New York. Qui sono le peggiori condizioni, non solo perchè si tratta di una grande città di oltre quattro milioni di abitanti; ma perchè ivi sbarcano e si fermano il maggior numero degli emigranti di tutte le razze e di tutte le nazionalità.

A New York, dice Bingham, approdano i predatori delinqnenti di tutte le nazioni come se fosse il loro feudo; e New York è il feudo degli Ebrei russi; della *Hunchakist* — associazione criminosa degli Armeni; della *Camorra* napoletana, della *Mafia* siciliana, della *Tongs* cinese; New York è il grande mercato delle *schiave bianche,* delle prostitute; la sede degli anarchici stranieri, che portano nei suoi *slums* l'incendio e l'assassinio; a New York sinanco si sono dati la posta e i cospiratori serbi e gli emissari del Governo russo per uccidervi l'editore di un giornale rivoluzionario. Ma a New York secondo lo stesso Bingham, il maggior pericolo viene rappresentato dagli italiani, che hanno nella *Mano nera* una vera ispirazione criminosa (*loco cit.* pagine 384, 385 e 391).

Pure guardando alla criminalità totale di New York e alle categorie varie dei reati, solamente nell'omicidio si trova che ha il primato (*Statistics of cities 1902 e 1905).* Ma queste pubblicazioni danno gli *arresti* per omicidio; i quali arresti, come da per tutto, vengono arbitrariamente classificati. Pure lo stesso Shipley in base alle statistiche del 1904-906 è indotto a conchiudere che New York per gli omicidi si trova in condizioni assai migliori di molte altre città americane, Lexington, Kansas, Lousville, St Louis, S. Francisco, Chicago, Cleveland.

Per una parte di New York, la circoscrizione giudiziaria denominata *Contea di New York*, che comprende una popolazione calcolata a 31 dicembre 1908 di 2,600,000 sopra un totale di oltre 4,300,000 si hanno notizie dettagliate e precise nelle pubblicazioni officiali giudiziarie (1).

I reati di cui si occupò il *District attorney 's office* nel 1908 furono i seguenti. (pag. 10-12).

Processati	Cifre assolute	Per 100,000 ab.
Assault (1°, 2°, e 3° grado)	1,209	45,4
Burglary » » »	1,263	47,4
Forgery » » »	207	7,4
Grand larceny» » »	2,277	85,3
Homicide	42	
Manslaughter (1° e 2° grado omicid:o)	40	
Murder (1° e 2° grado assassinio)	38	4,6
Tentato murder (1° grado)	3	
Rape	157	5,8
Robbery	353	13,2
Totale reati	7,877	295,2

Queste cifre dicono che non è affatto più elevata che nelle migliori città di Europa la delinquenza contro le persone e contro la proprietà nella Contea di New York (2).

Riesce di speciale interesse il seguente prospetto (*Report* pag. 54 e 78):

---

(1) *Annual Report of the Chief Clerk of the district at torney office County of New York for the Year ending December 31 1907* (New York. Martin Brown Company 1908); Idem pel 1908 (New York 1909).

Riporterò soltanto i dati del 1908, che per l'omicidio rimontano anche al 1900.

(2) Dalle distinzioni sull'omicidto si comprende che le cifre sopra esposte riguardano i reati pei quali vi fu processo e non i soli condannati.

CONTEA DI NEW-YORK ACCUSATI E CONDANNATI PER OMICIDIO
(MURDER E MANSLAUGHTER 1° e 2° grado Cifre assolute)

Anno	Condannati	Assoluti	Discharged	Dismissed by grand jury	Totale
1899	37				
1900	27	12	29	16	84
1901	25	17	15	31	88
1902	31	11	52	49	143
1903	42	08	34	32	116
1904	37	14	19	34	104
1905	32	13	12	53	110
1906	53	22	24	45	144
1907	39	10	13	47	109
1908	35	17	27	41	120
Totale	358	124	225	348	1018
Media annua	35,8	13,8	25,0	38,6	112,0

Guardando alla curva degli accusati noi troviamo delle oscillazioni fortissime non affatto proporzionate alla curva della immigrazione totale e di quella italiana. L' aumento dell'anno 1902 può attribuirsi all'aumento di tutti gl' immigrati e degli Italiani; ma l' aumento di questi ultimi continuò negli anni successivi e gli accusati di omicidio diminuirono fortemente. Il brusco rialzo del 1906 può anche attribuirsi all' aumento d'immigrazione italiana; ma segue fortissima la diminuizione degli omicidi nel 1907. I rimpatri eccezionali cagionati dalla crisi cominciarono nell'ultimo trimestre.

In base alla popolazione della Contea per 100,000 abitanti nella media del periodo decennale i condannati per omicidio furono 1,3 e il totale degli accusati 4,2; nel 1908, rispettivamente 1,3 e 4,5. Questa cifra pel totale degli accusati corrisponde presso a poco a quella data da Shipley (*Minerva* p. 977) ed è degna di fede.

Nei *Reports* ecc. pel 1907 (p. 45) e 1908 (p. 55 a 58) ci sono i nomi dei condannati per omicidio e a giudicare dai medesimi ci sarebbero stati quattordici Italiani condannati per *murder e manslaughter* nel 1907 sopra 39; e tredici sopra 35 nel 1908. Supponendo che nella Contea di New York gl'Italiani

fossero nella stessa proporzione che nella intera
città,—cui lo stesso Bingham ne assegna non meno
di 500,000 — si avrebbe nella media del 1907-1908
che gl'Italiani condannati per omicidio furono 4,0
per 100,000; cioè il triplo dello insieme degli abitanti e siccome non c'è da pensare che gli accusati
italiani siano sfuggiti alla condanna in proporzione
maggiore dello insieme degli abitanti—essendo invece assai probabile che siano stati condannati in
una misura maggiore ; e giacchè il numero degli
accusati sta a quello dei condannati come 3 ad 1,
si può ammettere che gl' Italiani accusati di omicidio siano stati nell'ultimo biennio 1907-908: 12,10
per 100,000 abitanti.

Questa quota è altissima; ma è assai lontana da
quella che loro assegnano e il Bosco e lo Shipley.
Non è il caso di supporre che la criminalità degli
Italiani in New York sia inferiore a quella delle
altre città; il caso inverso è sicuro. E' Bingham,
che ha proclamato : essere la grande metropoli il
feudo dei delinquenti di ogni razza; e degli Italiani
del mezzogiorno e della Sicilia in ispecie.

VI.

I dati esposti sinora assegnano indiscutibilmente
una superiorità nella più grave delinquenza agli
Italiani; cui seguono alcuni dei popoli, che costituiscono la massa degli *undesirables* — Austriaci e
Ungheresi sopratutto.

In generale la nuova immigrazione *undesirable*
mostra una decisa prevalenza nei reati contro le
persone, a base di violenza, mentre la vecchia e
particolarmente quella irlandese delinque maggiormente contro la società e contro la proprietà. Nella
prima fanno eccezione i Russi e gli Ebrei ; e di
ordinario si ha da fare con Russi di religione ebraica:
i disgraziati, che lasciarono l'Impero moscovita per
isfuggire agli scellerati *pogroms*.

Qualunque sia la delinquenza della nuova immigrazione *undesirable* è certo che essa suscita mag-

giore ripulsione nei Nord-americani , per orgoglio
di razza, per vero pregiudizio *chauviniste*, per ignoranza di certe condizioni, che fanno apparire diversa di quella che è la realtà e producono anche
in buona fede delle strane illusioni. Tra poco i
lettori vedranno su questo ultimo punto ciò che i
più competenti Nord-americani hanno detto sulla
illusione che si genera quando non si tiene conto
della età e del sesso dei delinquenti messi in rapporto con le analoghe condizioni demografiche della
popolazione totale.

Contro il pregiudizio della superiorità della propria razza, contro lo *chauvinisme*, che fa attribuire
agli stranieri una parte maggiore della criminalità
degli Stati Uniti, si è levato colla solita imparzialità il Koren.

A prima vista il rapporto tra la popolazione totale e il numero dei prigionieri darebbe ragione al
pregiudizio. Poichè i dati della comparazione sarebbero i seguenti (*Prisoners* ecc. pag. 18).

PERCENTUALE DEGLI STRANIERI BIANCHI

	Nella popolazione totale sopra i 10 anni 1906	Tra i detenuti 1904
Stati Uniti	19,5	23,7
North Atlantic. division	28,3	32,7
South Atlantic. »	4,1	6,5
North Central »	20,6	16,0
South Central »	4,9	10,5
Western »	24,6	25,1

Da questa prima comparazione risulta già che
nella North Central gli stranieri danno una percentuale di detenuti inferiore a quella che loro
spetterebbe e che tale percentuale è di poco superiore nella Western division. Ma in certi Stati della
North Atlantic, dove più numerosi sono gli stranieri bianchi, la loro percentuale di detenuti è inferiore a quella che loro spetterebbe come negli
Stati di New York, Connecticut e Rhode Island.
Supera quella media dell'intera Divisione in ordine
decrescente negli Stati del Maine, New Jersey,

Pennsylvania, New Hampshire, Vermont e Massachussetts.

Ma hanno una importanza di gran lunga maggiore i mutamenti che sono avvenuti tra i censimenti dei detenuti nel 1890 e nel 1904 e che si possono rilevare da questo prospetto (*Prisoners* ec. pag. 18):

PERCENTÚALE DEI DETENUTI NATIVI BIANCHI E STRANIERI

	1890		1904	
	Nativi	Stranieri	Nativi	Stranieri
Stati Uniti	71,8	28,2	76,3	23,7
North Atlantico Div.	65,6	34,4	67,3	32,7
South Atlantic »	89,6	10,4	93,5	6,5
North Central »	76,4	23,6	84,0	16,0
South Central »	83,9	16,1	89,5	10,5
Western »	67,2	32,8	74,9	25,1

E' evidente tra le due date il peggioramento notevole avvenuto tra gl'indigeni, tra i nord-americani bianchi. Il quale risulta anche maggiore pel fatto che dal 1890 al 1904 è aumentata sopratutto la immigrazione nuova degli *undesirables*. Se a questa, adunque, spettasse quella influenza deplorevole, che le viene attribuita nella genesi della delinquenza si sarebbe dovuto verificare il fenomeno inverso. Perciò il Koren giustamente e imparzialmente conchiude: « ha poco fondamento in realtà la credenza « popolare, che gli stranieri sono quelli che riem- « piono le prigioni degli Stati Uniti. Essi, però, « sembrano leggermente più inclini (*prone*) a com- « mettere i piccoli reati che i nativi. Probabilmente « ciò deriva dal fatto, che gli stranieri sono più « concentrati nelle città (*Prisoners* ec. 41).

Questo giudizio sintetico favorevole agli stranieri, acquista valore di dimostrazione matematica quando si esaminano tre speciali ed importanti condizioni demografiche, che devono servire per rendere comparibili i dati sulla proporzione dei detenuti colla popolazione; e cioè: 1° il numero complessivo degli individui di ciascuna nazionalità, che si è messo in rapporto col numero dei detenuti appartenenti

alla medesima; 2° le proporzioni dei sessi nella popolazione totale e tra i detenuti· 3° le proporzioni
di ogni gruppo di età nell'una e tra/gli altri.

1.° *E' errata la proporzione tra popolazione e delinquenti.* Il Koren avvertì che le comparazioni tra
tutti gli stranieri bianchi e il numero dei detenuti
appartenenti a ciascuna nazionalità sono errate e
fanno apparire maggiore del vero la delinquenza
degli stranieri; perchè il numero degli stranieri in
base al quale si sono ricavate le percentuali dei
detenuti è quello dato dal censimento del 1900 e
l'altro dei detenuti è quello del 1904. Ora dal
1900 al 1904 la immigrazione nuova è fortemente
aumentata; quindi la rispettiva percentuale dei detenuti diminuisce proporzionalmente a tale aumento.
(*Prisoners* ec. pag. 19 e 40).

Quale importanza può avere questa correzione
si può comprendere da questi dati: dal 1901 al 1904
la totale immigrazione negli Stati Uniti fu di 2,806,577
In questa cifra colossale la nuova immigrazione,
undesirable, nei suoi tre gruppi principali — italiana, austro-ungarica e russa — era rappresentata
da 738,289 Italiani, 668,486 Austro-ungarici e 473,838
Russi; cioè da circa due terzi di quella totale.

Gli emigranti Italiani in generale rimpatriano
nella proporzione del 50 per 100 (1); si può quindi
calcolare, che circa 360,000 Italiani si devono ag-
giungere a quelli censiti negli Stati Uniti nel 1900.
In base a questa semplice correzione, quindi, la
percentuale della delinquenza degli Italiani si deve
ridurre del 42 per 100, quasi della *metà!*

Nè si può sospettare che la riduzione possa su-
bire delle modificazioni in base ad una diversa distribuzione degl'Italiani immigrati dal 1901 al 1904; poichè risulta che essi si diressero sempre di preferenza

---

(1) Colajanni: *Manuale di demografia* pag. 400. I rimpatri dagli Stati Uniti furono il 53 % nel 1907 e il 295 nel
1908; ma questi furono la conseguenza eccezionalissima della
crisi terribile scoppiata nell'ultimo quatrimestre del 1907.

nella *North Atlantic Division* e sopratutto nello
Stato di New York, come si rileva dagli *Annual Reports* che pubblica il *Commissioner General of immigration* degli Stati Uniti e dall'analisi, che fece il
Sheridan della immigrazione del 1905-906 (*loco cit.*
pag. 409).

2. *Sesso ed età.* — Riunisco insieme, per amore
di brevità, queste due discriminanti della criminalità degli Italiani in America.

La loro evidenza risulta da queste premesse:

In Italia, come altrove, in una misura maggiore
o minore, le donne e i minorenni danno un minimo contributo alla delinquenza in generale e sopratutto all'omicidio e inversamente lo danno massimo i maschi e gli adulti. E' chiaro, adunque, che
un gruppo nel quale le donne rappresentano il 50
per 100 della popolazione totale e i minorenni sotto
20 anni sono pure dal 30 al 40 per 100 dello insieme del gruppo totale una delinquenza rappresentata ad esempio da 100 reati per 100 mila abitatanti, non è comparabile con un altro gruppo nel
quale le donne sono appena il 20 per 100 e i minorenni non raggiungono nemmeno la stessa percentuale e che abbia lo stesso quoziente di criminalità. Gli stessi 100 reati del secondo gruppo,
siccome nella popolazione totale i maschi, cioè i
candidati al delitto, sono più numerosi che nel primo
e che il quoziente di criminalità deve ottenersi dal
rapporto tra i reati e coloro che possono commetterli, è evidente che la delinquenza di questo secondo gruppo sarà minore. Se invece di 100 nel
secondo gruppo, ad esempio, i reati fossero 130,
sebbene a prima vista la delinquenza apparirebbe
maggiore, essa sarebbe uguale in realtà o molto
vicina a quella del primo.

Chiarisco meglio semplificando i termini del confronto, cioe supponendo che la discriminante si
riferisca ai soli minorenni e che questi non diano
alcun contributo al delitto. Nel primo gruppo noi
abbiamo che i 100 reati si devono dividere in una

popolazione di 100 persone delle quali 60 sono adulti maschi e 40 minorenni; il quoziente di criminalità sarebbe $x = \frac{100}{100-40}$ cioè 1,66. Nel secondo gruppo i minorenni sono 20 su cento e gli adulti 80; il quoziente di criminalità x verrebbe dato da $\frac{100}{100-20}$ cioè 1,25. La manifestazione statistica del reato nel secondo gruppo dovrebbe elevarsi a poco più di 132, perchè il suo quoziente di criminalità arrivasse a 1,66 come nel primo gruppo.

Un grande statistico degli Stati Uniti, Mayo Smith, rilevò il grave errore che si commetteva dai propri concittadini comparando la delinquenza degli indigeni con quella degli stranieri immigrati senza tener conto della correzione da fare nel proporzionare i delitti dei due gruppi della diversa composizione demografica del sesso e della età. « Il de-« litto, egli scrisse, è più frequente tra la popola-« zione straniera degli Stati Uniti, che tra i nativi. « Ma se teniamo conto del maggior numero di adulti « maschi tra gli stranieri, noi dobbiamo sospettare « che la loro maggiore criminalità sia dovuta molto « più alla diversa proporzione dei sessi e delle età « che alla nazionalità » (1).

L importanza di queste osservazioni si rileverà da questi dati di fatto. Secondo il censimento americano del 1900 negli Stati Uniti la percentuale della popolazione secondo tre gruppi di età era la seguente (2):

Nella popolazione	da 0 a 19 anni	da 20 a 59 anni	Sopra 60 ann
Popol. bianca totale	44,4 %	49,2 %	6,4 %
Fra i bianchi nati			
negli Stati Uniti	49,6 »	45,2 »	5,2 »
» bianchi stranieri	10,5 »	74,8 »	14,7 »

E' evidente la profonda differenza che c'è nella

---

(1) *Statistics and sociology*. New York 1893 pag. 25.
(2) *Supplementary analisis of census 1900*. Washington pag. 154.

composizione secondo i gruppi di età tra i nativi e gli stranieri in generale.

In quanto al sesso le differenze negli Stati Uniti erano le seguenti (*Abstract* ecc pag. 13):

	PER 100 ABITANTI	
	Maschi	Femmine
Nella popolazione totale	51,0	49,0
Tra i bianchi nativi	50,5	49,5
Tra i bianchi stranieri	54,4	45,6

Questa differenza tra i bianchi nativi e stranieri nella percentuale dei due sessi aggrava quindi quella tra i gruppi di età.

Le due differenze sono maggiori tra tutti gli stranieri per gl'Italiani. Ciò si può scorgere da questo prospetto sulle condizioni demografiche degli immigrati negli Stati Uniti. (Media del 1902-1907) (1)

	PER 100 IMMIGRATI		
	Maschi	Femmine	Sotto i 14 ann
Irlandesi	47,9	52,1	4,9
Tedeschi	60,5	39,5	16,3
Inglesi	62,2	37,8	14,3
Polacchi	70,6	29,4	9,7
Russi e Ruteni	72,0	28,0	6,7
Italiani del Sud	80,0	20,0	10,9
» del Nord	79,6	20,4	8,6

E' evidente la grande prevalenza dei maschi e degli adulti tra gl'immigrati delle diverse nazionalità ; è massima in quanto al sesso tra gl'Italiani. La scarsissima natalità irlandese spiega la percentuale più bassa tra loro degli individui sotto i 14 anni. In Italia (Censimento del 1901) c'erano 50,3 femmine e 49,7 maschi e 34,1% individui d'ambo i sessi da 0 a 15 anni.

Vediamo adesso se in realtà la delinquenza secondo i gruppi di età e secondo il sesso corrisponde all'ipotesi che si è posta per la determinazione del

(1) N. Colajanni : *Manuale di Demografia*. 2ª Ed. L. Pierro, Napoli 1909 pag. 455.

diverso quoziente di criminalità secondo i diversi gruppi di età e alla diversa percentuale dei sessi.

In quanto al sesso in Italia nel quinquennio 1896-900 su 100 condannati ci furono 82,13 maschi e 17,87 femmine. La differenza è maggiore per i gruppi di età: su 100 condannati ce ne furono appena 2,82 dai 9 ai 14 anni: 10,20 dai 14 anni compiuti ai 18 (1).

Sotto i 14 anni, poi, è del tutto trascurabile la percentuale dei condannati per omicidio. In Italia tra 2,624 condannati di ogni età per omicidio qualificato e 6,328 condannati per omicidio semplice ed oltre l'intenzione dal 1896 al 1900 sotto 14 anni ce ne furono soltanto 7 e *108*; cioè: 0,3 e 1,7 %. Su 100 condannati sotto 14 anni per omicidio qualificato e per omicidio semplice e oltre l'intenzione ce ne furono 0,03 0,43. Su 100,000 abitanti sotto l'età di 14 anni i condannati per omicidio qualificato furono 0,04 e per omicidio semplice 0,64. Invece per i principali gruppi di età che sono più numerosi tra gl'immigrati italiani negli Stati Uniti si ebbero (2):

Per 100,000 abitanti	Omicidio	
	qualificato	semplice
da 18 a 21 anni	3,55	12,70
» 21 » 25 »	4,02	13,27
» 25 » 30 »	4,54	9,13

Il primato nell'omicidio tra gl'Italiani negli Stati Uniti sarebbe, quindi, una conseguenza della prevalenza tra gli gl'immigranti della nostra nazionalità del sesso maschile e degli adulti sopra i 14 anni.

Un ultima ricerca a conferma delle precedenti.

---

(1) *Notizie complementari alle st. giudiziarie penali* 1896-900. Roma 1909 pag. XL e XLV.

(2) *Notizie complementari delle statistiche giudiziarie penali degli anni 1896 900.* Roma, 1909. Pag. XLIX a LI.

Tra i detenuti negli Stati Uniti le percentuali secondo i sessi furono (*Prisoners* pag. 16):

	1904 Maschi	Femmine
Totale popolazione bianca	94,5	5,5
» bianchi indigeni	95,9	4,1
» bianchi stranieri	91,5	8,5

Tra gli stranieri in genere, la partecipazione della donna al delitto è maggiore che tra i nativi, forse, per il maggior numero di adulte, tra i primi e per la posizione sociale privilegiata che occupano nel loro paese le donne americane; ma tra gli Italiani in ispecie è minima. La minore partecipazione della donna al delitto rende più marcata la differenza nella composizione della immigrazione italiana secondo il sesso e rinforza le mie conclusioni.

Secondo l'età i condannati arrestati negli Stati Uniti nal 1904 si distinguevano così (*Prisoners* ecc. pag. 49):

	Totale delinquenti	Major offenders	Minor offenders
da 10 a 19 anni	10,1	17,1	8,4
» 20 a 44 anni	69,2	74,2	59,0
» 45 in sopra	19,9	8,6	22,7

Anche questi dati rinforzano le precedenti osservazioni: perchè essendo minore negli Stati Uniti la partecipazione dei minorenni al delitto, in un gruppo sociale, qual' è quello degli immigrati italiani, nel quale sono poco numerosi gl' individui sotto i 14 anni, il quoziente degli adulti viene aggravato. La partecipazione dei minorenni è probabilmente inferiore a quella dell'Italia, perchè l'immigrazione prevalentemente di adulti, sposta la composizione per età della popolazione totale.

La diversa costituzione demografica s'impone tanto nella comparazione della delinquenza di due gruppi sociali diversi, che Commons ispirandosi al giusto criterio di Mayo Smith corresse le cifre del Censimento del 1890. Questo aveva dato 898 detenuti per ogni milione di bianchi nativi e 1,768 per ogni milione di bianchi stranieri; ma proporzionando i detenuti colla popolazione adulta (in età di votare) trovò 3,305 detenuti tra i bianchi nativi e 3,270 tra gli stranieri (*loc. cit.* pag. 168 e pagina 169).

Prima di Commons ci fu Hastings Hart a fare la stessa correzione nell' *American Journal of Sociology* (novembre 1896). Egli dimostrò come quattro e quattro fanno otto che non tenendo conto delle differenze nell'età gli stranieri davano il 41 % in più di delinquenti dei bianchi nativi; tenendo conto di tali differenze, invece, erano i bianchi nativi a dare il 50 % in più di delinquenti dei bianchi stranieri (1). Tenendo conto delle differenze nella proporzione dei sessi la posizione degli stranieri migliora ancora di più.

Perciò: tenendo conto del numero degli Italiani aumentato dopo il 1900 per proporzionarlo ai detenuti del 1904; del maggior numero degli adulti maschi nella popolazione italiana; riducendo questa ad una popolazione *normale*, secondo i sessi e secondo le età, la delinquenza degli italiani negli Stati Uniti si ridurrebbe del 75 o almeno del 70 per 100 di quella, che apparisce!

Rimane certamente grave la delinquenza degli Italiani negli Stati Uniti; ma assai meno grave di quello che vorrebbero farla apparire le impressionanti descrizioni delle cronache giornalistiche, anche per l'omicidio, come ho dimostrato indiscutibilmente in base ai documenti ufficiali nord — americani.

Con gratitudine e soddisfazione si deve constatare che alcuni stranieri, che non possono essere sospettati di tenerezza e di parzialità per gli Italiani hanno riconosciuto le esagerazioni che ai loro danni si diffondono. Così tra i tedeschi, Schultze

---

(1) Barrow : *Opera citata* pag. 204 e 205.

osserva : la delinquenza degli Italiani comunemente
creduta altissima non è, in realtà, tale ; e i nomi
terribili di *maffia* e *mano nera* , sfruttati da qualche delinquente abile, non corrispondono ad orga
nizzazioni realmente esistenti ed attive. In New
York dove vivono mezzo milione d'Italiani, il numero degli omicidi non é nè assolutamente , nè
relativamente molto alto ed é inferiore a quello
osservato tra gli irlandesi ». Altrettanto benevolo
si mostra il Caro (1). Il primo nella difesa ha anche ecceduto.

Lo stesso Steiner, pur non sottraendosi allo spirito
di esagerazione giudicando gli Italiani privi del
senso del rimorso e della vergogna, trova tutte le
attenuanti pel fatto che essi sono numerosi nelle
prigioni. « Mi si consenta dire enfaticamente, egli
dichiara, che le statistiche sono ingannatrici e che
non ostante il grande numero d' Italiani nelle prigioni, tra loro vi sono meno delinquenti di quello
che indicano le statistiche. Molti vi si trovano perchè hanno commesso degli atti che non sono puniti nel loro paese, molti altri vanno in prigione
perchè non conoscono il linguaggio delle Corti »,
(pag. 272 e 273).

Infine va notato che nel momento più triste pel
nome italiano, dopo l'assassinio di Petrosino , nel
più autorevole giornale inglese si mettevano le cose
a posto con rara imparzialità.

Il *Times* (16 marzo 1909 pag. 10) in una corrispondenza da New York protesta contro le esagerazioni americane esplose in occasione del nefando
delitto di Palermo e mette in evidenza le beneme-
renze degli italiani. « I loro reati di sangue osserva-

(1) E. Schultze : *Die Italiener in den Vereinigten Staaten*
(in *Zeitschrifts fur Socialwissenschaft* ottobre 1906), L. Caro
*Die Statistick der osterreichischen ungarischen und polnischen
Auswanderung nach den Vereinigten Staatem von Nord America* (in *Zeitschrift fur Volkwirtschafsoṣialpolitik und Ver-
waltung* 1907, fasc. 1°).

sono per lo più impulsivi e passionali; ma questi reati non erano affatto sconosciuti in America prima dell'arrivo degli italiani. Il ricordo della prima civiltà nelle regioni occidentali senza attingerlo alle pagine di Bret Harte fa testimonianza della frequenza degli omicidi, e nessuno che conosce le condizioni del Sud e che ha letto, ad esempio, le circostanze dell'assassinio recente dell'ex Senatore Carmack potrà assicurare che la vita vi è sicura. »

In *Latini e Anglo-Sassoni*, infine, avevo già ricordato qual'era la delinquenza dei Nord americani anche nel secolo XIX senza la benchè minima influenza degli Italiani, e in una forma recentissima e veramente scellerata sono oggi i nativi di razza bianca, che hanno acquistato in alcune grandi città dell'Unione una tristissima fama. La storia del *Baby-farming* supera in orrore quella della *Mano nera*.

Egli è vero che gl'Italiani negli Stati Uniti commettono una maggior quantità di gravi reati. Ma questi sono sempre i meno numerosi — pochissimo numerosi in senso assoluto, come gli omicidi — e la *quantità* dal punto di vista sociale ha una grande importanza.

Ora gl'Italiani negli Stati Uniti danno tra gl'indigeni e tra gli stranieri il minimo numero di vagabondi, di ubbriachi, di prostitute, come hanno riconosciuto Sargent, Forbey, Foster Carr, Bingham (1); come risulta dalle statistiche ufficiali. La massa di quelli tra gl'Italiani, che vengono considerati come i peggiori — Napoletani, Calabresi e Siciliani — il Bingham confessa che sono tra i *migliori cittadini* di New York (2).

---

(1) Bingham (*loco citato*) Per gli altri si riscontri l'articolo di Francis Oppenheimer: *The Truth about the Black Hand* (*Published by The National liberal immigration League.* Head quarters, 150. Nassau Street. New York).

(2) Of the 500,000 Italiens in New York City to day, 80 % are from the South, from Naples, Sicily and Calabria; and,

VII.

Ciò che si è detto sinora serve a ridurre le esagerazioni ingiuste, che corrono sulla delinquenza degli Italiani negli Stati Uniti. C' è dell' altro da osservare.

Gl'Italiani all'estero delinquono più o delinquono meno che in patria? Insomma.il mutamento dell' ambiente quale influenza esercita su di essi ? La ricerca ha uno speciale interesse perchè si possa stabilire *se* e *quanta* responsabilità spetta allo Stato e alla Società della grande repubblica americana nella criminalità italiana.

Che all'uno e all'altra ne spettino una assai considerevole scaturisce dai dati statistici non copiosi, ma sempre sufficienti, che si posseggono sulla delinquenza degl'Italiani all'estero (1).

Non servono le acute considerazioni di Augusto Bosco sulla delinquenza degli Italiani in Francia nè quelle del de Stoutz sulla criminalità degli Italiani a Ginevra perchè gl'immigrati nelle due vicine repubbliche appartengono prevalentemente agli Italiani del Settentrione, mentre quelli degli Stati Uniti appartengono prevalentemente agli Italiani del Mezzogiorno e della Sicilia.

---

the great bulk *of these people are among our best citizens.* (*Loco cit.* pag. 385).

(1) Come per gl'Italiani agli Stati Uniti corrono grandi esagerazioni sulla loro delinquerza presso le altre nazioni. Ho cercato di correggere tali esagerazioni in *Latini e Anglosassoni.*

Sulla delinquenza degl' italiani all' estero si riscontrino : G. Tosti : *La delinquenza in Francia e l' immigrazione degli italiani.* (Nel *Bollettino del Ministero degli esteri*, 1896); G. de Stoutz: *Quelques chiffres sur la criminalité des italiens à Genève.* (Genève 1898); G. Ciraolo : *La delinquenza degli italiani all'estero.* (Nell'*Italia colonia'e* 1901); A. Bosco: *La delinquenza in vari Stati d' Europa.* (Roma 1903 , pagina 41 e seg.).

Ora essendo considerevolissima la differenza nella criminalità tra gl'Italiani delle due grandi divisioni regionali non si possono trarre giuste conclusioni dalla comparazione tra la delinquenza degli Italiani in Francia e quella degli Italiani negli Stati Uniti.

Si presta meglio alla comparazione quella degli Italiani nella Città di Buenos Ayres. I dati che su di essa posseggo non sono molto dettagliati ; ma permettono sempre qualche induzione (1).

In base alla popolazione del 1904, ch'era di 950,891 abitanti, nel Penitenziario di Buenos Ayres nella media del 1897-906 erano accusati di :

Omicidio . . . . 18,4 per 100,000 abitanti.
Aggressione . . . 67,2 » » »
Furti . . . . . . 52,6 » » »
Altri reati contro la
proprietà. . . . 13,0 » » »

E' elevata la media degli omicidi; ma siamo nell'America del Sud dove sono più numerosi — più che il triplo nel Chilì — anche dove mancano o sono scarsi gl'Italiani. In compenso appare tenuissima la criminalità contro la proprietà.

In quanto alla nazionalità degli arrestati nello stesso periodo di tempo, mancano i dati pei singoli reati e non si hanno che quelli complessivi, che risultano dal seguente prospetto:

Nazionai. nel censim. del 1904	Percent. arrestati 1898-906	Percentuale Penitenziario nel 1906	Rimessi al Penitenziario a 31 Dic. 1906	Rimasti nel prigione prev. nel 1906	Mandati nelle
Argentini 55,0	36,90	39,64	40,61	36,61	
Italiani 24,0	32,02	28,28	30,60	30,11	

La criminalità degli Italiani apparisce alquanto superiore a quella degli Argentini; ma tra gli stra-

---

(1) Le notizie sul numero degli Italiani e degli stranieri in Buenos Ayres sono tolte dal *Recensement de la ville di Buenos Ayres, 1904* (Buenos Ayres, 1904 pag. 23 a 40); quelle sulla delinquenza dall'*Annuaire de la Ville de Buenos Ayres, 1906* di Casares et Martinez. (Buenos Ayres, 1907 pag. 258 a 275).

nieri soltanto i Francesi la danno inferiore. Il fenomeno si può spiegare colla diversa composizione
sociale delle singole nazionalità : si sa che le classi
più povere e lavoratrici commettono un numero di
reati di gran lunga superiore a quello delle classi
ricche ed agiate; ed a Buenos Ayres sono gl'Italiani
che rappresentano le classi più derelitte.

Ma c'è poi la discriminante demografica che in-
verte completamente le percentuali e mostra che
in Buenos Ayres gl'Italiani delinquono meno degli
Argentini e delle altre nazionalità.

Infatti nel numero totale delle donne — la cui
delinquenza è minima — e secondo le età le percentuali sono le seguenti :

Nazionalità	Su 100 donne	Donne su 100 abit.	Gruppi età su 100 abitanti		
			15-19 anni	20-39 anni	40-49 ann
Argentine	59,9	51,9	11,7	22,9	4,1
Italiane	20,3	40,3	6,5	45,0	2 ,4

Percentuale delle donne entrate nella casa di correzione di Buenos Ayres nei 1906		Percentuale delinquenti secondo l'età arrestati nel 1897-906	
Argentine	71,7	da 16 a 20 anni	17,95
Italiane	6,4	» 21 a 40 anni	62,15

E' evidentissimo che sommando i maschi e le
femmine entrati nelle prigioni e nella Casa di correzione nel 1906 — numeri assoluti: maschi 1358
femmine 1224 — le proporzioni mutano in favore
degli Italiani come risulta da questo prospetto in
cui si tiene conto soltanto della diversità di sesso,
senza badare alla diversa composizione della popolazione per gruppi di età :

Entrati nelle prigione e nell'asilo di correzione nel 1906

Argentini (maschi e femmine) 1416	Per 100,000 ab.	270,9
Italiani » » 489	»	» 213,9

Il miglioramento risulta notevole anche senza
tener conto della forte differenza che c'è tra Argentini e Italiani nei tre gruppi di età.

Ha maggiore interesse per questo studio la criminalità degli Italiani in Tunisia perchè quì la immigrazione è rappresentata quasi esclusivamente da Siciliani e sopratutto delle tre provincie di Palermo, Trapani e Girgenti, che danno nell'isola la più alta criminalità.

Sul carattere etnografico di tale immigrazione non c' è dubbio. Carletti, Regio Console a Tunisi nel suo Rapporto sulla *Tunisia e l'emigrazione italiana* (*Bollettino dell'Emigrazione*. 1903, N. 2) osserva: « La psicologia della nostra colonia è la psicologia della razza siciliana; la nostra colonia è l'immagine impiccolita della Sicilia, riportata sopra un quadro di modeste proporzioni in uno sfondo tu nisino » (1).

Studiando soltanto l' omicidio, il reato caratteristico dei Siciliani e quello di cui si sono maggiormente allarmati gli Stati Uniti, si trova che nel sessennio 1901-906 in media i Siciliani giudicati pel reato di omicidio furono 9, per ogni anno; e siccome gl'Italiani in Tunisia si avvicinano ai 100,000; così se ne può conchiudere che gli omicidi degli Italiani furono 9 per 100,000 ab.; se la Colonia la si vuole ridurre ad 80,000 quanto la considerava il censimento del 1900 la proporzione si eleva a 11 per 100,000 ab. Tenendo conto delle correzioni demografiche, relative al sesso e all' età degli immigrati fatte precedentemente per gli Stati Uniti, gli omicidi dei Siciliani in Tunisia per una popolazione *normale* si possono ridurre al *maximum* a 6 per 100,000 abitanti. Ma gli accusati di omicidio in media per tutta la Sicilia furono nel quinquennio 1900-905 ben 25,55 per 100,000 ab.; 34,02 per le sole tre provincie di Palermo, Trapani e Girgenti. L'ambiente Tunisino, adunque, migliora e non peggiora l'im-

---

(1) Il Carletti che fa grandi elogi dei Siciliani in Tunisia. avverte esplicitamente che non è siciliano. Elogi uguali loro fa il Loth (*Le peuplement italian in Tunisie et en Algerie.* A. Colin. Paris. 1905) che pure è allarmatissimo della italianizzazione delle due colonie.

migrazione italiana o meglio la siciliana, ch' è la più detestata negli Stati Uniti (1).

Perchè l'ambiente nord-americano, specialmente quello delle grandi città, peggiora gl'Italiani — se peggioramento c'è; ed ho giù dimostrato che c'è miglioramento—e dallo spirito della *mafia* fa assurgere all'organizzazione della *mano nera*? (2)

Le condizioni intrinseche facilmente generatrici della delinquenza, la miseria e l'ignoranza, che gl'Italiani portano seco negli Stati Uniti, quì trovano un ambiente, che ne intensificano l'azione. In questa intensificazione la responsabilità massima spetta agli Americani.

Una condizione che esercita una grande influenza è quella del sovraffollamento — *overcrowding—* nelle *Piccole Italie* (3) E noi sappiamo che Roosevelt pensa che sia dovere precipuo dello Stato di provvedere a questo grave malanno; e conosciamo pure per confessione degli stessi Nord Americani, come risulta dagli atti delle loro associazioni e dei loro congressi, e dal discorso dell'Amy A. Bernardy, che nulla o ben poco si è fatto pel passato per distrurre questi fo-

---

(1) Devo le notizie sulla criminalità degli Italiani in Tunisia al sig. Yvernès (lettera del 21 Ottobre 1907) direttore della statistica giudiziaria della Francia, cui rivolgo vivi ringrazia menti. A proposito dell' ambiente a coloro, che giurano ancora nella influenza delle razze e del clima ricorderò che gli Indigeni giudicati per omicidio in Tunisia furono appena 9 all'anno nel quinquennio 1901-905; cioè 0,52 per 100,000 abitanti: meno che tra gli anglo-sassoni!

(2) In Sicilia e specialmente a Palermo si guardano con terrore gli *Americani*, cioè i Siciliani reduci dagli Stati Uniti. Ad essi attribuiscono la recrudescenza della delinquenza ptù grave.

(3) Delle tristissime e spaventevoli conseguenze del sovraffollamento mi sono lungamento intrattemento nel mio *Manuale di Demografia* (2ª Ed. L. Pierro Napoli 1909). Sull'azione degli *slums* e del sovraffollamento (*overcronding*) come generatrice della delinquenza insistono giustamente tutti gli scrittori nord-americani. Si riscontri, ad esempio, ciò che dice il Barrow (*opera citata* pag. 206 a 208).

colai d'infezione, questi centri di cultura di tutti
i microbi, che insidiano l'organismo biologico e
sociale; e ch'è recente il vigoroso movimento che
mira ad attenuare, se non ad eliminare, tale inesauribile sorgente di criminalità.

A queste cause, per così dire generiche d'intensificazione si aggiungono quelle specifiche, che riguardano gli organi, che hanno il compito diretto
di combattere la delinquenza: la polizia e la magistratura.

La polizia, che dovrebbe dare la caccia ai delinquenti, nella maggior parte delle grandi città è
deficiente per numero, é male organizzata, è corrotta
sino alla più sfacciata complicità con quelli che
dovrebbero essere i suoi designati nemici.

La deficienza numerica della polizia venne deplorata vivamente per New York, da chi ne stava
a capo, dal generale Bingham nell'ultimo *Annual
Report of the Police Commissioner City of New
York for the Year ending December 1908* (1); e
l'aveva già deplorata nel suo articolo più volte
citato della *North American Review*. Ma la polizia
non è solo numericamente deficiente; ma è anche
male organizzata: ha uno scarsissimo numero di
agenti, che conoscono la lingua italiana e non ha a-
genti segreti. Su queste deficienze l'opinione del
pratico competente, il Bingham, si trova d'accordo
con il sociologo criminalista, lo Shipley.

Lo stesso Bingham sull'insieme della organizzazione e dell'azione della polizia nord americana
giustamente osserva : « Noi dobbiamo combattere
una criminalità medioevale con metodi anglo-sassoni. Contro di essa le nostre leggi sono fiacche.
Basta questo esempio. Il 26 novembre 1908 si arrestò a Ellis Island un Italiano. Il certificato penale
segnava una condanna ad otto giorni di carcere, ma
le autorità italiane (ministero dell'interno) avvisa-

---

(1) Brown Company. New York 1909.

rono che egli aveva commesso omicidii ed altri gravi
reati ed era stato condannato in contumacia a vita
dalla Corte di assise di Napoli. Il 15 dicembre egli fu
liberato, ed ora é a New York. (*Report of the Commissioner of police* ecc. pag. 23,24).

Sullo stesso argomento l' ex console italiano G.
Branchi,e che esercitò tale carica lungamen]e a New
York, in una lettera al *Times* (*The Times Weeckly*
16 agosto 1907) in risposta ad una intervista del
rev. Campbell, che accusava gli italiani e gli altri
europei meridionali di New York di costituire
associazioni segrete a scopo di assassinio, avverte
« che ci sono esagerazioni e calunnie; ma che in
ogni modo la maggiore responsabilità ricade sulla
polizia di New York, che non ha che due agenti
che conoscono l' italiano; che non penetra nelle
strade e nei quartieri abitati dagli Italiani, che non
li sorveglia e dà prova di una indifferenza e di una
trascuratezza inverosimili nell' adempimento dei
proprii doveri. L'America, egli conclude, non deve
lamentarsi che della propria organizzazione. »

Ma ciò che è più spaventevole è la corruzione
profonda e generalē della polizia delle grandi città,
prima tra tutte quella di New York.

Non posso dilungarmi a riprodurre ciò che William Stead scrisse della polizia di Chicago nel suo
pamphlet diffuso a centinaia di migliaia di copie,
If Christ kame to Chicago; nè ciò che il Norvins
scrisse sulla polizia di S. Francisco e di altre città
nei suoi articoli della *Revue* di Iean Finot *(Le bandi-
tisme politique aux Etats Unis.* 1905); nè la esposizione delle vergogne delle città (*The shame of cities*)
del Lincoln Steffen ecc. Non posso fare a meno,
però, di riprodurre poche frasi caratteristiche sulla
polizia di New York pubblicate da un ex capo della
medesima, nella più autorevole Rivista del nuovo
mondo : nella *North American Review* (*Police cor-
ruption and nation.* N.º di ottobre 1901). Egli scrive:
« Mentre la polizia raramente fa eseguire le leggi, le
« società private e gli individui hanno provveduto

« essi; e sono sempre riusciti a far punire i rei —
« *meno quando è intervenuta la polizia* ... »

« Gli agenti della polizia proteggono i ladri di ogni
« sorta e *intimoriscono e terrorizzano le loro vittime...*
« La polizia è complice coi ladri... I piccoli nego-
« zianti pagano un tributo ai *masnadieri della poli-*
« *zia...* Dovunque c'è una violazione delle ordinanze
« del municipio *si può essere sicuri che c'e la mano*
« *della polizia...* Il *policeman* può fare tutto ciò che
« gli piace: *egli è inviolabile...* La polizia praticamente
« è al *disopra della legge...* Chiunque va a New York
« *impara a violare la legge dai custodi della stessa*
« *legge...* La corruzione, i metodi, la condotta della
« polizia di New York **servono di modello a tutte**
« **le altre** ». (1).

Queste terribili constatazioni sono state rinnovate
più da recente da George Kibbe Turner nel *Mac
Clure' s Magazine* (giugno 1909) in un articolo dal
titolo eloquente : *Il dominio dei criminali professionali sulla Tammany di New York. Uno studio di un
nuovo periodo di decadenza nel governo popolare delle
grandi città* (2).

Ma chi lo crederebbe? Il Bingham, che per
condurre bene la lotta contro il delitto in New-
York voleva riformare e aumentare la polizia... venne
licenziato !

Il licenziamento pare che sia stato l'effetto di
una campagna condotta dal giudice Gaynor contro
le vergogne inaudite della polizia; ma chi è responsabile del suo improvviso e non motivato licenziamento è il *mayor* (sindaco) di New York, che
lo aveva nominato. Bingham narra la propria car-
riere ed esplicitamente lascia intendere nell' *Ham-*

---

(1) In *Latini e anglo sassoni* ho riprodotto estesamente tanto
i fatti e i giudizi di Stead, quanto quelli di questo ex capo
della polizia che mette sempre i punti sugli *i* con nomi e co-
gnomi (pag. 321 a 332).

(2) E' stato largamente riassunto nella *Rivista popolare* del
31 luglio 1909.

pton's Magazine che egli deve la destituzione allo zelo spiegato nel dare la caccia ai ladri, ai ruffiani, alle prostitute, a tutta la *mala vita (cadets, grafters, fixers)* di New York complice interessata e protetta dalla *Tammany Hall.* Egli difende sin che può ciò che c'è di onesto nella polizia di New York; ma riconosce del pari « che su 10,000 policemens ce ne sono circa 2000 senza scrupoli e capaci di tutto ; che non sia inferiore a *100 milioni di dollari* la somma dispersa in ruberie e corruzioni di ogni genere e spesa per assicurare l'impunità dei politicanti criminali e corrotti ».

Egli stesso dichiara « che nel primo anno dello esercizio della propria carica avrebbe potuto mettere insieme circa tre milioni di franchi. Il generale cita come per esempio dei casi, in cui gli vennero offerti venticinquemila, cinquantamila franchi, ecc., semplicemente da un solo proprietario di stabilimento. Egli fa notare che la grande potenza della nota *Tammany* si basa appunto sulla influenza che questa associazione ha sulla polizia, ed aggiunge che da questo lato New York è la città dove nella polizia esiste la maggiore anarchia. La maggior parte delle leggi fiscali sono state inventate per potere ricattare coloro, che le violano. La polizia e la magistratura alla dipendenza della *Tammany hall*, conchiude, sono la **causa principale del dilagare della delinquenza in New York »** (1).

Tutto ciò che si è rilevato sul conto della polizia pur troppo si applica alla magistratura, che a New

---

(1) *The policing of our lawless cities* (nell'*Hampton Magazine* di Settembre 1909). Fra i più recenti articoli, che si occupano della corruzione della polizia e della magistratura di New York e delle altre grandi città del Nord-America oltre quelli citati di Mc Clure e di Bingham sono ricchi di fatti solamente incredibili un altro articolo di Bingham: *The organized Criminals in New York* ed un altro di Kibbe Turner: *The Daughters of the Poor*, entrambi nel *Mc Clure 's Magazine* di Novembre 1909.

York è pure alla dipendenza della stessa *Tammany Hall* e nelle altre città da analoghe associazioni politico-criminose.

Il Kibbe Turner nel citato articolo del *Mc Clure's Magazine*, dà la dimostrazione di questa vergognosa e criminosa relazione tra magistratura e amministrazione; relazione, che riesce alla impunità dei delinquenti; relazione, dice il Kibbe Turner, che *mette la seconda città del mondo alla dipendenza di una popolazione di delinquenti, specialmente di ladri e di ruffiani.*

Egli continua: « Ma pare che i delinquenti e specialmente le « perdute e i loro sfruttatori non « contino soltanto sulla complicità della polizia, ma « anche su quella dei magistrati della Corte della « notte (*night court*) (1). Così nel 1908 su 147 casi de « nunziati al giudice Finn nel distretto (*precint*) di « polizia di Tenderloin non si ebbero che 4 sentenze « di arresto: una su 20 casi innanzi al giudice « Walsh ».

« Invece quando funzionano giudici severi come « Corrigan e Cornell, che condannarono due terzi « o un terzo degli accusati, le vie furono relativa- « mente liberate (*clear*) da tali malfattori ». (2)

« E' da notarsi che durante il 1906 due dei se- « dici magistrati di Manhattan e Bronx furono co- « involti in pubblici scandali; uno si dimise e l'altro « fu rimosso dalla Corte di appello di New York. « Così può avvenire che migliaia di questi crimi- « nali — che se fossero puniti, come disse il Com- « missario di polizia, farebbero diminuire assai il « male sociale, — riescono a non farsi colpire mai ».

« La stessa impunità godono i *saloons* per le con-

----

(1) La *Night Court*, istituita da tre anni circa è rappresentata da una specie di Pretore, che ha l' obbligo di sedere e giudicare dalle 8 alle 12 pomeridiane.

(2) Persona che conosce New York mi ha spiegato come qualche volta s'incontrino giudici onesti: sono quelli, che in certe circostanze vengono nominati dal governatore.

« travvenzioni alle leggi sulle *accise* di New York.
« Durante il periodo del 1° maggio 1907 al 30 aprile
« 1908 ci furono 2837 arrestati che andarono in-
« nanzi alla Corte; soltanto *cinque* ebbero lievi
« punizioni ossia 0,17 per 100 ».

« La stessa impunità viene assicurata alla orga-
« nizzazione dei ladri. I reati di furto e di rapina
« costituiscono crimini (*felonies*) e vanno innanzi
« al Grande Giury e alla Corte delle Sessioni gene-
« rali. Ma i giudici che presiedono sono tutti nomi-
« nati dalla *Tammany Hall.* C'è tutta una organiz-
« zazione per fare rimanere impuniti i delinquenti e
« ci sono degli specialisti, dei sensali (*fixer*), che si
« assumono come un affare qualunque il consegui-
« mento della impunità degli accusati. Il delitto
« professionale come tutte le altre intraprese, dalla
« grande tendenza moderna negli affari, è sospinto
« ad organizzarsi. E lo ha fatto. Il magistrato Cor-
« rigan depose innanzi alla *Commissione Page: che
« un attorney* (specie di pubblico ministero) *divenga
« quasi il rappresentante di tutti i buoni borsaiuoli
« (good pickpockets) in New Yorh non dovrebbe
« sembrare probabile*; *pure ciò é certo* »!

« Vi sono sette distinte linee di difesa, a cui ri-
« corre un criminale a New York per isfuggire alla
« punizione. La prima è la soppressione dei testi-
« moni o la fabbrica delle false testimonianze. Le
« altre sono l'uso del denaro o dell'influenza colla
« polizia o coi magistrati di tutte le specie e di
« tutti i gradi. In ogni sezione o collegio (*assem-
« bly district)* della città, vi sono parecchi agenti
« della organizzazione politica democratica, che
« funzionano (*watching*) continuamente a fare sfug-
« gire il delinquente alla giustizia in ogni stadio ».

« Uno dei mezzi di aiutare i criminali é l'uso
« largo della *condanna condizionale* (suspended sen-
« tence) ».

« Dieci anni or sono la sospensione della sen-
« tenza di un condannato per fellonia in New York
« era rarissima. Le cose sono ora mutate.

« Durante gli ultimi quattro anni il numero delle
« persone condannate per crimine (*violente assault,*
« *burglary* e *larceny*) e che godettero della condanna
« condizionale si é triplicato ».

« Nel 1908 fu applicata ad un terzo dei condan-
« nati. Intanto non si esercita alcuna sorveglianza
« su questi criminali, cui viene applicata la con-
« danna condizionale ».

« E tali reati che sono i più frequenti in New
« York sono più specialmente commessi dai delin-
« quenti professionali italiani o ebrei ».

« Intanto c'è un terribile circolo vizioso in New
« York: quanto più i criminali sfuggono al castigo
« tanto più sono i delinquenti e tanto più elementi
« ci sono a disposizione della *Tammany Hall* nelle
« votazioni; quanto più voti essi procurano ai suoi
« capi tanto più certa è la loro influenza nella mala
« amministrazione della giustizia! »

Kibbe Turner dà altri dettagli ed altre prove
sulla esistenza e sulle conseguenze di questo fatale
circolo vizioso e ricorda anche i tentativi vani per
romperlo. I tentativi s'infrangono pel fatto che
la *Tammany* ha nelle mani 750 milioni di lire —
Bingham dice un miliardo—all'anno che essa spende
nei servizi pubblici.

E New York, come la qualificò Massimo Gorki,
è la città di Mammone ! Perfettamente identico è il
giudizio di Bingham.

Della impunità nel maggior numero dei casi c'è
la prova nei documenti ufficiali. Nel *Report of the
Police Commissioner* ecc. (pag. 24) si legge: « *Black
Hand Cases* (casi di mano nera): riportati nel 1908:
424; arresti 215; condanne 30; liberazioni (*discharges*) 156; casi pendenti 23. Durata complessiva delle
condanne: 2 anni, 2 mesi e 5 giorni.... »

« *Esplosioni di bombe* : casi riportati 44; arresti 70;
condanne 9; liberazioni 58; pendenti 3. Durata complessiva delle condanne: anni 5, mesi 6 e giorni 10 ».

Altra documentazione ufficiale—più degna di fede,
perchè viene dal magistrato. Nel *Report of the Chief*

Clerck of the District Attorney Office County of New York a pag. 6 si legge che dal 1899 al 1908 sopra 49,734 processati ci furono solo..... 4538 condannati.

E perchè non si creda che questa vergognosa percentuale si riferisca ai reati lievi e alle contravvenzioni riferirò dallo stesso *Report* (pag. 54 e 78) i risultati per l' omicidio consumato (*Murder* 1º e 2º grado e *Manslaughter* 1º e 2º grado): accusati e giudicati dal 1900 al 1908 numero 1,018 ; condannati 321, cioè appena il 31,33 % ! E si tratta del reato più grave, contro il quale negli Stati Uniti la opinione pubblica è inesorabile.... (1).

E ci sono giornalisti a New York che invocano leggi eccezionali.... per l' Italia !

Il Bingham nel *Report* della polizia osserva che l' America del Nord sembra che sia la *terra promessa* dei delinquenti Italiani; poichè in Italia le leggi contro di loro sono rigorose e i delinquenti sono trattati come delinquenti. In America le leggi sono assai più miti e la mancanza di provvedimenti adequati di polizia non ne permette l'applicazione.

La verità, come si è visto per la sua stessa confessione, è più brutta : le leggi, che non sono miti

_____

(1) In Francia sono di moda gli studi sulla pena di morte come efficace rimedio contro l'omicidio. Riferisco perciò questi dati. Le sentenze di morte nel 1904 furono: nella North Atl. Division 38. South Atl. 12. North Central 22. South Central 6. Western 28. In tutto 106.

Gli omicidi nella North Atl. Div. furono 1,4 per 100,000 ab. Nel Western 4,8.

Gli omicidi più numerosi negli Stati Uniti furono: 147 Kentucky; 134 Tennessee; 120 Alabama; 138 Mississipì; 154 Luisiana; 150 Texas; 126 Pennsylvania; 75 California.

Nei primi sei Stati solo 4 furono condannati a morte: 3 nel Texas e 1 nel Mississipì. In Pennsilvania 23 furono condannati a morte e 14 in California sopra 75 omicidi. Solo 2 donne furono condannate a morte: 1 in Pennsilvania e 1 nel Vermont.

(*Special reports. Prisoners and juvenile delinquents in institutions* 1904. Washington 1904, pag. 37).

non si applicano perchè la polizia e la magistratura sono corrottissime — corrotte sino all'inverosime. Non spetta, adunque, ai Nord-Americani la massima responsabilità se gli Stati Uniti sono divenuti la terra promessa dei delinquenti italiani?

Rispetto a questa responsabilità io non formulo, che un solo augurio: che la polizia e la magistratura vengano riformati in guisa da rispondere al loro compito, alla loro missione; che i poteri pubblici tutti e la pubblica opinione aiutino dapertutto il risveglio che c'è tra gl'Italiani degli Stati Uniti, che prima a Chicago e poi a New York, a Filadelfia, a Boston, a Pittsburg hanno costituito le associazioni della *Mano bianca* per combattere la *Mano Nera* (1).

VIII.

Se questo studio avesse intenti partigiani potrebbe arrestarsi al paragrafo precedente, nel quale venne assodata la grave responsabilità dell'ambiente e del governo dei singoli Stati e delle più grandi città dell'Unione nella genesi della delinquenza degli Italiani; ma esso mira più in alto: alla ricerca della verità pel miglioramento, per la elevazione morale degli Italiani. Perciò mi trattengo in ultimo su alcuni fattori sociali, che hanno maggiore influenza nella genesi della criminalità e specialmente su quella dell'omicidio; e quest'ultimo paragrafo serve a stabilire la grande responsabilità dello Stato Italiano.

Se gl'Italiani arrivano negli Stati Uniti poveri, ineducati o male educati e analfabeti, e quindi predisposti alla delinquenza certamente di ciò non

---

(1) La Società italiana *La mano bianca* di Chicago fu incorporata secondo le leggi dello Stato dell'Illinois con atto del 6 Dicembre 1906. (Vedi *The italian* **White Hand** *society in Chicago*. Chicago Tipografia dell'*Italia* 101 E. Harrison **Street**, 1908).

sono colpevoli gli Stati Uniti; ma la madre patria.

Non posso intrattenermi della miseria, la cui azione diretta o indiretta, sulla criminalità è enorme eome dimostrai in *Sociologia Criminale* e come più da recente, riprendendo nella sua interezza la mia tesi, dimostra il Bonger (1). Per questa è indiretta e non grande la responsabilità dello Stato italiano. Ma questa responsabilità è diretta, immediata, grandissima per l'analfabetismo degli emigrati italiani.

In uno scritto pubblicato nella *Rivista penale* (L' *omicidio in Italia*. Gennaio 1901) dimostrai il rapporto innegabile tra analfabetismo ed omicidio cogli esempi dell'Italia, della Russia, della Spagna, dell'Ungheria. Ivi è elevatissimo l'analfabetismo; e ivi è più frequente che altrove l'omicidio. Il rapporto viene confermato nel confronto tra i due fenomeni nelle diverse provincie d'Italia. Quasi senza eccezione si può dire che scendendo dal Nord verso il Sud cresce la percentuale degli analfabeti e cresce quella degli omicidi. Si viene alle stesse conclusioni, salve alcune rare eccezioni, nello esame dei due fenomeni nei vari Governi della Russia. Viceversa in Europa gli Stati a minimo analfabetismo— Germania, Scandinavia, Francia, Inghilterra—sono quelli che danno il minimo contributo all'omicidio.

Negli Stati Uniti si ha una conferma piena del rapporto.

Cominciamo da un primo rapporto: quello tra l'analfabetismo complessivo e la delinquenza di tutti gli Stati Uniti.

Nel 1900 l'analfabetismo della popolazione totale al di sopra di 10 anni era di 10,7 %; discendeva a 4,6 pei bianchi nativi per salire: a 12,9 tra i bianchi stranieri; a 18,2 tra i Giapponesi; a 29,0 tra i Cinesi; a 44,6 tra i Negri.

Ora tra i delinquenti tutti per reati commessi nel 1904 l'analfabetismo era del . . . . . 17,9 %

---

(1) *Criminalite et conditions economiques.* Amsterdam. G. P. Tiere Edit. 1905.

Tra i delinquenti per reati contro la società. 12,6 »
  »   »   ›   »   la proprietà. 15,5 »
  »   »   »   »   le persone. 25,4 »
  »   »   »   per *assaults* . . 28,2 »
  »   »   »   per omicidio. . 28,6 (1)

Il Koren osserva che dovunque negli Stati della Unione prevalgono, Negri, Italiani, Messicani, pel loro elevato analfabetismo si eleva la cifra dei reati contro le persone (*Prisoners*, ecc. pag. 58). Ciò che si sa dell'analfabetissimo degli italiani immigrati negli Stati Uniti — che oscilla attorno al 50 per 100; un poco più alto del medio analfabetismo dei Negri ! (2);—della grande prevalenza dello omicidio nelle provincie italiane a massimo analfabetismo; e della prevalenza degli Italiani di queste provincie precisamente tra gl'immigrati nella grande repubblica autorizza ad indurre : che se gli Italiani negli Stati Uniti primeggiano, accanto ai Negri, nell'omicidio, ch' è il reato che li rende maggiormente invisi ai nativi, ciò si deve all'analfabetismo ch' è il carattere comune agli uomini appartenenti all'una e all'altra razza — ai bianchi d'Italia ed ai Negri di America.

_____

(1) *Prisoners* ecc. pag. 200.
(2) A vergogna dell'Italia riporto queste notizie. Negli antichi Stati a schiavi del Sud dal 1870 la spesa per le pubbliche scuole fu di dollari 917,411,089 , di cui 160,000,000 dollari, cioè oltre 800 milioni di lire italiane per le sole scuole dei Negri. Nel 1876-77 la spesa in tali Stati fu di dollari 11,531,073; nel 1906 907 di dollari 53.027,569. Nel primo anno gli alunni bianchi furono 1,827,139 : i negri 571,506. Nel secondo anno: alunni bianchi 4,671,135 ; negri 1,682,725 (*Report-Commissioner education 1908*, vol. 2° pag. 941). L'analfabetismo dei negri al disopra di 10 anni dal 1890 al 1900 discese da 57,1 % a 44,5 ; (*Abstract of the Tewelfth Census 1900* Washington 1904 pag. 16) cioè in una misura più che doppia della discesa che dell'analfabetismo degli italiani che in *20 anni* tra i censime..ti del 1881 e del 1901 per la popolazione sopra i 6 anni discese da 61,9 a 48,5 !... E la discesa fu minore nel mezzogiorno e in Sicilia...

Questo rapporto generale viene confermato negli Stati Uniti dalla distribuzione dei due fenomeni nelle grandi divisioni e ne singoli Stati di ciascuna divisione come risulta dalle seguenti cifre (1).

	ANALFABETI SU 100 AB. SOPRA I 20 ANNI			PER 100,000 ABITANTI	
	Bianchi nativi	Bianchi stran.	Colorati	deten. per omic. a 30 giugno 1904	Det. per omic. commes. dur. 1904
Stati Uniti	4,9	11,5	47,4	13,3	3,0
North Atlant. Div.	2,0	15,2	15,3	6,2	1,4
South »	» 11,5	11,3	51,1	21,3	4,1
North Central »	2,9	7,9	24,8	8,4	1,9
South »	» 11,7	18,8	52,5	24,5	6,2
Western »	» 2,4	7,7	13,4	23,3	4,8

Credo che non occorra alcuna considerazione per illustrare il rapporto tra omicidio e analfabetismo. Nella *North Atlantic Division* e nella *North Central* al minimo analfabetismo dei bianchi nativi, che costituiscono la grande maggioranza degli abitanti ed anche dei Negri corrisponde il minimo di omicidî; e viceversa nella *South Atlantic* e nella *South Central* al massimo analfabetismo corrisponde il massimo degli omicidî.

In queste due grandi divisioni c'è pure il massimo numero dei colorati (Negri quasi tutti): 35,8 per 100 nella *South Atlantic* e 30,3 nella *South Central*, che sono i più analfabeti.

Una eccezione verrebbe rappresentata dalla *Western division* dove l'analfabetismo è minimo ed è asrai elevato il numero degli omicidi.

Questa contraddizione non può spiegarsi colla presenza di numerosi Italiani e Negri, perchè vi sono in numero molto inferiore alla media dell'Unione

(1) I dati completi sulla percentuale dei bianchi nativi dei bianchi stranieri, degli Italiani dei negri e del rispettivo analfabetismo, dei detenuti per omicidio a 30 giugno 1904 e dei detenuti per omicidio commesso durante il 1904 saranno riprodotti in *appendice* per comodità di coloro, che vogliono controllare la esattezza delle mie induzioni.

e delle altre Divisione· Shiplev, come si sa, la spiega colla presenza dei Cinesi e dei Giapponesi. Ma se questa spiegazione vale per la California non vale per qualche altro Stato della *Western Division* dove sono pure numerosissimi gli omicidi (1).

Il confronto tra i due fenomeni nei singoli Stati di ciascuna divisione è altrettanto istruttivo.

Nella *North Atlantic Division* è massimo l'analfabetismo dei nativi nel Maine e Pennsylvania e sono alte le cifre dei detenuti per omicidio a 30 giugno 1904 nell'uno e nell'altro Stato; basso il numero dei condannati per omicidio commesso durante il 1904.

Nel Connecticut c'è la più alta cifra dei detenuti della *Division*, mentre vi è minimo il numero degli analfabeti nativi; però vi è elevatissimo quello dei bianchi stranieri: 26,1 per 100. Nel Massachussetts al minimo numero di analfabeti tra nativi e stranieri corrisponde il minimo numero di omicidi.

Nella *North Atlantic Division* il Vermont e il New Jersey costituirebbero una parziale contraddizione. Nello Stato di New York, la presenza dei numerosi Italiani, compensatrice degli scarsissimi Negri, non vale ad elevare il numero degli omicidi al disopra della media delle *Division*, anzi vi è alquanto inferiore, perchè inferiore è la media degli analfabeti.

Nella *South Atlantic Division* la South Caroline, la Georgia e la Florida col massimo numero dei Negri analfabeti danno le cifre più elevate di de-

---

(1) Tra i minorenni condannati per reati commessi nel 1904 c'è la stessa prevalenza, negli Stati a massimo analfabetismo. Infatti gli analfabeti tra i condannati minorenni furono: Stati Uniti 10,1 %, *North Atlantic Division* 9,3; *South Atlantic* 24,2; *North Central* 6,0; *South Central* 28,0; *Western* 4,6 (*Prisoners* ec, pag. 282). La statistica distingue gli analfabeti assoluti da coloro che sanno leggere ma non scrivere. Questi, però, sono una trascurabile quantità: 215 sopra 11,153 e non ne ho tenuto conto. Nella popolazione totale gli analfabeti sotto i 20 anni sono circa 7,5 % (*Abstract* ecc. pag. 17).

tenuti a 30 Giugno 1904 e di detenuti per omicidio
commesso durante il 1904. Nel Delavare e nel
Maryland ai minimo analfabetismo corrisponde il
numero minimo di omicidi della Divisione. Lo dà
elevato la West Virginia, quantunque non abbia numerosi gli analfabeti e i Negri; ma vi sono
assai numerosi i minatori — è il terzo per numero
di minatori nell'intera Unione. A questi minatori
lo Shepley attribuì la frequenza degli omicidi; aggiungo che le provincie di Caltanissetta e di Girgenti in Italia che danno la più alta cifra degli
omicidi denunziati sono quelli che hanno il maggior numero di minatori — nelle miniere di zolfo;—
così pure nelle regioni minerarie della Spagna. E'
una influenza criminogena specifica della professione?

Nella *North Central Division* l'analfabetis o supera nel Missouri di molto la media della Divisione
ed è altissima la quota degli omicidi. Iowa, Wi-
sconsin, Minnesota, Nebraska con analfabetismo
minimo danno minimo omicidio. Formano una con-
tradizione Ohio e Indiana con analfabetismo inferiore alla media della Divisione.

Più spiccata eccezione viene rappresentata dal
Kansas: analfabetismo minimo anche tra stranieri
e Negri; altissima cifra di detenuti per omicidio a
30 giugno 1904 e di detenuti per omicidio commesso
durante il 1904: rispettivamente 29,6 e 7,2 per
100,000 abitanti cioè più del triplo della media
della divisione. Il Koren spiega questa eccezione
col fatto che c'è una prigione federale. (*Prisoners* ecc.
pag. 23); ed io avvertii già che la spiegazione non
regge.

Nella *South Central Division* il massimo analfabetismo tra i nativi, gli stranieri e Negri della
Lousiana e dell'Alabama; e quello basso dei nativi
e degli stranieri, ma sempre altissimo dei Negri
nel Missisipì danno le cifre più alte degli omicidi
in tutta l'Unione.

In questi Stati la proporzione degli analfabeti

riesce assai più elevata perchè vi sono numerosissimi i Negri: 45,3 ner 100 nell'Alabama , 58,7 nel Missisipì, e 47,2 nella Luisiana !

Nell'alta cifra degli omicidi nei primi due Stati non contribuiscono gli stranieri che vi sono in minima proporzione : 0,8 0,5 per 100 ; sono più numerosi, specialmente gli Italiani nella Louisiana. Lo Stato di Oklahoma colla minima proporzione di analfabeti dà anche una cifra piccolissima di òmicidi : solo 0,2 detenuti per omicidio commesso durante il 1904 per 100,000 abitanti.

Gli altri Stati si avvicinano alla media della Divisione nei due fenomeni ; fa eccezione il Texas che ha un numero superiore alla media negli omicidi a 30 giugno ed ha numero di analfabeti inferiore alla media. Ma vi sono numerosi gli stranieri e questi sono più analfabeti che nell'intera Divisione 25,4 per 100.

Nella *Western Division,* infine, confermano pienamente il rapporto : New Messico e Arizona col massimo numero di analfabeti e di omicidi; Utah, ldaho, Washington ed Oregon col minimo numero degli uni e degli altri. Contraddicono più o meno il rapporto lo Stato di Montana, Wyoming e Nevada ed un po' la California

Se si tien conto della molteplicità straordinaria dei fattori dei fenomeni sociali, — fattori che volta a volta si sostituiscono, si elidono reciprocamente, si addizionano, reagiscono l'uno sull'altro, facendo fallire l'isolamento di ciascuno—si dovrà ritenere davvero straordinario il numero dei casi, che confermano il rapporto sopra enunziato tra analfabetismo e omicidio; rapporto che negli Stati Uniti trova altra conferma caratteristica nel fatto che i Negri e gl'Italiani, colla percentuale più alta di analfabeti, per lo appunto sono quelli che danno il maggior contributo all'omicidio.

Nulla di più umiliante per i bianchi che si credono di razza superiore ai negri il vedersi a questi ultimi perfettamente accomunati nelle manifestazioni dei due sopracennati importantissimi fenomeni sociali.

Il discredito morale che all'Italia e agli Italiani viene da tale constatazione è enorme, è incredibile; è tale che precisamente contro il dato di fatto sempre constatabile esattamente, l'analfabetismo, e contro l'altro col quale potenzialmente si connette, la tendenza omicida, si è levata l'opinione pubblica negli Stati Uniti, pur nascondendo altri moventi economici, ed ha indotto i legislatori a votare provvedimenti che chiudano le porte della grande Unione repubblicana agli immigranti analfabeti. Il *Lodge Bill* non troverà sempre *il veto* presidenziale e finirà col trionfare.

L'Italia dovrebbe evitare l'affronto; dovrebbe con vigoria di mezzi e d'intenti nello interesse morale ed economico del paese combattere efficacemente l'analfabetismo e cancellare la macchia che l'offusca. Questo bisogno, questo dovere altamente proclamai nel XV Congresso della *Dante Alighieri* (Settembre 1904) proclamando forte ed alto che la benemerita società doveva volgere lo sguardo non ai soli *irredenti* di olre Isonzo, ma anche ai milioni d'*irredenti*, che vivono oltre l'Atlantico (1). Il mio consiglio parve una deviazione poco patriottica a taluni; ma oggi con singolare compiacimento ho visto che il XIX Congresso della stessa *Dante Alighieri* tenutosi in Brescia, alle falde delle Alpi, me assente, ha proclamato lo stesso bisogno e lo stesso dovere : quello di combattere l'analfabetismo, che nuoce all'Italia in casa propria e la disonora nel mondo.

Debelliamo questo nemico ed avremo iniziato opera di vera redenzione morale degli italiani, distruggendo uno dei massimi fattori dell'omicidio, il crimine più grave ed antiumano ch'è caratteri-

---

(1) Vedi : La *Dante Alighieri e gli emigrati analfabeti*, Roma 1904. Presso la « Rivista popolare ». Cent. 75.

stico, per la sua intensa manifestazione, dei Negri
e degli Italiani ! (1).

---

(1) Non posso lasciare questo argomento senza ricordare
un'altra triste abitudine che hanno gl'Italiani, principalmente
del mezzogiorno e delle isole — quello di portare delle armi;
abitudine incivile che all'analfabetismo si connette e che con-
tribuisce a fare terminare in efferato omicidio qualunque con-
tesa che tra Inglesi termina con uno scambio di cazzotti.

Il giudice Thomas Davis in Newark condannando quattro
Italiani perchè trovati col coltello in tasca rivolse loro queste
parole che suonano giusto e severo ammonimento per gli
italiani in America e per quelli in patria :

« Voi foste sottoposti a giudizio penale, e riconosciuti rei
dalla giuria nei singoli addebiti. Ciascuno di voi fu implicato
in una rissa, in cui il coltello e il revolver furono largamente
usati.

« La frequenza di questi procedimenti dinanzi alle Corti di
questa contea, è ormai nota a tutti, e non v'è chi non la
commenti pubblicamente.

« Voi appartenete alia medesime razza, e per conseguenza
le azioni dei singoli si risolvono a discredito della vostra nazione, nonostante che molti di essi, dimoranti in queste con-
trade, siano rispettabili cittadini, e degni dell'ammirazione
comune.

« E' tempo, ormai, che i più notevoli cittadini, e le Società più importanti della vostra nazionalità italiana, insorgano
contro l'abuso delle armi. Molti delitti si coordinano appunto
all'abusivo porto di esse.

« La Corte non potrà non incoraggiare una simile crociata,
col punire severamente i colpevoli ».

APPENDICE I.

Percentuale degli italiani e dei Negri negli Stati Uniti.

Detenuti per omicidio e morti per omicidio

STATI	Italiani nel 1900 Cifre assolute	Percentuale ital.	Percentuale Negri	Detenuti per omic. per 100,000 ab. 30 Giugno 1904	Detenuti per omic. commessi durante il 1904 Per 100,000 ab.	Morti per omicidio (1) Per 100,000 ab. Città	Distr. rurali
Stati Uniti	484,207	0,6	12,1	13,3	3,0		
North Atlant. D.	352,065	2,3	1,9	5,6	1,4		
Msine	1,334	0,2	0,3	6,2	0,3	1,7	0,7
New Hamphire	947	0,2	0,2	5,4	1,2	1,1	1,3
Vermont	2,154	0,6	0,3	4,3	0,9	2,0	0,6
Massachussets	28,785	1,0	1,3	4,0	0,9	1,3	0,5
Rhode Island	8,972	2,1	2,2	5,2	1,3	3,0	2,1
Councteiut	19,105	2,1	1,8	7,6	1,3	0,7	0,3
New York	182,248	2,5	1,5	6,1	1,2	2,2	0,5
New Iersey	41,865	2,2	3,8	5,9	1,5	1,2	0,7
Pennsylvania	66,655	1,0	2,5	5,6	1,9	5,8	4,8
South Atlant. D.	10,509	0,1	35,8	21,3	4,1		
Delaware	1,122	0,6	16,6	8,4	5,2		
Maryland	2,449	0,2	19,8	9,9	1,6	4,6	2,2
Columria Dis.	930	0,3	31,3	—	—		
Virginia	781	0,04	35,7	13,5	3,2		
Wast Virginia	2,921	0,3	4,5	25,3	5,4		
NorthCarolina	201	0,01	33,3	13,1	2,9		
SouthCarolina	180	0,01	58,4	24,0	4,4		
Georgia	218	0,009	46,7	33,4	6,2		
Florida	1,707	0,3	43,7	52,0	7,0		
North Central D.	55,085	0,2	2,1	8,4	1,9		
Ohio	11,021	0,2	2,3	6,9	1,9		
Indiana	1,327	0,05	2,3	6,7	1,4	2,7	1,1

(1) *Mortality 1905* (Washington 1907 pag. 194-195) *Mortality 1906* (pag. 224-225). La media non risulta di 5 anni in tutti gli Stati. Nel Connecticut, ad esempio, pei distretti rurali é degli anni 1903 e 1905, nel Vermont per le città del 1904 e 1905 ecc. Per la California, Pensilvania Colorado e North Dacotn pel solo 1906. Pel Maine. Michigan, Rhode Island e New-Hampshire la media è del 1902-906. Quella del Morylan d' è pel solo 1906.

STATI	Italiani nel 1900 Cifre assolute	Percentuale ital.	Percentuale Negri	Detenuti per omic. per 100,000 ab. 3° Giugno 1904	Detenuti per omic. commessi durante il 1904 Per 100,000 ab.	Morti per omicidio per 100,000 ab. Città	Morti per omicidio per 100,000 ab. Distr. rurali
Illinois	23,523	0,4	1,8	8,9	1,9		
Michigan	6,178	0,2	0,9	6,8	1,1	1,04	1,0
Wisconsin	2,172	0,1	0,5	6,2	0,7		
Minnesota	2,222	0,1	0,8	5,2	0,7		
Iowa	1,198	0,05	0,6	4,9	0,9		
Missouri	4,345	0,1	5,2	10,1	2,7		
North Daotca	700	0,2	2,3	8,9	2,4		
South Dacota	360	0,08	5,2	6,6	2,4	—	2,2
Nebraska	752	0,07	0,9	4,6	1'7		
Kansas	987	0,06	3,7	29,6	7,2		
South Central D.	26,158	0,1	30,3	24,5	6,2		
Kentouky	679	0,03	13,3	24,7	6,5		
Tennessee	1,222	0,6	23,8	17,7	6,3		
Alabama	862	0,04	45,3	32,1	6,1		
Missisipi	845	0,05	58,7	26,8	8,3		
Louisiana	17,431	1,2	47,2	36,3	10,3		
Texas	3,942	0,1	20,4	29,4	4,4		
Ind. Territory	573	0,1	22,8	—	—		
Oklahoma	28	0,007	7,7	—	0,2		
Arkansas	576	0,04	28,0	14,4	7,1		
Western Divis.	40,210	0,9	5,3	23,3	4,8		
Montana	2,199	0,9	7,0	34,8	6,7		
Wyomyng	781	0,8	3,8	39,1	3,8		
Colorado	6,818	1,2	2,0	23,2	6,9	12,0	15,6
New Messico	661	0,3	7,7	42,0	7,2		
Arizona	669	0,5	24,4	59,1	18,2		
Utah	1,062	0,3	1,5	5,9	1,3		
Nevada	1,296	3,0	16,4	49,6	14,2		
Idaho	779	0,4	4,5	16,7	0,5		
Washington	2,124	0,4	4,2	14,0	2,9		
Oregon	1,014	0,2	4,6	9,9	1,8		
California	22,777	1,5	5,5	25,3	4,7	11,3	8,3

APPENDICE II.

Analfabetismo negli Stati Uniti per 100 abitanti da 10 anni in sopra

	Bianchi		Negri
	Nativi	Stranieri	
STATI UNITI	4,6	12,9	44,5
North Atlantic Div.	1,6	15,9	13,8
South " "	11,4	12,9	47,1
North Central "	2,3	9,4	21,7
South " "	11,2	22,8	48,8
Western "	2,7	8,5	13,1
Alabama	14,8	9,3	57,4
Arizona	6,2	35,3	12,7
Arkansas	11,6	8,0	43,0
Califormia	1,0	8,7	13,4
Colorado	2,7	8,1	13,0
Connecticut	0,8	16,3	11,5
Delaware	5,6	18,3	38,1
Districtof Columbia	0,8	7,0	24,3
Florida	8,6	11,6	38,4
Georgia	11,9	7,0	52,4
Idaho	0,9	6,0	14,5
Illinois	2,1	9,1	18,1
Indiana	3,6	11,4	22,6
Indian Territory	14,0	19,0	42,8
Iowa	1,2	7,1	18,5
Kansas	1,3	8,5	22,3
Kentuky	12,8	10,9	40,1
Louisiana	17,3	28,6	61,1
Maine	2,4	19,4	14,2
Maryland	4,1	13,4	35,1
Massachussetts	0,8	14,6	10,7
Michigan	1,7	10,3	10,9
Minnasota	0,8	8,4	7,9
Mississipi	8,0	10,7	49,1

	Bianchi		Negri
	Nativi	Stranieri	
Missouri	4,8	9,3	28,1
Montana	0,6	7,0	11,4
Nebraska	6,8	9,8	11,8
Nevada	0,6	7,5	23,0
New Hampshire	1,5	20,5	11,9
New Iersey	1,7	14,1	17,2
New Messico ,	29,4	34,8	19,1
New York	1,2	14,0	10,8
North Carolina	19,5	6,1	47,6
North Dakota	0,9	7,8	12,8
Ohio	2,4	11,1	17,8
Oklahoma	2,5	8,3	26,0
Oregon	0,8	4,1	8,8
Pennsylvania	2,3	19,9	15,1
Rhode Island	1,8	18,7	14,1
South Carolina	13,6	6,5	52,8
South Dakota	0,6	6,7	13,3
Tennessee	14,2	9,7	41,6
Texas	6,1	30,3	38,2
Utah	0,8	6,1	6,3
Vermont	2,9	21,4	14,6
Virginia	11,1	10,9	44,6
Washington	0,5	4,5	11,6
West Virginia	10,0	21,5	32,3
Wisconsin	1,3	11,1	11,4
Wyoming	0,7	8,2	17,2

APPENDICE III.

Morti per omicidio per 100,000 abitanti In Italia Regioni	1864-65	1887-88	1905-906	Diminuizione — o Aumento + tra 1864-65 e 1905-906	1887-88 e 1905-906
Piemonte	4,0	3,0	2,8	— 30,0 %	— 6.6 %
Liguria	4,2	3 0	3,1	— 26,1 »	+ 3,3 »
Lombardia	9,4	1,5	1,9	— 79,7 »	+ 26,6 »
Veneto	—	1,5	1,0	—	— 33,3 »
Emilia	4,2	3,0	2,0	— 52,3 »	— 33,3 »
Toscana	5,3	3,0	2,7	— 49,0 »	— 10,0 »
Marche	10,9	4,5	2,8	— 74 3 »	— 37,7 »
Umbria	15,2	4,5	3,8	— 75,0 »	— 15,5 »
Lazio	—	11,0	7,1	—	— 35.4 »
Abr. e Molise	13,9	8,5	4,1	— 66,1 »	— 51,7 »
Campania	11,7	10,0	6,3	— 46,1 »	— 37,0 »
Puglie	9,0	7,0	5,4	— 40,0 »	— 22,8 »
Basilicata	31,8	8,0	5,3	— 83,3 »	— 33,7 »
Calabria	17,2	10,5	7,1	— 58,7 »	— 32,3 »
Sicilia	19,0	9,0	8,3	— 56,3 »	— 7,7 »
Sardegna	11,3	5,5	6,3	— 44,2 »	+ 12,6 »
Regno	**10,5**	**5,0**	**4,0**	**—61,9**	**—20,0**

Omicidi volontari denunziati per 100,000 abitanti	Media 1879-83	1902-906	Diminuizione — Aumento +
Piemonte	8,21	4,84	— 41,0 %
Liguria	7,01	5,89	— 15,9 »
Lombardia	4,87	2,99	— 38,6 »
Veneto	6,37	2,31	— 63,7 »
Emilia	9,02	4,01	— 55,8 »
Toscana	13,76	4,96	— 63,7 »
Marche	15,70	5,70	— 63,6 »
Umbria	16,96	5,47	— 65,9 »
Lazio	27,85	11,61	— 58,3 »

Omicidi volontari
per 100,000 abitanti
in Italia

	Media 1879-83	1902-906	Diminuzione — Aumento +
Abruzzi e Molise	29,67	10,61	— 64,5 »
Campania	32,03	15,30	— 52,2 »
Puglie	18,61	9,23	— 50,4 »
Basilicata	32,03	13,44	— 58,0 »
Calabrie	35,79	12,27	— 67,7 »
Sicilia	37,36	24,09	— 35,0 »
Sardegna	34,24	18,28	— 46,6 »
Regno	**17,88**	**8,94**	**—50,0** »

Ho aggiunto in *Appendice* questi dati sul decorso dell'omicidio in Italia a prova della possibilità della sua rapida diminuizione. I risultati che si sono attenuti sinora sono considerevoli; eppure non si può dire che lo Stato e la Società abbiano fatto molto per ottenerli !

A chi osserva che la diminuizione si è ottenuta in conseguensa della emigrazione, che allontana dall'Italia gli adulti che più facilmente ammazzano rispondo : 1° che la diminuizione fu fortissima nei morti per omicidio tra il 1864 65 e il 1887-88 quando l'emigrazione ¡era piccolissima ; 2° che la diminuizione fu notevole anche in quelle regioni — Toscana, Marche, Umbria, Lazio — nelle quali la emigrazione si è sviluppata dopo il 1905-906.

DIRECTORY

FIFTH GRAND

MASQUERADE BALL

GIVEN BY

LODGE C. COLOMBO, I. V.
NO. 414
ORDER SONS OF ITALY

OUR MOTTO

LIBERTY--EQUALITY--FRATERNITY

MONDAY, FEB. 7, 1921

TO BE HELD AT

DUNN HALL

AT 8.00 P. M.

FIRE INSURANCE

FIELD & SON

OVER RUTLAND SAVINGS BANK

CLARK & ADAMS

REAL ESTATE
FIRE INSURANCE
26 1-2 Merchants Row   Rutland, Vt.

La Loggia C. Colombo Italia Vittoriosa No. 414
Ordine Figli d'Italia

Fu nata da una società di mutuo socorro organizzata il primo Marzo, 1914, da circa 40 membri presieduta dalla seguente Amministrazione:

Domenico Paolucci, Presidente
Gerardo Ricci, Vice-Presidente
Antonio Bizzarro, Tesoriere
Angelo Iannetti, Segr. Corrispondenza
Alfredo Francioni, Segr. Finanza

Consiglieri

Giuseppe Ragosta
Giovanni Di Lorenzo
Nicola Romano

Iniziata sotto l'ordine Figli d'Italia il giorno 11 Luglio, 1915, col nome di C. Colombo fondata sui principii di

Libertà — Uguaglianza — Fratellanza

Amministrazione
1915-1916

Domenico Paolucci, Venerabile

Giosofatto Romano, Ass. Ven.

Gaspare Crociata, Ex-Ven.

Giuseppe Ragosta, Oratore

Angelo Iannetti, Segr. Archivista

Giovambattista Romano, Segr. di Finanza

Antonio Bizzarro, Tesoriere.

Amministrazione
1920-1921

G. Raffaele Cioffi, Venerabile
Arduino. Secci, Ass. Ven.
Ludovico Ricci, Ex-Ven.
Mariano Magro, Oratore
Pellegrino Menduni, Segr. Archivista
Saverio Marro, Segr. di Finanza
Antonio Bizzarro, Tesoriere

CURATORI

Francesco Rovere   Carmine Polzella
Gennaro Bove   Giuseppe Romano
          Francesco Altrui

CERIMONIERI

Salvatore Cioffi   Gennaro Musella

SENTINELLA

Nicola D'Errico

DIRECTORY

A

Name	No. in Family	Address
Abatiello, Carmine		15 Forest st.
Mrs. Stella		
Abatiello, Raffaele	9	80 Strongs ave.
Mrs. Angelina		
Abrunzo, Antonio		30 Meadow st.
Mrs. Nunziata	4	
Adama, Saverio		116 Post st.
Albano, Vincenzo		104 Franklin st.
Mrs. Anna		
Altobelli, Luigi	6	32 Meadow st.
Mrs. Giuseppa		
Antonio		
Domenico		
Lorenzo		
Vittorio		
Altrui, Francesco		83 Strongs ave.
Amoriello, Giuseppe	7	290 West st.
Mrs. Vincenza		
Armando, Leonardo		West Rutland
Ansalone, Salvatore		
Antonini, Donato		44 Wales st.
Mrs. Brigida, vedova		
Roberto C., U. S. A.	3	
Aurello, Samuele		125 Post st.

B

Name	No. in Family	Address
Baccei, Aldrige		Center Rutland
Barbagallo, Giovanni		43 Union st.
Mrs. Vesnera		
Alfio	3	
Bandelloni, Lorenzo		Center Rutland
Barone, Domenico		93 Granger st.
Giuseppe		
Bartolino, Dante		Center Rutland
Bashi, Daniele		100 Granger st.
Belfiore, Luigi		288 West st.
Mrs.		
Bellomo, Antonio		276 West st.
Mrs.		
Antonino		
Giuseppe		39 Grove st.
Benincasa, Carlo		58 River st.
Mrs. Giovannina	6	307. West st.
Bifano, Giovanni		121 So. Main st.

Name	No. in Family	Address
Bizzarro, Anna		South st.
Antonio		
Maria		
Angelo	7	
Carlo		78 Travers pl.
Mrs. Clementina		
Filomena		
Gennaro		
Bolgioni, Rizzieri	4	12 Meadow st.
Mrs.		
Airi		
Bordone, Federico		83 Granger st.
Bove, Gennaro		33 Howe st.
Mrs. Filomena	2	
Luigi		7 Pine st.
Mrs. Teresina		
Alberto		
Gennaro		
Concetta	7	
Pietro		7 Williams st.
Mrs. Antonetta	7	
Pellegrino		5 Pine st.
Mrs. Filomena		
Teresa		
Pietro		
Luigi		
Vittorio		
Lora	7	
Breotti, Nobbio		74 Travers pl.
Mrs. Maddalena		
Nicola		
Rosina		
Luisa		
Antonetta		
Carmela	9	
Bruso, Giuseppe		52 Cherry st.
Bruzza, Francesco		6 Pine st.
Mrs.		
Ines		
Emma		
Maria	5	
Burreto, Maria		137 Post st.
Befi, Antonio		266 West st.
Antonino		

C

Name	No. in Family	Address
Carofano, Tommasi		73 Travers pl.
Luigi		
Mrs.		
Giuseppe		197 West st.
Mrs. Scafata		

RUTLAND SAVINGS BANK

RUTLAND, VERMONT

Incorporated 1850

RESOURCES

Loans and Securities	$8,356,207.51
Cash on Hand and in Banks on Interest	394,428.99
	$8,750,636.50

LIABILITIES

Due Depositors	$7,908,583.07
Surplus and Interest	817,184.49
State Taxes Accrued	24,868.94
	$8,750,636.50

Name	No. in Family	Address
Carofano, Giuseppe		78 Travers pl.
Pasquale		
Mrs. Caterina		
Paolo		
Francesca		
Caligiuri, Serafino		13 Forest st.
Mrs. Carmela		
Calistri, Vincenzo		35 Forest st.
Calvano, Giovanni		Meadow st.
Calvi, Vincenzo		6 Pine st.
Mrs.	4	
Cambrani, Salvatore		36 Strongs ave.
Cantone, Angelo		78 Travers pl.
Mrs. Barbera		
Pietro		
Pasquale		
Stella	10	
Cantone, Pietro		78 Travers pl.
Mrs. Gemma		
Cappabianca, Pasquale		176½ West st.
Caruso, Domenico		78 Franklin st.
Carosone, Giuseppe		Center Rutland
Carosone, Antonio		Center Rutland
Catozzi, Alfredo		51 E. Washington st
Alessandro		217 State st.
Fernando		103 Bellevue ave.
Ceccarelli, Arturo		Center Rutland
Celentano, Emanuele		36 Strongs ave.
Mrs. Vicenza	5	
Cereghino, Luigi		95 State st.
Mrs.		
Maria		
Rosina		
Vittoria	5	
Chiesi, Andrea		Center Rutland
Egisto		
Virginio		
Cristiano, Luigi		58 River st.
Cimieca, Giovanni		Barrett st.
Cillo Domenico		61 Howe st.
Mrs.		
Cioffi, Alfonso	6	61 Howe st.
Mrs. Maria		
Antonio	4	296 West st.
Angelo		51 Howe st.
Mrs. Caterina		257 West st.
Emilio	4	72 Granger st.
Francesco		72 Granger st.
Antonio		296 West st.
Mrs. Maria		
Giovannina		

Name	No. in Family	Address
Pietro	12	257 West st.
Giuseppe		
Mrs. Amalia		
Giuseppe		Park st.
Mariano		30 West st.
Mrs. Benedetta		
Raffaele G.		30 West st.
Mrs. Luigia		
Antonetta		
Clementina		
Manfreda		
Mario	6	
Raffaele		49 Forest st.
Mrs. Luisa		
Francesco		
Giuseppe	10	
Clementi, Raffaele		182 Granger st.
Coladonato, Giuseppe		Sheldon pl.
Salvatore		
Colutti, Giuseppe		119 Spruce st.
Corso, Giuseppe		66 River st.
Costa, Ida	6	149 State st.
Crapo, Napoleone		124 Strongs ave.
Cristello, Adriano		441 West st.
Emanuele		76 Strongs ave.
Wid. Giuditta		76 Strongs ave.
Guglielmo		441 West st.
Curcio, Orazio		75 Granger st.
Cuzzaniello, Guglielmo		78 Travers pl.
Cioffi, Salvatore		235 West st.
Mrs. Filomena		
Rocco		
Maria	8	
Cioffi, Pasquale		77 Travers pl.
Mrs. Giovanna	7	

D

Name	No. in Family	Address
Dalto, Antonio		151 South st.
Mrs.		
Angelo		
Pietro	4	
Denardo, Nicola		
Mrs.		Mill Village, R. F. D. 1
De Blasio, Pasquale		125 So. Main st.
De Lauri, Angelo		
Mrs.	6	50 Cherry st.
Del Bianco, Errico		Center Rutland
Oreste		298 West st.
Dellivenici, Domenico		298 West st.
Dellolive, Giacomo		11 Ripley pl.
Denerro, Giovanni		34 Meadow st.

Name	No. in Family	Address
De Palma, Michele		83 Granger
Mrs.	5	
De Palma, Paolo		74 Granger
Mrs. Sadie	2	
Di Domenico, Angelo		9 Royce st.
Mrs. Emma		
DeMasi, Giovanni		Post st.
D'Errico, Nicola		Spruce st.
De Asio, Nicola		266 West st.
Mrs.		
De Francesco, Giovanni		Brown st.
Mrs.	6	

E

F

Name	No. in Family	Address
Falco, Pietro		88 Franklin st.
Ferrara, Luigi		36 Strongs ave.
Mrs. Concetta		36 Strongs ave.
Ferrara, Giovanni		36 Strongs ave.
Mrs. Elisabetta	6	
Ferretti, Agnesa		60 Pine st.
Florio, Francesco		235 Columbian ave.
Mrs.		
Lazzaro		80 Franklin st.
Nicola C.		15 Kendall ave.
Rocco		235 Columbian ave.
Fontana, Giuseppe		67 River st.
Mrs. Pasqualina		
Formato, Aniello		127 Post st.
Foti, Giuseppe		167 South st.
Mrs.		
Franceschi, Alfredo		399 West st.
Mrs. Ida	3	
Francioni, Cosimo		56 Union st.
Mrs. Agnesa	4	
Wid. Mrs. Luisella		
Franzoni, Alba		110 Crescent st.
Alesandro		446 West st.
Almo		137 Church st.
Aristide		10 City Park
Mrs.	7	
Attilio		110 Crescent st.
Augusto		446 West st.
Emilio		264 West st.
Geno		10 City Park
Giuseppe		109 Robbins st.
Fregosi, Adolfo		38 Water st.

Name	No. in Family	Address
Amilcare		145 Library ave.
Mrs.		
Giorgianna		
Ida		
Ada	4	
Antonio		38 Water st.
Aspasio		
Ugo		
Nicola		
Ernesto		43 Union st.
Fucci, Cristiano		76 South st.
Mrs.		
Luigi		
Raffaele	7	
Luigi		165 Spruce st.
Raffaele		49 Forest st.
Fucilo, Benne		121 Spruce st.
Fusco, Giovanni		Center Rutland
Lorenzo		
Luigi		
Secondino		

G

Name	No. in Family	Address
Gallipo, Giovanni		City Park
Gallo, Salvatore		
Mrs. Carmine	9	158 South st.
Pasquale		78 Travers pl.
Mrs. Assunta	6	
Gariano, Giuseppe		88 Franklin st.
Garofano, Antonio		105 Temple st.
Mrs.		
Saverio		
Maria		
Pasquale	14	
Garruso, Domenico	6	78 Franklin st.
Gatti, Angelo		72 Cleveland ave.
Mrs.	7	
Genovesi, Caterina		Proctor Road
Emilio		143 Forest st.
Remo		Center Rutland
Gentili, Alcide		Center Rutland
Mrs.	5	
Ghio, Adolfo		447 West st.
Giglioli, Evaristo		41 Cleveland ave.
Mrs. Edita		
Edith		
Angelina		
Josephine		
Paul		
Guinetti, Pietro		243 Columbian ave.
Mrs. Lena		

H

I

Name	No. in Family	Address
Iani, Achille		21 Ripley pl.
Ianni, Luigi		393 West st.
Napoleone		389 West st.
Iachetta, Pietro		76 Travers pl.
Icolari, Giuseppe		97 State st.
Mrs. Giuseppina		
Angelina V. L.		
Vincenza C. G.		
Bianca B. L.		
Tittina A. C.		
Avida E. L.	7	
Ippolito, Salvatore		298 West st.
Mrs.	5	
Iannetti, Angelo		30 Meadow st.
Mrs. Romilda	8	
Ienco, Saverio		62½ River st.
Mrs.	4	

L

Name	No. in Family	Address
Laface, Domenico		62 River st.
Giuseppe		83 Granger st.
Michele		235½ West St.
La Grasso, Carlo		234 West st.
Landi, Raffaele		88 Franklin st.
Lanzillo, Alberto		116 Post st.
Alfonso		
Giuseppe		
Lanzillo, Tommasi		83 Strongs ave.
Mrs. Amalia	10	83 Strongs ave.
La Penna, Francesco e Vincenzo		113 Spruce st.
La Vecchia, Angelo		127 Post st.
Antonio		137 Spruce st.
Larino, Ruggiero		81 Spruce st.
Lemmo, Gaetano		82 Franklin st.
Lertola, Adolfo		78 Simond st.
Emma		34 Watkins st.
Guglielmo		34 Water st.
Lucarina, Carlo		72 Simond st.
Pilato		Center Rutland
Lavena, Vittorio H.		61 Evergreen ave.

M

Name	No. in Family	Address
Magro, Giuseppe		62 River st.
Mainolfi, Francesco		316 West st.
Mrs.	5	

Name	No. in Family	Address
Mainolfi, Pasquale		84 Granger st.
Mrs. Maichina		
Francesco	5	
Manfredo, Vito		107 Franklin st.
Elisabetta		
Maria		
Emma		
Concettina		
Equina		
Lucia		
Nicola		
Antono, U. S. N.	9	
Marcantonio, Salvatore		16 Killington ave.
Marchetti, Giuseppe		Center Rutland
Marconi, Giovanni		Center Rutland
Marino, Natale		74 Travers pl.
Mrs. Vincenza		
Mariotti, Vasco		71 Simond st.
Matteo, Domenico		71 Spruce st.
Francesco		62 River st.
Mazzariello, Pasquale		22 Meadow st.
Mrs. Giuseppa		
Viatrice	6	
Giovanni		74 Travers pl.
Melise, Giovanni		5 Pine st.
Mrs. Giovannina		
Pasquale		
Luigi		
Menduni, Pellegrino		1 Evergreen ave.
Mrs.	4	
Migliari, Gennaro		125 So. Main st.
Mono, Michele		Center Rutland
Montella, Pasquale		83 Spruce st.
Mrs.	7	
Marigliano, Achille		Center Rutland
Morisetti, Vittorio		122 Baxter st.
Marro, Saverio		76 Travers pl.
Mrs. Filomena		
Maria		
Michele		
Antonio		
Virginia		
Moscatiello, Raffaele		21 Meadow st.
Mrs. Filomena	8	
Luigi		235½ West st.
Antonio		75 Travers pl.
Mrs. Angelina		
Giuseppe	4	
Moscatiello, Nicola		
Bricida	5	
Marro, Nicola		266 West st.
Mrs. Elvira	8	

Name	No. in Family	Address
Marotta, Andrea		
Mrs. Angiolina	4	
Magro, Mariano		
Maiello, Michele		51 Howe st.
Mrs.	6	
Musetti, Daniele	3	Water st.
Musella, Gennaro		295 West st.
Musella, Raffaele		235 West st.
Mrs.	3	

N

Napolitano, Raffaele	5	65 River st.
Narciso, Francesco		
Mrs.		316 West
Nimera, Francesco		138 West st.
Notti, Giovanni		63 Howe st.

P

Padotiechia, Arolato		264 West st.
Pagnozzi, Giuseppe		61 Killington ave.
Palermo, Calogero		247 West st.
Palesi, Leno		8 City Park
Panarello, Domenico		80 Franklin st.
Paolucci, Francesco		129 South st.
Mrs. Giovanna	8	
Giovanni		20 Meadow st.
Mrs.	6	
Pasquale		40 Howe st.
Domenico C.		35 Howe st.
Giuseppe		
Olimpia		
Umberto	4	
Pietro		20 Meadow st.
Pedone, Domenico		121 Spruce st.
Ignazio		
Mrs.		
Luigi		36 Strongs ave.
Pellegrinelli, Cesare		206 Columbian ave.
Pellistri, Almo		132 State st.
Penti, Giovanni		61 Killington av.
Mrs.	4	
Persico, Francesco		58 River st.
Mrs.	5	
Perugia, Alessandro		Water st.
Mrs.	5	
Petrillo, Antonio		125 River st.
Mrs. Giovannina	4	
Piccini, Enrico		260 West st.
Picucci, Michele		
Mrs. Dolorata	4	

Name	No. in Family	Address
Piscopo, Pasquale		111 State st.
Mrs. Giovannina	7	
Fitaniello, Carmine		13 Forest st.
Mrs. Giorgina	6	
Pizanello, Geno		West cor. Simond st.
Ponte, Luciano		36 Strongs ave.
Polzella, Carmine		49 Forest st.
Mrs. Antonia	4	
Pratico, Pasquale		165 Spruce st.
Filippo		
Mrs.		68 River st.
Natale	5	164 Spruce st.

Pratto, Pasquale		36 Strongs ave.
Prozza, Michele		53 Union st.
Mrs. Andriana	4	
Palluotto, Pasquale		Howe st.
Mrs.	4	

R

Ragosta, Giuseppe		76 Travers pl.
Giovanni		32 Meadow st.
Mrs. Principia	2	
Ragucci, Arturo		274 West st.
Mrs.	4	

Name	No. in Family	Address
Ratti, Carlo A.		215 Grove st.
Mrs.	4	
Ravenna, Bernardo		383 West st.
Garibaldi		71 Simond st.
Menio		231 State st.
Regula, Pietro		80 Simond st.
Ribolino, Umberto		Center Rutland
Ricci, Antonio		193 West st.
Vito	4	80 Franklin st.
Luigi		193 West st.
Mrs. Maddalena	2	
Ludovico		7 Pine st.

THING CO.

TOO

FAIR HAVEN

LLS

LUDLOW

VALUES

TO ALL

Name	No. in Family	Address
Mrs. Francesca	2	
Michele		80 Franklin st.
Pasquale		149 Forest st.
Mrs. Margherita		
Giovannina		
Luisa		
Antonio		
Ricci, Giuseppina	7	149 Forest
Rizzica, Luigi		73 Travers pl.
Mrs. Filomena		
Giuseppe	4	
Rocco, Angelo		127 Post st.
Romano, Domenico		111 Maple st.

Name	No. in Family	Address
Giosofatto		2 East st.
Mrs. Felice	4	
Giovanni B.		34 Meadow st.
Giuseppe		12 Meadow st.
Wid.		
Nicola		17 Meadow st.
Mrs. Filomena	8	
Pasquale		20 Meadow st.
Mrs.	5	
Salvatore		91 First st.
Giambattista		Forest st.
Mrs.	7	
Rosario, Lemmo		77 Spruce st.
Mrs.	3	50 Cherry st.
Rotelli, Mariano		
Mrs.	4	
Roveri, Francesco	3	54 Union st.
Teresa		
Amelia		
Ruffano, Francesco		121 So. Main st.
Russo, Antonio		76 School st.
Mrs. Emma	3	
Ruggiero, Agostino		Center Rutland

S

Name	No. in Family	Address
Sabatasso, Giuseppe		36 Strongs ave.
Mrs. Giovannina		
Sacco, Andrea		74 Travers pl.
Vincenzo		77 Travers pl.
Mrs. Maddalena	3	
Salada, Francesco		Center Rutland
Salvioli, Enrico		131 State st.
Sangrigoli, Vincenzo		79 Spruce st.
Mrs.	5	
Santini, Sachelli		Center Rutland
Startori, Salvatore		67 River st.
Secci, Arduino J.		71 Williams st.
Mrs.	2	
Segale, Giuseppe		95 State st.
Mrs. Vittoria		
Selva, Michele		Center Rutland
Signori, Selina		7 Kingsley ave.
Simiele, Antonio		62 River st.
Startari, Salvatore		30 Meadow st.
Mrs.	5	
Socia, Roberto		142 Crescent st.
Sofia, Antonio		78½ Strongs ave.
Solari, Achille		Center Rutland
Floido		Center Rutland
Soldati, Alessio		145 Library ave.

Name	No. in Family	Address
Spero, Angelo		7 Pine st.
Spino, Vito		22 Meadow st.
Mrs. Asterina	5	
Stellato, Manuele		125 So. Main st.

T

Name	No. in Family	Address
Tartaglia, Salvatore		78½ Strongs ave.
Taverino, Enrico		433 West st.
Tenerami, Almo		104 Franklin st.
Terenzini, Cesare		Center Rutland
Manotti		Center Rutland
Riccardo		
Taverelli, Achilli	5	32 Water st.
Tesconi, Ernesto		12 Meadow st.
Adelina		
Battista		32 Water st.
Tosi, Oreste		46 Forest st.
Mrs.	5	
Trombetta, Francesco W.		87 West st.
Mrs. Mariuccia	4	

V

Name	No. in Family	Address
Vaccariello, Francesco		292 West st.
Mrs. Francesca	4	
Valente, Arcangelo		16 Meadow st.
Giovanni		
Michele		125 Post st.
Nazareno		305 West st.
Mrs. Maria Giovanna		
Luigi		
Luisa	5	
Antonetta		
Valente, Nicola		77 Travers pl.
Pietro		49 Howe st.
Varrecchione, Francesco		78 Travers pl.
Vitagliano, Costantino		61 Howe st.
Mrs.	4	
Generoso		116 Post st.
Mrs.	3	
Vitello, Vincenzo		43 Union st.
Mrs. Chiarina	4	
Vitta, Teresa		Loretto Home
Vitagliono, Costantino		Post st.
Mrs. Antonetta	2	

Z

Name	No. in Family	Address
Zampone, Pietro		
Zampelle, Alcide		34 Meadow st.
Andrea		Center Rutland
Zingale, Basilio		81 Spruce st.

Fratelli delli Loggia C. Colombo
Italia Vittoriosa No. 414
O. F. d'I.

Abrunzo, Antonio	30 Meadow st.
Altrui, Francesco	83 Strongs av.
Armando, Leonardo	West Rutland
Ansalone, Salvatore	
Amoriello, Giuseppe	290 West st.
Bizzarro, Antonio	
Breotti Nobbio	
Bove, Pellegrino	5 Pine st.
Bifano, Giovanni	121 So. Main st.
Bove, Luigi	7 Pine st.
Barbagallo, Giovanni	43 Union st.
Bove, Pietro	71 Williams st.
Belfiore, Luigi	West st.
Bruzza, Francesco	6 Pine st.
Bizzarro, Carlo	78 Travers pl.
Bolgioni, Rizzieri	12 Meadow st.
Bove, Gennaro	33 Howe st.
Befi, Antonio	266 West st.
Befi, Antonino	266 West st.
Carofano, Tommasi	73 Travers pl.
Corso, Giuseppe	66 River st.
Cioffi, Alfonso	51 Howe st.
Cioffi, Mariano	30 West st.
Cioffi, Raffaele	49 Forest st.
Cioffi, Angelo	51 Howe st.
Cioffi, Raffaele	30 West st.
Cioffi, Salvatore	235 West st.
Cioffi, Pasquale	77 Travers pl.
Cioffi, Giuseppe	257 West st.
Cillo, Domenico	61 Howe st.
Carofano, Giuseppe di Dom.	78½ Travers pl.
Carofano, Giuseppe di Tom.	197 West st.
Carofano, Luigi	73 Travers pl.
Carofano, Pasquale	16 Meadow st.

Name	Address
Caligiuro, Serafino	13 Forest st.
Cantone, Angelo	78 Travers pl.
Ceccarelli, Arturo	Center Rutland
Cristelli, Adriano	441 West st.
Carosone, Antonio	Center Rutland
Cappabianca, Pasquale	176 West. st.
Curcio, Orazio	75 Granger st.
D'Errico, Nicola	Spruce st.
De Palma, Paolo	74 Granger st.
De Asio, Nicola	266 West st.
De Blasco, Pasquale	So. Main st.
Di Domenico, Angelo	9 Royce st.
De Palma, Michele	83 Granger st.
D'Amante, Francesco	
Florio, Rocco	235 Columbian ave.
Florio, Lazzaro	80 Franklin st.
Foti, Giuseppe	167 South st.
Florio, Vito Antonio	235 Columbian av.
Fontana, Giuseppe	67 River st.
Fregosi, Amilcare	Center Rutland
Fusco, Secondino	Library ave.
Fusco, Luigi	Center Rutland
Franzone, Alessandro	
Ferrara, Giovanni	36 Strongs av.
Florio, Giovanni	Center Rutland
Gallo, Pasquale	78½ Travers pl.
Galigiuri, Serafino	13 Forest st.
Guglielmo, Domenico	176 West st.
Iachetta, Pietro	76 Travers pl.
Ippolito, Salvatore	298 West st.
Iannetti, Angelo	30 Meadow st.
Icolari, Giuseppe	97 State st.
Lanzillo, Tommasi	83 Strongs ave.
Lanzillo, Alfonso	116 Post st.
Larino, Ruggiero	81 Spruce st.
Landi, Raffaele	Franklin st.
Mosella, Gennaro	295 West st.
Mosella, Raffaele	295 West st.
Menduni, Pellegrino	1 Evergreen ave.

Name	Address
Marro, Saverio	76 Travers pl.
Mainolfi, Pasquale	84 Granger st.
Mainolfi, Francesco	316 West st.
Mazzariello, Pasquale	22 Meadow
Marro, Nicola	266 West st.
Marotta, Andrea	Forest st.
Maffeo, Pietro Agostino	Franklin st.
Moscatiello, Raffaele	21 Meadow
Marino, Natale	74 Travers pl.
Magro, Mariano	34 Meadow st.
Melise, Giovanni	5 Pine st.
Maiello, Michele	51 Howe st.
Mosetti, Daniele	34 Water st.
Montella, Pasquale	83 Spruce st.
Moscatiello, Antonio	73 Travers pl.
Magro, Vincenzo	
Narciso, Francesco	316 West st.
Paolucci, C. Domenico	35 Howe st.
Petrillo, Antonio	125 South st.
Palluotto, Pasquale	Howe st.
Piscopo, Pasquale	109 State st.
Polzella, Carmine	49 Forest st.
Paolucci, Giovanni	20 Meadow st.
Prozza, Michele	54 Union st.
Perugi, Alessandro	Center Rutland
Romano, Giosofatto	2 East st.
Romano, Giovambattista	Forest st.
Ragosta, Giuseppe	76 Travers pl.
Ragosta, Giovanni	32 Meadow st.
Ricci, Luigi	193½ West st.
Ragucci, Arturo	274 West st.
Ricci, Antonio	193½ West st.
Ricci, Ludovico	7 Pine st.
Rovere, Francesco	
Ratti, Primo	74 Travers pl.
Ruggiero, Agostino	74 Travers pl.
Riboli, Enrico	Center Rutland
Sabatasso, Giuseppe	36 Strongs ave.
Segale, Giuseppe	95 State st.
Scaravilla, Salvatore	67 River st.

Name	Address
Secci, Arduino	71 Williams st.
Startori, Salvatore	30 Meadow st.
Solare, Achille	Center Rutland
Soldato, Alesio	145 Library ave.
Sangrigoli, Vincenzo	79 Spruce st.
Tartaglio, Salvatore	78½ Strongs ave.
Varricchione, Francesco	78 Travers pl.
Vaccariello, Francesco	292 West st.
Valente, Nazareno	305 West st.
Valente, Michele	125 Post st.
Vitello, Vincenzo	43 Union st.
Zampone, Pietro	South st.

WEST RUTLAND

Ambrosi, Felice	Marble st.
Augustino, Francesco	Alberston rd.
Bandelloni, Lorenzo	Clarendon ave.
Basette, Emma	West Rutland
Bioti, Giuseppe	West Rutland
Campagna, Antonio	82 Baxter st.
Caruso, Andi	Albertsin rd.
Cristiano, Leonardo	Anderson st.
Costa, Luigi e Mrs.	Clarendon av.
Tommaso	129 State st.
De Crescenzo, Ernesto	Pleasant st.
DeGange, Santo	Roberts st.
De Mincio, Leonardo	Long st.
De Nofrio, Martino	Albertson rd.
Derigo, Carlo	Baxter st.
Drapala, Antonio	Pleasant st.
Duclo, Giuseppe	Marble st.
Dutelle, Giuseppe	Cor. High st.
Federico, Pietro	Pleasant st.
Fusco, Andrea	Pleasant st.

Name	Address
Gola, Luigi	29 Long st.
Grazeni, Antonio	Whipple Hollow
Izzo, Alberto	Pleasant st.
Benedetto	Pleasant st.
Teresa	Pleasant st.
Ianto, Felice	Pleasant st.
La Belle, Eduardo	Castleton rd.
Giuseppe	Main rd.
Librizzi, Leonardo	25 Baxter st.
Lipruna, Antonio	Pleasant st.
Logan, Giovanni	Woodwardville st.
Roda, Giovanni	Pleasant st.
Michele	Harrison ave.
Sabatino, Angelo	Pleasant st.
Filippo	
Vincenzo	
Selva, Giovanni	Harrison ave.
Vaccosello, Leone	Pleasant st.
Pietro	
Valente, Tommaso	Blanchard av.
Zapala, Antonio	Pleasant st.

Name	Address

PROCTOR

Baratta, Francesco	33 North
Bardi, Ario Dante	West st.
Attilio	West st.
Barsi, Biagio	West st.
Pietro	West st.
Berretta, Giulio	33 North
Bola, Giorgio	10 Taylor Hill

CLAUSON SHOE CO.

DRESS SHOES

FOR

MEN, WOMEN and CHILDREN

---

19 CENTER STREET,

Boni, Alessandro	10 Taylor Hill
Canapa, Rogeri	10 Cross st.
Cicone, Camcesso	2 Tony sq.
Nicola	
Cineta, Francesco	30 East st.
Dangi, Angelo	30 South street
Dazzi, Primo	West Rutland
De Leonardis, Francesco	27 Tony sq.
De Lucia, Giuseppe	2 Tony sq.

Name	Address
Forro, Rossilio	15 West st.
Luigi	
Franzoni, Ceccardo	25 North st.
Lea	
Pietro	
Fregosi, Alfa	Proctor
Gallipo, Arcangelo	16 Green sq.
Gatti, Luigi	Curry ave.

Name	Address
Gola, Giovanni	Patch Hill
Greso, Martino	29 Tony sq.
Irango, Andrea	2 East st.
Inangola, Giovanni	2 Tony sq.
Laruccia, G.	Post office bldg.
Lazzari, Adriano	Geno Hill
Giuseppe	Stiles st.
Luciani, Antonio	22 West st.
Maiello, Giovanni	33 Tony sq.

Name	Address
Nitrini, Spartaco	Stiles st.
Passani, Augusto	5 Green sq.
Pillistri, Aristide	8 Green sq.
Lena	
Nella	
Ratti, Amerigo	45 North st.
Amerigo J.	Cor. East & Williams
Cesare	East st.
Olga	45 North st.
Rocchi, Irene	Geno Hill
Andrea	
Saggi, Giovanni	Tony sq.
Michele	79 West st.
Serri, Francesco	26 North
Raffaele F.	
Sirria	
Francesco	12 Taylor Hill
Torri, Duilio	29 North
Trombetta, Giuseppe	1 Zadock ave.
Verta, Lula	Proctor
Vice, Paolo	25 Patch Hill
Zampelli, Achille	25 West st.

Program of the Ball

8:30 Opening of the Ball

Address by Mayor James C. Dunn

10:00 Award of Waltz Prize

10:15 Grand March

Committee of Arrangements

Giosofatto Romano, Presidente

Giuseppe Ragosta, Segretario

Giuseppi Icolari, Tesoriere

Domenico Paolucci

Giuseppe Segale

Giovanni Bifano

Patronesses

Mrs. Marvelle C. Webber

Mrs. James C. Dunn

Mrs. C. A. Simpson

Mrs. Carleton E. Wilson

Mrs. T. P. Mangan

Mrs. Charles E. Tuttle

Mrs. O. H. Coolidge

Miss Berenice R. Tuttle

Marsden Electric Company

Electric and Gas Supplies, Fixtures and Appliances
Lamps, Wiring

Berwick House Block

Rutland,   Vermont

GIUSEPPE CIOFFI

Importer of

Pure Italian Olive Oil

◻

Macaroni a Specialty

◻

Telephone 1284 - W

257 West Street,   Rutland, Vermont

Committee of Award
of Prizes

Mrs. D. B. Locke

Giosofatto Romano

Giuseppe Ragosta

Giuseppe Icolari

Giuseppe Segale

Domenico Paulucci

Giovanni Bifano

PRIZES

◙

$20 Gold Piece for Best Waltz

Silver Cup for Best Couple

Silver Cup for Best Single

Silver Cup for Funniest Couple

Silver Cup for Best Group

Order of Dance

1—One Step ....Noah's Wife

2—Fox Trot ....Bells

3—Waltz ....Sunny Tennessee

4—Fox Trot ....Feather Your Nest

5—One Step ....Pickaninny Rose

6—Fox Trot ....Chili Bean

7—Waltz ....Wishing Land

8—One Step ....Rose of My Heart

9—Fox Trot ....Avalon

10—Waltz ....Honolulu Eyes

11—Fox Trot ....Japanese Sandman

12—Waltz ....If You Could Care

Intermission

1—Fox Trot ....I'm a Jazz Vampire

2—Waltz ....Little Curly Head

3—Fox Trot ....Hop Skip and Jump

4—One Step ....Rosie

5—Fox Trot ....Grieving for You

6—Waltz ....Everybody Knows

7—One Step ....Humming

8—Waltz ....Hiawatha's Melody of Love

9—Fox Trot ....Margy

10—Waltz ....When I Looked in Your Eyes

Extras

ITALY AMERICA SOCIETY

ITALY AMERICA SOCIETY

"To create and maintain between the United States and Italy an International friendship based upon mutual understanding of their national ideals and aspirations and of the contribution of each to progress in science, art, and literature, and upon co-operative effort to develop international trade."

Headquarters

In the United States:

Italy America Society,
25 West 43rd Street,
New York City.

Italy America Society,
17 N. State Street,
Chicago, Ill.

In Italy:
Associazione Italo- Americana,
271 Corso Umberto I,
Rome.

Charles Evans Hughes—First President

Honorary Presidents:

The Italian Ambassador to the United States
The American Ambassador to Italy

Honorary Vice-Presidents:

Richard Washburn Child
Lloyd C. Griscom
Hamilton Holt
Robert Underwood Johnson
John G. A. Lieschman
Thomas J. O'Brien
Henry White

President:
Paul D. Cravath

Vice-Presidents:
James Byrne
Robert Perkins

Secretary:
Henry Burchell

Treasurer:
Thomas W. Lamont

Manager:
Irene di Robilant

Executive Committee

The Italy America Society in the United States is a non-political, non-partisan, non-sectarian organization, governed by American citizens, and supported entirely by membership dues and contributions.

SOMETHING ABOUT IT

The Italy America Society was founded in 1918 by
a group of American gentlemen headed by Charles Evans
Hughes for the purpose of acquainting the American
public not only with Italy's part in the World War, but
with the actual standing of that country in the various
aspects of its social and intellectual life.

The fact, that so many Italians have come to our
shores in recent years makes it necessary for every good
citizen to know more about their background, and to
realize how important has been Italy's contributions
towards the upbuilding of this democracy.

The Italy America Society endeavors to disseminate
accurate and reliable news regarding Italy.

By its public meetings it furnishes a vehicle through
which the message of distinguished Italians can reach the
world at large. It has grown considerably during its
five years of life and its membership, made up of American men and women interested in Italy, is truly representative of the best in the community.

Monthly Bulletins are sent to members as well as to
the leading Universities and Chambers of Commerce.

Members have the opportunity of meeting all
Italians of distinction who visit the United States, and
when in Italy enjoy the hospitality of our sister organization the Associazione Italo-Americana, and the Library
of American Studies located in one of Rome's historical
palaces.

The Society has established an exchange of lecturers,
and supports two Fellowships enabling an American
Student who has studied Italian to spend a year in Italy,

and an Italian with a good knowledge of English to study in the United States for an academic year.

Through the efforts of the Italy America Society, the drama, music, art, and literature of modern Italy are becoming familiar to the American public, whilst American ideals and writers are gaining appreciation in Italy.

The Society has an active branch in Chicago and committees in various cities of the Union. In the centre of America's greatest city it maintains an office open all the year around with a reference file consulted daily by members and institutions from all over the country, from which accurate and unbiased information can be secured regarding trade, travel and the intellectual life of Italy.

If you have been in Italy and have spent some happy hours there will you not join the Society and recall in your American life the pleasant memories you associate with things Italian?

If you plan to visit Italy or are interested in Italian culture will you not help us to make it known to all Americans, and when in Italy will you not want to meet Italians and interpret the ideals of this land of promise to them?

If you are merely interested in the welfare of mankind, in an international understanding and peace which can only be obtained through education and knowledge of one another, will you not join the Society, which will help you to understand the ideals and problems of a people whose destiny is of great interest to our future?

ITALY AMERICA SOCIETY
25 West 43rd Street
New York City

Membership is open to American citizens and Italian citizens resident in the United States.

Membership is open to men and women alike.

The Italy America Society does not concern itself with the internal politics of Italy and membership does not in any way conflict with the duties of good citizenship. It is realized that each member's first duty is to the land of his birth or adoption.

I hereby apply to become a...
(class of membership)

member of the ITALY AMERICA SOCIETY, and enclose subscription for membership dues.

Name ....

Address ....

Nationality ....

Business ....
(for record purposes only)

Proposed by ....

Classes of Membership

Annual	$ 10.00
Contributing (annually)	25.00
Commercial (annually for firms only)	100.00
Life	100.00
Fellows	250.00
Sustaining	500.00
Patrons	1,000.00

Make checks payable to THOMAS W. LAMONT, TREASURER, Italy America Society, 25 West 43rd Street, New York City

ITALY AMERICA SOCIETY ENDOWMENT AND FELLOWSHIP FUND

Persons who are interested in the objects of the Italy America Society, but do not wish to add to their obligations by becoming members may subscribe to the Endowment and Fellowship Fund which the members and their friends are gradually raising.

Subscribers will receive an artistic diploma and be classed as benefactors of the Society.

———————

I hereby enclose .... to be applied to the Endowment and Fellowship Fund of the ITALY AMERICA SOCIETY with the purpose of furthering intellectual relations between the United States and Italy.

Name ....

Address ....

Make checks payable to THOMAS W. LAMONT, TREASURER, Italy America Society, 25 West 43rd Street, New York City

WHAT THE SOCIETY HAS ACCOMPLISHED

Year 1918

The Society was organized, published its by-laws and endeavored by publications and meetings to afford recognition to Italy's part in the World War.

Was instrumental in having an American Regiment sent to Italy.

Year 1919

By a concert at the Metropolitan Opera House raised $45,000.00 for the Italian Red Cross.

Gave a public dinner in honor of Ambassador Macchi di Cellere.

Year 1920

Organized a nation-wide celebration of Italy America Day on the fifth anniversary of Italy's entrance into the war.

Gave a public dinner for Ambassador Romano Avezzano with an attendance of 900 persons.

Year 1921

Created the National Dante Committee, organizing a nation-wide recognition of the Six Hundredth anniversary of the death of Dante.

Sent a party of 180 American college students to visit Italy on a Summer Tour.

Gave public dinners in honor of Ambassador Rolandi Ricci, and General Diaz.

Contributed towards sending Prof. Kenneth McKenzie as Exchange Professor to Italy.

WHAT THE SOCIETY HAS ACCOMPLISHED

Year 1922

Enlarged the membership of the Society.

Donated a book entitled Modern Italy to all its members.

Organized an information file on Italian trade and educational institutions consulted by persons and institutions all over the United States.

Year 1923

Organized a Branch of the Society in Chicago.

Founded the first two Fellowships for the exchange of students.

Sent an American Professor to lecture at the University of Rome.

Organized a lecture tour covering every state in the Union for Princess Santa Borghese.

Year 1924

Was instrumental in having the Italian Dramatist, Luigi Pirandello, come to America, and financed the production of his play "Henry IV"

Conducted a nation wide survey of the status of the study of the Italian language in American colleges and high schools.

Assigned the Eleonora Duse Fellowships to an American student.

I PROBLEMI DELLA MADRE IN UN PAESE NUOVO

CECILE L. GRIEL, M. D.

I PROBLEMI DELLA MADRE IN UN PAESE NUOVO

Introduzione.

ESIDERATE qualche volta di essere laggiu, in Italia, dove la vita è tanto più semplice? In questo paese, sentite in parlare di lingua straniera, che non ha nessun senso pei vostri orecchi. Nella fretta e la confusione della vita Americana, siete spinta da parte e forse canzonata da folle spensierate che non vedono in voi che una forestiera. Voi Italiani, che siete così caldi ed amichevoli, pensate che siamo molto freddi e formali, in questo paese. Veramente, è un mondo nuovo molto strano, nel quale vi siete avventurata. Chiunque ha vissuto molto tempo nella vostra patria d'oro e dalle colline coperte di olivi, ove ogni cosa si riveste un pochino del calore e del colore del vostro cielo azzurro, sà e sente quanto questo paese debba sembraryi duro e strano.

Siete stata probabilmente molto sorpresa ed afflitta dalla mancanza di simpatìa e di interesse che noi Americani abbiamo dimostrato per voi e per la vostra famiglia. Quando noi siamo venuti da voi, in Italia, come viaggiatori o per stabilirci fra voi, ci avete sorriso e, sorridendo, ci avete dato benvenuto. Quando ci siamo provati a parlare il vostro linguaggio dolce e musicale, non ci avete canzonati, ma bensì vi siete sforzati ad intenderci e ci avete aiutati. Ci avete incoraggiate, col vostro: "La Signora parla abbastanza bene", quando eravamo riuscite a dire solo due parole, e queste molto male. Ed adesso, vi meravigliate e vi domandate perchè, in simile circostanza, noi non siamo pronti ad aiutarvi nello stesso modo.

Le nostre abitazioni sono strane quanto le nostre usanze e la nostra lingua. Avete lasciato la vostra casina bianca, tutta stuccata, nella Campagna, le vostre vigne, le vostre melagrane, ed i vostri giardini splendidi, coi muri verdèggianti dai muschi di centinaia d'anni, tutto tanto bello, e voi stessa facenti parte di tutto questo colore e di tutta questa vita. Nelle nostre città grandi, la gente vive così accalcata che non c'è più posto per giardini ove i bambini possano giuocare al sole. Ma questo non è il caso in America soltanto. Nelle città grandi, in Italia, la gente non ha più posto di quel che non ne abbia qui. Ogni volta che molte migliaia di gente àbitano nello stesso posto, questo deve per forza essere il caso, perchè non c'è abbastanza posto perchè ogni famiglia possa avere una casa separata. Se non vi piace di vivere nelle nostre città affollate, c'è moltissimo spazio nella campagna, dove potete avere la vostra casina e molta buona aria e sole.

Perfino i nostri abiti Americani non vi piacciono dapprima quanto quelli che portavate in Italia. Non sono di gran lunga nè così sciolti ne così comodi. Come è bello e diritto, il vostro corpo! Come noi Americani vi abbiamo ammirate, quando vi vedemmo portare, ben bilanciati in testa, un gran paniere oppure una vera montagna di soprabiti, senza che mai succedesse un accidente!

Voi dite che il più duro è di tirar su i vostri figli, qui in America. I più piccini vanno a scuola e presto imparano a parlare, a vestirsi ed a pensare come Americani. Essi giuocano là dove non li potete sorvegliare, con altri ragazzi che non ubbidiscono ai loro genitori come dovrebbero farlo. Presto essi pure cominciano a disubbidirvi, e non sapete più come fare. I vostri figli piu grandi vanno a lavorare in manifatture Americane, ove è d'uopo che sappiano l'Inglese per capire le direzioni che le sono date riguardo al lavoro che devono fare, e per poter parlare coi loro compagni di lavoro. Presto essi fanno le cose che vedono fare ai ragazzi ed alle ragazze Americani, e voi non siete sicura se queste cose stiano bene, o no.

Non potete più dirigere la condotta dei vostri figli, dicendo loro che i ragazzi e le ragazze per bene in Italia non si conducono come fanno loro. Essi rispondono subito che non conoscete le usanze Americane, che sono veramente preferibili a quelle Italiane. L'immigrazione in questo paese ha cambiato un poco la vostra posizione verso i figli vostri. Non siete più la sola che debba essere ascoltata circa molti soggetti. I figli vostri parlano con autorità su ciò che è l'abitudine o di moda nell'America, e voi non sapete se hanno ragione o torto. Cominciate a sentire un allontanamento fra voi ed i figli vostri, e questo vi addolora.

Non dovete permettere questo, dovete trovare un mezzo per rimediarci. Vi dovete americanizzare voi stessa, se non fosse per altro che per proteggere i figli vostri dai trabocchetti della vita. Sforzatevi di riuscire in questo, malgrado le condizioni dure che vi circondano, malgrado i molti piccoli bambini che avete, malgrado un marito che torna a casa stanco, e malgrado i vostri molti altri doveri. Dimostrate ai figli vostri che siete interessata nelle cose nuove che essi stanno imparando, e lasciate che essi vi insegnino le usanze Inglesi ed Americane.

Il Baby nato in America.

UESTO piccino è molto diverso da Luigi, Bettina ed Alfredo. Questi nacquero nella bella Italia, ma quest'ultimo è nato in America ed il suo nome è stato americanizzato. Ho veduto il suo certificato di nascita. Voi gli daste il nome di Salvatore, ed invece fu iscritto come "Sam". Ma questo sta benissimo, Mamma. Noi, semplicizziamo e scorciamo tutti i nomi, qui. Questo fa parte della fretta, (chiamata "Hurry-up") della vita Americana. Non passerà molto tempo prima che cominciate a capire come si fanno le cose qui in America. Il nuovo Baby vi aiuterà in questo, e perfino Luigi, Bettina ed Alfredo vi saranno di aiuto.

L'ultima vostra creaturina ha un nome americano ed è un Americano. Forse non avete neppur seguito l'usanza Italiana di avere una levatrice pel vostro parto. Un Dottore fu chiamato a casa vostra per aiutare lo sgravo. Questo pure è Americano. Noi siamo sempre per "la sicurezza avanti tutto" (Safety first), e sappiamo che, se ci fossero delle difficoltà durante il parto, un bravo Dottore dovrà essere chiamato in ogni modo. Per questo, noi donne Americane abbiamo un Dottore per questa importantissima funzione fin dal principio, per eliminare qualunque pericolo tanto per la Madre che pel nascituro. I Dottori sono educati piu scientificamente in questo genere di lavoro, e, se, per, caso, non potete pagare niente oppure soltanto pochissimo pei servigi resi dal Dottore, domandate al'Infermiera Municipale od alla Signora che vi fa visita tanto spesso, di accomodare la cosa per voi; esse possono farvi entrare in un ospedale, gratis, o con pochissima spesa. Il consiglio che vi do adesso potrà aiutarvi col prossimo piccino, anche se arriva troppo tardi per il piccolo Sam.

Questo nuovo Baby, Sam, è meraviglioso. Egli ha degli occhioni neri come la sua bella Mammina, ed i suoi capelli mostrano digià dei ricciolini soffici e bruni. Questo somiglia a suo Padre, non è vero? Le gambine del Piccinino sono forti e fossettate. Guardatelo, come si stende e si rimugina. È così che cresce. Non fasciamo i nostri Piccini, in questo paese. Abbiamo fiducia che cresceranno diritti e forti, lasciando fare la natura. Essa lo formò, ed essa farà crescere i suoi membri forti e diritti se li lasciamo liberi e diamo loro abbastanza spazio per crescere. Forse non crederete questo, ma provatelo e vedrete se non è vero! Sam è un Americano, e gli Americani sono strani e differenti.

Sam avrà migliori opportunità della maggior parte dei Babies Americani, perchè lo allatterete col proprio latte. Disgraziatamente, la più parte delle Madri Americane non possono o non vogliono allevare i loro Piccini. L'allevare un Bambino prende molto tempo, ma ne vale la pena, perchè non c'è nutrimento che possa prendere il posto del latte materno.

Non c'è che uno svantaggio possibile nell'allattare il vostro Baby. Forse lo lascerete puppare troppo spesso. Il preparare le bottiglie per nutrire i Piccini artificialmente, dà molto disturbo, e per questo regoliamo le ore con gran cura, e non gli diamo la bottiglia mai più spesso che una volta ogni due ore. Quando allevate il vostro Baby, è tanto facile di aprire il vestito vostro, e porre la bocchina dolce la dove può bere liberamente. La natura allora apre le fontane, e Baby beve e beve da ghiottoncino che egli è. Questo va benissimo per lui-e per voi pure-al momento. Baby si addormenta, e voi potete accudire alle vostre altre faccende per qualche tempo. Se lo allattate troppo spesso, però, egli non si addormenterà sempre. Gli verranno dei dolori di pancina, ed egli si istizzirà, piangerà e perderè la dolcezza abituale della sua disposizione ed allora lo potrete far puppare e ripuppare quanto vorrete, ma la stizza continuerà, ed egli continuerà a piangere. Provate il nostro metodo Americano di nutrirlo ad intervalli regolari di due ore, ed egli vi si abituerà in poche settimane e prospererà con questo sistema.

Sam incontrerà pure qualche svantaggio, per esser nato in America, e questi dovranno esser sormontati in qualche modo. In Italia, quando vivevate in campagna, o in un villaggetto, avevate un'abbondanza d'aria fresca per i bambini, il vostro clima e cosi dolce e caldo che essi potevano restar fuori tutto il giorno senza impensierirvene. Questo, in America, è impossibile. Probabilmente, siete in un appartamento di quatro stanze. La vostra famiglia è numerosa, e vivete gli uni addosso agli altri. Questo è male, ma non è la colpa di nessuno in particolare, e migliaia di Americani provano la stessa difficoltà, vediamo dunque come si possa sormontare.

Manteniamo i nostri Piccini in salute col bagnarli spesso e tenendo le finestre aperte notte e giorno nelle camere ove dormono parenti e figliuolini. Questo aiuterà molto; e qualche volta, quando Baby sarà di cattivo umore, la sera, e che non vuol addormentarsi, per quanto lo culliate o quanto spesso gli abbiate dato la poppa, provate a spogliarlo ed a dargli un altro bel bagno caldo. Mettetegli una camicetta pulita, per tenerlo fresco e lindo, e copritelo bene nel suo letticciuolo, lasciando tutte le finestre aperte. È quasi certo che dormirà come un angioletto fino alla mattina.

Non tenete il Baby troppo tempo nelle braccia, e non lasciate che Bettina lo faccia neppur lei. Questo è cattivo per la salute di Bettina, e non farà nessun bene nemmeno al Piccino. Insegnategli ad essere tranquillo e quieto ed a divertirsi da se. Non esitate ad andare a chiedere consiglio alla casa di ricovero (Settlement) la più vicina e lasciate che le infermiere municipali del vostro distretto vi aiutino quando siete in dubbio. Questo modo di procedere per tirar su i Piccini in un paese nuovo, è molto duro, ma siamo lieti che siete venuti fra noi, e desideriamo aiutarvi quanto piu possiamo. Voi avete molto da darci, e noi abbiamo molte cose da offrivi, a voi, a Luigi, a Bettina ad Alfredo ed al nuovissimo Cittadino Americano, Sam. Cerchiamo di capirci e di diventare amici. Noi chiediamo la vostra amicizia. Non volete la nostra?

9

Quando i Ragazzi Cominciano ad Andare a Scuola.

QUI in America, le leggi che regolano l'educazione obbligatoria sono rigide e stringenti. Appena i ragazzi hanno sei anni, sono costretti ad andare a scuola. In certe provincie d'Italia, potete mandarli a scola, o no, come vi pare. Ma questo non farebbe affatto per i ragazzi, quà. Quelli che sanno leggere e scrivere sarebbero più arguti, piu svelti di quelli che non sanno, e questo darebbe loro un vantaggio su questi ultimi. Questa è "l'America libera", con privilegi uguali per tutti, pei ricchi e pei poveri, e certo voi desiderate che i vostri Figli abbiamo il meglio di quel che l'America ha da offrire.

Qui, non soltanto siete costretti a mandare i ragazzi a scuola, ma siete felici di mandarceli, quando vedrete come imparano presto l'Inglese e come bene imparano a conoscere l'America, grazie al contatto cogli altri ragazzi, nati qui. Forse, dapprima vi dispiacerà quando, avendo parlato in Italiano a Bettina, essa vi risponderà in Inglese. Ma, veramente, e superfluo il temere che essa si scordi della lingua natìa. Il ragazzo non richiede che cinque anni per fissarsi in mente perfettamente un linguaggio; poi, a casa, le parlerete spesso Italiano, e così essa non se lo potrà scordare.

Se, poi, ve ne sentite, potete fare due cose in una; potete rendere i ragazzi molto felici, permettendo loro di

10

giuocare a scuola con voi qualche minuto ogni giorno, ed allo stesso tempo vi insegneranno l'Inglese. Non crediate che questo tempo, benché rubato alle faccende della giornata, sia perso: creerà una grande amicizia fra voi ed i ragazzi, perchè i bambini si deliziano a vedere la Mamma interessarsi a tutto quello che fanno. Questo è il momento per formare una grande bella, durevole amicizia coi figli vostri, un'amicizia che durerà tutta la vita.

Quando poi la piccola Bettina diventerà una vera Signorina Americana, più tardi, essa verrà a confidarvi tutti i suoi segreti, le sue piccole avventure, e tutti i suoi problemi, precisamente come viene da voi adesso coi suoi piccoli problemi infantili. Ecco come noi proteggiamo i flgli nostri, qui, in America. Ne facciamo amici nostri; i nostri compagni, "Chums" come li chiamiamo.

Giorgio pure si diletterà a giuocare a scuola con Mamma. I ragazzi non sono che uomini in piccolo; e gli uomini—soprattutto gli uomini Italiani si dilettano a tener conto delle donne loro. E stato dovere loro dal principio del mondo in quà, e Giorgio sarà presto o tardi, il protettore di una donna; certo, non potrebbe far meglio che cominciare col proteggere sua Madre! Piu tardi, vedrete che Giorgio verrà a narrarvi le sue prodezze sul campo del "Baseball" (pallone), ed a dirvi che gusto egli ha provato nell'essere a capo della sua classe. Poi, un poco più tardi, egli seguiterà a confessarvi come il cattivo ragazzaccio del vicinato abbia cercato d'indurlo a fumare, o a bestemmiare, o a buttar delle pietre ad un povero cane malato. E voi, buona Madre, l'incoraggerete con delle carezze materne, e vi sentirete tanto felice di aver fatto un amico, un "Chum" di vostro figlio! poichè potete essere sicura, che se avete piena conoscenza di tutto ciò che fa nella sua prima giovinezza, egli resterà

11

vostro amico e "Chum" anche più tardi, quando, inevitabilmente, si presenteranno i grandi e duri problemi della vita.

Questo è il solo mezzo che noi Americani abbiamo trovato per conservare i figli nostri buoni e sinceri. La vita è tanto complicata, qui, e non possiamo seguire i figli nostri ovunque vadano; per questo, l'incoraggiamo a restare onesti, autonomi, e sinceri. Così siamo certi che si sapranno guardare dovunque, da soli.

Vi meraviglierete quando vedrete quanto le maestre Americane amino i loro piccoli amici forestieri. Fanno tanto più che insegnar loro soltanto a leggere, scrivere, e gli altri studii! S'interessano **specialmente** a loro, ed hanno molto a cuore il loro benessere, e la loro salute. Esse li adottano come una specie di "sorella maggiore" ed i ragazzi s'indirizzano a lei spesso coi problemi casalinghi, e la maestra capisce e simpatizza. Se i bambini hanno bisogno di cure speciali o di speciali consigli medici, la Maestra se ne avvede molto presto. Essa lo dice alla Guardia malati della Communità (Settlement Nurse) o alla Signora visitante (Welfare Worker), oppure essa viene da voi personalmente. Sarete sorprese di vedere quanto esse siano gentili e cortesi; come, entrando in casa vostra, cercheranno di scoprire la causa del disturbo, e come cercheranno tutti i mezzi per aiutarvi a sciogliere in vostri problemi in questo mondo tanto grande e nuovo! Veramente, esse **possono** aiutarvi, e lo fanno volentieri.

Qualche volta, i ragazzi hanno bisogno di cure mediche; forse la spina dorsale della Bettina si e storta un pochino, e non ve ne siete avveduta: voi siete tanto occupata, ed l'ultimo bambino prende quasi tutto il vostro tempo! Ma la Maestra se ne avvede presto, quando la governante o guardia-malati della Communità, esamina i bambini, o quando è l'ora della ricreazione nella ginnastica. La Maestra vi manda allora alla farmacia, o la guardia-malati vi ci può accompagnare, se vi vergognate ad andare sola o se siete troppo occupata.

In questo paese, abbiamo certe idee stabilite circa l'igiene e lo sviluppo dei ragazzi. Sono state accuratamente studiate per adattarsi alle condizioni del nostro clima e della nostra vita. Viviamo in città, in strade e case affollate, e per questo siamo costretti a trovare altri mezzi par dar spazio ai nostri figliuoli, per permetter loro di esercitarsi e di giucare per diventare belli forti e sani.

Qui, dunque, abbiamo divisato una nuova scuola per giuocare, con maestre speciali per questi scuola-giardini (playgrounds). Ed adesso, Bettina, Giorgio e gli altri possono saltare correre, dondolarsi e giuocare quanto vogliono, e la Mamma non ha piu ragione d'impensierirsi, perchè sono costantemente guardati da occhi provetti, e cosi niente di male puo accadere loro. Quest' è una splendida idea Americana, questa scuola di giuoco-lavoro! Ma aspettate un po'! Non ne dobbiamo essere troppo alteri, e chiamarla Americana, senza essere interamente sicuri di ciò che avanziamo. Veramente, avete la stessa cosa a Roma; la Signora Montessori ve l'introdusse. Avete parecchie "Case di Bambini" in Italia, ma qui, ci serviamo più generalmente di quest'idea, e l'abbiamo connessa praticamente con i centri scuolastici in tutta l'estensione della nostra patria; questo è di grande aiuto alle Madri che hanno tanto da fare—tanto le Madri Americane che quelle nate in altri climi—è tanto meglio pei bambini che di stare a casa, in stanze auguste e zeppe, o di giuocare nelle strade sudicie ed affollate, ove i Piccinini sentono dire ogni sorte di cose che non sono per orecchie di bambini, e dove gli occhioni innocenti si posano su cose che dovebbero rimanere sconosciute all'infanzia. Questa mancanza di spazio pei giuochi dei fanciulli fa parte integrale della tragedia della vita nelle città grandi, ma ci sforziamo a ripararci al più presto che possiamo, e avremo delle "scuole di giuochi" dappertutto.

Tutte le Bettine ed i Giorgi e tutti i ragazzi in generale hanno bisogno di essere molto, molto forti e di stare molto, molto bene, per poter fare la loro parte del lavoro di questo mondo con efficacia ed intelligenza quando saranno cresciuti. Essi sono Americani, sia perchè sono nati qui, sia perchè vostro marito ha già ottenuto le sue seconde carte di cittadinanza. Questo rende tutta la famiglia cittadini di questi Stati Uniti di America. Nelle scuole, cominciamo il nostro compito di sviluppare tanto la mente che il corpo dei nostri ragazzi, con esercizii sagaci e ben diretti, cosicchè, quando l'opportunità si presenterà loro più tardi, quando saranno cresciute, saranno ben preparati, mentalmente e fisicamente, per il lavoro che avranno da fare. Le Scuole, le "scuole per giuochi", le Case della Communità (Settlements) e le Chiese vi possono aiutare benissimo a sciogliere queste difficoltà, e, che più è, non soltanto lo possono fare, ma lo fanno volentieri. Ma voi pure, potete aiutare, malgrado le faccende di casa, i pensionanti, il bucato, il lavoro di casa e tutto il resto. Forse, sarà un po' duro adesso, ma, quando i ragazzi saranno cresciuti (riposiamo molta fiducia), ed educati, aiuteranno essi pure, ed allora le cose anderanno meglio.

Noi crediam — come un grande aiuto a preservare una buona salute — nella pulizia personale, e, se siete di quest'opinione voi pure, aiuterete la Maestra, tenendo i ragazzi pulitissimi. Certuni fra noi abbiamo delle tinozze in questo paese, ed altri non ne hanno. Questa e una splendida abitudine per i piccini. Però, finchè non

14

siete sicura che sanno bene come lavare tutto il loro corpicino, orecchi, testa e tutto il resto, sarà bene che sorvegliate questo compito, e che vi assicuriate che sia fatto bene ed in regola.

Gli abiti per la scuola devono essere molto semplici. Le Madri Americane sono qualche volta abbastanza stolte per vestire troppo bene i figli loro quando vanno a scuola. Voi non farete questo errore. Gli Italiani portano sempre degli abiti semplici e comodi: seguitate questo buon sistema! Le figlie vostre sono tanto belle, cogli occhioni neri e la carnagione olivastra, ed adornate soltanto di una vita bianca e di una sottana scura'. Forse Bettina avra voglia di un bel nastro rosso, posato sui suoi bei riccioli neri,. . . . Brava! questo stà bene! Da colore alla stanza, e la rende allegra. Ma Bettina deve pure spazzolare bene bene i suoi capelli, per renderli lustri ed attraenti!

Giorgio strappa i suoi abiti tanto presto! Tutto i ragazzi fanno lo stesso! Certe Madri Americane tengono un paio di calzoni extra per la scuola, e la camicetta bianca è cambiata pure. Quest' è soltanto una buona abitudine da prendere. I ragazzi imparano qualunque cosa, se gli adulti insegnano loro con pazienza ed amore, e li **potete** avvezzare, e la Maestra vi aiuterà.

Quando i Ragazzi Vogliono Divertirsi.

CHE può fare, una Madre affaccendatissima, per divertire la sua famiglia, che cresce a vista d'occhio? Questa è una questione agitata da tutte le Madri di questo gran paese. I ragazzi non possono stare sempre in Scuola. Quando sono a casa, di due cose l'una: o fanno tanto fracasso da far impazzire la Madre loro, oppure si rincantucciano in un angolo oscuro dove non penetra l'aria fresca, e dove non fanno nessuno esercizio ne hanno niente che faccia loro alcun bene.

Questo è paese del "Hurry up", della "fretta e furia." Capite, no è vero? Vogliamo far tutto, quanto più presto e quanto meglio possiamo. Crediamo bene di spicciarci anche nel soddisfare i desiderii (bisogni) dei nostri figli appena hanno l'età del ragionamento. Abbiamo usato questo stesso sistema di prestezza anche nelle nostre idee per i giuochi e la ricreazione dei figli nostri. Utilizziamo ogni giuoco, facendolo contare per qualche cosa nel loro benessere futuro. Le ricreazioni devono avere un doppio scopo: il primo è di rendere i ragazzi felici e lieti—il secondo, di fortificare contro le lotte dell'avvenire la loro salute ed il loro carattere. Tutti i giuochi di scuola tendono a questo, ed i giuochi di casa devono avere la stessa direzione.

I ragazzi sono pieni di energia e di vitalità, e le loro risate sonore e la loro attività irrequieta a casa non è indizio di cattiverìa ne di malvagità. Sono soltanto ragazzi sani e normali che instintivamente cercano di esercitare i loro polmoni ed i corpi loro. Qualche volta, questo annoia molto gli adulti, ma non è per questo che si possa chiamare perverso. È l'ambiente che non stà bene. Se facessero questo frastuono nel parco o nella "corte dei giuochi" della scuola, non vi annoierebbe affatto. Ma in quattro stanze strette ed anguste, con un bambino piccino che sta dormendo in una di esse, la posizione diventa qualche volta molto dura. Per questo, vedremo adesso ciò chè si può fare pei giuochi in casa.

Una bambinella che conoscevo aveva perso sua Madre da qualche anno; ed il povero Padre non sapendo che cosa fare per divertir la sua bambina le aveva comprato dozzine Sopra dozzine di bambole, di giocattoli e di belle cose di ogni genere. Questa bimba aveva la propria serva, e tutto ciò che il denaro puo procurare, ... eccetto la gioia! La bambina era infelice, e rimase tale finchè non venne in quella casa, una "Mamma nuova", alla quale piaceva di far da cuoca e da pastic ciere da se. La piccola "nuova figlia", coi suoi bei vestitini, nastri celesti ecc, la guardava fare con invidia, e, notando ciò, la "nuova Madre" invitò "nastri celesti" ad aiutarla. Fu questa la prima volta che la bambina godette nel giuoco. La "nuova Madre" aveva risolto il problema, e "nastri celesti" si addobba adesso ogni giorno di una cuffiettina e di un grembiulino da lavoro ed aiuta la Mamma. Tutti i giuocattoli sono stati regalati a delle bambine che avevano troppe vere faccende da fare, troppo bucato, troppo da spazzare e da badare al baby, e non abbastanza occupazione nel campo imaginativo. Ecco tutto il segreto. Qualche volta desideriamo di giuocare pretendendo essere persone

17

cresciute (adulti) ed altre volte bramiamo tutti di ridiventare come bambini.

I ragazzi hanno tanta individualità, e noi, qui, amiamo molto l'individualità. Questo nuovo paese offre migliaia di opportunità nuove pei giovani e pei vecchi, ma la nuova generazione ha un vantaggio meraviglioso qui in America, se dimostra di essere "individuale" "originale". Per questo, studiamo i figli' nostri e cerchiamo di indovinare al più presto ciò. Ed i giuochi — i giuochi naturali — spesso spesso ce ne rivelano il segreto. La più parte dei ragazzi amano "pensare colle mani". Ai maschi piacciono i giuocattoli che sviluppano le tendenze costruttive. Perfino quando Giuseppe spezza i suoi giuocattoli o fa un buco nel suo tamburo nuovo, non ci arrabbiamo con lui. Scrolliamo le spalle, ridiamo, e pensiamo che dev'essere un ragazzo molto intelligente; e pronostichiamo che esso farà una bellissima carriera e diventerà un grande inventore o un grande tecnologo.

Cosi pure, la Bettina guarderà con ammirazione la splendida bambola che le portò "Santa Claus" e si maraviglierà del prodigio del suo cappello rosa e del suo vestito, cucito addosso, di seta e trina. L'ammirerà, si, ma non l'amerà veramente, ne ci si divertirà. Poi, un bel giorno, un terribile accidente rovinerà il bel vestito di seta, e la Bettina piangerà e sarà infelice. Allora la Mamma, se è intelligente, dirà: "Vieni, Carina, vediamo se non possiamo farle un vestito piu bello di ciò che il negozio non potesse!" E gli occhi ancora umidi di Bettina danzeranno dalla gioia. E, in un batter d'occhio, la vedrete tagliare e cucire per la bambolina sua, ed essa avrà fatto **due** cose senza neanche avvedersene: avrà imparato a giuocare per benino, tranquillamente, senza far rumore che svegli il baby, ed avrà imparato un'arte nuova, un mestiere che forse la renderà molto felice piu tardi.

Cerchiamo di fomentare nei figli nostri, per mezzo dei giuochi, ogni sorte di qualita morali per mezzo dei giuochi, e, qui pure, potete aiutarci. incoraggiate i ragazzi tutto il tempo e cercate di interdirli. S e potete insegnare personalmente a Bettina a far la calza o la trina come facevate laggiù, nel vostro "Paese", (fatelo.) siate certa di farlo. In questo momento, il far la calza è di voga. Ogni donna e ragazzina in questo paese cerca di farlo. Potete essere una grande amica della vostra figliuoletta se l'insegnate voi stessa e se l'incoraggiate ad aiutare essa stessa più che può.

Qualche volta, poi, i ragazzi desiderano altre forme di divertimento, e così parliamo un poco dei cinematografi. (Movies). C'è sempre un teatro cinematografico ad uno o due "blocchi" da casa nostra, e spesso siamo indotti ad andarci, precisamente perchè è tanto vicino. Poi, è molto buon mercato. Possiamo tutti darcene il lusso, ed è buono pure perchè è una forma di divertimento che tutta la famiglia può godere insieme. Questo è splendido.

Però, c'è qualche cosa che non va bene col cinematografo, giusto come c'è con tante altre cose che ci piacciono in questo mondo. Che terribili pitture danno la piu parte di questi teatri! In quasi ogni pittura, ci sono assassinii, rubamenti e mariolerie d'ogni specie. Poi, ci sono le pitture dell' "Ovest selvaggio" ("Wild and Wooly West"), in cui un impossibile vaccaro ("Cow-boy") dincorre a cavallo, furiosamente, un ladro che ha assassinato sua moglie e bruciato la casa sua. Esso galoppa e galoppa e spara e spara finchè i ragazzi

19

spettatori si eccitano tanto quasi da urlare forte. Più tardi, Giorgio, che non ha capito bene, pensa (quanto deve esser bello di essere quella sorte di vaccaro, e la sua imaginazione comincia a lavorare in una cattiva direzione. Noi adulti sappiamo bene che non c'e nessun Ovest di quella sorte, e certamente nessuni uomini come quei vaccari, ma Giorgio non lo sa. Poi i magnifici vestiti di seta a coda, gli automobili ed il lusso della bella ragazza di magazzino, che spende una vita di peccati! Noi vediamo nell'ultima filma che "La retribuzione del delitto è la morte." Ma la piccola Bettina non vede cotesto. Essa vedrà soltanto ciò che la sua imaginazione brama di vedere, i vestiti ricchi e soffici, i gioielli ed i bellissimi ambienti.

Capite il mio punto di mira, o Madre? I Cinematografi sono buonissimi per quelli fra di noi che non hanno che pochi denari da spendere pei piaceri loro, ma non sono sempre buoni pei ragazzi. Non è vero che qualche d'une di quelle pitture vi eccitano tanto? Potete appena dormire quando tornate a casa, e spesso spesso vi sognate di ciò che avete visto. I figli vostri si eccitano anche più di voi, e stanno svegli pensando a ciò che hanno visto, quando invece dovrebbero dormire. Vengono loro in testa un monte di idee cattive, e qualche volta si provano perfino ad imitare gli attori dei cinematografi! Questo può essere molto pericoloso per loro. Dovreste cercare di informarvi sul carattere delle pitture prima di condurci i figli vostri. Se vi è possibile, andateci con loro, e discutete insieme, dopo, tutto ciò che avrete veduto.

L'aria non è mai molto buona in un Cinematografo, e per questo i ragazzi non dovrebbero andarci molto spesso. Poi c'è sempre folla, ed e facile che i ragazzi ci prendano qualche malattia contagiosa. Cercate sempre di lasciar qualched'uno a casa per badare al baby, perchè egli non puo godere delle pitture ed è facilissimo che egli si ammali per via dell-aria viziata. L'aria buona e pura è necessarissima specialmente per i Babies.

L'arte di giuocare coi bambini è una grande arte, e le Madri Americane non l'hanno imparata troppo bene. I ragazzi imparano meglio a conoscere gli adulti della loro famiglia dai giuochi che in qualunque altra maniera. "Oh, ma che la Mamma sa giuocare davvero?" disse eccitatamente ultimamente, una ragazzina quando, ad una festa di matrimonio, sua Madre era stata sfidata a ballare la Tarantella dei suoi giorni antichi di Capri. Naturalmente, questa madre aveva perso moltissimo a non giuocare colla figliuoletta sua ed a non insegnarle, essa stessa, a ballare il bellissimo ballo di quel giardino del mondo, Capri. Mamma non sospettava neppure che, in tutto questo paese, nelle "corti da giuocò e nelle ginnastiche quel suo ballo natìo, fosse insegnato alle bambine Americane, e come sarebbe stata altera la sua piccola Margherita di poter insegnare alla sua compagna di scuola il ballo che sua Madre "ballava quando era una bambina in Italia."

Non affettate di essere troppo grandi, coi bambini. E molto piu facile di giocare che di lavorare, e la vita è tanto piena di lavoro! È per mezzo dei nostri ragazzi che rinnoviamo la nostra gioventù, non nel guardarli e scuotendo la testa in segno di disapprovazione, ma nell'entrare in tutto quello che fanno, con zesto e con vero entusiasmo. Il più gran complimento che una ragazza possa fare alla Madre sua, qui in America, è di chiamarla la sua Compagna ("my Chum") ed per il maschio è quando dà una forte manata sulla spalla della Mamma, esclamando: "Mamma — per Bacco, ma sei un buon compagnone!"

Compagni di giuoco dei Figli vostri.

"VOGLIO che i figli miei stiano, in casa, con me. Non mi piace che il mio Giorgio stia nella strada con quel ragazzo Irlandese. Non so che cosa facciano, laggiù. Non capisco ne il loro Inglese ne i loro giuochi, e non mi piace!"

Bettina è diventata cattiva, adesso. Balla con Maria Warshopsky. Quella è polonese, credo. Neppur questo mi piace. Essa si scorda di parlarmi Italiano quando torna a casa. Essa scoiattola troppo per tutto il vicinato colle sue amiche. Io non le conosco, e voglio che la mia Bettina stia a casa con me."

Si, cara Madre, questo è pur troppo vero. Voi vorreste tenere i figli vostri vicini, là dove sapete che sono al sicuro sotto le vostre ali. Nessun danno può accadere loro quando essi sono con voi, perchè dimorate al quinto piano d'altezza, sù per aria, lontana lontana dalle strade curiose. Qui, almeno, potete vedere ciò che essi facciano. Ma un'infinità di cose terribili possono succedere nelle strade americane, affollate. Questo tirar sù una famiglia in un paese nuovo è davvero durissimo. Pensate soltanto quanto tutto fosse facile semplice nel "Paese" in Sicilia o nella bella Campagna Romana! Tutti i ragazzi giuocavano insieme nella strada soffusa di sole, e voi vi sedevate sulla soglia della porta facendo la calza, o chiacchieravate colle commari radunate intorno alla fontana. La vita era così sicura e semplice. Tutti si conoscevano e tutti parlavano Italiano.

Ma le cose sono molto, molto differenti quà, Mamma cara. E dura, ma, alla fine, la vincerete. La DOVETE vincere. Il fatto stesso che abbiate avuto il coraggio di impaccare tutto ciò che avevate al mondo e di lasciare la vostra Patria diletta per venire quà a tentare la fortuna, questo fatto solo dimostra che **dovete** vincere. Un bel giorno Bettina, Luigi, Alfredo e perfino Baby Sam diventeranno Americani schietti. Allora, saranno tanto alteri della loro "Mammina" che ha avuto il coraggio di fare il lungo viaggio in terza classe e di portarli con se in America. Capiranno quanto tutto questo sia stato duro per voi, e vi benediranno per tutto quello che avrete fatto per loro.

Ed adesso, parliamo un po' di questa questione dei compagni di giuoco dei figli vostri. Sono essi veramente cattivi, oppure **sembrano** essi soltanto cattivi a voi, perchè non capite il modo americano di giuocare? Certo, vi sembra molto strano, quando pensate a vostro Giorgio, nato in Italia, nero di capello, che ha come compagno quella testa rossa di Jimmy Murphy, nato in Irlanda. Questo **è** strambo, o Mamma, ma è molto Americano. Un grande scrittore ha chiamato l'America una "grande caldaia," perchè tutte le razze e tutte le classi si confondono qui, e sono fuse in un nuovo metallo meraviglioso, che si chiama Americanismo. Questa è una delle cose che amerete di più, in questo paese nuovo, quando sarete arrivata a capirci. È splendido questo maniera di popoli diversi e di farne Americani!

Non è veramente un problema per voi, Madre, quando il vostro Giorgio giuoca con "Jim l'Irlandese" o Bettina si diverte con "Maria la Polonese". Il vero problema è di sapere che sorte di amico Giorgio frequenta, e che cosa dicono e pensano Bettina e Maria quando passeggiano per la strada a braccetto? E la sorte di amicizia che i figli vostri contrattano adesso, che sarà di grande importanza per voi più tardi, e non la nazionalità o la razza di questi amici.

Quale è il legame che attira ed unisce Compagni di giuoco? Mi pare che debba essere lo stesso di quello che ci attira, noi altri adulti. È la nostra comune preferenza per le stesse cose. Questa è la regola generale, e moltissime Maestre hanno trovato che le amicizie contratte nella scuola sono per lo più belle e pure e vengono dall'avere gli stessi gusti e dalla preferenza per gli stessi giuochi. Bettina e "Maria la Polonese" forse non parleranno che poche parole della stessa lingua Inglese; Ma parlano una lingua migliore di quella Polonese, Italiana o perfino dell'Inglese: Esse parlano il linguaggio dell'anima; capiscono a vicenda le loro speranze ed i loro sogni, perchè Iddio le ha fatte dello stesso materiale. Questa lingua dell'anima è veramente la piu bella lingua di tutte.

Le amicizie dei maschi sono per lo più affatto innocue, esse pure, malgrado che esso non chiami i suoi amici con nomignoli. Giorgio è probabilmente un bel ragazzo, normale e sano, che giuoca al pallone (baseball) col suo amico Irlandese. Giuocano alla trottola insieme, e gridano ed urlano e fanno tanto rumore quanto qualunque degli altri ragazzi che avete mai conosciuti. Eppure, tutto questo non è che un giuoco innocuo ed è buon esercizio, e non c'è nulla che vi debba impensierire.

Finchè sapete chi sono gli amici dei vostri figli che essi frequentano fuori di casa, e che cosa stanno facendo, siete salva. Le Madri Americane capiscono questo. Incoraggiano gli amici dei figli loro a venire in casa loro, ed imparano a conoscerli bene. Ben presto si avvedono se l'influenza esercitata da questi amici sia buona o no. Dovete imparare a studiare questo, voi pure. Non potete pretendere che i vostri ragazzi stiano sù in casa con voi sempre e sempre. Questo non e ammissibile in una città grande. Devono correre, giuocare ed andare a scuola, ed esser cosi lontani da voi e della vostra influenza la più gran parte della giornata.

Come ho gia detto, la più gran parte delle amicizie di gioventù (d'infanzia) sono semplici e reali. Ma spesso non sono cosi semplici quanto paiano. I ragazzi naturalmente inclinano verso un' esagerata ammirazione od affezione (Hero worship). Qualche volta, il gran ragazzo, già mezzo cresciuto, che osa fare cose viziose che Giorgio non avrebbe mai pensato fare da solo, implica Giorgio e lo trae in vie storte. È adesso che dovrete esser la migliore amica di Giorgio, dimostrandogli che non è affatto cosa eroica, il gettar delle pietre nella finestra del povero ciabattino, od il torturare un cane o un gatto. Giorgio forse non vedrà che male ci sia in questo. Forse penserà che il ragazzo piu grande è molto scaltro ed intelligente, perchè può scappare cosi presto e schivare il poliziotto. Ora, prendete a parte Giorgio, e spiegategli che nessuno è un eroe, quando è crudele o cattivo. Incoraggiatelo invece a giuocare coll'innocente "Jim l'Irlandese", che va in Chiesa ed adora la sua "Mudder" (Mammina).

E Bettina pure, corre certi pericoli nelle amicizie sue, pericoli che dovete cercare di vedere e di capire. Neppur Bettina potete tenere tutto il tempo sott'occhio ed è necessario che sappiate chi si lega in amicizia con essa. Le bimbe piccine sono per lo più buone imitatrici e seguono facilmente l'iniziativa di qualche ragazzina più grande che ha saputo eccitare il loro interesse. Forse sarà il modo di vestire di quella ragazza piu attempata, o il suo modo di parlare, o qualche cosa di misterioso che pare che sappia, e di cui essa sussurra continuamente. Le Bimbe piu piccole l'ammireranno e l'adoreranno e cercheranno di imitarla. Se questa ragazza fosse per avventura una della quale la compagnia fosse nociva, potrebbe far molto danno alla vostra piccola Bettina.

In questo paese, ove siamo tutti sempre tanto affrettati ed occupati, è d'uopo che siate in relazione costante con tutti i vostri figli. È un lavoro duro, e che prende molto tempo, ma è la vostra sola salva-guardia. Incoraggiate gli amici dei figli vostri a venire in casa vostra, e riceveteli bene quando vengono. Imparate l'Inglese, cosicchè possiate capire tutto quello che essi dicono. Speriamo che tutti i ragazzi desiderano di essere sinceri e veraci, ma pur troppo sappiamo che, certe volte, questo non è il caso. Come potete rischiare la moralità dei figli vostri, semplicemente perchè non sapete l'Inglese?

Certi ragazzi usano un linguaggio vile fra di loro. È rivoltante perfino agli orecchi gia accostumati ai diversi significati della vita; cosa non deve essere di tragico quando queste cose sono sentite per la prima volta da un ragazzo o da una bimba innocente, a faccia angelica? Vidi un giorno due bei angioletti biondi, che parevano non avere piu di sette a nove anni, seduti lungo una via, ascoltando un gruppo di giovanotti scostumati che parlavano frà di loro. Nessuno pareva badare a quei due cari bambini, seduti lì, ascoltando un linguaggio e vedendo cose che farebbero rabbrividire un vecchio!

Compagni sani sono il più grande aiuto che possiate avere pel benessere dei figli vostri. Giacchè siete costretta a perdere di vista, per delle lunghe ore ogni giorno, i figli vostri, fate amicizia coi loro amici, per poter sapere tutti i loro sogni e le loro speranze. Se questa amicizie sono buone e tendono ad elevare il carattere, incoraggiatele il più che potete. Fate anche di più, coll'amalgamarvi con loro. Conducete la piccola "Maria la Polonese" dai vostri amici Italiani ed alle vostre riunioni; cercate di far la conoscenza della Madre di Maria pure, ed aiutatevi a vincenda coi vostri problemi. Essa pure, è straniera in questo paese, si sente isolata quanto voi vi sentite sola. In questo modo, saluterete insieme il Nuovo Mondo, per mezzo dei Figli vostri, e sarete altere tutte e due di ciò che saranno diventati questi vostri figli, a cagione del vostro intendimento e della vostra buona direzione. Voi amerete questo Mondo Nuovo, che vi ha dato tante belle opportunità pel vostro Giorgio, per la vostra Bettina e per ogni altra Madre ed i suoi Figli che vengono da noi.

La Bimba di dieci anni.

Q UESTA mia bimbetta di dieci anni diventa ogni giorno di più come un indovinello per me. Non la posso più capire affatto. Era cosi buona e dolce; faccva sempre quel che le si diceva, ma adesso. . . . Be'! mi è diventata un rebus che non sò spiegare."

Si, Madre! La vostra Bettina Italiana e tutte le bambine di dieci anni nel mondo sono tutte degl'indovinelli, a dieci anni. Le Madri Americane hanno preciso le stesse difficoltà colle loro bambine, a quest'età. Ma non cessano di cercare la soluzione di questo rebus. Esse lavorano indefessamente, finchè, alla fine, non arrivino a scoprire quale ne sia la causa.

È un po' più facile per la Madre, Americana di nascita, di capire la sua bimba a quest'età. Essa stessa era un piccolo rebus di dieci anni, in questo paese, qualche tempo fa. E per questo, essa intende un pochino meglio come stanno le cose colla sua bambinella, che ha dieci anni oggi Essa parla Inglese, ciò che non avete ancora imparato a fare, e questo è un altro vantaggio che essa su di voi.

Tutte le Bambine di dieci anni nel paese, Bettina non esclusa, stanno ancora sognando. Il fatto stà che vivono in un mondo di sogni loro, che si fabbricano da se. Niente par loro importante, o reale, se non è contenuto nel loro piccolo mondo egoistico dei sogni loro. Ed è per questo che nulla, a questo periodo, ferma i loro pensieri o la loro attenzione.

Abbiamo fabbricato molti nomi coi quali cerchiamo di spiegare a noi stessi la ragione della condotta di tutte le Bettine a dieci anni. Le chiamiamo testarde, sognatrici, capricciose cattive, malvagie, troppo quiete, e troppo rumorose. Oh! abbiamo molti modi di sgridare le nostre bambinelle di dieci anni. Facciamo tutto, eccetto cercare di capirle e di simpatizzare con loro. Venite, cerchiamo di metter le nostre teste insieme e di vedere che cosa c'è, che cosa succede e che cosa vi è di male. Volete?

Prendiamo per esempio, una bambina sbadata come la vostra Bettina. — Pare impossibile, ma non sembra che vi riesca di correggere le sue abitudini. Essa è proprio disperata quando il suo vestitino nuovo si strappa il primo giorno che se lo mette. Essa supplica: "Non l'ho fatto a posta, Mamma!" e promette di far meglio la prossima volta. Ma essa non si ricorda la sua promessa. Il giorno seguente, essa è da capo nel suo mondo dei sogni, ove non vi sono chiodi brutti o punte sporgenti per strappare gli abiti. Il suo vestito è strappato da capo, ed è una povera bambina piangente che torna a casa, da Mamma.

Questa volta, poi, vede subito, scritto chiaro nella faccia di sua Madre, che non se la passerà liscia. Ci vede la promessa di una sculacciata, o dell'esser mandata a letto senza cena; e, per questo, essa butta le braccia intorno al vostro collo, e piange così dirottamente che siete per forza costretta a perdonarle, baciandola. Questo è bellissimo per tutte e due, per voi e per essa. La vostra Bettina è un miglior rebus che non siano molte altre bimbe di dieci anni. Le Madri di quelle bimbe spesso le sgridano troppo forte, e le bimbe si rincantucciano ostinatamente, e rifiutano di escire e di ammettere che sono dispiacenti.

Uno strano cambiamento si palesa in tutte le piccole Bettine a quest'età. Esse si trovano in una confusione mentale che precede l'età della pubertà. In questo periodo, esse sono facilmente trasportate dalle proprie emozioni, ed hanno bisogno di uno freno simpatico ma fermo, da mani esperimentate.

Se voi non formate bene il carattere della vostra Bettina a quest'epoca, rischiate di subirne le cattive conseguenze per tutta la vita. Gran parte della sua condotta attuale vi guiderà circa il modo di agire con essa in avvenire, quando il cambiamento dall'infanzia all'età matura sarà compiuto.

Oggi, la vedete ancora come una Baby. E sotto molti riguardi, essa lo è ancora. Essa giuoca con maschi della sua età, affatto inconscia della differenza di sesso. Essa non conosce nè sente nessuna differenza. Guardatela piuttosto, rotolare giù da una collina di margherite col fratello suo e con gli amici di questo, in una bella giornata d'estate. Guardate come le sue sottanine volano sù, al di sopra del suo capo, e le sue gambine ossute, nude quasi fino all'anca! Se ne preocupa essa? E essa conscia di nessuna mancanza di decoro? Essa? No!

E un altro giorno, guardatela arrampicarsi attraverso di una chiusura, coi maschi; eccola che va, colle sottanine riprese intorno alla testa se ve ne è bisogno. Sù, sù, essa monta, ed oltrepassa l'ostacolo. Eccola in cima alla chiusura, a quattro zampe, — e le sottane sopra la testa. Col petto in terra, essa si sdraia sulla terra. Scoppia dal ridere a vedersi così grulla. I maschi ridono con lei, e par loro che sia una buona camarata che valga quasi quanto uno di loro. Essi l'aiutano a rialzarsi, scuotono la polvere dalle sue vesti; Tiran giù le vesti stracciate e tiran sù le calze bucate. Poi, tutto lo sciame di inconsci compagnoni s'invola per andare a notare nel rivuletto, o per un'altra diversione, tutti insieme.

Nè Bettina nè i ragazzi coi quali essa giuoca sono consci del fatto che la Natura stà effettuando in loro il suo magno piano. L'attrazione dei sessi c'è, malgrado che i ragazzi innocentemente giuocano insieme, non ne siano consci. Bettina vi proverà questo in mille maniere che dovete imparare a riconoscere da voi. Essa è veramente per se stessa, a dieci anni, un rebus quanto lo è per voi, e non sa essa stessa perchè fa questo o quello. La vedrete cambiare atteggiamento molte volte in una giornata, e dovete provarvi ad intenderla in ognuna di quelle disposizioni. Così essa verrà sempre da voi per il consiglio del quale ha tanto bisogno ora.

Ed adesso, lasciate che studiamo il vostro piccolo rebus da un angolo totalmente diverso. Oggi, essa non guarderà neppure nè suo fratello nè i suoi amici. Non vorrà affatto escire con loro. Se il fratello cerca di deciderla ad andare con loro, essa gli risponderà con poco buona maniera che essa odia il modo ruvido col quale giuocano i ragazzi. Poi, insisterà che odia pure i ragazzi stessi. Il lusingarla od il canzonarla non avrà nessum effetto oggi. Essa è occupata.

Essa sta leggendo qualche cosa che l'interessa molto più delle preghiere o del sarcasmo di suo fratello. Guardiamo un po, gentilmente, Mamma, e vediamo che cosa è che interessa Bettina così tanto oggi. Racconti di fate, naturalmente. Lo dovevamo indovinare. Adesso, essa è nel suo mondo dei sogni, che, a dieci anni, le sembra tanto reale. Forse, in questo minuto, essa si vede come la Principessa Indiana che è tanto bella che centinaie di innamorati hanno sospirato e languito per la sua mano.

Se guardate, dopo qualche ora, da una fessitura nell'uscio, quando Bettina avrà finito di leggere il suo racconto delle fate, la vedrete probabilmente addobbata in una delle vostre vesti a coda lunga. Una tenda da finestra sarà avvolta artisticamente intorno alla sua testa, per raffigurare il lungo velo fluttuante che le principesse dei racconti delle fate portano sempre. Ai bracci avrà dei gioielli d'oro e di pietre preziose risplendenti rappresentati da anelli da tende e smeraldi di carta metallica. Oh, abbiamo tutte veduto le Bettine delle case nostre divertirsi un mondo. Che nutrimento mentale sono per loro, quelle storie delle fate! Qualched'une fra noi non hanno attaccato veruna importanza al travestimento di Bettina in Principessa Indiana, e l'abbiamo preso per un innocente divertimento. Ma in esso sta nascosto, però, un intento significativo.

La mia idea personale è che abbiamo molti libri molto migliori dei racconti delle fate per le Bettine di dieci anni. Questi non sono sempre innocenti. Qualche volta essi sono effettivamente immorali e nocivi di carattere. Noi adulti, li potremmo leggere, se fossimo stolti abbastanza per farlo, e non ci troveremmo che del divertimento. Ma i ragazzi hanno bisogno di qualche cosa del quale si possano servire nel vasto giuoco della vita che aspetta loro.

I racconti delle fate sono per lo più totalmente stranieri alla vita, senza realtà, stimulano soverchiamente l'imaginazione e sono stupidi nella loro assurdità romanzesca. Cercate di apprendere qualche cosa delle nostre librerie pei fanciulli, Madre. Andate a quelle librerie da voi stessa. La bibliotecaria è una delle migliori vostre amiche americane. Ditele l'età ed il temperamento di Bettina. Essa vi capirà, che parliate Inglese o Italiano. La sua educazione è stata diretta sulla soluzione di questi stessi problemi. Lasciate che essa vi indichi una buona listina di libri per il vostro piccolo rebus. Nel fare della bibliotecaria Americana un'amica vostra, Bettina sarà indirizzata nella via che la condurrà a leggere dei buoni libri e la sua vera educazione sarà incominciata.

Malgrado che Bettina sia ancora una bambina, in questo momento, Bettina si trova proprio sul limite della pubertà, ed è d'uopo che la guardiate adesso contro i pericoli. Ciò che osservate adesso, è il processo di "fra mezzo". Questo si dimostra in questi cambiamenti spontanei — dalla bimba inconscia e di propensità quasi maschili, alla quale piacciono i giuochi virili, — alla bambina sensitiva ed imaginativa che anela il suo "Principe "Charmant" ", che deve venire per portarla via nel suo castello di fate, sù in cima ad un monte. Questo castello è tanto, tanto lontano dall'appartamentino di cinque stanze ove ci sono sempre dei piatti da lavare! dove Mamma è sempre tanto stanca e sopraffatta dal lavoro, e dove il Baby piange ed è di cattivo umore tutto il tempo.

Adesso consideriamo un solo altro esempio delle molte disposizioni in cui forse troverete la vostra Bimba diecenne. Osservatela a giuocare colle sue bambole, un bel giorno, e voi vedrete un altro aspetto del vostro rebus di dieci anni. Fate che essa non vi veda, o non vi riescirà ottenere informazione alcuna. Forse, oggi giuocherà alla "Mamma" colle figlie sue bambole. Essa le bacia e le culla finchè non si siano addormentate. Più tardi, essa le veste e le cura come una vera Madre. Poi, il giorno seguente, essa dice che non giuocherà mai più colle bambole finchè vivrà. Esse non sono che giuocattoli per i Babies, ed essa le detesta.

Adesso, cara Madre Forestiera, questi cambiamenti repentini da una cosa all'altra, sono un segno sicuro del cambiamento graduale di Bettina, che esce dall'infanzia per diventare una donna. Un bel giorno, le gambine ossute e diritte si riempiranno e metteranno delle belle curve. Presto Bettina non si strapperà più il vestito. Essa prenderà pensiero degli abiti suoi e vi dirà ciò che le piace portare. Molti cambiamenti si succederanno, ed il vostro rebus si scioglierà da se. Un bel giorno, vedrete una bellissima ragazzina davanti a voi, invece della bimba che ci vedete adesso.

Non vedete, o Madre, che questo momento è una via di traversa pericolosa nella vita della Figlia vostra? Essa stessa vi da i segnali di pericolo. Voi sola, da buona e vera amica e da ingegnere provetto, dovete imparare a leggerli. Se mettete della dolcezza ed i raggi di sole della perfetta intelligenza, nella vita della piccola Bettina, oggi, vedrete una brava e bella donna emergere e svilupparsi da questa curiosa bambinella, che si leverà e vi chiamerà "Benedetta".

La Figlia vostra a quattordici anni.

H! avete un ricevimento! Mi dispiace tanto di esser venuta oggi. Ah, Bettina compie i suoi quattordici anni oggi, e volete che mi trattenga qualche minuto? Grazie. mi farà piacere di stare un pochino con voi. Ma, vi prego, non cantate "Sole mio" o "Addio Napoli" o nessuno di quelle belle canzoni Napoletane, che mi rendono pazza di desiderio per il golfo di Napoli e l'azzurro della baia del Vesuvio. Se le cantate, mi scorderò dicerto che non sono che una "Signora Dottoressa Americana", che, come tutti gli Americani, non ha nè sentimento per la musica ed il canto, ed è tanto fredda!

No, Madre, non siamo veramente freddi, ma non abbiamo il vostro modo latino tanto dimostrativo, di far vedere ciò che sentiamo. Sentiamo veramente, molto profondamente, e so che ci amerete quando ci conoscerete meglio.

Ma quante chiacchiere a proposito degli Americani, e non una parola sulla nostra piccola Bettina. Dovrebbe oggi occupare il centro del palco scenico, se mai in vita sua. Bettina, vieni quà! Lascia che ti guardi. In fede mia, sei una ragazzina cresciuta. Dove è la mia bambinella d'ieri? Come sei bella oggi. La tua veste della festa natalizia ti tocca la punta delle scarpe ed i tuoi capelli sono delicatamente appuntati sulla testolina. Veramente, Bambina mia, sei cresciuta e diventata donna durante la notte.

Prima che ce ne accorgiamo, ci saranno degli ammiratori, dei ricevimenti e degli innamorati. Poi, un bel giorno, vi saranno delle nozze. Ma vieni, Bettina, non arrossire e sogghignare. Sai bene che ciò che dico è vero. Noi, le Madri di oggi, eravamo bambine come voi, tempo fa, ed i nostri pensieri erano gli stessi dei vostri.

Non vi urtate, Madre, di ciò che dico alla vostra Bettina di quattordici anni. Non sono spudorata. So bene che non parlate di amore o di innamorati alle vostre ragazze in Italia, Ma adesso, siamo Americani. Lasciate che si cambi l'antico adagio: "Quando siete a Roma, fate come fanno i Romani" Qui, è cambiato in: "Quando siete in America, fate come fanno gli Americani". Questo è veramente un consiglio splendido. Le Madri Americane hanno sistemato la sorte di famigliarità da camerati colle figlie loro che rende possibile per loro il discutere ogni sorte di soggetto. Non esiste nessuna falsa modestia fra Madre e Figlia. Tutto può essere discusso quando esse sono sole, fra Madre e Figlia. La vostra bella Bettina dagli occhi neri avrà bisogno di quest'amicizia intima a partire da questo suo compleanno, coi suoi quattordici anni, finchè essa non avrà un focolare suo.

Prima di tutto, Madre, la vostra Bettina è nata Italiana. Questo vuol dire molto. Le ragazze dei paesi tropicali Latini sviluppano e maturano molto più presto di quello che non facciano quelle di parentaggio Inglese-Americano o Danese-Americano. Guardate al bel corpo ben formato di Bettina, e vedrete la differenza al primo colpo d'occhio. Bettina è cresciuta su larghe proporzioni. Essa è piuttosto grassotta ed ha il petto perfettamente sviluppato. Il lavoro di Madre Natura per essa e praticamente finito. Il corpo suo non ha che delle curve, tutti gli angoli e le goffaggini sono attenuate in una grazia arrotondata donnesca. Bettina e' un modellino perfetto di una donna, — esattamente come era sua Madre a quattordici anni, laggiù a casa sua, nel suo paese in Italia, quando essa aveva quattordici anni.

Voi eravate protetta molto meglio di quel che non lo sia vostra Figlia oggi. In Italia, eravate fra gli amici vostri. Probabilmente, non lavoravate fuori della casa "vicina". Vostra Madre, malgrado che non vi dicesse nulla del gran piano dell'esistenza nostra, pur vi guardava e vi proteggeva da ogni danno. Essa vi accompagnava ovunque, e non avete mai potuto rubare una sola parola con un bello straniero del villaggio vicino, dagli occhi bruni, malgrado che egli ci si provasse con tutto l'intento. Avete mai scambiato un tantino di sorrisetto, desiderando che gli occhi della Madre vostra non fossero così acuti? Probabilmente, lo avete fatto, e non c'era nessun male. Non è che naturale di desiderare di godersi la vita, di assaggiare di un po' di poesia e di sognare sogni di bellissima fantasia, quando uno ha quattordici anni . . . ed è adolescente.

Bettina è adolescente oggi. Adolescente vuol dire essere poetica, conscia di se stessa, un briciolino capricciosa ed un pochino vana. Vuol dire di contemplare timidamente l'avvenire, là dove la moglie e la madre si rizzeranno un dì. Guardiamo se possiamo seguire lo sviluppo di Bettina da quando Madre le sussurrò all'orecchio per la prima volta la revelazione dell'aurora della sua natura di donna. Tutte le ragazze si conducono quasi lo stesso a quest'età. Se avete osservato la condotta di qualched'une di loro a questo periodo, sapete abbastanza bene come si comportano tutte.

Il primo segno dell-aurora di abdolescenza e della giovine maturità muliebre e venuto molto lentamente. Pareva che cominciasse piuttosto colle gambe di Bettina che colla sua testa. Probabilmente, nel tornare da scuola un bel giorno, essa ha annunciato inaspettatamente che i suoi vestiti erano troppo corti. Non voleva che i ragazzi guardassero le sue gambe. Bene, questo era il primo segno, Madre. Quando ha scoperto, Bettina, che possedeva delle gambe? La settimana passata essa non lo sapeva. E cosi, avete allungato il suo vestitino, ed il primo passo nel cambiamento dall'infanzia alla pubertà era fatto.

Non molto tempo dopo, avete scoperto Bettina che stava spazzolando e spazzolando i suoi capelli e cercando di appuntarli sù, come una ragazza cresciuta. Quando le avete domandato, molto sorpresa, ciò che questo significava, essa vi ha risposto "Sono slanca di aver i capelli che mi cascano sul viso, e li voglio appuntare in modo che non caschino più". Questo veramente, voleva dire un mondo. Bettina aveva fatto il secondo passo verso maturità.

Poi, vi sono i suoi amici. Vi ricordate, non è vero, come, pochissimo tempo fa, Bettina giuocasse coi ragazzi, affatto inconscia di nessuna differenza fra essi e lei stessa? Ma questo si cambiò molto presto. Un giorno, vi siete accorta che Giuseppe, il piccolo compagno di giuoco della casa dall'altra parte della strada, è diventato molto cortese per Bettina. Adesso, porta i suoi libri, e le compra dei gelati e dei dolci. Il Fratello Alfredo, che non era adolescente, fu molto piccato di perdere i suoi due compagni di giuoco allo stesso tempo. Egli non poteva capire questa nuova deferenza per Bettina. Ed essa non si curava affatto di lui in questo momento.

Un bel giorno, Alfredo crescerà lui pure. Allora, si guarderà attorno per la sorella di Giuseppe o per qualche altra ragazzina per la quale egli possa comprare dei gelati e dei dolci. Ed egli non permetterà a sua sorella di immischiarsi nei suoi affari a questo punto. Ma oggi, parleremo di Bettina, e qualche altro giorno, discuteremo i nostri ragazzi.

E triste per noi Madri, la realizzazione del fatto che i nostri Piccini, che ieri ancora dipendevano da noi per tutti i loro bisogni, sono, oggi, uomini e donne cresciuti. Ma questo è la vita, Madre. I figli nostri **devono** crescere e staccarsi da noi. Probabilmente ogni Madre, in questo mondo intero sparge qualche lagrima quando si avvede di aver perso la sua piccinina (Baby Girl) e che si rende conto dell'accaduto.

Bettina e' felice nella sua nuova alterezza di adulta. E così voi vi proverete ad essere felice con e per essa. L'intendere tutto riguardo il benessere dei Figli proprii, è uno dei trionfi ed una delle difficoltà della Maternità. In quest'epoca importantissima, dovrete compiere ciò che sembra essere veri miracoli. Le ragazze sono cosi sensitive, a quattordici anni.

Se non vi siete comportata molto gentilmente e con gran cura, con Bettina, durante la sua prima giovinezza (infanzia), cosicchè essa sia certa che la capite, sarete certa di perdere la sua amicizia in questo frangente. Essa si ritirerà da voi e dagli interessi di famiglia come un fiore che chiude i suoi petali all'avvicinarsi della notte. E restera allontanata da voi, se non vi dimostrate molto affezionata e simpatica.

In America, dovrete fidarvi all'onore di Bettina più spesso di quel che non fosse necessario per vostra Madre di fidarsi al vostro onore. Non la potete seguire ovunque essa vada, nè potete forzare una giovane ragazza a restare troppo in casa. Sarete forzata di lasciarla cantare, giuocare, ballare ed esprimere se stessa quasi come se fosse nata in America.

Lasciate che Bettina abbia degli amici per bene. Incoraggiateli a venire a casa vostra. Mostrate di prendere molto interesse nei suoi amici ed imparate a parlare Inglese. Se no, non saprete che cosa essi dicano, facciano o pensino, in questa strana America. Ed è d'uopo che lo sappiate. Vostro Marito non puo guadagnarsi la confidenza di Bettina come lo potete voi, anche se parla Inglese. Nessuno può aiutare Bettina come lo potete voi, Madre, negli anni pericolosi che stanno per venire.

Tutto sarà giusto come questa festa di nascita. Come è felice ed allegra, Bettina, oggi. Essa è proprio radiante. Le gote sue sono come delle rose Italiene. Giuseppe l'ha rimarcato pure. Oh, si, lo rimarca. Ed altri uomini se ne avvedranno.

Nel primo eccitamente del sentirsi una bella e giovane donna, e di avere degli uomini che sono tanto pronti e che glielo dicono, Bettina è in pericolo di perder la testa. Ma non vogliamo neppur pensare che nessun danno possa sopraffare il nostro bello ed innocente fiore.

Per questo, Madre, la dovete guardare molto teneramente. Le dovete parlare adesso dei soggetti i più intimi della vita. Ditele che Dio le ha dato questo bel corpo perfetto, perchè Egli l'ha dotata, come tutte le altre donne, del sacro privilegio della Maternità. Presto o tardi, essa adempirà il suo destino. E per questo dovete preparare la sua mente e dirle gentilmente e dolcemente come sono meravigliosi i piani di Dio per le sue "Donnine". Ditele che il suo corpo è sacro e che è dovere suo di conservarlo puro, pulito e sano.

Poi, qualche giorno, il vero "Principe" del racconto delle fate verrà, e Bettina e Voi, Madre, ve ne rallegrerete insieme. E Dio, pure, vi guarderà di lassù e sorriderà. Egli sa e capisce tanto bene quanto sia stato duro per voi qui, tanto lontana dalla vostra cara Patria.

Vostra Figlia, nata in Italia, che lavora in Fattorie Americane.

LA Bimba sa che sua Madre è la sua migliore amica. Si attacca ad essa per salvezza, e sente che nessun gran male può accaderle finchè Mamma è vicina. Ma, quando lo sviluppo s'avvicina, le cose cambiano d'aspetto. Sfortunatamente, Madre e Figlia, a quest'epoca, della loro vita, non son sempre le amiche intime che dovrebbero. Questo avvèra in ogni paese. Ed è pur vero per voi, Madri Italiane, perfino laggiù, a casa vostra, nel vostro paese di poesie e di fiori. Ma è ancora molto più tragicamente ero per voi quì, trapiantate in un paese ove tutto vi sembra confuso e strano, ed in cui voi stesse state guardando, stralunate, questo mondo nuovo.

Bianca probabilmente prese a portar il busto e le scarpine a tacchi alti nella prima settimana che fu quà, e la guardaste, meravigliandovi, e scuoteste il capo, tristemente. Finalmente, in disperazione, vi accostaste sempre più alla vostra Colonia Italiana, faceste vieppiù parte della vita di questa Colonia, e cercaste di vivere il più possibile come vivevate nel vostro Paese, in Italia. Ma Bianca nòn fece così. Essa si procurò un posto in una grande Fattoria Americana, e prestissimo cominciò a considerarsi come Americana.

41

A partire da quest'epoca, voi e Bianca avete avuto sempre meno e meno in comune, e, finalmente, pare adesso che stiate per perderla del tutto. Ora, essa si rivolge quasi a chiunque piuttosto che a voi, sua Madre, per aiutarla nelle sue perplessità o nelle sue confidenze. Questa è una situazione bruttissima, ed è d'uopo che facciate tutto quello che potete per rimediarla. Può darsi che essa trovi amici eccellenti, la cui influenza sulla sua vita casalinga avvenire sarà forse la migliore imaginabile, ma può darsi invece che sia tutt'altro. E tutto un giuoco d'azzardo e non ne potete prendere il rischio. Dovete persuadervi che il vostro consiglio materno, dato ad intervalli, non basta per l'assoluta sicurezza della Figlia vostra. Quando essa lascia la casa, la mattina, altre influenze la circondano. Essa fa la conoscenza di persone nuove; incontra ogni specie di stranieri; amicizie eccitanti si sviluppano fra di essa ed uomini e ragazze della sua età. Non siete presente per consigliarla, e non sapete neppur ciò che accada. Un danno gravissimo può succedere, ed allora ne saprete tutti i dettagli, ma sarà troppo tardi. La vita, qui in America, è più dura per Bianca, che aveva già oltrepassato l'età dell'andare a scuola quando venne in questo paese, che non lo sia per le vostre figlie minori d'età. Essa è cresciuta in Italia, ed è più difficile per lei di capire le usanze Americane. Più i ragazzi sono giovani, e più è facile per loro di imparare linguaggi ed usanze (costumi). Se Bianca vuol imparare l'Inglese, deve studiare la sera, quando è gia stanca della sua giornata di lavoro. Anela di fare essa pure, tutte le cose strane che vede fare alle ragazze Americane, malgrado che non sappia come cominciare. Se la scoraggiate, le farà probabilmente senza farvelo sapere. È molto meglio di cercare a capire ed a soddisfare i suoi desiderii.

Tutte le città grandi hanno delle scuole serali, come complemento alle loro scuole regolari ed ai loro "Settlements".

In molte Manifatture, vi sono delle stanze ed impiegano maestre per delle classi serali o date a mezzogiorno, per donne e ragazze. Spessissimo, ci sono dei ricevimenti ed altri divertimenti, provveduti per le alunne di queste classi speciali. Questa è un'opportunità splendida per tutte le donne e ragazze forestiere, di imparare la lingua Inglese e le usanze Americane. Persone responsabili sono sempre a capo di queste classi, e dovreste incoraggiare Bianca a frequentarne una. Fate voi stessa delle frequenti visite alla scuola di Bianca, per rendervi conto dei suoi progressi, e delle amicizie che essa sta formando.

Se Bianca non impara l'Inglese in una classe regolare, lo imparerà più adagio nella fattoria ove lavora. Il linguaggio usato in molte Fattorie non è buon Inglese, ma spesso e'profano ed urta qualunque persona Americana distinta che lo senta. Se Bianca non impara di meglio, imiterà questo linguaggio delle Fattoria. La ragazza che stà a lavorare accanto ad essa, forse esclamerà: "Oh my God" "Oh, Lord" o qualche altra espressione profana quando è sorpresa o arrabbiata. Bianca crede che tutti gli Americani devono parlare cosi, ed essa si servirà di queste espressioni la prossima volta che si troverà fra Americani. Tutti dimostreranno la loro sorpresa ed il loro disgusto, ed essa si domenderà il perchè. Essa non sa che un'espressione simile non è mai usata da gente colta.

I forestieri sovente giudicano la gente dal linguaggio da essi usato. Se sentono una ragazzina usare dei termini profani, credono subito che essa sia una ragazza per bene. La ragazza, invece, forse non fa che imitare

43

il linguaggio che sente usare là dove lavora, senza rendersi conto che sia profano. Ma gli stranieri non lo sanno. Può darsi che Bianca sia quella ragazza della quale si stà parlando. Voi sapete bene che Bianca è una ragazza per bene e buona, ma gl'indifferenti non la conoscono altro che dal linguaggio che usa. Vi piacerebbe che Bianca passasse per volgare e rustica, semplicemente perchè non ha potuto attendere le classi ed imparare l'Inglese in una maniera regolare?

Bianca vi sfiderà quando insistete dicendo che essa non può formare amicizie con uomini finchè non sia, a giudizio vostro, in stato di maritarsi. Essa vi dirà che le ragazze Americane sono famigliari con uomini coi quali passeggiano liberamente da buoni camerati. Voi siete stata educata in Italia con delle idee molto differenti e vi è semplicemente impossibile di sentire che le maniere Americane stanno bene. Ma troverete impossibile di tenere le ragazze chiuse in casa, in America, come è solito per le Madri farlo in Italia. Il solo modo che abbiate di proteggere Bianca, è di cercare di capire le relazioni che esistono in questo paese fra giovani e ragazze.

Qui in America, Bianca si trova in una posizione molto differente di ciò che non fosse in Italia. Qui, le ragazze sono più libere, e chiunque non ci è avvezzo è atto a lasciarsi abbindolare per ignoranza. Il Padrone, in una Fattoria, cerca spesso a prevalersi di una ragaza forestiera che non capisce le maniere Americane. Egli dice loro che stà benissimo, qui in America, che un uomo passi il braccio intorno alla vita di una ragazza o che si prenda anche altri privilegi, ed essa suppone che deve essere così, e che egli le dice la verità. Voi dovete scoprire quale sia la vera situazione, e le dovete spiegare con molta cura che deve risentirsi contro queste attenzioni degli uomini. Allora, essa saprà cosa rispondere al Padrone od a qualunque altro uomo che cercherà di essere troppo famigliare con lei.

Le ragazze Americane sono educate a credere che le amicizie loro coi giovani non hanno nulla che fare con l'amore o col matrimonio. Le ragazze in questo paese remano, nuotano, prendono parte ai giuochi, studiano od hanno qualunque specie di associazione sana con uomini; i limiti imposti dalla modestia e del ritegno le sono insegnati in età tenera. Se esse si confidano liberamente colle loro Madri e che le possono parlare francamente circa i loro amici, e condurre questi amici a casa loro, nessun male può accadere. Una amicizia sana e l'abilità di conoscere e di discernere lo stato dell'animo loro, ne saranno i soli risultati.

Vi sono qualche volta, nelle fattorie, delle donne più attempate, e di mente leggera, che danno dei consigli molto malaccorti alle ragazze, ed offrono loro delle informazioni che nascondono suggestioni nefande. Dovrete avvisare Bianca di non le stare a sentire. Spesso sono esse che aiutano a far cascare le ragazze in posizioni critiche. Una Figlia dovrebbe sempre rivolgersi a sua Madre per i consigli dei quali ha bisogno, e la Madre dovrebbe esser sempre pronta a darle questi consigli nel modo il più completo ed il più sano.

Le ragazze Americane amano le loro stanzuccie. Bianca desidererà pure avere la sua stanzetta. Dovreste cercare di procurarle un posticino ove essa possa sedersi e sognare, o leggere o scrivere, o fare qualunque di quelle piccole cose che influiscono tanto sulla felicità di una ragazza giovane quì in America. Per lo meno, cercate di avere un posto, nella vostra casa tanto piena, ove essa possa ricevere i suoi amici. Questo vi sembrà essere senza importanza, ma renderà Bianca molto felice.

Adesso, cara Madre di Bianca, non vi pare che valga la pena di fare tutto questo per la Figlia vostra? Essa sarà tanto felice, quando si avvedrà che cercate realmente di intendere i problemi della sua vita. Potete essere sua Madre e sua sorella maggiore allo stesso tempo. Se succedete in questo, non avrete più bisogno di impensierirvi per Bianca. Essa non sarà che troppo felice di venire da voi quando avrà bisogno di simpatìa o di consiglio.

I risultati sul Figlio vostro degli anni che passano.

"A gioventù deve avere i suoi capricci". Questo è un vecchio adagio che è usato in quasi tutto il mondo, riguardo ai giovani. È tanto facile, per i genitori, di rinunciare a qualunque responsabilità per la condotta dei loro figli, col dire: "Che volete! esso è giovane! Deve avere i suoi capricci. Non vi tormentate per lui; Ne escirà benissimo alla fine!"

Ma non sapete, o Madre, che nessuno può seminare gramigna e raccogliere grano? Questo è vero per la vita umana come lo è per i campi. La Bibbia dice: "Seminate il vento, e raccoglierete la bufera." Un gran numero di uomini si sono dati alla disperazione più tardi, nella vita, perchè nessuno li ha frenati quando non erano che giovani scapestrati. I giovani pagano un prezzo tremendo, più tardi nella vita, per i peccati commessi da loro per l'ignoranza dei loro genitori.

Dio è tanto buono e la Natura è una cosi buona Madre protettrice, che, in qualche modo, molti, frà i nostri giovani arrivano a crescere fino ad una splendida virilità. Malgrado la nostra negligenza e la nostra facile indifferenza, essi riescono a diventare uomini validi. Ma la proporzione di questi giovani che sono negletti dai loro genitori, e, malgrado ciò, riescono a diventare uomini buoni, è molto piccola. Se le cose vanno male e che vostro Figlio, cara Madre Italiana, paga nella sua vita avvenire con torture ed infelicità per la gramigna che gli fu permesso di seminare, nè voi nè suo Padre potrete mai schivarne la responsabilità ed il biasimo.

Noi diamo troppa libertà ai nostri figli, in questo paese. Il biasimo di ciò che accade casca su di noi quanto su di voi, che siete nate in altri paesi. Le nostre Figlie ricevono molta attenzione e permettiamo spesso ai nostri figli di condursi come piu piace loro. È egli vero che i maschi sanno proteggersi meglio di quel che non lo sappiano le ragazze? Certo che no. Essi hanno bisogno della direzione dei loro genitori quanto le loro sorelle. Essi sono i futuri Padri della razza, esattamente come le sorelle loro ne sono le futuri Madri. Vostro Figlio, arrivato all'età della pubertà, non sarà onesto, buono o puro di ciò che non lo sia a quattordici anni.

I vostri Figli giovani stanno troppo tempo nelle strade. Certi d'uni fra loro non hanno lavoro definito. Altri vendono giornali o fanno da messaggieri. Ma tutti vedono e sentono cose, per le strade, che sono molto nocive. Molti ragazzi impiegati come messaggieri sono mandati in posti ove i parenti loro non anderebbero mai. Questi ragazzi accettano molto presto un'idea sbagliata della vita, che non si scorderanno mai più. Ragazzi che lavorano in commerci esercitati nelle strade, facilmente si abituano a spendere una gran parte dei loro guadagni in giuochi d'azzardo. E molto duro per loro di resistere a questa tentazione quando hanno il denaro in tasca.

Uno dei più grandi scrittori che abbiamo, che si chiama Marro, ed è Italiano, ci ha detto che i delitti fra ragazzi dai 12 ai 14 anni crescono enormemente di numero. I ragazzi di quest'età, che stanno molto tempo nelle strade, diventano molto scaltri. Essi cominciano a credere che il solo modo di diventare ricchi è di truffare, e di ingannare il prossimo. Da principio, non sono disonesti che in cose piccole, ma ben presto essi praticano la loro disonestà in cose importanti.

Non è veramente la colpa di questi ragazzi, se sono cresciuti così e che sono arrivati ad una vita viziosa. Furono negletti dai genitori loro proprio quando avevano bisogno della più gran cura. Ragazzi, come il vostro Alfredo, fra i dodici ed i quattordici anni, sono adolescenti e si trovano nel periodo il più difficile di tutta la vita loro. Si trovano precisamente fra l'infanzia e la virilità. Se Alfredo ve lo potesse spiegare, egli ammetterebbe che si trova in uno stato di mente curiosissimo. Le sue emzioni sono tutte frammiste fra quelle di un ragazzo e quelle di un uomo, ed egli ha adesso il più gran bisogno di suo Padre, del pastore della sua chiesa, o di un amico di età matura che gli possa parlare liberamente.

I ragazzi nascondono molto più di quel che non si creda i loro pensieri intimi. Un monte di cose gli si affacciano alla mente, ma non ne parlano con nessuno. Generalmente si sviluppano in età ancora tenera. Non c'è una epoca definitiva, per ragazzo, in cui esso diventi conscio del suo sesso, come esiste per la sorella sua. Certi ragazzi, a nove anni, capiscono cose che un altro ragazzo non sa a sedici anni.

Ecco una cosa interessante che maestri e lavoratori che hanno studiato i problemi di parenti di natività straniera, come voi, vorreberro che imparaste. È questa: Ragazzi dai dodici ai sedici anni, che hanno molta responsabilità a casa diventano uomini migliori di quel che non siano i ragazzi ricchi della stessa età. Perchè è vero questo? Potete voi indovinarlo? No? Bene! guardate nella stanza davanti di casa vostra, e ne vedrete la risposta. Ecco, vi è Alfredo. Esso ha quattordici anni, non è vero? Guardatelo camminare in sù ed in giù per addormentare il Baby. Che bella pittura!

Voi dite che questo lavoro non gli piace molto. Non suppongo che gli piaccia. Esso preferirebbe esser giù nella strada, giuocando alla palla o coi pattini a rotelle. Naturalmente. Non vi piacerebbe di più di andare a passeggiare nel bel sole d'estate o di andare all'opera, o di esser fuori comprando dei bei vestiari per voi stessa, che di essere dove siete? Tutti, perferiamo far le cose che ci piacciono.

Naturalmente, ragazzi devono avere un' opportunità di giuocare. Essi somigliano un tantino a degli animali selvaggi. Hanno bisogno di un modo senza rischio, che permetta loro di eliminare dal loro sistema lo "spirito dell'animale selvaggio". Essi hanno bisogno di essere domati, e diretti lungo la via, finchè non siano arrivati all'età virile. Ecco perchè i giuochi di sport, sotto la direzione di buoni maestri ed uomini savii, hanno tanta buona influenza nella vita dei ragazzi giovani, in una città grande. Ai ragazzi piace immensamente lo stare in compagnia. Essi si sentono il bisogno di correre, giuocare, di fare degli esercizii e di gridare. Senza nessun proponimento qualsiasi, essi fanno delle amicizie di ogni specie, semplicemente per avere la gioia della sociabilità e per giuocare. Ed è da quest'istinto del bisogno di giuocare che i maestri hanno sviluppato molti mezzi splendidi per lo sviluppo dei nostri ragazzi.

Le magnifiche attività sociali delle Chiese, Souole, Case co-operatrici (Settlements) vi possono essere di un aiuto meraviglioso, Madre. Prendetene conoscenza, ed incoraggiate Bettina ed Alfredo a frequentarli. Prima di tutto, è come l'avere due case in una. La casa vostra è troppo piccola per aver molto spazio per giuocare. Per di più, siete troppo occupata per poter dirigere i giuochi dei Figli vostri. E così, ecco la seconda parte di casa vostra,—il ginnasio gratis, le Case co-operative (Settlements) e le vostre Chiese, ove potete mandare i figli vostri per tutte queste cose.

La corte è grandissima. Alfredo potrà buttare la palla nel paniere ed esercitare i suoi muscoli sulle sbarre del "Gim". poi, gli sarà dato di poter far un bagno a doccia per rinfrèscarsi per benino dopo il riscaldamento del giuoco. Un Maestro dall'espressione felice—che non è che un ragazzo cresciuto esso stesso,—sarà il fratellone di Alfredo. Come l'amerà Alfredo, e come l'ammirerà, e, meglio di tutto, come si sforzerà ad imitarlo!

I ragazzi amano la forza, e spesso, essi ammirano i pugilatori semplicemente perchè sono forti ed hanno muscoli bene sviluppati. La forza è buonissima, ma un buon carattere, e la nobiltà, generosità e gentilezza maschili sono ancora migliori. Nella ginnastica, i ragazzi possono imparare tutti e due nello stesso tempo. I maestri di ginnastica sono belli uomini che gl'insegnano ad esser forti. I ragazzi ammirano tanto questi maestri che presto cercano di imitarli. E così scoprono che un uomo può essere bello, grande a forte e restare onesto e diritto di mente e carattere.

Bettina ed Alfredo vanno allo stesso Centro di ricreazione. Mentre Bettina impara a cucinare ed a cucire, Alfredo si dondola su di un trapezio o sta a giuocare alla palla col paniere (Basket-ball). Adesso, sono le nove e mezza—tempo di tornare a casa. Alfredo ha fatto il suo bagno a doccia, si è messo il suo cappotto ed il suo capello, ed è pronto ad andarsene. Esso è gia uscito, con un gruppo di amici. Parlano ancora di un punto che è rimasto indeciso durante la partita. Tutt'ad un tratto, Alfredo si arresta bruscamente: "Mi sono scordato affatto la mia sorella! Bisogna che la conduca a casa; ma che bella serata per fare alle palate di neve! Potrei battervi tutti, a quel giuoco lì!"

Ed esso torna indietro. Bettina sta seduta sull'orlo di una seggiola, cucendo una cuffina di raso rosa per regalo di Natale per qualched'una. Alfredo si arrabbia col lei perchè non è pronta per andare a casa. "Oh, vieni, dunque! È tempo che le bimbe siano a casa ed a letto. Se non ti ho riportata a casa alle nove e mezza, sono io che sarò sgridato, e tu stai qui e cuci, e cuci, e cuci!"

Alfredo è del tutto disgustato dalla lentezza di Bettina. A quest'età, i ragazzi, per lo più, non amano pazzamente le sorelle loro. Ma la cosa importante, è che egli prende cura della sua sorella. Egli la porta a casa, e si comporta bene e da vero uomo in questo. Sarebbe stato facile per lui di lasciare stare Bettina ed andarsene gaiamente cogli amici suoi. Quest'educazione, che gl'insegna a guardare la sorella sua, è eccellente per Alfredo. Adesso, è il tempo opportuno, in cui esso deve imparare a rispettarla ed a proteggerla. Se gli si insegna ad apprezzare donne e ragazze mentre egli è piccino, sarà un buon marito ed un buon Padre quando egli sarà cresciuto.

La più gran parte delle Madri Americane guastano i loro Piccini ad un'età molto tenera. Vi siete mai trovata seduta in un carro, vedendo un bambino vestito benissimo dare dei calci a su'Madre perchè essa non faceva ciò che voleva lui? Adesso, che razza d'influenza avrà questa Madre sul figlio suo quando avrà raggiunto l'età di quattordici o sedici anni? Fa rabbrividire il solo pensarci.

Pochissime Madri capiscono i figli loro. Li sculacciano quando sono molto cativi. Questo, però vuol dire per lo più, quando Mamma essa stessa è nervosa o di cattivo umore, e non quando il ragazzo è più cattivo del solito. Forse, egli avrà fatto la stessa cosa molte volte prima, e nessuno si è avvisato di correggerlo. Ma oggi, perchè la Mamma è stanca, la condotta di Alfredo la secca, ed egli si becca una sculacciata! Questo non gli piace affatto. Egli se ne risente amaramente, ed in questo, egli ha ragione. Se ieri non faceva male a far così, non deve essere male neppur oggi. Ma ieri, la Mamma ed il Babbo risero, quando egli fece la stessa marachella, e lo trovarono intelligente e buffo!

È molto difficile per voi, cara Madre forestiera, di tener d'occhio vostro figlio in quest'America tanto affaccendata. Ma c'è qualche cosa che potete fare. Mandate vostro figlio in un Centro di ricreazione in una scuola o in una Chiesa. Tutti questi Centri sono gratis. Incoraggiatelo ad andarvi, ed andate col lui qualche volta, tanto per fargli vedere che vi interessate ai piaceri suoi. Qui, vostro figlio svilupperà buoni abitudini di contegno e di responsabilità nella vita. Più tardi, potrete fidarvi di lui ovunque, ed egli sarà tanto fiero ed altero della Mamma sua, che lavora tanto, e che ha sacrificato tutto pel bene e per la felicità dei figli suoi.

Quando Alfredo sarà cresciuto, egli aiuterà a portare il fardello della famiglia, come un gigante, forte e sano. Ed allora, cara Madre, paziente, stanca, e voi pure, Padre rifinito dal lavoro, potrete sedervi in una seggiola dai cuscini morbidi e riposarvi della vostra lunga lotta. I figli, nati in Europa, ma tirati-su in America, quelli frà questi, che hanno avuto una buona educazione nella loro giovinezza, hanno molta reverenza per i loro cari genitori, nati nel mondo vecchio. E, malgrado che amino l'America, e che siano Americani, pur gli restan memorie tenere dei loro giorni d'infanzia, laggiù a traverso l'Oceano.

Quando la Malattia Entra la Casa.

 COSÌ, il povero Luigino è tornato a casa malato, dalla Scuola, questo dopo-pranzo. Che peccato. Mi dite che ha 'sette anni, e che ha giusto incominciato il lavoro della prima Classe? Sono tanto dispiacente per voi, Madre! Le vostre giornate sono abbastanza piene di lavoro così, senza avere i figli malati. Posso vedere il caro piccino? Oh, egli è sdraiato nella stanza da letto, e Bettina, Alfredo e Baby Sam lo stanno guardando? Benissimo; Entriamo, e vediamo un po' se possiamo scoprire quale sia il suo male.

Come sei saldo, Luigino caro. Non è troppo calda, questa stanza, e le coperte tue, non sono esse troppo pese? Si? Allora, lascia che apra un pochino la finestra, e che faccia escire un po' di quest'aria calda e viziata. Allora, un po' di buon' aria fresca potrà entrare. No, Madre cara, l'aria fresca non lo farà infreddare. Non prendiamo freddo da troppa aria fresca nè durante la giornata, nè durante la notte, neppure quando dormiamo. Tutt'al contrario, sono le finestre chiuse ermeticamente e l'aria caricata di gas avvelenato e di germi che svolazzano nelle nostre stanze, che ci fanno infreddare e ci rendono malati.

Ecco! sentite come tosse e stranutisce Luigino. Questo sembra che sia una brutta infreddatura che si dimostra. Lascia che senta la tua testina e che ti guardi in gola, Luigi.—Ne ero certa! febbre e mal di gola, e le amigdale tutte coperte con macchie bianche. Mio Dio! che sbaglio, di lasciar che gli altri ragazzi vengano in questa stanzetta angusta con un piccolo malato! Ma qui, ci sono tutti i quattro insieme, che respirano la stessa aria malsana, piena di bacteria dalla gola ammalata di Luigi, per di più di tutti gli altri germi. Corrono un gran rischio di ammalarsi, loro pure.

I bambini sono molto suscetttibili alle malattie degli altri. Se volete assicurarvi contro un'epidemia generale nella famigliuola intera,—e frà i vicini per di più,— dovete immediatamente mettere il bambino malato in una stanza separata. Questo è ciò che chiamiamo "La sicurezza avanti tutto", qui in America. (Safety first). Non perdete mai nulla per essere troppo guardinghi. Se la gola malata di Luigino si risolve in un'infreddatura moderata che si guarisce con uno o due giorni di letto, a casa, Hurrah per Luigino! Tanto meglio per lui. Ma se, al contrario, la malattia si dichiara seria, e che sia contagiosa, avrete preso una precauzione saggia, tenendone gli altri ragazzi lontani.

Un bambino malato e di cattivo umore basta per qualunque Madre al mondo, a curare e tenerne di conto. Vedete, Madre, cura e protezione, protezione e cura e cosi da capo, sono i soli mezzi che altri abbiano trovato per tirar sù i loro figli qui, in America. Avete pure sentito anche in Italia, il vecchio adagio che spesso si ripete qui: Un'oncia di prevenzione vale una libbra di cura.

L'esperienza ci ha insegnato parecchie lezioni, a noi Americani. Abbiamo per lo più pagato caramente ogni lezione ricevuta, ma le abbiamo imparate, e questo è molto. Ed adesso, il nostro piu gran desiderio, cara Madre forestiera ed afflitta, e di rendere la vostra via in questo mondo nuovo più facile di ciò che non fosse la nostra quando noi arrivammo dal mondo vecchio.

Abbiamo imparato che è molto, ma molto più facile di prevenire le malattie dei nostri figli, che di spendere il nostro denaro, che ci è costato tanto a guadagnare, per le note dei Dottori e per le medicine. Per di più, infreddature e malattie costanti indeboliscono i nostri figli, cosicchè doventano vieppiù suscettibili e resistono male alle malattie. Avrete sentito spesso delle Madri che dicono: "Il figlio mio sembra che si prenda tutte le malattie che ci siano al mondo." E così capite che cosa vuol dire. Vuol dire molto probabilmente che il ragazzo ha ciò che chiamiamo noi poca resistenza al male. Egli deve essere pallido delicato, magro e nervoso.

E così, Mamma cara, lasciate che si mandino via gli altri Piccini dalla stanza di Luigino. Adesso, Luigi, devi essere bravo, e non devi piangere perchè bisogna che tu stia solo. Lo sò, è duro, di star a letto e di non avere nessuno che vi parli e vi diverta, quando uno si sente tanto male. Mi dispiace tanto, Caro. Cerca di essere un bell'uomino ben paziente. Se stai tranquillo oggi, e che tu fai tutto ciò che la Mamma ti dice di fare, presto starai bene da capo, e te ne anderai nella bell'aria fresca dell'inverno.

Lo vedi, Luigi, come Baby Sam e forte e grasso e felice, oggi. Egli non è ancora che un piccolissimo bambinello, e un baby come lui chiappa molto piu facilmente di te, un bel ragazzo grande con un buon appetito, tutte le malattie alle quali egli è esposto. Non vorresti vedere le belle rosine delle sue gote appassire e tutto il suo corpicino dimagrire perchè saresti stato troppo egoista per starne lontano per qualche giorno? No! Lo sapevo! Naturalmente, no. E cosi, Madre cara, finchè non avremo visto quanto sia seria la malattìa di Luigi, badate che gli altri ragazzi stiano fuori della stanza, che non vadano a scuola, e che stiano lontani dai loro camerati (compagni) di giuoco, nella strada. Ogni Madre ama i proprii figli come voi amate i vostri. È dovere vostro di cercare d'impedire i figli loro di prendere una malattìa per stare troppo vicini ai figli vostri, quando avete questa malattia in casa.

Certe malattie dei bambini sembra che vengano a tutti, piu presto o piu tardi, — la rosola e la tosse cavallina, per esempio. Ma questo non da il diritto a nessuna Madre di essere negligente. Nessuna Madre esporrà volentieri i figli suoi ad una malattia che essa sà fare strage nella vicinanza. Certe Madri sgomberano qualche volta per la paura perchè un ragazzo sullo stesso pianeròttolo avrà la tosse cavallina o la febbre scarlattina. Non possiamo biasimarle per questo. Queste malattie dei ragazzi sono pericolose e costano molto cordoglio ai parenti.

Spesso certe Madri, Italiane come Americane cercano di proteggere i figli loro alle spese di altri. Questo non è giusto. Le Madri che vivono in appartamenti angusti ed affollati, si danno quasi alla disperazione la devono tenere un branco di ragazzi sani e rumorosi in casa perchè uno dei membri della famiglia è malato. E così permettono ai regazzi di andare a scuola o di giuocare nella strada con una folla di altri ragazzi. In questo modo, un'epidemìa di rosolìa o di tosse cavallina o di qualunque altra malattia contagiosa può spargersi nella vicinanza. Questo non è giusto, non è veri? Se esponete i figli della vostra vicina a qualche malattìa essa farà probabilmente lo stesso con voi, pei figli vostri. Tutte le Madri dovrebbero essere in simpatìa l'una coll'altra, e cercare di proteggersi a vicenda.

E per questo, Luigino avendo mal di gola, è d'uopo che egli resti solo nella camera finchè il dottore o la guardia della Casa Co-operatrice venga e non dica che esso può uscire. Gli altri ragazzi devono astenersi dall'attendere la scuola, finchè il dottore non abbia deciso quale sia il male.

E qui veniamo ad unà altra questione importante. Quando deve decidere la Madre che il bambino è abbastanza malato da dover chiamare il Dottore? Questa è una questione molto dura per qualunque Madre. I ragazzi si ammalano dal mangiar troppo o da altre cause così leggere che le Madri qualche volta trascurano molto e chiudono un occhio sui primi sìntomi. Per regola generale, la febbre, il mal di gola od un' eruzione sono segni di pericolo coi ragazzi. Eppure perfino questi possono esser causati da qualche leggiero errore. In questo caso, una buona dormita e l'andare a letto senza cena, sono tutte le medicine necessarie. Ma non ne potete esser sicura, e nessuno, fuori di un dottore perito può decidere se la malattìa sia seria o no. E per questo, dovete ancora una volta seguire la regola che dice che la prevenzione vale meglio della cura. Mettete il bambino a letto, a parte dagli altri ragazzi e mandate a chiamare il dottore o la guardia-malati della Casa di Co-operazione. (Settlement).

Non cercate di amministrare medicine da voi. La vita del figlio vostro è troppo preziosa per servire di esperimento. Se Luigino non è molto malato, i vostri rimedii casalinghi soliti non sono necessarii. Egli risanerà senza medicine di sorte. Ma, se sembra più malato l'indomani che non lo fosse quando è tornato da scuola il giorno avanti, mandate subito per la guardia della Casa di Co-operazione più vicina. Essa saprà precisamente ciò che sarà bene suggerire. Le potete affidare Luigi. È stata educata per questa specie di lavoro e lo ama. Essa conosce i dottori e le farmacie e sa molto bene cio che si deve fare e come si deve fare. Essa vi darà saggi consigli circa l'evitare l'infezione degli altri figli vostri e di quelli dei vicini.

Essa sarà una buona, cara amica od una sorella affezionata per voi nell'ora del bisogno; Luigi le vorrà tanto bene de prenderà le medicine cattive che essa gli darà, senza neppur fare delle boccaccie. Essa è tanto buona e gentile coi ragazzi. Vi sarà di somma gratificazione l'averla in casa vostra. E, cosi, non vi spaventate, cara Madre solinga, in un paese nuovo. Avete molti amici che desiderano di conoscervi e di aiutarvi. Certed'une delle nostre leggi ingieniche vi sembreranno forse molto dure quando affetteranno la casa vostra, ma ricordatevi che sono fatte per voi e per me e per tutti, affinchè si cresca tutti forti e robusti per poter combattere con successo la gran battaglia della vita. Questo ci aiuterà a conseguire ciò che l'America brama con tutto il cuore dare a tutti quelli che approdano alle sue rive.

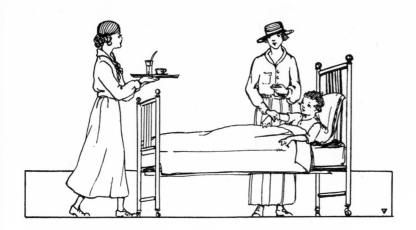

L'Associazione che si chiama Matrimonio.

COME, Amica mia, piangete? Siete infelice? No, non è niente? Piangete di gioia, mi dite? Oh, adesso, capisco. Vostra Figlia, Bettina, si è sposata ieri, e vi sentite isolata, senza di lei. Quanto comprendiamo le vostre legrime, noi tutte, Madre, Madre Italiane, Greche, Americane, o Cinesi! Qualunque Madre in tutto l'universo, che ha mai avuto una Figlia cresciuta e che si è maritata, e che lascia le braccia della Madre per quelle del Marito, tutte queste Madri sanno ciò che sentite oggi.

Vi sono certe specie di pene e certe specie di gioie che tutte le Madri hanno in comune. Le rendono tutte sorelle, non importa ove vivano o che lingua parlino. Quando le vostre Figlie lasciano il nido materno per un altro nido, in altra parte, questo è doloroso. Vostro nuovo Figlio puo essere il miglior uomo che esista sulla terra, ed il cuore forse vi canterà in seno per la grande felicità di Bettina, ma i lagrimoni pur vogliono scorrere, ed i vostri sorrisi sbiadiranno ogni tanto. È duro, il pensare che immenso cambiamento avverrà nella vita della vostra Bettina, dopo che l'anello nuziale sia stato posto e che i voti sacri siano stati pronunciati.

Non avete pianto, quando siete tutti andati all'Ufficio del Sindaco, nel Municipio, per assistere al matrimonio civile, non è vero? Lo so. Il senso di isolamento non si è palesato altro che quando il servizio di Chiesa è stato finito e che tutta la cerimonia era finita.

Non vi potete capacitare che Bettina è sposata. Pare impossibile che essa non sia più in casa vostra, nè sotto la vostra cura. Tutta la cara vita famigliare fra Madre e Figlia è cambiata, adesso. Bettina è adesso in un gran mondo nuovo che le appartiene, e dove la prima considerazione appartiene a suo Marito, e non più a sua Madre. Si, Mamma, tutto questo è ben duro. Non vi biasimo, se piangete. Noi, le Madri Americane, abbiamo pianto come voi piangete oggi. Molte, molte fra noi hanno sentito lo stesso isolamento ed abbiamo creduto che niente ci potrebbe mai consolare. Ma abbiamo imparato presto che le Figlie non lasciano veramente mai la Madre. Al contrario, molte Figlie non riguardano mai la Madre loro come la loro migliore (più cara) amica, finchè, essendo maritate esse stesse, si trovano faccia a faccia coi problemi della donna maritata.

E cosi, Madre, asciugate le vostre lagrime, e sparecchiate il resto del banchetto nuziale. Il bel vestito di seta deve esser riposto per un'altra occasione. Sono state nozze bellissime. Bettina aveva dei fiori ed un lungo velo bianco. I giovani hanno buttato una scarpa vecchia dietro alla carrozza per portarle buona fortuna, e l'amica intima di Bettina le ha buttato del riso avanti i piedi, per migliore fortuna. Le usanze Americane sembrano curiose, non è vero?

Bettina sarà di ritorno a casa in pochi giorni. Ed allora, verrà da Mamma per pregarla di aiutarla ad accomodare la sua casetta nuova. Voi l'aiuterete, al principio del suo nuovo viaggio nella vita, e le darete i consigli profondi di Madre e di Donna. Invece di essere Madre e Figlia, sarete due care amiche, che, tutte e due sapranno che cosa voglia dire la vita ed il matrimonio, e che si possono parlare a core aperto.

Oh, Madre, qualche cosa di meraviglioso e di nuovo è entrato nella vostra vita quando Bettina si è sposata. Adesso, essa verrà davvero da voi per essere consigliata, e seguira il consiglio che le darete. Essa vi vedrà tutt'ad un tratto, come siete realmente,—bella, buona ed eroica. Capirà adesso, perchè avete intrappreso ïl lungo viaggio attraverso l'Oceano cupo. Capirà che un amore profondo e lo spirito di associazione vi decise a viaggiare con tutta la familgia, e venire attraverso il mare, nel paese ove ïl Padre suo era venuto anni prima. Adesso capirà perchè vostro Marito rise e pianse allo stesso tempo in quel giorno meraviglioso quando vi incontraste di nuovo ad Ellis Island. Capirà perchè piangeste tutti e due—probabilmente, voi piangeste moltissimo, ed il Marito vostro si provò ad aver l'aria di non essere pazzo di gioia. Capirà perchè suo Padre vi tenne tanto tempo stretta stretta e perfino si scordò di rimarçare quanto erano cresciuti i ragazzi, in questi lunghi anni di separazione.

Si, cara Madre, molte cose si schiudono chiaramente alle nostre Bettine, quando l'amore le raggiunge e che loro stesse cominciano gli affari della vita in associazione, chiamata Matrimonio. Bettina amerà sua Madre, rifinita dal lavoro, con un amore nuovo a più profondo. Questo amore nuovo dirà, senza favellare,: "Oh, cara, cara Madre, adesso, capisco quanto sia stato duro per te e pel Babbo. Quanto voi due dovete avere sognato, e sperato e sofferto e quanti piani dovete aver fatto in tutti quei lunghi anni. Quanto vi dovevate amare a vicenda per aver avuto il coraggio di andar avanti, sempre avanti, quando le disgrazie sono sopragiunte. Che Dio mi dia forza di fare il mio dovere verso mio marito, verso la casa mia ed i figli miei, come voi avete fatto per i vostri."

Ma adesso, bastano le lagrime, Madre! sappiamo essere pratiche, tanto per cambiare! Mentre Bettina è in viaggio per la sua piccola luna di miele, potete accomodare la sua casina. Le ci vorranno molte cose, nella sua casa nuova, che a voi parranno assolutamente inutili. Probabilmente, siete voi che avete ragione. Voi non aveste che pochissime cose di lusso, quando vi sposaste, laggiù nella vostra terra natìa. Ma, se Bettina non somiglia straordinariamente alla sua Mamma nata in Italia, essa avrà acquistato delle idee nuove per l'ammobiliamento della casa, idee che ha imparate negli Stati Uniti. Essa vorrà probabilmente ornare il suo nuovo apparamento con della mobilia di velluto, delle pitture ad olio, dai gaii colori, con tappeti che covrono tutto l'impiantito, delle belle tende e della roba di cristallo. Se suo marito è un lavorante, forse non avrà messo molto denaro da parte, e per questo dovete consigliare Bettina di non comprare troppo ad un tratto.

Noi abbiamo una cattiva abitudine, negli Stati Uniti, quella di comprare le cose delle quali abbiamo bisogno, e poi di pargarle a rate. Noi chiamiamo questo "Il piano a rate (Installment Plan). Se compriamo una tavola che costi 15 dollari, noi paghiamo un dollaro la settimana, finchè non sia interamente pagata. Le cose comprate così costano sempre più di quelle per le quali si fa un pagamento immediato. Per lo più queste cose sono fatte male e di materia di infima qualità. Inoltre, soccombiamo spesso alla tentazione di comprare più di ciò che ci bisogna, quando non dobbiamo pagare immediatemente. Questo ci fa entrare in debiti che passiamo lunghi e penosi mesi a pagare, per aver preso delle cose delle quali non avevamo bisogno. Qualche volta, queste cose sono usate e perdono tutto il loro valore, molto prima di quando non sono interamente pagate.

Sò che gli Italiani sono, per lo più, più economi degli Americani, e che amano pagare subito per quel che comprano. Ma Bettina è molto Americana nelle sue idee, e dobbiamo aiutarla a capire che molte usanze chiamate Americane non sono tali affatto. Degli Americani stupidi e nocivi hanno il torto di insegnare un cattivo gusto alle donne forestiere. Essi danno cattivi esempii che i forestieri che capitano nel nostro paese credono essere veramente Americani e, come tali, cercano di copiare.

Bettina spesso sarà disorientata e confusa nella sua nuove vita maritale. Questa vita le darà molte libertà delle quali non ha mai goduto prima. Ma le darà pure molte difficoltà da sormontare. Se suo marito guadagna un salario settimanale, lo darà probabilmente tutto a sua Moglie. Essa dovrà condurre la casa e pagare tutte le note. È qui che potete aiutare moltissimo la nuova moglicina. Insegnatele a comprare per la sua casina, giusto come fate per la vostra. Gli Italiani s'intendono così bene a tirar il miglior partito possibile da poco denaro, nel maneggio della casa. In questi tempi, è' d'uopo saper come servire del cibo buono e sostanzioso ai nostri mariti affamati, colla minore spesa possibile. Insegnatele a fare i vostri piatti prediletti, come li avete imparati da vostra Madre, laggiù in Italia. Incoraggiatela a comprare soltanto quei cibi che sono veramente a buon mercato e sostanziosi.

Potete dire alla sposina come faceste a cominciare a metter da parte dei fondi quando voi e Pietro decideste di venire quà in America. Vostro marito guadagnava molto meno di ciò che il marito di Bettina non guadagni oggi. E per questo, malgrado il prezzo alto della roba, Bettina può metter da parte un pochino se ci si prova e se le insegnate a farlo regolarmente. Noi, che guadagnaimo un salario meschino, siamo forzati a mettere da parte in piccoli mucchi. Una Banca nella quale possiamo mettere un cinque o dieci soldi ogni giorno è un buonissimo modo di cominciare. Fa tanto piacere a vedere il mucchietto crescere.

Quest'abitudine di spendere con cura e di mettere da parte farà di Bettina una moglie felice ed una vera socia di suo marito. Egli guadagna il denaro, e sua moglie bada alla casa ed ai loro interessi e fa in modo che il denaro renda più che possibile. Piace alle donne Americane di considerarsi come le associate dei loro mariti. Esse sentono che, nello stato conjugale non c'è nè "tuo" ne "mio", ma che tutto è "nostro". Soci, negli affari, dividono fra loro tanto il male che il bene. E cosi pure, nell'associazione chiamata Matrimonio, tutto e diviso in comune. Il lavoro di ognuno è ugualmente importante alla loro felicità maritale ed al mondo in generale.

Voi scuoterete la testa in modo saggio, e riderete e simpatizzerete con essa, e racconterete a Bettina la vostra prima disputa con Pietro. Diceste allora che non volevate più mai vederlo, perchè egli non volle fare ciò che vi piaceva. Voi scappaste a casa vostra per raccontare l'accaduto a vostra madre, precisamente come Bettina è scappata da voi dopo la prima disputa con suo marito.

Voi le farete osservare con orgoglio la vostra bella e numerosa famiglia, quasi tutti cresciuti adesso. Le direte: "Pietro era un bel giovanotto, ed abbiamo tirato sù una bella famiglia. Eramo soci laggiù in Italia, e siamo sempre soci, qui in America. Abbiamo sofferto molto ed abbiamo dovuto lavorare forte. Ma Pietro è

65

sempre il mio socio, e gli voglio ancora più bene di quando ero la sua sposina novella. Il vero matrimonio consiste nel lavorare insieme, vivere insieme ed il soffrire insieme."

L'economia pei cibi

O H, vi chiedo scusa, Madre. Vi sono capita-
ta proprio al momento del pasto. Mi dispiace tanto. Verrò un'altra volta. No,
veramente? Volete che stia a mangiare
con voi e coi ragazzi? Siete certa? Ebbene,
mi farà tanto piacere, ed eccomi pronta.
Ma ho paura che non ci sarà cibo bastante
pei ragazzi, se desino qui così inaspettatamente. Misuriamo tutti le nostre spese di tavola,
in questi tempi! Tutto quel che si compra costa tanto!
Dite che ce n'è più che a sufficienza? Ah, sì, mi scordavo
per un momento che stavo parlando ad una padrona di
casa Italiana. Essa ha sempre un buon desinare, con
abbondanza per tutti. Dite che è un desinare molto
semplice, soltano la minestra e gli spaghetti? Oh, che
buon odore!

Odoro il vostro modo delizioso Italiano di cucinare
gli spaghetti. A noi Americani gli spaghetti ci piacciono
moltissimo, e gli facciamo spesso. Quanto ci potreste
insegnare, per prepararli bene a renderli deliziosi:
Noi, per lo più facciamo cuocere i nostri finchè non siano
morbidi, e diventano come una gelatina. Dovreste far
pubblicare nei giornali Inglesi il vostro modo di cuci-
narli, e così si potrebbe imparare da voi. Qualche volta,
quando ci sentiamo pronti a fare qualche prodigalità,
spendiamo un monte di denari nei piccoli restoranti

Italiani, quieti e comodi, semplicemente per avere un pasto di spaghetti come gli sapete fare voi altri Italiani.

Adesso, mi permettete di dirvi qualche cosa? Voi avete chiamato questo un pranzo semplicissimo, e ve ne siete scusata. Veramente è tutto quello che Dottori e persone che hanno studiato i problemi dei cibi chiamerebbero un pasto perfetto. Non lo sapèvate? Naturalmente, non lo sapèvate. Le Donne Italiane sono le migliori cuoche del mondo, — almeno così crediamo, — eppure non hanno studiato ne in classi ne in libri.

Il vostro pranzo è delizioso. La minestra è una suppa densa e ricca della quale noi Americani non sappiamo nulla. Eppure, vedete quanto questa minestra è nutritiva ed a buon mercato. Cavolo, fagioli, qualche fetta di cipolle, un pochino di cacio, ed ecco la minestra! Pochi soldi di ognuno di questi ingredienti compongono la porzione giornaliera, ed avete un cibo perfetto. Esso contiene tutte le qualita di ricchezza dei cibi che ci abbisognano in una giornata fredda d'inverno. Noi donne Americane dobbiamo imparare molto da voi. L'economia vi è naturale, e fate dei piatti meravigliosamente buoni e nutrienti con legumi ed altre cose che noi buttiamo via o che non conosciamo.

La salza di pomodori che fate, è una festa in sè stessa. Noi non la possiamo fare affatto. Pare che siate nate col ticchio particolare per fare questa salza. Poi, ci è quest'insalata che è una tentazione in se stessa. Voi sapete fare delle insalate con tanti legumi freschi che noi Americani sappiamo appena usare. Il vostro olio d'oliva è sempre buono, perchè sapete come e dove comprarlo. ed il vostro aceto non è mai acido. Semplicemente qualche goccia di un vino rosso un pochino aspro, e ci siamo!

Oh, cara Madre Italiana, noi donne Americane abbiamo risolto molti problemi casalinghi per la nostra vita, dei quali voremmo rendervi consapevoli. Ma voi pure, ci potete insegnare molto in compenso. Si dovrebbe unirci ed imparare a conoscerci, ad amarci e ad aiutarci a vicenda. I tempi sono duri per tutti, adesso. Lo capite, questo. La guerra ci ha fatto soffrire tutte ugualmente, ricche e povere, Italiane ed Americane. Malgrado che le battaglie siano attualmente finite adesso, dovremo ancora per molto tempo aiutare pel sostenimento dei nostri fratelli e delle nostre sorelle d'Europa. Questo c'insegna ad essere ancora più guardinghi di prima per tutti i cibi. I lavoranti non hanno mai avuto troppo, cominciando e dal vero principio della guerra, le nostre borselline sono state esaurite sempre più.

Non c'è che una consolazione per noi donne, Ed è questa. La guerra ci ha costituite tutte sorelle. L'angoscia ed il dolore sono stati comuni a tutte. Delle Madri Americane hanno dato i figli loro, i loro cuori ed i loro denari per l'Italia come per l'America e per le altre Nazioni Alleate. Le Madri Italiane hanno dato i figli loro, i loro cuori ed i loro denari pure per la stessa causa. E così, vedete, siamo realmente sorelle adesso. La guerra ha contribuito più al ravvicinamento ed al buon intendimento fra donne, all'apprensione dello stesso sentimento l'una per l'altra, di ciò che molti anni di vita comune nello stesso paese, sotto condizioni normali potrebbero mai aver conseguito.

Molte economìe, nei generi diversi, sono ancora richieste dal nostro Governo. L'economìa dei cibi è, senza dubbio, la più importante fra tutte. Noi donne Americane, troviamo questo molto difficile da conseguire. Siamo una Nazione molto spenditrice. Vogliamo avere troppo di tutto quello che produciamo. Questo è la verità per i ricchi, come lo è per i poveri. La vostra economìa consiste nell'usare gli stessi cibi nutrienti e semplici che avete sempre usati, tanto in America che in Italia, e nell'insegnarci a far lo stesso.

Non vi provate ad imitare delle nuovità fantastiche Americane proprio in questo momento. Aspettate un poco. Quando la prosperità sarà tornata, lasceremo che i ragazzi c'insegnino i piatti Americani che hanno imparato a fare nelle loro Scuole e nelle Case Co-operativa. Forse, vi piaceranno tanto per cambiare, ma noi Americani desidereremo pur sempre un buon, semplice pranzo di spaghetti come abbiamo oggi.

E così, adesso ci separiamo di nuovo per accudire al nostro lavoro del dopo pranzo. Tutti ci sentiamo benone. Guardate le belle gotine grasse e abuffanti e dal bel colore, di Baby Sam. Questo è il più bel complimento che possiate ricevere, o Madre. Prova indiscutibilmente come i ragazzi sino ben nutriti e che buon cibo preparate per loro.

Non una briciola è stata persa. Guardate la tavola. Questo è quello che ci ha predicato tutto il tempo il nostro Controllore dei cibi — cibi semplici, nessun pasto stravagante, poco zucchero e nessuno sciupìo. Ed ecco che ubbidite le leggi Americane perfettamente bene, e lo fate senza frastuono, e senza desiderare laudi per questo. Fate precisamente ciò che fanno centinaia di migliaia di donne Italiane in tutto il mondo, perchè avete imparato presto a gustare i pranzi semplici che vostra Madre v'insegnò a fare, laggiù nella vostra amata Italia.

E cosi, "Buon giorno," Madre. Me ne vado al mio lavoro. Spero che m'inviterete di nuovo, e che verrete alla Casa Co-operativa per rci tutto quello che potete per aiutarci nel nostro economizzare cibi. E noi per contro, vi diremo tutto ciò che sappiamo noi; E così noi

donne, lavorando insieme, aiuteremo sciogliere la parte
che riguarda la casa ed i cibi, di quei grandi problemi del
mondo, per dopo la guerra."

La Madre che lavora fuori di casa

UESTO è veramente un po' tardi, per fare una visita, lo so. Sono venuta diverse volte la mattina, ma non vi ho mai trovata a casa. Lavorate tutto il giorno nel magazzino finendo pastrani? Ed adesso, sono le nove, e siete ancora a lavorare, pulendo la casa e lavando i panni della famiglia. Cara mia, che giornata piena di lavoro. Dovete essere molto stanca. Dite che siete rifinita? Lo credo bene. Certe altre donne, che non lavorano forte come lo fate voi tutti i giorni, nella fattoria, si sentono stanche, quando viene la sera. E certe d'une fra loro, non hanno doveri di famiglia ne responsabilità di sorte che l'aspettano a casa, quando ritornano dal lavoro.

Quanto tempo è, dacchè avete cominciato a lavorare fuori di casa? Avete lavorato cosi' dalla morte di vostro marito in poi? Prima di questo, non facevate che un poco di rifiniture a casa, non è vero? Adesso, è necessario che guadagnate più denaro per mantenere i bambini. Era duro abbastanza, quando il marito vostro lavorava colla vanga e la picca, facendo soltanto quattordici dollari la settimana. Era soltanto perchè la vostra natura di donna Italiana vi suggeriva tutte le economìe e tutte le parsimonie, che vi era possibile di andar avanti allora. Noi Americani, abbiamo sempre osservato che gli Italiani possono comprare piu ròba con un dollaro di ciò che non

72

ci sia dato a noi di fare. E voi arrivate sempre a metter da parte qualche dollaro per le disgrazie che, presto o tardi, sono sicure di capitare ad ognuno di noi. Adesso, mi dite che non soltanto non riescite a metter nulla da parte, ma che non potete neppur andare avanti col denaro che guadagnate? Vi credo! Sappiamo bene noi tutti, quanto sia duro il poter far faccia a tutti gli impegni, in questi tempi.

Avete più bisogno di imparare l'Inglese adesso di quel che non l'abbiate mai avuto. Quando viveva vostro Marito, egli poteva tradurre qualunque cosa per voi, e poteva accudire agli affari di famiglia che richiedevano l'uso della lingua Inglese. Ma adesso, siete sola, e, se non potete aiutarvi da voi in Inglese, le cose potrebbero andar male per voi. Le Donne che non sanno l'Inglese sono sempre in pericolo imminente di accidenti, perchè non possono leggere le ammonizioni stampate, nè capire gli ordini. I principali (padroni) dovrebbero metter sù avvertimenti in lingue diverse per avvisare gli impiegati, o il capo-squadra (foreman) dovrebbe aver l'obbligo di avvertire le lavoratrici forestiere del come devono fare per evitare gli accidenti. Ma dovete sempre stare sulla vostra guardia. Una delle grandi manifatture ha provato che, dacchè le sue lavoranti hanno imparato a leggere un poco l'Inglese, non vi sono più che la metà degli accidenti che succedevano prima. E così, vedete, il leggere la lingua Inglese vi è di grande protezione. Vi insegnerà cose riguardanti il vostro lavoro che dovreste sapere, e vi guarderà dai pericoli.

Durante un anno — non un anno di guerra — vi furono 25,000 (Venticinque mila) accidenti e morti in uno degli Stati Uniti. Potete voi capire come questo possa succedere? Generalmente, proviene da sbadataggine e dall'essere per questo chiappato nel macchinario,

— o da bruciature. Un pànico promosso dalla paura che causa la vicinanza di qualche pericolo e un'altra causa di accidenti. Padronanza di se stesso, ed il potere di pensar forte e presto, per decidere quale sia la cosa da farsi immediatamente quando un pericolo vi minaccia, sono le due cose che al mondo salvano più vite. Procurate di ricordarvi questo, se, in qualunque momento, un pericolo vi minaccia nella vostra Fattorìa, e mettetevi a pensare forte e presto.

Non permettete a nessuno di forzarvi a lavorare troppe ore di filato. Molti accidenti hanno per causa lo strapazzo. Anche se siete obbligata a fare il vostro lavoro di casa quando tornate a casa la sera, questo è pure un cambiamento di occupazione e non è pericoloso. È quando lavorate allo stesso lavoro finchè non siate completamente esausta e nervosa che vi è pericolo che diventiate sbadata ed immemore, e così che causiate un accidente.

Ditemi tutto ciò che riguarda il vostro lavoro, Madre. Quante ore lavorate, e quali sono le condizioni nel magazzino in cui lavorate? Sapete bene, non è vero, che tutte le vostre sorelle — le Donne di tutto questo immenso paese — stanno lavorando e facendo piani per procurare le migliori leggi possibili per quelle frà noi che sono obbligate a lavorare nelle Fattorìe. Le Donne s'interessano di più, adesso, nel procurare buone condizioni di lavoro per le donne lavoratrici, che in nessum altra cosa. Avete veramente bisogno di sapere qualche cosa delle leggi che governano le industrie in questo paese. Certi fra i manifatturieri forse potranno essere disonesti o indifferenti per la vostra sicurezza e pel vostro benessere e potranno valersi della vostra derelizione o della mancanza in voi dello studio dell'Inglese. Il nostro Governo a Washington, stà passando molte

74

buone leggi speciali per la protezione delle Donne, perchè eglino sanno quanto voi altre siate impotenti. Andate all'Istituto Internazionale o dalle Young Women's Christian Association della vostra Città e domandate loro di spiegarvi tutte queste cose. Se avete bisogno di qualche d'una che prenda cura del Baby mentre state al lavoro, vi troveranno una buona casa ove persone responsabili s'incaricano di tener cura dei bambini durante il giorno (Day-Nursery) vicino a·casa vostra. La, il vostro bambino sarà ben tenuto e ben curato finchè non lo andiate a riprendere, la sera.

Vi sono molte amiche che pensano alle vostre inquietudini e che si sforzano di arrivare fino a voi per aiutarvi. La guerra ha reso la vita tanto dura per tutte noi che abbiamo il più vivo desiderio di aiutarci a vicenda. Molte, molte donne lavorano oggi quanto lavorate voi. Certe fra di loro, occupano i posti degli uomini che non sono ancora tornati dalla guerra. Altre hanno perso il marito loro, — come voi — e lavorano per mantenere le loro famiglie. La vita è cambiata per noi tutte.

Dovete prendere cura di voi, adesso, perchè la vita stessa dei vostri Figli dipende da voi sola. Voi siete il loro Padre nonchè la Madre loro e la vostra felicità dipende dalla vostra salute e la vostra salvezza. Avete bisogno di un poco di aria fresca quanto ne hanno bisogno i ragazzi. Non potete stare in buona salute senza di questo. Non potreste camminare nell'andare e nel tornare dalla fattoria, ogni giorno? So bene che andate sempre di fretta e che siete tanto stanca e che vi preme di essere il più presto possibile coi ragazzi, a casa. Ma un poco di aria e di cambiamento vi riposerà e vi farà tanto bene, anche se per questo dovete arrivare un poco più tardi a casa.

L'aria fresca è un tònico meraviglioso. Precisamente in questo momento siamo nella primavera e perfino nelle città affollate la primavera ci chiama. Guardate fuori della vostra finestra. Vedete quel grande albero, laggiù nel Parco? Sta diventando verde e stende nuovi rami. Questo è il modo della vita. Qualche volta, sembra tutto triste e bigio come le foglie morte in una foresta d'autunno. Poi, un bel giorno, la primavera viene, e tutta la vita diventa dolce e rinnovata, e le giornate tristi e fastidiose sono passate. E così, Madre, guardate in alto e lontano, verso la primavera. La guerra è cessata e gli altri affanni passeranno pure. malgrado che oggi paia essere tanto buio, seguitate a vivere bravamente giorno per giorno e ricordatevi che noi, Donne Americane, ci sforziamo a conoscervi e ad aiutarvi.

La Madre stessa.

HE bellissimo mazzo avete sulla tavola quest'oggi, Madre! Rose rosse e garofani bianchi. Sono incantevoli, e tutto nella casa ha un'aria di festa e d'allegria. È un compleanno, è la festa natalizia di qualche d'uno, non è vero? Quella di Baby Sam forse? No? Non ho indovinato lene. Ma allora di chi è la festa? Del marito vostro? La vostra propria festa, mi dite? Quaranta-cinque oggi? Bene, ho piacere di essere qui, per poter augurarvi tante altre buone feste, molti altri bei mazzi di fiori e moltissima felicità nel mondo nuovo. Presentate oggi all'occhio un bellissimo quadro. È magnifico, di vedere la Mamma vestita nel suo più bell'abito, e celebrando la sua propria festa. E che bella cosa, di avere tutti i figli a casa, congratulando la Mamma e facendo una festa del suo compleanno.

Questo è proprio Americano. Noi donne Americane rifiutiamo di invecchiare e non vogliamo star in un cantuccio per la sola ragione che siamo madri di famiglia e che i nostri capelli cominciano a diventare grigi. Facciamo il nostro dovere verso i mariti nostri e verso i nostri figli, ma vogliamo fare di più. Siamo d'opinione che le Donne hamno quattro posti importanti nella società. Prima di tutto, siamo mogli e madri, secondariamente, siamo donne di casa: terzo, siamo cittadini del paese in cui viviamo, ed, in fine, — ma questo è pure molto importante, — siamo DONNE. Le donne sono delle ragazze cresciute, precisamente come gli uomini sono dei ragazzi cresciuti. Abbiamo ancora molte delle stesse brame per la vita e la bellezza che ci agitavano quando eravamo ragazze giovani.

Qualche volta i nostri mariti si scordano che le madri che stanno lavorando alle faccende di casa ed ai molteplici doveri che le impongono la cura dei ragazzi, anelano pure un poco più di attenzione personale. I mariti spesso si scordano di vedere le loro innamorate dei tempi passati, e le loro spose, i cui occhi erano "i più belli del mondo, ed i cui capelli erano i più neri che si fossero mai veduti", nelle madri del presente, che lavorano indefessamente per loro, per la casa, e per la famiglia. Questo è molto duro per noi donne. Abbiamo bisogno di vedere il nostro lavoro apprezzato, e di essere l'oggetto dell'amore a dell'attenzione dei nostri di casa. Tutte le donne sono campagne in questo. L'età, nè i costumi diversi dei paesi ove si possono trovare, non ci cambieranno mai nulla.

Qualche volta, i nostri Figli se lo scordano pure, e cominciano a credere che la Mamma dovrebbe starsene sempre a casa, facendo la cucina, cucendo, pulendo a badando ai Piccini. Non apprezzano il sacrificio della Madre loro, finchè non venga una malattia, e che non c'è più nessuno per fare tutto questo lavoro. Allora, tutto va a rotta di collo, e cominciano a meravigliarsi come mai la Mamma potesse far andare tutto così bene in casa.

Un vecchio adagio Americano dice: "L'uomo lavora dall'alba all'alba, ma il lavoro della donna non è mai finito." ("Man works from sun to sun, but Woman's work is never done.") L'avete mai sentito? Non vi pare molto vero? Si, tutte le donne che tengono casa sanno quanto questo sia vero, ma non si lagnano mai. A che prò? Continuano e continuano e vanno sempre avanti, perchè sentono che è dovere loro di far così. E forse lo è. La vita è tanto piena di lavoro, e tutti dobbiamo fare la mostra parte.

E molto più facile per le Madri, di far la loro parte, se c'è un poco di sentimento, di poesia e d'amore nella loro vita. Un piccolo regaluccio, un mazzettino di mammole, il cui profumo forse le riporterà in pensiero a molti anni passati, ed attraverso migliaia di miglia di onde di mare, nel paese della loro gioventù e dei loro sogni. Tutte le donne hanno questo bisogno in cuore. Un pensiero gentile dei ragazzi crescenti, un poco di attenzione dal marito, rendono molto più facile di comportare le dure realtà della giornata. Riempiono la casa di felicità e procurano la contentezza di tutta la famiglia.

Il vostro Compleanno è una gran festa, Madre, e per questo, desideriamo tutti di aiutare a rendervi felice. Qualche volta, i figli vostri sembrano un poco egoisti. La vita piace tanto a loro, ed hanno tanti bisogni. Le ragazze hanno bisogno di tanti abiti nuovi, quando crescono. Ed i ragazzi hanno molti interessi, fuori di casa, che costano denari. Il Padre loro si priva, egli pure, di molte cose, per poter comprare ai ragazzi ciò che bramano e di cui hanno bisogno. Ma quando fù che la povera Mamma ha avuto un vestito interamente nuovo o che è andata ad un teatro od un concerto? Oggi, pensiamo a tutti i vostri sacrificii, e vogliamo che sappiate che, davvero, apprezziamo ciò che avete fatto.

Migliaia di Madri, portano le vite che le figlie loro hanno scartate perchè non le considerano più di moda, precisamente come avete fatto voi. Spesso, hanno aiutato le signorine figlie loro a vestirsi per il ballo o per ricevimento. Le Madri sono felici di vedere i proprii figli contenti. Ma, qualche volta, amano pure di divertirsi un pochino, esse pure. Soprattutto voi, Madri Italiane, amate la musica, non è egli vero? È parte del vostro retaggio Italiano.

Non dovete arrossire e dire che siete troppo vecchia per cominciare ad imparare l'Inglese ed i modi Americani. La vita, presa da un angolo nuovo, ci forza ad imparare molte cose nuove. Siete stata coraggiosa di venire in un paese nuovo e sconosciuto, coraggiosa quanto lo furono gli antichi esploratori, come il vostro Cristòforo Colombo. Eglino sono venuti a vedere che cosa ci fosse di buono nel mondo nuovo, lontano lontano, e così faceste voi pure. E così, escite ogni tanto di casa, Madre, e cercate di capire ciò che l'America intende per voi. Potete trovare tante buone cose, qui!

Molte donne Americane desiderano di conoscervi e di aiutarvi. Verremo da voi ad insegnarvi a conoscerci meglio, se ci permettete di venire. Tutte le Americane che conoscono l'Italia vi amano digià. Bella Italia! Queste parole cantano digià nelle nostre orecchio, come il vostro languido "Bel Canto" nella vostra migliore musica. Sappiamo ciò che l'Italia e gli Italiani hanno fatto per gli Stati Uniti, e siamo contenti che tanti Italiani siano venuti qui per vivere con noi.

Ed adesso, beveremo alla vostra salute, Madre. Alla vostra festa. Salute! Possiate vivere e lavorare felici e contenti! Colla mano nella mano, lasciateci sperare di formare un cìrcolo di amicizia fra le donne di tutti i paesi. Ed allora l'amore farà cascare tutte le barriere stupide delle lingue differenti. A rivederci, Madre. Grazie pel fiore del vostro prezioso mazzo di festa. L'apprezzo molto. A noi donne, ci piacciono tanto le piccole gentilezze e cortesie, non è vero Madre?

The Truth About the Black Hand

BY

FRANCIS J. OPPENHEIMER

(FRANCIS OPP)

Author of

"Jewish Criminality," "The Nationality of New York Criminals," "The Catholic Church in the United States," etc., etc.

Reprinted from THE DENVER REPUBLICAN, January 17, 1909

(By permission of the New York Herald Co.)

PUBLISHED BY

THE NATIONAL LIBERAL IMMIGRATION LEAGUE

HEADQUARTERS, 150 NASSAU STREET

NEW YORK CITY

THE TRUTH ABOUT THE BLACK HAND

BY

FRANCIS J. OPPENHEIMER

A DAY scarcely passes without some mention coming up in the news of the misdoings of the Black Hand Organization. Either it is a "Black Hand Robbery," or a "Black Hand Kidnapping," or a "Black Hand Dynamiter." Sometimes it is even a "Black Hand Murder!"

So constant and so recurrent are the reported activities of the Black Hand Society, that the whole country is kept in a state of continuous alarm. The question now has come up, somewhat paradoxically, it at first may seem, "Is there actually such a criminal organization of Italian immigrants and who operate together under the name, The Black Hand?" Prominent Italian citizens of New York City and elsewhere, and those also in a position to know most, scoff at the very idea, and deny, without exception, that such a society exists.

Police Sergeant Petrosino, who had charge of the Italian Bureau at Headquarters, up to the time of his death, recently said, "As far as they can be traced, threatening letters are generally a hoax. Some of them are attempts at blackmail by inexperienced criminals, who have the idea suggested to them by reading about the Black Hand in the sensational papers, but the number of threatening letters sent with the deliberate intention of using violence as a last resort to extort money, is ridiculously small."

Baron E. Edmund Mayor des Planches, Ambassador to the United States from Italy, and Dean of the Diplomatic Corps in Washington, who has made a comprehensive study at first hand of the conditions of his own countrymen in the United States, has repeatedly announced that he does not believe that there exists any such organization of criminal Italians operating from any central agency, as is sometimes understood by the Black Hand.

Dr. G. E. Di Palma Castiglione, in charge of the Labor Information Office for Italians, New York City, and who is in touch every hour of every day in the week with Italian immigrants of all classes, says, "As an organization, the Black Hand is non-existent. Its sole existence is, in fact, confined to a literary phrase."

If the name of Cæsar was a name to "conjure with," so is the phrase "Black Hand" a phrase to "conjure with," for it has so compelling and so terrible a symbolism, that one must almost believe it exists, when he has only heard the words. The "Black Hand" is a myth, in so far as it suggests the existence of a society of criminal Italian immigrants.

Those Italians who have been arrested for being members of the "Black Hand" organization, really are independent malefactors who may happen to be socially intimate. They might frequent the same saloon, or they occasionally might play a game of cards together, but these meetings are a matter of chance, and accidental, and not the result of the workings of any criminal organization. In a real grave crime, two or three, or even more of these independent desperadoes might band together, and not because each one of them is a member of the "Black Hand," but because this (as every detective throughout the United States knows), is a common practice among criminals of all nationalities, Americans as well as Italians.

Prominent Italians have often made the complaint that every immigrant of dark complexion, unless he wears a Turkish fez when arrested, is put down in the police blotter as being an Italian, and that by this slip-shod method of keeping police records, the Italians are made responsible for a variety of crimes committed by other nationalities. An anecdote has been going the rounds of the Italian bankers and journalists of New York City, of how a rather well known Italian merchant of Hoboken, Mr. Bianchetti by name, had been pictured and featured in the papers as being one of the assassins involved in the murder of King Humbert. His photograph was copied in a great many of the sensational papers throughout the United States, and the story eventually found its way to the merchant's native land. Mr. Bianchetti, said to have been an honest and a much respected man, lost all his friends and his business failed. He died within a year of the publication of the story, his acquaintances said, of a broken heart.

The Black Hand is not a tree spreading its roots deeper and deeper in the national life of the United States, as so many hysterical if well meaning sociologists, have affirmed. Nor is it, as others fancy, some closely woven "system." "The Black Hand" is only a general name appropriated by individual Italian criminals, under which to hide their own individual crimes.

Some contend that the "Black Hand" has scarcely ever been heard of in Italy. An English student of criminology, A. F. Griffiths, has many facts and many arguments to prove that the "Black Hand" really is a Spanish organization, and that it had its origin in Spain. His contention is, that individual criminal Italians in this country, on hearing the phrase "Black Hand," were fascinated by its harrowing symbolism; that they adopted it, and from the United States it spread to Italy. Criminals employ this phrase, because they know that it acts on the receiver of a threatening letter the same as the name of a well-known journalist acts on the judgment of a hesitating editor.

The one criminal organization in Italy is the "Camorra," and the "Camorra" exists only in the city of Naples. Its influence is not to be found even in the Neapolitan suburbs. The poverty of Naples is responsible for the existence of this organization, for the "Camorra" is in a sense a sort of criminal collection agency. In fact, the only way a "usurer's" debt may be collected in Naples is through the co-operation of the "Camorra." Most of these Neapolitans are so poor that a judgment from the courts against them is scarcely of any value. The "Camorra" has its chiefs and officers, and it is made up of ruffians, "cadets" and all kinds of irresponsible adventurers. These criminals who belong to the "Camorra," of course, do far more desperate things than collect debts.

The number of Neapolitans who come to the United States is very, very small. According to official statistics, the total number of persons who emigrated from Italy, from 1904 to 1907, was 2,690,174. (1) Only 17,837 of these 2,690,174 were natives of Naples! It would not be just to humanity, to fancy that each and every one of these sixteen thousand was a member of the "Camorra" simply because he was born in the city that has the "Camorra." Furthermore, it is certain that a great proportion of these Neapolitans sailed for Brazil, Australia and Canada.

Many writers guess that the "Mafia" is another criminal organization, which operates in Sicily. The "Mafia" is not a society but it is rather a spirit of revenge, a primitive seeking for justice, and is much like the Kentucky feuds. It is less cruel and less unrelenting, perhaps, than the feuds of our Southern mountaineers, for it at least stops at the grave.

The usual impression given by unscientific writers, is that America gets "the scum of Italy." It is very significant that

---

(1) See: " Statistica della Emigrazione Italiana per l'estero negli anni 1904-1905 e negli anni 1906 e 1907."—Roma, Tipografia Nazionale—via Umbria.

3

only 1½ per cent. of Italian immigrants are debarred by the immigration officers.(1) Usually, the Italian immigrants are strong, healthy and able bodied, their ages varying from twenty to thirty-five years. Seldom do any of these sons of "sunny Italy" become paupers. Mr. James Forbes, chief of the Mendicancy Department of the Charity Organization Society, says he has never seen or heard of an Italian tramp. Detective-Sergeant Petrosino gave out a statement just before his death, that he had never seen or heard of an Italian girl "on the street," and in her special report to the United States Government, Miss Emily Fogg Meade praises the chastity of the Sicilian women in Hammonton, N. J.

The Italians do drink considerably of their own light wines. Nevertheless they are a very temperate race. John Foster Carr states: "With the exception of the Russian Jews, the Italians are by far the most sober of all nationalities." James J. Starrow, Boston, in a recent article said, "The Italian drunkard hardly exists."

"So far from being the scum of Italy's paupers and criminals, our Italian immigrants are the very flower of her peasantry," Mr. Carr says. "They bring healthy bodies and a prodigious will to work. They have an intense love for their fatherland, and a fondness for old customs, and both are deepened by the hostility they meet, and the gloom of the tenements they are forced to inhabit. The sunshine, the simplicity, the happiness of the old ways are gone, and often you will hear the words, "Non c'e piacere nella vita!" (There is no pleasure in life here!) But yet they come, driven from a land of starvation to a land of plenty. Each year about one-third of the great host of the industrial recruits from Italy, breaking up as it lands into little groups of twos and threes, and invading the tenements, almost unnoticed, settle in the different colonies of New York.

"New York tenement houses are not adapted to life as it is organized in the hill villages of Italy, and a change has come over every relation of life. The crowded living is strange and depressing. Instead of work accompanied by song in orangeries and vineyards, there is silent toil in the canons of a city street. Instead of the splendid and expostulating carabiniere, there is the rough force of the New York policeman to represent authority." (2)

The report of the Commissioner-General of Immigration contained in 1904 a study of "foreigners" who were at that time in-

---

(1) See: "*Annual reports of the Commissioner - General of Immigration,*" Government Printing Office, Washington, D. C.

(2) See: "The Coming of the Italian." *Outlook*, Feb. 24, 1906.

4

mates of the various public institutions of the United States. The total number of alien inmates was 44,985. Of this number, but 3,266 were Italians! In the penal institutions there were confined 1,318 Italians as against 9,825 other aliens. In the insane asylums there were 718 Italians, as against 19,764 other aliens, and in the various charitable institutions, there were 1,230 Italians, as against 15,393 other aliens. Of the 1,318 Italians in the penal institutions, only 75 were charged with grave crimes. (1)

In the report made by Miss Emily Fogg Meade for the Federal Government, published in the Bulletin of Labor, May, 1907, there is given a sympathetic account of the rise and growth of the Italian colony at Hammonton, Atlantic county, N. J. This town, which now numbers between 4,000 and 5,000, has been built up almost entirely by Sicilians. Miss Meade praises the Sicilians as fruit growers and as farmers, and in general has nothing but enthusiasm for their temperance and industry. She indicates that in 1906-7 they paid $4,936 in taxes in Hammonton, and that there were 1,370 names of Sicilians and Southern Italians on the tax register. In her report, it is stated, that nearly every week or so, a house is built between Rosedale and Winslow by some Sicilian.

"The Italians of Hammonton show themselves to be a social people, with simple, natural tastes," she says. "Their love of home and children is healthful. They are ignorant, primitive, childlike, but their faults will largely be mended by contact with good American customs. Their courtesy, gentleness and love of outdoor life and simple pleasures are actual contributions to American life. The country environment seems to develop their better qualities and they take a normal part in the life of the community." Miss Meade's general summing up of the colony of Sicilians at Hammonton, is as follows: "In Hammonton are found the results of twenty years' contact of a typical American population with the lowest class of Sicilian immigrants. It is a safe conclusion that what the Italian has been able to accomplish in Hammonton he can achieve elsewhere under similar circumstances."

If journalists would confine themselves to studying the especial characteristics of Italian criminals, it would be legitimate, but so many unscientific and uninformed if not frenzied sociologists, strike not only at the Italian criminals, but at the Italians generally, and for no reason other than that they are Italians. The publication of their hastily prepared and sensational articles is an offense against American common sense as well as

(1) See: " *Report of the Commissioner-General of Immigration for the fiscal year ending June 30, 1904,*" Washington, D. C., page 49 and following.

against our Italian American citizens. Articles of this kind powerfully contribute to the creating of an unfair feeling all over the country against a large part of our Italian American citizens. These stories circulate not only in the larger cities, where the readers are more sophisticated, but they filter their way into the interior of the country, into the smaller towns, and are there read by the gentle and peaceful farmers.

These simple-minded readers believe firmly that every word printed in these articles is correct, that the stories are founded on fact. In every Italian they meet, even if it be only some poor worn out laborer with his dinner pail, who "grieves not" and "never hopes," "Stolid and stunned, a brother to the ox," they see a member of the "Black Hand." And if in those portions of the country, there are no Italians, the residents take steps to prevent any from ever settling there.

"America is still in need of settlers," Mr. Castiglione, Labor Information Office for Italians, said. "Large tracts of its territory must yet be cultivated. In arousing a feeling of disgust and fear against Italians, many sections of the country are prevented from using them in the development of their natural resources. Last year the Legislature of the State of Virginia passed a resolution the aim of which was to prevent the settlement of Italians in Virginia. In South and North Carolina, there was also some legislation enacted to this effect, and yet in St. Helena, North Carolina, there is a flourishing settlement of Italians. They have their own church, co-operative store, bake oven, blacksmith shop, and they have designed and built all these buildings themselves. The Southern newspapers, time and time again, have featured and written up the virtues and progress of the Italian colonists at St. Helena, N. C.

"Is it fair to the Italians, is it useful for this country that this feeling should be created and spread among its citizens?" Mr. Castiglione continued. "Of the 2,000,000 Italians in the United States many have taken out their papers and become naturalized Americans. Many of them have settled here with their families, are sending their children to American schools, are investing their savings in American real estate, in American industries, in American business enterprises. In the city of New York alone conservative estimates place the value of real estate owned by Italians at $20,000,000."

The large majority of the Italians work hard to build the subways, to ditch the sewers, and they are obliged, because of the limited wages they receive, to live in unhealthful slums where many of them contract tuberculosis and other dreaded diseases.

6

Others are building the railroads, excavating the canals. To do this primitive, but necessary work, these misunderstood laborers are compelled to live in cheerless labor camps, far from any civilization (and ignorant of its luxuries), sleeping in shanties, sustaining themselves on poor food. Still others are working in the mines, and the quarries, risking their lives every moment of every day, with practically no protection of the law in case of accident. Many Italians devote themselves to agriculture. Prosperous agricultural colonies are to be found in Vineland, N. J.; Tentitown, Ark., and St. Helena, N. C. In California, Italians own and cultivate the largest and finest vineyards. In Louisiana many Italians are working in the cotton fields and on sugar plantations.

"Is it fair to arouse a national feeling against these people, and only because some of them (and the statistics eloquently show how proportionately few) are criminals?" Mr. Castiglione asks. "This is unfair not only to the Italian race, but it is even injurious to the best interests of America and Americans."

Consigli
Agl' Immigranti

ADVICE FOR IMMIGRANTS.
KEEP THIS! IT WILL BE USEFUL TO YOU.

Exchange your money for American money before leaving Ellis Island. The Money Exchange there is under Government control and no one else can give you more.

Take your baggage checks to the Baggage Room at the Barge Office, where the officials will see that honest expressmen deliver your baggage to your address.

Do not give or show your money to people you do not know well. Do not change banknotes for strangers who may give you worthless money in exchange.

Guides may be engaged on Ellis Island at the desk in the passageway near the ferry to take you direct to your address in or around New York City. Look for men with "Immigrant Guide and Transfer" on their caps.

Do not stop over in New York City if you are going to other places. Ask to be taken to the station at once or you may fall into the hands of greedy persons who will swindle you.

It is seldom necessary to take a cab, as nearly all parts of the city can be reached by street cars, where the fare is five (5) cents.

Do not go with strangers into saloons and restaurants where you may have your money and valuables stolen.

If you need information ask a policeman, who is easily known by his uniform. A metal badge does not always mean that the person is an official; swindlers use such badges to mislead and cheat.

If you are asked to pay large fees in advance for a job and high railway fares, ask for information before accepting the terms. There are Free Labor Agencies which will help you find work or direct you to reliable employment agents.

If you meet with an accident while at work, do not deal with a lawyer who comes to see you, and do not agree to pay him half of the money he says he can get for you. Find out about the lawyer before you accept his offer.

Use the Post Office "Money Orders," which you can buy at any Post Office, if you have to send money back to the old country or from one part of the United States to the other. This is the safest, cheapest and quickest way to send money.

Deposit your savings in regular savings banks, as these are under the control of State officials. Ask whether a private bank in New York State is supervised by the State before you put your money in it. Do not patronize any bank not under government supervision.

2

CONSIGLI AGL' IMMIGRANTI.

Conservate questo Notiziario!
Esso vi sarà utile.

Se al vostro arrivo avete moneta da cambiare, cambiatela prima di lasciare Ellis Island—Là, i Cambiavalute sono sotto il controllo del governo, e nessuno potrà darvi dippiù.—

Consegnate gli Scontrini (checks) dei vostri bagagli nel Baggage Room (ufficio dei Bagagli) che trovasi nel Barge Office, dove impiegati appositi cureranno che onesti carrettieri trasportino il vostro bagaglio a domicilio.

Non consegnate, nè mostrate la vostra moneta a persone sconosciute.

Non cambiate la vostra moneta per biglietti di banca esteri, con sconosciuti che vi potranno ingannare, dandovi in cambio biglietti di nessun valore.

Avendo bisogno di Guide, esse possono essere procurate ad Ellis Island nel passaggio, vicino al Ferry. Dette Guide vi condurranno direttamente al vostro domicilio, oppure, se è vostro desiderio, in giro per New York.

Dette Guide si riconoscono, per aver scritto sul cappello: "Immigrant Guide and Transfer."

Non vi fermate in New York, se siete diretti in altri luoghi. Domandate di essere condotti alla stazione direttamente, altrimenti correte il rischio di cadere nelle mani di persone che vi potranno derubare.

È raro il caso che abbiate bisogno della carrozza, per essere condotto al vostro domicilio, giacchè quasi tutte le parti della città possono essere raggiunte da trams, e la corsa è di 5 soldi.

Non andate con stranieri in Restoranti e Bars, dove la vostra moneta ed oggetti di valore possono venirvi derubati.

Se avete bisogno d'informazioni domandate ad un policeman, il quale è facilmente riconoscibile dalla sua divisa.

Una placca di riconoscimento, non indica in tutti i casi che la persona, che ne è in possesso, sia realmente un ufficiale. Molti imbroglioni usano tale disfintivo per ingannare ed imbrogliare.

Se venite richiesto di pagare anticipatamente larghi compensi per ottenere lavoro in posti dove occorre spendere molto per recarvici, domandate informazioni, prima di accettare tali proposte.

Vi sono delle Agenzie di Collocamento al lavoro, gratis, le quali vi aiuteranno a trovar lavoro, oppure a dirigervi a delle Agenzie oneste, e di tutto riposo.

Se avete la disgrazia di capitare in qualche accidente, mentre vi trovate sul lavoro, non contrattate con alcun avvocato, che eventualmente possa venire a trovarvi, e non vi accordate di cedergli la metà della moneta, che egli vi promette di ottenere come compenso per voi. Ottenete informazioni sull'avvocato, prima di accettare il contratto.

Usate i Vaglia Postali, i quali potete ottenerli in qualunque Ufficio Postale, se dovete mandare denari al vostro paese, oppure in qualunque parte degli Stati Uniti.

Questo metodo è il più uro, il meno dispendioso, ed è la niera più sbrigativa di mandare m eta.

Depositate i vostri rispar ii nelle Casse di Risparmio (Savings Banks) giacchè esse sono sotto il Controllo dello Stato. Infoimatevi, se avete intenzione di depositare i vostri denari, in qualche banca privata, se essa sia sotto la tutela dello Stato, prima di decidervi a farlo.

Non patrocinate alcuna Banca, che non sia sotto la tutela dello Stato.

IL FIENO DA BUONI GUADAGNI NELLO STATO DI NEW YORK

INVEST YOUR MONEY IN AMERICA. Each year thousands of dollars are sent by immigrants to their home countries for the purchase of land. When the alien goes home, the freedom and opportunities in the United States have become so dear that he is restless to return. In the United States land is cheaper than in Europe, it is more productive, taxes are very low and the prices for produce are much higher. If you have your own farm, you are independent, and if hard times come you can produce the necessities of life. When the factories close down you will not suffer. The cities are now crowded with idle men who cannot properly provide for their families; the farmers are prosperous.

Buy your land through one of the many bureaus conducted by the State officials and thus be secure against possible swindling. Beware of "home lots," workingmen's homes and payments on installment schemes.

Call or write for information, advice or direction, which is furnished *free* by the North American Civic League for Immigrants, 127 Madison Avenue, New York City, Tel. 8310 Madison.

FARMING IN NEW YORK STATE.

New York State to-day offers great opportunities for the man who wants to buy a farm, or who wants to work on a farm, with the intention of some day owning one himself.

The chance for a man with small capital to own a farm in New York State is greater to-day than, perhaps, anywhere else either in this country or in Europe.

To the foreign-born farmer who has come to this country, New York

4

UNA MODERNA STALLA PER VACCHE NELLO STATO DI NEW YORK

"Investite i vostri denari in America."

Ogni anno migliaia e migliaia di dollari degli immigranti sono mandati in paesi di provenienza dei medesimi allo scopo di comprarvi terreni. Poi quando l'immigrato ritorna al suo paese, si avvede che la libertà e le opportunità che godeva in America, hanno trasformato il suo essere, ed anela tornare in America nuovamente.

Negli Stati Uniti i terreni sono molto più a buon mercato che in Europa, ed il terreno è più produttivo. Le tasse sono miti ed i prezzi delle derrate molto alti.

Se voi possedete la vostra terra (farm) voi siete indipendente, e se tempi difficili sopravvengono, avete di che produrre il necessario per la vita. Le città sono attualmente popolate di un numero considerevole di persone disoccupate, che non possono convenientemente provvedere per le loro famiglie—mentre i farmaiuoli (proprietarii di terreni) sono prosperi.

Comprate le terre per mezzo di uno dei tanti appositi Uffici, che lo Stato mantiene appositamente, e così essere sicuri contro possibili frodi. Attenti per la compra dei così detti "Lotti di Terreno."—"Case per i lavoratori" e "Pagamenti ratuali."

Andate oppure scrivete per informazioni, consigli ed assistenze, le quali sono fornite gratis, alla "The North American Civic League for Immigrants" 127 Madison Ave., New York City Telefono 8310 Madison Square.

L'Industria Agricola nello Stato di New York.

Attualmente lo Stato di New York offre grandi opportunità, a chi vuole comprare una farm, oppure voglia lavorarvi, col proponimento di divenirne padrone eventualmente.

L'opportunità per la persona con un piccolo capitale di possedere una farm nello Stato di New York, è maggiore oggi, che non lo sia in qualunque altra parte degli Stati Uniti e dell'Europa.

Al contadino, o farmaiuolo, il quale è venuto in America, lo Stato di New

5

State offers a most cordial welcome, and guarantees him as good an opportunity to buy a farm and to make a living as any place in the country. FOR THE MAN WHO HAS MONEY TO BUY A FARM. Good land in New York State can be bought for from $25 to $50 per acre at the present time. This includes comfortable houses, large barns and all necessary farm buildings. The land is all under cultivation, and, with intelligent and careful farming, will raise large yields of the best crops.

the crops of Germany, Hungary, Poland, France, Northern Italy, Bohemia, Holland, Sweden, etc., are grown, and there is no reason why an experienced farmer from any of these countries should not be able to make a good living raising the same crops in New York State.

SECOND: Farms can be bought in New York State to-day at far less than their actual value, and in some cases less than the cost of the buildings alone. In a number of cases these cheap farms, within two or three years after being purchased by

L'INDUSTRIA DEI LATTI CINU E LA PRIMA NELLO STATO DI NEW YORK

Here are some of the many advantages New York State has for the farmer, which apply equally as well to the farmer from a foreign country as to the man born here.

FIRST: New York State raises successfully any crop suited to its climate, which, with proper cultivation, will pay handsome returns. New York State produces more hay, dairy products, potatoes and buckwheat than any other State in this country, and she ranks among the first in oats, corn, fruits, poultry, etc.

In New York. State practically all

industrious farmers, have yielded crops bringing more than the cost of the entire farm. As one example, a farm was bought in Central New York several years ago for $15 per acre, and two years after its purchase it produced a crop of peas worth $50 per acre.

THIRD: The New York farmer has his market right at his door. The large number of cities in New York State, with the 6,000,000 in New York City and vicinity, assure the New York farmer of a ready sale for what he raises, at the best prices obtainable.

York offre un ospitale benvenuto, e gli garantisce l'opportunità di acquistare una farm, dargli il mezzo di guadagnarsi la vita, a condizioni non inferiori a qualunque altro posto d'America.

Per chi ha la possibilità di comprare una farm.

Buon terreno nello Stato di New York può presentemente essere acquistato dai 25 ai 50 dollari per Acre. Tale prezzo include case confortabili, arghi granaii, e tutti i necessarii stabili per una farm. Il terreno è tutto coltivabile, e con intelligente cultura, si possono ottenere abbondanti prodotti.

Qui appresso enumeriamo alcuni dei tanti vantaggi che lo Stato di New York offre ai farmaiuoli, i quali sono estensibili tanto agli Americani, quanto agli Stranieri.

1.° Nello Stato di New York cresce abbondantemente ogni prodotto adequato al suo clima, i quali, ben maneggiati danno degli splenditi risultati.

Lo Stato di New York produce più eno, prodotti latticinii, patate ed vena, che qualunque altro Stato d'America. Ed esso è annoverato tra i primi Stati per la produzione del grano, granone, frutta, pollame etc.

Quasi tutti i prodotti della Germania, Hungary, Polonia, Francia, Nord Italia, Boemia, Olanda, Svezia etc. nascono nello Stato di New York, e perciò non vi è ragione di dubitare che un esperto coltivatore, proveniente da tali paesi non possa essere abile a procacciarsi un buon sostentamento continuando la lavorazione del terreno nello Stato di New York.

2.° Farme possono essere acquistate oggi giorno nello Stato di New York, molto al disotto del loro valore reale, ed in molte circostanze a prezzi al disotto del valore degli stabili contenuti in esse.

In un numero di casi, queste farme a buon mercato, dopo 2 o 3 anni, hanno apportato all'industrioso coltivatore, una rendita superiore al prezzo d'acquisto della farma stessa. Come per esempio, una farm che parecchi anni fa, venne acquistata, nel centro dello Stato a $15 l'Acre, dopo 2 anni produsse un ricolto di piselli, valutato a $50 per Acre.

3.° Il farmaiuolo di New York ha il mercato di sbocco, proprio alla sua porta. Il gran numero di Città

PASCOLI UBERTOSI PER LE PECORE

7

There is an increasing demand all the time for the products of the New York farms; and the foreign-born farmer who does not know the language of this country will get just as good prices and can sell his products just as quickly as anyone else.

FOURTH: New York has an excellent climate the year around. Near neighbors are at hand everywhere. Good roads are found all through the State. The schools are the best in the world. Taxes, too, are very low in New York State.

FIFTH: New York State has a law which protects farmers against losses of many kinds. This is particularly beneficial for the farmer who has come here from abroad.

LA CITTA DI NEW YORK E UN INESAURIBILE
MERCATS PER LA VERDURA

CROP YIELD PER ACRE.

Crop	Yield per acre
Corn	32-100 bu.
Wheat	21-30 bu.
Oats	34-50 bu.
Barley	25-35 bu.
Rye	16-25 bu.
Buckwheat	19-25 bu.
Potatoes	75-350 bu.
Hay	1½-5 tons
Tobacco	1150-1800 lbs.
Cabbage	20-40 tons
Brussels sprouts	3000 qts.
Cauliflower	5000-6000 heads
Celery	10,000-25,000 bunches
Asparagus	2000-3000 bunches
Gooseberries	200-400 bu.
Currants	100-150 bu.
Cranberries	175-350 crates
Snap Beans	75-120 bu.
Lima Beans	70-100 bu.
Beets	400-700 bu.
Carrots	500-750 bu.
Cucumbers	150,000
Grapes	3-10 tons
Horse Radish	3-5 tons
Onions	450-900 bu.
Parsnips	600-900 bu.
Green Peas	100-150 bu.
Peppers	30,000-50,000
Raspberries	50-100 bu.
Blackberries	50-100 bu.
Quince	100-300 bu.
Spinach	200-250 bu.
Strawberries	100-350 qts.
Tomatoes	10-15 tons
Turnips	800-1000 bu.

CROP YIELD PER TREE.

Apples	25-40 bu.
Peaches	5-10 bu.
Plums	6-8 bu.

8

LAVOCATORI NEI CAMPI DI GRANONE

nello Stato, con i 6,000,000 di popolazione nella Città di New York e d'intorni, assicurano il coltivatore di un'immediata vendita dei suoi prodotti, ed ai migliori prezzi ottenibili.

Vi è una continua domanda di prodotti dalle farme di New York, ed i coltivatori stranieri, che non comprendono la lingua del paese, sono sicuri di fare tali prezzi, e vendere le loro derrate così presto, come i coltivatori americani.

4.° Lo Stato di New York, gode di un eccellente clima tutto l'anno. Si è in contatto di proprietarii di altri poderi continuamente. Lo stato è intersicato da buonissime vie di comunicazione. Le scuole sono le migliori nel mondo. Le tasse sono molto basse nello Stato di New York.

5.° Lo Stato di New York ha una legge che protegge il coltivatore con-

tro molte perdite di diversi generi. Questa è particolarmente favorevole per i coltivatori che vengono dall'estero.

Quello che potete produrre in ogni Acre di terreno nello Stato di New York.

PRODOTTI PER OGNI ACRE

Granturco	da 32 a 100	bu.
Grano	" 21 a 30	bu.
Orzo	" 34 a 50	bu.
Avena	" 25 a 35	bu.
Segala	" 16 a 25	bu.
Frumento	" 19 a 25	bu.
Patate	" 75 a 350	bu.
Fieno	" 1½ a 5	tons.
Tabacco	" 1150 a 1800	lbs.
Cavoli	" 20 a 40	tons.
Brussels Spronts	" 3000	qts.
Cavoli di fiore.	" 5 a 6000	capi.
Sedani	" 10 a 25000	mazzi
Asparigi	" 2 a 3000	mazzi
Gooseberries	" 200 a 400	bu.
Currants	" 100 a 150	bu.
Cranberries	" 175 a 350	craters
Fagiuoli	" 75 a 120	bu.
Fagiuoli grandi.	" 70 a 100	bu.
Barbabietole	" 400 a 700	bu.
Carote	" 500 a 750	bu.
Cocomeri	" 150,000	
Uva	" 3 a 10	tons.
Radici	" 3 a 5	tons.
Cipolle	" 450 a 900	bu.
Parsnips	" 600 a 900	bu.
Piselli	" 100 a 150	bu.
Peperoni	" 30000 a 50000	bu.
Raspberries	" 50 a 100	bu.
Blackberries	" 50 a 100	bu.
Quince	" 100 a 300	bu.
Spinacci	" 200 a 250	bu.
Fragole	" 100 a 350	qts.
Pomidori	" 10 a 15	tons.
Turnips	" 800 a 1000	bu.

FOR THE MAN WHO WANTS TO WORK ON A FARM.

There is a great demand for farm labor in New York State, commencing April 1st of each year. For a man foreign born, who knows anything about the handling of milch cows or general farming there is a ready job waiting at good wages.

Farmers in New York State pay from $25 up per month, with board, for farm help; and where the men board themselves, they receive correspondingly higher wages.

A good many men have started in working on a farm in New York State, apply to

COMMISSIONER OF AGRICULTURE,
ALBANY, N. Y

EDUCATION FOR NEW YORK STATE.

Learn to speak English as quickly as possible. You make more money get on faster, and are less likely to be injured at work if you learn English Get the addresses of night schools i your neighborhood or get a number o. your countrymen together and ask fo a teacher in English.

All children between six (6) an fourteen (14) years of age ar

COMODA CASA COLONICA

State, saved up their wages until they had enough money to buy a farming outfit of their own, when they obtained a farm to work on shares. In several years they have been able to buy the farm. A number of men who want to sell their farms will take a small payment down and let the rest be paid for from the products of the place, especially where a man has worked for them and they know that he is honest and reliable.

For detailed information regarding farms for sale and also regarding places to work on farms in New York

obliged by law to attend school and must know how to read and write before they can stop going to school There are free public schools all ove the country. Parents who do no send their children to school are liable to arrest and fine.

WORK.

Bring the birth certificates of al your children coming to this country This is the quickest and surest way of getting their "work papers."

Children under fourteen (14) wil not be allowed to work in a factory or store.

10

PRODOTTI PER OGNI ALBERO

Mele ....25—40 bu.
Pesche .... 5—10 "
Susine .... 6— 8 "

Pel lavoratore che voglia lavorare in un podere (farm)

Vi è una grandissima richiesta di lavoratori di terreno nello Stato di New York, cominciante dal 1.° di Aprile di ogni anno. Per qualunque straniero, che sappia qualche cosa riguardo l'allevamento degli animali, mungere il latte, ed il lavoro di campagna in generale, vi è subito un posto pronto ad attenderlo e buon salario.

I padroni di poderi pagano ai lavoratori da $25 al mese in su, oltre il vitto e letto. In casi ove i lavoratori pensano da loro per il mangiare e dormire i salarii sono corrispondentemente più alti.

Una grande quantità di persone incominciarono coll'andare a lavorare sul podere posseduto da un altro. Risparmiarono i loro salarii, tanto da comprarsi gli utensili per lavorare, indipendentemente, ottenendo una farm, i cui profitti venivano divisi col proprietario, e dopo pochi anni divenirne i padroni assoluti.

Molti si contentano di vendere i loro poderi, accettando solamente poco contante in acconto, ed il resto gradatamente col ricavo delle vendite dei prodotti della farma stessa, specialmente in casi, quando la persona che acquista ha già lavorato per il padrone che vende, ed ha dato prova della sua onestà ed esattezza.

Per dettagliate informazioni riguardante terreni da vendere, oppure per ottenere posti da lavoratori nelle farms nello Stato di New York, dirigersi al *Department of Agriculture* Albany, N. Y.

L'Istruzione nello Stato di New York.

Imparate a parlare l'Inglese il più presto possibile. Esso vi abilita a guadagnare più moneta—a crearvi più speditamente, e c'è minore probabilità di capitare in qualche accidente, mentre sul lavoro. Procuratevi l'indirizzo di una scuola serale nel vostro vicinato, oppure riunitevi una quantità di paesani volenterosi e fate domanda per un istruttore d'Inglese.

Tutti i ragazzi tra i 6 ed i 14 anni d'età sono obbligati dalla legge di attendere alla scuola, e debbono apprendere a leggere e scrivere, prima che sia loro permesso di lasciare la scuola.

Vi sono scuole gratis ovunque in America. I genitori i quali non mandano i loro figli alla scuola, possono venire arrestati e sottoposti a multa.

LAVORO

Conducete con voi dai vostri paesi d'origine i certificati di nascita dei vostri figli. Questo è il metodo più sicuro e spicciativo per procurare loro il permesso pel lavoro (work papers).

Ragazzi aldisotto dei 14 anni di età non possono lavorare nelle fattorie o

UNA VIGNA NELLO STATO DI NEW YORK

11

D'ISINFEZIONED ALBERI DI MELE A
MEZZO DELL' INNAFFIAMENTO

that every one may be protected from
disease.

Report any case of disease in your
family to the Board of Health at 55th
Street and 6th Avenue, New York
City, and you will be told what to do.

NATURALIZATION

All men and women over eighteen
(18) years of age meaning to make
their home in the United States

GRAPPOLI D'UVA

Children between fourteen (14)
and sixteen (16) cannot go to work
until they get a paper from the Board
of Health showing that they are fourteen (14) years old, have been to
school and are strong enough to work.

Children under sixteen (16) in factories can only work between eight
A. M. and five P. M., never at night,
and only eight (8) hours a day for six
days a week. In stores nine (9)
hours and until seven P. M.

Work cannot be done at home by
families living in tenements unless
the landlord has a license allowing
home work in his building, from the
Commissioner of Labor, 381 Fourth
Avenue, New York City.

In America there are many occupations in which an immigrant cannot
make a living unless he has applied
for his first citizenship papers or becomes a citizen. One of the first
things to do is to ask about these occupations and to apply for citizenship
papers.

HEALTH.

The Board of Health of each city
prints laws about the care of the
health, the streets and houses in order

12

nei negozii.

Ragazzi tra i 14 ed i 16 anni di età non possono andare a lavorare, se non ottengono una carta dal Board of Health, nella quale è dichiarata, che hanno raggiunto il 14° anno di età hanno frequentato la scuola e sono forti abbastanza per lavorare. Ragazzi sotto il 16° anno d'età possono lavorare in fattorie tra le 8 ant. e le 5 p.m., mai di notte, e solamente 8 ore al giorno per 6 giorni della settimana. Nei negozii 9 ore, sino alle 7 di sera.

Nessun lavoro può essere fatto a domicilio dalle famiglie, senza che il padrone della casa, abbia ottenuto dal Commissioner of Labor 381 Fourth Ave., New York, una regolare licenza.

In America vi sono molti impieghi, che l'immigrato, non può occupare, ammenochè non abbia fatto domanda per la prima carta di naturalizzazione, oppure sia divenuto cittadino. Perciò una delle cose più importanti è d'informarsi di tali impieghi e fare immediatamente domanda per la Carta di Cittadinanza.

UN ALBERO DI PESCHE DI FOUR ANNI NELLO STATO DI NEW YORK

should declare their intention to become citizens, so that they may be naturalized.

No knowledge of English is necessary for the first paper, which can be got from the Federal Court or the Courts of Record of the State in the district in which you live, soon after you land and costs one dollar ($1.00).

To get the full naturalization paper, you must live in the country five years, speak English, understand something about the government of the country, as well as prove by two friends that you are of good moral character, take the oath of allegiance to the United States Government and renounce the country of your birth. This second paper costs four ($4.00) dollars.

FREE BUREAUS OF INFORMATION.

The State of New York has established at 40 E. 29th St., New York City, and at 165 Swan St., Buffalo, N. Y., a Bureau of Industry and Immigration for the benefit of immigrants who are in the State of New York.

The Bureau is open daily from 9 A. M. to 3 P. M., Wednesday evenings 7 to 10, and Sunday mornings 9.30 A. M. to 1 P. M. Services are entirely free, and may be obtained by telephoning 7779 Madison, in New York City, or calling in person at the office.

The Bureau is a place where any immigrant may bring his troubles, and where he may receive information, advice, direction and assistance which will enable him to help himself.

Many immigrants, new in this country, ignorant of the English language, without friends or employment, are without the means to secure a hearing or the rights now guaranteed to them by the laws of this country.

UNA MODERNA FATTORIA PER POLLICULTURA

14

IGIENE

Il Dipartimento di Salute Publica di ogni città, stampa continuamente per uso publico le Leggi, per la cura della Salute, pulizia delle strade, case etc. in modo che ognuno possa previnirsi da malattie.

Qualunque caso di malattia nella vostra famiglia se verrà riportato al Board of Health, 55 Strade e 6sta Avenue, nella città di New York—otterrete tutte li informazioni del caso per come potervi regolare.

Per Naturalizzarsi Cittadini Americani

Uomini e donne al disopra dei 18 anni di et á, che intendono di stabilirsi negli Stati Uniti, debbono dichiarare la loro intenzione a voler divenire cittadini, in modo da essere naturalizzati.

Nessuna conoscenza dell'Inglese è necessaria per la prima carta, la quale può essere ottenuta dalla Corte Federale, oppure dalla Corte dei Records dello Stato, nel distretto in cui voi abitate, appena dopo sbarcato, e costa uno scudo.

Per ottenere la piena Carta di Cittadinanza, bisogna aver dimorato negli Stati Uniti 5 anni, parlare Inglese, e comprendere qualche cosa delle leggi del paese. Provare per mezzo della testimonianza di due testimoni di essere di buon carattere morale. Prendere il giuramento di ubbidienza al Governo degli Stati Uniti, e rinunziare al paese d'origine. La Seconda Carta costa 4 dollari.

Ufficio d'Informazioni Gratis.

Lo Stato di New York ha stabilito al No. 40 E. 29th St., New York City, ed al No. 165 Swan St., Buffalo, N. Y. un Bureau of Industries and Immigration per la protezione degli immigranti, nello Stato di New York.

L'Ufficio è aperto giornalmente dalle 9 a.m. ale 5 p.m. Mercoledì sera dalle 7 alle 10, e Domenica mattina dalle 9.30 a.m. alle 1 p.m. Qualunque cooperazione viene resa gratis, ed essa può ottenersi a mezzo del te-

FIENO COCCOGLITICCIO

lefono 7779 Madison, nella città di New York, oppure presentandosi in persona all'ufficio.

Il Bureau, è un luogo, ove qualunque immigrante può indirizzarsi per ottenere informazioni, consigli, direzione· ed assistenza.

Molti immigranti, appena arrivati, ignoranti della lingua inglese, senza amici e lavoro, e senza mezzi sono impossibilitati procurarsi i mezzi per ottenere che i loro reclami venissero ascoltati, e potessero godere dei benificii e dei diritti che le leggi del paese loro accordano.

15

The Bureau does not furnish employment, but it directs men and women to reliable places where employment is furnished, and where they will be treated honestly and dealt with in a business-like way.

The Bureau protects immigrants, and is in a position to help those who have been exploited, cheated, mistreated, misinformed or intimidated by either their own countrymen or Americans with whom they have to deal, as employment agents, padrones and commissaries in labor camps, employers, foremen and contractors, private bankers, notaries public, emigrant home and boarding houses, steamship ticket agents, advocates, military and inheritance agents, loan and land companies, medical companies, benevolent societies.

Call or write in any language to these addresses:

New York State Bureau of Industries and Immigration, 40 East 29th St., New York City. Created by the State to look after your interests and furnish reliable information of any kind.

New York State Department of Agriculture, 23 Park Row, New York City. Provides a bureau where full and reliable information may be had, free of charge, concerning agricultural opportunities in the State of New York.

United States Division of Information, 17 Pearl Street, New York City, maintains this division to give information concerning all matters of interest to immigrants; where land may be purchased or rented, where farm work or work of other kinds may be obtained, and general information concerning all of the United States. You may write to this division for information, no matter where you locate in the United States. No charge is made for any service rendered.

IL PRIMO ABBEVERAGGIO

Il "Bureau" non fornisce lavoro, ma dirige uomini e donne presso agenzie oneste, ove il lavoro viene accordato, e dove saranno trattati onestamente, e lealmente.

Il "Bureau" protegge gl'immigranti, ed è in posizione di aiutare tutti quelli che sono stati sfruttati, derubati, maltrattati, mal informati oppure intimiditi sia dai loro compatrioti, oppure da americani nel corso delle loro transazioni, sia da Agenti di Collocamento al lavoro, padroni, Commissarii nei lavori di campagna, foremen, e contrattori—banchieri privati, notaii publici, emigrant home, and boarding houses, agenti marittimi, avvocati, agenti militari, agenti di eredità—Compagnie per la vendita dei lotti di terreno—Compagnie di prestiti—Compagnie mediche —Società filantropiche.

Recatevi personalmente a questi indirizzi, oppure scrivete in qualunque lingua.

New York State Bureau of Industries and Immigration, 40 E. 29th Street, New York City.

Stabilito dallo stato per proteggere i vostri interessi e fornirvi informazioni di qualunque specie.

New York State Department di Agricultura

23 Park Row, New York City

Questo Departimento provvede un Bureau, ove complete ed autentiche informazioni possono ottenersi, gratis, in riguardo delle opportunità agricole nello Stato di New York.

La United States Division of Information.

17 Pearl Street, New York City

Mantiene una "Divisione" per dare informazioni, concernenti fatti d'interesse agli immigranti; dove terreno può essere acquistato od affittato—dove lavoro di campagna, o qualunque altro lavoro può essere ottenuto, e qualunque altra informazione, concernente gli Stati Uniti.

Potete scrivere a questa Divisione per informazioni, da qualunque luogo voi risediate.

Nessun compenso è richiesto per sarvigi resi.

LE ANITRE SI VENDONO A CARO PREZZO IN NEW YORK

17

MONEY.

The American coins are:

a copper coin called a cent...................	1 cent
a nickel coin called a nickel	5 cents
a silver coin called a dime.................	10 cents
a silver coin called a quarter..............	25 cents
a silver coin called half a dollar.	50 cents
a silver coin called a dollar	100 cents

There are paper bills for one dollar, two dollars, five dollars, ten dollars and twenty dollars, and gold coins for two dollars and a half ($2.50), five dollars ($5.00) and twenty dollars ($20.00).

The dollar has the following value in foreign money:

$1.00 is about	5	kronen in Austria........10 kronen equals......$2.03
1.00 " "	3⅘	crowns in Denmark......10 crowns " 2.68
1.00 " "	5	francs in France........10 francs " 1.93
1.00 " "	4⅛	marks in Germany.......10 marks " 2.38
1.00 " "	4	shillings in Gt. Britain... 1 pound " 4.86
1.00 " "	5	drachmas in Greece......10 drachmas " 1.93
1.00 " "	5	lire in Italy.............10 lire " 1.93
1.00 " "	2½	florins in Netherlands....10 florins " 4.02
1.00 " "	5	crowns in Norway.......10 crowns " 2.68
1.00 " "	1	milreis in Portugal...... 5 milreis " 5.40
1.00 " "	2	rubles in Russia.........10 rubles " 5.15
1.00 " "	5	pesetas in Spain.........25 pesetas " 4.82
1.00 " "	20	piastres in Turkey......100 piasters " 4.40
1.00 " "	3⅘	crowns in Sweden.......10 crowns " 2.68

18

MONETA.

I denari di metallo in America sono:

di rame, chiamato un cent, 1 cent
di nickel, chiamato un nickel, 5 cents
di argento, chiamato un dime, 10 cents
di argento, chiamato un quarter, 25 cents
di argento, chiamato mezzo dollaro, 50 cents
di argento, chiamato un dollaro, 100 cents.

Vi è moneta di carta per 1 dollaro, 2 dollari, 5 dollari, 10 dollari, 20 dollari. Di oro per 2 dollari e mezzo ($2.50) 5 dollari ($5.00) e venti dollari ($20.00).

Il dollaro corisponde ai seguenti valori esteri:
$1.00 è circa 5 kronen in Austria, 10 corone sono uguali $2.03
$1.00 è circa 4⅘ crowns in Danimarca, 10 corone sono uguali $2.68

$1.00 è circa 5 francs in Francia, 10 franchi sono uguali $1.93
$1.00 è circa 4⅛ marks in Germania, 10 marki sono uguali $2.38
$1.00 è circa 4 shellini in Gran Brettagna, 10 dramma sono uguali $1.93
$1.00 è circa 5 drachmas in Grecia, 1 sterline sono uguali $4.86
$1.00 è circa 5 li erin Italia, 10 lire $1.93
$1.00 è circa 2½ fiorini in Olanda, 10 fiorini sono uguali $4.02
$1.00 è circa 5 corone in Norvegia, 10 corone sono uguali $2.68
$1.00 è circa 1 Milreis nel Portogallo, 5 milreis sono uguali $5.40
$1.00 è circa 2 rubli in Russia, 10 rubli sono uguali $5.15
$1.00 è circa 5 pesetas in Ispagna, 25 pesetas sono uguali $4.82
$1.00 è circa 20 in Turkia, 100 piastres sono uguali $4.40
$1.00 è circa 3⅘ corone nella Svezia, 10 corone sono uguali $2.68.

19

In the Italian Quarter of New York

By Joseph H. Adams
Brooklyn, N. Y.

Published by
The Congregational Home Missionary Society
287 Fourth Avenue, New York

ITALIAN CHILDREN IN REAR YARD OF A MULBERRY STREET HOUSE

IN THE ITALIAN QUARTER OF NEW YORK

By Joseph H. Adams, Brooklyn, N. Y.

THE old original Italian quarters in New York City were commonly called "hotbeds of crime." For years the worst of the inhabitants of southern Italy and Sicily have poured into certain districts of New York and the surrounding country, and these aliens have given the police no end of trouble for the past fifty years or more. The "Black Hand," the "padrone," and the "vendetta" gangs have opposed government and tried to overthrow social and moral standards, and it was not until strenuous laws and immigrant regulations had been enforced that this undesirable tide was in a measure checked. The fines imposed on the steamship companies and the very rigid rules carried out by the Government officials at Ellis Island have resulted in the careful sifting that debars a greater portion of the really undesirable and criminal classes.

Years ago the Italian quarter properly extended three blocks each way from the famous "Mulberry Bend," the turn in Mulberry Street just above Worth that has since been obliterated on one side by that garden spot and playground, Mulberry Bend Park. For years the buildings which stood on this now beautiful spot were known as the blackest holes in New York City, and the number of crimes committed around these two blocks was a disgrace to a community of civilized people in which progress is the distinctive characteristic. This locality, from Five Points up to Canal Street, was so infested with the worst and most desperate thieves and murderers years ago that it would cost a detective's life to try to ferret out suspected criminals. When the old buildings were razed to the ground and the foundations brought to view, numerous passages were found that led from cellar to cellar, and several under the streets, so that pursued criminals could make good their escape through these hidden "undergrounds." When excavating and leveling the ground for the present park, it was a common thing to unearth bones and scraps of clothing—grim reminders of some uncanny burial in a back yard, the result of a murder or "unknown death." The police records show that many who were enticed to the buildings in this section from 1830 to 1870 were lost forever, swallowed up without a trace, but the ghastly disclosures of years later bore the mute evidence of past crimes.

Since the great influx of the Italian race into this country a quarter of a century or more ago, the limited district about Mulberry Street east and west as far up as Bleecker Street and between the Bowery and the boundary line, Centre Street, could not hold the ever increasing population, and since about 1870 the Italians have spread to the four quarters of the city, to Brooklyn and the Bronx, and to the colony known as "Little Italy," located on the East Side between Ninety-fifth and One Hundred and Tenth Streets from Second Avenue to the East River. The largest Brooklyn settlement is known as "Paradise Park," a low stretch of land thickly populated with Italians, extending from Blythebourne to Bath Beach between New Utrecht and Twelfth Avenues. This section gives the police of Brooklyn considerable trouble, for here stabbing affrays, "Black Hand" tragedies, and murders often occur. A large percentage of the criminal court cases involve Italians, as their quarrelsome, hot-headed ways and their thirst for revenge to right some trivial offense or imaginary insult lead them into all sorts of trouble, which frequently results in a murder. The educated Italians refrain from disturbance of any kind, and try as far as possible to quell any riotous feeling among the others; but in the lower classes that colonize here instead of spreading out over the country, there will always be more or less trouble until education and mission work shall appeal to the better side of their natures and give them something better to look forward to than chicanery, cheating, and intrigue, combined with stealth to carry out some blackmailing scheme against their more prosperous neighbors.

The Italian quarter is the home of the padrone system, and while the police and the Gerry Society have been instrumental in breaking the backbone of this oppressive system with regard to the children, who in former years were compelled to beg, steal, and prey upon the public by soliciting alms under false pretenses, no amount of pressure brought to bear on the heads of these alien companies has been effective in breaking them up. In fact, the strongholds are either unknown, or if in sight, the head of the padrone colony cannot be found. He "is not known," and it is the policy of those around him to know nothing about him, his whereabouts, or his business. The monstrosity, the oppression, and the penalties of this system are little known to the average New Yorker, but many sad-eyed Italians are selling fruit in the streets, working on bootblack stands, grinding organs, gathering rags, and looking over the dumps at the river wharves, half or more of whose earnings are going into the pockets of the lazy padrones who "toil not, neither do they spin," but live in luxury and on the fat of the land through the sweat of their feebler countrymen's brows.

From certain alleys, every morning bright and early one may see the procession of push cart fruit venders —slaves of the system, for the padrones are large wholesale purchasers of fruits, nuts, and flowers. Each peddler has his cart and number, and is accountable for his load of fruit. Not one can escape the lash of the system, and should any of them get into trouble and be arrested word is quickly passed and a bondsman soon appears. The padrone knows his men and watches over them, protecting them—not for themselves, but for his profit. Then at night when the returns come in each peddler is given a daily pittance, and the balance goes to the lord and master, for does he not own the cart, the fruit, and—yes, the soul and body of the peddler, who is kept in constant fear and slavery through the system?

The writer has stood near a runway leading to a basement and watched no less than fifty organ grinders coming home after a day's work, each with the same sort of instrument, and having entered the basement the organs were wheeled in and arranged along the wall in rows. The "agent," a trusted confederate of the padrone, received the money collected by the Italians, who were then handed checks for the day's work. Sometimes a lively scene follows small receipts, and on several occasions in the same night the writer has seen tired organ grinders thrown down and their clothing searched. One suspect who had saved some small change in his coat lining was treated to a sound thrashing, and he dared not cry out, for the system is so subtle and sure that any offense against the "head one" would bring on disastrous results.

Huddled in the closely confined quarters of the Italian tenements there are numerous artificial flower establishments, where great numbers of Italian women, girls, and children work about long tables, deftly fashioning bits of cloth into bunches of gaily-colored flowers for cheap decorations of all sorts. Many of these girls have come over from Italy ostensibly to visit aunts and uncles, who vouched for their maintenance and support, but who in reality sell them or their labor into the padrone system. Little does the child or young woman know that she is to enter a form of slavery. She is put to work in a flower shop or at sorting rags to work out her passage

REAR OF A TENEMENT IN THE ITALIAN QUARTER

money, and it takes oh, so long! because she does not know how much she earns—that is kept from her. She is ill fed and poorly clad, and thrown into company with women of sensual habits, lewd thoughts, and demoralizing language, and the evils of such a life constitute one of the curses of the densely populated Italian districts—the padrone system again, which is felt in every section and in every household.

Throughout the Italian neighborhoods one will find basements in which rags are sorted—rags of all kinds and colors, some filled with disease-breeding germs and all more or less filthy —for it is almost alone the Italian who will gather rags, bottles, and discarded household articles from the ash cans and streets, sort and classify them, and get something out of them. The Italians of the tenements and slums—those of them who will work —always select that class of occupation which offers the least resistance. They will stand all day in the street beside a push cart full of fruit or peanuts, but few are found in factories or shops where skilled labor is employed, or in any of the callings where other nationalities abound. To be sure, the Italians will work on subways, tunnels, ditches, foundations, and other work where the highest wages are paid and the least brain work required. As a gang foreman of a squad working on the tracks of a railroad once said, "They make good mud-slingers, but very poor soldiers." The remark was made after one of their number accidentally stepped on a third rail and was instantly killed. The fifty or more laborers lost their heads completely. They were panic-stricken, and ran in all directions like a scattering flock of sheep, and it was not until the foreman shot his pistol into the air that they stopped and returned, knowing well that trouble would be the result of disobedience, for the padrone got them their jobs and he grafts some of their wages. The tribute paid to these usurers is the blood money wrung from these simple-minded sons and daughters of Italy, held under the lash through ignorance of our laws and customs, whereas they could become independent and properly self-supporting under other conditions, and if educated, many of their troubles could easily be overcome.

There is not an article of food or clothing that does not find its way into the Italian section at one time or another, and some of them are manufactured right in the neighborhood for local consumption. Mulberry and Elizabeth Streets from Worth to Grand or to Bleecker Street present an interesting panorama, closely re-

A GENERAL STORE IN THE ITALIAN QUARTER

AN ITALIAN SPAGHETTI SHOP

sembling the quarters in Rome, Florence, and Milan, where the general melange of humanity and its needs are jumbled together in an indescribable mass. The street and sidewalks teem with merchandise on stands, push carts, old wagons, and trays strapped to the shoulders of the venders, for no reasonable chance to attract the purchaser is lost by the Italian merchant. As one passes along the street, it is impossible to take everything in at one trip. The shops overrun with all sorts of wares, both foreign and domestic, and mingled with the Italian names of the wares and their prices, are the gaudy signs of cheap domestic manufactured soaps, washing powders, and matches. Bread is seldom sold within the buildings except on rainy days. It is placed outside on wooden trays, and even hung up on the sides of the buildings on nails to attract attention. Tubs of pickles that can be smelled a block away, dried fish, baskets full of dandelion plants fresh from the country, barrels of vegetables, tubs of dried beans and peas, cocoanuts, bananas, strings of red peppers and garlic, blad-

AN ITALIAN BREAD VENDER

ders full of imported lard, bundles of kindling wood, dried beans, and links of sausages, are but a few items of the miscellaneous assortment that one sees in traveling but a few steps. Here and there throughout the district one will see the spaghetti shop, outside of which the "green" article fresh from the macaroni press, is hanging on long poles and moving in the breeze as a field of ripe wheat is swayed by the wind. Inside the shop are also rows of poles loaded with spaghetti, and with the often filthy conditions prevailing in these shops and the dirt blown on the macaroni hanging outside, one can imagine the great amount of filth, disease germs, and other impurities the Italians take into their systems through this one medium alone.

One of the greatest benevolent features that can be carried out in the Italian sections is to teach the children the English language and get them into the schools and keep them off the streets and away from the demoralizing influence of the men and women of questionable character who inhabit these localities. In recognition of the advantages of education to these alien children of the old quarters, The Board of Education of New York City maintains a school especially for Italian children, the same as it does for the Negro children, in the district lying between Twenty-third and Thirty-fourth Streets and Sixth and Ninth Avenues, and in this school particular attention is given to teaching not only the English language, but the advantages of freedom, its proper meaning and limitations, and in so doing it is calculated that in time the oppression practiced on the ignorant classes will be done away with. There are certain conditions, however, that cannot be met in this way, but these are well within the scope of mission work, for that which is already being done is meeting with good success. The city is doing what it can in the way of schools, dispensaries, and tenement house regulation, but the personal side of the inhabitants of these quarters can only be reached by the mission workers, and those directly interested in the personal conditions and habits of these people.

[NOTE.—This leaflet and several others published by The Congregational Home Missionary Society give a very accurate description of the conditions under which thousands of Italians in New York City are living. What is true of New York applies to almost every large city in the United States.

The Society is at present maintaining twenty-nine missions among Italian immigrants in this country, and the calls for several times that number are constant. We are hoping that the time will come when there will be sufficient funds in our Treasury to enable us to do the work needed to bring these foreign fellow citizens to a truly American standard of living and religious faith.]

ITALIAN CHILDREN OF CAMP MEMORIAL MISSION, NEW YORK CITY

ITALY AMERICA DAY
CELEBRATION

in recognition of

THE FIFTH ANNIVERSARY
OF ITALY'S ENTRANCE INTO THE WAR

THE HIPPODROME
SUNDAY EVENING, MAY TWENTY-THIRD
1920

UNDER THE AUSPICES OF
THE ITALY AMERICA SOCIETY

ITALY AMERICA DAY
CELEBRATION

THE HIPPODROME
SUNDAY EVENING, MAY TWENTY-THIRD
1 9 2 0

AN AMERICAN TRIBUTE
TO ITALY

WILLIAM HOWARD TAFT
Former President of the United States

AM glad to hear that the citizens of the United States of Italian origin and American friends of Italy are about to celebrate the anniversary of Italy's entering the great war for world liberty. Her sons did such noble work for that cause at most critical periods, that all who love her may well rejoice to commemorate her great war days.

Italy and the United States have always been close friends. We cherish her great names. Mazzini, Garibaldi, and Cavour do not belong to Italy alone, but to all who value liberty and progress.

We in this country earnestly hope and pray that Italy in the burdensome work of reconstruction and in her recovery from the effects of her enormous sacrifices, may steadily march toward the normal, and win renewed happiness for her people.

ALFRED E. SMITH
Governor of New York State

EN of Italian birth have for centuries been leaders in the development of our civilization. Their accomplishments in the fields of art, letters and science have been great and far-reaching. We of America have benefited by their genius and their scientific attainments, as well as by their physical brawn and perseverance.

To our admiration for Italy's glorious past is

3

now added a profound respect and an undying gratitude for the magnificent part which she played in the war for liberty. Under conditions of staggering difficulty, she showed an idealism, a heroism, a sacrifice and an achievement that have challenged the world's admiration.

In the difficult tasks of reconstruction, Italy is again in the forefront of courageous endeavor. No other country has equalled her in the promptness and vigour with which she has met the problems of economic rehabilitation. In boldness, in thoroughness, in fairness, and in effectiveness, her fiscal measures point out the way to the nations of the world.

WILLIAM G. McADOO
Former Secretary of the Treasury

TALY'S notable contributions to civilization have earned for her the enduring admiration of mankind. Her achievements in the recent war for liberty and democracy in which America engaged with such enthusiasm and effect have strengthened the ties of international friendship which have long existed between the two nations, and assure a future development and co-operation which will benefit them and the world.

WILLIAM GUGGENHEIM

TALY has admirably weathered the greatest storm that has ever threatened the civilization of the world, and during the trials and tribulations that have necessarily followed, she commands our deepest sympathy, which will not be denied her.

America's unbounded admiration and respect

4

for Italian greatness as fostered by Mazzini, Cavour, and Garibaldi, and fully sustained by those momentous events of the World War, suffer no diminution with the march of time.

FRANK W. TAUSSIG
Professor of Economics, Harvard University

AM impressed with the vigour and courage with which Italian statesmen have grappled with their extraordinarily difficult financial problems. Difficulties of the same sort exist in every country of Europe, and in few, if indeed in any, have they been faced with so much frankness and courage. The laws on the taxation of war profits and on the taxation of property seem to me deserving of great recognition. They indicate a determination to face the need of great sacrifices. Drastic measures are indispensable, and it is very greatly to the credit of Italian statesmen that they have not hesitated to enact them.

HERBERT HOOVER

IVE years ago today the great Italian nation, loyal to the spirit of nobility and sacrifice that marks the traditions of her glorious past, cast in her lot with the nations which were fighting against the oppressive ambitions of the central powers. Hundreds of thousands of American citizens followed her every move with eager sympathy, quick to glory with her in her proud achievements, in the bravery and daring of her soldiers, and sad in her noble losses. When later years brought the youth of this nation to European soil to fight a common enemy, the fortunes of war decreed that many of our best

5

should render their services in Italy itself, whether on the flying fields, in the ambulance service, or through the auxiliaries of the Red Cross or Y. M. C. A. There, common service and common sacrifice welded enduring bonds of admiration and friendship.

The people of Italy and America have both made their struggle for national unity and freedom. They have been moved by the same impulses and the same ideals. Their ties of sympathy have ever been close, but their recent association has brought new understanding to both. I am convinced that this new understanding will grow with time and will bring into the relations of these two great nations a new era of co-operation and affection.

LIEUT COLONEL THEODORE ROOSEVELT

IT gives me great pleasure to add my word of endorsement in behalf of what Italy did during the war for civilization. Her service was signal and her loss great. I sincerely trust that this country may make it evident by its attitude on all questions that the heartiest understanding exists and will continue to exist between the two countries.

HONORARY COMMITTEE

CHESTER H. ALDRICH

B. D. ATTOLICO

GEORGE F. BAKER, JR.

JAMES BARNES

HARRY S. BLACK

WALTER LAWRENCE BOGART

JAMES BROWN

CHARLES S. BURCH

BERTRAM DE N. CRUGER

JAMES H. DARLINGTON

SHERMAN DAY

JOHN ADAMS DRAKE

BENJAMIN FAIRCHILD

AUSTIN B. FLETCHER

DANIEL CHESTER FRENCH

THOMAS STAPLES FULLER

ALBERT EUGENE GALLATIN

LLOYD C. GRISCOM

AUGUSTUS V. HEELY

MYRON T. HERRICK

RICHARD H. HUNT

WILLIAM FELLOWES MORGAN

FRANK A. MUNSEY

FREDERIC NEWBOLD

SAMUEL L. PARRISH

GUIDO PEDRAZZINI

ROBERT P. PERKINS

EDOARDO PETRI

FRANKLIN A. PLUMMER

KERMIT ROOSEVELT

THEODORE ROUSSEAU

LINDSAY RUSSELL

HOWARD VAN SINDEREN

CHARLES STEELE

H. K. TWITCHELL

WHITNEY WARREN

GEORGE W. WICKERSHAM

LOUIS WILEY

7

ITALY AMERICA DAY
CELEBRATION

▽

This theatre, when filled to its capacity, can be emptied in five minutes. Choose the nearest exit now, and in case of need, walk quietly (do not run) to that exit, in order to avoid panic.

▽

PROGRAM

I.   Marcia Reale
     The Star Spangled Banner
     ORCHESTRAL SOCIETY
     Conductor: Max Jacobs

II. Overture from William Tell . . . *Rossini*
     ORCHESTRAL SOCIETY

III. Address
     HON. CHARLES EVANS HUGHES
     President of the Italy America Society

IV. Aria — Selected
     JOHN CHARLES THOMAS

V.   Address
     HON. BAINBRIDGE COLBY
     Secretary of State of the United States

VI. The Battle of Vittorio Veneto
     COL. ASINARI de BERNEZZO
     Military Attache of the Italian Embassy

8

VII. Aria: "O Don Fatale," from Don Carlos *Verdi*
MME. ELEANORA DE CISNEROS
of La Scala, Milan, and the Chicago Opera Company

VIII. Address
HON. F. H. La GUARDIA
President of the Board of Aldermen

▽

INTERMISSION

▽

IX. "Dance of the Hours" . . . . . *Ponchielli*
ORCHESTRAL SOCIETY

X. Address
HIS EXCELLENCY THE ROYAL ITALIAN AMBASSADOR
BARON ROMANO AVEZZANA

XI. Aria from Pagliacci . . . . *Leoncavallo*
NICOLA ZEROLA

XII. Address
COMM. GIANNI CAPRONI

XIII. Aria from Samson et Dalila . . *Saint Saens*
MME. DE CISNEROS

XIV. Address
HON. MEDILL McCORMICK
United States Senator from Illinois

XV. March from Aida
ORCHESTRAL SOCIETY

▽

Steinway Piano Used

9

PATRONESSES

BARONESS ROMANO AVEZZANA

MRS. EMMA G. AUCHINCLOSS
MRS. DONN BARBER
MRS. J. STEWART BARNEY
MRS. FELICE BAVA
MRS. GORDON KNOX BELL
MRS. ARTHUR BENINGTON
MRS. R. BERTELLI
MRS. HERMANN M. BIGGS
MRS. CHARLES A. BRISTED
MRS. HENRY W. CANNON
MRS. WINTHROP CHANLER
MRS. GEORGE S. CHAPPELL
MRS. RICHARD W. CHILD
MRS. LUIGI COSTA
MRS. HENRY IVES COBB
MRS. PAUL D. CRAVATH
MRS. WILLIAM DARRACH
MRS. PAOLO DE VECCHI
MRS. JOSEPH DI GIORGIO
MRS. WILLIAM B. DINSMORE
MRS. CHARLES H. DITSON
MRS. DAVID DOWS
MRS. JOHN ADAMS DRAKE
MRS. JOHN R. DREXEL
MRS. STUART DUNCAN
MRS. J. CLIFTON EDGAR
MRS. ERNESTO G. FABBRI
MRS. EDWARD R. FINCH
MRS. GEORGE BARTON FRENCH
MRS. JOHN J. FRESCHI
MRS. R. HORACE GALLATIN

MRS. EDWIN GOULD
MRS. GEORGE JAY GOULD
MRS. MABEL GRAVES
MRS. WILLIAM GUGGENHEIM
MRS. JOHN HENRY HAMMOND
MRS. J. HORACE HARDING
MRS. E. HENRY HARRIMAN
MRS. J. BORDEN HARRIMAN
MRS. McDOUGALL HAWKES
MRS. SAMUEL N. HINCKLEY
MRS. HAMILTON HOLT
MRS. EDWARD M. HOUSE
MRS. CHARLES EVANS HUGHES
MRS. WALTER B. JAMES
MRS. ROBERT U. JOHNSON
MRS. OTTO H. KAHN
MISS ANNIE KING
MRS. Le ROY KING
MRS. W. DE LANCEY KOUNTZE
MRS. C. GRANT LA FARGE
MRS. J. HENRY LANCASHIRE
MRS. FREDERICK N. LONGFELLOW
MRS. CHARLES M. MacNEILL
MISS ELISABETH MARBURY
MRS. MARKOE
MRS. EDWARD McVICKAR
MRS. HUGH MINTURN
MRS. JOHN PURROY MITCHEL
MRS. PIERPONT MORGAN
MRS. WILLIAM FELLOWES MORGAN
MRS. LOUIS GOUVERNEUR MORRIS

10

PROGRAM GIRLS

MISS MARGARET SMITH
Chairman

MINNETTE BARRETT

GERALDINE BERGH

PEGGY BOLAND

IONE BRIGHT

MARGARET GREEN

EVELYN HERBERT

FLORENCE HOSTETTER

MARGARET HURTER

ETHEL INTROPODI

VERA ROYER

DOROTHY TIERNEY

BETTY WALES

EMILY REBIE WALKER

CONSTANCE WALLACE

ROSE WINTER

THE ITALY AMERICA SOCIETY

▽

"*. . . To create and maintain between the United States and Italy an international friendship based upon mutual understanding of their national ideals and aspirations and of the contributions of each to progress in science, art and literature, and upon coöperative effort to develop international trade.*"

▽

THE war had gone on too long before Americans awakened to the part which Italy was playing in the conflict, and to the gallantry and resourcefulness of her troops in the defence of civilization. Her services and her sacrifices had been minimized by propaganda of the enemy; her one defeat had been magnified; her purposes and her national aspirations had been distorted and belied.

It was because of the need of an awakening to the true facts of Italy's participation in the war that Americans, acting spontaneously, founded the Italy America Society. As the first permanent organization working for the mutual interests of Italy and the United States, the new Society was eagerly welcomed by Americans the country over as affording a means of expressing their friendship for Italy and their appreciation of her brilliant war record. For this reason, the Italy America Society, under the vigorous and determined management of its Executive Committee, leaped forward in membership and in accomplishment as no other organization of its kind has ever done in so short a space of time.

The Society was founded in March, 1918. Within two months, it had focussed on Italy the attention of the entire United States. It organized a nation-wide celebration of Italy America Day on the third anniversary of Italy's entrance into the war, enlisting for the purposes of the celebration the services of the Governors of States and the Mayors of the larger cities. It held a notable meeting at the Metropolitan Opera House, at which the Secretary of War was one of the principal speakers, and to which the President of the United States and His Excellency the Royal Italian Ambassador lent their patronage. It sent the proceeds of the meeting, somewhat more than $45,000, to the Italian Red Cross.

According to the statement of the Secretary of War, the Italy America Society secured from our Government the policy of sending American troops to Italy. It directed the attention of American bankers to Italy's need of American capital. It secured the publication of thousands of pieces of news which were accepted by the press for no other reason than that the news was demanded by a revived public interest in Italy and a new recognition of her importance as one of our allies.

Again in 1919 a national celebration was held, this time with the Secretary of the Navy as one of the principal speakers.

In 1919, as in 1918, the Society was host to many distinguished Italians who visited this country, bringing them into personal contact with leaders of American opinion, and securing wide currency for the messages which they brought to America. Among them were the late Count Macchi Di Cellere, Admiral Ugo Conz, Captain Bevione of the Italian Bureau of Information, and a group of Italian press correspondents. During the present year, the Italy America Society has given a banquet in honor of His Excellency the Royal Italian Ambassador.

UT all these are matters that lie behind us. The important fact is that the future holds greater opportunities and greater obligations than did the past. Heretofore, the Society's work has necessarily been confined largely to developing a sound organization. The real work—the actual building of close financial, commercial, and intellectual contacts with Italy—lies ahead.

The Society has planned a comprehensive program of activities. It is developing a trade service to furnish information to American business men with regard to industrial and commercial developments and opportunities in Italy. Its Committee on Economic Relations is undertaking an exhaustive study of economic and industrial questions for the mutual benefit of the two countries.

The Society is attempting to secure the publication of adequate and accurate news regarding Italy in this country, and is combating any tendency toward misrepresentation of Italy in the public press. It is publishing a series of booklets on various phases of Italian achievement, and is arranging reciprocal art exhibits and exhibitions of industrial products. Through its Committee on Education, it is working toward an exchange of professors and students between Italy and America. Other activities are planned.

The Italy America Society will welcome as members those who have the foresight to read clearly the benefits that must accrue both to Italy and to America through closer relations between the peoples and governments of the two countries.

COMMITTEE ON ARRANGEMENTS

WILLIAM GUGGENHEIM, Honorary Chairman
RICHARD WASHBURN CHILD, Chairman

ARTHUR BENINGTON
JOHN FOSTER CARR
MISS EMILY H. CHAUNCEY
MISS MARGHERITA DE VECCHI
DOUGLAS L. DUNBAR
JOHN J. FRESCHI
WILLIAM J. GUARD

F. M. GUARDABASSI
MRS. ETHEL CRYDER HIGGINS
FRANCIS H. MARKOE
RONALD H. PIERCE
CARLO CIULLI RUGGIERI
IRWIN SMITH
WHITNEY WARREN

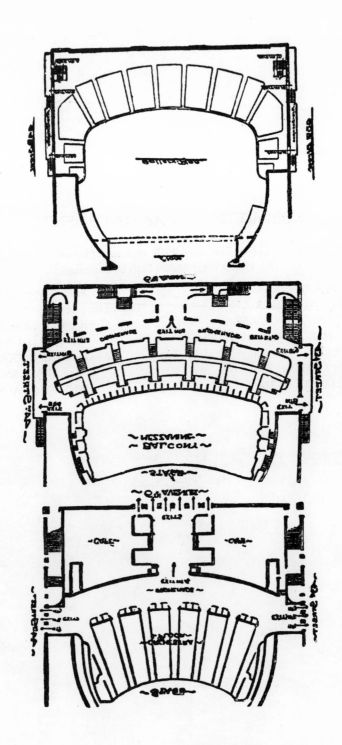

TONTITOWN, ARK.

CELEBRATING THE ITALIAN AMBASSADOR'S VISIT TO TONTITOWN

The World's Ideal Vineyard

Has at Last an Outlet for Her Wonderful Productions Through the New Railroad Which Has Just Been Completed

WONDERFUL DREAM COME TRUE

Kansas City & Memphis
Railway Company

Arkansas, Oklahoma & Western Railroad
Monte Ne Railway

Rogers, Ark.

Connections with All Frisco Lines

History of Tontitown
The Gem of the Ozarks
In the Center of the Great Southwest

The romantic story of the founding of Tontitown reads like a novel.

Father Bandini, an Italian priest, and a man of noble birth, gave up his career at home in order to devote his life to the bettering of the conditions of his countrymen in America.

To the people of Arkansas, to the sociologist, Father Bandini needs no introduction. Less than a dozen years ago, when his people who had been colonized in an unhealthy climate were dying of malaria fever in the swamp lands, this noble man with great personal courage and sacrifice, enduring hardships and even hunger, brought his people to this plateau of the Ozarks and founded Tontitown, naming it after Enrico Tonti, the first explorer of Arkansas.

After their first hardships the people recovered their health, and in a few years this spot that was nothing but a wilderness, became a prosperous community, the former waste places made veritable gardens.

He founded the incorporated village of Tontitown on broad non-sectarian lines. His church and home are open to all good people, irrespective of their religious views.

By his great personal efforts and self-denial he has endeared himself to the people and is their "Guide, Philosopher and Friend." Tontitown was brought to the attention of the world at large only recently through a magazine article in *Everybodys,* and other references in the daily press.

The visit of the Italian Ambassador of Washington, to the colony, primarily Italian than

FATHER BANDINI
Founder of Tontitown

mixed, and now intensely American, added to the interest in and knowledge of its people, their aims and success in attaining them.

This little colony of thrifty, happy, hard-working people, which had been growing and advancing steadily f o r years, increasing in population a n d prosperity through their own efforts, and the yield of their crops of peaches, pears, apples, berries, and the never-failing grape, being suddenly brought before the world, and people who were really astonished at the progress, the excellent school a n d church facilities of the little village, so far away from a railroad, a number of American and Irish Catholic families joined the settlement, and Tontitown made its appearance on the geographical maps as a thrifty village.

Through the suddenly acquired fame of Tontitown, by the magazine articles, hundreds of Catholics from all parts of the world asked Father Bandini to admit them to the colony. He steadily refused, until now, for the reason that so many of the applicants had just enough money to purchase their lands, and as there was no railroad, they were, of necessity, limited to the culture of vines as the only certain crop, which would necessitate them not being able to support themselves for three or four years, to become charges on their more successful neighbors.

He said to all applicants that when the railroad they had dreamed of and worked for so many years, was built into Tontitown he would admit them, for then they could market quick crops, such as strawberries, raspberries and vegetables.

"The Railroad"

The Kansas City and Memphis Railroad

An ambitious dream of ten years has just been realized by the people of Tontitown, on the completion of the Kansas City and Memphis Railroad, which was finished August, 1911.

This road is being built from Siloam Springs, Ark., to Memphis, Tenn. It has just reached Tontitown and thereby opening up a wonderfully productive country. Tontitown is now within nine hours of Kansas City and ten hours of St. Louis.

Heretofore additions to the colony have been discouraged for the reason that there was only a limited available market for the berries and other quick money crops, and the newcomers could hardly support themselves while waiting for the grapes and orchards to come into production.

Now the new road assures an outlet for any and all its products, be they fruits or vegetables, and there is nothing to hinder its full development, and accessories are invited.

The farming tracts are adjoining the town and each tract of five acres, properly cultivated, will produce enough for the maintenance of the family.

Railroad and Commercial Facilities

There are now in operation wine presses, a cider and vinegar factory, cheese factory and evaporator and drying factory.

The completion of the Kansas City and Memphis Railroad, which runs direct to Tontitown, has realized the dream of Father Bandini, and the community is furnished with a means of disposing of its products and is also placed in direct communication with the large commercial centers.

The completion of this road places Tontitown in the very center of a rich farming community, and to Tontitown all the farmers within a large radius of square miles will bring their produce for sale and for shipment. There is a grand opening here for all kinds of stores, and the most liberal inducements will be offered to those who wish to embark in business.

An Abundance of Winter Feed

Climatic Conditions and Fertility of Soil are to be Found Nowhere so Favorable to the Support of Human Life as at Tontitown

Finer Wheat Than That of Which the Dakotas Boast Grows Near Tontitown

Products

Tontitown is situated in the heart of the fruit-producing belt.

For seven months in the year the fruit and vegetable growers of this country are shipping stuff to market. This district has long been famous for its fruits, particularly apples and strawberries.

But the list of profitable possibilities includes also fancy cherries, plums, pears, grapes, melons, tomatoes, asparagus, celery, onions, early radishes, lettuce, etc. Because of the nearness of large market cities these products reach the consumer in fresh condition, commanding the best prices. All surplus is readily disposed of on the ground to the canning factory and evaporator. This insures the growth against waste in any form.

Crops of all sorts may be made to overlap, so that every day is profitable, with no dreary "dead" season. This is intensive farming under the very best conditions, requiring least outlay of capital.

Poultry keeping is also profitable on this plateau. With right management eggs may be supplied to Northern markets at the minimum of cost in winter, when prices are at the maximum, and this applies also to winter hatched broilers. With our mild, open winters each of these operations is a dollar maker. A small acreage in fruit may be made to yield richly in garden truck as well, and will also sustain its poultry flocks, its pigs and milch cow—all supported from the products of the owner's bit of ground.

One article manufactured in Tontitown has almost worldwide fame, inasmuch as the "Old World" flavor of her Chanti wine, made from the grapes, grown and cared for almost like children by the colonist, each family having its own vine-

Tontitown Grapes

Better Corn Than Illinois Can Produce

yard, and many hundreds of gallons of wine are shipped to all parts of the world every year. There are two million grape vines in bearing at present.

Live Stock—It is very evident that a country abounding in such a variety of nutritious and heavy yielding food, forage and pasture crops, must be a grand live stock country.

While cattle, sheep, horses and mules all flourish, it is one of the greatest hog countries, as swine plagues are unknown.

Live Stock

Tontitown's Own Products

The cheese made from goats' and cows' milk, such as only the people from the old country know how to make, with its delicious flavor, is quite a money-making product.

The real thing in Italian Spaghetti is made by hand at Tontitown, and is just as good as the imported article.

Opportunity Cannot Knock at the Door of the Man Who Has No House. Lay the Foundation for a Home in Tontitown.

Strawberry Pickers

Buy Tontitown Dirt, to Own Real Estate is the Best Object a Saving Man can Have

The Tontitown Apple Crop

Gathering the Apple Harvest

Packing the Apples for Shipment

Tontitown Products

An Industrious Scene at Tontitown

A Tontitown Onion Field

Schools and Educational Advantages

Considering the very short time Tontitown has been founded—about 12 years—its educational advantages are truly remarkable. Its large Academy, which cost about $4,000.00 is a worthy tribute to the intelligence of the inhabitants. At

Father Bandini's Home

Bandini and the Sisters. Surely, for a town that a dozen years ago was unheard of, this is up-to-date American progression.

The Tontitown School

present there are about one hundred pupils enrolled, and some of them have come from as far away as Chicago in order to receive the educational advantages which the Sisters are so well fitted to impart. The training of the young mind under their careful tuition and the spiritual guidance which Father Bandini gives, finds them well-fitted

Hotel Golden Lion

The "Hotel Golden Lion," at Tontitown, is a magnificent building on the Italian style, modern and up-to-date, with its own water works and lights, built at a cost of $8,000.00.

The Tontitown Band

for life's battle. At the present rate of progress, it will be only a short time before this Academy will be heard of as a college possessing all the advantages of a modern university.

Plans are under way for the founding of a free hospital, under the supervision of Father

Tontitown Products

Catholic Church

The Catholic Church at Tontitown is the only church in North Arkansas which has regular services.

In no Other Place are the Climatic Conditions and the Fertility of the Soil so Favorable to the Support of Human Life

---

The Price of Three Cigars Daily Lays the Foundation for a Home in Tontitown

A Great Boom in Tontitown, Ark.

An investment with the Kansas City and Memphis Development Co., while the Tontitown boom is still in its infancy, is a sure cure for poverty, and a word to the wise is sufficient. The next focus of great activity is at Tontitown, Arkansas

Our Proposition is Backed by Men of Enthusiasm, Capital, Energy and Skill

The Kansas City and Memphis Development Co. is composed of railroad men who realize the great industrial and commercial possibilities of this beautiful Ozark country. They are willing to spend millions building railroads to open to the world a path to nature's own gardenspot.

Inducements to Merchants— Great Future for the Merchant in Tontitown

Now that the railroad is in, the farmers within a large radius of square miles will trade at Tontitown instead of going on miles beyond for shipping facilities, as heretofore.

We want all kinds of stores—hardware, drugs, grocery stores and dry goods.

Also men in all lines and professions will do well.

In this new town, it does not require much capital to get a start.

COME AND START WITH THE STARTERS AND STAY WITH THE STAYERS.

A Wine Ranch

Big Spring Near Tontitown

Climate and Health Conditions

Health conditions are perfect in the town and surrounding country. This plateau is 1,250 feet above sea level. Malaria is unknown. There are no mosquitoes. The average temperature for the year is 58 degrees. The summer days are never excessively hot, while the nights are always delightfully cool. In winter there is no extreme cold, in fact, work in the fields and gardens may be carried on practically every day in the year.

The water of the Ozark country is one of the remarkable waters of the world for promoting health. It is very pure with practically no solids. It comes from numerous springs, and for that reason this country is noted as a health resort. There are many resorts near Tontitown, among them the famous Eureka Springs, Sulphur Springs and Monte Ne.

Every farm can have its own adequate supply of water, and on many of them are found everliving springs, cool and refreshing.

A Convenient Spring

The Call of Tontitown

To the intensive farmer, to the run-down city man who is tired of the dirt and the smoke, the congested condition of the city; to those who are working for a pittance—come to Tontitown—it calls you, heed the call "back to nature"—and come to us and settle down in "God's Country," far from the maddening crowds; you will be assured of health and happiness.

Come where the land will yield from $50.00 to $100.00 an acre if planted in strawberries, blackberries, onions, beans or potatoes, and where the never-failing grape crop will yield from $300.00 to $500.00 per acre.

Come and be one of us in the land where the maringo grows; no mosquitoes, but plenty of hunting, fishing, sunshine and flowers, and where a good healthy living can be had almost for the asking.

Many thousands got rich from small investments at Seattle, San Francisco and Galveston. It's knowing how, and where, and when to act that makes fortunes.

Remember, prompt action is the important element in money making.

The Kansas City and Memphis Development Company

Offer Lots on the Following Terms:

Business Lots, 25x140, inside, at $200.00; 5% down and 10% per month afterward.

Business Lots, 25x140, corner, at $225.00; 5% down and 10% per month afterward.

Residence Lots, 50x140, corner, at $225.00; 5% down and 10% per month afterward.

Residence Lots, 50x140, inside, at $200.00; 5% down and 10% per month afterward.

Five-acre tracts or any number of acres can be secured for you through the Kansas City and Memphis Development Co. Although we do not offer them for sale, they can be had on easy terms from $150.00 per five acres to $500.00, dependent upon location and quality.

TAKE ADVANTAGE OF OUR PLAN OF CITY BUILDING AND SECURE A HOME IN TONTITOWN.

NOTHING IS MORE WORTHLESS THAN A BUNCH OF RENT RECEIPTS.

Opportunity cannot knock at the door of the man who has no house. Lay the foundation for a home in Tontitown.

A Word in Conclusion

Testimonials

Tontitown, Ark.

As a resident of Tontitown, I am able to state that grapes in this locality yield from $300.00 to $500.00 per acre and are of fine flavor and quality. Apples yield from 200 to 300 baskets to the acre.

JAMES NEAL.

This is only one of the many testimonials from Tontitown. A good living and the proper maintenance of a family with fine school advantages are assured to any who come to Tontitown.

Upon the signing of the contract and making the first payment, the Trustee will contract to deliver deed to property, upon completion of payments as per contract. FIVE PER CENT DISCOUNT FOR FULL CASH PAYMENTS.

All payments should be made to J. E. Felker, Trustee and Cashier of Bank of Rogers, Rogers, Ark.

For References:

Bank of Rogers, Rogers, Ark.,
and
Kansas City & Memphis Railroad Co.,
Rogers, Ark.

Make a Personal Investigation of our Proposition and You will Say that Half has Never Been Told.

Come Where the Berry Crop Yields From
$50.00 to $100.00 an Acre

THE enterprising man needn't have fear,

OF any investment offered here,

NATURE has provided beauty, health and wealth,

TONTITOWN, "The Gem of the Ozarks" in the World's Fruit Belt,

IN the land of all-year sunshine, yet sufficient rain,

TONTITOWN offers many openings for the enterprising brain,

OPPORTUNITY is knocking at the door of every man,

WHO wants to get rich, but can't find a plan,

NOW is the time, Tontitown is the place and the land.

Come Where the Grape Crop Yields From
$300.00 to $500.00 an Acre

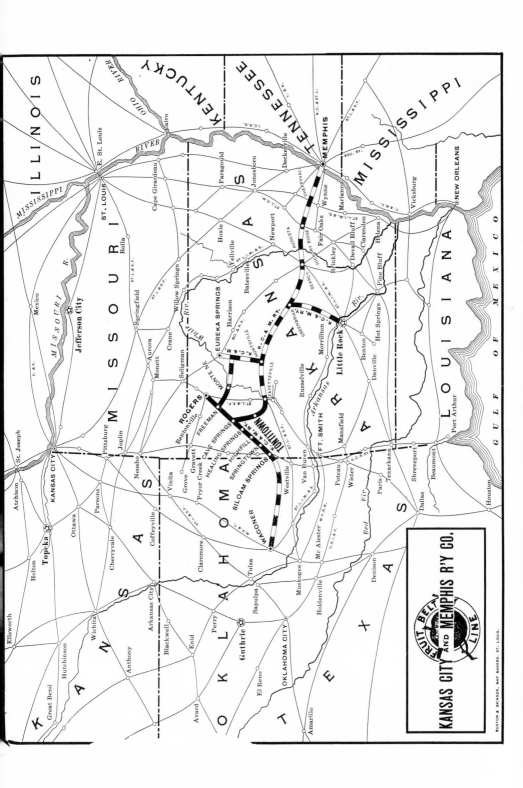

KANSAS CITY AND MEMPHIS R'Y CO.

FRUIT BELT LINE

BUXTON & SKINNER, MAP MAKERS, ST. LOUIS.

The Italian American Experience

An Arno Press Collection

Angelo, Valenti. **Golden Gate.** 1939

Assimilation of the Italian Immigrant. 1975

Bohme, Frederick G. **A History of the Italians in New Mexico.** (Doctoral Dissertation, The University of New Mexico, 1958). 1975

Boissevain, Jeremy. **The Italians of Montreal:** Social Adjustment in a Plural Society. 1971

Churchill, Charles W. **The Italians of Newark:** A Community Study. (Doctoral Thesis, New York University, 1942). 1975

Clark, Francis E. **Our Italian Fellow Citizens in Their Old Homes and Their New.** 1919

D'Agostino, Guido. **Olives on the Apple Tree.** 1940

D'Angelo, Pascal. **Son of Italy.** 1924

Fenton, Edwin. **Immigrants and Unions,** A Case Study: Italians and American Labor, 1870-1920. (Doctoral Thesis, Harvard University, 1957). 1975

Forgione, Louis. **The River Between.** 1928

Fucilla, Joseph G. **The Teaching of Italian in the United States:** A Documentary History. 1967

Garlick, Richard C., Jr. et al. **Italy and Italians in Washington's Time.** 1933

Giovannitti, Arturo. **The Collected Poems of Arturo Giovannitti.** 1962

Istituto di Studi Americani, Università degli Studi di Firenze (Institute of American Studies, University of Florence). **Gli Italiani negli Stati Uniti** (Italians in the United States). 1972

Italians in the City: Health and Related Social Needs. 1975

Italians in the United States: A Repository of Rare Tracts and Miscellanea. 1975

Lapolla, Garibaldi M. **The Fire in the Flesh.** 1931

Lapolla, Garibaldi M. **The Grand Gennaro.** 1935

Mariano, John Horace. **The Italian Contribution to American Democracy.** 1922

Mariano, John H[orace]. **The Italian Immigrant and Our Courts.** 1925

Pagano, Jo. **Golden Wedding.** 1943

Parenti, Michael John. **Ethnic and Political Attitudes:** A Depth Study of Italian Americans. (Doctoral Dissertation, Yale University, 1962). 1975

Protestant Evangelism Among Italians in America. 1975

Radin, Paul. **The Italians of San Francisco:** Their Adjustment and Acculturation. Parts I and II. 1935

Rose, Philip M. **The Italians in America.** 1922

Ruddy, Anna C. (Christian McLeod, pseud.). **The Heart .of the Stranger:** A Story of Little Italy. 1908

Schiavo, Giovanni Ermenegildo. **Italian-American History:** Volume I. 1947

Schiavo, Giovanni [Ermenegildo]. **Italian-American History:** The Italian Contribution to the Catholic Church in America. Volume II. 1949

Schiavo, Giovanni [Ermenegildo]. **The Italians in America Before the Civil War.** 1934

Schiavo, Giovanni E[rmenegildo]. **The Italians in Chicago:** A Study in Americanization. 1928

Schiavo, Giovanni [Ermenegildo]. **The Italians in Missouri.** 1929

Schiro, George. **Americans by Choice:** History of the Italians in Utica. 1940

La Società Italiana di Fronte Alle Prime Migrazioni di Massa. (Italian Society at the Beginnings of the Mass Migrations). New Foreword (in English) by Francesco Cordasco. 1968

Speranza, Gino. **Race or Nation:** A Conflict of Divided Loyalties. 1925

Stella, Antonio. **Some Aspects of Italian Immigration to the United States:** Statistical Data and General Considerations Based Chiefly Upon the United States Censuses and Other Official Publications. 1924

Ulin, Richard Otis. **The Italo-American Student in the American Public School:** A Description and Analysis of Differential Behavior. (Doctoral Thesis, Harvard University, 1958). 1975

Valletta, Clement Lawrence. **A Study of Americanization in Carneta:** Italian-American Identity Through Three Generations. (Doctoral Dissertation, University of Pennsylvania, 1968). 1975

Villari, Luigi. **Gli Stati Uniti d'America e l'Emigrazione Italiana.** (The United States and Italian Immigration). 1912

Workers of the Writers' Program. Work Projects Administration in the State of Nebraska. **The Italians of Omaha.** 1941